CORROSION OF GLASS, CERAMICS AND CERAMIC SUPERCONDUCTORS

CORROSION OF GLASS, CERAMICS AND CERAMIC SUPERCONDUCTORS

Principles, Testing, Characterization and Applications

Edited by

David E. Clark and Bruce K. Zoitos

University of Florida
Department of Materials Science and Engineering
Gainesville, Florida

np

NOYES PUBLICATIONS
Park Ridge, New Jersey, U.S.A.

Library of Congress Catalog Card Number: 91-16010
ISBN: 0-8155-1283-X
Printed in the United States

Published in the United States of America by
Noyes Publications
Mill Road, Park Ridge, New Jersey 07656

10 9 8 7 6 5 4 3 2 1

Library of Congress Cataloging-in-Publication Data

Corrosion of glass, ceramics and ceramic superconductors : principles, testing, characterization, and applications / edited by David E. Clark and Bruce K. Zoitos.
p. cm.
Includes bibliographical references and index.
ISBN 0-8155-1283-X
1. Ceramic materials--Corrosion. 2. Glass--Corrosion. 3. Ceramic superconductors--Corrosion. I. Clark, David E. II. Zoitos, Bruce K.
TA455.C43C68 1991
620.1'404223--dc20 91-16010
CIP

MATERIALS SCIENCE AND PROCESS TECHNOLOGY SERIES

Editors

Rointan F. Bunshah, University of California, Los Angeles *(Series Editor)*
Gary E. McGuire, Microelectronics Center of North Carolina *(Series Editor)*
Stephen M. Rossnagel, IBM Thomas J. Watson Research Center *(Consulting Editor)*

Electronic Materials and Process Technology

DEPOSITION TECHNOLOGIES FOR FILMS AND COATINGS: by Rointan F. Bunshah et al

CHEMICAL VAPOR DEPOSITION FOR MICROELECTRONICS: by Arthur Sherman

SEMICONDUCTOR MATERIALS AND PROCESS TECHNOLOGY HANDBOOK: edited by Gary E. McGuire

HYBRID MICROCIRCUIT TECHNOLOGY HANDBOOK: by James J. Licari and Leonard R. Enlow

HANDBOOK OF THIN FILM DEPOSITION PROCESSES AND TECHNIQUES: edited by Klaus K. Schuegraf

IONIZED-CLUSTER BEAM DEPOSITION AND EPITAXY: by Toshinori Takagi

DIFFUSION PHENOMENA IN THIN FILMS AND MICROELECTRONIC MATERIALS: edited by Devendra Gupta and Paul S. Ho

HANDBOOK OF CONTAMINATION CONTROL IN MICROELECTRONICS: edited by Donald L. Tolliver

HANDBOOK OF ION BEAM PROCESSING TECHNOLOGY: edited by Jerome J. Cuomo, Stephen M. Rossnagel, and Harold R. Kaufman

CHARACTERIZATION OF SEMICONDUCTOR MATERIALS—Volume 1: edited by Gary E. McGuire

HANDBOOK OF PLASMA PROCESSING TECHNOLOGY: edited by Stephen M. Rossnagel, Jerome J. Cuomo, and William D. Westwood

HANDBOOK OF SEMICONDUCTOR SILICON TECHNOLOGY: edited by William C. O'Mara, Robert B. Herring, and Lee P. Hunt

HANDBOOK OF POLYMER COATINGS FOR ELECTRONICS: by James J. Licari and Laura A. Hughes

HANDBOOK OF SPUTTER DEPOSITION TECHNOLOGY: by Kiyotaka Wasa and Shigeru Hayakawa

HANDBOOK OF VLSI MICROLITHOGRAPHY: edited by William B. Glendinning and John N. Helbert

CHEMISTRY OF SUPERCONDUCTOR MATERIALS: edited by Terrell A. Vanderah

CHEMICAL VAPOR DEPOSITION OF TUNGSTEN AND TUNGSTEN SILICIDES: by John E.J. Schmitz

(continued)

Ceramic and Other Materials–Processing and Technology

SOL-GEL TECHNOLOGY FOR THIN FILMS, FIBERS, PREFORMS, ELECTRONICS AND SPECIALTY SHAPES: edited by Lisa C. Klein

FIBER REINFORCED CERAMIC COMPOSITES: by K.S. Mazdiyasni

ADVANCED CERAMIC PROCESSING AND TECHNOLOGY–Volume 1: edited by Jon G.P. Binner

FRICTION AND WEAR TRANSITIONS OF MATERIALS: by Peter J. Blau

SHOCK WAVES FOR INDUSTRIAL APPLICATIONS: edited by Lawrence E. Murr

SPECIAL MELTING AND PROCESSING TECHNOLOGIES: edited by G.K. Bhat

CORROSION OF GLASS, CERAMICS AND CERAMIC SUPERCONDUCTORS: edited by David E. Clark and Bruce K. Zoitos

Related Titles

ADHESIVES TECHNOLOGY HANDBOOK: by Arthur H. Landrock

HANDBOOK OF THERMOSET PLASTICS: edited by Sidney H. Goodman

SURFACE PREPARATION TECHNIQUES FOR ADHESIVE BONDING: by Raymond F. Wegman

FORMULATING PLASTICS AND ELASTOMERS BY COMPUTER: by Ralph D. Hermansen

Preface

During the last several decades significant progress has been made in understanding corrosion processes in ceramic materials. Development of new theoretical approaches and improvements in analytical techniques have served to unify many of the concepts developed in early corrosion studies. The objective of this volume is to provide scientists and engineers with a compilation of state of the art understanding in ceramic corrosion.

Section I, Theory, Testing and Characterization, begins with a discussion of the general theoretical concepts of ceramic corrosion, explores the applicability and limitations of theory in predicting corrosion and concludes with a discussion of testing strategies and options for characterization of corroded ceramics.

Section II covers glass corrosion and presents thorough discussions of the two most prominent approaches to glass corrosion--geochemical and thermodynamic. This is followed by seven applications chapters which discuss specifics of corrosion in nuclear waste, optical, bioactive, halide and geologic glasses, as well as corrosion of glazes and enamels and conservation issues in the preservation of corroded ancient glasses.

Corrosion of ceramic materials is covered in Section III. This section is comprised of five applications chapters discussing corrosion in glass-ceramics, masonry, cutting tools and TZP ceramics. The fifth chapter discusses high temperature, non-aqueous corrosion of engineering ceramics.

Finally, Section IV includes six chapters which cover corrosion of the recently developed ceramic superconductors. Three of these chapters discuss fundamental aspects of corrosion in superconductors, while two chapters are devoted to processing and treatment methods for reducing corrosion rates in these materials. The last chapter suggests important research areas that should be further explored.

It is our hope that this text will serve to document the current status of understanding in this field, reveal areas of deficiency and suggest directions for future study.

Department of Materials Science and Engineering
University of Florida
Gainesville, Florida

David E. Clark
Bruce K. Zoitos

Acknowledgments

No undertaking of this magnitude would be possible without the support and assistance of numerous individuals. The editors express their deepest gratitude to the contributors and their staffs for their cheerful cooperation through numerous "visions and revisions." The logistical assistance and guidance of George Narita, Vice President of Noyes Publications, is gratefully acknowledged. We also thank Ms. Rebecca Schulz, graduate student in the Materials Science and Engineering Department at the University of Florida, for assembling the text. The quality of typography and layout of this book are the direct result of her diligence and perseverance.

Finally, the editors thank those closest to them for providing the supportive environment necessary to survive the tribulations of such a project.

Contributors

P. Bruce Adams
Precision Analytical
Watkins Glen, NY

Vasantha R.W. Amarakoon
NYS College of Ceramics
at Alfred University
Alfred, NY

Aaron Barkatt
The Catholic University
of America
Washington, DC

Shyama S. Bayya
NYS College of Ceramics
at Alfred University
Alfred, NY

James P. Bennett
U.S. Bureau of Mines
Tuscaloosa Research Center
Tuscaloosa, AL

Temel H. Büyüklimanli
University of Florida
Gainesville, FL

Gregory T. Chandler
University of Florida
Gainesville, FL

David E. Clark
University of Florida
Gainesville, FL

Richard A. Eppler
Eppler Associates
Cheshire, CT

Rodney C. Ewing
The University of New Mexico
Albuquerque, NM

Dennis S. Fox
National Aeronautics and
Space Administration
Lewis Research Center
Cleveland, OH

Bernd Grambow
Kernforschungszentrum
Karlsruhe GmbH
Karlsruhe, Germany

Larry L. Hench
Advanced Materials Research Center
University of Florida
Alachua, FL

Hamid Hojaji
Vitreous State Laboratory
The Catholic University of America
Washington, DC

Yasuro Ikuma
Kanagawa Institute of Technology
Atsugi, Kanagawa, Japan

Nathan S. Jacobson
National Aeronautics and Space Administration
Lewis Research Center
Cleveland, OH

Carol M. Jantzen
Westinghouse Savannah River Co.
Savannah River Laboratory
Aiken, SC

Michael J. Jercinovic
The University of New Mexico
Albuquerque, NM

Tapio T. Lepistö
Institute of Materials Science
Tampere University of Technology
Tampere, Finland

Alexander Lodding
Chalmers University of Technology
Gothenburg, Sweden

Tapio A. Mäntylä
Institute of Materials Science
Tampere University of Technology
Tampere, Finland

Walter J. McCracken
GE Aerospace
General Electric Company
Philadelphia, PA

Karen A. Michael
Vitreous State Laboratory
The Catholic University of America
Washington, DC

Richard J. Parry
Pilkington Technology Centre
Lathom, Lancashire, England

Rebecca L. Schulz
University of Florida
Gainesville, FL

Catherine Simmons
Advanced Materials Research Center
University of Florida
Alachua, FL

Joseph H. Simmons
University of Florida
Gainesville, FL

James L. Smialek
National Aeronautics and Space Administration
Lewis Research Center
Cleveland, OH

Pallavoor N. Vaidyanathan
Microfabritech®
University of Florida
Gainesville, FL

Pamela B. Vandiver
Conservation Analytical Lab
Smithsonian Institution
Washington, DC

William B. White
The Pennsylvania State University
Materials Research Laboratory
University Park, PA

George G. Wicks
Westinghouse Savannah River Co.
Aiken, SC

Masahiro Yoshimura
Research Laboratory of
Engineering Materials
Tokyo Institute of Technology
Midori, Yokohama,
Kanagawa, Japan

Bruce K. Zoitos
University of Florida
Gainesville, FL

NOTICE

The views, opinions and findings contained in this book are those of the individual authors. The Publishers, editors and authors assume no responsibility nor liability for errors or any consequences arising from the use of the information contained herein. Mention of trade names or commercial products does not constitute endorsement or recommendation for use by the publishers, editors or authors.

Final determination of the suitability of any information, procedure, or product for use contemplated by any user, and the manner of that use, is the sole responsibility of the user. This book is intended for informational purposes only. Expert advice should be obtained at all times when implementation is being considered, particularly where hazardous materials or processes involving glasses, ceramics or ceramic superconductors are encountered.

Contents

SECTION I
THEORY, TESTING AND CHARACTERIZATION

1. THEORY OF CORROSION OF GLASS AND CERAMICS 2
William B. White
Introduction 2
Chemical Mechanisms of Dissolution and Corrosion 3
Kinetics 7
Rate Equations for Crystalline Ceramics 7
Rate Equations for Glasses 13
Surfaces 16
Characterization of Surface Layers 16
Role of Surface to Volume Ratios 18
System Variables 19
Open and Closed Systems: Rates of Transport 19
pH as an Open System Variable 20
Eh as an Open System Variable 21
Effects of Temperature 21
The Arrhenius Rate Law 21
Dissolution in Supercritical Fluids 25
Corrosion by the Vapor Phase 25
Summary and Conclusions 26
References 26

2. PREDICTING CORROSION 29
P. Bruce Adams
Introduction 29

Prerequisites . . . 31
Corrosion Models . . . 31
Composition . . . 32
Surface Condition . . . 32
Other Data Sources . . . 34
Standard Test Procedures . . . 35
Interferences . . . 36
Target Environment Conditions . . . 37
Analytical Approach . . . 38
Precision and Accuracy . . . 38
The End User . . . 39
Methodology . . . 39
Ranking Based on Test Results . . . 39
Estimation of Results . . . 40
Duplicating the Environment . . . 41
Modeling . . . 41
Applications and Discussion . . . 41
Composition and Product Development . . . 41
Preservation, Storage and Use . . . 42
Health and Safety . . . 44
Fundamental Studies . . . 47
Summary . . . 48
References . . . 49

3. CORROSION TESTING AND CHARACTERIZATION . . . 51
D.E. Clark and B.K. Zoitos
Introduction . . . 51
Methodology . . . 52
Laboratory Tests . . . 56
Field Tests . . . 75
Characterization of Corrosion . . . 78
Solution Analysis . . . 79
Solid/Surface Analysis . . . 81
Structure . . . 81
Infrared Spectroscopy . . . 81
Raman Spectroscopy . . . 81
Nuclear Magnetic Resonance . . . 82
X-Ray Photoelectron Spectroscopy . . . 82
Extended X-Ray Absorption Fine Edge Structure . . . 84
Microstructural Characterization . . . 84
Transmission Electron Microscopy . . . 86
BET Gas Adsorption Analysis . . . 87
Small Angle X-Ray Scattering . . . 87
Compositional Analysis . . . 87

Secondary Ion Mass Spectrometry 88
X-Ray Photoelectron Spectrometry 88
Auger Electron Spectroscopy 88
Electron Microprobe Analysis 88
Rutherford Backscattering 89
Nuclear Reaction Analysis 90
Data Analysis and Reduction 91
Summary 96
References 97

4. CHARACTERIZATION OF CORRODED CERAMICS BY SIMS 103
Alexander Lodding
Introduction 103
Main Principles of Functioning Instrumentation 104
Experimental Aspects of SIMS 105
Secondary Ion Emission and Quantitation 107
Applications to Corrosion Layers on Glasses 111
Applications to Impurities and Corrosion Products in Ceramic Superconductors 115
Summary 120
References 120

SECTION II
CORROSION OF GLASS

5. GEOCHEMICAL APPROACH TO GLASS DISSOLUTION . . . 124
Bernd Grambow
Introduction 124
Short Overview on Concepts of Glass Dissolution 125
Alkali Extraction vs Matrix Dissolution 125
Limitations of the Interdiffusion Theory 126
Factors Affecting Glass Matrix Dissolution 127
Formation of Altered Surface Layers 127
Progress in the Geochemistry of Rock/Water Interactions Relevant to Glass Corrosion 128
The Rate Limiting Step in the Glass/Water Reaction 131
Implications for the Mechanism of Initial Selective Alkali Depletion 133
Kinetic Equations for Glass Matrix Dissolution 134
General Rate Equation for Glass Dissolution 134
Influence of Silica Transport on the Corrosion Rate 137
Comparison of Silica Transport and Interdiffusion Mechanism 139

Geochemical Modeling of Glass Dissolution 140
Reaction Path Modeling of Surface Layer and Solution Composition 140
Reaction Kinetics–Modeling with GLASSOL 142
Validation Process for the Model 146
Summary 148
References 148

6. THERMODYNAMIC APPROACH TO GLASS CORROSION . . 153
Carol M. Jantzen
Introduction 153
Kinetic Versus Thermodynamic Approach to Glass Durability 155
Thermodynamic and Structural Basis of Glass Hydration 156
Glass Hydration Mechanisms 156
Thermodynamic Representation of Glass Hydration Mechanisms 159
Ion Exchange 160
Matrix Dissolution 160
Surface Layer Formation 160
Redox Species 162
Effects of Solution pH and Eh 162
Structural Basis of Hydration Thermodynamics 165
Relation of ΔG_{hyd} to Nonbridging Oxygen Bonds 165
Relation of ΔG_{hyd} to Ionic Potential (Z/r) and Ionic Field Strength (Z/r^2) 170
Experimental Techniques 174
Parameterization of the Factors Affecting Glass Durability 174
MCC-1 Testing 174
MCC-3 Testing 176
Product Consistency Test (PCT) 177
Controlled Environmental Testing 178
Glass Dissolution: A Function of Thermodynamic Equilibria 179
Variations in Glass Composition 179
Statistical Determinations 179
Laboratory and In-Situ Examples 186
Variations in Surface Layer Formation 188
Variations in Leaching Solutions/Environments 191
Glass Dissolution: A Function of Kinetic Test Parameters 192
The Kinetic (SA/V)•t Parameter 192

The Kinetic Temperature Parameter 194
Kinetic vs Thermodynamic Contributions 196
Glass Dissolution : A Function of Environmental Conditions 201
Quantification of a Pourbaix Diagram for Glass 202
Determination of the Pourbaix Diagram Stability Fields 203
Conclusions 209
References 210
Appendix A 216

7. NUCLEAR WASTE GLASSES: CORROSION BEHAVIOR AND FIELD TESTS 218
G.G. Wicks
Introduction 218
Background–High Level Waste (HLW) 218
High Level Waste Inventory 218
High Level Waste Characteristics 219
High Level Waste Disposal Strategy 219
Advantages of Glass 222
Processing Considerations 224
Vitrification Facilities and Practical Operating Experience 224
SRS Vitrification Process 224
DWPF 226
Technical Performance 227
Flexibility 227
Thermal Stability 227
Mechanical Integrity 227
Radiation Stability 232
Chemical Durability 232
Chemical Durability of Nuclear Waste Glass–Testing and Evaluation 233
Leaching Tests 233
Laboratory Intercomparison Tests 233
Basic Glass Corrosion Studies 233
Repository Relevant and Field Testing of Waste Glass Systems 234
Chemical Durability of Nuclear Waste Glass–Performance 234
Integrated Study Approach 234
Parameters of Importance 235
Time 235
Temperature 238

pH . . . 239
Nuclear Waste Glass Structure and Corrosion Processes . . 241
Waste Glass Structure . . . 241
3 Stage Corrosion Process for Nuclear Waste Glasses . . . 243
Stage I: Interdiffusion . . . 243
Stage II: Matrix Dissolution . . . 243
Stage III: Surface Layer Formation . . . 243
Modeling of Corrosion in Nuclear Waste Glasses . . . 246
In-Situ Testing and Performance of SRS Waste Glass . . . 249
Materials Interface Interaction Tests [MIIT] . . . 250
Compositional Correlations . . . 250
Post-Test Analyses of SRL 165/TDS Waste Glass . . . 253
Precipitated and Glass Reaction Zone . . . 253
α_0-Outermost Precipitated Salt Layer . . . 253
α_1-Precipitated Glass Layer . . . 255
β_0-Major Depletion Zone . . . 255
β_1-Gradient Zone . . . 255
β_2-Diffusion Zone . . . 255
Brine Analyses . . . 258
Summary . . . 258
References . . . 260

8. THE CHEMICAL AND ENVIRONMENTAL STABILITY OF OPTICAL GLASSES . . . 269
R.J. Parry
Introduction . . . 269
Optical Glasses . . . 269
Refractive Index and Dispersion . . . 270
The Cold Working of Optical Glasses . . . 272
Durability of Optical Glasses . . . 273
Durability Related Defects in Polished Optical Glass . . . 273
Staining . . . 274
Latent Scratch . . . 274
The Need for Durability Testing . . . 274
The Durability Testing of Optical Glasses . . . 275
Weathering . . . 275
Acid Resistance . . . 278
Alkaline Resistance . . . 283
Summary of Test Methods . . . 285
The Effect of Composition on Durability . . . 285
Silicate Glasses . . . 285
Borate Glasses . . . 288
Phosphate Glasses . . . 289
Glasses in General . . . 290

The Effect of Micro-Organisms 291
Fungi 291
Bacteria 293
Summary 294
References 295

9. SURFACE MODIFICATION OF BIOACTIVE GLASSES AND CERAMICS 298
Larry L. Hench
Introduction 298
Bioactivity 298
Five Types of Glass Surfaces 299
Kinetics of Silicate Glass Surface Reactions 301
Type III Glass Surfaces 303
Type IIIA Bioactive Glass and Ceramic Surfaces 304
Bonding Mechanisms of Bioactive Ceramics 305
Compositional Effects on Bioactivity 307
Soft Tissue Bonding 309
Effect of Composition on Interfacial Strength 310
Conclusions 311
References 311

10. CORROSION OF HEAVY-METAL FLUORIDE GLASSES 315
Catherine Simmons
Introduction 315
Glass Preparation 316
Test Methods 317
Aqueous Corrosion Processes 317
Reactions of HMF Glasses with Water 317
The Effect of pH 318
Surface Condition 320
Transform and Crystal Layer Formation 321
Formation of Hydroxyl Species 323
FTIR Analysis 324
XPS Analysis 326
Coatings 327
Summary 327
References 328

11. CORROSION OF GEOLOGICAL AND ARCHAEOLOGICAL GLASSES 330
Michael J. Jercinovic and Rodney C. Ewing
Introduction 330
The Occurrence of Natural Glasses 331

Geologic Systems 334
General Alteration and Corrosion Concepts as Applied to Natural Glasses 335
Alteration of Low Silica Glass 337
Introduction 337
Palagonite: General Properties and Composition 338
Mechanism and Rate 344
Secondary Mineralization 348
Effects of Higher Temperatures or Long Exposure Times 352
Alteration of High Silica Glass 354
Introduction 354
Mechanism and Rate 355
Influence of Elevated Temperatures 356
Implications 357
Nuclear Waste Disposal 357
Archaeology 358
Planetary Geology 359
The Geochemistry of Seawater, Groundwater, and Mineral Deposits 362
Geochemistry of Seawater 362
Influence on Groundwater Composition 362
Mineral Deposits 364
Summary 365
References 366

12. CORROSION OF GLAZES AND ENAMELS 372
Richard A. Eppler
Introduction 372
Corrosion Processes 373
Ion Exchange 373
Hydroxyl Attack 373
Kinetics of Corrosion 374
Corrosion by Hydrofluoric Acid 379
Effect of Coating Composition 380
Resistance to Ion Exchange 380
Resistance to Hydroxyl Attack and Stability Diagrams . . . 381
Tests for Corrosion Resistance 385
Lead and Cadmium Release from Ceramic Coatings 386
Kinetics of Lead Release 387
Procedures for Controlling Lead and Cadmium Release Under Production Conditions 387
Typical Results 388
Formulating a Coating for Low Lead Release 388

Prediction of the Lead Release of a Ceramic Coating ... 388
Summary ... 390
References ... 391

13. **CORROSION AND CONSERVATION OF ANCIENT GLASS AND CERAMICS** ... 393
Pamela B. Vandiver
Introduction ... 393
Contributing Factors ... 395
Case Studies in Corrosion ... 395
Low-Fired Paleolithic Ceramics ... 395
Neolithic Plasters ... 398
Near Eastern Bronze Age Aqueous Glass Corrosion ... 399
A Roman Glass Vessel from Samaria ... 404
Glazes on Pottery and Tiles ... 415
Treatments ... 423
Historical Perspective ... 423
Future Prospects ... 424
Summary ... 425
References ... 427

SECTION III
CORROSION OF CERAMICS

14. **CORROSION OF GLASS-CERAMICS** ... 432
Walter J. McCracken
Introduction ... 432
Corrosion of a Model System ... 435
Experimental ... 436
Results and Discussion ... 437
Summary ... 452
References ... 452

15. **CORROSION OF CERAMIC CONSTRUCTION MATERIALS IN ACIDIC ENVIRONMENTS** ... 455
James P. Bennett
Introduction ... 455
Standard Tests ... 457
Published Data ... 458
Chemical and Physical Properties of Ceramic Materials ... 461
Acidproof Brick ... 461
Specialty Materials ... 462
Silica ... 463

Silicon Carbide 464
High Alumina 465
Acid Exposure Tests and Results 465
Acid Corrosion 465
HCl 467
HNO_3 471
H_2SO_4 471
Volume Expansion 474
Ambient Exposure 475
Autoclave Exposure 476
Summary 478
References 479

16. DEGRADATION OF CERAMIC CUTTING TOOLS 481
P.N. Vaidyanathan
Introduction 481
Brief History of Cutting Tool Materials 482
Degradation Mechanisms 483
Abrasive Wear 483
Plastic Deformation 484
Fracture 485
Chemical Degradation 485
Chemical Dissolution 486
Diffusion Wear 486
Complications Due to Cutting Fluids 486
Taylor's Tool Life Equation 487
Ceramic Cutting Tool Materials 487
Tungsten Carbide/Cobalt Composites 487
Steel Machining 488
Aluminum Machining 488
Nickel-Based Alloys 488
Titanium Alloys 488
Non-Metallic Composites 488
Aluminum-Oxide-Based Composites 488
Steel Machining 488
Nickel-Based Alloys 489
Non-Metallic Composites 489
Silicon-Nitride-Based Composites 489
Steel Machining 489
Cast Iron Machining 489
Nickel-Based Alloys 489
Combating Degradation 489
Selection of Tool Materials 489
Coatings–The Interim Solution 490

The Future 490
References 490

17. DEGRADATION OF TZP CERAMICS IN HUMID ATMOSPHERES 492
T.T. Lepistö and T.A. Mäntylä
Introduction 492
Zirconia Ceramics 493
Partially Stabilized Zirconia (PSZ) 494
Tetragonal Zirconia Polycrystals (TZP) 495
Fully Stabilized Zirconia (FSZ) 495
Environment-Induced Phase Transformation 495
Effect of Environmental Factors on Phase Transformation 496
Temperature 496
Environment 498
The Effect of Material Parameters on Phase Transformation 500
Stabilizing Agent and Stabilizer Content 500
Grain Size 502
Density 502
Grain Boundary Phase and the Shape of the Grains 503
Effect of Aging on Material Properties 504
Strength 504
Fracture Toughness 505
Wear Resistance 506
Electrical Conductivity 506
Proposed Methods to Avoid Phase Transformation 507
Proposed Mechanism for Degradation 508
Corrosion Mechanism 508
Destabilization Theory 509
Stress-Induced Transformation 510
Summary 511
References 512

18. HIGH TEMPERATURE CORROSION OF ENGINEERING CERAMICS 514
Nathan S. Jacobson, James L. Smialek, and Dennis S. Fox
Introduction 514
Test Methods 516
Thermodynamics of Corrosion 517
Kinetics of Corrosion 525
Microstructural and Mechanical Property Effects 529
Composites 540

Protective Coatings 540
Summary and Conclusions 543
References 543

SECTION IV
CORROSION OF CERAMIC SUPERCONDUCTORS

19. CORROSION OF CERAMIC SUPERCONDUCTORS: AN OVERVIEW 548
Aaron Barkatt, Hamid Hojaji and Karen A. Michael
Introduction 548
Weathering in Moisture-Containing Environments 551
Decomposition upon Contact with Liquid Water 555
Inherent Mechanism of Attack by Water: Ionic Mobility and Matrix Dissolution 557
Effects of Starting Materials 558
Acid/Base Aspects of the Interaction of Ceramic Superconductors with Aqueous Environments 559
Oxidation/Reduction Aspects of the Interaction of Ceramic Superconductors with Aqueous Environments . . 561
Effects of Presence of Moisture During Synthesis 566
Interaction of Superconducting Oxides with Nonaqueous Solvents 567
Effects of Microstructure: Porosity and Grain Boundaries 571
Protection Against Environmental Attack 573
Summary 574
References 575

20. ROLE OF PROCESSING IN THE CORROSION OF CERAMIC SUPERCONDUCTORS 583
Gregory T. Chandler
Introduction 583
Reactivity with Water 584
Degradation Reactions 584
Effect on Superconducting Properties 585
Effects of Other Solvents 586
Sample Preparation and Characterization 587
Effect of Particle Size on the Stability of $YBa_2Cu_3O_{7-x}$ in a Humid Environment 588
Density and Weight Changes 588
Change in Microstructure 593
Change in Superconducting Properties 593
Effect of Sintering Temperature on the Stability

of $YBa_2Cu_3O_{7-x}$ in a Humid Environment 597
Proposed Model for the Role of Processing in the Corrosion of Ceramic Superconductors 597
References 600

21. CORROSION OF $YBa_2Cu_3O_{7-x}$ IN HIGH AND LOW HUMIDITY ENVIRONMENTS 601
Yasuro Ikuma and Masahiro Yoshimura
Introduction 601
$YBa_2Cu_3O_{7-x}$ in Liquid Water 602
Reaction of $YBa_2Cu_3O_{7-x}$ in Saturated or High Humidity 603
Reaction of $YBa_2Cu_3O_{7-x}$ in Low Humidity 606
Summary 611
References 612

22. SURFACE SCIENCE TECHNIQUES FOR ANALYSIS OF CORROSION OF THE CERAMIC SUPERCONDUCTORS ... 615
Temel H. Büyüklimanli and Joseph H. Simmons
Introduction 615
X-Ray Photoelectron Spectroscopy 615
Application of XPS to 123 Compounds 616
Measurement of Surface Contamination Using XPS 617
Summary and Conclusions 628
References 629

23. REDUCING CORROSION OF CERAMIC SUPERCONDUCTORS WITH SOL-GEL COATINGS 632
S.S. Bayya and V.R.W. Amarakoon
Introduction 632
Moisture Degradation 633
Methods of Surface Passivation 634
Sol-Gel Coating of Powders 635
Experimental Procedure 637
Characterization 639
Results and Discussion 639
Conclusions 645
References 646

24. SUMMARY 648
D.E. Clark, R.L. Schulz, and B.K. Zoitos
Introduction 648
Zeta Potential and Its Role in Glass Corrosion 650
Background 650

Measurement Techniques 650
Microelectrophoresis 650
Streaming Potential 650
Electrokinetic Sonic Amplitude (ESA) 651
Glass Corrosion 652
Surface Passivation 655
Stress Corrosion 659
Glass and Ceramic Electrodes 660
Electrochemical Corrosion 662
References 663

INDEX 666

Section I.

Theory, Testing and Characterization

1

Theory of Corrosion of Glass and Ceramics

William B. White

Materials Research Laboratory and
Department of Geosciences
The Pennsylvania State University
University Park, PA 16802

INTRODUCTION

The glasses and ceramics discussed in this chapter are limited to oxide, fluoride, silicate, borate, phosphate, and related compositions that make up common ceramic materials. These materials are electrically insulating and contain few free carriers, so that chemical attack on their surfaces is mainly by acid-base type of reactions rather than electrochemical corrosion involving redox reactions. Crystalline ceramics and non-crystalline glasses of similar composition often have quite different corrosion behavior showing that structure at the atomic scale also plays a role.

The subject is addressed from a generic point of view, with specific examples intended only as illustrations. The object is to set down some of the most important reactions and their controlling parameters.

The literature comes from diverse sources written by researchers with very diverse motivations. Interest in glass corrosion comes from the necessity of understanding stability of glass containers, sheet glass, and other glasses in the presence of aqueous solutions (1). In general, the solutions are highly undersaturated, may be strongly acidic or basic, and the concern is with times in the range of a few years. In contrast, the literature on the dissolution (corrosion) of crystalline ceramics is relatively sparse because in the environments in which most of these materials are used, attack by aqueous solutions is not a problem. To geochemists, dissolution of crystalline minerals and rocks is the central process of rock weathering, diagenesis of sedimentary rocks, uptake of ions into groundwater from aquifer wall rock, and reactions of geothermal waters (2).

Here the fluids are often near saturation and the time spans of interest range from days to tens of thousands of years.

Much of the recent literature on glass corrosion comes from the nuclear waste research community (see Wick's chapter). Glasses designed for the immobilization of high level nuclear waste are chemically complex but their stability in a variety of fluids, over a temperature range of more than 100 degrees Celsius and, over time scales of thousands of years, is central to hazard assessment models for nuclear waste repositories (3). Finally, art history and archeology may be added to the list of disciplines where rate of weathering of glassy or crystalline objects is important either from the point of view of preservation of historically or aesthetically important artifacts or from the point of view of dating them (4) (see Vandiver's chapter).

CHEMICAL MECHANISMS OF DISSOLUTION AND CORROSION

Some ceramics are monophasic; that is they are made up of a single crystalline phase. Alumina or periclase refractories, and hot-pressed spinel or magnesium fluoride optical ceramics are examples. Likewise, most glass can be considered to be a monophasic ceramic. From a point of view of dissolution or corrosion behavior, a monophasic ceramic behaves as a single chemical compound although glass usually does not. Other ceramics are polyphasic; they are made up of more than one crystalline phase. Examples are the ceramics that have been proposed as nuclear waste immobilization materials, for example the Synroc family of titanium-rich ceramics. Still other ceramics are made up of crystalline phases plus glass (see McCracken's chapter).

To a first approximation the dissolution of the individual crystalline phases can be treated independently. This means that one is dealing with the dissolution mechanisms and kinetics of single chemical substances which is a great simplification. The bulk rate, R_{net}, of dissolution would then be the sum of the dissolution rates of the individual phases, R_i, normalized by the weighted surface areas, SA_i.

$$R_{net} = 1/SA\ \Sigma SA_i R_i \qquad (1\text{-}1)$$

If the dissolution of the bulk ceramic is examined in more detail, however, various second order effects appear:

(a) Dominant grains may buffer the reacting solutions and so modify the dissolution behavior of minority grains.

(b) There may be solvent-induced chemical reactions between phases.

(c) Some phases in the ceramic or their reaction products may serve to encapsulate other phases.

Microencapsulant ceramics pose a different sort of problem. These materials utilize a low solubility inert phase to surround and protect other more soluble phases. The bulk dissolution rate of such materials is determined by transport of fluid through the available interconnected pore spaces.

In the most general terms, there are several chemical mechanisms for the corrosion of crystalline and glassy materials:

(a) Congruent dissolution by simple dissociation. "Congruent" implies that the ratios of constituent elements in solution is the same as that in the dissolving solid. Take as an example the hot-pressed monophasic ceramic MgF_2 used as an optical window material in the visible and near-infrared regions. MgF_2 dissolves in water by a simple dissociation reaction:

$$MgF_2 \Leftrightarrow Mg^{+2} + 2F^- \tag{1-2}$$

This reaction is independent of pH and at equilibrium reaches a saturation limit which can be calculated from the free energies of formation of the chemical species involved. The surface of the crystalline grains retreats (not necessarily uniformly) as the crystals dissolve. No protective layers are formed and a concentration profile across the crystal surface is a single stair-step (Figure 1-1A).

(b) Congruent dissolution by chemical reaction with the solvent. Dissolution is by acid-base or hydrolysis reaction. No solid reaction products are formed and the surface retreats without formation of surface layers. An example would be the dissolution of high temperature MgO refractories in acidic solutions:

$$MgO + 2H^+ \Leftrightarrow Mg^{2+} + H_2O \tag{1-3}$$

Here, however, MgO is unstable in the presence of liquid water so there is a competing reaction:

$$MgO + H_2O \Leftrightarrow Mg(OH)_2 \tag{1-4}$$

which in turn dissolves in acidic solutions:

$$Mg(OH)_2 + 2H^+ \Leftrightarrow Mg^{2+} + 2H_2O \tag{1-5}$$

At equilibrium, all of these reactions must be satisfied simultaneously. Direct attack by hydrogen ions is often dominant in low pH regimes while hydrolysis by direct reaction with water is more common in near neutral regimes. Not only reaction rates, but reaction mechanisms may be pH dependent. The dissolving surface, however, retreats uniformly again with a stair-step concentration profile as in Figure 1-1A.

What is immediately apparent in even these simple examples is that there is a complicated solution chemistry, often made more complicated by speciation and complexation reactions within the solution. These interactions can be dealt

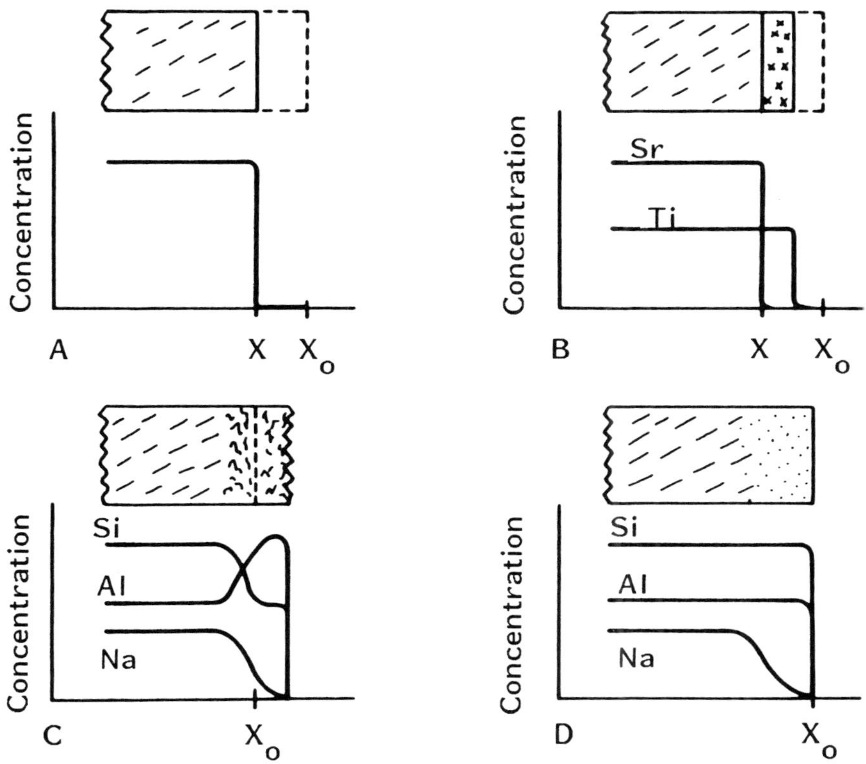

Figure 1-1. Sketches showing dissolution mechanisms for ceramics and glasses. A) Congruent dissolution with uniform retreat of dissolving interface. B) Incongruent dissolution with reaction product remaining in place. C) Corrosion of surface by chemical reaction between solvent and solid. D) Leaching of sodium aluminosilicate glass.

with most efficiently by computer programs such as WATEQF, MINEQL, or EQ3 which address the speciation and saturation of aqueous solutions. For the dissolution process itself, reaction path programs such as PHREEQE or EQ6 are helpful. See Nordstrom et al. (5) for a critical comparison.

(c) Incongruent dissolution with formation of crystalline reaction products. During incongruent dissolution there is a reaction of the original bulk solid with the coexisting liquid to form a new solid phase of different composition. The concentration of dissolved species do not occur in the same ratio as in the parent solid. In all cases there is a less soluble reaction product which may or may not accumulate on the dissolving surface to form a protective (to some extent) barrier layer. An example of incongruent dissolution is:

$$SrTiO_3 + 2H^+ \Leftrightarrow Sr^{2+} + TiO_2 \qquad (1\text{-}6)$$

Well crystallized strontium titanate reacts with aqueous solution to release strontium ions into solution but the titanium remains behind as the highly insoluble TiO_2. The retreating interface is often sharp. If the reaction products remain behind, the Sr concentration drops abruptly to zero at the reaction interface but the Ti concentration remains high out to the original surface (x_o in Figure 1-1B). This can be an equilibrium reaction with the strontium concentration at equilibrium being defined by the pH of the solution.

(d) Incongruent dissolution with formation of non-crystalline layers. A different type of reaction occurs when the original bulk phase is a framework structure such as an aluminosilicate. Although the reaction product has a different chemical composition than the original material, structural entities such as partially hydrolysed, polymerized silica and alumina tetrahedra remain. This product is often non-crystalline and has been called the "gel layer." The chemically modified surface may remain intact, even advance rather than retreat because of the lower bulk density of the gel layer. Sodium aluminosilicate is a prototype example.

$$\underset{\text{(Bulk)}}{NaAlSi_3O_8} + H^+ \Leftrightarrow Na^+ + \underset{\text{(Surface)}}{Al_2Si_2O_5(OH)_4} + H_4SiO_4 \qquad (1\text{-}7)$$

Although the reaction product is written with the kaolinite composition, the actual material on the dissolving surface may not be crystalline. There is a pronounced change in chemical composition through the "gel" layer in addition to density and textural changes (Figure 1-1C).

(e) Ion Exchange. Ion exchange reactions are those in which the mobile ions are leached from a more resistant matrix while leaving the matrix material more or less intact. Again sodium aluminosilicate can serve as an example.

$$\underset{\text{(Bulk)}}{NaAlSi_3O_8} + H_3O^+ \Leftrightarrow \underset{\text{(Surface)}}{(H_3O)AlSi_3O_8} + Na^+ \qquad (1\text{-}8)$$

The original interface is preserved (more or less) with the same Si/Al ratio. There is a continuous loss of Na through the leached layer.

There is no sharp distinction between breakup of the dissolving surface with concurrent formation of a gel layer and ion exchange reactions. Likewise there is usually some congruent dissolution as an additional background reaction. For albite this would be:

$$NaAlSi_3O_8 + 4H_3O^+ \Leftrightarrow Na^+ + Al^{3+} + 3H_4SiO_4 \qquad (1\text{-}9)$$

The net corrosion rate is a sum of rates due to the various chemical reactions and the various reaction mechanisms.

A key consideration is whether or not the solid phase can be in equilibrium with liquid water. If it is, dissolution in closed systems will approach an equilibrium saturation value and fresh solution must be introduced to keep the dissolution reaction going. If the solid is not stable in the presence of water, the hydrolysis reaction will continue until either the liquid is used up (for small quantities of solution in contact with large pieces of solid) or until the solid phase is entirely broken down (if large volumes of liquid are available). There is a further complication when the material is a glass. Glasses are everywhere metastable with respect to an assemblage of crystals with the same bulk composition. In competition with incongruent dissolution and ion exchange reactions is the recrystallization of the host material.

KINETICS

Rate Equations for Crystalline Ceramics

The uptake of ions in solution as a crystalline material dissolves often follows a reaction curve of the sort shown schematically in Figure 1-2A. If the crystalline material can exist in equilibrium with an aqueous solution, there is some concentration of each ionic species for which the solution is saturated. Assuming no material initially in solution, at time zero the crystal dissolves very rapidly. As dissolution proceeds the concentration of dissolved species builds up, the solution rate slows down and gradually approaches zero as the concentration of dissolved species approach their equilibrium concentrations. There is, therefore, a characteristic reaction time necessary for the system to approach some arbitrarily defined fraction of the saturation concentration.

Models for the dissolution of crystals and ceramics took as a starting point the traditional theory of reaction kinetics (6). Heterogeneous reactions occur at the interface between the dissolving crystals and the aqueous solution. The steps involved in any such reaction are:

(a) Transport of reactants to the solid surface.
(b) Adsorption of the reactants.
(c) Chemical reaction between an aggressive adsorbed species and some point of attack on the crystal surface. The detailed atomistic chemistry can be quite complex.
(d) Desorption of the reaction products.
(e) Transport of products into the bulk solution.

For a chain of individual reaction steps, the slowest step limits the overall rate. In general, control has been ascribed to either transport (steps a and e) or to surface reaction kinetics (step c).

In a transport-controlled reaction, the surface chemical kinetics are fast so that migration of ions through the fluid boundary layer becomes rate-limiting. In classical Nernst theory, the change in concentration is simply the flux, j, of

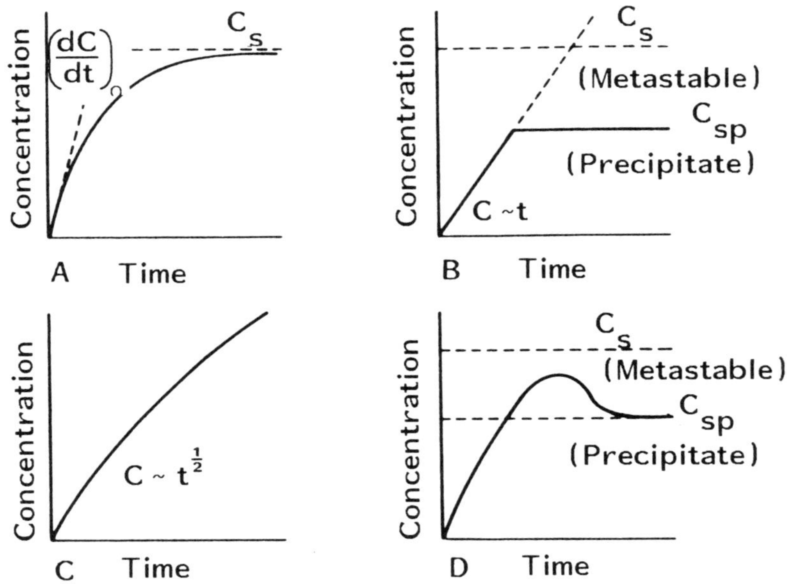

Figure 1-2. Rate curves for uptake of ions in solution according to various dissolution mechanisms. A) Dissolution when solid can come into equilibrium with liquid at saturation concentration, C_s. B) Linear kinetics truncated by saturation with respect to a precipitate phase with rapid precipitation kinetics. C) Diffusion-controlled kinetics with a parabolic rate law. D) Dissolution with precipitation of a product phase with sluggish precipitation kinetics.

material across the boundary layer which in turn is controlled by the layer thickness, δ, and the diffusivity, D, of such ions in the liquid. The driving force is supplied by the concentration gradient:

$$dC/dt = (SA/V)j = (D \bullet SA/\delta V)(C_s - C) = k_T(C_s - C) \qquad (1\text{-}10)$$

where k_T is a mass transfer coefficient. The integrated form of the equation is:

$$C(t) = C_s(1 - e^{K_T t}) \qquad (1\text{-}11)$$

Empirically, it is found that dissolution of substances with solubilities greater than 10^{-4} moles/liter is usually mass transport limited while dissolution is surface reaction controlled for substances with solubilities less than 10^{-5} mole/liter. Most crystalline ceramics and glasses of practical utility are quite insoluble so that dissolution is rarely limited by transport through the fluid boundary layer. Single crystal and polycrystal alkali halide optical materials and some water soluble detector (e.g. triglycine sulfate) and transducer (e.g. potassium dihydrogen phosphate) materials are obvious exceptions.

Providing that the dissolving solid will eventually come into equilibrium (saturation) with the solution whether by congruent or incongruent dissolution, the bulk rate equation is written:

$$dC/dt = (SA/V)k_s(C_s - C)^n \qquad (1\text{-}12)$$

The exponent, n, is the order of the reaction and k_s is a reaction rate constant. An alternate expression used for ionic solids is:

$$Rate = (V/SA)(dC/dt) = k(1 - \Omega)^n \qquad (1\text{-}13)$$

where Ω is the saturation ratio defined as:

Ω = (ion activity product)/(solubility product constant)

If n = 1, first order kinetics, equation 1-12 integrates to:

$$C(t) = (SA/V)C_s(1 - e^{-k_s t}) \qquad (1\text{-}14)$$

the same form as obtained from transport theory but with a much different definition of k_s. The rate constants in this formulation have units of 1/time and can be replaced by a characteristic reaction time ($\tau = 1/k_s$) which is a measure of the response time of the system. First order reactions are found for those systems where one of the concentration variables is dominant and where there is a one to one atomic reaction at the surface, for example the exchange of an alkali ion by a hydrogen ion. Higher order reactions are often observed. The competing rates of both forward and back reactions are incorporated in equations 1-12 and 1-13. An initial rapid rate, expressing mainly the forward reaction, decreases so that the concentration-time function curves over to become asymptotic to the horizontal line marking the saturation concentration.

Many ceramics, glasses, and minerals cannot exist in equilibrium with aqueous solution under ambient temperatures and pressures. For these substances, the back reaction is not usually observed. The trend toward metastable equilibrium is cut off by the precipitation of new, lower solubility, product phases. The product phases are sometimes hydrated and these can be regarded as the reaction products of the original phase with water. The system is then described by two reactions and two rate equations. The forward reaction is the dissolution of the primary phase. Instead of a back reaction, there is a precipitation reaction which, because it removes ions from solution, is often treated as a back reaction.

When dissolution of the metastable phases takes place far from equilibrium, the forward reaction rate is little influenced by the build up of reaction species in solution and obeys zero-order kinetics. That is:

$$dC/dt = (SA/V)k_+ \tag{1-15}$$

This equation leads to a linear concentration - time curve as shown in Figure 1-2B. Obviously, this equation describes only part of the process because the concentration increases without limit at large times, a physical impossibility. The precipitation reactions can be taken separately so that the net concentration of ions in solution is given by:

$$dC/dt = (SA/V)k_+ - (SA'/V)k_-(C - C_s) \tag{1-16}$$

where SA′ is the effective surface area of the precipitates, k_+ is the forward reaction rate constant and, k_- is the precipitation rate constant. (C - Cs) is the supersaturation with respect to the precipitate phases. If the precipitation kinetics are fast relative to dissolution kinetics, the concentration curve is cut off abruptly as shown in Figure 1-2B.

The dissolution rates of many materials varies with pH. Hydrogen ion activity can be factored specifically into the rate equations:

$$dC/dt = (SA/V)k_+'a_{H+}^n - (SA'/V)k_-(C - C_s) \tag{1-17}$$

The exponent, n, is unity for most crystalline materials in the acid regime. In alkaline solutions, n varies from -0.2 to -0.6.

There is a very large literature on the dissolution kinetics of insoluble ionic salts such as carbonates, sulfates, and phosphates. Most of these materials can exist in equilibrium with aqueous solutions and approach saturation following a rate equation that is some variant of equation 1-12 as the reaction proceeds under closed system conditions.

The dissolution of oxides often follows equation 1-17. In the example of Fe_2NiO_4 spinel (Figure 1-3), dissolution is congruent and the rate is linear in strongly acidic solutions. As the pH increases, the rate decreases as given by equation 1-17 and in addition a precipitation reaction of iron hydroxides appears. What appears in the solution chemistry is that the Ni^{2+} concentration continues to increase linearly but at a slower rate while the concentration of iron in solution increases even more slowly because of the precipitation reaction, giving the impression of non-stoichiometric dissolution.

The dissolution of anhydrous silicate minerals such as olivines, pyroxenes, quartz, and feldspars, particularly, have been examined in detail by the geochemical community because of their importance in the weathering of igneous and metamorphic rocks. There is a large amount of literature on the dissolution of feldspars, for example.

Rimstidt and Barnes (8) showed that the dissolution kinetics of quartz follow a variant of equation 1-17.

$$\left(\frac{\delta a_{Si}}{\delta t}\right)_{P,T,M} = \frac{SA}{M\gamma_{Si}}\left[k_+ a_{SiO_2} a_{H_2O}^2 - k_- a_{Si}\right] \tag{1-18}$$

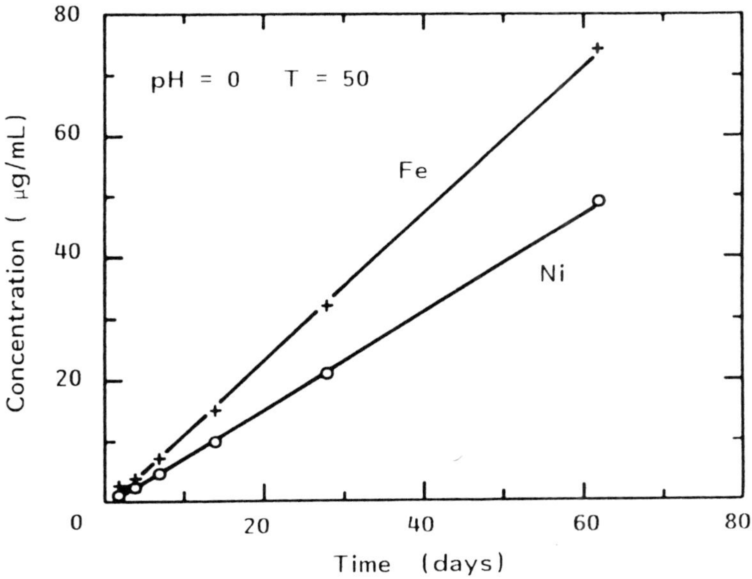

Figure 1-3. Dissolution of nickel ferrite in strongly acidic solution. Data from Nachlas (7).

M is the mass of solvent, Si is an abbreviation for H_4SiO_4, γ is the activity coefficient for H_4SiO_4 (near unity because H_4SiO_4 is neutral and dilute), a_{SiO2} is the activity of silica (= 1 for pure quartz), and a_{Si} is the activity of dissolved silica. The rate constants are functions of temperature but the system is independent of pH. Later investigators have shown that dislocations emerging from the crystal surface accelerate the dissolution rate (9). It was also found that the reaction is pH independent only below pH = 6; above this value, the rate increases as $a_{H+}^{-0.5}$ (10).

Lasaga (6) has reviewed the older data. In general, silicates dissolve by a surface reaction mechanism and often show linear dissolution with strong pH dependence of the sort described by equation 1-17. Lasaga has made an intercomparison both by the rates themselves and by considering the length of time required for a one millimeter cube of the crystal to completely dissolve (Table 1-1).

When the dissolution kinetics of silicate minerals were first examined, the concentration-time curves appeared to be paraboloid and it was assumed that the rate-controlling feature was a barrier layer on the surface, analogous to the layers that had been observed on glasses. However, when later workers looked for the surface layers with increasingly sophisticated characterization methods, the layers were not found. The initial rapid rate appears to be due to dissolution of small

Table 1-1. Rate of Release of Silica at 25°C and pH = 5

Mineral	Composition	Rate (moles $cm^{-2}day^{-1}$)	Lifetime* (years)
Quartz	SiO_2	8.64×10^{-13}	34×10^6
Muscovite	$KAl_3Si_3O_{10}(OH)_2$	2.56×10^{-13}	2.7×10^6
Forsterite	Mg_2SiO_4	1.2×10^{-12}	0.6×10^6
K-Feldspar	$KAlSi_3O_8$	1.67×10^{-12}	0.52×10^6
Albite	$NaAlSi_3O_8$	1.19×10^{-11}	80,000
Enstatite	$MgSiO_3$	1.0×10^{-10}	8,800
Diopside	$CaMgSi_2O_6$	1.4×10^{-10}	6,800
Nepheline	$NaAlSiO_4$	2.8×10^{-9}	211
Anorthite	$CaAl_2Si_2O_8$	5.6×10^{-9}	112

* Mean time required for the complete dissolution of a 1-mm crystal. All data from Lasaga (11).

particles attached to the surface of mineral grains and solution attack on high energy sites on the crystal surface such as the emergence points of dislocations, grain boundaries, and cleavage cracks (12). When these are consumed, the rate slows down to that of the bulk material but the overall rate curve gives a misleading impression of parabolic kinetics. When the effect of changing surface activity is taken into account, most silicate minerals so far examined dissolve by a linear kinetics mechanism.

It has been generally agreed that most of the mineral dissolution reactions are acid-base reactions and that at the heart of the matter is the reaction of a proton, hydronium ion, or water molecule with the cations on the surface of the crystal. Dissolution rates are highly sensitive to the presence of active sites for such attack. Stumm (13) and his colleagues have proposed an atomistic model in which the rate of dissolution is given by:

$$\text{Rate (moles } m^{-2}sec^{-1}) = kX_aP_jS \qquad (1\text{-}19)$$

where k is a rate constant in units of sec^{-1}. X_a represents the mole fraction of dissolution active sites. P_j is the probability of finding a specific site in the coordination arrangement of the precursor complex, and S is the surface concentration of sites in units of moles m^{-2}. The surface complexes - chelates or metal-proton complexes - are taken as the intermediate steps in the rate-limiting detachment of cations from the surface into solution. Application of the model leads to macroscopic rate equations of the form given above. Computation of the parameters of the theory from first principles or determination of surface concentrations by direct measurement are not possible at present.

Rate Equations for Glasses

The analysis of the corrosion and dissolution behavior of glasses had its roots in the container industry. Container glasses must be optimized for low rates of extraction of the component elements by whatever liquid is placed in the container. Early theory was cast almost entirely in terms of leach rate. There is a preferential dissolution of more soluble ions from multicomponent glasses by a process of ion exchange controlled by the diffusion of H^+ (or H_3O^+) into the glass coupled with the diffusion of leachable ions out of the glass.

Glass leaching is usually described in terms of Fick's laws in one dimension:

$$dC/dt = d/dx(D\ dC/dx) \tag{1-20}$$

where D is the diffusivity of the leachable ion. For a planar interface with a constant diffusivity, the solution of equation 1-20 leads to the observed parabolic rate law, with rate constant, k_p, which is often given in a simple empirical form:

$$\text{Rate (moles cm}^{-2}\text{)} = k_o + k_p t^{1/2} \tag{1-21}$$

The constant term, k_o, is usually not zero and has been attributed to rapid initial surface exchange between hydrogen (or hydronium) ions and exchangeable ions on or near the fresh glass surface. Like the corresponding models for crystalline silicate dissolution, the above equation describes only the forward reaction rate. The reaction continues without limit and if the supply of liquid is sufficient, the reaction ends only when the entire glass specimen has been leached (Figure 1-2C).

The build-up of ions in solution, however, may exceed the solubility of secondary reaction products which may then precipitate. Because of sluggish precipitation kinetics for silicate and aluminate reaction products, there can be considerable overshoot so that the concentration curve appears as shown in Figure 1-2D.

A good example of the comparative behavior of glasses and crystals of the same composition is provided by $NaAlSi_3O_8$, which occurs naturally as high quality crystals of the mineral albite and which can be made into an aluminosilicate glass. Figure 1-4 shows the comparison in corrosion behavior of the three elements that make up the material. The uptake of Si into solution is identical within experimental error regardless of whether the source is crystal or glass. The parabolic shape to the curve results from continued upward drift of pH rather than from diffusion-controlled kinetics. The removal of Na from crystalline albite is such that the Na/Si ratio is about 1:3 as required from the formula, thus confirming a congruent dissolution of the crystal. Sodium uptake from the glass, however, is much larger than that of silicon suggesting that sodium is being leached from a depth in the glass in addition to its dissolution

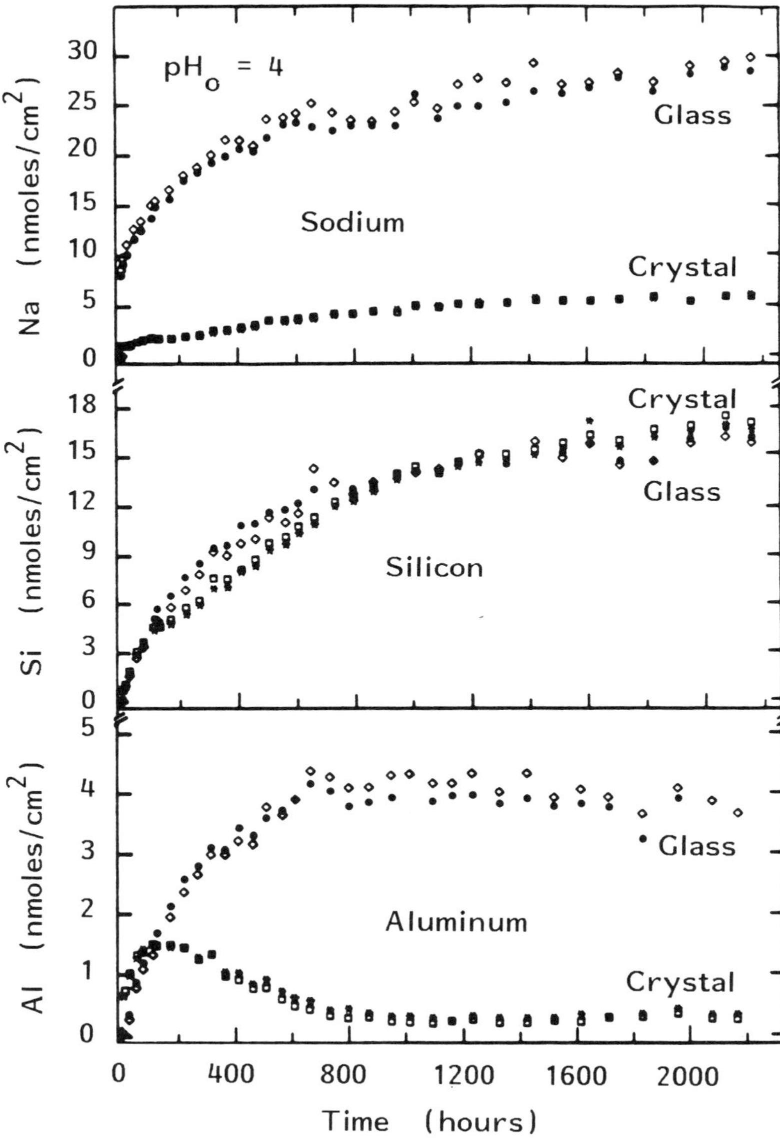

Figure 1-4. Comparison of the uptake of ions by dissolution of crystalline albite, $NaAlSi_3O_8$, and albite composition glass. Solid and open symbols show reproducibility of duplicate runs. Data extracted from Zellmer and White (14).

at the surface. The removal of Al from the glass occurs to a point where the Al/Si ratio is a little less than 1:3, the expected value from the chemical formula. This concentration of aluminum far exceeds the solubility of both aluminum hydroxides and the aluminosilicate layer silicates, so the retention of aluminum in solution must be metastable. However, Al extracted from the crystalline albite rises to a maximum a short time after the commencement of the experiment and then falls back to a low level. The results suggest much more rapid precipitation kinetics for the crystal compared to the glass of the same composition.

The albite example illustrates the continuing and unresolved problem of how to deal with the reaction layers that form during glass dissolution. Preferential extraction of leachable ions forms a leach layer that acts as a diffusion barrier to further extraction. Precipitation of network-forming ions such as silicon, aluminum, and boron not only forms an additional gel layer which may or may not act as a diffusion barrier, it also acts as a substrate for the adsorption and perhaps chromatographic separation of precipitating metal ions. The gel layer is metastable and ultimately recrystallizes. The near surface environment of a dissolving or corroding glass is thus quite complicated as illustrated in Figure 1-5.

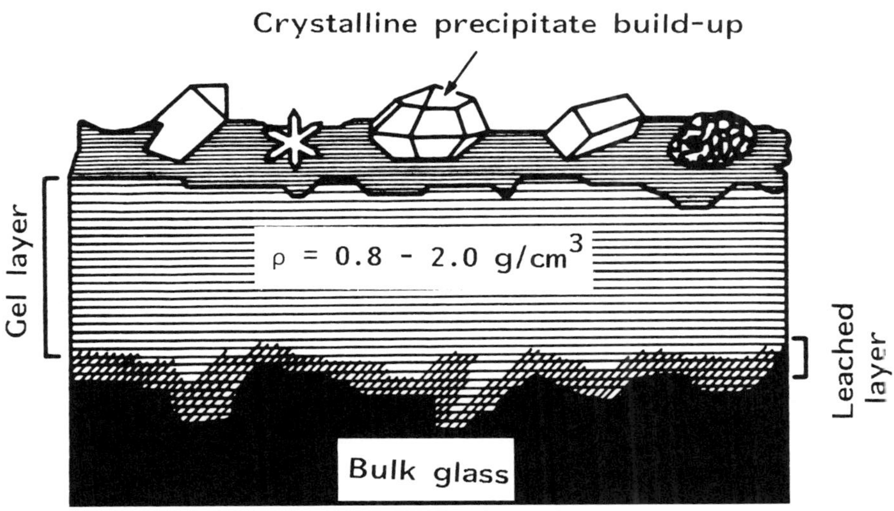

Figure 1-5. Sketch showing a corroded glass surface with ion-depleted leach layer, overlying precipitate or "gel" layer with ultimately recrystallizes into reaction product phases on the surface. From Mendel (15).

Leaving aside the additional complication of the gel layer, the ongoing dissolution process is composed a diffusive part with mobile ions moving outward through the depleted leach layer and the dissolution of the leach layer itself. It has long been recognized (16) that the parabolic kinetics seen at the early stages of glass dissolution shift asymptotically into linear kinetics at long times as would be expected if there was a steady state balance between diffusion and the surface reaction controlled dissolution of the glass surface. Empirically this is expressed by an equation of the form:

$$\text{Rate}(\text{g}\bullet\text{m}^{-2}) = At^{1/2} + Bt \tag{1-22}$$

Equation 1-22 forms the basis for many models of glass dissolution, usually accomplished by deriving an analytical form for the empirical constants A and B (see Eppler's chapter).

For silicate glasses, the cumulative uptake of silica in solution (in absence of a gel layer) is the best indicator of the degree of dissolution of the glass. One description of the process is the Wallace-Wicks equation for the dissolution behavior of nuclear waste glasses (17) (see Wicks' chapter).

$$dC/dt = [kC_s(1 - SA/V\ C_s - C)]/[1 + k\ q/D\ C] \tag{1-23}$$

where C is the concentration of silica in solution, C_s is the saturation concentration of silica, k is a rate constant, q is an empirical constant, and D is the diffusivity of silica through the leached layer.

The recognition seems to have dawned, long overdue, that the build-up of leached layers and gel layers on dissolving silicate glasses are transient phenomena and that ultimately the dissolution of glass is also a surface reaction controlled process. Current work on glass dissolution is often expressed as variants of the Grambow equation (see Grambow's chapter):

$$\text{Rate} = k_+(1 - a_s/K) \tag{1-24}$$

where k_+ is a forward reaction rate constant, a_s is the activity of dissolved silica, and K is the equilibrium constant for the dissolution reaction. The silica activity, a_s, is sometimes replaced by the reaction quotient, Q, in which case equation 1-24 is essentially identical to equation 1-13 that describes the surface-reaction controlled dissolution of crystalline ceramics in terms of a saturation ratio.

Phosphate glasses are generally less resistant to corrosion in aqueous solutions than silicate glasses. Studies of the dissolution of a family of alkali-calcium-phosphate glasses (18) shows that no leached layer is formed. These glasses dissolve uniformly by a surface reaction mechanism.

SURFACES

Characterization of Surface Layers

Dissolution mechanisms and rates for crystalline ceramics and glasses depend quite critically on the initial surface and on the evolution of leached layers, gels layers, and reaction product layers as the dissolution process proceeds. Characterization of these surfaces has brought into play nearly every surface characterization tool known to materials science (19). For those glasses that develop substantial surface layers, characterization by scanning electron microscopy, photoelectron spectroscopy, and diffuse reflectance infrared spectroscopy has proved useful. Depth profiling techniques such as secondary ion mass spectroscopy and sputter-induced optical emission spectroscopy both in conjunction with ion milling allow measurement of diffusion profiles through the glass surface layers (see Lodding's chapter). Success in measuring hydrogen concentrations in the hydrated layers has also been obtained by resonant nuclear reaction methods and by Rutherford backscattering.

It is generally agreed that the rate controlling diffusion step in the leaching of glass in the diffusion of hydrogen ions, hydronium ions, or water molecules into the glass surface. Models for dissolution based on this premise give good agreement with data (20). Comparison of the actual diffusivities of hydrogen-bearing species diffusing in compared with alkali ions diffusing out strongly support the diffusion of hydrogen-bearing species as the rate controlling step (21). Quantitative comparison of H/Na ratios within the diffusion profiles by sputter-induced optical emission spectroscopy (22), by resonant nuclear reaction (23), and by Rutherford backscattering (21) gives values in the range of 1.7 to 3.0, eliminating hydrogen ions but supporting either water molecules or hydronium ions or some combination of the two. Most of these kinds of experiments have been done either in near-neutral or in intermediate pH regimes. There is a cross-over between hydrogen ion as the reactive species at low pH and direct hydration by neutral water molecules in the near-neutral pH regime and it may be that no experiments have yet been done in sufficiently acidic system to observe hydrogen ion diffusion.

For the most part, gel layers, where they form, are not rate controlling. The gel layer is often permeable, patchy, and is sometimes observed to be peeled free of the glass surface. The chemistry within the gel layer is extremely complex as is the morphology of the layers. As indicated by the sketch in Figure 1-5, the gels are often non-planar and may have a complex internal structure with great variation in chemical composition within the layer. These layers occur on glasses with complex bulk composition such as the nuclear waste glasses and geochemical reaction path models seem to be adequate for a description of the reaction products found in the layer (24,25).

Demonstration of the absence of controlling surface layers in the dissolution of crystalline silicates and other ceramic materials is a much more difficult proposition. It has long been assumed that there would be a build up of hydrated layers on feldspars and other silicate crystals analogous to the layers that form on glasses of similar composition. Careful scanning Auger spectroscopic examination of feldspar surfaces reveals a chemical reaction layer only a few nanometers thick (26). Very careful hydrogen depth profiling using resonant nuclear reactions showed a hydration layer on a scale of less than 50 nm on a diopside, $CaMgSi_2O_6$, crystal that was interpreted to be the zone of active surface reaction (27). Such a zone of intermediate hydrated complexes would be consistent with the atomistic model described by equation 1-19.

Role of Surface to Volume Ratios

Dissolution rates are usually expressed in terms of the change in concentration of the solution as a function of time or as the mass of material released from a unit area of the dissolving solid as a function of time. Thus the various rate equations contain a scaling term usually expressed as SA/V, the surface area of the dissolving solid divided by the volume of solution. The relation between the uptake of dissolved ions and the dissolution rate is:

$$dC/dt = (SA/V)\ \mathrm{Rate} \tag{1-25}$$

Although the rate at which a dissolution reaction occurs is a property of the system as discussed in previous sections, the time required for the system to reach saturation or any prescribed fraction of saturation is directly proportional to the SA/V ratio. Expressed somewhat differently, the time for a particular experimental result to be achieved is scaled by the SA/V ratio. One must keep this relationship in mind when comparing results of experiments with quite different SA/V ratios (28) (see Clark/Zoitos chapter).

Dissolution reactions of fine powders with high surface area appear to take place faster than coarse grained or bulk materials although the rate, which by definition is normalized to surface area, may be the same. Likewise, reactions between thin liquid films and bulk solids reach equilibrium faster than the dissolution of the same solid into a large volume of liquid. For this reason, circulating pore fluids in permeable cements, ceramics, or rocks are likely to quickly come to equilibrium.

Several caveats must be observed concerning SA/V ratios. (a) Many laboratory experiments are carried out with modest amounts of solid in large volumes of solution (small SA/V) whereas many applications of the laboratory results are to situations of large surface and modest amounts of liquid (large SA/V). (b) Surface area is notoriously difficult to measure. Many porous materials are now known to be fractal, that is the magnitude of the surface area depends on the scale of the "ruler" with which one measures it. (c) Surface areas may change with time as dissolution proceeds. SA/V should be written SA(t)/V and not merely as a constant in the rate equation.

SYSTEM VARIABLES

Open and Closed Systems: Rates of Transport

In a thermodynamic sense, a closed system is one in which the total mass is fixed; an open system is one in which one or more components may be transported across the system boundaries. Dissolution in closed systems proceeds until the solution reaches saturation or until the source material is completely broken down or dissolved as described by the appropriate rate equation. Open system is used with two meanings in the literature of glass and ceramic dissolution. In one sense, the activities of certain components of the reaction chemistry can be controlled by external reactions. Examples are systems in which pH is buffered or in which the activity of dissolved silica is controlled by the presence of silicate minerals. In the second sense, open systems are those in which there is a continuous movement of liquid through the system. This includes laboratory experiments in which the leach solution is periodically replaced by fresh solution, various flow-through experiments (see Clark/Zoitos chapter), and, in the natural environment, systems with continuously flowing or percolating water.

The characterizing parameter is the transport rate/dissolution rate ratio. If this ratio is large, the dissolving glass or ceramic will be continuously bathed in fresh solution and the dissolution rate will be controlled by the initial slope of the rate curve, $(dC/dt)_o$, as indicated in Figure 1-2A. If the ratio becomes very small, the concentration of ions in solution will be given by the saturation values and the removal of mass from the system will be determined by the flow rate of the solution.

There is a crossover when the ratio is near unity as illustrated in Figure 1-6 for a nuclear waste glass. At high flow rates (short residence times) the glass surface is simply corroded by the highly undersaturated solutions and all elements dissolve at about the same rates. The leach rate becomes maximum and constant. At slower flow rates there is a preferred release of boron and sodium because these elements tend not to accumulate in precipitate phases. Aluminum has the slowest release rates because it tends to reprecipitate in the gel layers. At the lowest flow rates, saturation effects are the main controlling factors. Similar results are found for other nuclear waste glasses and ceramics (29).

The combination of high surface area to volume ratios and slow flow rates in porous media leads to the conclusion that the leaching of concretes and grouts, the evolution of sedimentary rock basins, and perhaps the extraction of radionuclides from high level nuclear waste repositories can be modeled using only saturation limits and models for the fluid flow field. The Pigford model (30) for nuclear waste repositories is based on this premise. Under certain circumstances the rate of extraction of elements from a mass of solid is nearly independent of the detailed chemical kinetics of the dissolution process.

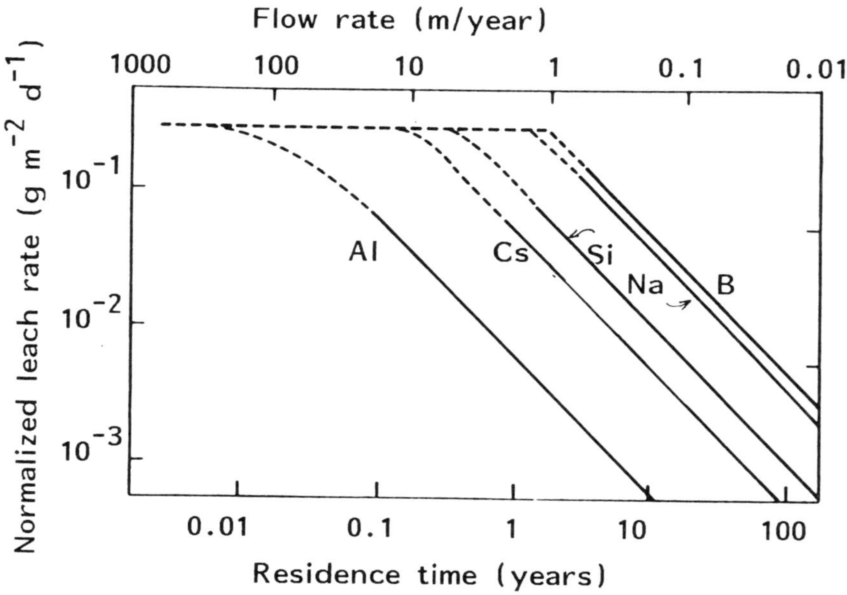

Figure 1-6. Normalized leach rates for elements extracted from a nuclear waste glass scaled to flow rate and residence time. Catholic University data drawn from Mendel's report (15).

pH as an Open System Variable

Dissolution experiments can often be broadly classified into "free drift" and "pH stat" types. In a free drift experiment, usually a closed system, the surface reaction of dissolving crystals or glass consumes hydrogen ions leading to a gradual increase in pH. Because the dissolution process is driven by the hydrogen ion activity, the concomitant loss of hydrogen ions causes a continuous decrease in dissolution rate in addition to changes in rate caused by buildup of leach layers or the onset of back reactions due to saturation effects. For certain glasses reacted with water, the pH can drift into a strongly alkaline regime which promotes breakdown of the silica framework and the rate of dissolution again increases. In pH stat experiments, either pH is buffered to maintain a constant value, or an automatic titrating system adds acid to the system to maintain pH at a constant value. pH stat experiments allow more accurate determination of rate constants; free drift experiments are a better representation of changing mechanisms and competing rate processes.

Eh as an Open System Variable

Most of the literature on dissolution kinetics concerns ceramics and glasses composed of silicates, phosphates, and other compounds of alkali and alkaline earth elements where redox potentials are not relevant. Although Eh (or pe) plays the same role in redox reactions that pH does in acid-base reactions, unlike the H^+ ion, there is no free electron in solution. Thus most solutions are poor Eh buffers and Eh is much more treacherous as a system variable (31) (see Jantzen's chapter). Eh does play a critical role in the dissolution of ceramics or glasses containing variable valence ions. The solubility of the ions depends strongly on their valence state. In principle, rate equations for dissolution reactions of variable valence materials should contain an electron activity term analogous to the hydrogen ion activity term in equation 1-17; however, few examples are available in the literature. Iron is a case in point. Compounds of Fe^{2+} are often relatively soluble whereas Fe^{3+} forms insoluble hydroxide precipitates. Figure 1-7 shows the dissolution of Fe_2NiO_4 in acid solution with the solution exposed to the atmosphere with its high oxygen pressure (labeled "oxidizing") compared with the same solution in the presence of hydrogen gas with a palladium catalyst used to expedite the reaction of hydrogen with the solution. The reducing conditions maintain the dissolved iron in the ferrous state and prevent build-up of a ferric hydroxide layer. The dissolution reaction, as determined from the metal ratios, remains congruent.

Redox reactions are of greatest importance in the dissolution of metal-containing waste forms. As a rule of thumb (to which iron and manganese are exceptions) the lower valence states tend to form insoluble oxides while the higher valence states are more soluble. Uranium dioxide used in reactor fuel pellets is highly insoluble but in oxidizing environments uranium oxidizes to U^{6+} which in the form of the uranyl, UO_2^{2+}, ion becomes moderately soluble (33). ^{99}Tc in its tetravalent state forms low solubility oxide compounds but readily oxidizes to the highly soluble Tc^{7+}.

EFFECTS OF TEMPERATURE

The Arrhenius Rate Law

The rate of most dissolution reactions increases exponentially with temperature obeying the Arrhenius equation:

$$\text{Rate} = A_o e^{-E_a/RT} \qquad (1\text{-}26)$$

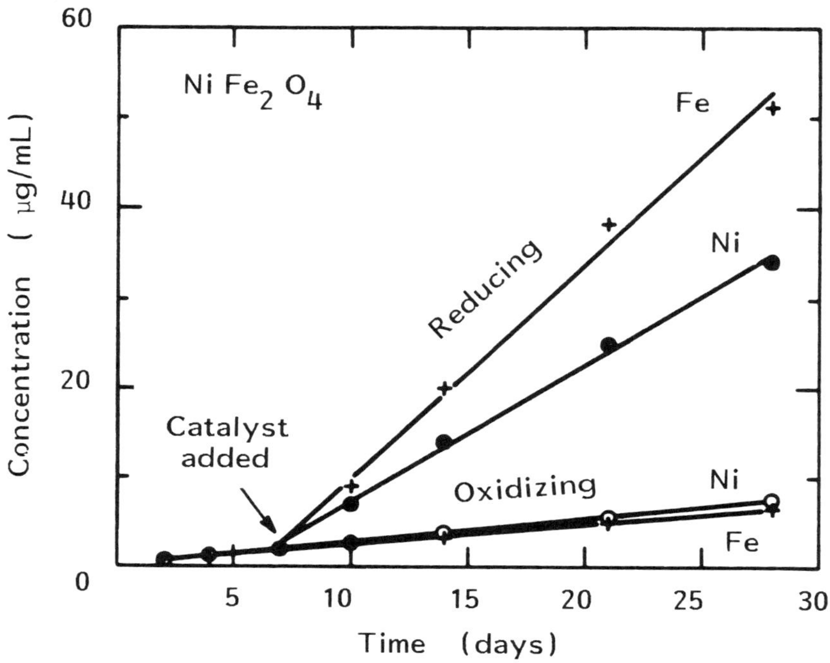

Figure 1-7. Dissolution of Fe_2NiO_4 spinel in acid solution (pH = 1, T = 70°C) under oxidizing (ambient atmosphere) and reducing (hydrogen gas at one atm.) conditions. From reference (32).

The pre-exponential factor, A_o, has a complex theoretical form and is usually taken as an empirical factor. The activation energy E_a is taken as a measure of the barrier height that must be surpassed for reaction to occur.

Diffusion processes provide low energy barriers and E_a takes on values in the range of a few kJ per mole. Surface reactions have a more substantial barrier, typically on the order of 80 kJ/mole, whereas recrystallization processes have activation energies of several hundred kJ/mole. Experimental values for E_a can be obtained by examining the dependence of rate, rate constant, diffusivity, or uptake of ions in solution as a function of temperature (see Clark/Zoitos chapter). These values can be used as evidence for the dissolution mechanism.

Because values for the activation energy of reaction limited dissolution tends to be about the same independent of the chemistry of dissolution, the dependence of dissolution rate on temperature appears to be very similar for many diverse types of reactions. Figure 1-8 shows some data on the release of ions into solution for a nuclear waste ceramic. The elements extracted include an alkali ion (Cs), an alkaline earth (Sr) and a rare earth (La). These elements are partitioned between several crystalline phases in the ceramic. In spite of these complications, the Arrhenius plots are linear with roughly the same slope.

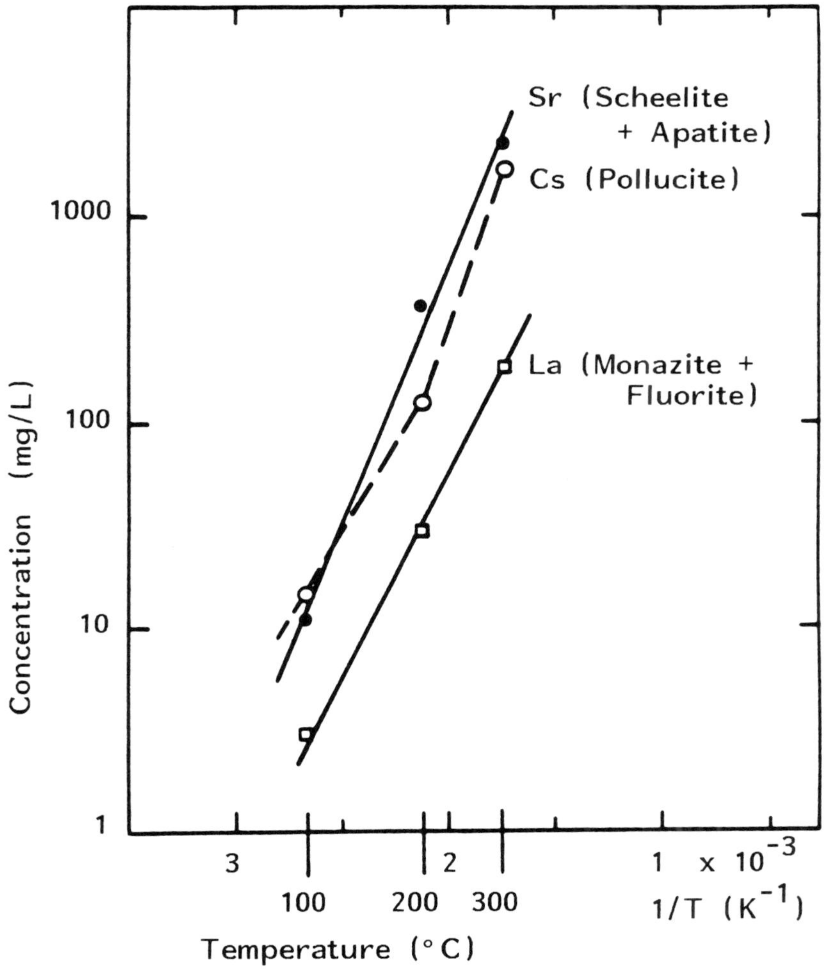

Figure 1-8. Dissolution of a nuclear waste ceramic as measured by the concentration in solution after 4 weeks at a pressure of 35MPa at the temperatures plotted. The ceramic is polyphasic with the source for each element given in parenthesis on the plot.

Figure 1-9 compares the temperature dependence of the dissolution of silica from natural tektite glasses with the temperature dependence of the dissolution rate of silica glass and crystalline quartz based on the rate equations of Rimstidt and Barnes. Again the Arrhenius plots are linear with roughly the same slope. The intercepts, of course, vary widely but the similar slopes reflect the similar values of the activation energies for all sorts of acid-base reactions.

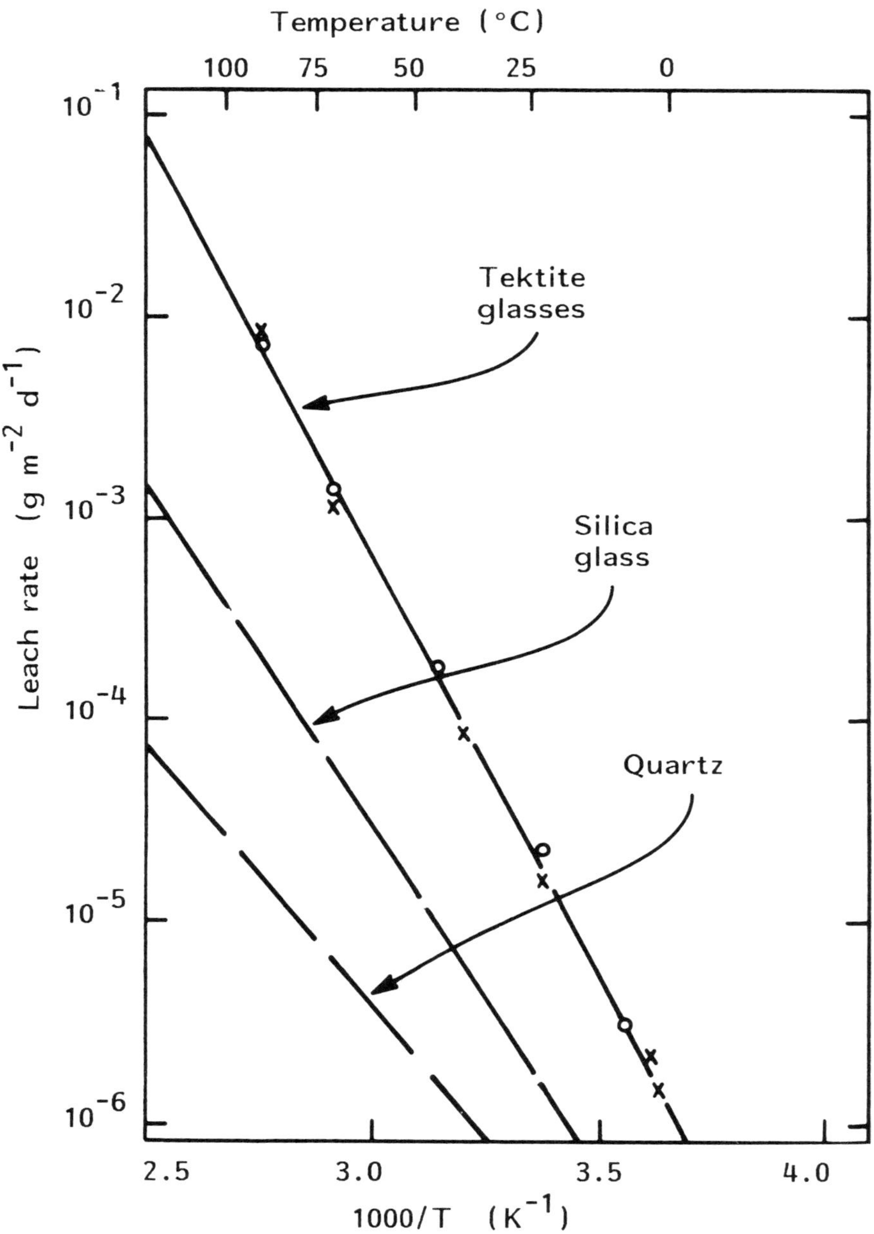

Figure 1-9. Rate of silica loss from crystalline quartz, silica glass, and from tektite glasses as a function of temperature. Quartz data and silica glass data calculated from equations of Rimstidt and Barnes (8) based on initial rates (no silica in solution). Tektite data are from Barkatt et al. (34). Crosses refer to indochinites; circles to australites.

Dissolution in Supercritical Fluids

The vapor pressure of water exceeds one atmosphere at 100°C and although the liquid-solid reactions can be followed to temperatures above 100°C, a pressurized system is required to maintain the liquid. The critical point of water is at 27.5 MPa and 373°C. Above the critical point of water, the distinction between the liquid and the vapor phase disappears.

Dissolution curves of crystalline ceramic materials usually continue through the critical point without a break. However, P-T-X phase diagrams for the system must be taken into consideration. Changing P and T can drive the system into new phase regions and then entirely new dissolution reactions and corresponding new kinetics regimes appear.

For glasses and other metastable phases, however, reactions in hydrothermal regions may be quite different. One major new factor that occurs is the change in mechanism of solid-solid recrystallization reactions. Recrystallization in dry systems or at low temperatures is controlled by solid state diffusion. Rates are low and crystallization often occurs only on geologic time scales making glass indefinitely "stable" on the human time scale. In hot liquids and more so in supercritical fluids, the crystallization mechanism shifts to one of dissolution and recrystallization. There is a great increase in recrystallization rate which typically has an activation energy of 300 to 400 kJ/mole, thus a much steeper slope on the Arrhenius plot. Short exposure to hydrothermal fluids often drives a system to the equilibrium assemblage of solid phases and the dissolution kinetics becomes that of the crystalline reaction products rather than the glass. Thick reaction layers of complex chemistry form rapidly on glasses reacted under hydrothermal conditions (35).

Corrosion by the Vapor Phase

Glasses can be hydrated and recrystallized by direct reaction with water in the vapor phase (36-38). Ionization in the vapor phase is negligible so that there is no question that neutral H_2O rather than H^+ or H_3O^+ is the active agent. Chemical profiles through the hydrated layer show a buildup of alkalis and other easily leachable elements on the surface. Unlike dissolution in liquid water where these elements are carried away in solution, vapor phase hydrolysis provides no transport mechanism and thus the hydration profiles for vapor hydrated glasses and water leached glasses are the reverse of each other. Hot water vapor, typically at 260°C, also promoted the recrystallization of the glass surface with the development of many small crystals of the phases appropriate to the glass composition.

The substantial corrosion of glasses by water vapor underscores the importance of hydrolysis of the network-forming elements as part of the overall dissolution process. This aspect of the mechanism is rarely included explicitly in discussion of the rate-controlling mechanisms.

SUMMARY AND CONCLUSIONS

Glasses and crystalline ceramics composed of oxides, silicates, borosilicates, and phosphates dissolve in aqueous solutions by a combination of hydrolysis of the framework-building elements near the surface and by acid-base reactions with the cations. Diffusion of mobile ions across depleted leach-layers may be rate controlling for some glasses. Rates usually increase linearly with hydrogen ion activity and exponentially with temperature.

Intense research activity in the decade of the 1980's, mainly carried out by the nuclear and hazardous waste communities in the case of complex glasses and waste immobilization ceramics and cements and by the geochemical community in the case of crystalline silicates has lead to some convergence in the description of the kinetics. There is considerable agreement on the forms of rate equations and on the relative importance of the various chemical reactions and chemical mechanisms of dissolution on a macroscopic scale. Understanding of corrosion mechanisms on an atomic scale is rudimentary although substantial current research is focused on this problem.

ACKNOWLEDGEMENTS

Experimental work on low temperature reactions of ceramics and glass on which this review was based is supported by the National Science Foundation under Grant No. DMR-8812824.

REFERENCES

1. Clark, D.E., Pantano, C.G., and Hench, L.L., *Corrosion of Glass*, Books for Industry, New York (1979).
2. Lasaga, A.C. and Kirkpatrick, R.J., *Kinetics of Geochemical Processes*, Mineralogical Society of America, Reviews of Mineralogy, Vol. 8 (1981).
3. Lutze, W., in *Radioactive Waste Forms for the Future* (W. Lutze and R.C. Ewing, eds), pp. 1-159, North-Holland, Amsterdam (1988).
4. Sayre, E.V., Vandiver, P.B., Druzik, J., and Stevenson, C.,eds. *Materials Issues in Art and Archaeology*, Mater. Res. Soc. Symposium Proc. 123, 321 pp. (1988).
5. Nordstrom, D.K., Plummer, L.N, Wigley, T.M.L., Wolery, T.J.,Ball, J.W., Jenne, E.A., Bassett, R.L., Crerar, D.A., Florence, T.M., Fritz, B., Hoffman, M., Holdren, G.R., Lafon, G.M., Mattigod, S.V., McDuff, R.E., Morel, F., Reddy, M.M., Sposito, G. and Thrailkill, J., in *Chemical Modeling in Aqueous Systems* (E.A. Jenne, ed), pp. 857-892, American Chemical Society Symposium Series 93 (1979).

6. Lasaga, A.C., in *Kinetics of Geochemical Processes* (A.C.Lasaga and R.J. Kirkpatrick, eds), pp. 1-68, Mineralogical Society of America, Reviews of Mineralogy, Vol. 8 (1981).
7. Nachlas, J.A., M.S. thesis, The Pennsylvania State University, University Park, PA (1983).
8. Rimstidt, J.D. and Barnes, H.L., *Geochim. Cosmochim. Acta* 44: 1683-1699 (1980).
9. Blum, A.E., Yund, R.A. and Lasaga, A.C., *Geochim. Cosmochim. Acta* 54: 283-297 (1990).
10. Knauss, K.G. and Wolery, T.J., *Geochim. Cosmochim. Acta* 52: 43-53 (1988).
11. Lasaga, A.C., *J. Geophys. Res.* 89: 4009-4025 (1984).
12. Grandstaff, D.E., *Geochim. Cosmochim. Acta* 42: 1899-1901 (1978).
13. Wieland, E., Wehrli, B. and Stumm, W., *Geochim. Cosmochim. Acta* 52: 1969-1981 (1988).
14. Zellmer, L.A. and White, W.B., in Fifth International Symposium on Water Rock Interactions, Extended Abstracts, pp. 652-655, International Association of Geochemistry and Cosmochemistry, Reykjavik, Iceland (1986).
15. Mendel, J.E., *Final Report of the Defense High-Level Waste Leaching Mechanisms Program.* Battelle Pacific Northwest Laboratory, Richland, WA, Report PNL-5157 (1984).
16. Rana, M.A. and Douglas, R.W., *Phys. Chem. Glasses* 2: 179-205 (1961).
17. Wallace, R.M. and Wicks, G.G., *Mater. Res. Soc. Symposium Proc.* 15: 23-28 (1983).
18. Bunker, B.C., Arnold, G.W., and Wilder, J.A., *J. Non-Cryst. Solids*, 64: 291-316 (1984).
19. Clark, D.E. and Yen-Bower, E.L, *Surface Sci.* 100: 53-70 (1980).
20. Harvey, K.B., Litke, C.D., and Boase, C.A., *Phys. Chem. Glasses* 27: 15-21 (1986).
21. Bunker, B.C., Arnold, G.W., Beauchamp, E.K. and Day, D.E., *J. Non-Cryst. Solids* 58: 295-322 (1983).
22. Houser, C.A., Herman, J.S., Tsong, I.S.T., White, W.B. and Lanford, W.A., *J. Non-Cryst. Solids* 41: 89-98 (1980).
23. Tsong, I.S.T., Houser, C.A., White, W.B., Wintenberg, A.L., Miller, P.D., and Moak, C.D., *Appl. Phys. Lett.* 39: 669-670 (1981).
24. Grambow, B. and Strachan, D.M., *Mater. Res. Soc. Symposium Proc.*, 26: 623-634 (1984).
25. Strachan, D.M., Krupka, K.M., and Grambow, B., *Nucl. Chem. Waste Management* 5: 87-99 (1984).
26. Hochella, M.R., Jr., Ponader, H.B., Turner, A.M., and Harris, D.W., *Geochim. Cosmochim. Acta* 52: 385-394 (1988).
27. Petit, J.C., Della Mea, G, Dran, J.C., Schott, J., and Berner, R.A., *Nature* 325: 705-707 (1987).

28. Ethridge, E.C., Clark, D.E., and Hench, L.L., *Phys. Chem. Glasses* 20: 35-40 (1979).
29. Barkatt, A., Macedo, P.B., Sousanpour, W., Barkatt, A., Boroomand,M.A., Fisher, C.F., Shirron, J.J., Szoke, P., and Rogers, V.L., *Nucl. Chem. Waste Management 4*: 153-169 (1983).
30. Chambre, P.L., and Pigford, T.H., *Mater. Res. Soc. Symposium Proc.* 26: 985-1008.
31. Lindberg, R.D. and Runnells, D.D., *Science* 225: 925-927 (1984).
32. Freeborn, W.P. and White, W.B., *Mater. Res. Soc. Symposium Proc.* 26: 719-726 (1984).
33. Wang, R. and Katayama, Y.B., *Nucl. Chem. Waste Management 3*: 83-90 (1982).
34. Barkatt, A., Boulos, M.S., Barkatt, A., Sousanpour, W., Boroomand,M.A., Macedo, P.B., and O'Keefe, J.A., *Geochim. Cosmochim. Acta* 48: 361-371 (1984).
35. McCarthy, G.J., Scheetz, B.E., Kormarneni, S., Smith, D.K., and White, W.B., in *Solid State Chemistry: A Contemporary Overview* (S.L. Holt, J.B. Milstein, and M. Robbins, eds) pp. 349-389, Amer. Chem. Soc. Adv. in Chemistry Ser. 186 (1980).
36. Bates, J.K., Jardine, L.J., and Steindler, M.J., *Science* 218: 51-54 (1982).
37. Bates, J.K., Seitz, M.G., and Steindler, M.J., *Nucl. Chem. Waste Management 5*: 63-73 (1984).
38. Bates, J.K., Steindler, M.J., Tani, B., and Purcell, F.J., *Chem. Geol.* 51: 79-87 (1985).

2

Predicting Corrosion

P. Bruce Adams

Precision Analytical
Watkins Glen, New York

INTRODUCTION

The technology and theory for predicting corrosion are powerful tools which seem to be on the brink of producing dramatic breakthroughs. But accurate predictions have sometimes been inhibited by an inadequate understanding of corrosion processes as well as an inability to deal mathematically with the many factors which impact the outcome of a glass-solvent reaction. None the less, predicting the corrosion behavior of materials has always been an important component of technology. It has been a vital, yet often unrecognized, contributor to product development as well as to basic science.

When corrosion theory fails to predict the behavior of glass, it is usually not the theories that are at fault. Rather, it is a blind reliance on these theories, coupled with a lack of appreciation for those factors that are either poorly-controlled or ignored. So the primary purpose of this chapter is to suggest a pathway for arriving at an optimum prediction process.

Very often the question that the corrosion engineer must answer is simply, "How will product A be corroded by target environment X ?" The answer may not always be easy to obtain. Testing procedures can be difficult to control. Target environments can be difficult to define. The principles of glass corrosion don't always seem to fit.

The literature often refers to "modeling" of corrosion, but the term connotes more than it can always deliver. Figure 2-1 shows that reality may sometimes present an obscure picture that may not fit a preconceived model (1). In this instance, the reactions have not been confined to the interface between the glass and the solvent; instead the growth of rod-like crystalline structures have created tubules which penetrate the parent glass. The fact is that there is no one single unifying concept that applies in all cases.

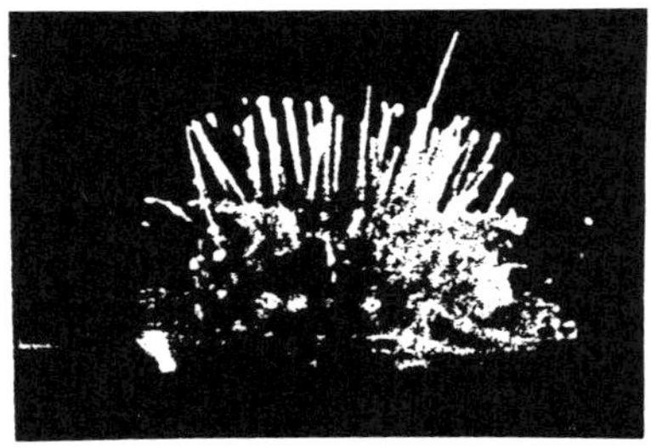

Figure 2-1. Microphotograph of tubules protruding from a corroded glass surface into the bulk glass. From reference (1).

The answers become more difficult when the questions become more complex. E.g. "What **composition** will perform to a given standard of expectation in target environment A?" The glass composition factor complicates efforts to obtain an answer. The model which fits one glass composition may not fit another. Figure 2-2 shows that there can be a dramatically different response to corrosion as a function of pH. The relationship between glass composition and corrosion behavior must be recognized and accommodated.

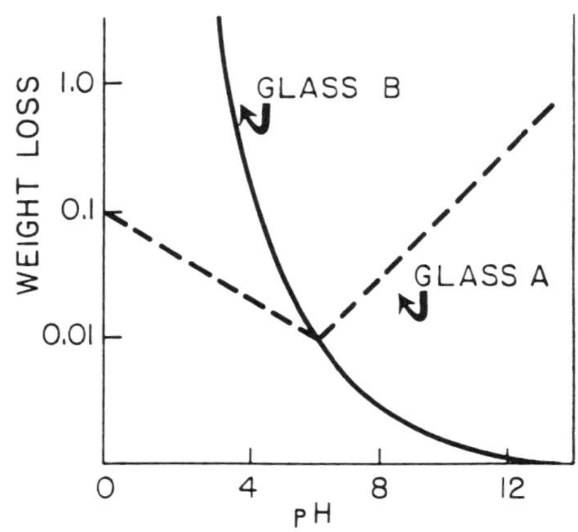

Figure 2-2. Corrosion rate vs. pH for two dissimilar glasses.

The list of complications sometimes seems to be unending. Prediction is never an exact process. The most important question is therefore, "How good is the prediction?"

It is a given that corrosion results will often be relatively imprecise, and perhaps inaccurate, with respect to the target environment. Figure 2-3 shows that some results can exhibit a considerable "range of uncertainty", in this case about two orders of magnitude (2). These results are not atypical of those which are often the best that can be obtained under very closely controlled conditions. Accuracy and precision must never be overlooked. They are the essence of prediction.

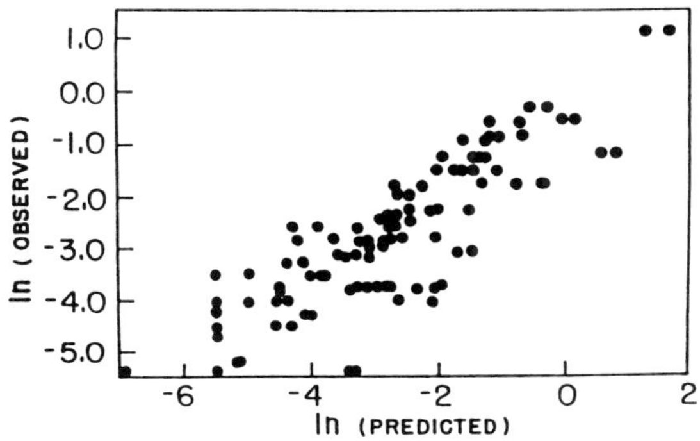

Figure 2-3. Correlation between observed and predicted results obtained by reaction of a variety of glasses.

This chapter is intended to be a guide to defining a process for predicting corrosion. The first section following the Introduction discusses some Prerequisite areas of information that are required as a knowledge base to begin the process. Following that is a section discussing various approaches, or Methods, for predicting corrosion. The section on Applications is intended to provide a sample of the many ways in which corrosion prediction has been used. Finally there is a Summary.

PREREQUISITES

There is a huge amount of information at the disposal of the corrosion engineer. Familiarity with this material is essential.

Corrosion Models

Certainly, the first prerequisite is an understanding of the classical theory of glass corrosion as well as the various supplementary and unifying concepts. This book provides the opportunity to acquire that understanding. The following is a brief summary of the important concepts.

Alkaline attack is described as a first order reaction whereby the glass structure uniformly disappears into the reacting solution; corrosion rate is proportional to time, increases with pH and is subject to the Arrhenius temperature equation. Acid attack is described as a diffusion-controlled leaching reaction wherein glass modifiers are selectively extracted; corrosion rate is proportional to the square root of time, increases only slightly with decreasing pH and is subject to the Arrhenius expression (3).

The fact that different metal oxides, as glass components, exhibit independent and different solubility as a function of pH is recognized. Stability diagrams provide a framework for explaining these effects (4).

Reaction layers form at the solvent-surface interface. Although they have long been recognized, better techniques for dealing with them have been developed in recent years.

For many glass-solvent systems, the corrosion reactions go to a stable and steady state at the reaction interface as reaction products saturate the system. Therefore, the thermodynamic approach can be helpful when the target environment is, or can be represented as, thermodynamically stable. This approach provides a useful tool for ranking various materials under conditions of thermodynamic stability (5) (See also Chapter 6).

When it is appropriate to assume that the system is dynamic, the kinetic model may be able to accommodate the many variables that are often involved in glass surface-solvent reactions. The mathematics should handle the time, temperature, and pH relationships as well as the effects of transport processes through reaction layers, solution changes and back reactions. Since corrosion experiments can seldom duplicate the kinetics of real performance, this approach offers the best opportunity to obtain quantitative predictions from experimental data, as recent efforts have demonstrated (6).

Composition

The effects of specific oxides on chemical durability have long been documented to a considerable degree. Figure 2-4 shows one such set of data (7). There is a vast amount of similar information relating to specific glass composition systems that is available from the literature. Note that one of the most useful methods of examining the composition effect is the thermodynamic approach (5).

The physical composition can be equally important in many instances. Most ceramics, whether they be multiphase or not, as well as some glasses which contain minor undissolved phases, react with a solvent so that one phase is preferentially attacked.

Surface Condition

The composition and topography of the surface and near-surface must be

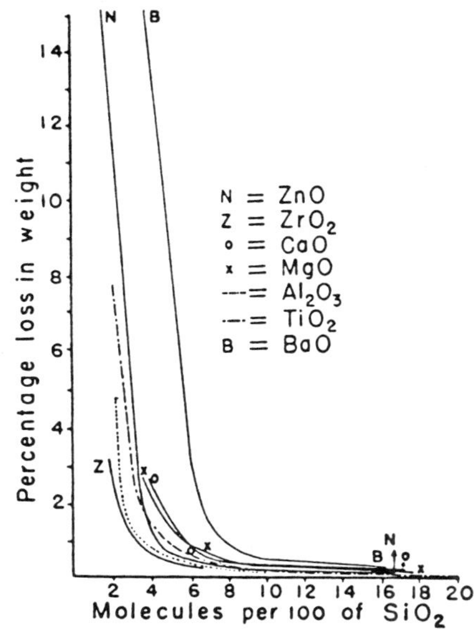

Figure 2-4. Effect on weight loss on the substitution of various oxides for sodium oxide in a glass composition. From reference (7).

considered. This is the zone of reaction and as such is far more important than the bulk of the material. The nature of the surface may dictate the type of test to be performed. Tests performed on one type of surface must not be blindly translated to infer how another type of surface will respond. Figure 2-5 is an example of differing reactivity for two different surfaces on the same glass (8). In this case the fractured surface has exposed the bulk glass composition compared to an alkali-depleted as-formed surface. Thus the fractured surface is more reactive.

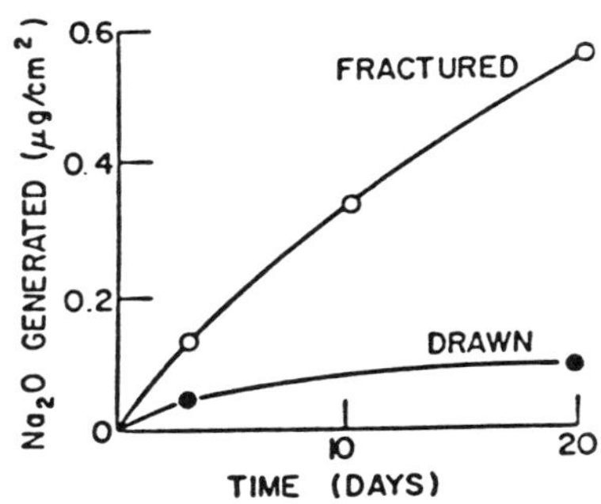

Figure 2-5. Comparison of alkali removed by weathering the same glass with dissimilar surfaces.

Other Data Sources

In most cases, it is unproductive to try to design a prediction protocol until available data has been reviewed. Information is available from many general sources, typically (3,7,9-11) of the nature of that which is shown in Tables 2-1 through 4 (9,10). (Note that Tables 2-1--3 reflect specific test data whereas Table 4 contains information that is deduced from other data). Journals and other publications provide more current and specific information reported by many researchers, the scope of which is suggested by Tables 2-2--5.

Table 2-1. Qualitative Ranking of Glasses Into Four Categories

TYPE	PRINCIPAL USE	WEATHERING	WATER	ACID
Soda Lime-1	Lamp Bulbs	3	2	2
Borosilicate-1	General	1	1	1
Borosilicate-2	Pharmaceutical	1	1	1
96% Silica-1	High Temp	1	1	1
Fused Silica-2	Optical	1	1	1
Glass Ceramic-1	Nose Cones	-	1	4
Glass Ceramic-2	Cooking Ware	-	1	2

Durability Key: 1 = excellent; 2 = good; 3 = fair; 4 = poor

There are also many private sources which maintain extensive records, such as glass manufacturers, government agencies and universities. Some data will be isolated bits or groups of information; others will be well organized systematic studies. They well may eliminate the need for a proposed study. At the least, they will help to delineate the protocol for a successful test program.

Table 2-2. Quantitative Results in Terms of Loss in Weight After 24 Hours at 95°C, mg/cm^2

GLASS TYPE	IN 5% NaOH	IN 5% HCl
Soda Lime-1	2.0	0.02
Soda Lime-2	2.0	0.01
Lead-2	4.0	0.02
Lead-2	4.0	0.01
Aluminosilicate-1	2.0	0.3
Aluminosilicate-2	1.0	0.4
Borosilicate-1	5.0	0.005
Borosilicate-2	4.0	0.005
High silica-1,2	2.0	0.001

Table 2-3. Quantitative Results in Terms of Na_2O Leached Out in 4 Hours at 90° C, μg/cm²

GLASS	IN H_2O	IN 0.02N H_2SO_4
Soda Lime-1	4	5
Soda Lime-2	1	1
Lead-1	3.0	10.0
Lead-2	1.0	5.0
Borosilicate-1	0.1	0.3
Borosilicate-2	0.1	0.3
High silica-1,2	<0.01	<0.01

Table 2-4. Estimations of the Upper Limits of Elements Extractable from Glass Bottles After 1 Year at 25°C.
(ppm in Solution)

GLASS TYPE	ELEMENT EXTRACTED	INTO STRONG ALKALI	INTO ACID OR WATER
Soda Lime-1	Si	10,000	10E+2
	Na	1,000	10E+2
	Fe	10	10E-1
Soda Lime-2	Si	1,000	10E 0
	Na	100	10E+1
	Fe	1	10E-2
Borosilicate-1	Si	1,000	10E 0
	Na	100	10E 0
	Fe	1	10E-3
Borosilicate-2	Si	1,000	10E-1
	Na	100	10E 0
	Fe	1	10E-3
High silica-1	Si	1,000	10E-1
	Na	1	10E-3
	Fe	1	10E-3
High silica-2	Si	1,000	10E-1
	Na	0.01	10E-5
	Fe	0.1	10E-3

Standard Test Procedures

Agencies such as ASTM and ISO promulgate standard test procedures. For meaningful interlaboratory tests, such standard tests are usually essential. For independent studies, they will often be worth adopting as is. In any event they will serve as an important resource to be studied for the rationale behind the

Table 2-5. Distribution of Publications on Corrosion in U.S. Literature for 1985-87. (Total -- 275 Publications)

GLASS PRODUCT CATEGORY	PER CENT
Ancient and Natural Glass	1
Flat Glass and Tubing	1
Optical, OWG	1
Solar, Dental, Dishware	1
Coatings	2
Biological	2
"Fundamental Studies"	10
General Composition Studies	12
Fibers, Fiber-Composites, Cement Reinforcement	14
Non-silicate and Non-Oxide Glasses	16
Nuclear Waste Glasses	40

details of procedure. A further discussion of standard tests is contained in Chapter 3 of this book.

Interferences

It is essential to have a broad knowledge of the interferences that can influence test results and their interpretation. For the most part, these "interferences" are simply factors which are not fully understood or controlled. A thorough reading of this book should enable the corrosion engineer to limit such errors. Some of the most common are briefly discussed below to underscore their importance.

"Static" tests in which the solution is left in contact with the test specimen may create a dynamically changing solvent. "Flow" tests, which are intended to simulate an invariant solvent, may create a dynamically changing reacted surface. If potential changes in the composition of the solvent and/or the reacting surface are not considered, the interpretation of results will be skewed. This aspect of corrosion prediction is discussed in detail in Chapter 5.

Different glass compositions exhibit different activation energies. Thus, it is not always valid to compare glasses in one test and then predict how they will compare in the target environment unless this effect is known and considered.

Different solvents at the same pH may produce different results. That is, there are other factors in addition to pH which influence the rate of reaction. This creates a particularly vexing problem where it might be more convenient to generalize with regard to pH.

Situations where one or more parameters change during the course of the reaction may increase the corrosion rate in some instances and decrease it in other instances. Unfortunately, real life target environments very often involve cycling conditions of temperature or solvent composition. For instance, in the weathering of some glasses, cycling which produces wetting and drying will cause the corrosion rates to decrease since a silica rich layer is formed which interferes with ion exchange. In the weathering of other glasses, corrosion rates remain constant or increase as alkaline weathering products destroy the previously reacted layer, thus exposing bulk glass.

Target Environment Conditions

It is perhaps redundant to say that the proposed environment of use, or the target environment, must be defined -- or at least the limits of one's understanding of it must be appreciated. Not only is there a need to know the temperature(s), the time(s), and the pH(s), but one must also know whether and how it changes, and, hopefully, what the real chemistry is.

It may be relatively simple to define the target environment; or it may be very difficult. For instance, it will be easy in a production situation such as the extraction of alkali from a glass part that is being washed in a process step; it will be somewhat more difficult for the extraction of radioactive material from a nuclear waste glass stored in an underground repository; and it will be extremely difficult for the extraction of lead by food from lead-glazed dishes (see Chapter 12).

In the first instance above, the glass parts are nearly always the same, the temperature is within a manageable range and easily controlled, the time is easily measured and the solvent will probable change little (or at worst the changes can be easily identified). Therefore, the target environment will be well defined.

In the instance of the nuclear waste disposal glass, the target environment is fairly easily identified. The temperature range, if not the specific temperatures, is reasonably well known. The solvent composition, that is, the chemical environment, is a little more uncertain, since there has been no definite decision by our society as to exactly which repository to chose. However, the time variable puts a real strain on the "models". Most test programs produce results in terms of days or weeks. Data from relatively short term tests must then go into a model which will be able to predict for a 10,000 - 1,000,000 year target environment reaction time.

In the lead extraction example, it is very difficult to define the "target environment," since there is a whole universe of environments. The pH of foods ranges from acidic to slightly basic, temperatures range from those in a refrigerator to boiling and the times of exposure range from a few minutes to several days. Further, the bulk compositions (chemical and physical) cover a wide field, as do the compositions of decorations and glazes. The "model" has to allow all of these variables to be translatable to the almost infinite number of target environments which comprise the real food consumption situations for a large number and variety of people, such as the population of the United States.

Thus, the construction of the test protocol will be quite different depending not only on what the target environment is but also how well it can be defined. Furthermore, a thorough analysis of the target environment and the problems associated with it may highlight some other key questions such as:

- What property do I want to predict? Do I want to measure a property change of the glass surface or do I want to measure a property change of the environment?
- How precisely and accurately will I need to know that property?

Analytical Approach

Although it may not be possible to make a final decision until the methodology is selected, it is essential to consider the various analytical approaches that are available and appropriate. These fall into two categories: techniques that are applicable to solution analysis, and techniques that are applicable to bulk material analysis.

Solution analysis methods include all those that are now utilized for the chemical analysis of glass and ceramics, wherein the material is first dissolved. Methods such as atomic absorption spectroscopy, gravimetry, electrochemistry, and fluorescence spectroscopy, are all suitable for the analysis of discrete samples of solution representing a particular set of test conditions. These or other less direct methods, such as pH measurement and conductivity, may be utilized for the *in situ* continuous monitoring of reacting corrosion solvents.

Material analysis methods likewise include all those that are now utilized for the chemical analysis of glass and ceramic materials in the solid state, as well as many that are used for physical analysis or characterization. Optical and electron microscopy (including scanning), X-ray energy dispersion, electron spin chemical analysis and Rutherford backscattering are typical of those techniques that can be applied to study changes in the reacted bulk material as well as to study the composition of newly-formed reaction layers. Those which can produce a profile of composition will be most helpful in understanding reaction mechanisms. On the other hand, relatively simple approaches which may reveal nothing about the mechanism of degradation, such as weight change, specular reflection, visual appearance and change in thickness, can provide very explicit information regarding the practical performance changes resulting from corrosion. Frequently, the most appropriate measurement is that of a specific property that is directly related to product performance.

Precision and Accuracy

The precision and/or accuracy of the final prediction is dependent on the precision and/or accuracy of the three prediction components. Those components are the test results, the target environment and the model. For instance, even if test results are obtained with relatively high precision, e.g.+/- 5 %, the accuracy of prediction may still be several orders of magnitude distant from reality if the target environment or the model are poorly defined.

A mathematically quantitative definition of precision and accuracy may not be possible, or necessary. In many applications, a prediction within a few orders of magnitude may be more than adequate. In others, much higher accuracy and/or precision may be required. The important point is that the test designer know what can be done and be certain that those who interpret the results understand the limitations of the three components of prediction.

The End User

Who is going to use the results? Will they be used by highly-trained technical personnel to design a process? Or will they be used by the layman to decide when and where a product should be available? Depending upon the answers to question such as these, the practical requirements on the prediction process will be quite different. Nor should one assume that the technical user will be any more discriminating than the layman unless he is made aware of the relevant factors.

METHODOLOGY

At this point, it should be clear what type of experiments, if any, will be useful. One should know the target environment with as much specificity as possible, know what factors are important and whether they will vary in the target application scenario, know what the required precision and accuracy must be in order to effectively deal with the questions relating to how the results will be used, and know what information is available to build on. Then, and only then, can the test protocol be well defined.

There are various levels of method complexity from which to chose. The simplest is to compare two glasses under a specific set of test conditions and then, with knowledge of how one behaves in the target environment, predict the behavior of the other in the same environment. The most complex involves testing the material under known conditions and then extrapolating the results so obtained to project the results under the target environment conditions.

Ranking Based on Test Results

Comparing test results is the simplest and probably the most widely used, often involving a standard test. It can be fairly accurate if the compositions are relatively close, or if there is a sure understanding of the effects of composition differences.

For example, assume that there was a desire to substitute a glass for Borosilicate-1 shown in Tables 2-1--4. If the criteria for acceptance were release of alkali into solution, then a test of the new glass under the conditions of Table 2-3 would provide at least a part of the answer. But suppose that the glass chosen was Borosilicate-2 and that the solution with which it was to be in contact was sensitive to contamination by barium or calcium ion. Then the data in Table 2-3 would be insufficient. It would be necessary to determine whether Borosilicate-2 contained either BaO or CaO and, if so, how much. It would be necessary to determine how much Ca^{++} is deleterious, what the specific temperature and time of storage would be, and the chemistry and pH of the solution. Depending on knowledge of these variables, further work might or might not be necessary.

When such an approach doesn't provide the answers with the required accuracy and precision, it may be necessary to seek another level of certainty for the prediction.

Estimation of Results

A better estimation can often be made with a few simple calculations based on knowledge of quantitative relationships. This need not be a complex exercise, and can produce useful results.

For example, if the problem were to determine an upper limit for the amount of sodium that could be introduced into ten milliliters of 0.1 N sulfuric acid stored for 9 months at 40°C in a 100 ml flask made of Borosilicate-1 glass that has previously been used to contain 5% hydrochloric acid for one year, then one could proceed as follows:

Assume that:

(a) the upper limit for extractables from code 7740 by "strong" acid in 1 day at 25°C is 10^{-4} mg Na_20 /cm^2 (12).

(b) the extraction of alkali from code 7740 glass increases by 1.5 X for every 10°C temperature increase (12)

(c) the extraction of alkali from code 7740 glass increases by $t^{½}$ for time, t (3),

(d) the flask has an internal area of 100 cm^2 exposed to 100 ml of solvent and

(e) the glass contains 4% Na_20.

Then it is possible to make the following computation:

$$Na_2O \text{ conc.} = (a) \times (b) \times (c) \times (d) \times (e), \tag{2-1}$$

where (a) through (e) are factors related to the corresponding assumptions above.

$$= (0.0001 \text{ mg/cm}^{2)} \times (1.5[40° - 25°]/10°) \times (270 \text{ days}^{½}/1\text{day}^{½})$$
$$\times (100 \text{ cm}^2/100 \text{ ml}) \times (4\ \%/100\ \%)$$

$$= 0.000148 \text{ mg/ml}$$

$$= 0.15 \text{ ppm } Na_20$$

Considering that the result may be high because the flask has been used, i.e. alkali has been preleached at the surface, and that the results may be low because sodium oxide is preferentially removed, this predicted result probably represents accuracy within one order of magnitude. In many cases, such a semi-quantitative prediction is more than adequate.

Duplicating the Environment

Sometimes the simplest and most practical approach is to duplicate the target environment.

Modeling

Modeling is the "purest" approach if and when it can be accomplished in the ideal manner: (1) definition of the target environment, (2) selection of the model that relates a statement of that environment to possible test conditions, (3) definition of the test conditions, (4) collection of the test data, (5) computation of the predicted results under the target environment conditions. It is rare, if ever, that the corrosion process can be totally modeled. Almost every effort involves many compromises.

However, in reality all prediction involves modeling. The model simply becomes more complex as the model approaches the conditions more exactly.

APPLICATIONS and DISCUSSION

There is virtually no glass or ceramic product for which chemical durability is unimportant. It may only seem unimportant in some instances because corrosion resistance so often exceeds requirements to the point that it does not appear to impact performance. But it is always a "silent" partner insuring that the product will do its job.

Composition and Product Development

The development of such diverse products as chemically-resistant laboratory ware, optical and ophthalmic lenses, optical wave guides, lighting products, electronic substrates, food service ware, biological implant materials, nuclear waste glasses, etc., have all involved a great deal of product development wherein chemical durability has to meet certain requirements. Sometimes this requirement is critical for production processing; sometimes for the conditions of use. Thus, predicting corrosion performance is an essential component of product development activity.

For example, in the preliminary development of a viable composition for cookware, thousands of routine screening corrosion tests must first be made to evaluate product variables associated with composition and heat treatments. Once this preliminary screening is done and final candidate compositions have been chosen, a more definitive test program can be started to assure that the product will meet the rigorous demands of the kitchen. Table 2-6 shows examples of some tests designed to simulate various aspects of cookware use in the laboratory compared to service tests in the kitchen (13). A variety of tests must be employed to "model" a target environment. In this case, each test "models" only one variable, or at best a few variables which are associated with the target environment. It takes a campaign of tests to demonstrate that the product will survive the rigors of household use. Such a campaign is typical of the process involved in predicting the performance of almost any glass or ceramic product.

Table 2-6. Laboratory and Service Tests for Cookware

TEST DESCRIPTION	TEST LIMIT	LAB VALUE	SERVICE VALUE
Dishwasher	Decal Off	1500 Cycles	Severe
0.3% Detergent	Decal Off	48 Hours	1 Year
		100 Hours	>2 Years
10% Citric	Decal Dull	>24 Hours	>2 Years
Lead Release	2 ppm	----------	---------
Abrasion/ Detergent	Decal Off	16 Hours	>2 Years

Preservation, Storage and Use

Very often, it becomes necessary to predict how a glass product already in use will react to a new environment or why it reacts to a given environment in some particular manner.

The need to preserve ancient glass items is an aspect of chemical durability prediction that has taken on importance in recent years. As some old glass items find their way from the confines of musty castles or are dug up from their graves in damp soil, their appearance starts to change dramatically. Some take on a "crizzled" appearance as shown in Figure 2-6 (14). This is caused by dehydration of the glass which had been previously saturated with water absorbed over years of exposure to a damp environment. Postulating the mechanism of degradation permits the evaluation of that model by testing it on experimentally produced compositions and the definition of a method to preserve such items. This can be as simple as protecting the item from the air-conditioned museum atmosphere.

Facades and other exterior surfaces that are constructed of ceramic materials often react noticeably with their environment. Laboratory tests can be designed to try to assess the effects of various atmospheres on different materials and to

Figure 2-6. Crizzling of ancient glass resulting from hydration and dehydration (left half: As received; right half: after hydration).

rate their ability to mimic target environments. Figure 2-7 (15) is an example of such an effort. It demonstrates that "modeling" a corrosive atmosphere by

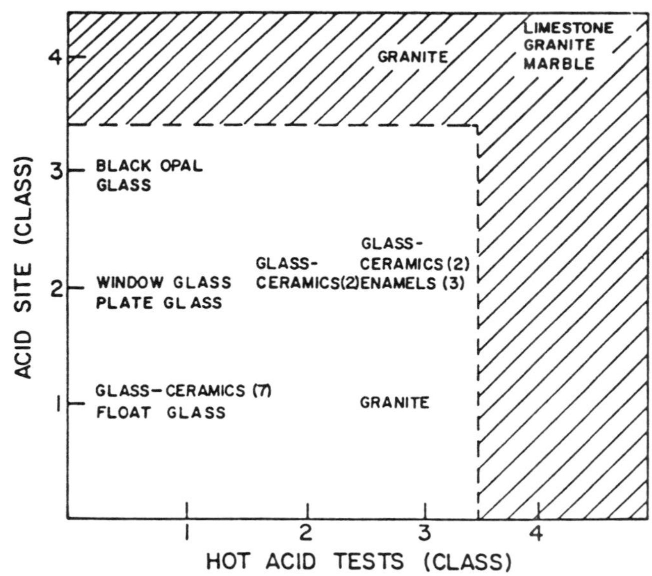

Figure 2-7. Comparison of laboratory (hot acid) tests and atmospheric (acid site) tests on a variety of glass and ceramic building materials. From reference (15).

using a laboratory solvent may work, sometimes. In this case, exact correspondence between the acid site exposure test and the laboratory HCl hot acid test would be exhibited for all materials if they fell on a straight 45 degree line.

Thus, the black opal glass shows more severe (class 3) degradation in the site test than in the lab test (class 1); on the other hand, some granite showed less severe (class 1) degradation in the site than in the lab test (class 3).
So some materials, such as glass-ceramics (essentially the same class in both tests) may be more reliably ranked by a lab test than other materials such as granite. Additionally, those materials which show very severe degradation (class 4) in both site and lab tests are demonstrably unsuitable for an environment of the type studied, acid in this case. It is clear that the atmospheric site tests come closer than the laboratory test to simulating the actual target environments. In the interpretation of such data, the other variables such as those associated with seasonal changes, transient atmospheric pollutants, the daily weather, and the design and direction of the exposed building faces must all be considered.

Designing a cleaning procedure for a specific purpose, such as preparing a glass substrate to receive a thin metal coating for an electronic device, is often a "cut and try" process to find out what works; if the process subsequently fails to do the job, another step or two is added. The result is often a cleaning procedure that contains many steps that are unnecessary, inappropriate, inadequate and/or counterproductive. A logical procedure can be designed by (1) defining the contaminants, (2) identifying solvents and methods that will remove the contaminants, (3) eliminating those solvents and methods that will simply contribute another undesirable contaminant and (4) eliminating those solvents and methods that will degrade the glass surface. This latter consideration, glass degradation, has to be specific to the composition, which can be unique and quite diverse, especially when considering the many special technical glasses, such as those illustrated by Table 2-7 (16), and comparing them with the more conventional commercial glasses, illustrated by Tables 2-1--4. The importance of understanding the interaction between solvent and glass is apparent to anyone who has ever tried to transfer a cleaning process from one glass to another.

Understanding the corrosion mechanism of multiphase ceramics often presents significant challenges that call for a great deal of creativity. For example, the corrosion of silicon carbide by combinations of acids and/or alkali can interfere with the life and function of wear-resistant seals, even though the corrosion rates may be very low, i.e. on the order of 0.00X mil per year (17). Electrochemical methods have therefore been used to study the effects of solvent concentration, silicon in solution and electrochemical potential on these very low rates. Figure 2-8 shows the results of one such study.

Health and Safety

During recent years, there has been a growing awareness of the contribution

Table 2-7. Some Technical Glass Compositions

	PERCENTAGE IN VARIOUS GLASSES							
	A	B	C	D	E	F	G	H
SiO_2	16	--	27	45	--	39	30	--
B_2O_3	10	20	--	--	7	--	10	29
Al_2O_3	--	--	--	--	18	--	10	5
Na_2O_3	--	--	1	6	14	4	--	--
K_2O	1	--	1	14	--	3	--	--
Li_2O	2	--	--	--	--	--	--	--
CaO	8	--	--	6	--	--	--	--
BaO	15	--	--	--	--	--	50	--
La_2O_3	13	30	--	--	--	--	--	--
ZrO_2	5	--	--	--	--	--	--	--
TiO_2	19	--	--	--	--	--	--	--
ThO_2	--	18	--	--	--	--	--	--
Ta_2O_5	--	5	--	--	--	--	--	--
Nb_2O_5	--	20	--	--	--	--	--	--
WO_3	--	7	--	--	--	--	--	--
ZnO	11	--	15	--	--	--	--	56
PbO	--	--	71	30	30	54	--	--
Pb_2O_5	--	--	--	--	33	--	--	--
V_2O_5	--	--	--	--	--	--	--	10

that lead glazed dishware can make to human lead intake via foods prepared or served in such items. Figure 2-3 shows one component of the process required to predict lead intake. It illustrates the difficulty in obtaining truly reproducible test results from one laboratory, especially when confronted with the uncertainties associated with a complex food chemical in contact with a product which often shows considerable variability in the very-near surface reacting zone. However, by quantifying and allowing for the uncertainties, such an approach

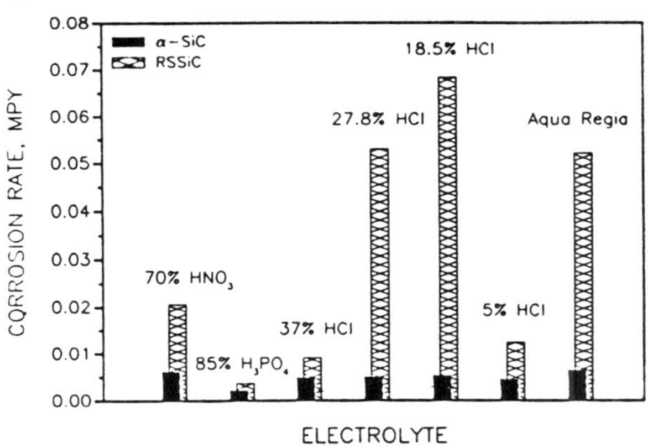

Figure 2-8. Comparison of the corrosion resistances of sintered α-SiC and reaction sintered SiC in various electrolytes. From ref. (17) by permission American Ceramic Soc.

enables the specification of safe lead release levels for ceramic food service products.

Chemical durability has become much more widely recognized by the technical community in the last 20 or 30 years because of the very essential role it plays in learning how to dispose of nuclear waste. In recent years roughly 40% of the publications, representing probably much more than 40% of the actual corrosion research, have been devoted to this topic.

All of this interest and activity has produced quality as well as quantity. Recognizing that there is a level of uncertainty, i.e. precision and accuracy, that in some instances can be defined in terms of tens of thousands of years, it is still now possible to design and melt glasses that will contain nuclear waste in a safe and reliable manner for tens of thousands of years (see Chapter 5). Application of the kinetic and thermodynamic concepts has played an important role in adding confidence to these predictions (Figures 2-9,10). Nonetheless, there are

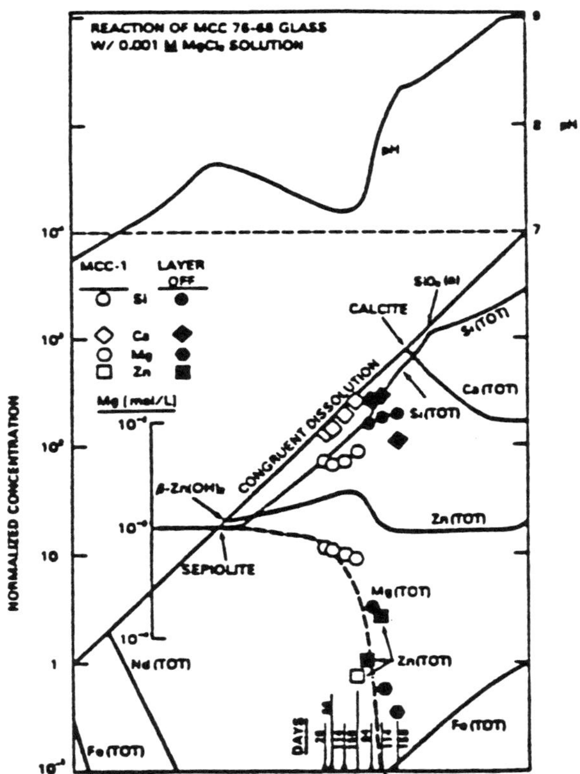

Figure 2-9. Predicted reaction product composition as a function of total corrosion for a nuclear waste glass. From reference (6).

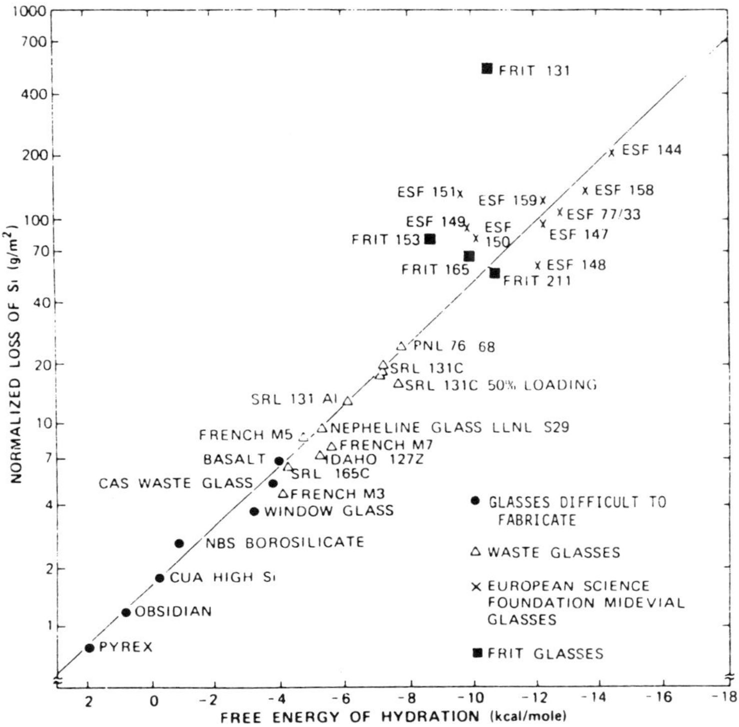

Figure 2-10. Predicting corrosion resistance from glass hydration energy. From ref. (5).

still areas to be explored so that the best possible host glass for a given waste can be defined for whatever site(s) are eventually chosen.

The problems that remain derive from various steps in the prediction process. For instance: (1) It will be difficult to define a geological environment as it may exist 10,000 years from now; it is extremely difficult to have to proceed, as researchers now do, in the face of a society which will not, or cannot, identify where they want the waste to go (2). The model will always contain some uncertainties, partly because of the uncertainties in knowing the target environment, but also because of the virtually endless number of variables that must be accommodated (3). Precision and accuracy of test results can be controlled by designing standard tests, but they must be recognized. So, prediction again comes down to precision and accuracy.

Fundamental Studies

Interesting and useful results may come from fundamental studies. On the one hand, we seek to understand mechanisms so that we can better predict how glass will behave. On the other hand, we seek to better understand glass composition and structure, so that we can better define its ultimate chemical stability and predict how products will function. Although it does not seem likely that we will ever define a fundamental chemical durability constant for glass, it is not unlikely that chemical durability may one day be related to several fundamental constants, or characteristics, of glass.

Isodurs, defined as lines of constant durability were used to identify the stability throughout a ternary system, as in Figure 2-11 (18). The data were obtained by testing a variety of experimental glass compositions. The durability result is recorded at the corresponding composition location on the ternary diagram. Lines of constant durability were then interpolated. Figure 2-11 shows them at 10 x 10^4, 100 x 10^4 and 1000 x 10^4. Thus it provides a visual "contour" map of durability changes as a function of three component composition variables.

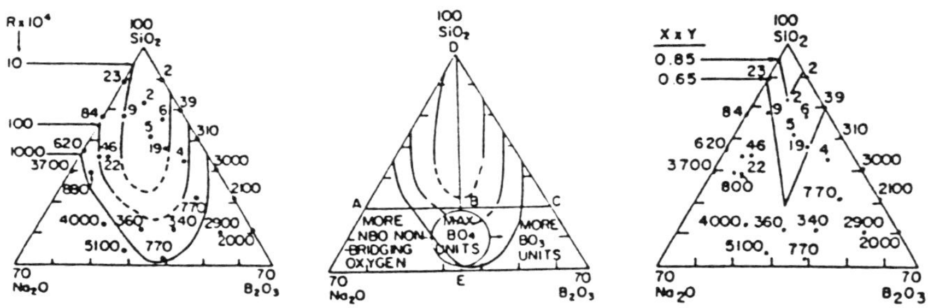

Figure 2-11. Left: defining composition contours that show invariant corrosion resistance (isodurs); Center: structural interpretation of isodurs; Right: calculated connectivity invariants. From reference (18).

As further shown in Figure 2-12, the "map" may be relatable to known or postulated glass or ceramic substructures, such as BO_3 vs. $B0_4$ in this case, or to mathematically-derived inferences about structure, such as the connectivity constant, reflecting bond strength.

SUMMARY

There is never one best set of tests or experiments that can be identified for predicting the corrosion characteristics of a glass or ceramic material. There are however, many mistakes that can be made and a lot of pathways that are less than optimum. The hope is that this chapter will encourage the engineer or researcher to stop and consider alternatives; and to never be sure that he has all the answers.

The prediction process steps are quite obvious, albeit often difficult to fulfill:

(1) Define the target environment
(2) Select the right model
(3) Identify "tests" which will fit the model
(4) Produce the data
(5) Plug the data into the model and hope for the best.

Although neither this chapter nor this book can pretend to provide all the answers as to how best to make the process described above be most effective, the following may be helpful:

(1) Decide what it is you need to know
(2) After you have defined the target environment, consider how close you really came
(3) Review the literature relating to similar test procedures and similar test results
(4) Select the simplest approach to the problem consistent with what you need to know and how well you know the target environment
(5) When the results come in, evaluate your prediction in the light of what you know you don't know, i.e. define the precision and accuracy of your prediction.

You could be off by 5% or you could be off 1000 X; you should know which it is and whether it is adequate for your purpose.

REFERENCES

1. Adams, P.B., in: *Scientific Basis for Nuclear Waste Management* (G. J. McCarthy ed.), Vol. I, pp 123-129, Plenum Press (1979).
2. Boyd, Y., Corning, New York, private communication.
3. Holland, L., *The Properties of Glass Surfaces*, Chapman and Hall (1964).
4. Paul, A., *Chemistry of Glass*, Chapman and Hall (1982).
5. Plodinec, M.J., Jantzen, C.M., and Wicks, G.G. *Advances in Ceramics* 8: 491-495, American Ceramic Society (1984).
6. Grambow, B., PhD Dissertation, Freie Universitaet Berlin (1984).
7. Morey, G.W. *The Properties of Glass*, 2nd ed., Reinhurst (1954).
8. Walters, H.V., and Adams, P.B., *J. Non-Cryst Solid* 183-199 (1975).
9. McLellan, G.W., and Shand, E.B., *Glass Engineering Handbook*, 3rd ed., McGraw-Hill (1984).
10. Britton, M.G., Adams, P.B., and Lonergan, J.R., in: *Materials and Processes, Part A* (J.F. Young, and R.S. Shane ed.), 3rd ed., pp 736-783, Marcel Dekker (1988).
11. Volf, M.B., *Technical Glasses*, Sir Isaac Putnam and Son (1964).

12. Adams, P.B., in: *Ultrapurity* (M. Zeif ed.), pp 293-351, Marcel Dekker (1972).
13. Morrissey, J.W., Corning, New York, private communications.
14. Brill, R.H., in: *Crizzling- A Problem of Glass Conservation. Proceedings of the Stockholm Conference* (1975).
15. Walters, H.V., and Adams, P.B., *Am. Cer. Soc. Bull.* 1224-1227 (1982).
16. *Annual Book of ASTM Standards*, ASTM C912-79, Sec. 15, pp 295-301, Am. Soc. Test. Mat. (1990).
17. Divakar, R., Seshadri, S.G., and Makuteswara, S., *J. A. Ceram. Soc.* 780-784 (1989).
18. Adams, P.B., and Evans, D.L., in: *Borate Glasses: Structure, Properties, Applications* (L.D. Pye, V.D. Frechette, and N.J. Kreidl ed.), pp 525-537, Plenum Publishing Corp. (1978).

3

Corrosion Testing and Characterization

D.E. Clark and B.K. Zoitos

University of Florida
Department of Materials Science & Engineering
Gainesville, Florida

INTRODUCTION

Over the last 100 years several thousand articles in the form of theses, dissertations, journal papers, reports and books have been written about glass and ceramic corrosion. The majority of these have appeared in the last 15 years in response to decisions to encapsulate nuclear waste in these materials. Most glasses and ceramics perform well in a variety of benign and harsh environments. For this reason, they are used to fabricate many products that directly affect our standard of living. These include containers for food, beverages and pharmaceuticals, windows for gamma, UV, IR and visible radiation protection; lenses for ophthalmic glasses, microscopes and telescopes; structural and architectural building materials, and replacement parts for humans such as crowns, teeth, middle ear bones and hip prostheses. The most demanding product, in terms of chemical durability, will be nuclear waste encapsulants. Production of these materials will begin in the United States in 1992 or 1993 (see Wicks' chapter). Once fabricated, they will be expected to endure for thousands of years without releasing harmful quantities of radionuclides to the environment.

The products and applications listed above rely on properties in addition to chemical resistance. In optical applications, transparency is the most important property. Under certain exposure conditions, surface films may form and substantially reduce light transmission. Although the extent of reaction may be extremely small (less than a few microns thick), the product may be rendered useless. In contrast, the same extent of corrosion on bioactive ceramics may not only be beneficial, but necessary, for the purpose of adhering to the surrounding tissue. Chemical durability is not the primary basis for selecting materials for either application. However, it does dictate the usefulness of the products in service.

The examples cited above help to illustrate a very important point. Different products can tolerate (even require) various degrees of corrosion damage. The degree of damage is dependent on both the mechanisms and extent of degradation. There is not a universal test or characterization technique that can be used to evaluate acceptable corrosion resistance for all glass and ceramic products. Thus, the main objective of this chapter is to describe a variety of tests and characterization techniques that are used in assessing chemical durability. We emphasize the importance of a systematic and integrated approach in order to better understand the corrosion processes.

METHODOLOGY

There are two basic components in all corrosion tests: the solid specimen and the test environment. Tests differ only in the manner in which these two components are combined and varied.

Figure 3-1 illustrates schematically the four steps involved in assessing chemical durability. These are: 1) material selection and sample preparation, 2) exposure, 3) characterization, and 4) data reduction and analysis. Although the stages are independent, the entire procedure must be well coordinated to provide answers to specific questions regarding chemical durability.

Generally, the specimen is in the form of either a bulk solid or crushed powder. Well polished (600 grit or finer) bulk samples are required if surface analyses are to be performed after the test. For example, infrared reflection spectroscopy (IRRS) and secondary ion mass spectroscopy (SIMS) are routinely used to characterize corroded surfaces (1-3). Rough surfaces result in diffuse scattering of the reflected IR beam making peak identification and changes due to corrosion difficult to interpret (see Figure 3-2) (4). Likewise, errors in depth profiling with SIMS may be significant when the area density of polishing grooves is large and their depth is about the same as the reaction layer thickness. Surface roughness can also affect solution analyses since the concentration of species released from the solid is directly related to the exposed surface area. The effects of surface roughness on boron leaching are shown in Figure 3-3 (5). As the surface roughness increases so does the amount of boron extracted at any given time. This is attributed in part to greater surface areas on the rough samples even though their geometric dimensions are equivalent.

There are generally two types of tests: laboratory and field. By far, the most widely used are laboratory tests because exposure conditions are easier to

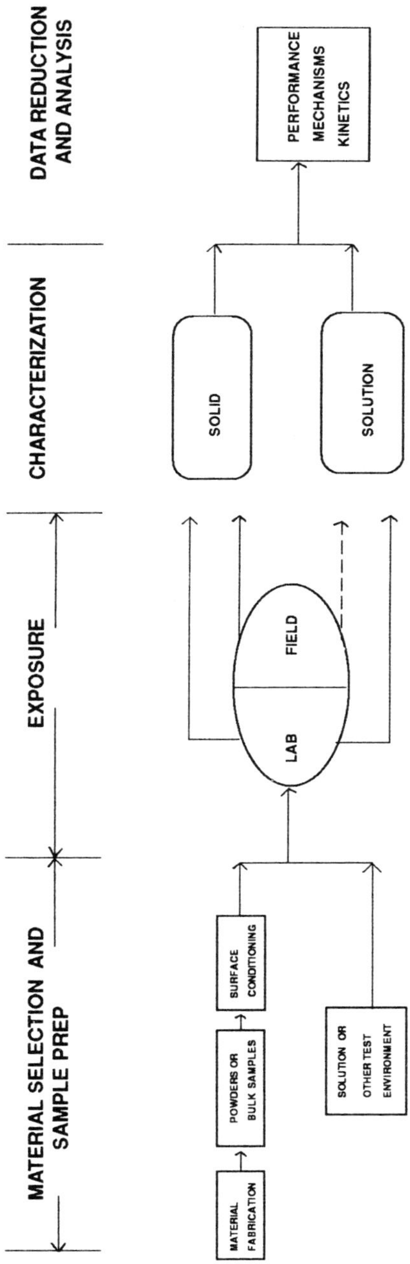

Figure 3-1. Test methodology.

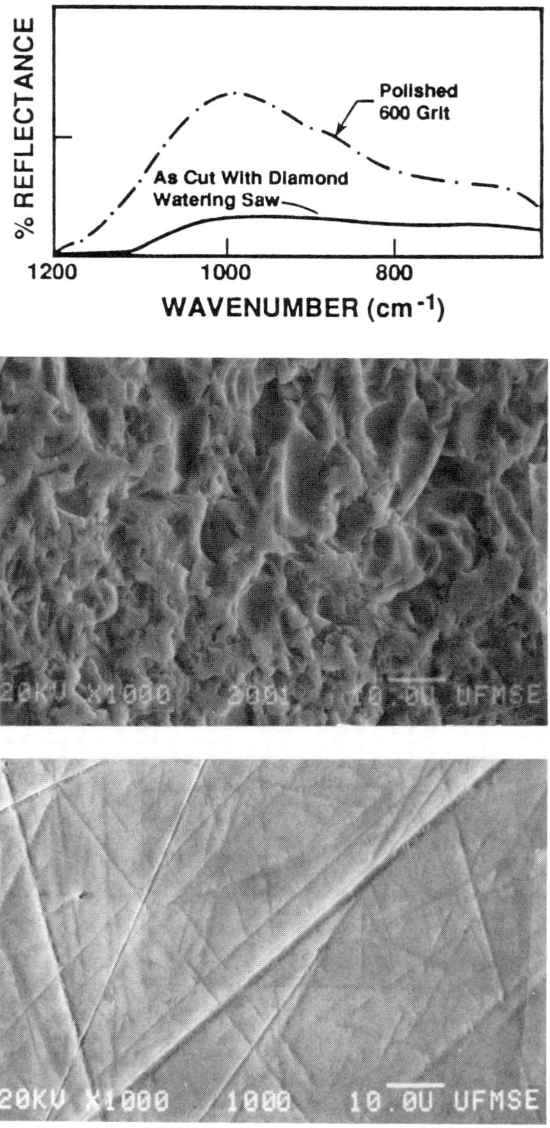

(a)

(b)
"As cut"

(c)
600 grit polish

Figure 3-2. Infrared reflectance spectra and scanning electron micrographs of a simulated nuclear waste glass with different surface finishes.

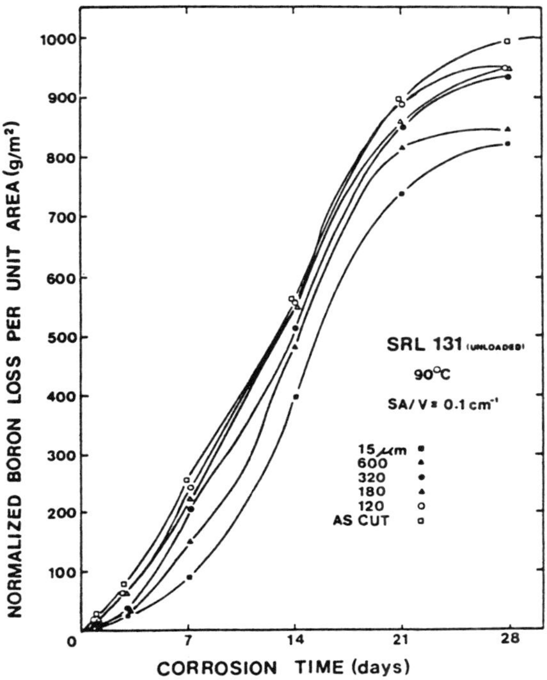

Figure 3-3. Normalized boron loss vs. corrosion time. Test was conducted at 90°C and the analysis was performed at room temperature.

control and, in most instances, these tests are less expensive. Due to better control and fewer variables, the data are easier to evaluate. One additional advantage of laboratory tests is that the solution in contact with the specimen can be analyzed and used together with surface analyses to provide a more complete picture of the corrosion processes. Unfortunately, laboratory tests do not always provide a realistic or accurate representation of the corrosion processes that occur during service. Field tests are required to provide this information. In field tests, specimens are placed in their use environment (or a close simulation) for specified times, removed and analyzed. In some cases the analysis may be a simple visual inspection. However, in most cases sophisticated surface analyses are required. The results from these tests can be compared to laboratory "scoping" tests to determine how successful the latter is in predicting performance under service conditions. Only in a few cases are field tests controlled, and in these the control is limited. For example, in a burial test the temperature may be maintained constant while other variables are allowed to fluctuate naturally. Thus, the number of potential interactions of the specimen with its environment is large. Laboratory tests involving limited variables are needed to help understand the complicated corrosion processes that occur in the field under more realistic conditions.

The end result of testing and characterization is data that can either be used to compare the performance of a material to that of a standard or compare the relative performances of different materials. If the test method is properly designed the data may also be used to determine the mechanisms and kinetics of corrosion and from these predict the long-term corrosion behavior (see Adams' chapter). The precision and accuracy of the data will depend substantially on the sample preparation, the extent to which the test procedures are controlled and the sensitivity of the characterization techniques.

LABORATORY TESTS

Test conditions vary considerably, as shown in Figure 3-4 and Table 3-1 (6-14). Examples of laboratory tests include crushed powder in contact with water; glass containers filled with water where the container itself is the sample; solid plates polished on all sides and immersed in solution; glass lenses exposed to controlled humidity (either cyclic or constant); hydrothermal and numerous others described in other chapters in this book. Recently, direct-current polarization measurements have been used to determine the corrosion rate of α-SiC in various acids (13). Application of electrochemical techniques to ceramics is new and, although the results appear promising, more research needs to be performed before their value can be ascertained. Some of the ASTM standards pertaining to ceramic and glass corrosion are listed in Table 3-2.

Where the specimen is exposed to humidity and reactive gases such as CO_2, the test is referred to as a weathering test. These tests are usually designed to minimize "runoff" and thus the products of corrosion remain on the surface (Figure 3-5) (see Parry's chapter and (15-19)).

Lab tests are usually not designed to eliminate CO_2 and other atmospheric gases. These gases are present in deionized and distilled water at room temperature and result in a pH = 5.66 at 25°C. As the temperature of the water is raised to about 85°C, the solubility of CO_2 becomes negligible and does not appreciably affect the pH (20). Certain groundwaters, do however, contain significant concentrations of carbonate and other ions that can affect both the pH as well as the corrosion products that form.

Prior to exposure, the material is thoroughly cleaned in either aqueous or organic solutions. The choice of cleaning solutions depends on the reactivity of the material as well as the objective of cleaning. For example, it might be desirable to not only remove the contaminants from the surface, but to chemically polish the surface at the same time. ASTM C 912-79 (reapproved 1984) (21) provides some useful guidelines for designing a cleaning procedure for specific materials. In general, combinations of ultrasonic rinses in high purity water and absolute ethanol are used for cleaning durable glasses prior to testing.

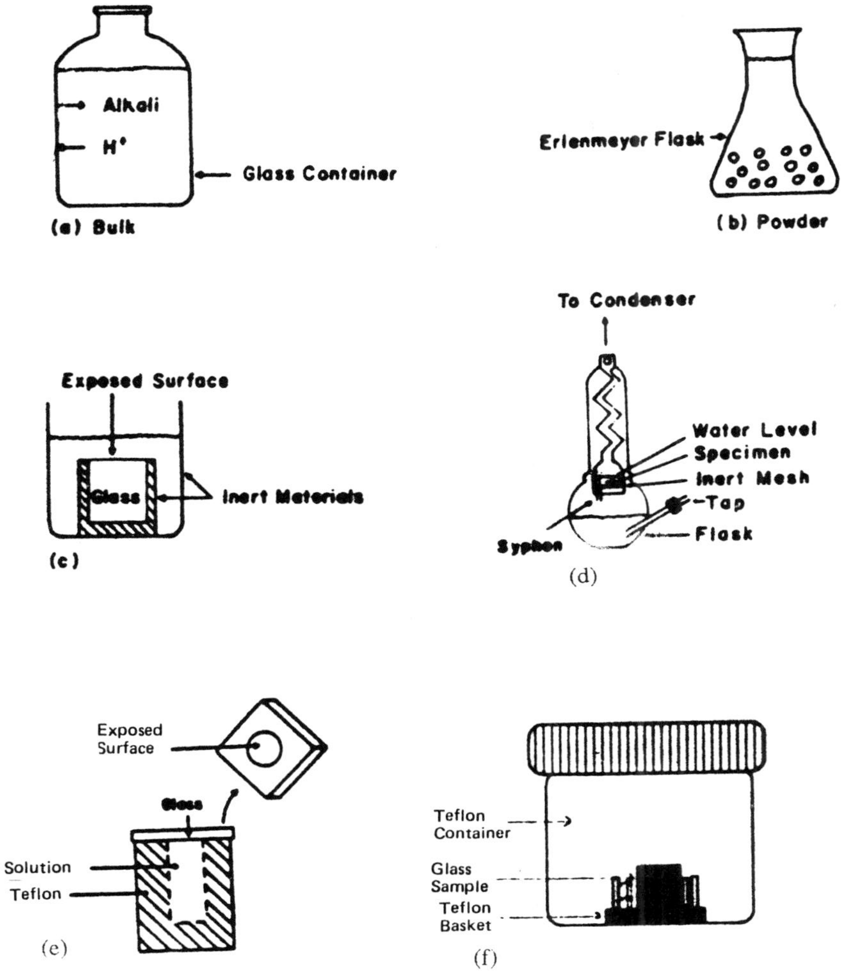

Figure 3-4. Corrosion test methods. Conditions: (a,b) ASTM C-225; T = 121°C, t = 1 hr (6) the glass container and Erlenmeyer flasks with glass powder are placed in an autoclave or steam sterilizer; (c) IAEA; T = 25°C, t = variable (8); (d) Soxhlet; T ≈ 100°C, t = variable (55); (e) Sanders-Hench; T = variable, t = variable, SA/V = 0.77 cm^{-1}. (f) MCC-1; T = variable (usually 90°C), t = variable, SA/V = $0.1cm^{-1}$. Adapted from (7).

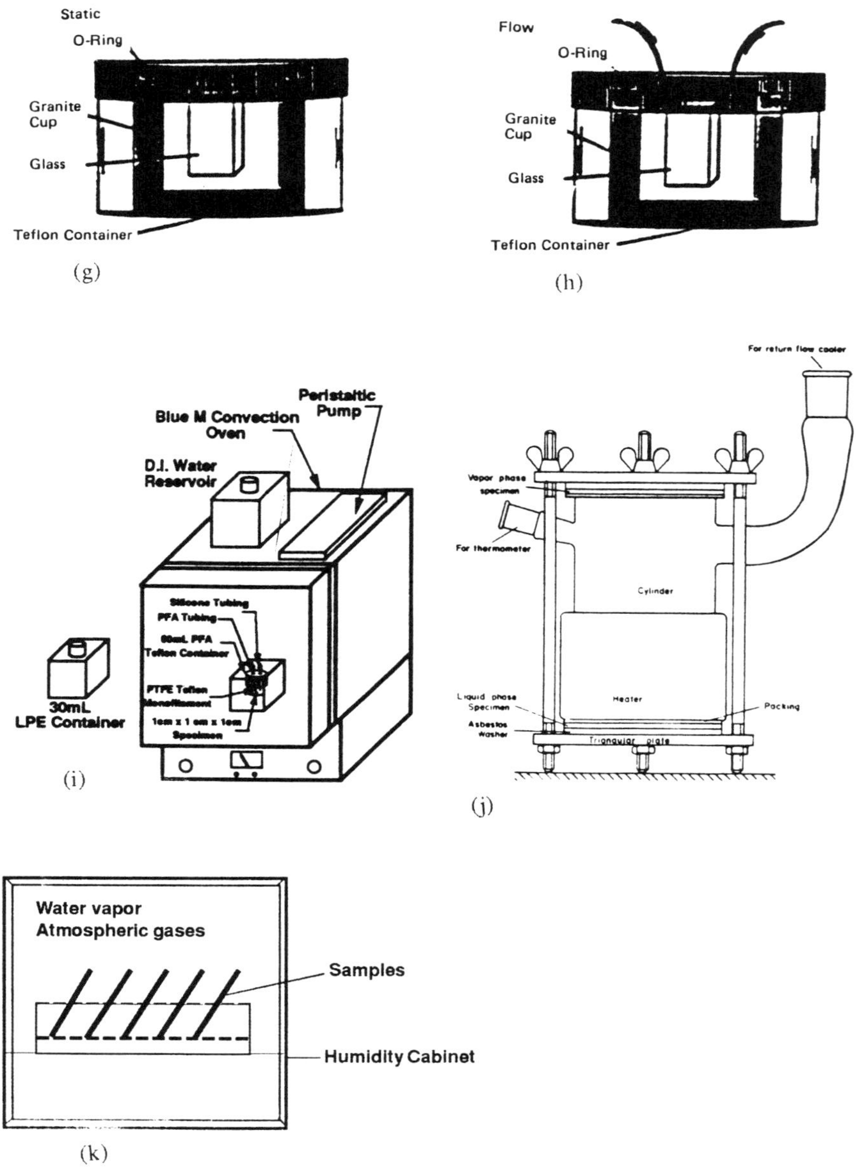

Figure 3-4 cont. (g,h) Rock cup; T = variable, t = variable, SA/V = $0.1cm^{-1}$; (i) Schematic illustrating all the major components of a flow test; (j) corrosion cell used to evaluate vapor and liquid corrosion; (k) weathering chamber; T = variable, t = variable, R.H. = variable.

Table 3-1. Comparison of US and German Glass Durability Test Methods Involving Crushed Glass.

Test Parameter	MCC-3	PCT	ASTM C-225-85
Particle Size	Both <325 and 100-200 mesh	100-200 mesh (0.11mm)	40-50 (0.356mm)
$wt_{solid(g)}$	4	4	10
Use standard glass	NO	YES	NO
V_{soln}	40	40	50
SA/V (cm^{-1})*	19.55 100-200 mesh	19.55	12.13
Time (days)	28,56,91 182,273,364, and longer	7	0.04
(SA/V)•t (cm^{-1}•days)	>547.4 for 100-200 mesh	136.85	0.50
T(°C)	40,90,110, 150,190	90	121
Leachant	ASTM Type I Water** or ground-water	ASTM Type I Water** or ground-water	ASTM Type I Water**
Filtration	<0.45 mm	<0.45μm	None
Leachate Analyses	pH and elemental	pH and elemental	Titration (Si conc being considered)
Agitation	10-14 cycles per minute roll or shake	Shake at start	None

Table 3-1 continued

Sample Crushing	Grind under alcohol	Analytic Grinder (WC blade)	Steel Mortor & Steel Jar Mill use magnet
Sample Washing	Wash with water & alcohol to remove fines	Wash with water & alcohol to remove fines	Wash with acetone to remove fines
Vessel Type	Teflon or Stainless Steel	Teflon or Stainless Steel	Predigested Pyrex Flask

Footnotes

* All surface areas calculated from an average glass density of 2.76 g/cc. The (SA/V)•t parameter is an indication of the reactivity of the test method. The higher the glass surface area (SA), the lower the volume of leachant (V), and the longer the test duration, the more the sample is reacted and the more stringent is the test of the glass durability.

** ASTM D-1193 Specifications for Reagent Water

	Type I	Type II	Type IV
pH	NA	NA	6.2 - 7.5
Total matter (mg/L)	0.1	0.1	2.0
Max soluble SiO_2 (mg/L)	not detectable	not detectable	no limit

Table 3-1 Courtesy of Carol Jantzen, Westinghouse Savannah River Co., Aiken, SC.

Table 3-2. ASTM Standards.

DESIGNATION	TITLE
C 912-79 (Reapproved 1984)	Standard Practice for Designing a Process for Cleaning Technical Glasses (1988 Vol. 15.02 Glass; Ceramic Whitewares)
C 225-85	Standard Test Methods for Resistance of Glass Containers to Chemical Attack (1990 Vol. 15.02 Glass; Ceramic Whitewares)
C 584-81	Standard Test Method for Specular Gloss of Glazed Ceramic Whitewares and Related Products (1988 Vol. 15.02 Glass; Ceramic Whitewares)
C 1084-87	Standard Specification for Heat-Treated Flat Glass Kind HS, Kind FT Coated and Uncoated Glass (1988 Vol. 15.02 Glass; Ceramic Whitewares)
C 724-81 (Reapproved 1987)	Standard Test Methods for Acid Resistance of Ceramic Decorations on Architectural Type Glass (1988 Vol. 15.02 Glass; Ceramic Whitewares)
C 373-72 (Reapproved 1982)	Standard Test Method for Water Absorption, Bulk Density, Apparent Porosity, and Apparent Specific Gravity of Fired Whiteware Products (1988 Vol. 15.02 Glass; Ceramic Whitewares)
C 424-80 (Reapproved 1985)	Standard Test Method for Crazing Resistance of Fired Glazed Whitewares by Autoclave Treatment (1988 Vol. 15.02 Glass; Ceramic Whitewares)
C 556-81	Standard Test Method for Resistance of Overglaze Decorations to Attack by Detergents (1988 Vol. 15.02 Glass; Ceramic Whitewares)
C 279-79	Standard Specification for Chemical-Resistant Masonry Units (1988 Vol. 04.05 Chemical-Resistant Materials; Vitrified Clay, Concrete, Fiber-Cement Products; Mortars; Masonry)
C 622-84	Standard Practice for Corrosion Resistance of Refractories to Molten Glass Using the Basin Furnace (1988 Vol. 15.01 Refractories; Carbon and Graphite Products; Activated Carbon)

Table 3-2 continued

DESIGNATION	TITLE
C 987-83	Standard Test Method for Alkali Vapor Attack on Refractories for Glass-Furnace Superstructures (1988 Vol. 15.01 Refractories; Carbon and Graphite Products; Activated Carbon)
C 126-86	Standard Specification for Ceramic Glazed Structural Clay Facing Tile, Facing Brick, and Solid Masonry Units (1988 Vol. 04.05 Chemical-Resistant Materials; Vitrified Clay, Concrete, Fiber-Cement Products; Mortars; Masonry)
C 267-82	Standard Test Method for Chemical Resistance of Mortars, Grouts, and Monolithic Surfacings (1988 Vol. 04.05 Chemical-Resistant Materials; Vitrified Clay, Concrete, Fiber-Cement Products; Mortars; Masonry)
C 395-85	Standard Specification for Chemical-Resistant Resin Mortars (1988 Vol. 04.05 Chemical-Resistant Materials; Vitrified Clay, Concrete, Fiber-Cement Products; Mortars; Masonry)
C 287-82	Standard Specification for Chemical-Resistant Sulfur Mortars (1988 Vol. 04.05 Chemical-Resistant Materials; Vitrified Clay, Concrete, Fiber-Cement Products; Mortars; Masonry)
C 413-83	Standard Test Method for Absorption of Chemical-Resistant Mortars, Grouts, and Monolithic Surfacings (1988 Vol. 04.05 Chemical-Resistant Materials; Vitrified Clay, Concrete, Fiber-Cement Products; Mortars; Masonry)
C 980-88	Standard Specification for Industrial Chimney Lining Brick (1988 Vol. 04.05 Chemical-Resistant Materials; Vitrified Clay, Concrete, Fiber-Cement Products; Mortars; Masonry)

Table 3-2 continued

DESIGNATION	TITLE
C 1012-87	Standard Test Method for Length Change of Hydraulic-Cement Mortars Exposed to Sulfate Solution (1988 Vol. 04.01 Cement; Lime; Gypsum)
C 911-87	Standard Specification for Quicklime, Hydrated Lime, and Limestone for Chemical Uses (1988 Vol. 04.01 Cement; Lime; Gypsum)
C 492-85a	Standard Test Method for Potential Expansion of Portland Cement Mortars Exposed to Sulfate (1988 Vol. 04.01 Cement; Lime; Gypsum)
C 3987-85	Standard Test Method for Shake Extraction of Solid Waste with Water (1988 Vol. 11.04 Pesticides; Resource Recovery; Hazardous Substances and Oil Spill Responses; Waste Disposal; Biological Effects)
C 738-81	Standard Test Method for Lead and Cadmium Extracted from Glazed Ceramic Surfaces (1988 Vol. 15.02 Glass; Ceramic Whitewares)
C 650-83	Standard Test Method for Acid Resistance of Ceramic Tile to Chemical Substances (1988 Vol. 15.02 Glass; Ceramic Whitewares)
C 735-81	Standard Test Method for Acid Resistance of Ceramic Decorations on Returnable Beer and Beverage Glass Containers (1988 Vol. 15.02 Glass; Ceramic Whitewares)
C 777-84	Standard Test Method for Sulfide Resistance of Ceramic Decorations on Glass (1988 Vol. 15.02 Glass; Ceramic Whitewares)
C 676-74 (Reapproved 1980)	Standard Test Method for Detergent Resistance of Ceramic Decorations on Glass Tableware (1988 Vol. 15.02 Glass; Ceramic Whitewares)

Table 3-2 continued

DESIGNATION	TITLE
C 675-85	Standard Test Methods for Alkali Resistance of Ceramic Decorations on Returnable Beverage Glass Containers (1988 Vol. 15.02 Glass; Ceramic Whitewares)
C 895-87	Standard Test Method for Lead and Cadmium Extracted from Glazed Ceramic Tile (1988 Vol. 15.02 Glass; Ceramic Whitewares)
C 872-83	Standard Test Method for Lead and Cadmium Release from Porcelain Enamel Surfaces (Cited 1988 Vol. 15.02 Glass; Ceramic Whitewares-However it is missing from vol.)

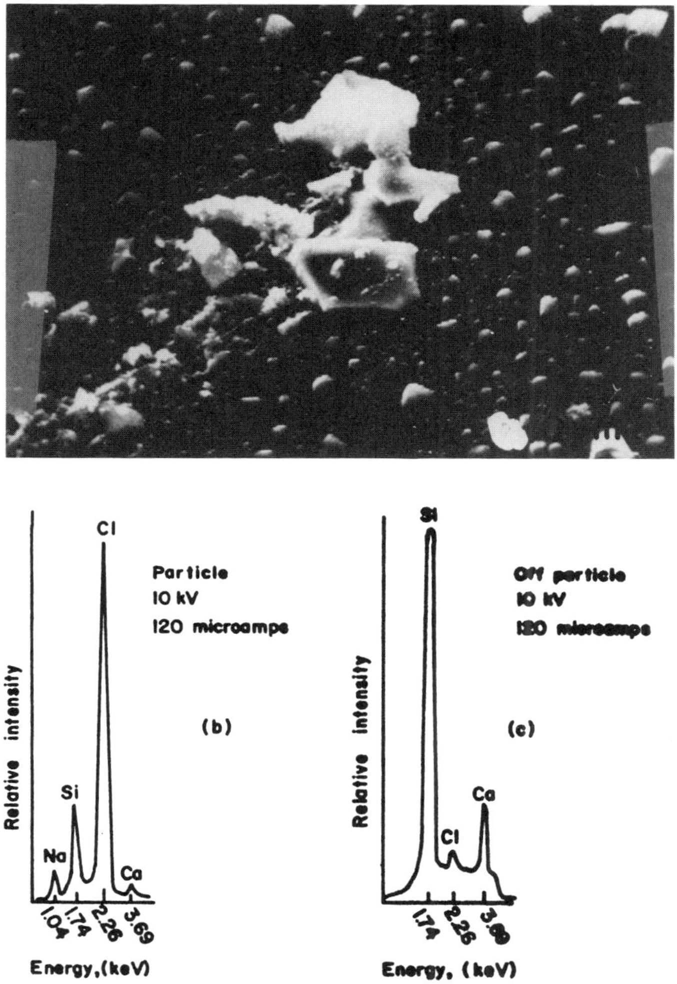

Figure 3-5. Scanning electron micrograph and energy dispersive x-ray spectra illustrating the formation of weathering products during exposure of a soda-lime-silicate glass to high humidity (100°C, ≈100% R.H., 20 days). Adapted from (18).

Other references related to glass cleaning are Adams (22), Hench (23) and Campbell and Adams (24).

If the relative performance of a material is required, the test procedure may be as simple as exposing a well-prepared sample to a standard solution for specified times. The Materials Characterization Center static test (MCC-1) shown in Figure 3-4f is designed to evaluate relative performance. Other standard tests such as those developed by the ASTM are shown in Figures 3-4a and b and are used for quality assurance. Materials that perform within acceptable limits under accelerated conditions are expected to perform satisfactorily in service. Design and use of such tests are usually based on years of field observations and correlation of these observations with laboratory test results. It should be realized that accelerated tests can only be used to predict performance when the mechanisms of corrosion are the same as those that occur under service conditions. Increasing temperature for the purpose of accelerating corrosion does not always meet this requirement (see Wicks' chapter).

The use of powders for evaluating corrosion has been the subject of much debate (25-27). A major concern is the presence of agglomerated fines which increase the true surface area and result in higher apparent extraction rates. Tables 3-3 and 3-4 summarize corrosion data for bulk and powder specimens where all other parameters are equal. The pH changes were nearly the same in both tests as were the Si leach rates. The most significant differences are observed with the early stage leaching of the more soluble elements: B, Na, and Li. After 28 days, the leach rates for these elements were nearly equivalent in both tests. In spite of these differences, it appears that powders provide as useful solution data as bulk samples.

Although the obvious experimental variables such as temperature, glass composition and leaching solution chemistry have been extensively studied, the role of glass surface area to solution volume ratio (SA/V) in corrosion studies has not been widely recognized until recently. Several papers which have been published on this subject illustrate that the experimental results are dependent on SA/V when other parameters are held constant (28-35).

The SA/V can be experimentally controlled in a laboratory in the range of about 0.001cm^{-1} to 100cm^{-1} when using monolithic samples. Higher SA/Vs do not produce sufficient solution for analysis even when using an ICP (inductively-coupled plasma spectrometer). Although smaller SA/Vs provide sufficient solution, the elemental concentrations in solution may be too small to detect with ICP. Alternatively, powders can be used to increase SA/V to values greater than 100cm^{-1}, but it is difficult to perform surface analyses on powders. The use of powders to achieve high SA/V is the basis for most solubility/saturation tests.

It has been shown by numerous investigators that both solution and surface analyses are required in order to obtain a thorough understanding of the corrosion processes and the effects of variables on these processes. For this reason monolithic samples are usually used in corrosion studies instead of powders. A combination of powders and bulk samples can be used to provide a very high SA/V. This condition leads to rapid saturation of the solution with respect to certain species from the glass (e.g. silica). Due to saturation effects bulk samples exposed to high SA/V solutions exhibit less surface degradation than do similar samples exposed to low SA/V solutions. Figure 3-6 shows the effect of solution concentration on glass corrosion using IRRS (36). Two identical bulk samples

Table 3-3. Total Mass Loss per Unit Area and Normalized Elemental Loss per Unit Area for the Corrosion of a SRL 131-29.8% TDS-3A Sample ($1cm^2$ x $0.2cm^2$). In deionized water at 90°C with SA/V = $0.1cm^{-1}$. Solution analyses performed at room temperature.

"BULK"

SAMPLE		CORROSION TIME	pH	MASS LOSS (gm/cm^2)	Si (gm/cm^2)	B (gm/cm^2)	Na (gm/cm^2)	Li (gm/cm^2)	Fe (gm/cm^2)	Al (gm/cm^2)
1	A	3 days	9.46	4.57	4.34	4.67	2.71	4.84	0.0	2.7
	B		9.45	5.37	4.74	4.67	2.81	4.30	0.0	3.3
2	A	1 week	9.55	8.92	6.63	9.35	3.92	8.60	0.0	3.3
	B		9.60	7.96	7.14	9.97	4.12	8.60	0.0	4.0
3	A	2 weeks	9.66	9.84	8.16	11.5	4.92	11.3	0.0	4.7
	B		9.64	9.92	8.67	11.8	5.12	11.3	0.0	4.7
4	A	3 weeks	9.69	10.1	10.2	12.1	5.92	16.7	0.0	4.7
	B		9.70	11.1	9.69	13.4	6.93	17.7	0.0	5.3
5	A	4 weeks	9.70	11.2	10.7	12.8	6.83	18.3	0.0	5.3
	B		9.64	11.0	11.2	14.0	7.03	19.9	0.0	5.3

Deionized water had an initial pH = 5.68 and resistivity = 10MΩ

Table 3-4. **Total Mass Loss per Unit Area and Normalized Elemental Loss per Unit Area for the Corrosion of Freshly Fractured -100 Mesh +200 Mesh SRL 131-29.8% TDS-3A Sample.** In deionized water at 90°C with SA/V = 0.1cm^{-1}. Solution analyses performed at room temperature.

"POWDERS"

SAMPLE		CORROSION TIME	pH	Si (gm/cm^2)	B (gm/cm^2)	Na (gm/cm^2)	Li (gm/cm^2)	Fe (gm/cm^2)	Al (gm/cm^2)
1	A	3 days	9.44	3.78	7.79	4.32	6.45	0.0	3.3
	B		9.45	4.13	8.72	4.42	6.45	0.0	3.3
2	A	1 week	9.56	6.12	13.1	6.32	10.7	0.0	4.0
	B		9.60	5.61	11.8	6.02	8.60	0.0	3.3
3	A	2 weeks	9.62	7.56	14.0	6.73	13.4	0.0	4.7
	B		9.61	8.16	14.3	7.03	13.4	0.0	4.0
4	A	3 weeks	9.69	9.69	14.6	7.63	15.1	0.0	5.3
	B		9.72	9.18	15.3	7.83	15.1	0.0	5.3
5	A	4 weeks	9.76	11.22	14.9	8.43	18.8	0.0	6.0
	B		9.71	12.24	16.2	8.94	18.8	0.0	6.0

Deionized water had an initial pH = 5.68 and resistivity = 10MΩ

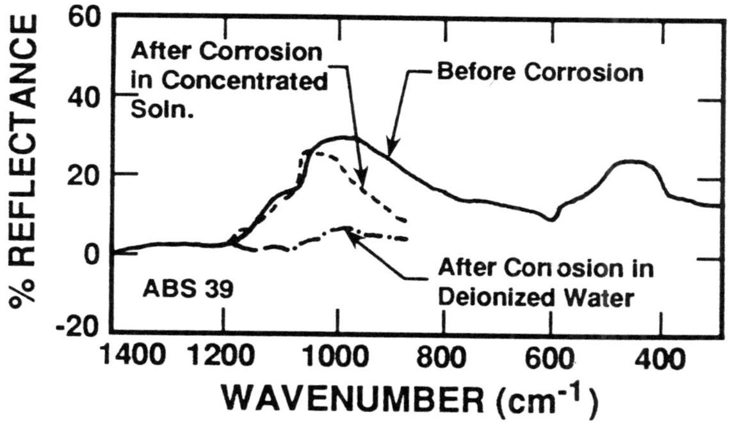

Figure 3-6. Infrared reflection spectra illustrating the reduced corrosion of an alkali borosilicate glass in a concentrated solution vs. deionized water. Concentrated solution means with respect to the elements in the glass. Test was conducted at 90°C, SA/V = $0.1cm^{-1}$, 28 days. Adapted from (36).

of an alkali borosilicate glass (ABS) were immersed in separate solutions: deionized water and deionized water with a high surface area powder of the ABS glass. In both cases the SA/V with respect to the bulk sample was $0.1cm^{-1}$. As can be seen from the spectra, the sample exposed to deionized water only has undergone more extensive surface degradation.

Although the initial, kinetically limited elemental leach rate is unaffected by SA/V, the rate of increase in concentration of species in solution is greater at high SA/V, as long as there are no changes in the dissolution mechanisms. Buckwalter et al. (29) found experimentally that the normalized mass loss for Si shows the greatest increase with SA/V, while the normalized mass losses of K, Na and B increase to a lesser degree. The reason for the increased leach rate for the Si is probably due to solution pH. It has been shown that the dissolution of Si increases significantly as the pH is increased. The solution pH increases as SA/V increases when their test parameters are held constant. A small, solid sample immersed in a large volume of solution will result in a slower change in solution pH than will a large sample immersed in the same volume of solution. Thus, the Si leach rate is expected to increase with SA/V.

In static corrosion experiments, where the same solution remains in contact with the glass over the duration of the experiment, the manner of achieving a specified SA/V is unimportant. For example, a SA/V of $0.1cm^{-1}$ can be obtained by immersing a $1.0cm^2$ surface area sample into $10cm^3$ of water, or by immersing a $6cm^2$ sample into $60cm^3$ of water. The relative changes in solution chemistry (i.e. solution concentrations and pH) will be equivalent in both cases.

SA/V can be varied over a wide range of values by changing either SA, V or both without affecting the solution analyses results. The absolute quantity of material released into solution will be greater for the high surface area sample at any given time until solution saturation is achieved. Once solution saturation has been achieved, the elemental concentrations in solution should be the same regardless of the SA/V. However, when the elemental concentrations are converted to normalized mass losses (g/m^2), these values will be inversely proportional to the SA/V under conditions of saturation. It has also been shown by Pederson et al. (32) and Barkatt et al. (31) that the solution concentration of a given element at a particular value of SA/V x t is a constant regardless of how that value of SA/V x t was obtained. In other words, with regard to solution concentration, a high SA/V test run for a short time produces the same result as a low SA/V test conducted for a long time. Chandler and Wicks (30) have also found this relationship to be valid, but only for particular elements such as Si, Na, B, Al and U. This concept may be used to provide an accelerated leach test where data obtained in the laboratory over a few days can be used to predict concentrations obtained over much longer periods of time under different SA/V values. It has been pointed out by Pederson et al. (32) that the use of SA/V for accelerated leach testing may produce problems when alteration phases are formed that have solubilities considerably less than those of the glass. In such cases the elements lost into solution may precipitate onto the glass surface or other parts of the corrosion cell, thus preventing solution saturation from occurring. Pederson et al. conclude that the dissolution of silicon is most important since saturation of that component reduces leaching rates of other elements whose concentrations are far from saturation. Events such as silica colloid formation and certain alteration phase formation may also modify the SA/V x t relationship.

The normalized boron release versus time is shown in Figure 3-7. In this case a higher SA/V yields a greater boron release after about 7 days of exposure. These differences are probably due to more rapid pH changes in the higher SA/V solution and to the fact that boron must be leached from a greater depth in the low SA/V solutions to yield equivalent solution concentrations.

Machiels and Pescatore (33) have studied the effects of SA/V on glass leaching from a theoretical point of view. The importance of SA/V in static as well as flow tests is modeled for network dissolution and diffusion-controlled reaction. A computer code based on the theoretical model has been tested for PNL 76-68[1] glass leached under static conditions and various SA/V values. There is a good correlation between experimental data and computer-generated plots for Na and Si normalized mass losses. In our own experience the role of SA/V in flow experiments is not as straightforward as in static experiments. For

---

[1]Simulated nuclear waste glass developed by Batelle Pacific Northwest Laboratories.

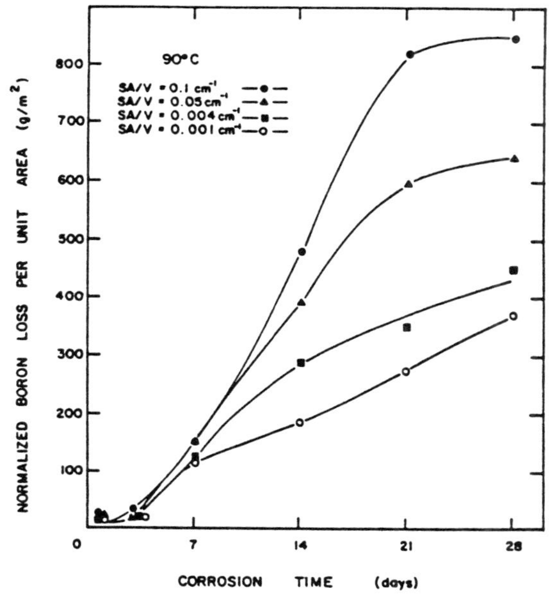

Figure 3-7. Normalized boron loss per unit area vs. corrosion time for an alkali borosilicate glass. Test was conducted at 90°C in deionized water. Analyses were performed at room temperature.

instance, the results of the experiment can be affected in a flow experiment by the method by which the SA/V is obtained. Reducing the volume of the leaching vessel to increase the SA/V will produce a smaller contact time with the sample than would be obtained by increasing the SA/V and maintaining the same volume to produce the equivalent SA/V.

In order to evaluate the effects of flow on wasteform leaching, MCC-4 was developed by the Materials Characterization Center (7). This test is one in which the solution passes through the leaching vessel once (i.e. single pass), and is similar to the test developed by Coles et al (37). Strachan et al. (38) have reported increased leach rates for silicon and strontium at a flow rate of $6mlh^{-1}$ compared to static testing. Similar results have been found by Adiga et al. (39).

Flow tests offer certain advantages over static tests in that more solution analyses can be obtained with time and solution chemistry can be better controlled. A certain amount of time is required for steady state conditions to be achieved in a flow test (Figure 3-8a) (40). Shorter times are required for more rapid flow rates, smaller vessel volumes, higher SA/V and higher temperatures. Steady state is defined as the condition where the concentration, C, of the species exiting the vessel does not change with time ($\delta c/\delta t = 0$). Figure 3-8b illustrates the steady state concentrations, C_{ss}, for several elements

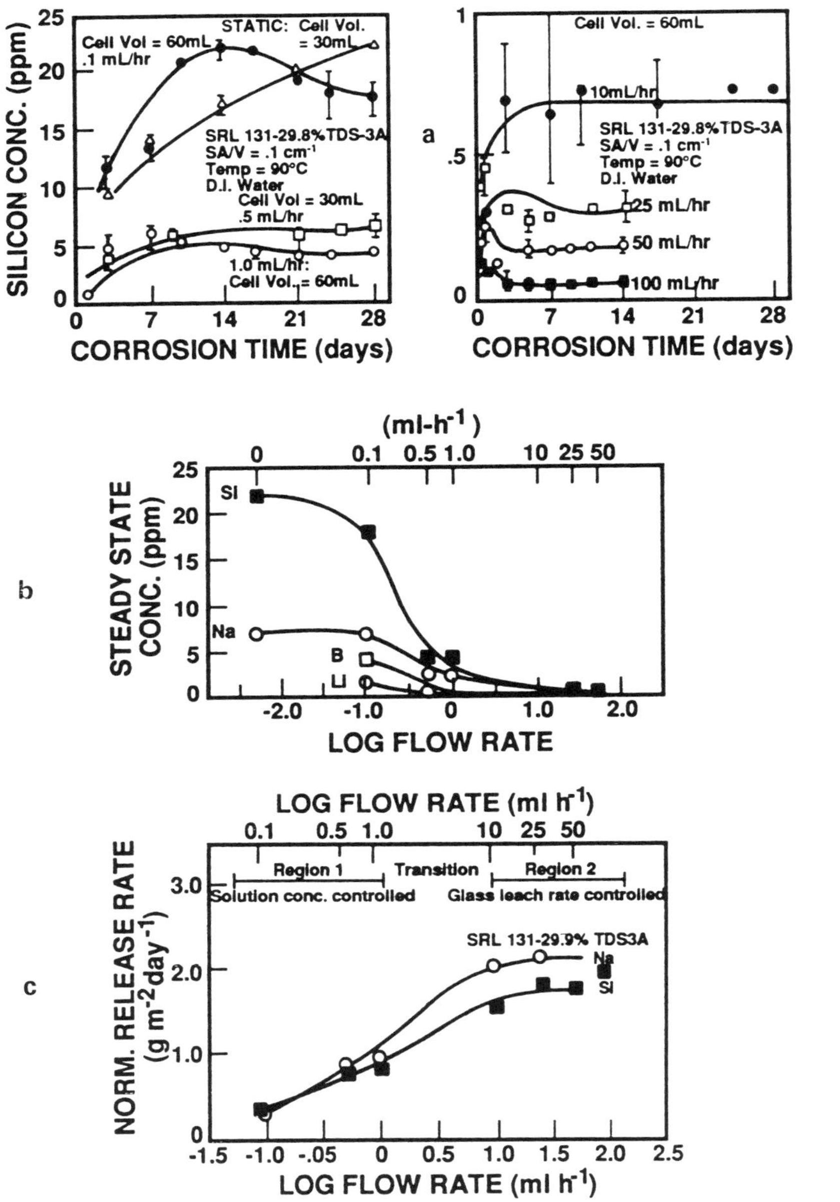

Figure 3-8. Flow test conducted on an alkali borosilicate glass at 90°C at various flow rates and times. Analyses were performed at room temperature. Adapted from (40).

over a wide range of flow rates. The corresponding normalized mass losses for these elements are shown in Figure 3-8b. At slow flow rates the normalized leach rate will be controlled by solution concentration effects, while at high flow rates the normalized leach rate will be controlled by the glass. This affect is shown in Figure 3-8c. It is possible that during testing, leached species are deposited onto the walls of the leach vessel. In static leach tests, the deposits can be recovered after the tests by acid stripping. These data are then added to those obtained from solution to give a more accurate analysis. Depositions that occur during flow tests cannot be accounted for until the end of the test, which may be several months.

Perhaps the single most important variable that controls chemical durability is solution pH (41-46). For this reason many of the standardized and industrially used tests require that the material be exposed to a wide range of solution pH. As seen in Figure 3-9 (44), alkaline and acidic solutions result in a larger extent of corrosion (%wt loss) than do more neutral solutions. The corrosion trend (U-shaped curve) is associated with the solubilities of the major oxide constituents in the material. Note the high value of elemental mass loss at pH 1.1 and pH 12 compared to pH 7 in Figure 3-10. Although the solubility of SiO_2 is not high at low pH, the solubilities of most of the other oxides that comprise about 40 wt% are high. Their release into solution results in a concomitant dissolution of SiO_2. In alkaline solutions the solubility of SiO_2 is high and its dissolution results in the simultaneous removal of most other oxides since it is the major network former.

In addition to buffer solutions, the pH of the leachant may be controlled by constant monitoring and adjusting with acids and bases. Such controlled pH tests are discussed in references 12 and 46.

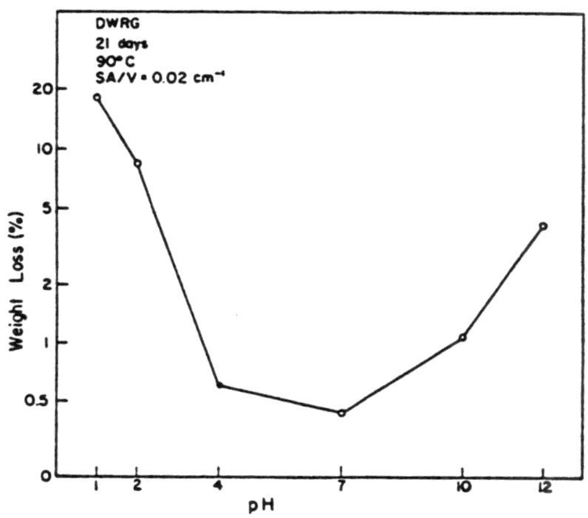

Figure 3-9. Total mass loss vs. pH after corroding an alkali borosilicate glass in pH buffer solutions at 90°C for 21 days.

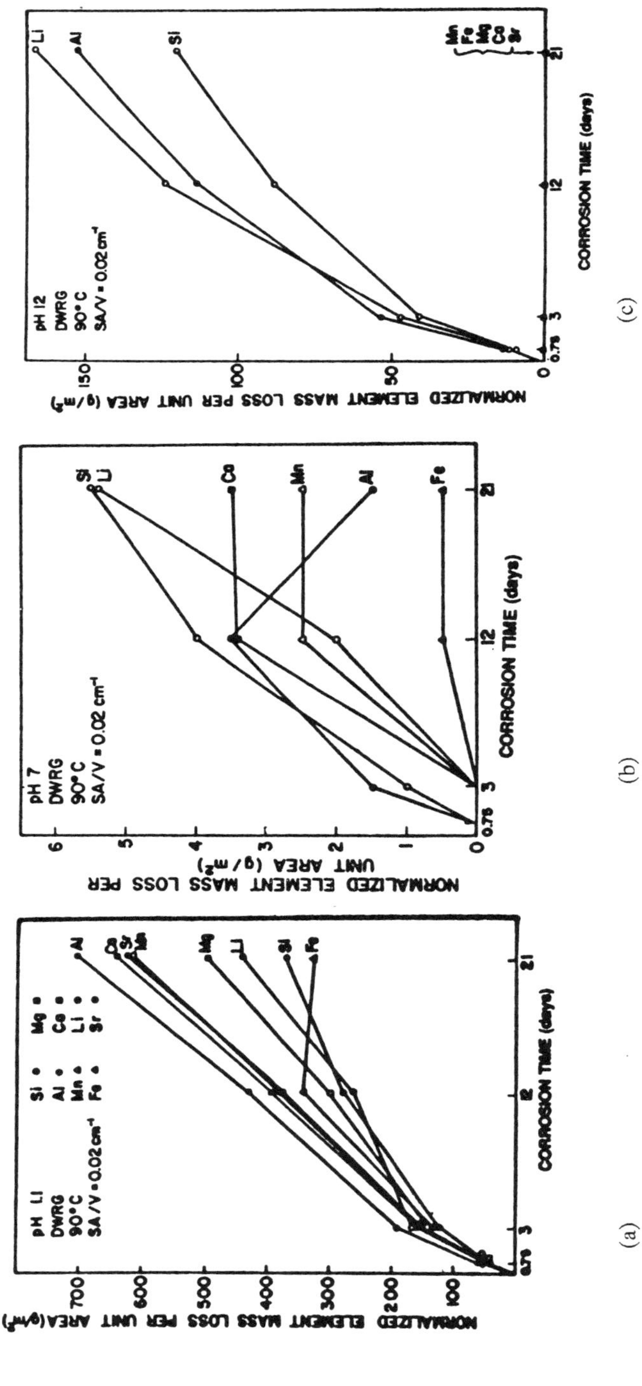

Figure 3-10. Normalized elemental mass loss per unit area vs. corrosion time for a simulated nuclear waste glass. (a) pH = 1.1; (b) pH = 7, and; (c) pH = 12. Analyses were performed at room temperature.

In many applications, materials may remain in constant contact with the same solution for long periods. During this time selective leaching of species can alter the solution pH which in turn affects the solubilities of other species in the material. In most silicate and borate glasses, alkali ions such as Li^+, Na^+, and K^+, are selectively removed resulting in an increase in solution pH as shown in Figure 3-11. When materials are allowed to remain in unbuffered static solutions for long periods extensive network dissolution (uniform or local) may occur.

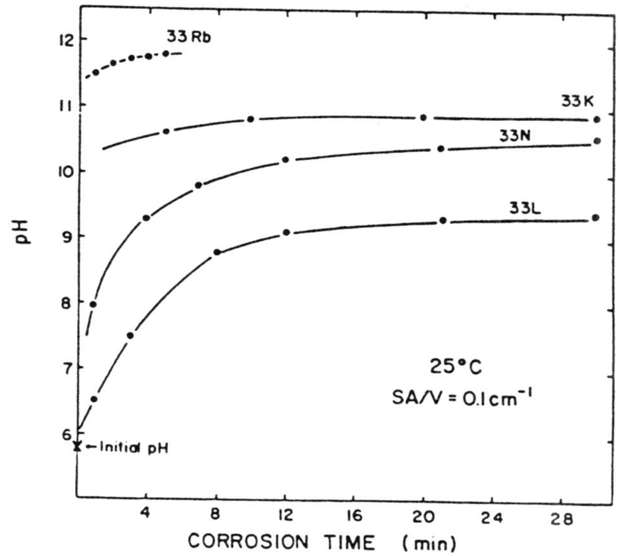

Figure 3-11. pH vs. corrosion time for alkali silicate glasses corroded in deionized water.

FIELD TESTS

The purpose of field tests is to simulate as closely as possible those environmental conditions that the material will experience during service and then to evaluate its performance. In some cases the field test may involve the use of the material in a typical application for specified times prior to its removal and evaluation. In other cases, defective products may be returned to the manufacturer for analysis and a determination of why performance was "different" from that "expected". The results of each of these tests helps to provide a data base that can then be used for product improvement or assist in the specification of the environmental limitations within which the product can be expected to perform satisfactorily. In most applications field tests cannot be designed to exactly simulate service conditions. The reason for this is that ceramic materials are expected to last for many years in a variety of changing environments. Therefore, the results from field tests are generally used in combination with laboratory test results to obtain a more comprehensive understanding of degradation mechanisms.

There are many field tests involving ceramics reported in the literature (47-53). However, one of the most comprehensive and systematic field tests is that for simulated nuclear waste glass reported by George Wicks in this book. This test is referred to as a materials interface interactions test (MIIT) and is presently in progress at the Waste Isolation Pilot Plant (WIPP) near Carlsbad, New Mexico. This test methodology is based on a prior experiment begun in 1982 at the Stripa Mine in central Sweden. Stripa is an inactive iron ore producing mine set in a granite geology. In order to understand the leaching behavior of nuclear waste forms under practical repository conditions, a joint project was initiated between the University of Florida (UF), Savannah River Laboratory (SRL), and the Swedish Nuclear Fuels Safety Division of the Nuclear Fuel Supply Co. (SKBF/KDS). The objectives of the burial test were to:

- Evaluate the leaching behavior of potential nuclear waste materials in the presence of groundwater, canister metals, overpack materials and geologic materials under a more (realistic) setting than can be obtained in the laboratory.
- Develop a characterization methodology for comparing the field data with laboratory data.
- Assess corrosion mechanisms and correlate these with those found in laboratory tests.

The materials to be tested were formed in the shape of pineapple slices shown in Figure 3-12 and stacked together to create different types of interfaces (i.e. glass/glass, glass/metal, glass/mineral). The assemblies were then placed into boreholes (≈ 345 meters below the surface), heater rods were inserted to maintain the desired temperature and the holes were sealed. Assemblies were harvested after 1 month, 3 months, 12 months and 24 months for analyses. The results of this field test provided some very useful information regarding the performance of materials under simulated repository conditions. Figure 3-13 illustrates depth of leaching vs. burial time determined from SIMS profiles. This depth of leaching is based on the boron which is leached from the greatest depth within the glass. After two years, the depth of leaching is only about 0.5µm for the burial sample compared to approximately 6µm from the laboratory leached sample. When all of the data from SIMS, IRRS and SEM are evaluated together, the results suggest that the mechanism of leaching in field tests are the same as those in laboratory tests and that the rates of degradation in the field are much lower than those predicted by laboratory tests. This latter finding can be explained by the lower SA/V used in the laboratory tests. As a result of these field tests more relevant laboratory tests have been developed that provide results more comparable to those obtained in the field (see Figure 3-4g,h,i).

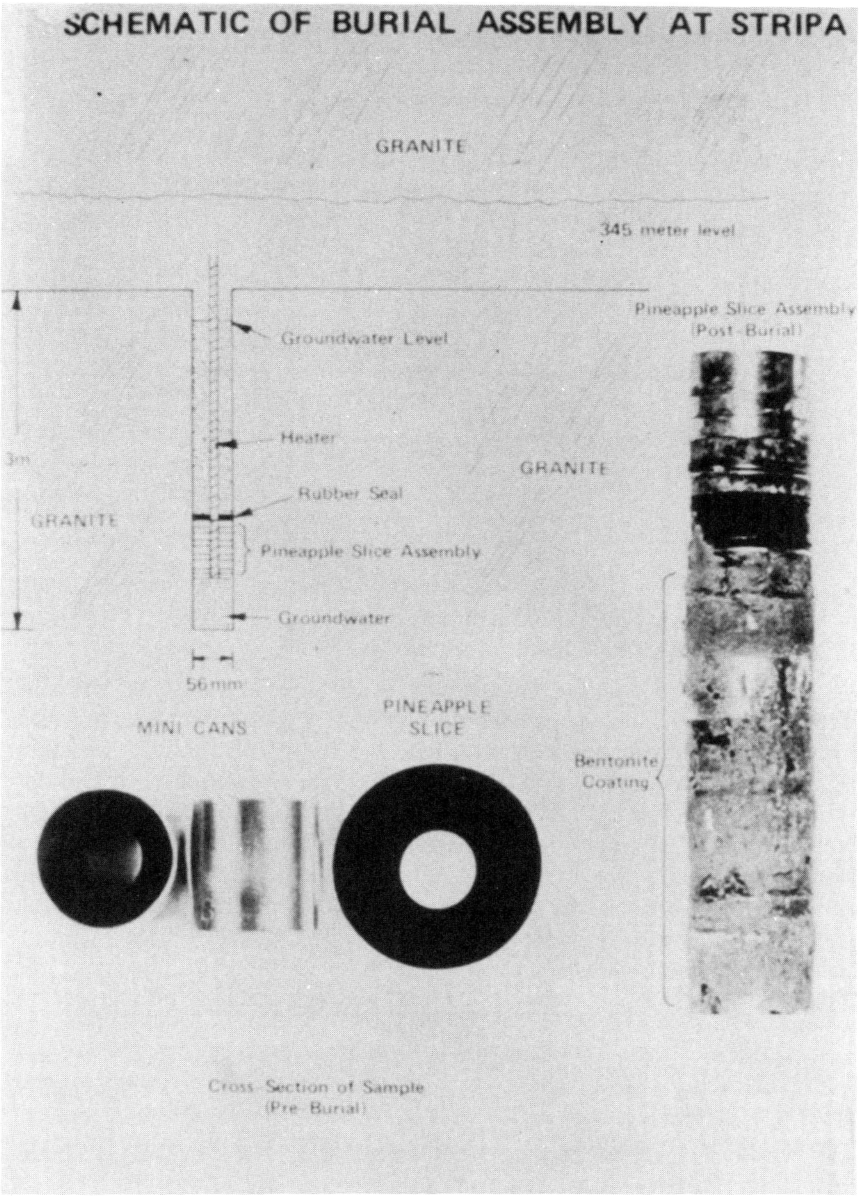

Figure 3-12. Diagram illustrating the positions of the samples in the Stripa mine during burial. The pineapple slice sample configuration is also shown along with a photograph of the 1 month, 90°C assembly immediately after removal from the borehole. The 90°C temperature was maintained with a heater element inserted in the center hole of the assembly. Adapted from (51).

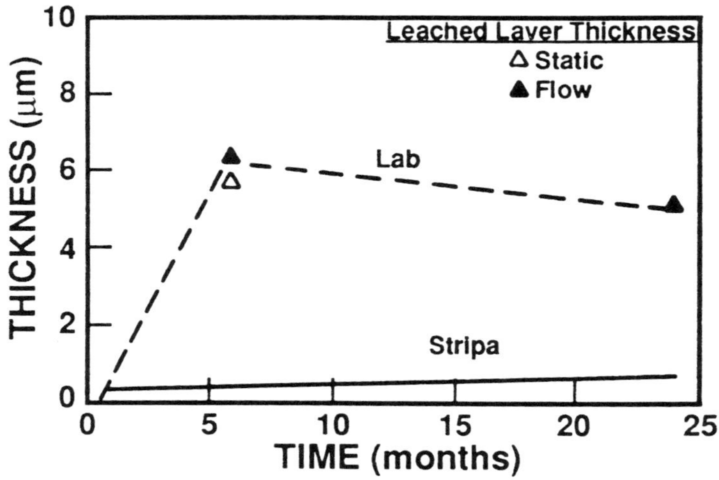

Figure 3-13. Depth of corrosion layer vs. exposure time for laboratory and burial samples (simulated nuclear waste glass). Depths were determined from SIMS profiles. All tests were conducted at 90°C. Laboratory samples had SA/V = $0.1 cm^{-1}$.

CHARACTERIZATION OF CORROSION

Typically, the output of a corrosion test is a solid specimen which has undergone corrosion and the solution with which it was in contact during the test. It is the task of characterization to assess aspects of corrosion by measuring specific parameters of these two test products. In many practical situations only a particular aspect of the corrosion process may be important, for instance contamination of a product by a corroded glass container, loss of strength in ceramic construction materials or loss of superconducting properties in oxide superconductors. However, if the objective of the study is to identify the mechanisms of corrosion, an integrated approach to characterization is required. Such an approach should utilize overlapping characterization techniques, each of which will provide a separate "puzzle piece" and when considered together will provide a broad view of material degradation in a corrosive environment. The objective of this section is to present an overview of the analytical techniques that are available and to provide examples of their use in understanding and characterizing the corrosion process.

Broadly considered, characterization methods may be classified as solution techniques or surface/solid techniques. Solution measurements are performed to measure the concentration of elements which have been leached from the ceramic into solution. Surface and solid studies may address specimen structure (atomic-level), microstructure or composition. An overview of techniques discussed in this chapter is given in Table 3-5. Technical aspects of surface techniques have

Table 3-5. Overview of Characterization Methods.

Solution

pH, specific ion electrodes
Colorimetry
ICP, ICP-MS

Solid/Surface

Structure
Infrared Spectroscopy
Raman Spectroscopy
Nuclear Magnetic Resonance
X-ray Photoelectron Spectroscopy
Extended X-Ray Absorption Fine Edge Structure

Microstructure
Transmission Electron Microscopy
BET Gas Adsorption Analysis
Small Angle X-ray scattering

Composition
Secondary Ion Mass Spectroscopy
X-ray Photoelectron Spectroscopy
Auger Electron Spectroscopy
Electron Microprobe Analysis
Rutherford Backscattering
Nuclear Reaction Analysis

been covered in detail in a number of references (54-60). The overall choice of a set of characterization techniques depends on the purpose of the study or test.

Solution Analysis

A number of techniques are available for determining ion concentrations in leachate solutions. pH and specific ion electrodes are used to measure hydrogen ion activity as well as activity of other ions. Colorimetric methods, utilizing ammonium molybdate for silica and curcumin, 1:1 dianthrimide, carminic acid or quinalizarin for boron detection, were among the earlier techniques of solution analysis (61,62). Colorimetric methods have largely been replaced by inductively coupled plasma spectrometry (ICP), because of its accuracy, ease of use and low detection limits (on the order of parts per billion).

ICP works by introducing a vaporized sample into a plasma torch at a temperature of approximately 7000K. At this temperature the atoms undergo excitation and, upon returning to ground state, emit characteristic optical radiation (63). Since each element has a specific set of emission lines (64) it is possible to use a spectrometer to observe individual line intensities. Comparison of these line intensities to those of a set of standards permits determination of specimen concentration.

Solution analyses have been conducted in numerous glass corrosion studies. In the case of borosilicate glass, the amount of boron appearing in solution has been used to assess the amount of glass reacted, since boron is not typically solubility restricted (65). Solution analysis played a key role in the early work of Grambow, who showed that the solution concentration of various metal cations as a function of pH in glass-leachate systems behaves as though the cation were in equilibrium with a solid precipitate phase (66). Clark and Hench (67) measured solution concentrations of lithium, aluminum and silicon over time to show changes in leaching mechanisms as shown in Figure 3-14. This figure shows that initially lithium is released from the glass, indicating ion exchange. Later, the proportions of each element in the solution approach those of the parent glass (indicating congruent dissolution) and finally, aluminum disappears from solution, indicating that precipitation has taken place.

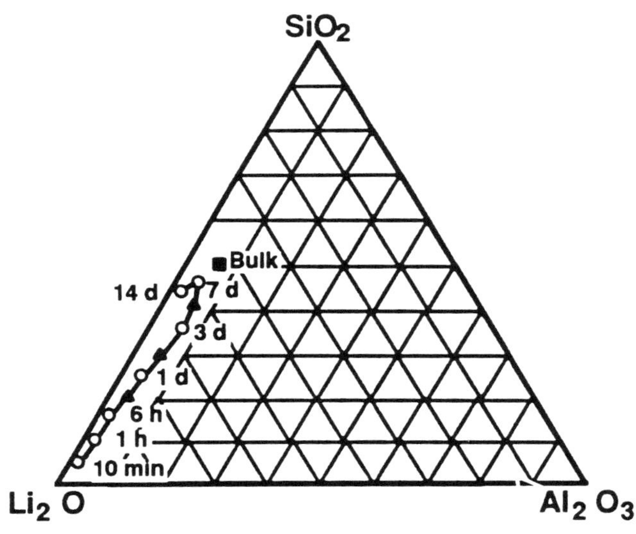

Figure 3-14. Evaluation of solution composition over time during leaching of Li_2O-Al_2O_3-SiO_2 glass. Adapted from (67).

Recent developments in solution analysis include the combined use of ICP and mass spectrometry (MS) (68). In this particular case, mass spectrometry was used to solve a unique problem within the Materials Interface Interaction Testing Program (described by George Wicks in Chapter 7 of this volume). Changes in solution concentration alone were not sufficient to monitor glass leaching, since all major elements (silicon, boron, sodium, lithium) occurred in the the glass as well as the surrounding geology. By noting that the glass had a different $^{7}Li/^{6}Li$ ratio than the surrounding brine, it was possible to determine the amount of lithium that had leached from the glass and thus make upper and lower estimates on the amount of glass which had reacted.

Solid/Surface Analysis

Structure

Perhaps the most fundamental level on which to approach glass corrosion is that of atomic-level structure. In this approach glass structure before and after leaching is examined and related to mechanisms of corrosion. Among the more commonly used methods for structural analysis are infrared and Raman spectroscopies, nuclear magnetic resonance (NMR), x-ray photoelectron spectroscopy (XPS) and extended x-ray absorption fine structure (EXAFS).

Infrared spectroscopy. Infrared spectroscopy is based upon the interaction of infrared radiation with vibrations of the dipole moment of atomic bonds. Of particular importance in silicate glasses are vibrations of siloxane bonds (Si-O-Si) and non-bridging oxygen bonds (Si-O:R, R = Na, Li, K, etc.). Water-based structures such as OH may also be examined using IR. Early work showed that reflectance peaks in the vicinity of 800 - 1200 cm^{-1} are due to contribution from bridging and non-bridging oxygen atoms (69). Upon leaching, the low wavenumber end of these peaks diminishes in intensity due to loss of alkali through ion exchange and the high wavenumber band shifts to even higher wavenumbers (70). Chen and Park (71) examined changes in infrared absorption spectra of blown glass films weathered under ambient room conditions. They observed a loss in intensity of the non-bridging oxygen peak, while the intensity of the bridging oxygen peak remained constant. In addition, they observed the growth of peaks in the 1300-1650 cm^{-1} region which was attributed to formation of a $Na_3H(CO_3)_2{\cdot}2H_2O$ reaction product. Aines (72) used diffuse reflectance IR to examine water and hydroxyl content in the leached layers of SRL 131, SRL 165 and ATM-1c simulated nuclear waste glasses. By examining reflectance peaks at 4500 cm^{-1} (OH stretching) and 5200 cm^{-1} (H_2O stretching and scissoring) and utilizing previously measured molar absorptivities for water they were able to measure the quantity of water and OH in liters/mol-cm within leached layers.

Raman spectroscopy. Raman spectroscopy is based on the coupling of monochromatic electromagnetic radiation with the polarizability tensor of structures in the glass. In this way Raman is capable of detecting vibrational modes due to bridging and non-bridging oxygen atoms (73) as well as larger silicate ring structures (74,75). Exarhos and Conway (73) used Raman spectroscopy to examine in situ structural changes during the leaching of sodium trisilicate fibers. They found that after seven days the spectrum changed from one characteristic of the unleached glass to one which was a combination of the spectrum of vitreous silica and silicic acid ($0.8H_2O{\cdot}SiO_2$). It was also noted that the number of non-bridging oxygen sites increased, presumably as a result of attack on the Si-O-Si network. Bunker, et al. (76) used Raman to examine the structure of several compositions of sodium borosilicate glass prior to and after leaching at pH 1, 9 and 12. The resulting spectra are shown in Figure 3-15. From this information, Bunker was able to identify Q^4, Q^3 and Q^2 units (silicate units with 4, 3 and 2 bridging oxygen bonds), fourfold rings made up of two neutral silicate and two anionic borate groups, and fourfold silicate rings. In addition, he was also able to obtain measurements of Si-OH/Si concentration ratios.

Nuclear magnetic resonance. In recent years, nuclear magnetic resonance (NMR) has been used to examine the local environment of atoms which possess a net magnetic moment. This type of spectroscopy is sensitive to local atomic environment and is based on the coupling of the magnetic moment of the atomic nucleus with externally applied magnetic fields. This method was used as part of the study by Bunker et al. to corraborate Raman results discussed previously. NMR spectra of ^{29}Si supported observations of Q^4, Q^3 and Q^2 groups and further indicated that Q^3 and Q^2 groups in the leached glasses involved exclusively silanol groups rather than Si-O:Na. ^{11}B NMR indicated that prior to leaching boron existed in charge-compensated anionic tetrahedral sites and neutral trigonal sites. In situations where boron was selectively leached from the glass, residual boron was found to exist solely in a threefold coodination, associated with B-OH bonds. At high pH values, where no selective boron leaching was observed, boron structures were identical to those in the original glass. ^{17}O NMR was used to examine the incorporation of oxygen from ^{17}O-labelled water into the leached glass. Bunker observed that ^{17}O was incorporated into Si-O-Si bonds for high alkali glasses and almost not at all for low-alkali content glasses. Bunker also observed the incorporation of ^{17}O into Si-O-B as well as non-bridging oxygen sites.

X-ray photoelectron spectroscopy. X-ray photoelectron spectroscopy (XPS), also known as electron spectroscopy for chemical analysis (ESCA), is based on the photoelectric effect and is capable of providing chemical state information (bonding, oxidation state) as well as quantitative compositional information. Veal et al. (77) used XPS to monitor changes in the oxygen 1s line upon addition of CaO to $Na_2O{\cdot}2SiO_2$. They found that this line consists of contributions due to bridging and non-bridging oxygen atoms, and upon addition of CaO the non-bridging oxygen peak grew at twice the rate that the bridging oxygen peak decreased, consistent with the production of two non-bridging oxygen sites and destruction of one bridging oxygen site per CaO molecule added. Having established this calibration, they used XPS to examine the

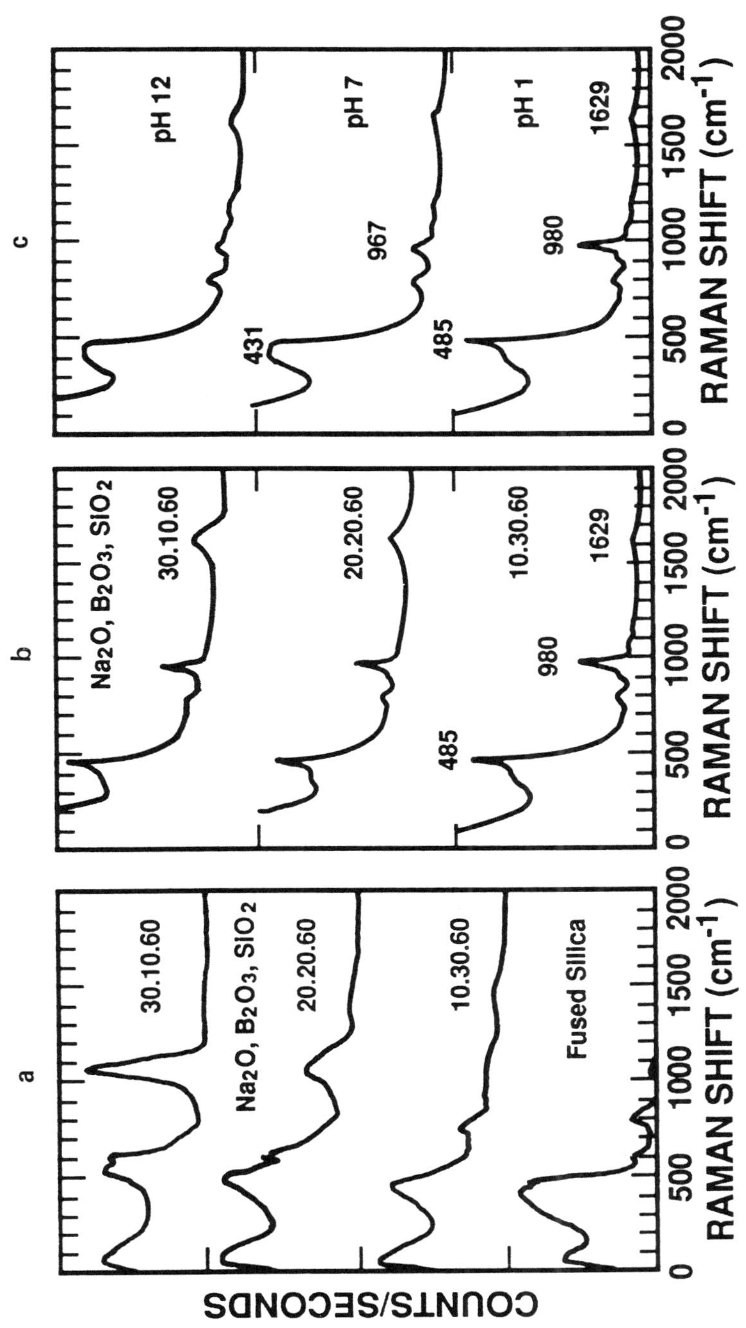

Figure 3-15. Raman spectra of sodium borosilicate glass: (a) unleached; (b) leached at pH 1; (c) 10.30.60 glass leached at pH 1, 7 and 12. Adapted from (76).

changes in oxygen bonding due to leaching. They found that upon leaching, the non-bridging oxygen contribution was significantly attenuated. By examining the O/Si ratio in the leached and unleached surfaces they concluded that extensive silanol condensation had occurred, and the surface was composed of nearly pure SiO_2.

Komizono and Clark (78) used XPS in conjunction with curve fitting techniques to examine the relative contributions of Fe^{2+} and Fe^{3+} to the Fe $2p_{3/2}$ line obtained from a leached sample of 165/TDS simulated nuclear waste glass. By assuming the Fe^{2+} and Fe^{3+} contributions to be centered at 708.3 and 710.6 eV, respectively, Gaussian distributions were fit to the background-corrected spectrum. The area of each contribution was taken to be proportional to the concentration of that specie. Similar work by Karim, et al. (79) led to the identification of oxidation state of Pu and Np in leached glass. Pu^{4+}, Pu^{3+} and Np^{4+} were found in the unleached glass. After leaching, no Pu signal was found and it was determined that neptunium was still in the +4 state.

Extended x-ray absorption fine edge structure. Greaves et al. (80-84) used extended x-ray absorption fine edge structure (EXAFS) to examine the local symmetry and radial distribution function of uranium, iron and lead in a variety of glasses. Normally EXAFS is not considered a surface technique; however, a special glancing-angle EXAFS arrangement was used in this case to limit beam penetration to between a few atomic layers and 1 micron, depending on angle.

In studies of uranium-containing glass (80,82) it was found that the coodination of uranium in the unleached glass was similar to that of uranyl fluoride fluorammonium, with U coordinated to two oxygens at 1.85A and four oxygens at 2.3 A, as shown in Figure 3-16c. The unleached glass showed no evidence of U-U coordination. EXAFS of leached surfaces show a broadening of the U-O subshell distributions, indicating a higher degree of disorder, as well as new peaks indicating U-U coordinations centered at 3.3 A and 4.8 A. (Figure 3-16a,b).

Similar studies of lead-iron-phosphate glasses (79,81) showed that Fe and Pb modifiers occupy very different environments in the glass. This information was used to explain the increased corrosion resistance of this glass with increasing iron content based on blocking (by iron) of a percolation network of lead diffusion pathways.

Microstructural Characterization

A limited number of studies have addressed the microstructure of leached glass. Corrosion product morphology and secondary mineral phase formation have been examined using transmission electron microscopy (TEM). Small-angle x-ray scattering (SAXS) measurements have been performed to measure particle size in leached glass, and porosity in leached glass has been studied using gas adsorption techniques (BET).

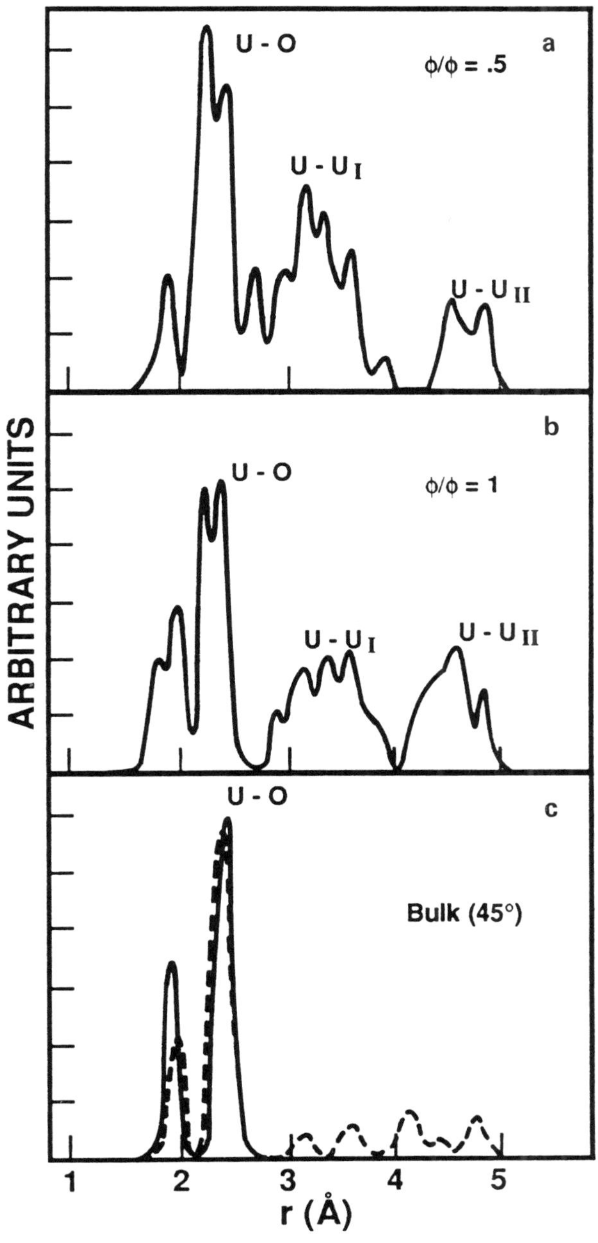

Figure 3-16. Partial radial distribution function of uranium in leached uranium-containing glass (a) at the outer surface; (b) at an intermediate depth; (c) bulk. Adapted from (80).

Transmission electron microscopy. TEM is a highly versatile technique, capable of microstructural analysis at the nanometer level, compositional analysis (using energy dispersive x-ray spectroscopy) and determination of crystal structure (using selected area electron diffraction). Use of TEM has been somewhat limited by difficulties in obtaining suitably thin (50-100 nm) specimens. This has been addressed by using ultramicrotomy as described by several authors (85-87). Abrajano et al. (88) conducted analytical electron microscopy on leached samples of non-radioactive and uranium-doped nuclear waste glasses prepared by ultramicrotomy. In the non-radioactive glass, six layers were identified as comprising the altered surface (Figure 3-17). Detailed analysis of these layers permitted identification of various mineral, clay and amorphous phases present in each. A similar study of the uranium-doped glass demonstrated the existence of a U-Ti phase within layers 3-5. This finding has significant implications for radioactive waste disposal, since it demonstrates a mechanism for the immobilization of uranium in spite of alteration of the glass.

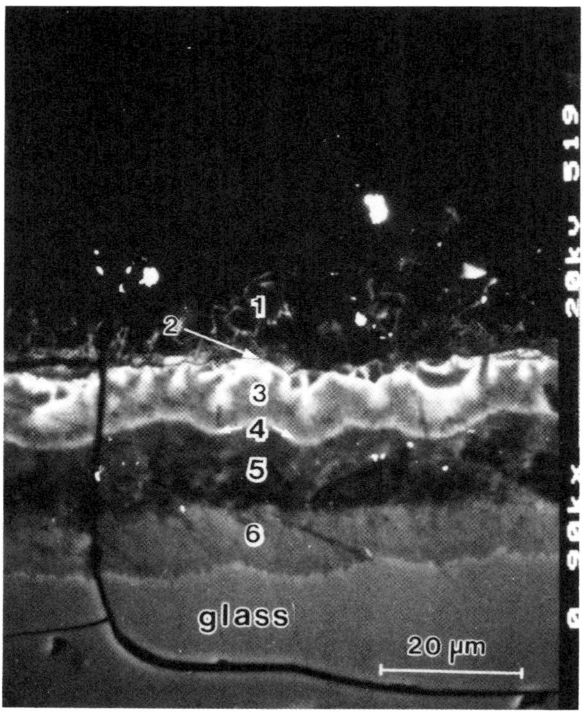

Figure 3-17. Transmission electron micrograph of leached nuclear waste glass. Photo courtesy of John Bates, Argonne National Laboratory. Reprinted by permission of John Bates and Clays and Clay Minerals (88).

Bunker et al. (76) prepared corroded glass for TEM by scraping leached material from the glass, grinding it in butanol and dispersing it onto a TEM grid. While this did not permit observation of layered structures or preserve spatial relationships among structures, it did provide sufficiently thin flakes for analysis. In leached binary alkali silicate and sodium borosilicate glasses porous structures were found which were qualitatively similar to aggregations of colloidal silica.

BET gas adsorption analysis. Porosity and surface area have been addressed in only a limited number of studies. These studies utilized BET gas adsorption analysis to measure these properties. Walker (88) measured the increase in surface area of bioglass leached 10 days in TRIS buffer as a function of sodium content of the parent glass. For high sodium content glass (24 mol% Na_2O) surface area was found to increase by a factor of 31,300 after leaching. Bunker et al. (76) used BET to measure the surface area of leached alkali borosilicate glass and determined the specific surface area to be 400 m^2/g with pores $\leq$30Å diameter, consistent with TEM results. Fillet, et al. (90) examined the relationship between drying technique, pore size and surface area of corroded R7T7 simulated nuclear waste glass. For samples dried conventionally, by controlled rate dehydration, dehydration using low surface tension liquids and hypercritical drying the specific surface area varied between 173 and 320 m^2/g, pore radius varied form 20 - 100 Å, and porous volume fraction from 0.312 to 0.991. It was concluded that hypercritical drying was the most likely method to preserve the original layer texture.

Small angle x-ray scattering. Small angle x-ray scattering (SAXS) is used to determine particle size in heterogeneous material. Tomozawa and Capella (91) used SAXS to probe the microstructure of leached $Na_2O{\cdot}3SiO_2$ and $4BaO{\cdot}28Na_2O{\cdot}68SiO_2$ glasses. Based on a mathematical analysis of the scattering curves, they found that the particles in the leached $Na_2O{\cdot}3SiO_2$ glass had a radius of gyration of 2-5 nm, corresponding to a diameter of 5-13 nm. The size of this microstructure increased slightly with increasing hydration time. For the $4BaO{\cdot}28Na_2O{\cdot}68SiO_2$ glass, the leached material had a radius of gyration ranging from 2-8 nm, corresponding to a spherical diameter of between 5 and 20 nm.

Compositional Analysis

Many of the techniques of compositional analysis are "surface sensitive", meaning that information is obtained from a limited, shallow depth (frequently ~Å) at the surface of the sample. However, corroded surfaces may be several microns deep, with compositions which vary drastically over depth. For this reason, many of surface sensitive techniques are combined with sputtering which progressively exposes fresh, deeper surfaces. This is usually the method of choice for techniques such as secondary ion mass spectrometry (SIMS), XPS and Auger electron spectroscopy (AES). Electron microprobe analysis has been used to examine fracture surfaces and polished cross sections without the necessity of sputtering. Finally, Rutherford backscattering (RBS) and nuclear reaction analysis (NRA) have been used to non-destructively obtain composition profiles in layers up to 2 μm in thickness.

Secondary ion mass spectrometry. SIMS is one of the most sensitive and accurate techniques available for measuring the composition vs. depth within leached glass. This method is covered in detail in the following chapter and will not be discussed here.

X-ray photoelectron spectrometry. XPS has been discussed as a probe of glass structure. Additionally, XPS is capable of quantitative compositional analysis. Pederson, et al. (92) used XPS combined with argon ion sputtering to measure composition profiles in leached 76-68 and 79-150 glasses. After accounting for depth-dependent sputtering rates, they compared element depletions within leached layers with solution concentrations as determined by ICP. For 79-150 glass it was found that sodium depletion within the layer agreed very well with the amount of sodium which had appeared in solution. In addition, they found that the aluminum concentration within the surface increased a normalized rate which was approximately equal to that of silicon depletion, suggesting that no aluminum had been lost from the surface. This suggests that the 79-150 composition underwent primarily selective leaching under these conditions. For the 76-68 glass, only about one-third of the solution concentration of sodium could be accounted for by the depth profile, indicating that that this glass had undergone a significant amount of matrix dissolution. Similar studies of R7T7 glass leached in brine have been conducted using XPS combined with argon sputtering (93).

Auger electron spectroscopy. Auger electron spectroscopy (AES) involves bombarding a specimen with energetic electrons to eject core electrons. Higher level electrons are de-excited and fill the resulting vacancy, transferring a characteristic quantity of energy to another high level electron which may then be ejected from the sample. Referred to as Auger electrons, the energy of these electrons indicates the element of origin, while the current is proportional to concentration. AES is frequently combined with argon ion sputtering to obtain depth profiles. Such applications are described for leached soda-lime-silica glass and PNL 76-68 simulated nuclear waste glass in reference (94) and bioactive glasses and glass-ceramics (95). Because AES utilizes an energetic electron beam, caution must be taken in determinations involving highly mobile ions such as alkalis (96). This problem may be ameliorated by using a liquid nitrogen cold stage to counteract beam heating effects.

Electron microprobe analysis. Electron microprobe analysis (EMP) is based on energy analysis of x-rays emitted during electron bombardment of a material. EMP has a lateral resolution of approximately two microns and thus may be used to examine the composition of cross-sections of leached layers which are several microns thick. Similar to AES, heating due to the electron beam may cause migration of mobile species. In addition to the use of a liquid nitrogen cold stage to remedy this problem, it is also possible to broaden the beam to reduce the degree of heating; however, this is done at the cost of resolution (97). Clark and Purdy (98) used EMP to examine sections of weathered chert arifacts. Based on intensity of x-ray lines from various elements, they were able to identify four distinct regions within the sample cross-section, as shown in Figure 3-18.

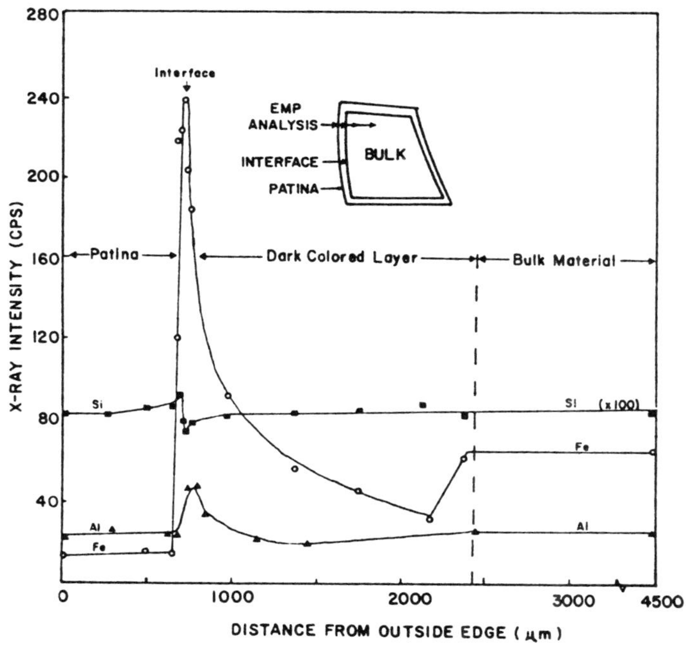

Figure 3-18. X-ray intensity of Si, Fe and Al in alteration rind of weathered chert, determined using electron microprobe analysis. Adapted from (98).

Rutherford backscattering. Rutherford backscattering (RBS) is based on measurements of energetic ions elastically scattered from the outer 1-2 μm of a material. By proper analysis of the energy distribution of the scattered particles (99) it is possible to deduce the mass and concentration profile of the elements responsible for scattering. RBS has the additional advantage of being non-destructive. Sales, et al. (100) found that during the early stages of corrosion of Frit 21 glasses surfaces formed which were enriched in U, Sr, Gd, Nd, and Ta and depleted in Na, Cs, and Si. These enrichments were attributed to the formation of insoluble metal complexes such as $Ti(OH)_4$, $SrCO_3$, etc. For leaching times greater than six hours, the RBS profile approached that of an uncorroded surface, an effect attributed to removal of the layer as a result of pH increase at longer leaching times. This is shown in Figure 3-19, which shows RBS spectra as a function of time for Frit 21 with 5 wt% SrO. RBS was used

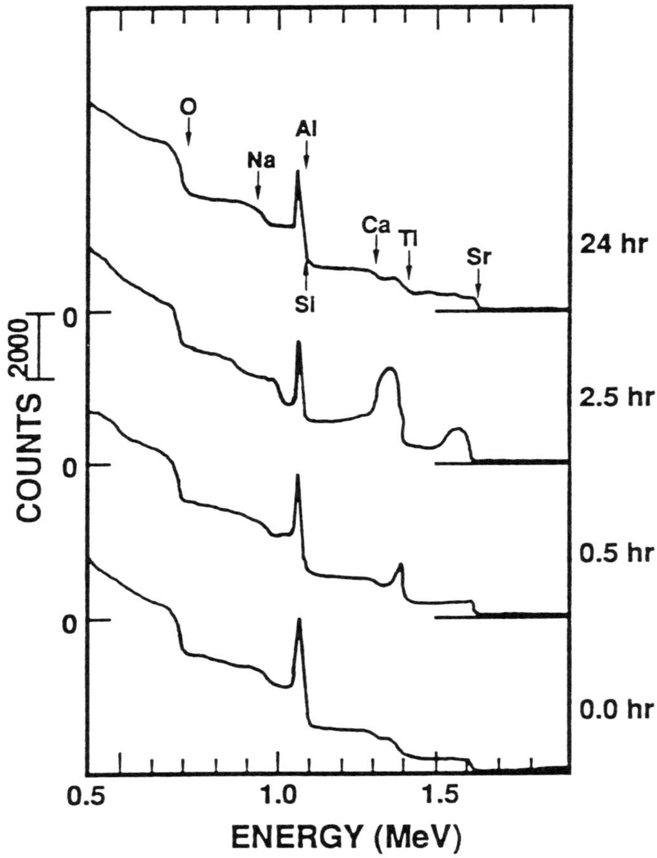

Figure 3-19. RBS spectrum from corroded and uncorroded Frit 21 doped with 5wt% SrO. Numbers on right indicate corrosion time. Adapted from (100).

by Bunker, et al. (101) to examine K and Na profiles in (1-x)Na_2O·xK_2O·3SiO_2 glasses. They determined that while Na was completely removed from the leached region, only a fixed fraction of K was depleted. The concentration of K remaining was found to be dependant on solution pH, with more K remaining in glasses leached in basic solutions.

Nuclear Reaction Analysis. Similar to RBS, nuclear reaction analysis (NRA) is also a non-destructive depth profiling technique which utilizes an ion beam. NRA is used primarily for profiling light elements and is particularly useful for measuring hydrogen profiles. Hydrogen profiling is based on the $^{1}H(^{15}N,\alpha\gamma)^{12}C$ nuclear reaction. As the energy of the impingent $^{15}N^{2+}$ is increased, it penetrates to a greater depth within the sample. 4.4 MeV gamma rays resulting from the reaction are then measured using a NaI(Tl) detector (102). A similar method, utilizing the $^{23}Na(p,\alpha)^{20}Ne$ reaction, is used to profile Na.

Schnatter, et al. (103) used NRA to examine hydrogen profiles in a variety of leached soda-lime-silica glasses. They found that it was important to cool samples to below O °C before and during measurement to prevent hydrogen (H_2O) loss. For samples treated in this manner, hydrogen was found to replace leached sodium in a ratio of 3:1, suggesting H_3O^+ as the active specie in ion exchange.

DATA ANALYSIS AND REDUCTION

Solution data is usually obtained by atomic absorption spectroscopy (AAS), inductively-coupled plasma spectroscopy (ICP), direct-coupled ion spectroscopy (DCP), colorimetry, ion chromatography, selective ion electrodes and pH electrodes. In nearly all cases the concentrations of elements in solution are measured and reported in parts per million (ppm). This is equivalent to milligrams of element per liter of solution (mg/l) in dilute solutions typically encountered in leachates. Some investigators convert the elemental loss to oxide loss and report these data in milligrams of oxide loss per gram of material tested (mg/g). In addition to causing confusion, these types of data make it difficult to compare experimental results obtained at different laboratories unless specimen surface area is specified.

In the ASTM method C 225-85 for determining the resistance of glass containers to chemical attack, the solutions are titrated. For acid resistance, the volume (ml) of 0.020N nitric acid consumed in the test is reported. For water resistance, the volume (ml) of 0.020N sulfuric acid required for titration is reported. In all test methods discussed in this chapter (except) where an increase in temperature is used to accelerate the leaching, all analyses are carried out at room temperature.

Recently, there has been a significant effort in the corrosion community to standardize methods of reporting data as well as conducting the tests. Solution data is typically converted to a normalized mass loss $(NL)_i$ in g/m^2 using (for static tests) (7):

$$(NL)_i = \frac{C_i}{f_i(SA/V)} \tag{3-1a}$$

where: C_i = concentration of element i in mgl^{-1}
f_i = mass fraction of element i in the specimen
SA = surface area of the specimen in m^2
V = volume of leachant in m^3.

For example, in a typical test of a glass specimen containing 36wt.% K, SA = $2.5x10^{-4}m^2$, V = $2.5x10^{-5}m^3$, C_i = 325ppm after 0.5 hours of exposure.

$$(NL)_K = 325/(0.36)(10)\ gm^{-2} \tag{3-1b}$$

$$(NL)_K = 90gm^{-2} \tag{3-1c}$$

The average leach rate for K in deionized water over 28 days is:

$$LR_{AV} = 90gm^{-2}\ /0.021\ days \tag{3-2a}$$

$$LR_{AV} = 4.3\ x10^3\ gm^{-2}day^{-1} \tag{3-2b}$$

In contrast, the average leach rate of typical nuclear waste glasses after 28 days of exposure in deionized water is less than 1 $gm^{-2}day^{-1}$. The average leach rate usually decreases with time due to either saturation effects or the formation of protective surface films. The instantaneous leach rates for each element can be determined by plotting $(NL)_i$ versus time. The slope of this curve $\delta(NL)_i/\delta t$, at any given time is equal to the instantaneous leach rate, which is equal to the average leach rate only if $(NL)_i$ is linear with time. When the specimen is dissolving congruently, all elements should be released into the solution in the same proportion as they are present in the solid. Under this condition, $\delta(NL)_i/\delta t$ will be the same for all elements. However, it is more common to observe different $\delta(NL)_i/\delta t$ for each element with the more mobile and soluble elements exhibiting the highest instantaneous leach rates (Figure 3-10).

Total mass loss can also be obtained by weighing the sample before and after the test. It is given by (for both static and flow tests) (8):

$$L_m = \frac{W_i - W_f}{SA} \tag{3-3}$$

where: L_m = mass loss/unit area (g/m^2)
W_i = initial weight of specimen (g)
W_f = final weight of specimen (g)
SA = surface area of specimen (m^2)

The average total leach rate and instantaneous leach rate can be determined in the same manner as for the leach rates based on normalized mass loss. Care must be exercised when using equation 3-3 since significant alteration of the solid can occur with little weight loss due to precipitation. However, under conditions of congruent dissolution, and without precipitation, the leach rates based on normalized mass loss and total mass loss should be equivalent. Figure

3-20 illustrates these concepts for a glass that dissolves congruently. In this example, $\delta(NL)_i/\delta t$ is also constant indicating congruent dissolution over the entire exposure time.

Increases in temperature cause increases in corrosion rates. As long as the mechanisms of corrosion are not changed the total rate, L_m, can be expressed by:

$$L_m = Ke^{-E/RT} \tag{3-4}$$

where: K = constant
E = activation energy for corrosion, kcal/mol
R = gas constant, 2kcal/mol
T = absolute temperature, K

$$\ln L_m = \ln K - E/RT \tag{3-5}$$

By plotting $\ln L_m$ versus 1/T, the activation energy can be determined for the corrosion process. From the Arrhenius plot shown in Figure 3-21 for a 33mol% K_2O - 67mol% SiO_2 glass, the activation energy was calculated as 20kcal/mol. Using these data the reaction rate due to congruent dissolution can be calculated for any temperature between 25°C and 90°C.

Other equations based on solution data have been derived by various investigators to describe the leaching process under various conditions. Sanders and Hench (104) use the following equation for binary glasses to describe the mechanisms of leaching:

$$\alpha = \frac{Y_{soln}}{X_{soln}} / \frac{Y_{glass}}{X_{glass}} \tag{3-6}$$

where: Y = moles of SiO_2
X = moles of R_2O

When $\alpha = 0$, the alkali oxide, R_2O, is being selectively removed from the glass. When $\alpha = 1$, the glass is dissolving congruently and all species are being released into solution in the same ratio that they are present in the glass. $\alpha = 1$ for the 33 mol% K_2O - 67 mol% SiO_2 glass shown in Figure 3-20. α may change with time due to the formation and subsequent dissolution of surface films (105).

Since solution data is typically obtained in ppm, α can be obtained from:

$$\alpha = \left(\frac{ppmSiO_2}{2ppmR}\right)\left(\frac{MWR}{MWSiO_2}\right)\left(\frac{Pm}{1-Pm}\right) \tag{3-7}$$

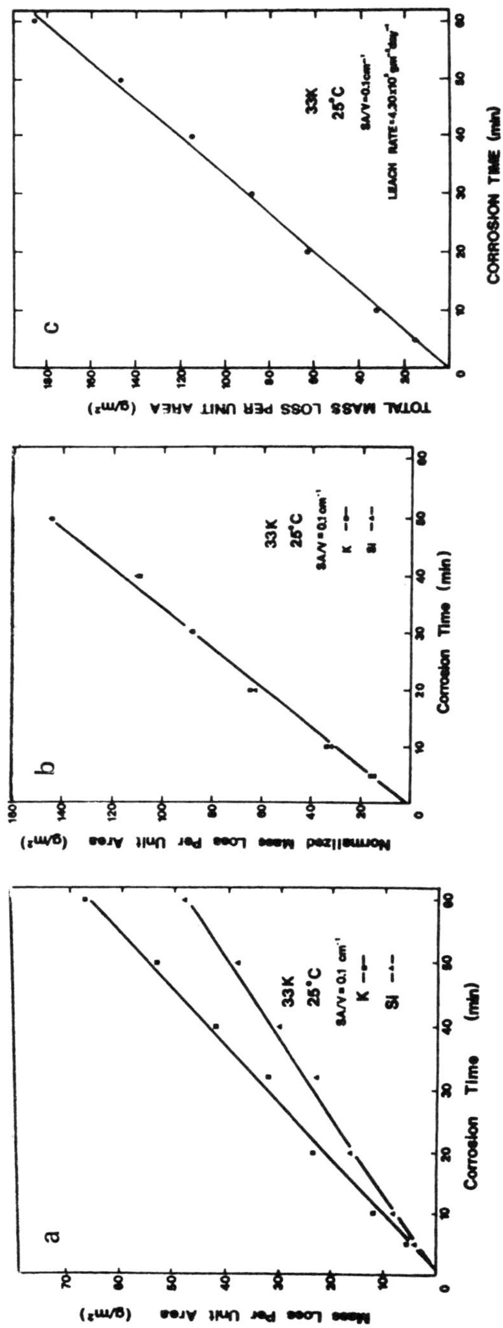

Figure 3-20. Mass loss vs. corrosion time for a 33 mol% K_2O-67 mol% SiO_2 glass. (a) Mass loss per unit area; (b) Normalized mass loss per unit area (NL)i, and (c) Normalized total mass loss, (Lm).

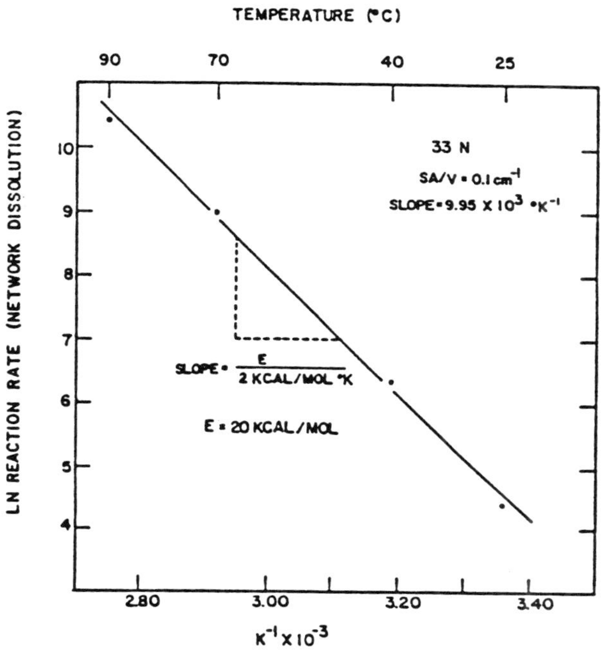

Figure 3-21. Arrhenius diagram for a 33 mol% Na_2O - 67 mol% SiO_2 glass corroded at various temperatures. The corrosion (leach) rates used for this diagram were for network dissolution.

where MW is the molecular weight of the indicated species and Pm is the mole fraction of R_2O in the glass. It should be noted that the value of α only provides information concerning the mechanisms of corrosion, not the extent of corrosion.

Regarding flow, it can be shown that the leach rate can be expressed by (106):

$$\frac{dC}{dt} = \frac{SA}{V}(LR) - \frac{F}{V}C \tag{3-8}$$

where: C = solution concentration of a particular species, gm^{-3}
SA = sample surface area in m^2
V = solution volume, m^3
F = flow rate, m^3d^{-1}
LR = leach rate $gm^{-2}d^{-1}$.

This equation is obtained by considering mass balance of elements entering solution from the glass and leaving the system as a result of flow. By collecting solution data over time and calculating dc/dt, it is possible to determine glass leach rates. This equation is generally applicable and capable of accounting for time-varying leach rates and flow rates, as well as the effect of mixing within the leaching vessel.

It is important that the test methodology provide results that are both accurate and precise and many standard tests are now designed with this in mind. Accuracy refers to how close the test values (the arithmetic means when two or more values are obtained) is to the real value of durability. Precision refers to the closeness of the test results to each other (the standard deviation). At least two test results are required to specify precision. A set of durability measurements may be very precise, but inaccurate. Scatter in data leading to either imprecision or inaccuracy may arise from two general sources: variations in the test procedures and variations in the analyses. With the availability of calibration standards and other advances in analytical equipment, it appears that the largest source of error is related to variations in test procedures. For this reason round robin tests are most likely to yield high precision within each of the participating laboratories and more of a spread in accuracy between the laboratories.

A recent round robin Product Consistency Test (PCT) was conducted by the Materials Characterization Center (MCC) involving seven laboratories and four glass compositions (107). The PCT is a seven day test where glass powders are exposed to pure water at 90 ± 2°C and the leachate is analyzed for specific elements. One objective of the round robin was to evaluate the tests' ability to discriminate between the durabilities of similar glass compositions. The results indicate that the ability to discriminate between two glasses varied with laboratory and analyte. In all laboratories the ability to discriminate improved as the number of replicate tests was increased. Furthermore, B followed closely by Na and K (then Al, Li, Si) provided the best discrimination within each of the laboratories. It appears that laboratories experienced in conducting chemical durability tests will be able to perform the PCT with sufficient accuracy and precision to discriminate between glasses with small compositional differences.

SUMMARY

In this chapter we have attempted to familiarize the reader with various testing and characterization methods used for assessing the chemical durability of ceramics. Although the major emphasis has been on glass, the principles and procedures are, in most cases, applicable to all types of ceramics. A good review of the chemical corrosion of technical ceramics can be found in reference 108. Much progress has been made in the development and refinement of laboratory and field test methodologies over the last 15 years. Especially important, advances have been made in the application of characterization techniques for studying ceramic corrosion. Consequently, we now have a much better understanding of the degradation mechanisms than we did a decade ago. This new understanding will enable the materials engineer to design ceramic products with more controlled and predictable environmental stability.

An integrated/comprehensive approach to corrosion involving both solution and surface analyses and well controlled test procedures is the goal of scientists and corrosion engineers. Although much has been done to achieve this goal, there are still several areas of research that deserve more aggressive efforts. These include: 1) standard surface analyses procedures and data reduction, 2) more thorough microstructural analyses of corroded surfaces and, 3) more emphasis on mass balance between solution and surface analyses.

ACKNOWLEDGEMENTS

The authors thank Westinghouse Savannah River Company under subcontract No. AX-715234 for partial financial support during the preparation of this chapter.

REFERENCES

1. Sanders, D.M., Person, W.S., and Hench, L.L., *Appl. Spectrosc.* 28[3]: 247-255 (1974).
2. Clark, D.E., Zoitos, B.K., Wicks, G.G., and Molen, K.A., in: *Testing of High Level Waste Forms Under Repository Conditions* (T. McMenamin, ed.), EUR 12017EN, 140-151 (1989).
3. Lodding, A.R., Engström, E.U., and Odelius, H., in: *Testing of High Level Waste Forms Under Repository Conditions* (T. McMenamin, ed.), EUR 12017EN, 127-139 (1989).
4. Clark, D.E., Ethridge, E.C., Dilmore, M.F., and Hench, L.L. *Glass Technol.* 18[4]: 121-124 (1977).
5. Hench, L.L. and Clark, D.E., *Surface Properties and Performance Prediction of Alternative Waste Forms*, Annual Report submitted to U.S. Nuclear Regulatory Commission (1981).
6. Jantzen, C.M., and Bibler, N.E. *The Product Consistency Test for the DWPF Wasteform*, to be published in proceededing of the 2nd International Seminar on Radioactive Waste Products, Julich, Federal Republic of Germany, 1990.
7. *Materials Characterization Center, Nuclear Waste Materials Handbook-Test Methods*, DOE/TIC-114000, Pacific Northwest Laboratories, Richland, WA.
8. Hespe, E.D., *Atom. Energy Rev.* 9: 195-207 (1971).

9. Clark, D.E. and Ethridge, E.C., *Am. Ceram. Soc. Bull.* 60[6]: 646-649 (1981).
10. Gould, J.H. and Moss, R.I., *Am. Ceram. Soc. Bull.* 61[12]: 1307-1310 (1982).
11. Malow, G., in: *Scientific Basis for Nuclear Waste Management V* (W. Lutze, ed.) 25-36 (1982).
12. Chemical Durability Committee of the Society, *Glass Technol.* 7[2]: 35-41 (1966).
13. Divakar, R., Seshadri, S.G., and Srinivasan, M., *J. Am. Ceram. Soc.* 72[5]: 780-784 (1989).
14. Hench, L.L. and Clark, D.E., *Surface Properties and Performance Prediction of Alternative Waste Forms*, Annual Report submitted to the U.S. Nuclear Regulatory Commission, NUREG/CR-3472, Vol. 1 (1983).
15. Simpson, H.E., *J. Am. Ceram. Soc.* 42[7]:337-343 (1959).
16. Simpson, H.E., *Am. Ceram. Soc. Bull.* 30[2]: 41-45 (1951).
17. Walters, H.V. and Adams, P.B., in: *Glass Surfaces* (D.E. Day, ed.) 183-199, North-Holland (1975).
18. Clark, D.E., Pantano, Jr., C.G., and Hench, L.L., *Corrosion of Glass*, Books for Industry, Inc., New York (1979).
19. Chen, H. and Park, J.W., *Phy. Chem. Glasses* 22[2]: 39-42 (1981).
20. Hench, L.L. and Clark, D.E., *Surface Properties and Performance Prediction of Alternative Waste Forms*, Annual Report submitted to the U.S. Nuclear Regulatory Commission (1984).
21. American Society for Testing and Materials, ASTM C 912-79 (Reapproved 1984), ASTM Standards 286-291 (1984).
22. Adams, P.B., in: *Glass Surfaces* (D.E. Day, ed.) 275-276, North-Holland (1975).
23. Hench, L.L. and Ethridge, E.C., in: *Surface Contamination*, Vol. 1, 313-326, Plenum Press, New York (1976).
24. Campbell, D.E. and Adams, P.B., *J. Testing and Evaluation* 14[5]: 260-265 (1986).
25. Sykes, R.F., *Glass Technol.* 6[6]: 178-183 (1965).
26. Chemical Durability Committee, *Glass Technol.* 7[6]: 181-182 (1966).
27. Gebhardt, F., Graff, U., Merker, L., and Vermoehlen, H., *Glastech. Ber.* 54[8]: 257-264 (1981).
28. Ethridge, E.C., Clark, D.E., and Hench, L.L., *Phy. Chem. Glasses,* 20[2]: 35-40 (1979).
29. Buckwalter, C.Q., Pederson, L.R., and McVay, G.L., *J. Non-Cryst. Solids* 49: 397-412 (1982).
30. Chandler, G.T., Wicks, G.G. and Wallace, R.M., in: *Advances in Ceramics* (D.E. Clark, W.B. White and A.J. Machiels, eds.) Vol. 20 455-463, Am. Ceram. Soc., Inc. (1986).
31. Barkatt, Aa., Macedo, P.B., Sousanpour, W., Barkatt, A., Boroomand, M.A.,

Szoke, P., and Rogers, V.L., in: *Scientific Basis for Nuclear Waste Management VI*, (D.G. Brookins, ed.), Vol. 15, 227-234, North-Holland (1983).

32. Pederson, L.R., Buckwalter, C.Q., McVay, G.L., and Riddle, B.L., in: *Scientific Basis for Nuclear Waste Management VI*, (D.G. Brookins, ed.), Vol. 15, 47-55, North-Holland (1983).
33. Machiels, A.J. and Pescatore, C., in: *Scientific Basis for Nuclear Waste Management VI*, (D.G. Brookins, ed.), Vol. 15, 209-216, North-Holland (1983).
34. Kamizono, H., Clark, D.E., and Lodding, A.R., *J. Nucl. Sci. and Tech.* 26[4]: 441-448 (1989).
35. Fillet, S., Vernaz, E., Nogues, J.L., and Jacquet-Francillon, N., in: *Advances in Ceramics* (D.E. Clark, W.B. White and A.J. Machiels, eds.) Vol. 20, 443-453, Am. Ceram. Soc., Inc. (1986).
36. Clark, D.E. and Hench, L.L., in: *Scientific Basis for Nuclear Waste Management VI*, (D.G. Brookins, ed.), Vol. 15, 113-124, North-Holland (1983).
37. Coles, D.G., Weed, H.C., Jackson, D.P., and Schweiger, J.S., in: *Radioactive Waste in Geologic Storage* (S. Fried, ed.) ACS Symposium Series No. 100, 93-114, Am. Chem. Soc. (1979).
38. Strachan, D.M., in: *Advances in Ceramics* (G.G. Wicks and W.A. Ross, eds.) Vol. 8 12-18. Am. Ceram. Soc. Inc., (1983).
39. Adiga, R., Barkatt, Aa., and Clark, D.E., in: *Advances in Ceramics* (D.E. Clark, W.B. White and A.J. Machiels, eds.) Vol. 20, 487-494, Am. Ceram. Soc., Inc. (1986).
40. Hench, L.L., Clark, D.E., and Harker, A.B., *J. Mater. Sci.* 21: 1457-1478 (1986).
41. Paul, A., *Chemistry of Glasses*, Chapman and Hall, New York (1982).
42. El-Shamy, T.M., Lewins, J. and Douglas, R.W., *Glass Tech.*, 13[3]:81-87 (1972).
43. Wood, S. and Blachere, J.R., *J. Am. Ceram. Soc.* 61[7-8]: 287-292 (1978).
44. Hench, L.L. and Clark, D.E., *Surface Properties and Performance Prediction of Alternative Waste Forms*, Annual Report submitted to the U.S. Nuclear Regulatory Commission, NUREG/CR-3472, Vol. 2 (1986).
45. Isaid, J.O. and Priestly, D., *Phys. Chem. Glasses* 26[6]: 221-222 (1985).
46. Bunker, B.C. and Arnold, G.W., in: *Scientific Basis for Nuclear Waste Management VI*, (D.G. Brookins, ed.), Vol. 15, 151-158, North-Holland (1983).
47. Tait, J.C., Hocking, W.H., Betteridge, J.S., and Bart, G., in: *Advances in Ceramics* (D.E. Clark, W.B. White and A.J. Machiels, eds.) Vol. 20, 559-565, Am. Ceram. Soc., Inc. (1986).
48. VanIseghem, P., Timmermans, W., Debruyn, W., Dresselaers, J., and Neerdael, B., in: *Advances in Ceramics* (D.E. Clark, W.B. White and

A.J. Machiels, eds.) Vol. 20, 649-656, Am. Ceram. Soc., Inc. (1986).
49. Wicks, G.G. and Molecke, M.A., in: *Advances in Ceramics* (D.E. Clark, W.B. White and A.J. Machiels, eds.) Vol. 20, 657-667, Am. Ceram. Soc., Inc. (1986).
50. Hench, L.L., Lodding, A.R., and Werme, L., in: *Advances in Ceramics* (G.G. Wicks and W.A. Ross, eds.) Vol. 8 310-322, Am. Ceram. Soc., Inc. (1983).
51. Clark, D.E., Zhu, B.F., Robinson, R.S., and Wicks, G.G., in: *Advances in Ceramics* (G.G. Wicks and W.A. Ross, eds.) Vol. 8 324-336, Am. Ceram. Soc., Inc. (1983).
52. Walters, H.V. and Adams, P.B., *Am. Ceram. Soc. Bull.* 61[11]: 1224-1227 (1982).
53. Waksman, D., Roberts, W.E., and Byrd, W.E., *Durability Build. Mater.*, 31[1]: 1-21 (1985).
54. Czanderna, A.W., *Methods of Surface Analysis*, Elsevier (1975).
55. Kane, P.F. and Larrabee, G.B., *Characterization of Solid Surfaces*, Plenum Press (1974).
56. Hench, L.L., in: *Characterization of Materials in Research, Ceramics and Polymers* (V. Weiss and J.J. Burke, eds.), pp. 211-251, Syracuse University Press (1975).
57. Lee, C.T. and Clark, D.E., *Appl. Surf. Sci.* 20: 397-412 (1985).
58. Woodruff, D.P. and Delchar, T.A., *Modern Techniques of Surface Science*, Cambridge University Press (1986).
59. Feldman, L.C. and Mayer, J.W., *Fundamentals of Surface and Thin Film Analysis*, North-Holland (1986).
60. Various authors, *Materials Research Society Bulletin*, 12(6): 22-105 (1987).
61. Subcommittee AII of the International Commission on Glass--Chemical Durability, *Glass Tech.* 16(4): 80-83 (1975).
62. Subcommittee AII of the International Commission on Glass--Chemical Durability, *Glass Tech.* 14(1): 14-19 (1973)
63. Fassel, V.A. and Kniseley, R.N, *Anal. Chem.* 46(13): 1110A-1120A (1974).
64. Winge, R.K., Peterson, V.J. and Fassel, V.A., *Appl. Spectrosc.* 33(3): 206-219 (1979).
65. Sheetz, B.F., Freeborn, W.P., Smith, D.K., Anderson, C., Zolensky, M. and White, W.B., *Mat. Res. Soc. Symp. Proc. Vol. 44* (C.M. Jantzen, J.A. Stone and R.C. Ewing, eds.), pp. 129-134, Materials Research Society (1985).
66. Grambow, B., *Mat. Res. Soc. Symp. Proc. Vol. 11.* (W. Lutze, ed.) pp. 93-102, North-Holland (1981).
67. Clark, D.E. and Hench, L.L., *Nuc. Chem. Waste. Mgt.* 2: 93-101 (1981).
68. Sasoon, R.E., Gong, M., Adel-Hadadi, M., Brandys, M., Barkatt, A. and Macedo, P.B., *Testing of High-Level Waste Forms Under Repository Conditions*; Commission of the European Communities Report EUR 12017 EN, (T. McMenamin, ed.) pp. 204-210, CEC (1989).

69. Sanders, D.M., Person, W.B. and Hench, L.L., *Appl. Spectrosc.* 26: 530-545 (1972).
70. Clark, D.E., Ethridge, E.C., Dilmore, M.F. and Hench, L.L., *Glass Tech.* 18(4): 121-124 (1977).
71. Chen, H. and Park, J.W., *Phys. Chem. Glasses* 22(2): 39-42 (1981).
72. Aines, R.D., Weed, H.C. and Bates, J.K., *Mat. Res. Soc. Symp. Proc. Vol. 84* (J.K. Bates and W.B. Seefeldt, eds.), pp. 547-558, Materials Research Society (1987).
73. Exarhos, G.J. and Conaway, W.E., *J. Non-Cryst. Solids* 55: 445-449 (1983).
74. Galeener, F.L. *J. Non-Cryst. Solids*, 49: 53-62 (1982).
75. Ravesz, A.G. and Walrafen, G.E., *J. Non-Cryst. Solids*, 54: 323-333 (1983).
76. Bunker, B.C., Tallant, D.R., Headley, T.J., Turner, G.L. and Kirkpatrick, R.J., *Phys. Chem. Glasses* 29(3): 106-120 (1988).
77. Veal, B.W., Lam, D.J., Paulikas, A.P. and Karim, D.P., *Nucl. Tech.* 51: 136-142 (1980).
78. Komizono, H. and Clark, D.E., *Nippon Seramikkusu Kyokai Gakujutsu Ronbunshi* 97(4): 494-97 (1989).
79. Karim, D.P., Lam, D.J., Diamond, H., Friedman, A.M., Coles, D.G., Bazan, F., and McVay, G.L., *Mat. Res. Soc. Symp. Proc. Vol. 6* (S.V. Topp, ed.), pp. 67-73, North-Holland (1982).
80. Thornley, F.R., Barrett, N.T. Greaves, G.N. and Antonini, G.M., *J. Phys. C: Solid State Phys.* 19(25): L563-L569 (1986).
81. Greaves, G.N., Gurman, S.J. and Jenkins, R.N., *J. Phys. Colloq.* C8(2): 787-791 (1986).
82. Barrett, N.T., Antonini, G.M., Thornley, F.R., Greaves, G.N. and Manara, A., *Mat. Res. Soc. Symp. Proc. Vol. 84* (J.K. Bates and W.B. Seefeldt, eds.), pp. 571-576, Materials Research Society (1987).
83. Greaves, G.N., Gurman, S.J., Gladden, L.F., Spence, C.A., Cox, P., Sales, B.C., Boatner, L.A. and Jenkins, R.N., *Philos. Mag. B* 58(3): 271-283 (1988).
84. Greaves, G.N., Barrett, N.T., Antonini, G.M, Thornley, F.R., Willis, B.T.M. and Steel, A., *J. Am. Chem. Soc.* 111: 4313-4324 (1989).
85. Jercinovic, M.J., Kaser, S. and Ewing, R.C., *Testing of High-Level Waste Forms Under Repository Conditions*; Commission of the European Communities Report EUR 12017 EN, (T. McMenamin, ed.), pp. 183-191, CEC (1989).
86. Lutze, W. and Ewing, R.C, *Testing of High-Level Waste Forms Under Repository Conditions*; Commission of the European Communities Report EUR 12017 EN, (T. McMenamin, ed.), pp. 91-105, CEC (1989).;
87. Vernaz, E. and Godon, N., *Testing of High-Level Waste Forms Under Repository Conditions*; Commission of the European Communities Report EUR 12017 EN, (T. McMenamin, ed.), pp. 81-90, CEC (1989).
88. Abrajano, T.A., Bates, J.K., Woodland, A.B., Bradley, J.P. and Bourcier, W.L., Secondary Phase Formation During Nuclear Waste Glass

Dissolution, to be published in *Clay and Clay Minerals*.
89. Walker, M.M., Masters Thesis, University of Florida, Gainesville, Florida (1977).
90. Fillet, S., Phalippou, J., Zarzycki, J. and Nogues, J.L., *J. Non-Cryst. Solids* 82: 232-238 (1986).
91. Tomozawa, M. and Capella, S., *Comm. Am. Ceram. Soc.* C24-C25, February (1983).
92. Pederson, L.R., Thomas, M.T. and McVay, G.L., *J. Vac. Sci. Technol.* 18(3): 732-737 (1981).
93. Roggendorf, H., Conradt, R. and Schmidt, H., *Mat. Res. Soc. Symp. Proc. Vol. 127* (W. Lutze and R.C. Ewing, eds.), pp. 89-96, Materials Research Society (1989).
94. Clark, D.E. and Yen-Bower, E.L., *Surf. Sci.* 100: 53-70 (1980).
95. Berger, G. and Atzrodt, V., *Phys. Stat. Sol.* 85: K9-K13 (1984).
96. Pantano, Jr., C.G., Dove, D.B. and Onoda, Jr., G.Y., *J. Vac. Sci. Technol.* 13(1): 414-418 (1976).
97. Clark, D.E., Hench, L.L. and Acree, W.A., *J. Am. Ceram. Soc.* 58(11-12): 531-532 (1975).
98. Clark, D.E. and Purdy, B.A., *American Antiquity* 44(3): 517-524 (1979).
99. Matzke, Hj. *Fresenius Z. Anal. Chem.* 319: 801-808 (1984).
100. Sales, B.C., Boatner, L.A., Naramoto, H. and White, C.W., *J. Non-Cryst. Solids* 53: 201-226 (1982).
101. Bunker, B.C., Arnold, G.W., Beauchamp, E.K. and Day, D.E., *J. Non-Cryst. Solids* 58: 295-322 (1983).
102. Della Mea, G., Dran, J.C., Petit, J.C., Bezzon, G. and Rossi-Alverez, C., *Nuc. Inst. Meth. Phys Res.* 218: 493-499 (1983).
103. Schnatter, K.H., Doremus, R.H. and Lanford, W.A., *J. Non-Cryst. Solids* 102: 11-18 (1988).
104. Sanders, D.M. and Hench, L.L., *J. Am. Ceram. Soc.* 56[7]: 373-377 (1973).
105. Clark, D.E. and Yen-Bower, E.L., *Surface Science*, 100: 53-70, North-Holland (1980).
106. Zoitos, B.K. and Clark, D.E., in: *Materials Stability and Environmental Degradation* (Aa Barkatt, E.D. Verink, Jr. and L.R. Smith, eds.) Vol. 125 169-176 (1988).
107. Piepel, G.F., Jones, T.E., Eggett, D.L., and Mellinger, G.B., *Product Consistency Test Round Robin* Conducted by the Materials Characterization Center-Summary Report, PNL-6967/UC-70 (1989).
108. Fisher, D.J., in: *High Temperature Chemistry of Inorganic and Ceramic Materials* (F.P. Glasser and P.E. Potter, eds.) Special Pub. No. 30, 1-10 (1976).

4

Characterization of Corroded Ceramics by SIMS

Alexander Lodding

Physics Department
Chalmers University of Technology
Gothenburg, Sweden

INTRODUCTION

As complete a knowledge as possible of the distribution and kinetics of a wide range of elements, even those present in low concentrations, is of primary importance in many branches of materials research. Advanced study of problems such as phase formation, doping, stoichiometry, atomic transport or corrosion, requires analytical equipment capable not only of high sensitivity of element detection, but also of good spatial resolution in specimen morphology. A powerful technique combining these capabilities is offered by Secondary Ion Mass Spectrometry (SIMS) (1,2). Sensitivity in the sub-ppm range, applying to nearly all elements in the periodic table, is here paired with sub-micron resolution in elemental depth profiling or surface mapping. Due to its particular set of assets, SIMS has in recent years undergone a rapid expansion in multidisciplinary applications (3).

As regards SIMS investigations of ceramics and other electrically insulating materials, major progress has been achieved recently due to elimination of particular artifacts of charging and by the development of reliable quantitation (4-10).

The aim of this chapter is to discuss the characterization of ceramics and related materials by means of profiling and imaging SIMS, as illustrated by applications on corroded glasses and superconducting ceramics.

MAIN PRINCIPLES OF FUNCTIONING INSTRUMENTATION

Secondary ion mass spectrometry is based on a) bombardment (sputtering) of the specimen surface by energetic primary ions, and b) mass analysis of the ionized sputter products.

One may distinguish between "static" and "dynamic" SIMS. In the former mode, a relatively low intensity of primary ion bombardment is employed and only the outermost atomic layers of material surfaces are studied for chemical information. In the dynamic mode, sensitivity as well as lateral resolution is furthered by intensive sputtering, and appreciable penetration into the specimen is effected during analysis. The present discussion of analytical applications is concerned exclusively with dynamic SIMS.

The main types of information are a) mass spectra (point analysis), b) elemental profiling (in depth or along surface), and c) imaging (surface mapping by mass-resolved ion microscopy). The primary information, i.e. secondary ion intensities at different atom mass numbers, can under favorable circumstances be quantified to the respective element concentrations.

The typical renderings of two types of modern SIMS instrumentation may be seen in Table 4-1. "Type A" is a good multi-purpose machine, designed foremostly for high sensitivity and good in-depth resolution. "Type B" is chiefly an imaging instrument, characterized by very high lateral resolution. As the three aspects of performance are mutually competitive (1), the two designs of SIMS complement each other. The applications to be discussed in this chapter involve mainly Type A instrumentation (see Figure 4-1), but examples of Type B imaging applications are also shown.

Table 4-1. Two Mutually Complementary Instrumental Designs of SIMS

	TYPE A	TYPE B
CLASS	DIRECT-IMAGE MICROANALYZER	SCANNING MICROPROBE--MICROSCOPE
MAKE	CAMECA IMS-3F/4F (1)	UC-HRL (1,11)
MASS SEPARATION	MAGNETIC-ELECTROSTATIC	QUADRUPOLE
MASS RESOLUTION $M/\Delta M$	$\cong 2 \times 10^4$	$\cong 5 \times 10^2$
SECONDARY ION ENERGY ACCEPTANCE WINDOW	≅ 0.5 - 130 eV	≅ 0.2 - 15 eV
PRIMARY IONS	O_2^+; O^-; Ar^+; N_2^+; Cs^+	Ga^+; In^+
Prim. ion acceleration	≅ 2 - 18 kV	≅ 25 - 45 kV
Prim. ion current density	$\cong 1 - 10^2$ mA/cm^2	$\cong 10^2 - 10^3$ mA/cm^2
Prim. ion probe diameter	0.2 - 250 µm	0.02 - 0.05 µm
Lateral resolution	--> ≅ 200 nm	--> ≅ 20 nm
Depth resolution	--> ≅ 0.5 nm	--> ≅ 2 nm
Element detection limits (typical)	10^{-3} - 10 ppm	1 - 10^4 ppm

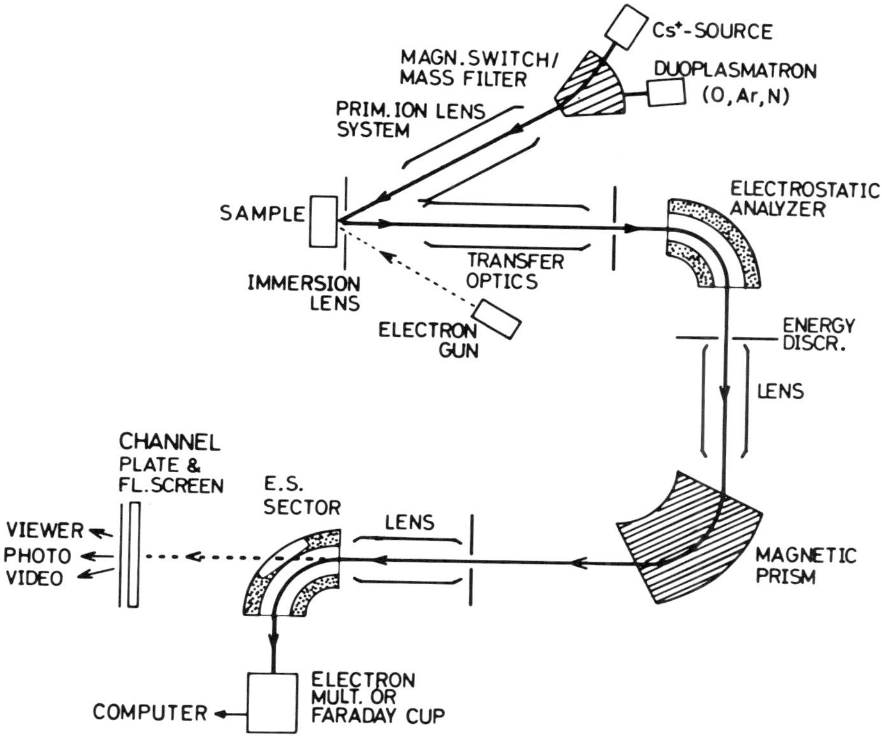

Figure 4-1. Schematic drawing (1) of Cameca Direct-Image Ion Microanalyzer.

EXPERIMENTAL ASPECTS OF SIMS

As primary ions impinge on the surface of an insulating specimen, charged secondary species, particularly electrons, are emitted. A charge develops on the surface. The effect on the ion optics includes defocusing and/or deflection of the primary beam, often the disappearance of the secondary ion signal, and sometimes spark discharge. Also, the specimen potential is affected, with serious consequences on the accuracy of measurement.

To reduce these effects, the insulator surface is routinely coated with a conductive layer, usually ca 10nm Au. With the Type A ion probe instrument, best results have hitherto been obtained when using O^- primary ions. When analyzing positive secondary ions, the negative primaries enter at a relatively high angle of incidence; also the secondary electrons are to some extent turned back to the specimen with this combination of polarities. If the aperture swept in the conductive layer by sputtering is not too wide, surface charging remains moderate, shifting the effective specimen potential only by some 100eV or less.

Even so, the shift may be sufficient to inhibit the reproducibility and quantitation of the measurement. This is illustrated in Figure 4-2, where the kinetic energy distributions of different ionic species, sputtered from a glass, are schematically plotted (1). It can be seen that the relative ionic yields of elements vary considerably with energy. The mass spectrometer receives ions within a pre-selected pass-band (EWW in Figure 4-2). As the specimen potential changes due to charging, the elemental energy distributions drift past this "window," and the relative sensitivities of element detection are affected. As any quantitation can only be meaningful within a definite kinetic energy range of secondary ions, the shift introduces a serious complication.

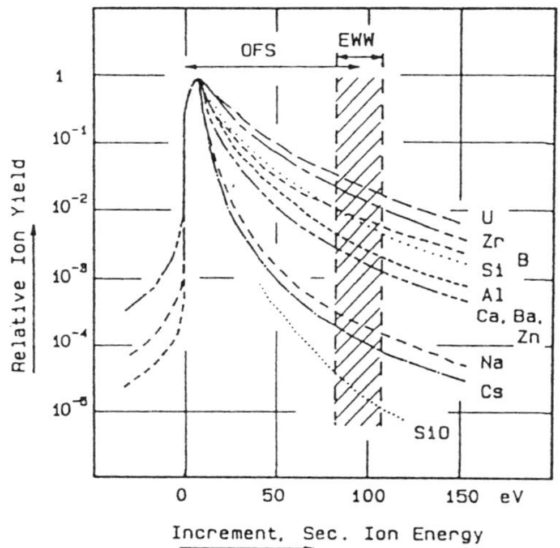

Figure 4-2. Kinetic energy distributions of different ions sputtered from a glass. Primary ions: 15 keV O^-.

For a few Type A ion probes a voltage-compensating software routine has therefore been devised. Periodically the mass range is scanned, a pre-selected sharp reference peak is recognized, and sample voltage is adjusted to return the peak to the correct potential with respect to the mass spectrometer. This allows very stable and reproducible recording of spectra and in-depth profiles for insulator specimens, which would otherwise not easily be amenable to quantitative SIMS study.

In some laboratories, positive secondary ions are studied by using positive primaries rather than O^-, and neutralizing surface charging by means of an electron flush beam. So far such a procedure (12) has presented problems of ion yield and stability, and appears to be less extensively employed than the O^- routine.

None of the above-mentioned precautions is sufficient to ensure good results in the study of negative ion spectra. However, a special type of electron gun

(13), developed for the Type A instrument, is, in combination with O^- primary ions, capable of maintaining a self-regulating constant specimen potential, and has in recent years made possible excellent profiles and images in negative secondary ion spectra from insulators, including glasses (14).

In the Type B instrument the charge effects may be to some extent alleviated by a relatively thick Au coating, and particularly in the negative spectrum quite stable SIMS imaging of insulators can often be performed (7). However, minimization of charging usually necessitates relatively low primary ion intensities, which introduces further limitations in the already relatively poor counting statistics of the quadrupole type analyzer. Also, the shifting of the energy window position is here a serious impediment to reproducibility and quantitation. Nevertheless, by means of the Type B instrument excellent mass-resolved images of insulator surfaces have been obtained at a unique lateral resolution (7,8,10).

SECONDARY ION EMISSION AND QUANTITATION

A SIMS spectrum is characteristic of the analyzed solid but is also conditioned by factors such as type, energy and direction of the incident ions, by the ambient atmosphere, and by the technicalities of the mass spectrometer. The intensity of an individual mass peak is proportional to the concentration of the corresponding isotope or element according to:

$$I_{LM}/c_L \propto b_M I_p (S_\eta) \gamma_L \qquad (4\text{-}1)$$

where I_{LM} denotes the ion current of element L at the peak of mass number M with isotope abundance b_M, c_L is the concentration of L, I_p is the total primary ion current impingent on the analyzed area, S is the sputtering yield (secondary particle per incident ion), η the transmission of the spectrometer, and γ_L the positive or negative ionizability of L (ions per secondary particle).

The key entity in sensitivity and quantitation is the elemental ion yield γ_L, which may vary within wide ranges not only from one element to another, but also depending on the nature of the sputtered matrix ("chemical effect"). The yield of the positive ions is usually furthered by the presence of electro-negative reactive species (e.g. oxygen, sulphur, chlorine); that of negative ions by the presence of species that lower the work function for electron emission. The chemical effect favors high and stable secondary emission from ionic and compound media, including most ceramics. The material classes of interest in this chapter can therefore be considered as particularly suitable for sensitive SIMS analysis.

In regards to quantitation, in most practical cases the instrumental and matrix dependent factors in Equation 4-1 can be bypassed when considering the yield of element L only relative to a reference peak R (corresponding to a known concentration c_R), which leads to a considerably simplified expression:

$$[(I/c)_L/(I/c)_R](b_R/b_M) = (\gamma_L/\gamma_R) = (RSF)_L \quad (4\text{-}2)$$

the last term denoting the "Relative Sensitivity Factor" of element L.

Table 4-2 lists representative RSF-values in the positive spectrum for selected elements in a nuclear waste glass and in an oxide superconductor. The relative ionizabilities have been obtained in careful measurements on chemically well-defined standards (4,9,15). It is found in the calibration studies that the variations in RSF, between different glasses in the alkali-boro-aluminosilicate group, or between different superconductor oxides of interest, are relatively small, roughly within one order of magnitude (9,15).

As mentioned above (see Figure 4-2), the relative ionizabilities are energy dependent. The RSF values in Table 4-2 apply with OFS = 75eV and EWW = 50eV. This relatively high choice of energy offset is routinely used in many quantitative measurements on ceramics and glasses. The discrimination of low energy secondary ions is useful, as the sensitivity of elemental detection is strongly dependent not only on ionizability, but also on the elimination of spectral contaminations on the element mass numbers (1). These overlay mass spectra are mainly composed of molecular ion species from the matrix. Such "intrinsic" spectra from most classes of ceramics and glasses are relatively complicated and may mask many impurities of interest. However, the energy distributions of molecular ions are as a rule steep (e.g. SiO^+ in Figure 4-2), and thus an elimination of low energy species by relatively high offset (OFS) considerably simplifies the spectrum, and favors the detection of monatomic elemental ions. This is illustrated in Figure 4-3 a,b and 4-4 a,b, showing typical spectra from a glass and an oxide superconductor with a low offset compared with corresponding spectra where energy filtering has been employed.

Table 4-2. Relative Elemental Sensitivity Factors (RSF) in SIMS, Measured for Two Classes of Oxide Based Materials and Relating to Positive Secondary Ions in the 75-125 eV Kinetic Energy Range. Reference Element: Si.

Alkali Borosilicate Nuclear Waste Glass (15)				Oxide Superconductors Cu & Group IIa-Va Compounds (9)			
Na	1.2	Al	4.4	Na	100	Tl	1,2
K	1.6	Y	2.3	Cu	0.4	Y	22
Cu	0.15	La	1.8	Mg	6	C	0.1
Mg	1.85	C	0.15	Ca	24	Si	1.0
Ca	2.45	Si	1.0	Sr	30	Pb	0.2
Sr	1.65	Pb	0.35	Ba	26	Sb	0.2
Ba	1.15	Sb	0.5	Zn	0.15	Bi	0.15
Zn	0.13	Ti	2.4	Al	20	Ti	20
B	1.15	Fe	1.65	In	8	Fe	1.4

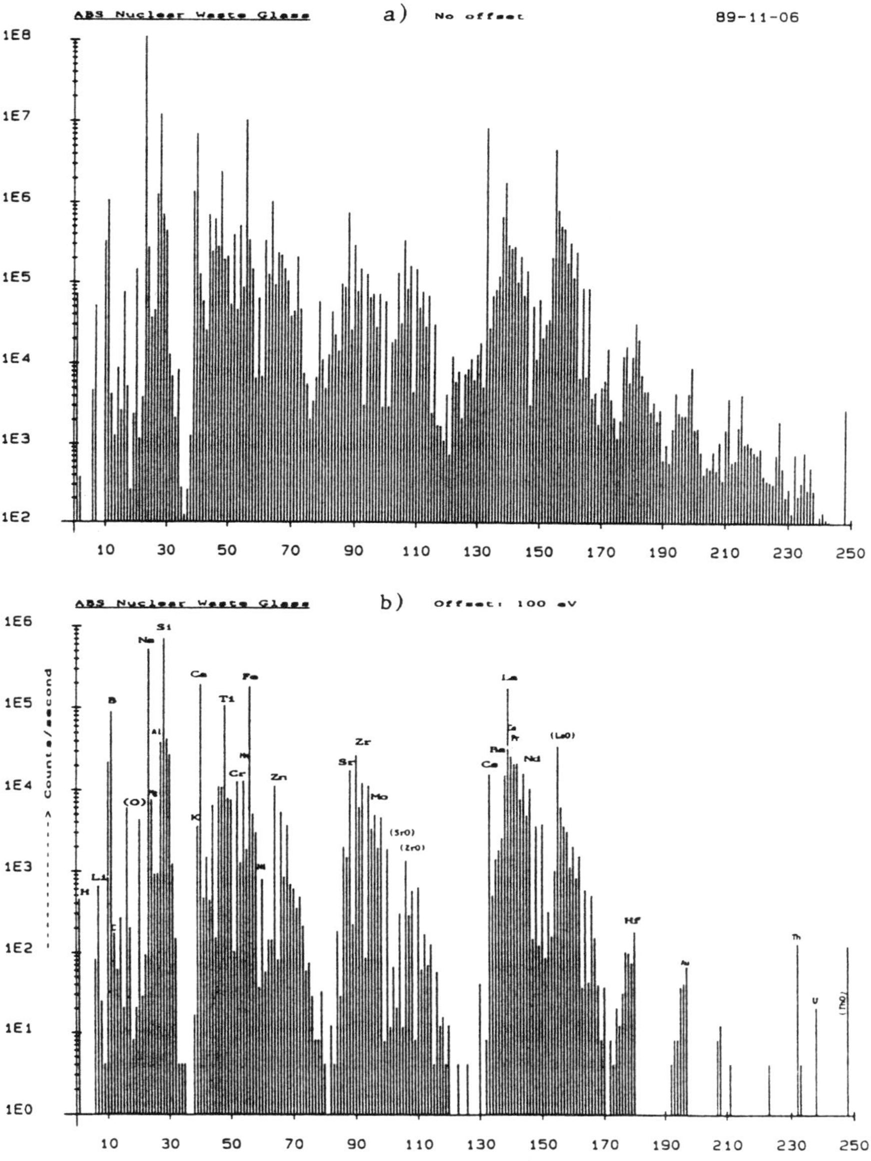

Figure 4-3. SIMS positive ion spectra of an alkali borosilicate glass with simulated reactor waste and fission products (14). a) No energy filtering of secondary ions. b) Only ions in energy range 75-125eV admitted into analyzer; relative enhancement of monatomic ions at the cost of molecular species.

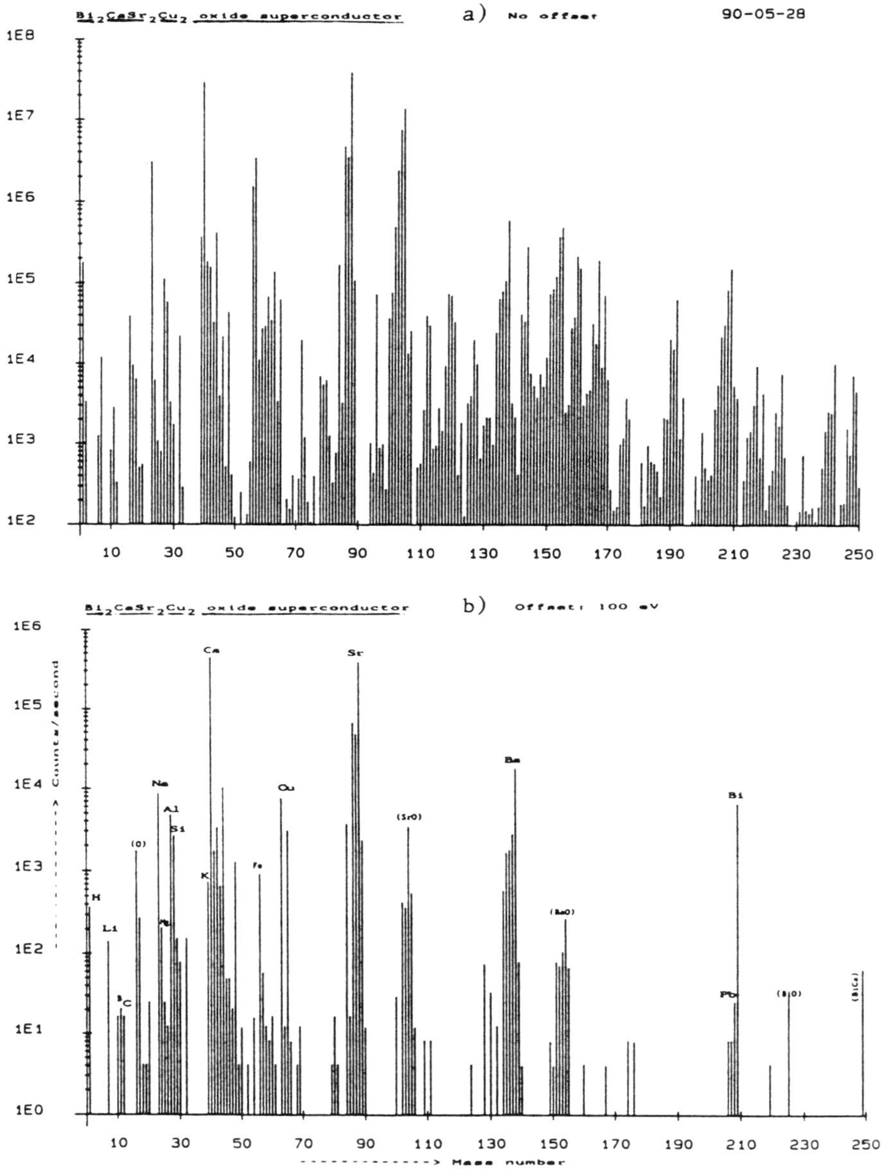

Figure 4-4. SIMS positive ion spectra of a sintered $Bi_2CaSr_2Cu_2O_8$ high T_c superconductor. a) No energy filtering of secondary ions. b) Only ions in energy range 75-125eV admitted into analyzer; relative enhancement of monatomic ions, supression of molecular (oxide, dimer, trimer, etc.) species.

Spectral contamination may, of course, also be suppressed if the spectrometer offers high mass resolution. However, the energy filtering procedure is readily applicable throughout the spectrum and as a rule causes smaller loss in the secondary ion signal. Another advantage of high energy offset is a decrease of element differences in ionizability (16).

To maintain good mass resolution, only ions ejected from a central portion of the sputtered crater are admitted for mass analysis. In order to obtain the correct depth scale through layers with different sputter rates, two or more in-depth profiles are recorded for each specimen, employing a stylus profilometer to measure the respective crater depths.

APPLICATIONS TO CORROSION LAYERS ON GLASSES

Due to recent technical advances, as discussed above (4,6), SIMS in three dimensions has now developed into a powerful tool in the characterization of glasses. Illustrated in Figures 4-5 and 4-6 are applications in the field of radioactive waste management (14-18; see also the chapter by G.G. Wicks in this volume). The waste form studied by SIMS contains small quantities of simulated reactor waste elements and fission products. The glass is exposed to leachants in laboratory or in realistic geological environments (17,18), which results in partial dissolution and surface precipitation, in loss of certain elements and in penetration of other elements from the environment.

The main aims of SIMS application to waste forms are: 1) to assess the depth of leaching as well as the loss or influx of each element after different times of exposure, in order to facilitate the prediction of long-term performance; and 2) to identify different corrosion phases and leaching zones, as a means of recognizing and understanding the mechanisms of corrosion.

In Figure 4-5 are plotted the depth profiles of 19 elements in a waste form glass after 2 years interaction with brine in a burial environment (14; G.G. Wicks, this volume). Figure 4-5a shows "raw data," i.e. the secondary ion counts at the mass peaks of the different elements. These profiles are not quantified, but allow several distinctive subzones in depth to be qualitatively recognized: such as the β layers, depleted in most glass components but penetrated with Mg and H from outside; or the outermost precipitation layers, enriched i.e. in Si and B. The dV profile in Figure 4-5a shows the increment in sample potential supplied by the automatic compensation for the build-up of surface charge on the insulating specimen.

Figure 4-5b presents the quantified profiles, in mole percent of the sum of all cations (excluding hydrogen). In the principal interaction layer, β_0, one may note nearly total depletion of Li, Na, Cs, Sr, Ba, Mn; there is also distinct loss of the network formers, Si and B; and major influx of chiefly Mg and H. Also typical is the peak of K in the "gradient layer" β_1 at the reaction front and the substitution of H for Li in the "diffusion layer" β_2. Certain elements (Al, Zr, Ti) are apparently "enriched" in the β layers; actually these elements may be considered as remaining nearly conserved in the glass (4,14), an assumption that offers a means to determine the net elemental balance, i.e., the molar loss or influx of every element partaking in the interactions.

This may be effected via the representation in Figure 4-5c, where all profiles are normalized to that of the "inert" element Al. The elemental loss or influx calculation for the β layers is made by integration of the respective decrements from the bulk concentration. In Table 4-3 such evaluation is listed (here concerning only the principal elements) for the glass as profiled in Figure 4-5. One may observe the nearly quantitative substitution of Mg and H for the leached out network formers and alkali elements.

Table 4-3. Element Balance in Depleted Layers on a Waste Form Glass after 2 Years Leaching in Brine, in mmol/m^2 (14). The "total" accounts for the balance of 19 cations.

	Total	Si	B	Alkali	Mg	H
Loss	405	165	20	190	-	-
Influx	350	-	-	-	170	160

Among the profiles in Figure 4-5, in particular those of O, Cl, and H show relatively low yields of positive secondary ions. It should be mentioned here that quantitative depth profiling in glasses may now also be performed for elements in the negative ion spectrum (14). This is particularly relevant when studying species with high positive ionization potential and high electron affinity (1,2).

Figure 4-6 shows examples of SIMS ion imaging (both polarities; positive above, negative below in figure), only recently developed as an efficient technique for insulating specimens (14,15). In these applications, as in the profiling procedure, the Type A instrument of Table 4-1 was employed, equipped with a new type of electron gun for charge neutralization (13). The images in Figure 4-6 show the corner areas, ca 60x60μm^2, of craters sputtered to ca 2μm depth through the leached layers and into the bulk of the specimen profiled in the preceding figure. The distributions of elements in the respective sub-layers are clearly evident in the "taper" representation of the crater walls. Note, for example, the narrow K peak at the reaction front, or the depletion of Si and B in the β layers and the influx of Mg, H, F, Cl, S and C.

Gate scan of the ion intensities across the imaged crater walls (from the surface outside the crater to the bulk glass at the crater bottom) yields semi-quantitative element profiles of a type similar to those in Figure 4-5a (except for the "tapered" depth scale), but shows here also the "raw data" distributions of species in the negative spectrum.

This new routine of elemental imaging by SIMS offers considerable promise concerning the characterization of various reaction phases on corroded glass.

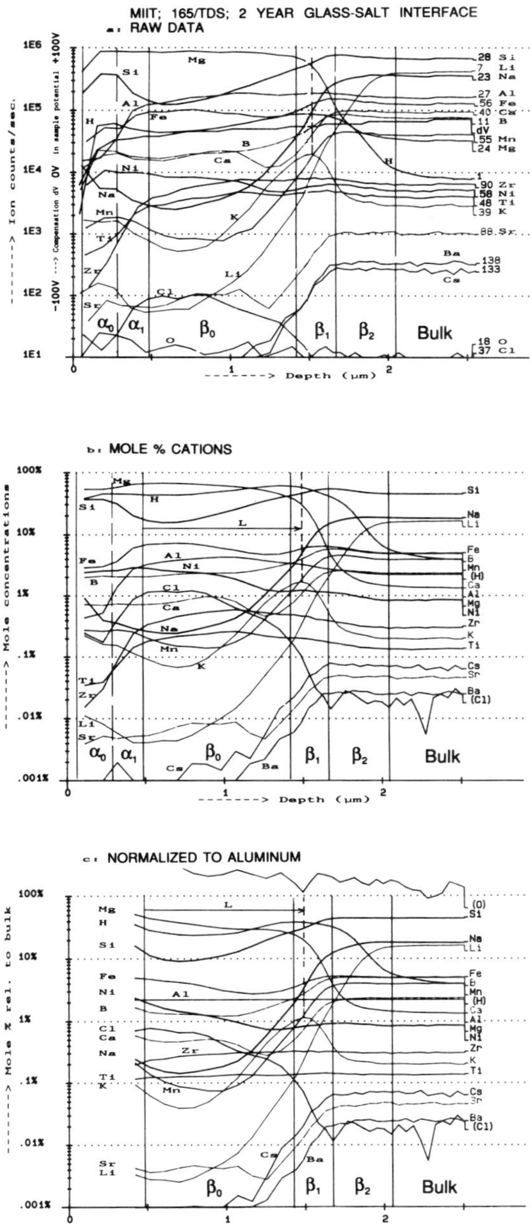

Figure 4-5. SIMS in depth profiles of elements in corroded layers on glass after 2 years leaching in brine. a) Secondary ion counts, not quantified. b) Mole concentrations (relating to sum of cations). c) Normalized to Al (element balance profiles).

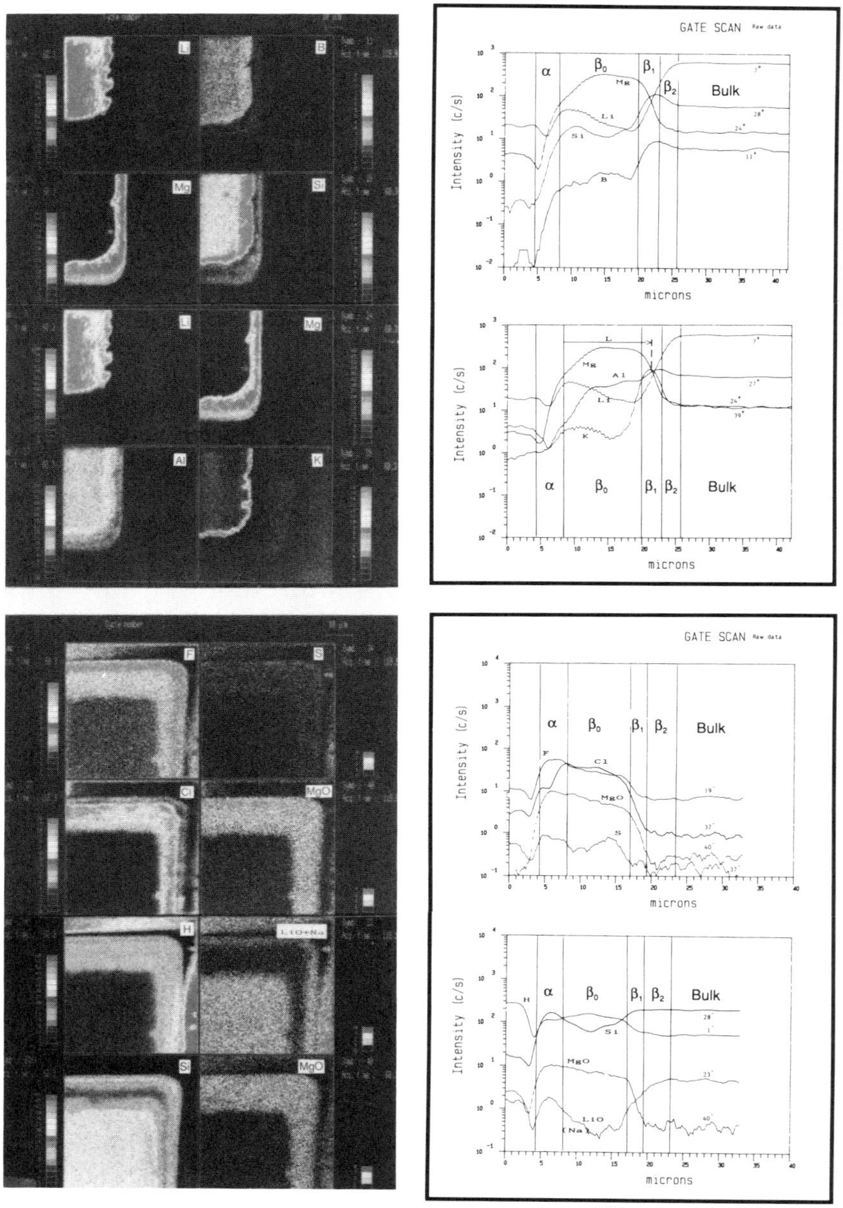

Figure 4-6. Left: Elemental images of sputter craters through corroded layers on glass. Right: Ion intensity scans across imaged crater walls (14).

APPLICATIONS TO IMPURITIES AND CORROSION PRODUCTS IN CERAMIC SUPERCONDUCTORS

The main uses of SIMS in the study of high-T_c superconductors and related oxides are based on the particular capability to identify stoichiometric phases and localize dilute but functionally important impurities. The illustrations chosen in this chapter show applications of two kinds: quantitative profiling in thin films of superconducting oxide material (5,6); and high resolution imaging of sintered and corroded material.

Figure 4-7 shows the in-depth variations in the concentrations of eleven elements in a film of $Ca_2Sr_2Bi_2Cu_3O_x$, deposited on sapphire substrate by the laser ablation technique. The quantitation was made by means of calibrated relative sensitivity factors (see Table 4-2). The assignments of RSF values were found (9) to apply (with an accuracy on the order of ±15% for the major elements) in different classes of high T_c materials, including the present examples. In the profiles the concentrations are quoted relative to the sum of cation constituents; consequently the approximate figure of 100% for oxygen corresponds to x≈9 in the stoichiometric formula.

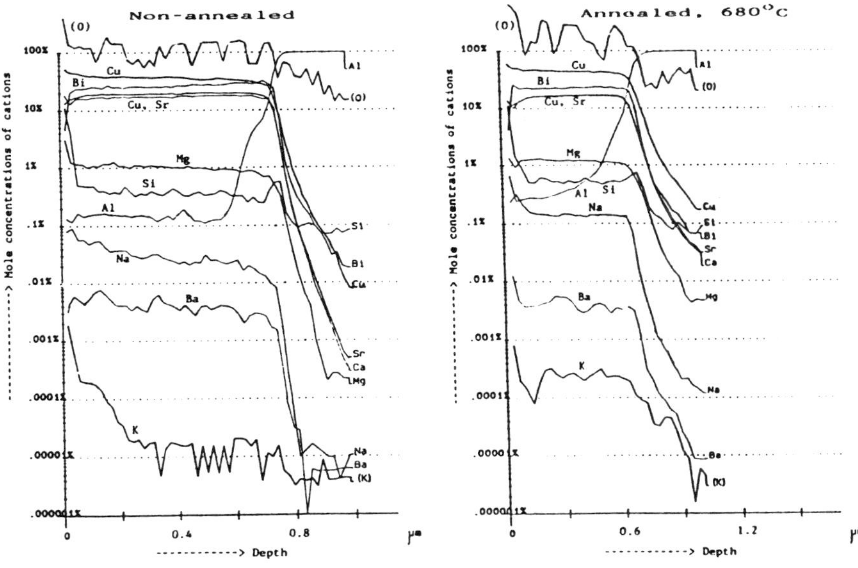

Figure 4-7. In-depth profiles of elements in alyer of $Ca_2Sr_2Bi_2Cu_3$ oxide superconductor on sapphire. Left: as produced (by laser ablation) Right: annealed after deposition, 24 hours at 680°C.

The oxygen profiles are seen to show considerable fluctuations due to poor counting statistics. The reason is that the small 18-O isotope (0.2%) had to be exploited, as the 16-O bombarding species was implanted during analysis. In similar measurements under preparation (19), 18-O^- is to be used as the primary ion.

The application in Figure 4-7 demonstrates the effects of annealing. One may note, i.a., the distinct increase in alkali impurity from the anneal environment; the in-diffusion of Al from the Al_2O_3 substrate; and a certain in-depth homogenization of the stoichiometry, achieved by annealing. The relatively high concentration suggested here for Cu (ca 44% instead of the normal 33%) is beyond the limits of experimental uncertainty and implies a shortcoming in the deposition process. Another artifact of production, as seen in Figure 4-7, is the relatively high level of Si impurity, including a typical enhancement peak at the substrate interface.

The other example of SIMS usefulness in the study of high-T_c relevant compound oxides puts the emphasis on high resolution elemental imaging, as achieved with the Type B instrument of Table 4-1 (10,11). Its outstanding lateral resolution, ≈20nm, has been exploited on thin film superconductors (20) as well as sintered ones (6,10). The illustrations presented in Figures 4-8 and 4-9 are taken from the latter group of materials.

The investigated specimen of $YBa_2Cu_3O_{7-x}$ was prepared from powders of oxide (Y and Cu) and carbonate (Ba), after calcination in a conventional furnace at 950°C and pressing at 430MPa to flat faced pellets. The subsequent sintering took place for 1 hour at 950°C, followed by slow cooling and annealing in flowing O_2. The superconducting fraction of the product was about 30%, with $T_c \approx 90K$ ((10); see also chapter by G. Chandler, this volume).

Investigation by quantitative SIMS ((10); Type A instrument) revealed a wide range of impurities, including, in concentrations of 0.1 mole-% or more, contaminants like Al, Si, Mg and Zr from the furnace. Particularly, significant amounts of carbon were found, some 0.1 mole-% in the interior and 0.7% at the surface. A few weeks' aging in the ambient air of the laboratory resulted in enhanced C levels: ca 0.3 mole-% in the bulk and as much as 2.5% locally at the surface. The naked eye aspect of the specimen also changed, from uniform dark gray to patchy with black and moss-green areas. Superconductivity progressively abated.

Figure 4-8 shows SIMS micrographs of a 30x30μm^2 area on the specimen about a week after fabrication. The topological aspect of the surface is most clearly observed in the integrated ion image (ISI; Figure 4-8f). One may roughly distinguish a fairly homogeneous area of packed crystallites in the top right portion of this image; a group of crystals, some 1-6μm in size, scattered over the more homogeneous region from bottom left to top right; and a large bottle-shaped crystal at the bottom right. The high intensity of the total ion image does full justice to the optimal lateral resolution and reveals i.a. some periodic grooving on the surface of the big crystal (possibly due to faceting effects in sputtering) with sharp details less than ca 50nm in size.

The likewise very sharp O^- image in addition shows crystal edge contrast, with a very similar topology to that of Cu^-. Closer examination reveals a

Figure 4-8. High resolution SIMS imaging of mixed phases on a sintered $YBa_2Cu_3O_{7-x}$ high T_c superconductor. a-e: Distribution micrographs of O^-, YO^-, Ba^+, Cu^-, and CN^-. f:ISI (total-ion topography). Scale: 30x30 μm^2.

decoration of crystallite faces, probably due to CuO, a decomposition effect (21).

The distributions of YO^-, Ba^+ as well as Y^- in the top left cluster are fairly homogeneous, except in its center and bottom edge portions ("cauliflower" aspect in the ISI, integrated secondary ion, image), which are depleted in Y and Cu. The homogeneous region may (after comparative study by quantitative SIMS) be assumed to be of the nominal $Y_1Ba_2Cu_3$ composition. The "cauliflower" areas are dominated by Ba, mainly as ringlike decorations on small crystallites, suggesting formations of $BaCO_3$. This is further implied by the CN^- distribution image. Carbon also appears to exist in relatively thick layers on several middle size crystals.

The bottle-shaped crystal is conspicuously void of Ba and emits an enhanced signal of Y^-. From a phase compatibility diagram of the Y-Ba-Cu oxide system (21) it appears probable that this crystal represents the (relatively rare) $Y_2Cu_2O_3$ phase. Finally, in a few isolated spots (e.g. the triangular tooth shaped crystal ca 7μm to the left of the top end of the "bottle"), Y, but not Ba and Cu, is absent, which suggests the well known decomposition phase (21) $BaCuO_2$.

Evidently the surface as imaged in Figure 4-8 exhibits a mixed array of phases, including some not closely compatible with the superconducting 1-2-3 stoichiometry.

Figure 4-9 shows element distribution images of a similarly produced specimen after more prolonged aging in room atmosphere. The surface is seen to be characterized by two main types of morphology. In areas which seem quite devoid of Y the "cauliflower" aspect dominates. Scrutiny reveals stacked arrays of rod-like crystallites, possibly $Cu_2Ba_2O_3$, with their faces decorated with $BaCO_3$, other forms of carbon, and/or CuO. In the other topography, islands of an Y-rich phase again occur as rod-shaped crystals, locally entangled or merging into more compact areas. They may be identified as the "green" phase, Y_2BaCuO_5, visible under the microscope or by naked eye as local green patches on the surface. It appears that aging causes gradual decomposition of the superconducting stoichiometry into its most compatible phases, viz. (22) $BaCuO_2$, Y_2BaCuO_5 and CuO. In addition, carbonate layers tend to dominate crystallite surfaces.

Figure 4-9f shows the O^- distribution at a particularly high ion optic magnification (only 10μm across the image). A sharp view is given of a boundary region between an array of "green" crystallites (right half), a larger crystal of the same stoichiometry (bottom left corner) and a "cauliflower" streak (left). The position of the magnified area in relation to the $40x40\mu m^2$ images is marked at the edge of Figure 4-9e.

It is obvious that high resolution SIMS is a particularly informative technique in the study of complicated microstructures. In the same degree, high sensitivity quantitative SIMS is eminently applicable in three-dimensional all-round elemental microanalysis. Combined, the performances of the two techniques open a promising future in advanced materials characterization.

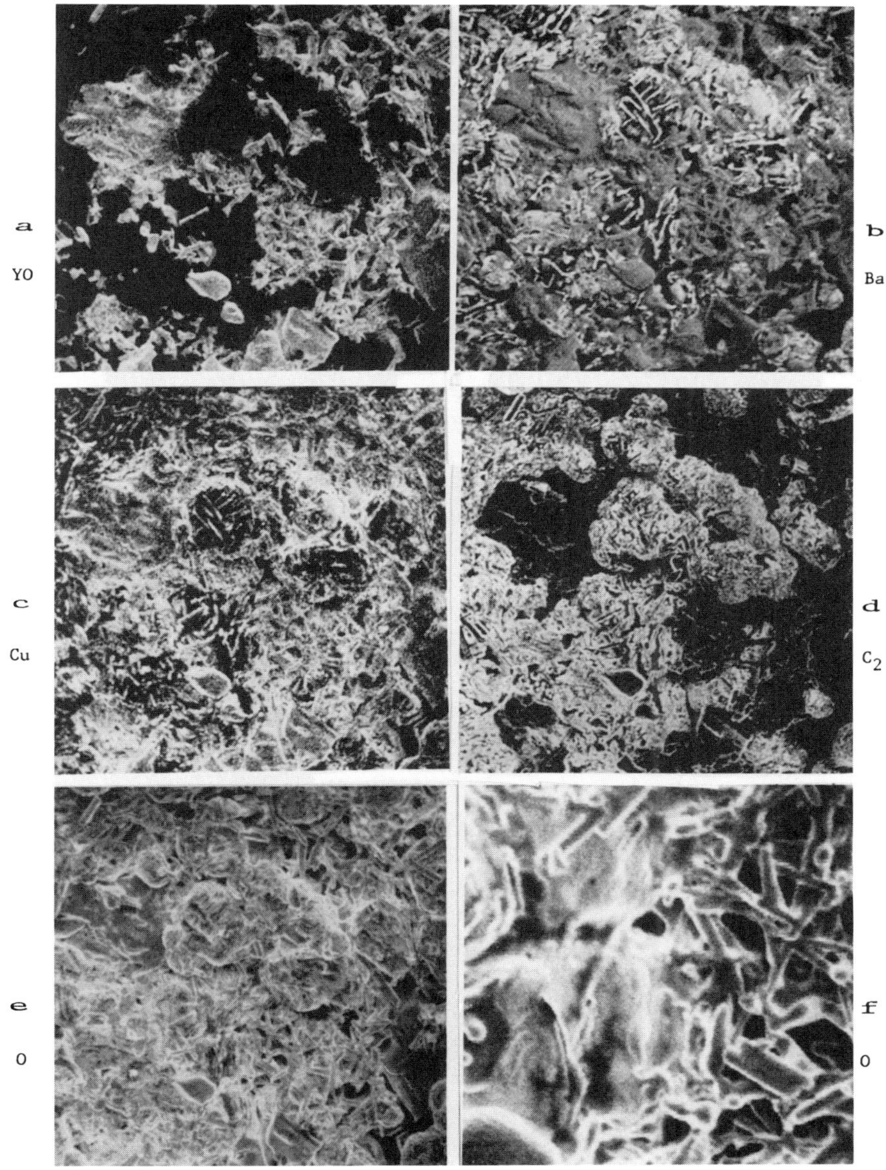

Figure 4-9. High resolution SIMS imaging of corroded surface of a sintered $YBa_2Cu_3O_{7-x}$ high T_c superconductor. Distributions: a: YO^-, b: Ba^+, c: Cu^-, d: C_2^-, and e, f: O^-. Scale, a-e: 40x40 μm^2, f: 10x10 μm^2.

SUMMARY

The technique of secondary ion mass spectrometry (SIMS) has asserted itself as uniquely applicable in the microanalytical and microstructural investigation of ceramics. In the profiling mode SIMS offers fast, complete and quantitative analysis with an outstanding (sub-ppm) detection sensitivity and with a depth resolution in the order of 10nm. In the imaging mode, elemental distributions can be mapped with lateral resolutions down to some 20nm. The particular experimental problems typical for the SIMS of insulating and/or multiphase materials have of late been elucidated and essentially solved.

In investigations of corroded layers on ceramics, SIMS supplies complete elemental balance in three dimensions as well as the identification of individual sub-layers and phases, thus contributing significantly to the knowledge of the various mechanisms of corrosion. Further, SIMS has proved to be a particularly efficient tool in the study of the distribution and kinetics of impurities and stoichiometric phases, with high relevance in ceramics processing. Extensive applications have in recent years been performed on glasses for nuclear waste management, on bioceramics and biomineralized tissues, and on high T_c superconductors. The use of SIMS in the characterization of numerous classes of ceramic materials is rapidly expanding.

REFERENCES

1. Lodding, A., in: *Inorganic Mass Spectrometry* (F. Adams, R. Gijbels, and R. van Grieken ed.), Ch. 4, pp 125-171, Chem. Analysis Ser. 95, Wiley & Sons (1988).
2. Benninghoven, A., Rüdenauer, F.G., and Werner, H.W., *Secondary Ion Mass Spectrometry*, Chem. Analysis Ser. 86, Wiley & Sons (1988).
3. *SIMS VII* (A. Benninghoven, C.A. Evans, K.D. McKeegan, H.A. Storms, and H.W. Werner ed.), Wiley & Sons (1990).
4. Lodding, A., Odelius, H., Clark, D.E., and Werme, L.O., *Mikrochim. Acta Suppl. 11*, pp 145-161 (1985).
5. Engström, E.U., Fischer, P.M., Lodding, A., Odelius, H., and Södervall, U., "Quantitative Applications of SIMS on Interdisciplinary Materials," to be published in *Surf. & Interf. Analysis.*
6. Lodding, A., Engström, E.U., and Odelius, H., in *Fortschrittliche REM Analyse in der Hochtechnologie* (V. Thien ed.), Vol. 13, pp 351-369, Deutscher Verb. Materialprüfung (1988).
7. Lodding, A., Norén, J.G., and Petersson, L.G., in: *SIMS VI* (A. Benninghoven, A.M. Huber, and H.W. Werner ed.), pp 865-871, Wiley & Sons (1988).
8. Lodding, A., Fischer, P.M., Odelius, H., Norén, J.G., Sennerby, L., Johansson, C.B., Chabala, J.M., and Levi-Setti, R., "Secondary Ion Mass

Spectrometry in the Study of Biomineralizations and Biomaterials," *Analyt. Chim. Acta*, 241(2), 299-314 (1990).

9. Engström, E.U., Lodding, A., and Odawara, O., in Ref. 3, pp. 705-708 (1990).
10. Chabala, J., Chandler, G., Clark, D.E., Engström, E.U., Levi-Setti, R., and Lodding, A., "Quantitative and Imaging SIMS of Sintered YBa_2Cu_3 Oxide Material," to be published in *Scanning Microsc.*
11. Levi-Setti, R., Crow, G., and Wang, Y.L., *Scanning El. Microsc.* 85-II: 535-542 (1985).
12. Tait, J.C., Hayward, P.J., Hocking, W.H., Betteridge, J., and Doern, D., "Analysis of Aluminosilicate Glass Samples from the In-Situ WIPP Salt Burial Test," *Nuclear Waste Management III*, (G. Mellinger, ed.) Amer. Ceramic Soc. 363-376 (1990).
13. Slodzian, G., Chaintreau, M., and Dennebouy, R., in *SIMS V* (A. Benninghoven, R.J. Colton, D.S. Simons, and H.W. Werner eds.), pp 158-160, Springer Ser. Chem. Phys. 44 (1986).
14. Lodding, A., Clark, D.E., Engström. E.U., Odelius, H., Schuhmacher, M., Zoitos, B.K., and Wicks, G.G., "SIMS Applications on Nuclear Waste Forms," **Ceramics Today - Tomorrow's Ceramics** (P. Vincenzini, ed.) pp 3121-3129, Elsevier (1991).
15. Lodding, A., Engström. E.U., and Schuhmacher, M., "Ion Optic Imaging of Element Distributions in Glasses," to be published in *Scanning Microsc.*
16. Engström, E.U., Lodding, A., Odelius, H., and Södervall, U., *Mikrochim Acta I*: 471-477 (1987).
17. Lodding, A., Engström, E.U., Clark, D.E., Werme, L.O., and Wicks, G.G., in: *Adv. in Ceramics* (D.E. Clark, W.B. White, and A.J. Machiels eds.), Vol. 20, pp 567-581 (1986).
18. Lodding. A., Engström, E.U., Clark, D.E., and Wicks, G.G., "Quantitative Concentration Profiling and Element Balance in SRL Glass after Two Years in WIPP," *Nuclear Waste Management III*, (G. Mellinger, ed.) Amer. Ceramic Soc. 317-333 (1990).
19 Ivanov, A., and Engström, E.U., to be published in *Surf. & Interf. Analysis.*
20. Levi-Setti, R., Chabala, J.M., Chang, R.P.H., Hansley, D.L., Ketterson, J.B., Li, D.Q., Wang, Y.L., and Wang, X.K., in Ref. 3, pp 693-696 (1990).
21. Frase, K.G., and Clarke, D.R., *Advanced Cer. Mater.* 2-2B: 295-300 (1987).

Section II.

Corrosion of Glass

5

Geochemical Approach To Glass Dissolution

Bernd Grambow

*Kernforschungszentrum Karlsruhe,
Postfach 3640, 7500 Karlsruhe, Germany*

INTRODUCTION

Studies of the reaction of silicate glasses with aqueous solutions began with modern chemistry. Most of the theoretical work has been concentrated on the short-term diffusion-controlled alkali leaching as an initial step of the glass/water reaction. Archeological and geological glass dating, in particular the need to dispose of nuclear waste in a vitrified state, have attracted significant attention to the long-term aspects and the mass balance of the glass/water reaction. Large experimental data bases have been generated (1) and models have been developed to describe the corrosion behavior of glasses under conditions which may occur in a future nuclear waste repository. Similarities between glass corrosion processes and rock/water interactions suggest the application of geochemical models to glass corrosion. Modeling is a necessary step for using experimental results to predict the long-term performance of any nuclear waste glass, but it is also useful for the dating of historical, archeological or geological samples or for the corrosion behavior estimation for glass compositions and conditions where experimental data are missing.

Geochemical models have already been applied successfully to many corrosion tests of nuclear waste glasses and to the alteration of basaltic glasses in the geochemical environment (2,3,4,5) (see Jercinovic's chapter). Glass corrosion behavior is described in terms of a mechanism which is based on glass network breakdown. In addition, saturation effects, surface layer formation, diffusion processes, and changes in solution composition are taken into account. The theoretical treatment of the glass/water reaction is based on transition state theory in combination with reversible and irreversible thermodynamics. Recent progress in the field of rock/water interaction has been applied to the theory of glass dissolution.

Computer programs such as PHREEQE (6)/GLASSOL or EQ3/6 can be used to model the glass/water reaction. Input variables are glass and solution compositions, time, temperature, glass surface area, solution volume, and flow rate. Rate and stability constants are required, and can be determined by experiments.

SHORT OVERVIEW ON CONCEPTS OF GLASS DISSOLUTION

The following short overview on concepts of the glass/water reaction develops the need for a general theory of glass corrosion. For a more detailed review the reader is referred to the literature (7,8).

Alkali Extraction vs. Matrix Dissolution

The reaction of alkali silicate glasses with aqueous solutions is usually interpreted as a result of a combination of two independent processes: initial diffusion-controlled extraction of alkali ions out of the glass matrix, and the dissolution of the glass matrix itself (9-12). The initial reaction shows a square root of time dependence for the alkali release, and it is described as an ion exchange (interdiffusion Na^+/H_3O^+ or Na^+/H^+) process (12) or as a result of water diffusion into the glass network (13). As the rate of release decreases with increasing depth of alkali depletion in the outer glass surface, matrix dissolution will become the dominant reaction. The coexistence of matrix dissolution and diffusion-controlled ion exchange or transport of water molecules was described mathematically by the solution to Fick's law for moving boundary conditions (12,14-16).

In the mass balance of the overall reaction, the relative contribution of selective alkali release and matrix corrosion depends on a diffusion coefficient D (m^2/sec) and on the rate of matrix corrosion, a, (m/sec). For times $t << a^2/D$ the overall reaction is dominated by diffusion-controlled release of alkali ions, for example, according to interdiffusion theory, the quantity Q (mol/m^2) of released alkali is given (17) by:

$$Q = C_o * 2\sqrt{(D*t/\pi)} \quad (5\text{-}1)$$

where C_o is the concentration of alkali in the glass. Matrix dissolution becomes dominant at $t >> a^2/D$ and the amount of leached alkali is given by:

$$Q = C_o(a\ t + D/a) \quad (5\text{-}2)$$

The diffusion coefficient in these equations is a possibly concentration-dependent alkali/hydrogen interdiffusion coefficient in the dry glass. The value of this coefficient is governed by the less mobile ion, which is the hydrogen-containing specie (18,19). The term D/a may be interpreted as the depth of selective alkali depletion (or of water penetration) and the term 'a t' describes the depth up to which glass network breakdown has occurred.

Limitations of the Interdiffusion Theory

In various cases, the diffusion coefficient of H_3O^+ or alkali ions in the bulk glass is a factor of 1000 or more lower than a diffusion coefficient which is obtained from interpreting solution data (15,20). The deviation was explained by the formation of a transformed surface layer with high ionic mobilities (21). Alternatively it was suggested (20) that the high diffusion coefficients are due to the penetration of water into the glass network. If diffusion in such a transformed surface layer is the rate limiting step, then the process by which this layer initially is formed (network hydration) must be faster than subsequent diffusion. The hydration reaction is based on the hydrolytic cleavage of glass network bonds. Thus, the nature of this process is similar to the processes which govern network dissolution. Transport processes can control the rate of hydration either by limiting the access of hydrolytic reactants or by retarding the release of mobile hydrolysis products which could participate in a back reaction. Several recent observations support that network dissolution and alkali ion exchange are governed by similar processes. The reactivity of the glass matrix was correlated to diffusion coefficients in the dry glass (17). In some experiments with highly reactive glasses a linear dependence of the release of alkali ions was observed without the etching of the glass surface which is usually associated with such constant rate processes (22). Transmission electron microscopy of leached alkali-silicate glass surfaces has shown that glass surface layers are highly porous and contain phase-separated water and silica spheres (23). High amounts of molecular water in the gel layer were determined by infrared transmission spectroscopy (24,25). The presence of free water molecules is probably a result of the condensation of silanol groups by the reaction (26) 2 -Si-OH -> -Si-O-Si- + H_2O. Silanol groups may have been produced prior to this reaction by ion exchange or by network hydrolysis. It has been shown by leaching alkali silicate glasses in $D_2{}^{18}O$ that the ratio of the surface uptake of oxygen to that of hydrogen is greater than one (27). This was interpreted as an indication for the predominant role of hydrolysis and condensation of glass network bonds for the release of alkali. Alkali leach rates were found to be a factor of 0.70 lower in D_2O than in H_2O (28). This isotope effect was explained by a proton transfer step in the hydrolysis of the silicate matrix preceding the release of alkali.

Factors Affecting Glass Matrix Dissolution

The reactivity of the glass matrix is correlated with the thermodynamic stability of a hypothetical mixture of glass component oxides and silicates with respect to hydrolysis and complexation in solution (hydration theory) (29). An improved model for calculating the Gibbs free energy of hydration has recently been incorporated into the geochemical code EQ3/6 (31,32). This model replaces the initial mechanical mixing model by an ideal solid solution model. For a wide range of glass compositions, e.g. window glasses, medieval glasses, basalt glasses and nuclear waste glasses, there appears to be a linear relationship of the calculated Gibbs free energy of hydration and the logarithm of the corrosion rate (30) (see Jantzen's chapter). As for other linear free energy relationships (LFER) (33) the physico-chemical basis for this process is not fully understood. The relationship shows that equilibrium properties of ideal solid phases can be used to estimate the dynamic properties of quite complex glass compositions in a general and predictable way.

In static leach tests it is observed that the corrosion rate of the glass matrix decreases with time because of saturation effects of silicic acid in solution, and this decrease can be described mathematically by a first order dissolution reaction (34,35). If a limiting silicic acid concentration in solution has been accumulated, the reaction rate may slow down by as much as a factor of 1000. The observations of saturation effects causes a problem for the theory of the glass/water reaction, because there is no solubility limit for glasses. A saturation effect must be related to the presence of a solubility-controlling solid phase. It has been shown that the overall affinity of the glass/water reaction remains large for long periods of time (32). Hence, the glass phase does not limit elemental solubility. The decrease of the reaction rate at saturation was explained by the condensation of surface silanol groups which may stabilize the glass surface (34) or by the precipitation of a stable non-permeable high silica surface film (35,37). Alternatively, it was suggested that the solubility-controlling phase is the gel layer, formed during alkali ion exchange in the initial stages of the glass/water reaction (31). Some authors (37-39) consider leaching to cease when silica is saturated in solution, but from a theoretical point of view this seems to be unlikely, because glasses are thermodynamically unstable with respect to an assemblage of alteration products (2,40). Tests performed at temperatures well above 100°C show that glass corrosion continues with a low rate despite silica saturation (2, 41-43). The understanding and quantification of corrosion processes under silica saturated conditions is a prerequisite for the long-term prediction of the performance of nuclear waste glasses under conditions typical for a repository.

Formation of Altered Surface Layers

The glass/water reaction leads in most cases to the formation of surface layers. Three major processes control layer formation, (1) hydration of the glass surface, interdiffusion (Na^+/H^+) and ion exchange, (2) reprecipitation of crystalline or amorphous reaction products growing from supersaturated solutions, and (3) pseudomorphic replacement of the glass phase by less-soluble reaction products remaining on the surface during reaction. Initially layer formation is governed by hydration/ion exchange. Whether processes (2) and/or (3) do occur depends on the glass composition. If they do, a multi-layered structure typically forms, with the layer formed by process (1) being closest to the pristine glass, and layers formed by process (2) at the interface to the solution (see Wicks' chapter). The various layers are affected in different ways by the ratio of the sample surface area to the solution volume (SA/V). The growth of the hydrated layer is very little affected by changes in the SA/V ratio. Layer thickness is typically in the range of a few tenths of a nanometer. Layers formed by hydration/ion exchange contain diffusion profiles for hydrogen, water and alkali elements and there is a smooth transition from the outer surface to the unaltered glass phase. It is either within the whole volume of this layer or at its outer surface, where the actual glass dissolution reaction occurs. Therefore, this layer is sometimes called 'reaction zone'. The growth of layers which are either precipitated from the solution or are formed by replacing the original glass phase is strongly affected by S/V. Layer thickness ranges from a few nm (at very high S/V) to mm size. It is often possible to mechanically remove these layers. If the layer-forming material is amorphous, it is often called a 'gel layer'. Crystalline phases may grow within the amorphous layer. At high temperatures (i.e. 200°C), a large fraction of the layer will become crystalline. As an example, Figure 5-1 shows various crystalline phases formed on the surface of a borosilicate nuclear waste glass during corrosion in NaCl-dominated brines. The cubic crystals are analcime. The formation of layers by processes (2) and (3) can both be described in terms of the thermodynamics of mineral solubility. The composition of any layer formed by solubility constraints depends both on glass and on solution composition. If the glass contains sparingly soluble constituents such as Fe or Zr, these elements will be concentrated at the surface. There is also a clear pH dependence of layer composition which reflects the pH dependence of solubility. For example, under neutral or acid conditions, Ca is released into solution, whereas under alkaline conditions Ca remains on the glass surface.

Progress in the Geochemistry of Rock/Water Interactions Relevant to Glass Corrosion

From the recent progress in the field of water/rock interactions, we can improve our understanding of the surface chemistry of glasses, because much emphasis has been on the prominent role of mineral surfaces in the weathering processes of rocks. The formation of sparingly soluble alteration products such as clays or zeolites during mineral dissolution is interpreted in terms of local

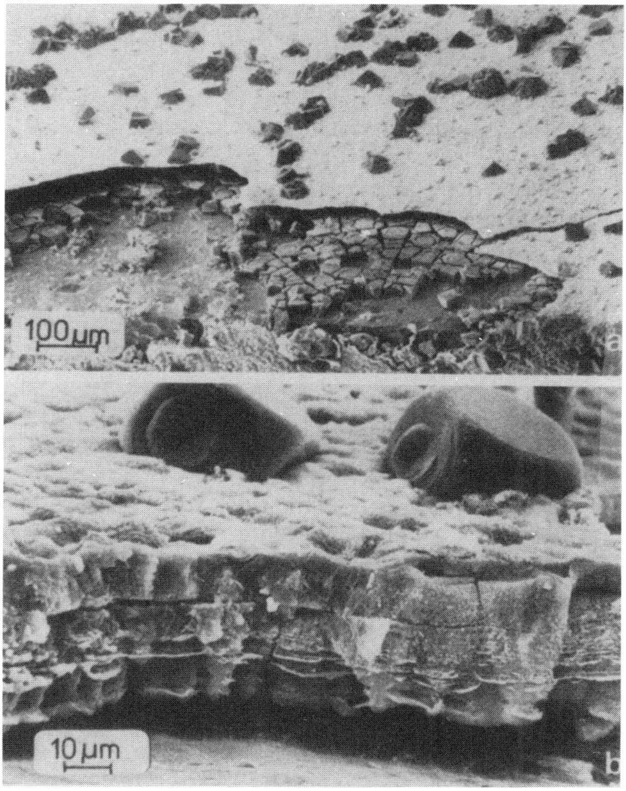

Figure 5-1. SEM-micrograph of corroded surface of borosilicate glass C31-3EC after 10 days in NaCl solution at 200°C; (a) analcime layer with an underlying amorphous layer, the pristine glass is at the bottom; (b) cross section of the surface layer (Malow, Lutze, Ewing, 1984 (79)).

equilibrium concepts. Thermodynamically stable or metastable phases form if the solution becomes oversaturated. During the dissolution process, initially formed phases may become instable with respect to other phases if, for example, the pH changes. Consequently the alteration product will start to dissolve. Reaction path models were developed which allow calculation of a paragenetic sequence of alteration products as a function of the amount of glass dissolved. As for glasses, the rate of dissolution of minerals often decreases with time. Initially, this effect was explained by the formation of protective surface layers (44,45). However, for alkali feldspars accurate measurements (46-48) have shown that surface layers are often not formed or are not protective. Only the dissolution of rather soluble minerals may be controlled by transport processes through surface layers. Alternatively, the decrease of reaction rate was attributed to initial surface defects or to solution chemistry (46).

Models were developed (49-54) for the corrosion of minerals, which considered activated surface configurations and solution chemistry. The overall hydrolysis reaction was described by a network of reversible and irreversible reactions. The reaction progress variable ξ was used as a measure of the overall reaction (55). Equilibrium thermodynamics was applied to reversible reactions (reactions among solution species, precipitation of insoluble phases etc.). The rate-limiting reaction step is a surface reaction. Initially, the extent of corrosion increases linearly with time, but close to saturation, the rate is proportional to the chemical affinity (50), A, or to a certain power of the affinity (53) of the rate-limiting reaction. The chemical affinity is a measure of the deviation of the system from saturation.

An important step in the understanding of the mineral/water reaction is the application of transition state theory to the mineral surface. According to this theory the rate limiting reaction is the irreversible decomposition of an activated complex (56). The reaction rate is given by the concentration, C^*, of this complex and a frequency factor, f. The concentration C^* is given by the mass action equation for the formation of the activated complex from the reactants. For corrosion of minerals the activated complex is a surface configuration of atoms in a partially destroyed silicate network. This complex cannot be observed directly and, therefore, the stoichiometry of this surface complex must be deduced from the reaction behavior of the solid. Based on transition state theory and irreversible thermodynamics Aagaard and Helgeson (50) derived a general rate equation for the reaction of minerals with aqueous solutions:

$$r_j = k_j * a_i^{-\nu_{i,j}} * (1\text{-exp}(-A_j/\sigma_j RT)) \qquad (5\text{-}3)$$

In this equation, r_j is the rate of the reaction j (various reactions in series are considered, where j denotes the rate limiting reaction), and k_j designates the steady state rate constant. The variable a_i denotes the activities of the reactants i and $\nu_{i,j}$ their stoichiometric factors in the rate limiting elementary reaction. The affinity of this reaction is A_j and is given by the ratio of the ion activity product IAP_j in solution to the equilibrium constant for this reaction (at saturation $A_j=0$, $r_j=0$). The stoichiometric reaction coefficient σ_j is the ratio of the rates of various reactions to the overall reaction. The coefficient is unity when only one irreversible reaction is considered. The gas constant is denoted R and T is the absolute temperature. The rate constant k_j is given by (50):

$$k_j = \alpha_j f K^* /(\sigma_j^* \gamma^*) \qquad (5\text{-}4)$$

where α_j is the transmission factor (usually close to unity), f is the frequency factor (f = kT/h; k = Boltzman constant, h = Plank constant) for the decomposition of the activated complex, K^* is the stability constant of the activated complex, and γ^* is its activity coefficient.

THE RATE LIMITING STEP IN THE GLASS/WATER REACTION

On hydration and/or dissolution various bonds (Si-O-Si, Si-O-Na, Si-O-B etc.) are hydrolysed. These bonds differ in their energy of hydration. The low stability of the Si-O-Na bond may be a major cause for the initial selective alkali release (ion exchange, possibly interdiffusion or water diffusion controlled) from silicate glasses. Nevertheless, for most glass compositions the rate of this process decreases to a steady state value which is given by the rate of matrix dissolution. At steady state all glass constituents are released with the same rate (congruent dissolution), as long as their solubility limits in solution have not been exceeded and sorption and coprecipitation processes can be neglected. Under these conditions the glass constituents are not released according to the hierarchy of bond energies, but a common activated surface complex exists which controls the release of soluble glass constituents by its formation and detachment.

The nature of the activated surface complex can be deduced from the effect of silica concentration on the matrix corrosion rate. The term "saturation" is ambiguous when used for glass, because there is no overall equilibrium between the glass phase and the solution. "Saturation" for glasses may become comprehensible when the term is referred only to a local surface equilibrium ("two dimensional saturation"): For every desorbing surface complex another complex is adsorbed from solution (forward rate = back reaction rate).

In contrast to saturation of other glass constituents, such as Ca, Fe, Nd, etc. in solution, the saturation of silica has a major effect on the rate. It can be concluded that the desorption of silica tetrahedra (as silicic acid) from the glass surface limits the rate of release of soluble glass constituents. At saturation, condensation of silanol groups stabilizes the glass network against aqueous attack. Hence, silica is the dominant constituent of the activated surface complex. The release of silicic acid from the glass matrix depends on the preceding hydrolysis of the four Si-O- bonds to the glass matrix. In the most simple case the precursor of the activated surface complex may be considered as a (-$Si(OH)_3$) group attached to the glass surface. The rate of corrosion could be calculated from first principles if the concentration of this complex would be known. Obviously, the concentration of this complex will depend critically on the population of hydrolysed bonds within the reacting zone at the glass surface.

In Figure 5-2 the meaning of the term "reaction zone" is illustrated. The reactions within the reaction zone of some nanometers depth will depend on glass composition, on the degree of silica saturation in solution adjacent to the glass, on the precipitation of silica-bearing phases, and on the surface pH. If the detachment of (-$Si(OH)_3$) is the rate controlling reaction for matrix dissolution, the other reaction will occur under local equilibrium conditions. This implies that there will be a dynamic equilibrium among hydrolysed and nonhydrolysed bonds in the reaction zone. Also, since glass network dissolution only becomes

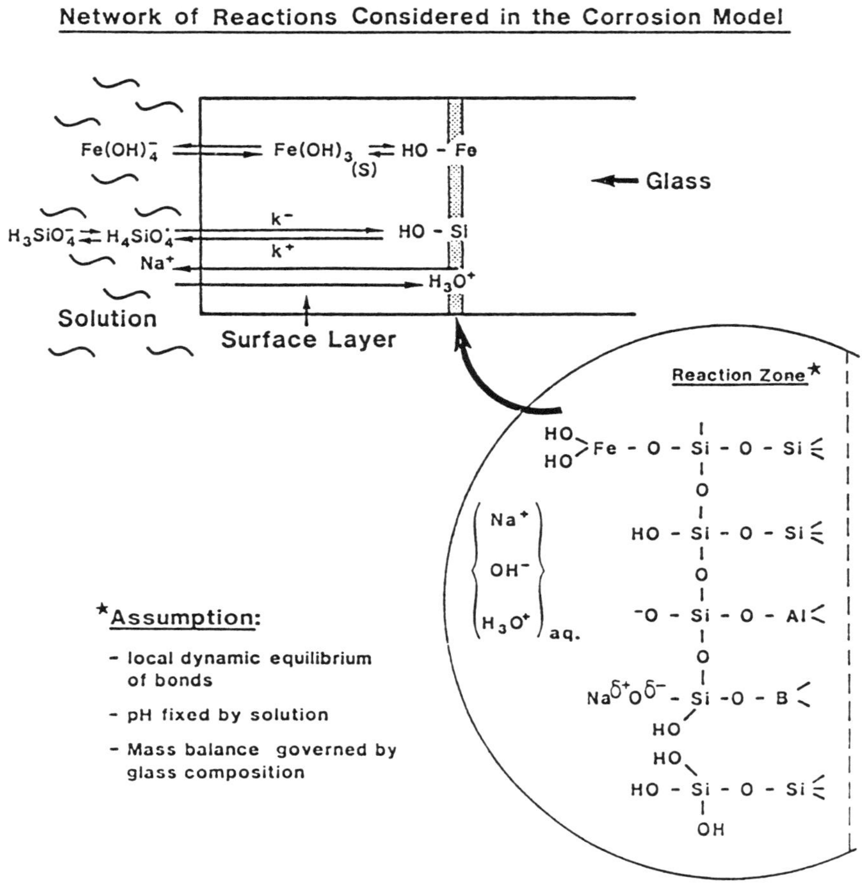

Figure 5-2. Illustration of the reactions which occur during glass corrosion.

rate controlling after the initial water-diffusion/ion exchange is at steady state, an ion-exchange equilibrium will be established between the reaction zone and the cations in solution.

The occurrence of dynamic surface equilibria appears to be the basis for the above-mentioned linear relationship between the glass corrosion rate and the Gibbs free energy of hydration of the glass phases. The Gibbs free energy of hydration of the glass phase is typically set equal to the Gibbs free energy of a solid solution of simple crystalline phases. This energy does not provide the driving force (chemical affinity) for the reaction, but it is related to the concentration of the activated surface complex. The more negative the hydration free energy, the more the surface equilibria are shifted to the side of the products (formation of hydrolysed bonds), and the higher is the concentration of activated surface complexes and the corrosion rate.

A more profound quantitative treatment of the surface equilibria would require the determination of the stability constants of the various surface reactions instead of the use of the Gibbs free energy of bulk phase hydration. If available, these constants could be taken directly from site binding models. Such models are developed to calculate surface processes (sorption or the formation of an electrical double layer) from surface ionization and complexation constants (57). The practical use of these models is limited, because the required constants are known only for a few pure oxides and hydroxides (SiO_2, TiO_2, $Fe(OH)_3$ etc.).

Implications for the Mechanism of Initial Selective Alkali Depletion

As described above, glass matrix hydrolysis is proposed (27) as an important precursor to alkali release initially as well as in later stages of reaction. The hydration equilibria do not apply only to the outer surface plane of the glass and, hence, to glass dissolution, but to the entire subsurface layer, which is penetrated by water molecules. The activity of silanol groups and that of other hydrolysed bonds will decrease with the penetration depth of water, because the activity of water decreases. The implications of this effect on water mobility are important. The inverse dependence of network crosslink density on the activity of silanol groups implies also an effect of this activity on the diffusion coefficient of water molecules in the transformed subsurface layer. It can be concluded that an effective diffusion coefficient for water transport in the transformed zone at the glass surface depends on the Gibbs free energy of the glass/water reaction and, hence, on glass composition and on the depth of water penetration.

As proposed in the model of Smets et al. (13), the release of glass constituents depends on the penetration of molecular water. Once water has diffused into the subsurface layer, glass constituents, such as the alkali ions, can be mobilized to an extent which is governed by the depth of water penetration. Other constituents, i.e. Si or Al, remain behind. The more negative the individual contribution to the overall Gibbs free energy of reaction $\Delta G_{k,R}$ of a given solid solution endmember representing glass constituent k is, the higher the mobility of k. Consequently, species such as B or Na are leached to a significant extent from the subsurface layer. At low pH, glass constituents which are insoluble at higher pH, such as the rare earth elements, can be released. For example, at low pH, elements such as Nd can be released like the alkali elements. Experimental evidence ('diffusion' profiles) for the selective release of Nd has been described in the literature (58). Since the behavior of B during leaching is more similar to the behavior of Na than to that of Si, one can conclude that the traditional separation of glass constituents into network formers, such as B or Si, and network modifiers, such as Na, provides no valuable categories for the leachability of glass constituents.

KINETIC EQUATIONS FOR GLASS MATRIX DISSOLUTION

General Rate Equation for Glass Dissolution

The transition state rate law for mineral dissolution (Eq. 5-3) can be applied to glass dissolution by assuming that an activated complex forms by the reaction of water molecules with surface silanol groups. On this basis a general rate equation for glass dissolution was derived (Eq. 5-5) which describes for a given reaction progress, ξ, the effect of glass composition and pH by the free energy of hydration ΔG_R, the effect of temperature T described by the activation energy E_A and the saturation effect described by the affinity A^*:

$$r_m = X \exp(-\Delta G_R(\xi)/RT) \exp(-E_A/RT) \ (1-\exp(-A^*(\xi)/RT)) \qquad (5\text{-}5)$$

r_m is the rate of glass matrix dissolution normalized to surface area. The proportionality constant X is given by the equation:

$$X = \alpha \ f/ \ \gamma^* a_{H2O} \ B \exp(1+\Delta S^*/R) \qquad (5\text{-}6)$$

The definitions of the symbols are the same as in Eq. 5-3 and 5-4. If the activation energy E_A is known, X can be derived from a plot of experimental rate data and calculated ΔG_R values. The relationship between the various terms in Eq. 5-5 is illustrated in Figure 5-3 with the help of an energy profile along the reaction coordinate. Eq. 5-5 may be expressed in an abbreviated form as:

$$r_m = k_+ \ (1-\exp(-A^*/RT)) \qquad (5\text{-}7)$$

where k_+ is the rate constant for the forward rate of reaction which depends on glass and solution composition and on temperature. In Eqs. 5-5 and 5-7, A^* denotes the affinity of the rate limiting reaction and describes the effect of saturation on the rate. It is given by the equation:

$$A^* = -2.3RT \log \ (IAP^*/K^*) \qquad (5\text{-}8)$$

where IAP^* is the ion activity product and K^* the stability constant for the rate limiting reaction and the term IAP/K is called the saturation index, because it describes the deviation of the system from saturation. It is important to distinguish the affinity of the rate limiting reaction in Eq. 5-7 from the affinity of the overall glass/water reaction. The affinity for the overall reaction may remain high when the affinity of the rate limiting reaction approaches zero. In the most simple case, the rate limiting reaction is:

$$SiO_2 + 2H_2O \rightarrow H_4SiO_4 \tag{5-9}$$

For this reaction the saturation index is $IAP^*/K^* = a_{H4SiO4}/K^*$ for ($a_{H2O} = 1$) and the rate is given by:

$$r_m = k_+ (1 - (a_{H4SiO4}/K^*)) \tag{5-10}$$

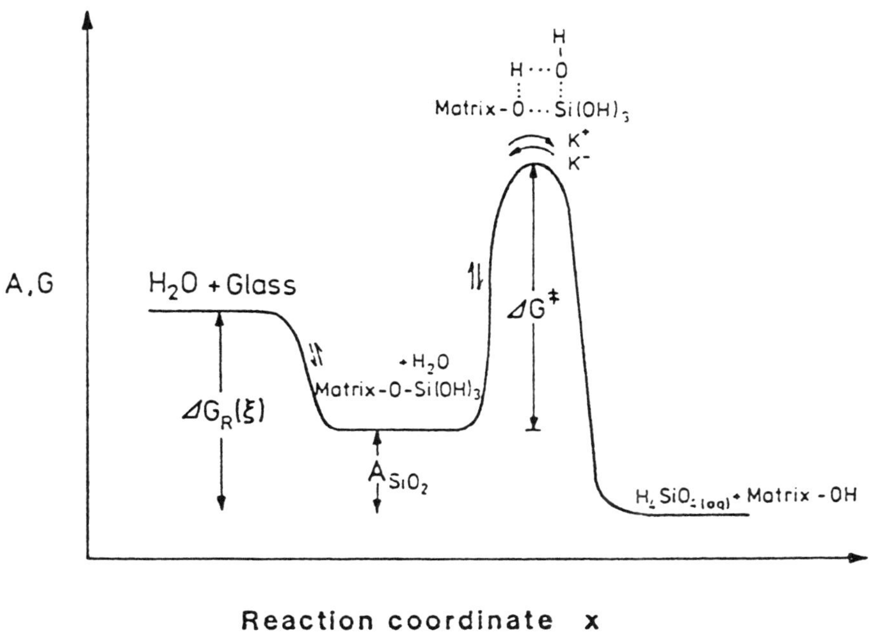

Figure 5-3. Energy profile along the reaction coordinate for detachment of H_4SiO_4 from the glass surface.

Although this equation has been derived for multicomponent glasses, it is similar to the equation of Rimstidt and Barnes (59) for the dissolution of silica polymorphs. It describes the effect of the silica solution concentration on the glass corrosion rate as a first order reaction and implies the existence of a 'saturation concentration' of silica. If not otherwise stated, in the present context the term 'silica concentration' is used as an abbreviation synonymous with the correct term 'activity of ortho silicic acid.' Depending on silica transport processes (see below), Eq. 5-10 applies either to bulk or to interfacial Si-concentrations. Initially (far from saturation), the reaction rate equals the forward rate constant, whereas the reaction should stop at a hypothetical state of true saturation ($A^*=0$). For glass dissolution the affinity of the overall reaction cannot become zero (no true equilibrium), because saturation only involves the reacting surface and soluble elements might still be released from the bulk glass phase. A final affinity of reaction and a respective final rate will remain under conditions saturated with silica. The mechanism of this process and the rate limiting step is not yet clear. If the affinity of the rate limiting step (desorption of silica groups) becomes zero at saturation, it is likely that the rate limiting step will change.

Equation 5-10 appears to be rather simple but the time dependence of the release of soluble glass constituents may become quite complicated, if additional reactions must be considered which consume or produce silicic acid (hydrolysis of silicic acid, precipitation of silicates, etc.). Only when none of these reactions occur can the time dependence of the solution concentration of silica be calculated from the integrated form of Eq. 5-10:

$$a_{H4SiO4} = a_{sat} \ (1 - \exp((-t \ k_+ \ S/V)/K^*)) \qquad (5\text{-}11)$$

where S/V is the ratio of the sample surface area to the solution volume. In all other cases, Eq. 5-10 must be integrated numerically, simultaneously solving the mass balance and mass action equations with all other reactions.

Using the kinetic interpretation of solubility phenomena (equal forward and back reaction) one may infer from the rate limiting detachment of silicic acid that the solid phase, which controls the silicic acid concentrations in solution under saturated conditions, is a silica polymorph. In contrast, Bourcier (31) has suggested that the affinity term of Eq. 5-7 is related to the composition of the glass surface layer ('gel layer'). Hence, glass corrosion rates would slow down when the solution is saturated with respect to the surface layer. There are complications related to the composition of the layer. The stoichiometry of the activated complex would become quite complicated if saturation effects with respect to a multicomponent layer would be rate limiting. Moreover, the composition of the gel layer changes considerably with time, water flow rate and surface area to solution volume ratio. For example, in a soxhlet test with iron-containing glasses the 'gel' layers will be enriched with large quantities of Fe whereas almost no silica remains in the layer (60). The solubility of the 'gel' would, therefore, be close to the solubility of iron oxyhydrates. The affinity term would become zero and the glass corrosion rate should decrease, when iron is saturated. Saturation of Fe is often the case, even under the fast flow conditions of a soxhlet test, but reaction rates remain constant (60). On the other hand iron oxyhydrates form a large distinct reaction product layer (not a 'gel') on the glass surface and the actual corrosion reaction will occur beneath this layer. It is likely that at the interface between the product layer and the pristine glass a silica rich 'gel' layer exists which is formed by alkali ion exchange/water diffusion. However, even in this case the composition of the gel layer will be a complicated function of environmental constraints. For example, under low pH conditions rare earth elements are typically released with the same mechanism (similar diffusion profiles were observed) as the alkali ions (58). Hence, at low pH the surface layer is depleted in rare earth elements, whereas at high pH the gel does contain these elements.

Influence of Silica Transport on the Corrosion Rate

In the general case, transport processes may contribute to or control the glass/water reaction rate. This will occur when the transport of silica from the glass/water interface to the bulk solution is slower than the rate of silica release from the glass. Slow transport would result in higher activities of silicic acid at the glass surface than in the bulk solution.

Transport of silica may either be limited by the growing surface layer, or, in case this layer is absent or very thin, by the Nernst film (hydrodynamic boundary layer of stagnant water) adjacent to the surface layer. There has been a discussion as to whether or not the surface layers are protective, i.e. decrease the rate of corrosion (61-66). In some cases, it could be shown experimentally that the surface layer had no effect on the leaching process (i.e. by removing the surface layer and showing that the rate was the same before and thereafter), whereas in other cases the reaction rate was decreased in the presence of a surface layer (63). In Figure 5-4 the various cases of potential silica transport gradients at the glass surface are described. In the cases a and b there is no gradient and the reaction rate will slow down only if silica saturation is approached in bulk solution (case b). In contrast, due to the silica gradients for cases c and d, saturation may be approached in the surface layer (case d), although the silica concentration in bulk solution is low.

The effect of a concentration gradient on the leach rate r_m can be calculated. The material dissolved from the glass must be transported into the bulk solution. Hence, the rate of transport $r_{t,Si}$ of silica can determine the rate r_m. The rate $r_{t,Si}$ per unit of surface area is given by a modification of Fick's first diffusion law

$$r_{t,Si} = D/L\ (a_s - a_b) \tag{5-12}$$

where D is the diffusion coefficient for silica, a_s and a_b are the activities of ortho silicic acid at the glass surface and in bulk solution, and L is the thickness of the transport barrier (surface layer or Nernst film). The reaction does not stop under saturated conditions, and release of soluble species can continue even without a concentration gradient of silica across the surface layer. Under this condition the release of soluble species is identical to the mass transfer of silica from the glass phase to a solid alteration product, and the rate of mass transfer r_{tr} equals the final rate: $r_{tr} = r_{fin}$.

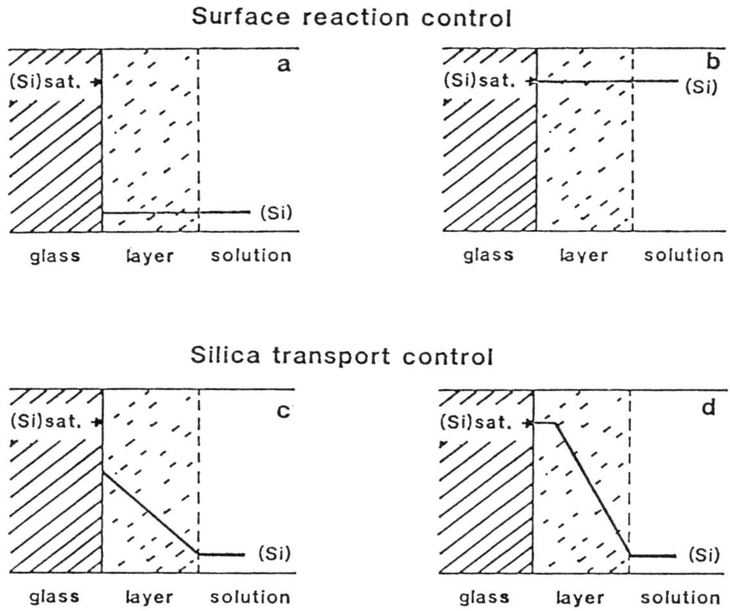

Figure 5-4. Various cases of silica transport gradients within a growing surface layer. (a), (b) no gradient, silica concentration in solution equals silica concentration in the aqueous phase within the surface layer, (c), (d) high concentrations of silica in the surface layer despite low concentrations in solution. (b) and (d) denote cases for saturation at the interface of the pristine glass.

Since $r_{t,Si} + r_{fin} = r_m$, Eq. 5-12 can be combined with Eq. 5-10 ($a_{H4SiO4} = a_s$) to give:

$$r_m = k_+ \, (1\text{-}\{a_s/K^*\}) = D/L \, (a_s - a_b) + r_{fin} \tag{5-13}$$

This equation is similar to the rate equation used in the model by Wallace and Wicks (37) (see Wicks' chapter), but it differs in the final rate term and in using activities of ortho silicic acid instead of using concentration values of total silica in solution.

The second relation of Eq. 5-13 can be used to calculate the activity of silicic acid at the glass surface from the activities in the bulk solution:

$$a_s = (a_b \, D/L + k_+ - r_{fin})/(D/L + k_+/K^*) \tag{5-14}$$

Equation 5-14 may now be introduced into the first relation of Eq. 5-13 to derive the rate expression:

$$r_m = k_+ \, (D/L \, (K^* - a_b) + r_{fin})/((D/L) \, K^* + k_+) \tag{5-15}$$

By comparing $(D/L)K^*$ with k_+ this equation provides a criterion for whether or not transport processes control the reaction rate. Since K^* equals the activity for silicic acid at saturation a_{sat} as long as $a_{H2O} = 1$, and since $(a_{sat} - 0)/L$ is the maximum possible concentration gradient at the surface, the term $(D/L)K^*$ may be understood as the maximum possible rate of transport for a given layer thickness L.

If this maximum rate of transport is much faster than k_+, the measured reaction rate is not influenced by transport and Eq. 5-15 approaches Eq. 5-10 (because $((D/L)K^* + k_+)$ equals about $(D/L)K^*$ and $k_+/((D/L)K^*)$ about 0). In contrast, if $(D/L)K^*$ is smaller than k_+, the rate of reaction may be controlled by transport of silica. It can be shown by analytical integration of Eq. 5-15 that under certain conditions the leach rate decreases with the square root of time (67).

Comparison of Silica Transport and Interdiffusion Mechanism

Silica transport in the surface layer and Na^+/H^+ interdiffusion in the transition zone between the surface layer and the pristine glass may yield the same square root of time dependence for the release of soluble elements from the glass. A plot of solution concentrations versus time alone does not unambiguously reveal the reaction mechanism. If the diffusion coefficient is determined from solution analyses, then its value may vary by orders of magnitude depending on whether water or alkali ions or silicic acid were assumed as the moving species. The reason is that with silica transport control the maximum transport gradient is governed by the solubility limit of silica and is three to four orders of magnitude smaller than for transport of H_20 or Na. Only by a combination of surface layer analyses and solution analyses can interdiffusion and silica transport processes be distinguished. If the alkali release is controlled by interdiffusion, diffusion profiles for alkali ions are generated in the dry glass without a sharp interface between the glass phase and a surface region. If on the other hand the alkali release is limited by the silica transport, measurable concentration profiles for alkali ions are not expected because the diffusivities in the altered surface layer are too high. Despite silica transport control, no silica gradient may be detected, because the measured concentration profiles of silica at the glass surface are governed by precipitated or remaining silica and not by the concentration of soluble silica in the aqueous phase in this layer. Silica transport controls the rate only if matrix dissolution is the fastest process, whereas interdiffusion will be the dominant reaction if matrix dissolution is slower.

When silica transport control is effective there may be concentration gradients of species other than silicic acid, including H^+. This may lead to different silica solubilities in the bulk solution and at the glass/water interface as a result of pH-variations. Precipitation of reaction products in the surface layer may be observed, though the bulk solution is not yet saturated. This effect is accounted for by attributing different reaction progress values and thus different concentrations to the surface and the bulk solution.

GEOCHEMICAL MODELING OF GLASS DISSOLUTION

A computer model was developed for the theoretical simulation of the glass/water reaction. The modeling is performed in two steps. First, the reaction path is calculated with the help of a geochemical code such as PHREEQE (6), EQ6 (68), or DISSOL (69). The term "reaction path" describes the mass transfer of dissolved glass constituents to solution and solid alteration products as a function of reaction progress, i.e. the amount of glass dissolved per unit volume of solution. The time dependence of glass corrosion is not considered.

The time dependence of the reaction is calculated after the reaction path has been determined with fair accuracy. One single reaction path model is valid for various calculations of reaction kinetics, as long as only the solution volume, the sample surface area or the flow rate of solution is changed. New reaction path calculations are required when changes in temperature, glass or initial solution composition are made, or when additional system components such as bentonite or steel corrosion products are added to the reaction system.

Reaction Path Modeling of Surface Layer and Solution Composition

The mass transfer from the glass into solution and/or solid reaction products (surface layer or elsewhere in the system) can be described by modeling with the help of geochemical codes. The codes EQ6, PHREEQE, SOLMNQ,and DISSOL (67,70-72) have already been used to describe glass corrosion and the formation of alteration products. The EQ6 code has several advantages when compared with PHREEQE. For example, one can consider the formation of solid solutions and the effect of high ionic strength brines. The reaction path is constrained by the initial glass and solution composition, the solubility of alteration products, the stability of solution species, and environmental parameters such as temperature, the composition of the gas phase (CO_2) and solution volume. Calculated results depend significantly on the thermodynamic data base for the solid phases, for solution species and for activity coefficients. In particular the thermodynamic data are uncertain for glass corrosion at temperatures well above room temperature and/or in highly saline solutions (ionic strength > 3M), or for trace elements, such as the constituents of radioactive waste. Different results are obtained if only the thermodynamically most stable phases are allowed to form or if the formation of metastable phases is considered. The formation of metastable phases is often observed in natural water systems and is a general result of low precipitation rates of the thermodynamically more stable phases. If only thermodynamically stable phases are considered, the calculated solution concentrations of glass constituents are often many orders of magnitude lower than the measured concentrations (32).

At present, a realistic reaction path cannot be determined from first principles but requires empirical knowledge on metastable phase formation. Accurate modeling requires a combination of forward and inverse methods (73). In the inverse mode, analytically determined solution concentrations of leachates resulting from glass corrosion tests are used as code input and saturation states are calculated for all solid phases contained in the data base. Typical results show both sub- and supersaturated phases. Supersaturation can be considered as strong experimental evidence for the slow formation of a given phase, if colloids are absent in the solution and if the thermodynamic data for the phase are correct. In the forward mode, the reaction path is calculated, i.e. a reaction model is applied to certain initial conditions (temperature, pH, pCO_2, Eh, glass and solution composition). The model includes a set of assumptions concerning the formation of metastable phases and the congruency of the dissolution reaction. Modeling can only attempt to match experimental data if the calculations either exclude the supersaturated phases which were determined by the inverse method or provisions are taken for considering precipitation rate laws. Typically, leach solutions are supersaturated with respect to stable phases such as goethite (FeOOH) or the clay mineral chlorite. Such phases must be excluded from the reaction path model because, otherwise, the computer code would calculate unrealistically low solution concentrations for elements such as Fe and Mg. Using the remaining phases of the data base, the code chooses for solution control those phases exceeding solubility limits at certain steps of a simulated dissolution reaction. The identification of the true reaction path for the dissolution of a given glass under a given set of experimental conditions is quite complicated and requires frequent comparison of the results from the inverse method (with reference to solution data of a variety of different experiments) and forward method. Experimental identification of alteration products is often not possible because, in many cases, these products are amorphous or only partly crystalline and constitute solid solutions of poorly known chemical composition. The reaction path model must be compatible with the subsequent calculations of reaction kinetics. According to Eq. 5-10, an activity of H_4SiO_4 which is higher than the saturation value would result in the cessation of the reaction (negative rates are not possible for the glass/water system). One can prevent the activity of silicic acid from exceeding saturation by using a SiO_2-modification as a precipitating phase with a solubility measured experimentally. The stability constant K_{Si} for this phase can be calculated by identifying a_{H4SiO4} for glass leaching under saturated conditions with K_{Si} and rearranging Eq. 5-10 and using the experimentally measured final rate r_{fin}:

$$K_{Si} = K^* \ (1 - r_{fin}/k_+) \tag{5-16}$$

As an example, Figure 5-5 shows the results of reaction path calculations for the nuclear waste glass R7T7 in deionized water at 90°C. The calculated solution concentrations are normalized to glass composition and are plotted versus the amount of glass reacted per unit volume of solution. Phases predicted to precipitate along the reaction path are $ZrSiO_4$, $Fe(OH)_3$(am), $Nd(OH)_3$, gibbsite ($Al(OH)_3$), analcime ($NaAlSi_2O_6$), montmorillonite, and SiO_2.

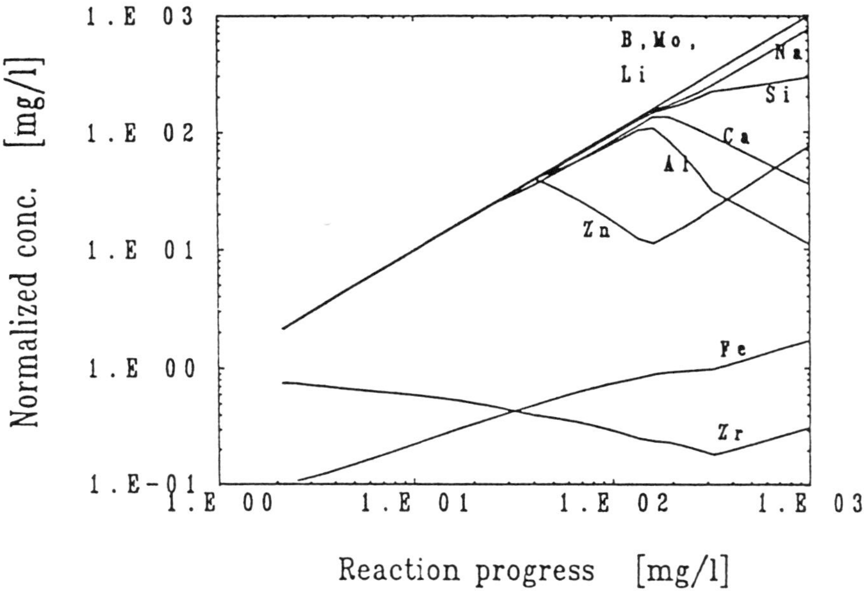

Figure 5-5. Results of reaction path calculations for the nuclear waste glass R7T7 in deionized water. Calculated solution concentrations are normalized to the glass composition.

The reaction path calculations are more complicated, when silica transport processes in the layer must be considered. The comparatively high concentrations of ortho-silicic acid in the surface layer refer to higher values in the reaction progress per unit volume. Two reaction progress scales are necessary: one for the bulk solution and another one for the aqueous phase in the surface layer. The link between both scales is established using Eq. 5-14.

Reaction Kinetics—Modeling with GLASSOL

In the case of EQ6, the code provides the possibility to describe the time dependence of the reaction using a transition state rate law. However, the code does not consider reaction rate control by the transport (diffusion) of reactant or product species. Hence, the code cannot describe the effect of growing surface layers as a transport barrier on the reaction rates. In many cases it is sufficiently accurate to neglect this effect but, in particular in tests with flowing leachants, surface layers may be an important transport barrier. In the case of the PHREEQE code, the calculation of reaction rates is impossible. Therefore, the code was modified to provide an output which can be used for the calculation of reaction kinetics. A code, GLASSOL, was developed to interpret this PHREEQE output in terms of kinetics. The effect of the growth of a surface layer as a barrier to silica transport is considered. In the present communication, the calculations of the time dependence of the glass corrosion reaction are only calculated with the PHREEQE/GLASSOL combination. With GLASSOL, Eq. 5-10 is solved numerically for time. For each step along the PHREEQE reaction path, a time value is determined. The reaction rate and the time increment between two steps of the reaction path model are calculated using the activities of ortho-silicic acid from the output of PHREEQE and a set of kinetic parameters determined experimentally. Because the activities of silicic acid calculated by PHREEQE are constrained by the sum of all reactions consuming silica (i.e. precipitation of silicates, sorption or hydrolysis of silicic acid), the effect of such reactions on the dissolution kinetics of the glass matrix is considered implicitly in GLASSOL. When using GLASSOL for the prediction or for comparison with data, it is possible to test the potential effect of various reaction scenarios on the rate of glass corrosion.

In cases when diffusion in a growing surface layer is described, the surface layer thickness, L, is calculated from the amount of glass reacted at each reaction step. An option is available that allows determination of a factor for the potential shrinking of the layer or for the layer porosity. This factor is calculated from a weighted difference between the molecular volume of idealized reaction products and the glass solid solution end members (see above).

Table 5-1. Parameters Used in GLASSOL to Model the Reaction of the Glass R7T7 Under all Experimental Conditions at 90°C

k_+	1.5	$g \cdot m^{-2} d^{-1}$
$\log K^*$	-2.9341	
A^*	6.0	J/mol
r_{fin}	0.0025	$g \cdot m^{-2} d^{-1}$
D	10^{-13}	m^2/sec

Since the model has been extensively applied (74) to the above-mentioned radioactive glass R7T7, it is worthwhile to use some leach data of this glass to illustrate the procedure of modeling. The required parameters are summarized in Table 5-1. The forward rate constant was identified with the rate in short-term tests performed in deionized water at low ratios of the sample surface area to the solution volume (SA/V = 10 m^{-1}). The saturation constant for silica K^* is calculated from the total silica concentration in leachates from experiments where saturation has been reached rapidly. Such tests were performed at high SA/V values (SA/V = 1000 m^{-1}). The results of the latter experiments were also used to determine the final rate of reaction under silica saturated conditions. After the determination of these parameters, the diffusion coefficient could be determined from the best fit of the model to experimental data, in particular the data from flow experiments. Leach tests were performed with a variety of initial solution compositions and flow rates. For the present context results for deionized water yield sufficient insight into the modeling of nuclear waste glass corrosion. For leaching R7T7 glass in static tests the MCC-1 test procedure was used (75). Solution data and calculated curves are contained in Figure 5-6. The calculated time dependence of glass corrosion is based on the reaction path model described in Figure 5-5. The calculated B-curve deviates by less than 20% from the average value for the Si, B, Na, Li, Mo and Cs data. For the release of Al the deviation between theory and experiment is larger (as much as a factor of 2).

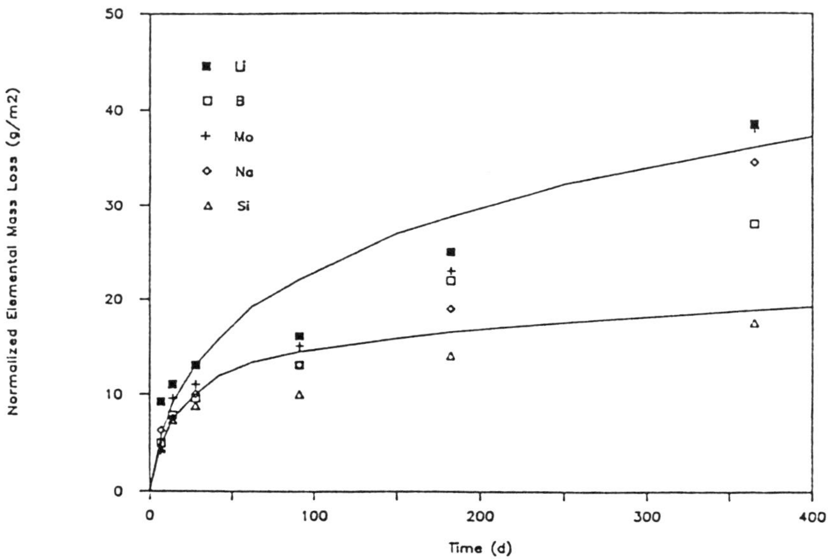

Figure 5-6. Comparison of experimental and calculated normalized solution data representing the release of various elements from the nuclear waste glass R7T7 during corrosion in static tests (90°C, SA/V = 10m^{-1}, deionized water).

Due to the dominant role of SiO_2-saturation, the groundwater flow rate and the accessible solution volume play an important role for the corrosion of nuclear waste glass. Depending on the flow rate and saturation concentration the release of glass constituents is independent of or proportional to the flow rate (38).

The effect of flow rate in dynamic leach tests has been considered in GLASSOL by the equation:

$$dC_G/dt = r_m\ S/V - C_G\ F/V \tag{5-17}$$

where F is the volumetric flow rate, C_G is the amount of glass reacted (expressed in units of concentration in solution), and r_m is the corrosion rate as calculated by equation 5-15. The consideration of the effect of diffusion for leaching in flowing solutions requires special attention. As with the static case, GLASSOL calculates a corresponding time value for every reaction step of the PHREEQE reaction path model. However, after some time, steady state is reached, which means that the reaction proceeds without further increase in solution concentration. Moreover, as the leach rate decreases due to diffusion in the growing surface layer, the steady state value decreases. The code GLASSOL uses only solution concentrations smaller than or equal to the steady state concentrations, and it follows the decrease of the steady state value with increasing layer thickness.

The flow test experiments were performed with the non-radioactive glass R7T7 using the MCC-4 test procedure (75). The effect of flow rate was calculated, using the same reaction path model that was used for static leaching of R7T7 glass. Calculated and experimental B solution concentrations (flow test 2.9 ml/d) are shown in Figure 5-7. The data show the typical maximum in solution concentration (38). It occurs after a reaction time of about 14 days which is similar to the residence time of solution in the leaching vessel. The curve deviate by less than 20% from the measured data.

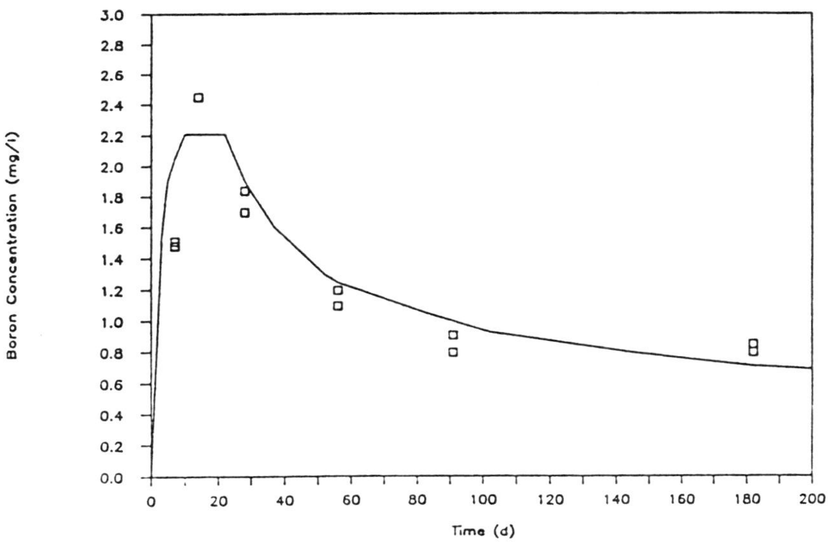

Figure 5-7. Comparison of experimental and calculated solution concentrations of boron, resulting from corrosion of R7T7 glass in tests with flowing water (90°C, SA/V = $10m^{-1}$, deionized water at 2.9ml/d).

To allow a more detailed insight into the effect of flow rate on the leaching mechanism, some additional curves are shown in Figure 5-8. These curves show the concentration of total silica in the bulk solution, in the aqueous phase within the surface layer, and the respective saturation concentration in the surface layer. The Si-saturation concentration in Figure 5-7 is not constant, because the pH varies with time in the surface layer as in the bulk solution. If transport of silica is hindered in the surface layer, a gradient is built up between the bulk solution and the surface layer (Figure 5-8). The bulk solution is far below Si-saturation but the surface layer becomes saturated with silica after about 200 days. The rate of reaction decreases with time. Consequently the steady state bulk solution concentrations of silica and of soluble elements will decrease.

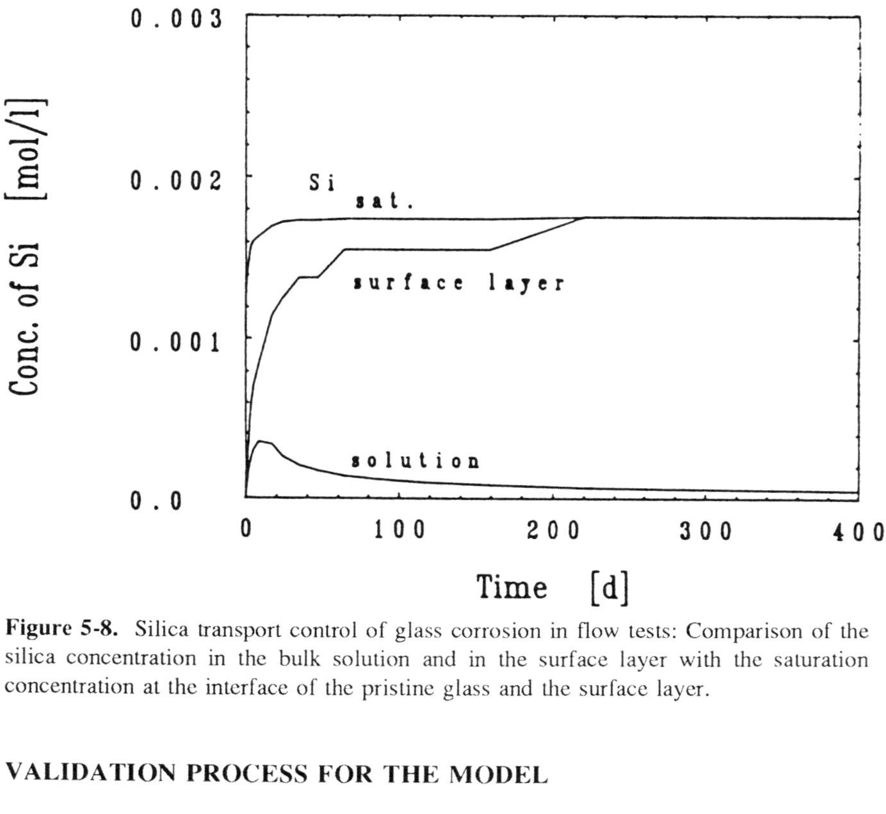

Figure 5-8. Silica transport control of glass corrosion in flow tests: Comparison of the silica concentration in the bulk solution and in the surface layer with the saturation concentration at the interface of the pristine glass and the surface layer.

VALIDATION PROCESS FOR THE MODEL

In particular, for the corrosion of nuclear waste borosilicate glass, models will be used to predict long-term materials performance. There must be sufficient confidence in the degree of understanding and quantification of the corrosion mechanism. The above described model is based on a combination of theoretically sound concepts with semi-empirical to purely empirical aspects. The general treatment of mass transfer by reaction path modeling and of kinetics by a first order dissolution reaction has a sound theoretical basis. On the other hand, the meaning of the term 'silica saturation' is still not well understood and the so called 'long-term rate' for silica saturated conditions can only be considered as an empirical upper limit to estimate long-term corrosion. A model which "explains" a single set of data cannot be considered confirmed, because different models may fit the same data and the lack of sufficient precision may not allow the distinction between "good" and "better" or "right" and "wrong" models. The validity of the model was assessed by comparing calculated results to an extensive experimental data base. This data base includes different glass compositions and a wide range of experimental conditions (Table 5-2).

Table 5-2. Summary of Conditions Used for Experimental Validation of GLASSOL at 90°C, Typical Solution Volume: 30 mL

Glass	Solution	final pH	SA/V m-1	flow mL/d	time d	system
R7T7	DI,GW,SI	7-9	10-10000	0-3	3-365	BE,MAG
SRL-131	DI,BA,SI	7-10	10	0-15	28-1400	
PNL-76/68	DI	7-9	10	0-14	28-365	
SM58	DI	7-9	10-100	0	1-365	
SAN60	DI	7-9	10-100	0	1-365	
MW	DI	7-9	10, 1320	0	7-365	

GW=granite water, SI=silicate water, DI=deionized water, BA=basalt water, BE=bentonite, MAG=magnetite

For the condition of each corrosion test, the model was used to calculate solution concentrations and pH, as well as normalized elemental mass loss and surface layer thickness. The temperature of all tests was 90°C. For each glass composition a separate set of kinetic parameters (k_+, K_{sat}, r_{fin} and D) were determined from the experiments. The parameters remained unchanged within the studied range of SA/V values, flow rates and groundwater compositions. The same list of potential alteration products (a detailed description of this list is given in ref. (67)) was used for all glass compositions.

For the soluble glass constituents such as Na, Li, B, Mo and Cs, and for Si the calculated results usually deviated by less than 30% from the results of solution or surface analyses. Due to limited thermodynamic data, discrepancies were larger for sparingly soluble elements such as Fe, Mg and Al. A detailed comparison between the data and the model is given in the references (1,3,67,76,77,78).

The fit of soluble element and silica release data shows that some general aspects of the reaction path (i.e. of solution chemistry and of mass transfer relations) and of the kinetic constraints are represented by the model in a reasonable manner, even without considering details of the reaction. A better agreement of experimental and modeling data will probably be obtained if some of these 'details' are considered, i.e., solid solution formation, degree of crystallinity of phases, alkali ion exchange, sorption phenomena in the surface layer, and the formation of colloids or polynuclear complexes in solution.

SUMMARY

In various fields of research and technology the corrosion of glasses must be described quantitatively. Examples are the dating of archeological and geological glass samples by the measurement of the thickness of corrosion rims and the prediction of the durability of nuclear waste borosilicate glasses. Corrosion of multicomponent glass compositions cannot be described in great detail by unqualified reference to ion-exchange and matrix dissolution mechanisms. A geochemical model is reviewed, which describes network dissolution, changes in solution chemistry, silica saturation and the formation of surface layers as potential transport barriers as a function of glass and solution composition and of environmental parameters such as temperature, pH, solution volume, surface area, and flow rate. The model is discussed in the context of the current understanding of the mechanism of glass corrosion and recent progress in the geochemistry of rock-water interaction. The model has been validated by application to large experimental data bases for the corrosion of nuclear waste glasses and has also been tested with data for natural glasses. Examples of these applications are given in the paper.

REFERENCES

1. *JSS-Project Phase IV Final Report*, JSS-Report 87-01, SKB, Stockholm, (1987).
2. Grambow, B., in: *Mat. Res. Soc. Symp. Proc.* (C.M. Jantzen, J.A. Stone, and R.C. Ewing ed.), Vol. 44, pp 15-24, Materials Research Society (1985).
3. Grambow, B., Hermansson, H.P., Björner, I,K., and Werme, L.O., in: *Mat. Res. Soc. Symp. Proc.* (L.O. Werme ed.), Vol. 50, pp 187-194 (1986).
4. Grambow, B. et al., in: *Advances in Ceramics*, (D.E. Clark, W.B. White, and A.J. Machiels ed.), Vol. 20, pp 465-474, American Ceramic Society (1986).
5. Grambow, B., Jercinovic, M.J., Ewing, R.C., and Byers, C.D., in: *Mat. Res. Soc. Symp. Proc.* (L.O. Werme ed.) Vol. 50, pp 263-272 (1986).
6. Parkhurst, D.L., Thorstensen, D.C., and Plummer, L.N., *Water-Resources Investigations 80-86*, U.S. Geological Survey (1980). (Extension by P. Offermann, Hahn-Meitner-Institut, Berlin (1983) and by W. Howden and

K. Kruppka, Pacific Northwest Laboratories (1984).
7. Newton, R.G., *Glass Technology* 26: 21-38 (1985).
8. Hench, L.L., Clark, D.E., and Harker, A.B., *J. Materials Science* 21: 1457-1478 (1986).
9. Schröder, H., *Glastechn. Ber.* 26: 91-97 (1953).
10. Zagar, K., and Schillmöller, A., *Glastechn. Ber.* 33: 109-116 (1960).
11. Douglas, R.W., and Elshamy, T.M.M., *J. Amer. Cer. Soc.* 50: 1 (1967).
12. Doremus, R.H., *J. Non-Crystalline Solids* 19: 137-144 (1975).
13. Smets, R.M.J., and Lommen, T.P.A., *Phys.Chem.Glasses* 23: 83-87 (1982).
14. Boksay, Z., Bouquet, G., and Dobos, S., *Phys. Chem. Glasses* 8: 140-144 (1967).
15. McGrail, B.P., Kumar, A., and Day, D.E., *J. Amer. Cer. Soc.* 67: 463-467 (1984).
16. Smets, B.M.J., and Tholen, M.G.W., *Phys. Chem. Glasses* 26: 60-63 (1985).
17. Doremus, R.H., in: *Treatise on Materials Science and Technology* (M. Tomozawa and R.H. Doremus ed.), Vol. 17, pp 41-69, Academic Press (1979).
18. Landford, W.A., Davis, K., Lamarche, P., Laursen, T., and Groleau, R., *J. Non-Crystalline Solids* 33: 249-266 (1979).
19. Scholze, H., *J. Am. Ceram. Soc.* 60: 186 (1977).
20. Bunker, B.C., Arnold, G.W., and Beauchamp, E.K., *J. Non-Crystalline Solids* 58: 295-322 (1983).
21. Doremus, R.H., Mehrotra, Y., Lanford, W.A., and Burman, C., *J. Mater. Science* 18: 612-622 (1983).
22. Schäfer, J., and Schaeffer, H.A., presented at U.S.C.V. Conference in Brussels (1984).
23. Bunker, B.C., Headly, T.J., and Douglas, D.C., in: *Proc. Mat. Res. Soc. Symp.*, Vol. 32, Elsevier Science Publishing Co., Inc. (1984).
24. R.D. Aines, Weed, H.C., and Bates, J.K., in: *Mat. Res. Soc. Symp. Proc.* (J.K. Bates and W.B. Seefeld ed.), Vol. 84, pp 547-558 (1986).
25. Ernsberger, F.M., *Glastechn. Ber.* 56: K(2), 963-968 (1983).
26. Scholze, H., *GLAS - Natur, Struktur und Eigenschaften*, Springer Verlag Berlin, 2. Aufl. (1977).
27. Baer, D.R., Pederson, L.R., and McVay, G.L., *J. Vac. Sci. Technol.* A2(2): 738-743 (1984).
28. Pederson, L.R, *Phys. Chem. Glasses*, (1986).
29. Paul, A., *J. Materials Science* 13: 2246-2268 (1979).
30. Plodinec J., Jantzen, C.M., and Wicks, G.G., *A Thermodynamic Approach to Prediction of the Stability of Proposed Radwaste Glasses*, DP-MS-82-66, Savannah River Laboratory (1982).
31. Bourcier, W.L., Knauss, K.G., and Merzbacher, C.I., *Proc. Sixth Intl. Symp. Rock-Water Interaction*, pp 107-110 (1989).
32. Bourcier, W.L., *Geochemical Modeling of Radioactive Waste Glass*

Dissolution using EQ3/6: Primary Results and Data Needs, UCID-21869, Lawrence Livermore National Laboratory (1990).

33. Brezonik, P.L., *Principles of Linear Free Energy and Structure-Activity Relationships and Their Application to the Rates of Chemical Reactions in Aquatic Systems*, Workshop on Aquatic Chemical Kinetics, Swiss Federal Institute of Technology, ETH, EAWAG, March (1989).
34. Grambow, B., *Glas Techn. Ber.* LVIK: 566-571 (1983).
35. Barkatt, A., Macedo, P.B., Gibson, B.C., and Montrose, C.J., in: *Mat. Res. Soc. Symp. Proc.* (C.M. Jantzen, J.A. Stone, R.C. Ewing ed.), Vol. 44, pp 3-14 (1985).
36. Strachan, D.M., in: *Advances in Ceramics* (G.G. Wicks and W.A. Ross ed.), Vol. 8, pp 12-18, The American Ceramic Society (1983).
37. Wallace, R.M., and Wicks, G.G., in: *Scientific Basis for Nuclear Waste Management* (D.G. Brookins ed.), Vol. 6, pp 23-28, North-Holland (1983).
38. Mendel, J.E. et al., *Final Report of the Defense High-Level Waste Leaching Mechanism Program*, PNL-5157, Pacific Northwest Laboratory (1984).
39. Zavoshy, S.J., and Chambre, P. L., *Mat. Res. Soc. Symp. Proc.* (C.M. Jantzen, J.A. Stone, and R.C. Ewing ed.), Vol. 44, pp 311-322 (1985).
40. Dibble, W.E., Jr., and Tiller, W.A., *Clays & Clay Minerals* 29(5): 323-338 (1981).
41. Fullam, H.T., *Solubility Effects in Waste-Glass/Demineralized Water Systems*, PNL-2614, Pacific Northwest Laboratory (1980).
42. Conradt, R., Roggendorf, H., and Scholze, H., *Mat. Res. Soc. Symp. Proc.* (C.M. Jantzen, J.A. Stone, R.C. Ewing ed.), Vol. 44, pp 155-162 (1985).
43. Freude, E., Grambow, B., Lutze, W., Rabe, H., and Ewing, R.C., *Mat. Res. Soc. Symp. Proc.* (C.M. Jantzen, J.A. Stone, R.C. Ewing ed.), Vol. 44, pp 99-106 (1985).
44. Wollast, R., *Geochim. et. Cosmochim. Acta* 31: 635-648 (1967).
45. Helgeson, H.C., *Geochim. et Cosmochim. Acta* 35: 421-469 (1971).
46. Petrovic, R., *Geochim. et Cosmochim. Acta* 40: 1509-1521 (1976).
47. Lagache, M., *Geochim. et Cosmochim. Acta* 40: 157-161 (1976).
48. Berner, R.A., and Holdren, G.R., *Geochim. et. Cosmochim. Acta* 43: 1173-1186 (1979).
49. Helgeson, H.C., in: *Geochemistry of Hydrothermal Ore Deposits* (Barnes ed.), pp 568-610, John Wiley & Sons (1979).
50. Aagaard, P., and Helgeson, H.C., *Am. J. Sci.* 281: 237-285 (1982).
51. Helgeson, H.C., Murphy, W.M., and Aagaard, P., *Geochim. et Cosmochim. Acta* 48: 2405-2434 (1984).
52. Lasaga, A.C., *Reviews in Mineralogy*, Vol. 8, Mineralogical Soc. America (1981).
53. Lasaga, A.C., *Journal of Geophysical Research* 89(B6): 4009-4025 (1984).
54. Dibble Jr., W.E., and Tiller, W.A., *Geochim. et Cosmochim. Acta* 45: 79-92

(1981).
55. Prigogine, I., and Defay, R., *Chemical Thermodynamics*, p. 543, Longmans, Green (1954).
56. Eyring, H., *J. Chem. Phys.* 3: 107-120 (1935).
57. Davis, J.A., James, R.O., and Leckie, L.O., *J. Colloid and Interface Science* 63(3): 480-499 (1978).
58. Scholze, H., Conradt, R., Engelke, H., and Roggendorf, H., in: *Scientific Basis for Nuclear Waste Management V, MRS Symposia Proceedings* (W. Lutze ed.), Vol. 11, pp 173-180, North Holland (1982).
59. Rimstidt, J.D., and Barnes, H.L., *Geochimica et Cosmochimica Acta* 44: 1683-1699 (1980).
60. Grambow, B., PhD Thesis, Freie Universität Berlin (1984), English trans. by E.I. Du Pont de Nemours and Co., Aiken, SC; DP-tr-78 (1985).
61. Malow, G., *Scientific Basis for Nuclear Waste Management V2, MRS Symposia Proceedings* (W. Lutze ed.), Vol. 11, pp 25-36, North Holland (1982).
62. Chick L.A., and Pederson, L.R., in: *Mat. Res. Soc. Symp. Proc.* (G.L. McVay ed.), Vol. 26, pp 635-642, Elsevier Science Publishing Co., Inc. (1984).
63. Grambow, B., and Strachan, D.M., in: *Mat. Res. Soc. Symp. Proc.* (G.L. McVay ed.), Vol. 26, pp 623-33, Elsevier Science Publishing Co., Inc. (1984).
64. Clark, D.E., Hench, L.L., and Wicks, G.G., in: *Advances in Ceramics* (G.G. Wicks and W.A. Ross ed.) Vol. 8, The American Ceramic Society (1983).
65. Barkatt, A., Simmons, J.H., and Macedo, P.B., *Nuclear and Chemical Waste Management 2*, pp 3-23 (1981).
66. Lanza, F., Conradt, R., Hall, A.R., Malow, G., Trocellier, P., and Van Iseghem, P., in: *Radioactive Waste Management and Disposal: Proceedings of the Second European Community Conference* (R. Simon ed.) pp 196-211, Cambridge University Press (1985).
67. Grambow, B., *Report to JSS-Project Phase IV*, JSS-Report 87-02, SKB, Stockholm (1987).
68. Wolery, T.J. *EQ3NR, A Computer Program for Geochemical Aqueous Speciation-Solubility Calculations*, UCRL-53414, Lawrence Livermore National Laboratory (1983).
69. Fritz, B., *Sci. Géol. Mém. Univ. Louis Pasteur*, 65: 197 (1981).
70. Bruton, C.J., in: *Mat. Res. Soc. Symp. Proc.* (M.J. Apted and R.E. Westermann ed.), Vol. 112, pp 607-619 (1988).
71. Savage, D., *Nuclear and Chemical Waste Management* 6: 14-39 (1986).
72. Crovisier J.L., Honnorez, J., and Eberhart, J.P., *Geochim. Cosmochim. Acta* 51: 2977-2990 (1987).
73. Plummer, L.N., in: *Proc. of the First Canadian/American Conference on Hydrology*, pp 149-177, National Water Well Assoc. (1984).
74. Materials Characterization Center, *Nuclear Waste Materials Handbook*,

DOE/TIC-1140, Pacific Northwest Laboratory (1981).

75. Test methods submitted for the *Nuclear Waste Materials Handbook*, Report PNL-3990, Battelle Pacific Northwest Laboratory (1983).
76. Grambow, B., and Strachan, D.M., *A Comparison of the Performance of Nuclear Waste Glasses by Modeling*, PNL-6698, Pacific Northwest Laboratory (1988).
77. Van Iseghem P., and Grambow, B., in: *Mater. Res. Soc. Symp. Proc.* (M.J. Apted and R.E. Westermann ed.), Vol. 112, pp 631-640 (1988).
78. Zwicky, H.U., Grambow, B., Magrabi, C., Aerne, E.T., Mohos, M., Werme, L.O., Bradley, R., and Barnes, B., in: *Mater. Res. Soc. Symp. Proc.* (W. Lutze and R.C. Ewing ed.), Vol. 127 (1989).
79. Malow, G., Lutze, W., and Ewing, R.C., *J. of Non-Crystalline Solids* 67: 305-322 (1984).

6

Thermodynamic Approach to Glass Corrosion

Carol M. Jantzen

Westinghouse Savannah River Co.
Savannah River Laboratory
Aiken, SC 29808

INTRODUCTION

The durability of glass determines its applicability, whether for use as a container for liquids, a window glass, an optical glass, or for solidification of nuclear waste. Corrosion of glass is the impairment of its properties by mechanisms such as weathering or by contact with water. The properties of glass are of vital interest in its application to the immobilization of high-level liquid nuclear waste (HLLW). The waste is vitrified in borosilicate glass to protect mankind from exposure to radioactivity (1). The glass will be poured into stainless steel canisters and subsequently emplaced in a geologic repository. The intrusion of groundwater into, and passage through, a repository is the most likely mechanism by which radionuclides may be removed from the nuclear waste glass and carried to the biosphere. Thus it is important that nuclear waste glasses be stable in the presence of groundwater over geologic time scales.

The effects of weathering on ancient glasses have been noted and the effects of aqueous attack on other industrial glasses have been well studied. The mechanism of glass corrosion and weathering are generally well understood but the time required to produce deleterious effects may be anywhere from a few hours to millennia. The rate of reaction depends on the glass composition, the environment, and the application (2). Here-to-fore these inter-dependencies have not been quantified and mathematical models could not be developed to reliably predict the durability of a given glass in a specific environment.

The usual tests for chemical durability involve exposure of a glass to water, to some other aqueous solution, or to a humid atmosphere for a given length of time at a given temperature. The practical aspect of the problem, then is to design a test which accelerates the corrosion process in a realistic manner. The ideal test would permit a direct prediction of glass performance to be made. Since this is rarely possible, tests are designed which rate glasses in the same durability sequence as in a real environment (2). The results of these tests can then be used for performance prediction by comparing unknown glasses with those whose real service performance is known.

It is impossible to conduct a laboratory experiment to directly demonstrate the stability of any man-made glass on geologic time scales. Because of the long radioactive half-lives of fission products and actinides in high-level nuclear waste, it must be isolated from the biosphere for 10^3 to 10^5 years. The existence of natural glasses, such as obsidians, basalts, and tektites, which are millions of years old, demonstrates that glasses can be formulated to survive for geologic times. The concept of using natural glasses as analogs for man-made glass durability was first proposed by Ewing (3). Laboratory comparisons of the durability of rhyolite glasses (4) and basaltic glasses (5) to nuclear waste glasses demonstrated that waste glasses, for example, could be fabricated to be as durable as some natural glasses. These mechanistic studies did not, however, provide a basis for predicting the glass compositions for which this would be true.

To quantify the relative durabilities, the performance of nuclear waste glasses relative to natural and ancient glasses (whose long-term performance is known) was investigated (6-12). A mathematical approach based on equilibrium thermodynamics (13) was applied to over 300 experimental laboratory results. Although equilibrium is rarely achieved in short term laboratory tests, the use of equilibrium thermodynamics furnishes a quantitative frame of reference for the relationship between any solid species and the environment on historic and geologic time scales.

Application of the thermodynamic model has enabled the relative durabilities of various glasses to be determined as a function of glass composition (8,9,12). Using this approach, the relative durability of natural glasses (tektites of 1-3 x 10^6 years and obsidians of 30 x 10^6 years) and ancient glasses (4th-9th century) has been shown to bracket the durability of waste glasses (6-9,12). Since the long-term performance of the natural glasses and the ancient glasses is known, the long-term stability of nuclear waste glasses has been interpolated.

During these studies, the thermodynamic approach of Newton and Paul (13) was expanded and shown to be applicable to a wide variety of glasses (6-12) in various environmental conditions (14). The thermodynamic approach was combined (10,11) with the thermodynamic approach used by Pourbaix (15) and Garrels (16) to describe the effects of natural aqueous environments on metallurgical and mineralogic species. The construction of Pourbaix diagrams for glass dissolution has conceptually and quantitatively unified the dissolution behavior of all glasses as a function of the conditions imposed in natural environments.

In this chapter, the quantification of glass durability from the thermodynamic aspect will be reviewed. The relationship between the thermodynamic and kinetic aspects will be demonstrated. The role of glass composition, glass structure, and the effect of varying test parameters will be discussed. The role of natural environmental conditions on glass durability will be described in terms of Pourbaix diagram construction.

KINETIC VERSUS THERMODYNAMIC APPROACH TO GLASS DURABILITY

The durability of a glass is a function of both its kinetic rate of approach to equilibrium and its final thermodynamic equilibrium state in an aqueous environment (17,18). Kinetic models have been applied to the time dependent corrosion of many glasses (19-22), nuclear waste glasses (23,24), and crystalline silicates (21,25). These models mathematically describe glass dissolution mechanisms of ion exchange, diffusion, and protective layer formation in the form of time dependent master equations. Such models are primarily and necessarily empirical in their derivation and basis. Although kinetic models describe the leaching behavior of a given glass, they cannot predict which of a given group of glasses will be most durable.

Alternatively, chemical thermodynamics has been used to predict the stability of vitreous (15,17-18,26) and crystalline silicates (15,16,27,28) in aqueous environments. Application of chemical thermodynamics does not require determination of the time dependent kinetics of the leaching processes. However, this approach compares the thermodynamic stability of the reactants (glass or crystalline silicates in solution) to the product species formed during leaching or weathering, e.g. silicic acid, cations or complexes in solution, surface precipitates, and/or solids. Such an ontological approach to glass durability was originally formulated by Newton and Paul (13,17,18) and applied to (1) the weathering of medieval window glasses and (2) to the relative stabilities of these glasses to Roman window glass, modern container glass, Pyrex® laboratory ware, and natural flint glass (13).

Newton and Paul's approach predicts glass durability from glass composition based upon thermodynamic hydration equations which represent the glass dissolution mechanisms of ion exchange and matrix dissolution. Determination of glass durability as a function of glass composition is significant to the development of durable commercial glasses and to the development of durable glasses for nuclear waste disposal.

To quantify the relative durabilities of a wide range of man-made and nuclear waste glasses, the thermodynamic approach of Newton and Paul was expanded to include (1) additional cationic species, (2) thermodynamic contributions from both the forward glass dissolution mechanisms, e.g. ion exchange and matrix dissolution, and the precipitation or back reactions which largely govern surface layer formation, and (3) the effects of increased pH on glass durability. The thermodynamic approach of Newton and Paul was applied to over 300 experimental laboratory results and shown to be applicable to a wide variety of glasses (6-12) in various environmental conditions (10-12,14).

The relationship between the thermodynamic and kinetic aspects of glass durability has been shown to be a function of the relative contributions of glass composition and imposed test conditions (29). The kinetic contribution to glass dissolution was shown to be primarily a function of the test geometry. Test parameters such as exposed glass surface area (SA), solution volume (V), test duration (t), and test temperature (T) have been shown to alter the kinetic contribution to glass dissolution. The thermodynamic contribution represents the long-term glass stability as a function of the glass composition and structure (bonding).

THERMODYNAMIC AND STRUCTURAL BASIS OF GLASS HYDRATION

Glass Hydration Mechanisms

During the glass dissolution process, there is a short period of alkali proton or hydronium ion-exchange followed by matrix dissolution and/or solution precipitation reactions (30,31). Solubility of the leached glass species, especially that of silica, is the dominant factor controlling the chemical durability of glasses in aqueous solutions. Because silica is the major network former of silicate glasses, a low solubility of silica is found for highly durable glasses. Silica solubility is a strong function of pH and increases significantly above pH values of ~9. This is the reason that most silicate glasses are poorly durable in basic solutions.

Glass dissolution can mechanistically be modeled as a combination of ion exchange and matrix dissolution reactions (8,17,19). The reaction zone, as originally described by Douglas and El-Shamy (22), has been determined (30, 31) to contain two reaction zones. One zone occurs at the leached layer solution interface where equilibrium is considered to be between the glass surface sites and the ions in solution. The second zone occurs at or near the leached layer glass interface where a counterion exchange occurs (Figure 6-1). The relative size of the two zones in glasses exposed to identical conditions varies according to the type of glass, e.g. highly durable glasses have thicker "glass-leached layer zones" and thinner "leached-layer solution zones". The opposite occurs for

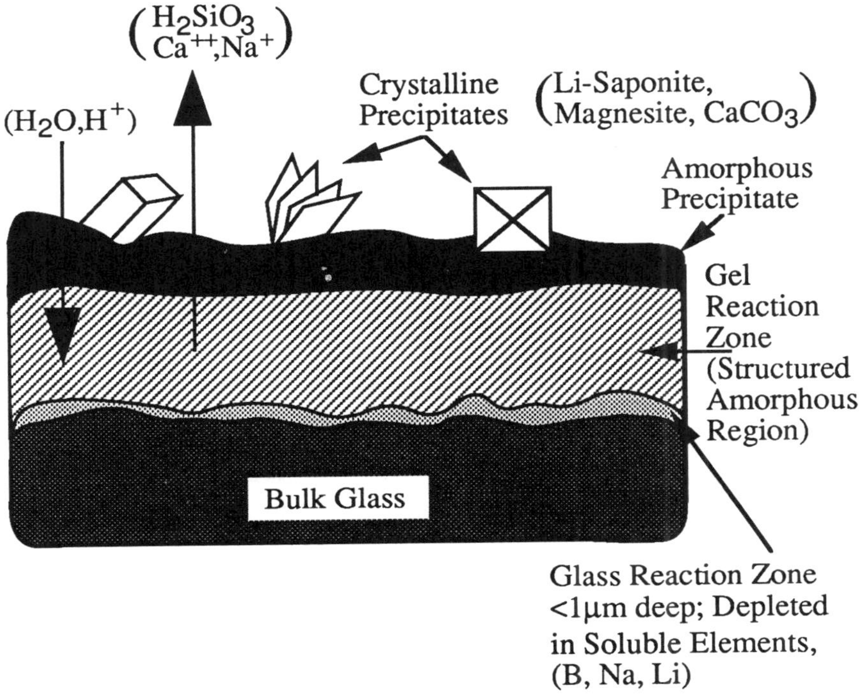

Figure 6-1. Schematic of glass dissolution mechanisms in aqueous solution, (ion exchange and matrix dissolution), coupled with hydrated amorphous surface layer formation ("back reactions" involving precipitation from solution) After Refs. 29,31,32.

glasses having low durability. With time, the dissolution rate is modified by formation of hydrated amorphous surface layers and/or precipitation of secondary metal hydroxo and/or metal silicate complexes (29,30-32; see Figure 6-1). These "back reactions" are caused by adsorption and/or changes in solubility of the metal cation and/or silicate species in solution (32).

The reactions at the glass surface-solution interface have been described in terms of two chemical reactions (19,22).

(a) Cations are released into solution as a result of ion exchange with protons from the solution:

$$\equiv \text{Si-OM} + H_2O \Leftrightarrow \equiv \text{Si-OH} + M^+ + OH^- \quad (6\text{-}1)$$

where M^+ is a monovalent metal cation.

(b) Silica is released into solution as the siloxane bonds in the glass matrix are attacked by the hydroxyl ions from the solution:

$$\equiv Si\text{-}O\text{-}Si\equiv + OH^- \Leftrightarrow \equiv Si\text{-}OH + \equiv Si\text{-}O^- \quad (6\text{-}2a)$$

$$\equiv Si\text{-}O^- + H_2O \Leftrightarrow \equiv Si\text{-}OH + OH^- \quad (6\text{-}2b)$$

A non-bridging oxygen, designated as NBO or $\equiv Si\text{-}O^-$ is formed in reaction (6-2a). This NBO can interact with water to form additional silanol bonds on the glass surface and, thereby, release another hydroxyl ion. In turn, this hydroxyl ion can serve as the source of hydroxyl for further breaking of siloxane bonds on the glass surface via reaction 6-2a. An increase in the activity of the hydroxyl ion in the solution will, therefore, favor increased removal of silica and suppression of reaction 6-1 as the activity of the hydrogen ion concentration is decreased (22).

The simplest case, the hydration of vitreous silica as expressed by the overall reaction (19,21):

$$SiO_2 + H_2O \Leftrightarrow H_2SiO_3 \quad (6\text{-}3a)$$

This represents a forward dissolution reaction and a reverse condensation reaction which are responsible for the formation of silica in the solution and in the leached layer, respectively. This hydration reaction has been quantitatively described (21) as a reversible first order dissolution reaction which has an associated equilibrium thermodynamic free energy. This thermodynamic equilibrium can be used to describe the dynamic exchange equilibria at the interface of the leached layer with the aqueous solution (22).

For the reaction of vitreous silica with water, the equilibrium constant, K, is:

$$K = \frac{(aH_2SiO_3)}{(aSiO_2)(aH_2O)} \quad (6\text{-}3b)$$

Since silica and water are assumed to be in their standard state at 25°C, then $aSiO_2$ and aH_2O are unity and the equilibrium constant is equal to aH_2SiO_3. The relationship between the equilibrium constant and the free energy of the exchange reaction (when the reactants and the products are in the standard state) is, therefore,

$$\Delta G^\circ = - RT \ln K, \quad (6\text{-}4)$$

where ΔG° = standard state free energy of change (cal mol^{-1}), R = gas constant (cal K^{-1} mol^{-1}), T = absolute temperature (K).

Thermodynamic Representation of Glass Hydration Mechanisms

In 1977 Paul (18) predicted the durability of glasses from their chemical composition by assuming that the glass/water reactions could be described in terms of the summation of the thermodynamic free energies for hydration reactions of the individual structural components, e.g. silicate or oxide groups, in a glass. This thermodynamic parameter, the free energy of hydration, was later correlated to alkali release for a series of poorly durable medieval glasses (13).

The thermodynamic approach of Newton and Paul (6-12,14,17,18,29) assumes that the glass is a mechanical mixture of silicate and oxide components. In particular, the cation hydration equations assume inosilicate groups (single chains of O:Si ratio of 3:1) and framework silicates (O:Si ratio of 2:1). Species not forming known silicate structures are treated as oxides. The overall free energy of hydration of a glass is assumed to be an additive function of the free energies of the hydration reactions of the individual silicate and oxide component end members. The formalism is:

$$\Delta G_{hyd} = \Sigma x_i \bullet (\Delta G_{hyd})_i \qquad (6\text{-}5)$$

where $(\Delta G_{hyd})_i$ is the free energy change of the thermodynamically most stable hydration reaction of component i at mole fraction x_i.

The use of additive functions of the compositional components of glasses to predict thermodynamic and physical properties is not restricted to glass durability studies. Bottinga and Weill (33) developed an empirical method of estimating the viscosity of magmatic silicate liquids based on the Arrhenius mixture rule. Similar approaches have been used to calculate the heat capacity of molten silicates (34) and the thermal expansion of oxide glasses (35) as additive functions of composition. Similarly, the Gibbs free energy of formation for complex silicate minerals has been calculated as additive functions of the simple oxide and hydroxide components (36).

Newton and Paul (13,17,18) did not apply their approach to borosilicate glasses. In order to evaluate borosilicate based nuclear waste glasses, Jantzen and Plodinec (16-18) chose a consistent set of hydration reactions which included a reaction for boric oxide hydrating to boric acid, and contributions from redox active ions. The thermodynamic data base of Paul (17,18) and Garrels and Christ (16) were used as the thermodynamic input for the hydration reactions.

Although Newton and Paul's approach mechanistically modeled glass ion exchange and matrix dissolution reactions it did not account for glasses where hydrated amorphous surface layers formed. The hydration thermodynamic approach of Jantzen and Plodinec (6-9,12,14) modeled the relative contribution of all three mechanisms as a function of the glass composition. If a stable hydration product was observed experimentally on a leached glass surface [the hydration reaction for this product usually had a higher free energy of formation than the theoretical $(\Delta G_{hyd})_i$], the free energy of formation for the observed hydration product, $(\Delta G_{hyd})_{obs}$), was preferentially used (8,9). The free energies of the various hydration reactions chosen by Jantzen and Plodinec are given in Table 6-1 and discussed below. The calculational details are given in Appendix A.

Ion Exchange. Network breaking glass structural species (e.g. alkali, alkaline earth, Fe^{2+}, Mg^{2+}, etc.) and intermediate glass structural species (e.g. Pb^{2+}) associated with the unpolymerized glass $(SiO_4)^{-4}$ tetrahedral network are considered to be released to the solution by ion exchange with protons in the aqueous solution. The polymerized $(SiO_4)^{-4}$ associated with these species is released to the solution or forms silanol bonds on the glass surface by attack of the OH^- on the nonbridging oxygen (NBO) bonds associated with each silicate group. Note that for inosilicates, two NBO bonds are available for OH^- attack. Model equations expressing these mechanisms are as follows:

$$M_2SiO_3 + 2H^+ \Leftrightarrow H_2SiO_3 + 2M^+ \tag{6-6a}$$

or

$$MSiO_3 + 2H^+ \Leftrightarrow H_2SiO_3 + M^{+2} \tag{6-6b}$$

where M is a monovalent cation in Equation 6-6a and a divalent cation in Equation 6-6b. Note that these equations are equivalent to those given in Equation 6-1.

Matrix Dissolution. Dissolution of the polymerized $(SiO_4)^{-4}$ glass network, not associated with the cationic silicate species, is expressed by the following equation:

$$SiO_2 + H_2O \Leftrightarrow H_2SiO_3 \tag{6-7}$$

Note the similarity of this equation to Equation 6-2. A similar reaction can be written for structural B_2O_3 hydrating to H_3BO_3 as shown in Table 6-1.

Surface Layer Formation. Recent investigations have shown that metal hydroxides and hydrated silicates form poorly soluble surface films on glasses during static leach testing (29-31,38-40, see Wicks'chapter). In some cases, these surface layers protect the underlying glass from further attack by water. In the thermodynamic formalism, dissolution of Fe_2O_3, Al_2O_3, Nd_2O_3, TiO_2, and other species which form hydroxides or hydrates that participate in surface layer precipitation reactions are written as follows:

$$M_2O_3 + 3H_2O \Leftrightarrow 2M(OH)_3 \tag{6-8}$$

where M^{3+} in equation (6-8) is a trivalent cation.

Table 6-1. Calculated Free Energies of Hydration for Glass Components[1].

Glass Components	Hydrated Species	$(\Delta G_{hyd})_i$ (kcal/mole)	$F=Z/r^2$	Z/r	CN[b]
Cs_2SiO_3	Cs^+, H_2SiO_3	-46.820	0.32	0.56	VI
			0.28[c]	0.53[c]	XII
K_2SiO_3	K^+, H_2SiO_3	-41.735	0.47	0.68	VI
			0.35c	0.60c	XII
$BaSiO_3$	Ba^{+2}, H_2SiO_3	-30.570	0.88	1.32	VIII
			0.71[c]	1.19[c]	XII
Na_2SiO_3	Na^+, H_2SiO_3	-28.815	0.82	0.91	VI
			0.65[c]	0.81[c]	VIII
$SrSiO_3$	Sr^{+2}, H_2SiO_3	-24.400	1.13	1.50	VIII
Li_2SiO_3	Li^+, H_2SiO_3	-22.740	1.49	1.21	VI
$CaSiO_3$	Ca^{+2}, H_2SiO_3	-16.116	1.71	1.85	VI
			1.39[c]	1.67[c]	VIII
$MnSiO_3$	Mn^{+2}, H_2SiO_3	-14.871	3.75[c]	2.19[c]	VI_H
			3.56	2.67	VI_L
$FeSiO_3$	Fe^{+2}, H_2SiO_3	-14.609	2.70[c]	2.32[c]	VI_H
			4.20	2.89	VI_L
$NiSiO_3$	Ni^{+2}, H_2SiO_3	-14.347	3.37	2.60	VI
$MgSiO_3$	Mg^{+2}, H_2SiO_3	-13.888	3.12	2.50	VI
B_2O_3	H_3BO_3	-9.930	75	15	IV
Al_2O_3	$Al(OH)_3$	-7.73	13.6[c]	6.38[c]	IV
Al_2O_3	$AlO(OH)$	-1.63	8.06	4.92	VI
Al_2O_3	$Al(OH)_3$ (am)	+3.04			
UO_2	$U(OH)_4$	+8.38	3.43	3.70	VIII
SiO_2	H_2SiO_3[a]	+5.59	34.6	11.76	IV
Fe_2O_3	$Fe(OH)_3$	+15.50	9.23	5.26	IV_H
TiO_2	$TiO(OH)_2$	+15.99	10.74	6.55	V
$ZrSiO_4$	$ZrO(OH)^+, H_2SiO_3$	+45.10	6.25	5.00	VI

a Due to choice of thermodynamic data base, $(\Delta G_{hyd})_{SiO2}$ is given by Paul (17,18) as +5.59 kcal/mole. However, a more realistic value of +3.7 kcal/mole (37) or +3.76 kcal/mole (28) may apply for vitreous silica

b Where H and L refer to ions in high spin or low spin states

c Preferred value of F or Z/r

---

1 $(\Delta G_{hyd})_i$ calculated from room temperature thermodynamic values. Values of crystalline silicates are substituted when data for amorphous species are not available. Equations assume that silicic acid dissociation at pH values of >9.5 has not occurred. Additional terms representing silicic acid dissociation should be added for leachates which exceed this value.

For example, in the original ΔG_{hyd} thermodynamic basis set (6) the hydration reaction for Al_2O_3 was written so that amorphous $Al(OH)_3$ was the hydration product (Table 6-1). Subsequently, three Al_2O_3-SiO_2 glasses known to have Al^{3+} in tetrahedral coordination (41) were leached. The glass monoliths were dried at 90°C for 1 hour (8) and the glass surfaces were then analyzed by x-ray diffraction. The $Al(OH)_3$ polymorph, gibbsite, was found to be the major phase on the glass surface along with poorly crystallized pyrophyllite [a clay in the smectite-montmorillonite family of composition $Al_2Si_4O_{10}(OH)_2$]. The free energy for the hydration of Al_2O_3 to gibbsite is -7.73 kcal/mol. versus +3.04 kcal/mol for hydration to amorphous $Al(OH)_3$. Also, when silica is consumed in the hydration to clay, an additional free energy contribution of -7.82 kcal/mol of Al_2O_3 needs to be added. The residual moles of SiO_2 are assumed to hydrate to H_2SiO_3 as in equation 6-7.

In glasses where Al^{3+} is known to be structural, or where surface layer analysis indicates the presence of gibbsite or clay, then the negative rather than the positive free energy of hydration for Al_2O_3 given in Table 6-1 should be used. In cases where the coordination of aluminum is not known or surface analyses are absent or are inclusive, then the surface is assumed to be amorphous $Al(OH)_3$ and the positive free energy of hydration for Al_2O_3 should be applied.

Redox Species. The approach of Jantzen and Plodinec (6-8) also accounts for the effects of glass redox. Experimentation on oxidized and reduced waste glasses which contain large amounts of iron and manganese has shown that glasses made under oxidizing conditions are more durable than those made under reducing conditions. In applying the data of Table 6-1, the redox state of iron can be a critical parameter, since Fe^{3+} (as Fe_2O_3) improves the calculated durability and participates in surface layer formation. However, Fe^{2+} (as $FeSiO_3$) reduces the durability.

Effects of Solution pH and Eh. Activity-pH diagrams provide the fundamental correlation between ion concentrations in solution (minimum solubility) and the free energy of a hydration reaction (16,42). Activity-Eh diagrams provide the fundamental correlation between redox sensitive ion concentrations in the solution as a function of both the oxidation potential of the solution and the free energy of the redox sensitive hydration reaction. Solubility concentrations differ from the activities by a factor, γ, known as the activity coefficient (16). If γ is approximately equal to one, then the ion activity approximately equals the ion concentration. The factor γ is dependent on the ionic strength of the solution and is a function of the multiple hydration, ionization, and complexation reactions that the ion participates in.

The interrelation of activity, pH, and Eh is shown schematically in Figure 6-2 for reactions which are:

a independent of Eh
b independent of pH

c independent of both Eh and pH
d dependent on both Eh and pH.

The activity of vitreous SiO_2 and vitreous B_2O_3 are independent of Eh, hence the activity (concentration) dependence can be accurately represented in activity-pH diagrams (Figures 6-3a and b). The influence of pH on the activity of vitreous silica can be thermodynamically calculated from the equilibrium constants for the hydration of SiO_2 to silicic acid and the equilibrium constants for silicic acid dissociation (15,18) (Figure 6-3a). The activity-pH diagrams, therefore, provide the fundamental correlation between silica concentration in solution[1] and the free energy.

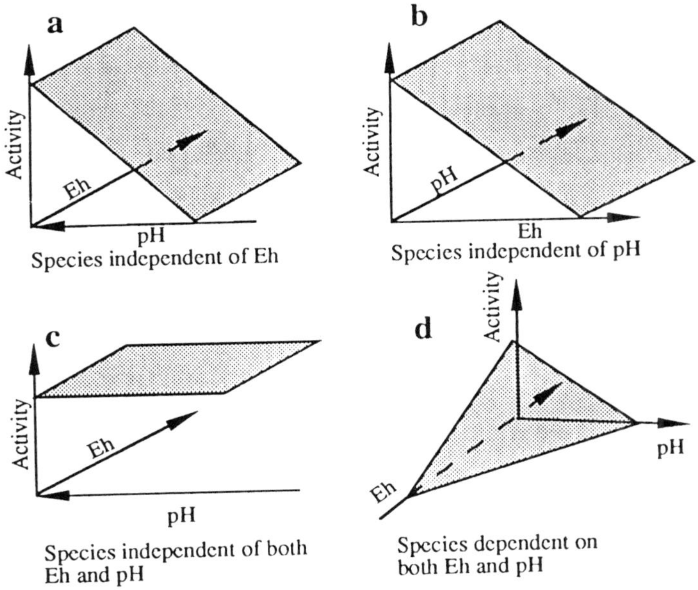

Figure 6-2. Three dimensional representation of the relations between the activity of a species with solution Eh and pH (after Krumbein and Garrels, Ref. 43). Note the different behavior of species which are dependent or independent of solution Eh and/or pH.

[1] This relation is supported by the data of Grambow (47,48) which suggests that dissolution of silicate and borosilicate glasses can be described by the activity diagrams for the dissolution of amorphous silica, and by the data of Reimus (49) which indicates that the aqueous solution activities for a range of nuclear waste glasses are all approximately the same because many of the dissolved species are at their solubility limit.

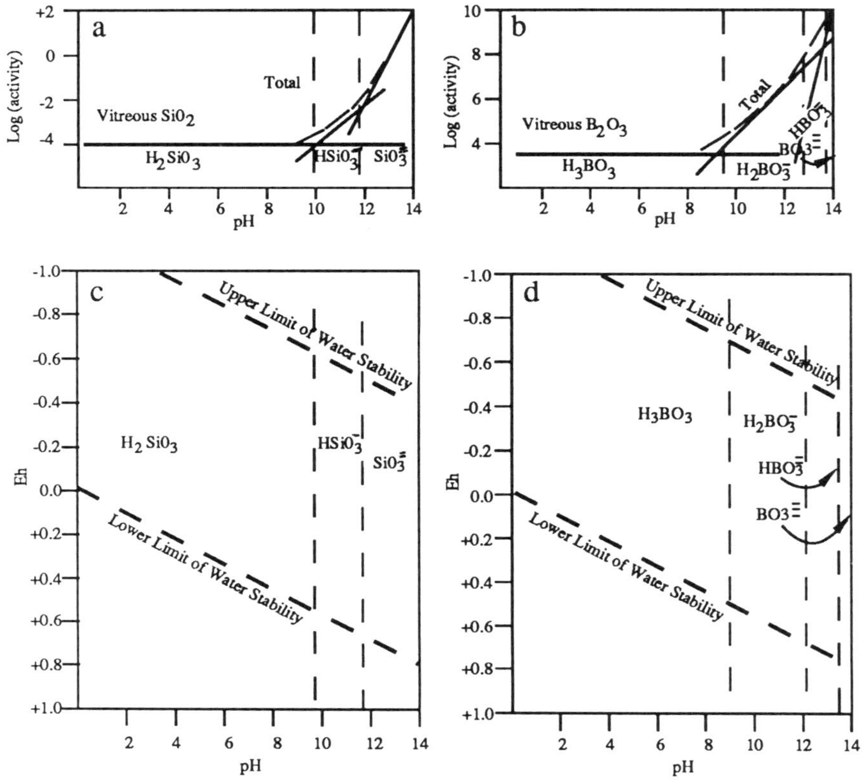

Figure 6-3. (a) Activity-pH relations for vitreous SiO_2 (after Pourbaix, Ref.15), (b) activity-pH relations for vitreous B_2O_3 (calculated), (c) pH-potential diagram for vitreous SiO_2 showing the "domains of dominance" of the aqueous species (after Pourbaix, Ref. 15), (d) pH-potential diagram for vitreous B_2O_3 showing the "domains of dominance" of the aqueous species (after Pourbaix, Ref. 15).

The vertical dashed lines in Figure 6-3a define the "domains of dominance" of the aqueous species. The "domains of dominance" can also be represented thermodynamically on the corresponding Pourbaix (Eh-pH) diagram for vitreous silica (Figure 6-3c). Because the equilibria are Eh independent, the "domains of dominance" for each of these aqueous species parallel the Eh axis (see Figure 6-2a).

The influence of pH on the activity of pure vitreous B_2O_3 (Figure 6-3b) is shown for comparison with that of vitreous SiO_2 (Figure 6-3a). The activity-pH data shown for vitreous B_2O_3 (Figure 6-3b) was calculated from the equilibrium constants for the hydration of vitreous B_2O_3 to boric acid and the equilibrium constants for boric acid dissociation determined from the data of Garrels and Christ (16). The "domains of dominance" of the aqueous species were taken from the Pourbaix diagram for B_2O_3 shown in Figure 6-3b. The equilibria appear as vertical lines parallel to the Eh axis because the dissociation reactions for H_3BO_3 are Eh independent (see Figure 6-2a). Since the sum of the activities of each species in solution determines its minimum solubility, the greater activity of vitreous B_2O_3 (Figure 6-3b) compared to vitreous SiO_2 (Figure 6-3a) may account for the consistently higher boron releases, relative to silicon releases, observed during borosilicate glass dissolution (47).

The thermodynamic approach (6-8,13) assumes that the silicate and borate components of a glass hydrate to silicic and boric acid. For poorly durable glasses where the alkali released from the glass drives the solution pH to greater than 9.5, the solubility of silica and borate increases rapidly due to dissociation as demonstrated in Figure 6-3. Jantzen (8,9,11) demonstrated that an additional contribution to the hydration free energy was needed to account for the dissociation of silicic and boric acid at high pH, e.g. the curvature in the activity-pH diagrams shown in Figure 6-3. The equations below were calculated from the dissociation constants of silicic and boric acid as a function of pH. The additional contributions to ΔG_{hyd} are as follows:

for H_2SiO_3

$$\Delta(\Delta G_{hyd}) = 1.364\left[-\log\left(1 + \frac{10^{-10}}{10^{-pH}} + \frac{10^{-21.994}}{10^{-2pH}}\right)\right] \tag{6-9}$$

for H_3BO_3

$$\Delta(\Delta G_{hyd}) = 1.364\left[-\log\left(1 + \frac{10^{-9.18}}{10^{-pH}} + \frac{10^{-21.89}}{10^{-2pH}} + \frac{10^{-35.69}}{10^{-3pH}}\right)\right] \tag{6-10}$$

Equation 6-9 applies for all silicate-based glasses including borosilicate glasses while Equation 6-10 applies only to boron-containing glasses. At pH values <7 the contribution to the ΔG_{hyd} term from equation 6-9 is minimal. At pH values <9 the contribution to the ΔG_{hyd} term from equation 6-10 is minimal. The adjusted free energy term was statistically shown to be more highly correlated with glass durability than the uncorrected hydration free energy for over 150 glasses tested (8,9). The pH-adjusted ΔG_{hyd} term is, therefore, preferentially used and glass durability can be calculated from glass composition alone or, more accurately, from glass composition and the pH of the aqueous environment.

Structural Basis of Hydration Thermodynamics

Relation of ΔG_{hyd} to Non-bridging Oxygen Bonds. Although the thermodynamic approach of Newton and Paul (13) assumes that glass is a mechanical mixture of inosilicate, framework silicate, and oxide components, no correlation between glass composition and glass structure was made. The relationship between bonding, composition, and durability of crystalline and vitreous solids has been examined for over 40 years. Glass scientists such as Stevels (48,49) have attempted to relate the proportion of non-bridging oxygen atoms (NBO) in a glass network to durability. Similarly, geologists have tried to classify the relative durability (weathering classification) of silicate mineral species on the basis of structural silica-tetrahedra linkages (50,51). Newton attempted a glass durability classification based on network-building SiO_2 versus network-breaking cations (M_2O) and network modifying (MO) species.

Recent studies have demonstrated that the rate-limiting step in silica-water reactions is breakage of the structural Si-O bonds (21). This has been confirmed by the Extended X-ray Absorption Fine Structure (EXAFS) by Greaves (53) which demonstrates that ion exchange occurs along percolation channels that exist in the glass and that the channels are characterized by the ionic bonding exhibited between NBO and the MO or M_2O species (Figure 6-4). Since the reaction mechanism is directly related to the composition of the solid (27), and hence to the number of NBO bonds (48,49) and their strength (21), it is not surprising that the dynamic exchange reactions at the glass-solution interface can be described in terms of the thermodynamic equilibria (22). This implies that the thermodynamic parameters are representations of the structural energetics of the hydration process and that the relation between the glass structure and durability is a function of the glass lattice energy. The bond strength is a component of the lattice free energy and an exact expression for the free energy of a glass has been derived and shown to be dependent only upon the concentration and energy of the bonds present in the structure (54).

White and Minser (55) derived a formula for calculating the number of non-bridging oxygen atoms in natural glasses such as tektites and obsidians. This formula was based on the glass composition. Jantzen and Plodinec (8) modified this equation to include additional species found in nuclear waste glasses and applied the formulation to over 150 natural, ancient, medieval, and nuclear waste glasses. The modified formula is as follows:

$$\begin{aligned} NBO = {} & 2(BaO + CaO + Cs_2O + FeO + K_2O + Li_2O + MgO \\ & + MnO + Na_2O + NiO + SrO - Fe_2O_3 - Al_2O_3) \\ & + 4(TiO_2 + ZrO_2 + UO_2)/\text{oxide mole sum.} \end{aligned} \tag{6-11}$$

This formula assumes that every mole of mono- and di-valent oxide component contributes two non-bridging oxygen atoms while every Fe_2O_3 and Al_2O_3 component create two bridging oxygen atoms. The formation of two NBO per mole of $M_2O + MO$ is consistent with the assumption of two NBO per inosilicate group while Fe^{3+} and Al^{3+} are assumed to form tetrahedral $(AlO_4)^{-5}$ and $(FeO_4)^{-5}$ groups. Tetravalent ions such as Ti, Zr, and U are assumed to create four non-bridging oxygen atoms.

Partitioning of Al^{+3} and Fe^{+3} between structural components of a melt and the different structural roles of Fe^{+3} and Fe^{2+} is especially critical to the calculation of the non-bridging oxygen atoms and the hydration free energy. The structural

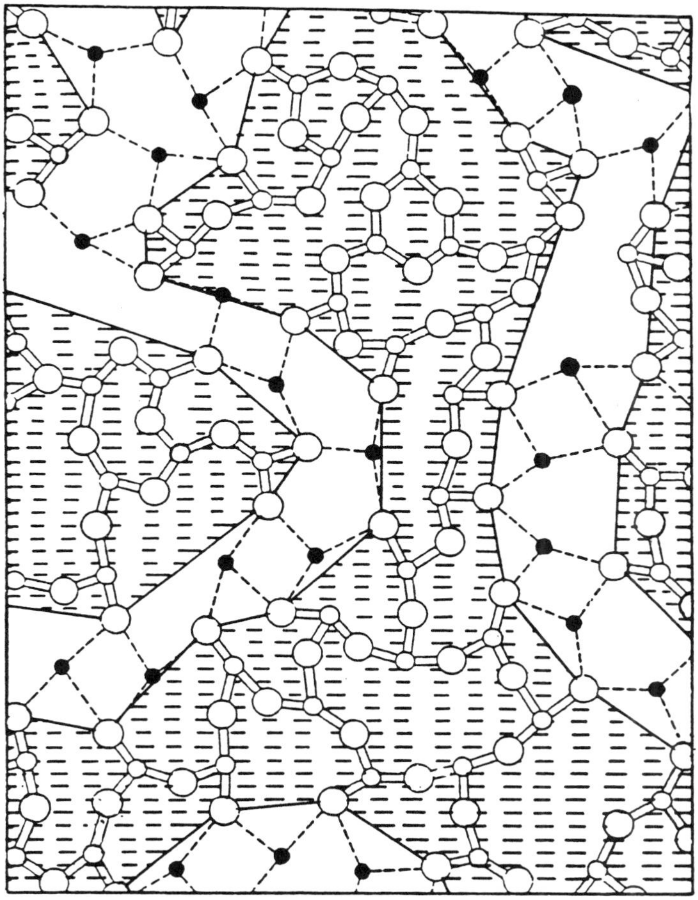

Figure 6-4. A modified random network (MRN) for a glass of nominal composition $M_2O_3(G_2O_3)_2$, where M represents the modifying cations and G represents the network forming cations. Covalent bonds are shown by the solid lines and ionic bonds by the dotted lines. The shaded regions are defined by the boundary disclinations which run through the G-O (non-bridging) bonds. The shading of the covalent bonds between G-O bonds highlights the percolation channels of M_2O_3 that run through the network (from Greaves, 53).

role of iron is controlled by an involved relationship between the degree of polymerization of the melt, types of cations present, amount of iron present, and its oxidation state (56). Although Al^{3+} is considered an intermediate (57) or conditional (58) glass forming ion, the addition of Al^{+3} to silicate melts increases the degree of polymerization of the melt by creating more network forming complexes. This transfers a divalent cation from a network breaking position to a charge balancing position within the aluminosilicate network (59). The monovalent alkali metals are more efficient at charge balancing within an aluminosilicate network (58,59). In iron silicate melts, the alkalis stabilize Fe^{3+} in tetrahedral coordination as a network former. If the M^+/M^{2+} ratio is less than that required to form tetrahedral Fe^{3+}, then mixed coordination of Fe^{3+} is found. Secondly, if only M^+ cations are available and $\Sigma M^+ = \Sigma Fe^{3+} = \Sigma Al^{3+}$, then Al is tetrahedral and Fe^{3+} acts as a network modifier. Thirdly, if M^+ is sufficient to stabilize either Fe^{3+} or Al^{3+} as a network former, but additional M^{2+} cations are available, then tetrahedral iron is preferentially charge balanced by alkali while tetrahedral alumina is preferentially charge balanced by divalent cations. The alkali-ferric iron associations are more stable than the alkali-alumina associations (60). In summary, in iron-free silica glasses, Al^{3+} would be expected to occur as a tetrahedrally coordinated cation, provided that either monovalent or divalent cations are present for charge balance complexing. When iron is present, the Fe^{3+} associations require cations with lower Z/r^2 ratios than the Al^{3+}.

Jantzen (8) correlated the calculated ΔG_{hyd} for over 150 glasses with the number of non-bridging oxygen (NBO) bonds calculated from Equation 6-11 (Figure 6-5a). Many of the glasses had complex chemistries (10 to 15 component species) and, hence, all the Fe^{3+} and Al^{3+} species in the glasses were considered tetrahedral. The calculated NBO term was regressed against the ΔG_{hyd} as the dependent variable (Figure 6-5a). The following expression was obtained:

$$\Delta G_{hyd} = -19.23\ \text{NBO} + 4.77 \qquad R^2 = 0.72 \tag{6-12}$$

Since this correlation involves the glass only and not the solution, no $\Delta(\Delta G)_{hyd}$ is added to the ΔG_{hyd} term during the calculation. A similar but weaker correlation ($R^2 = 0.47$) was observed when the dependent NBO term was regressed against the release of structural silicon from the 150 glasses during a static 28 day leach test (Figure 6-5b). This correlation does not allow for the effects of solution pH.

The correlation of ΔG_{hyd} with NBO and with the release of silicon from over 150 glasses is significant in that it experimentally validates the relationship between the thermodynamic hydration model and the bond energetics of multicomponent complex glasses. The hydration equations given in Table 6-1 represent the structural energies necessary to hydrate the metal ion silicates species of complex glasses. The correlation of NBO and silicon release validates

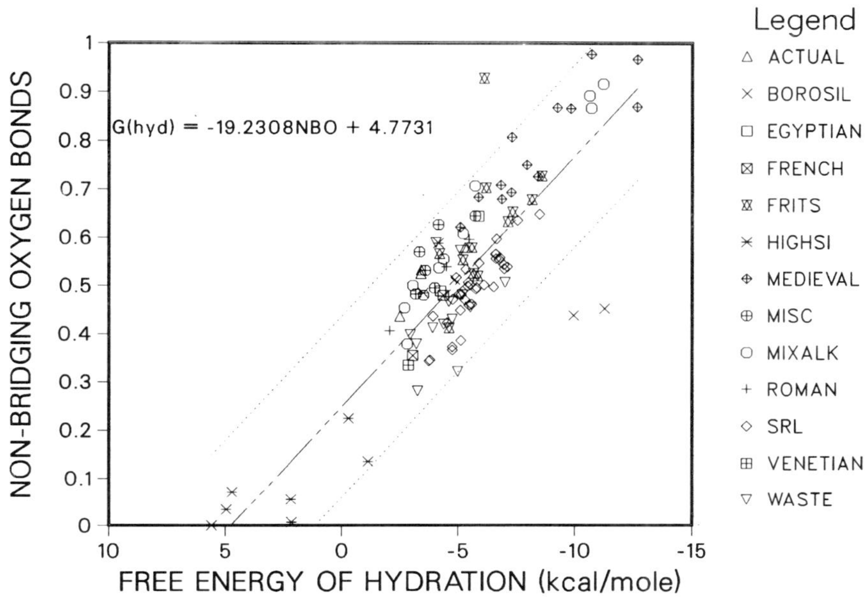

Figure 6-5a. Relation between the calculated number of non-bridging oxygen bonds (NBO) and the calculated free energy of hydration (after Ref. 8).

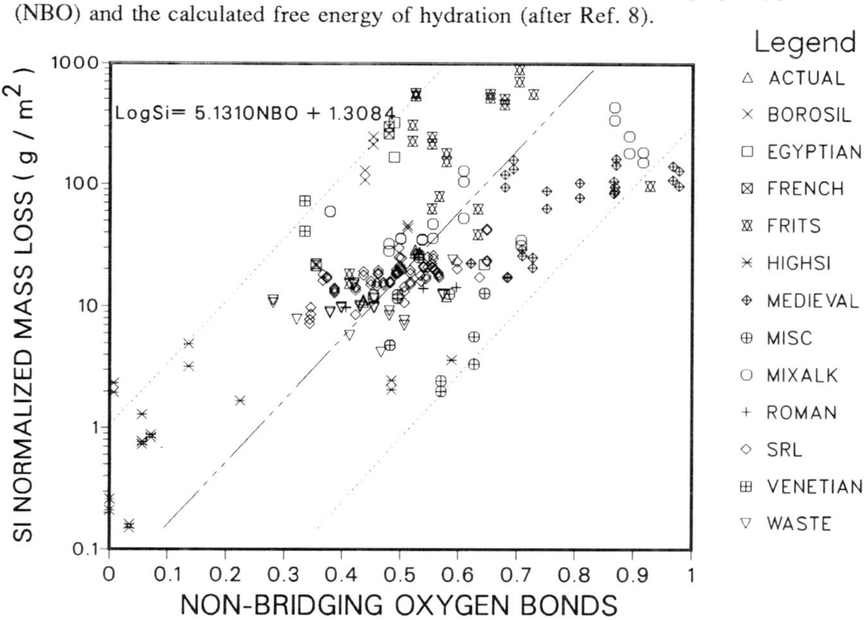

Figure 6-5b. Relation between the number of non-bridging oxygen bonds (NBO) and the amount of elemental silicon released by over 150 glasses during static leach testing in deionized water.

the relationship between glass durability and glass structure. This relationship is less well defined due to the effects of solubility, precipitation, and pH in the solution.

Relation of ΔG_{hyd} to Ionic Potential (Z/r) and Ionic Field Strength (Z/r^2)

Jantzen (8) also demonstrated that the thermodynamic free energies of hydration of the silicate and oxide glass components in Table 6-1 correlated with both the ionic potential (Z/r) and the ionic field strength (F) of these species. The correlation of the ΔG_{hyd} with the ionic potential implies that the structural components chosen are related to bond strengths in glass. Indeed, recent investigations (62) of glass structure have indicated that glasses are mixtures of silicate and oxide component clusters, e.g. the strained mixed cluster model or the quasi-molecular complex model. The latter model maintains that the size and type of the molecular complex is a function of the size and ionic charge of the associated cation (62).

The ionic character of fused silicates is widely accepted (64) and the ionic field strength criterion applies as discussed previously. For this criterion to apply to glass structures the ions are considered rigid spheres and the ionic field strength is:

$$F = Z/r^2 \tag{6-13}$$

Jantzen (8) listed the $(\Delta G_{hyd})_i$ of the component silicate and oxide species starting with the most negative free energies and ending with the most positive free energies. The calculations were based on the ionic radii of Whittaker and Muntus (64). The preferred values of Z/r and Z/r^2 were based on the coordination and spin of a given ion. By ordering the relative hydration potentials of the species based on $(\Delta G_{hyd})_i$, or on the ionic field strength, she observed groupings of the glass component species chosen into network-formers, -breakers, and -modifiers. The trend for the network breaking ions, for example, is for a higher Z/r as the $(\Delta G_{hyd})_i$ becomes more negative. Note that due to the highly covalent nature of the Si and B bonds, the ionic field strength for these species is not considered to be a valid number.

Table 6-1 shows a modification of the original Jantzen (8) table and includes the best known coordination (65) and spin of a given ion in silicate minerals. The data in Table 6-1 for the preferred Z/r and Z/r^2 ratios was plotted graphically against $(\Delta G_{hyd})_i$ in Figure 6-6. The values for the covalent Si and B were omitted. The data indicates that both the ionic potential and the ionic field strength are directly correlated to the calculated $(\Delta G_{hyd})_i$ for the ionically bonded species. The values used by assuming that Al^{3+} and Fe^{3+} are in tetrahedral coordination, as assumed in the calculation of NBO in the previous section, may be in error as indicated by the deviations of the points for Al^{3+} and Fe^{3+} in Figure 6-6.

Assumption of Al^{3+} and Fe^{3+} in VI fold coordination fits the observed trend in Figure 6-6 more accurately.

According to Scholze and Kreidl (66), the smaller the radius of the monovalent cation (corresponding to a higher field strength) the more attraction the NBO experiences from the cation. In terms of the thermodynamic free energy of hydration, this manifests itself as an increase in durability in the series $\Delta G_{hyd}Li^+ > \Delta G_{hyd}Na^+ > \Delta G_{hyd}K^+ > \Delta G_{hyd}Cs^+ >$ and $\Delta G_{hyd}Mg^{+2} > \Delta G_{hyd}Ca^{+2} > \Delta G_{hyd}Sr^{+2} > \Delta G_{hyd}Ba^{+2}$. Similarly, the ion selectivity concepts on which glass pH electrodes have been formulated depend of the competing interactions of cations in the glass and in the solution. Eisenman (67) represented these interactions as ion exchange reactions between the ion of interest in solution and the counterion in the glass. The free energies of these reactions termed "solvation energies" were found to be dependent upon the chemical composition of the glass. The solvation energies were also found to be dependent on the effective anionic field strength (67). The Z/r ratio has also been shown to correlate highly with the equilibrium constants of metal and metal oxide hydrolysis equations (68) as well as with the isolectric point of solids (IEPS). The IEPS determines the pH of the minimum solubility of hydrous metal, oxide, and silicate species (69).

Das (70) has provided evidence that glass dissolution by electrophilic and nucleophilic attack of OH^- on the glass network is highly correlated to the ionic field strength of the major cations in glass (Figure 6-7). The ionic radii determine the number of oxygen atoms with which the cations are coordinated in the glass structure. Conversely, then this represents the number of non-bridging oxygen bonds which are readily available for hydration.

The work of Feng (71) also demonstrated that the Z/r^2 ratio was related to the corrosion rate of a complex glass during powder durability tests in deionized water (Figure 6-8). The same relative position of the ions are noted in Figures 6-6, 6-7 and 6-8. Figure 6-6 indicates that glasses with a larger proportion of high Z/r^2 ions are more durable (more positive ΔG_{hyd}) while Figures 6-7 and 6-8 indicate that glasses with a larger proportion of high Z/r^2 ions exhibit lower corrosion rates (70,71). This is in keeping with the known correlation of more positive ΔG_{hyd} and lower glass corrosion rates (6-14,24).

The correlation of the $(\Delta G_{hyd})_i$ with lattice energies for the glass component species may also explain why the crystalline free energies can be applied to the vitreous state. The free energy of a crystalline material is equal to

$$\Delta G^\circ = \Delta U + p\Delta V - T\Delta S \qquad (6\text{-}14)$$

where G = free energy of formation, U = internal energy, V = volume, S = entropy, p = pressure, and T = temperature.

It is the internal energy term, U which represents the lattice energy and is a function of the ionic and electronic charges and the interionic distances (72,73).

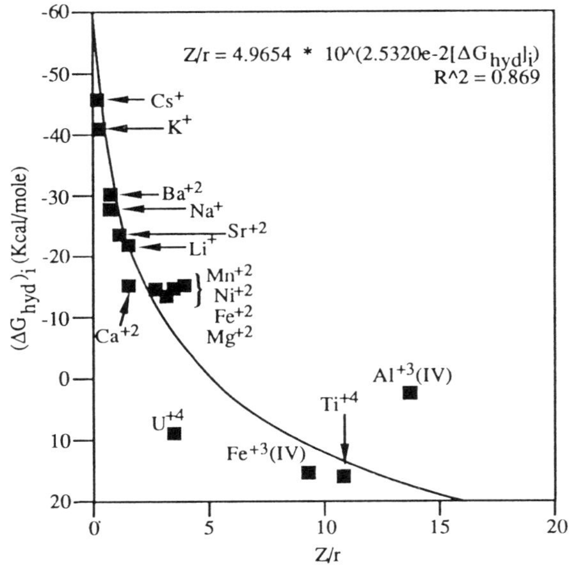

Figure 6-6a. Relation between the Z/r and the $(\Delta G_{hyd})_i$ for the silicate and oxide species.

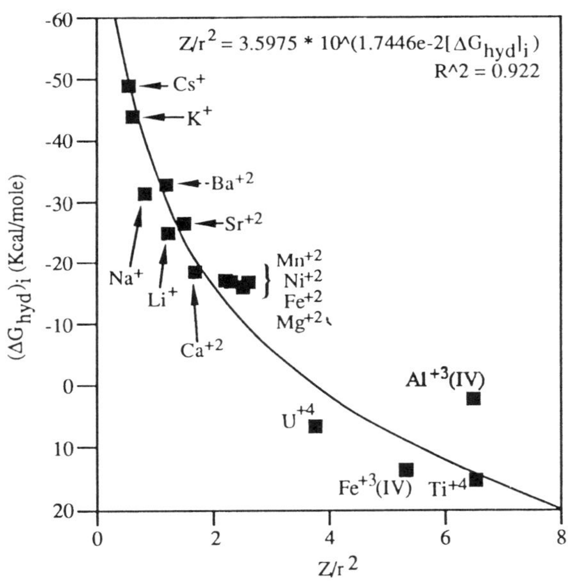

Figure 6-6b. Relation between the Z/r^2 and the $(\Delta G_{hyd})_i$ for the silicate and oxide species.

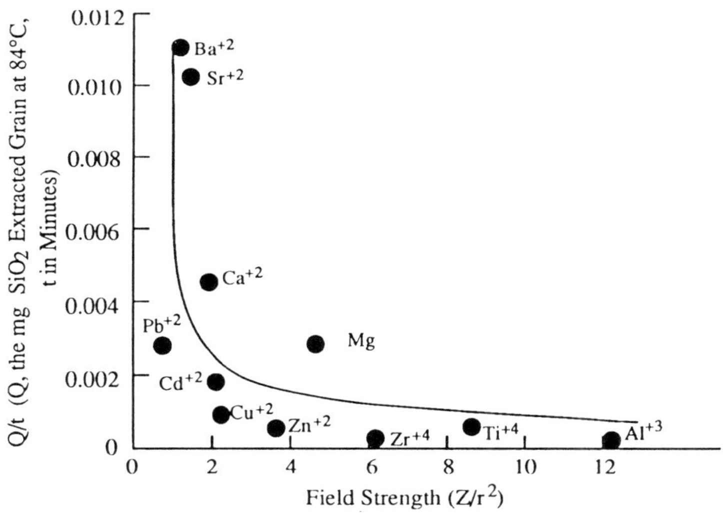

Figure 6-7. Effects of Field Strength, Z/r^2 of the cations in glasses during powder durability testing (after Das, Ref. 70).

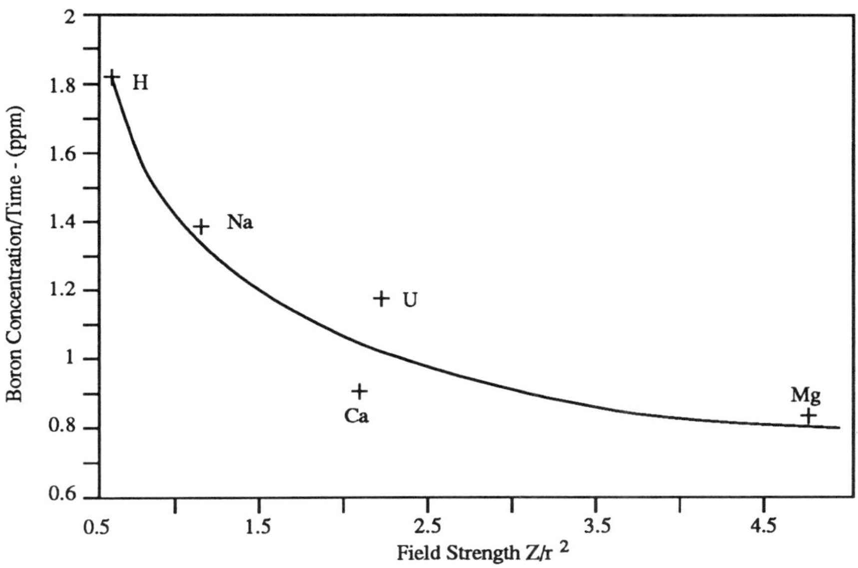

Figure 6-8. Effects of Field Strength, Z/r^2 of the cations in complex glasses during static powder durability testing (after Feng, Ref. 71).

Therefore,

$$\Delta G_{cryst} - \Delta G_{glass} = \Delta U_{cryst} - \Delta U_{glass} + p\Delta V_{cryst} - p\Delta V_{glass} + T\Delta S_{cryst} + T\Delta S_{glass} \tag{6-15}$$

For any given silicate system, the PΔV term can be neglected and the TΔS term is usually much less than ΔU so that $\Delta G \approx \Delta U$. The $\Delta(\Delta U)$ term between a glass and a crystal is small (74) since both are dependent on the electronegativity, mean value of the cation charge, and the cation radius (75). Since $\Delta(\Delta U)$ is small then $\Delta(\Delta G)$ is small and $\Delta G_{cryst} \approx \Delta G_{glass}$.

The correlation of the ionic potential (Z/r) and the ionic field strength (Z/r^2) with the $(\Delta G_{hyd})_i$ supports the conclusion that the calculated hydration equations are the thermodynamic representations of the structural energetics of the hydration process. Since the hydration thermodynamic approach assumes that glass structure is a primary function of glass composition, glass structure is not considered as an additional parameter affecting glass durability.

EXPERIMENTAL TECHNIQUES

Parameterization of the Factors Affecting Glass Durability

During laboratory experimentation, the following parameters which are known to affect glass durability (17) were varied:

(1) frequency of replenishing/changing the solution, e.g. time duration (t) of the experiment
(2) exposed surface area (SA) of the solid
(3) volume (V) of the leaching solution
(4) temperature (T) of leaching in °C
(5) type of leaching solution
(6) glass composition

Various test geometries including crushed glass and monolithic glass tests were performed in order to vary these parameters. The test methods are described below.

MCC-1 Testing. The Materials Characterization Center MCC-1 test procedure (76,77) is a monolithic glass durability test for determining the relative corrosion of glasses, glass-ceramics, or ceramics. Monolithic shapes are used so that a geometric surface area can be measured (Figure 6-9). The leachant can be deionized water or simulated groundwater and the test temperatures are 40°, 70° or 90°C. The minimum test duration is 28 days. Leachates are not filtered to remove colloids and/or particulates. Leachates are analyzed for pH and concentration of the glass species of interest. A standard glass is not used and,

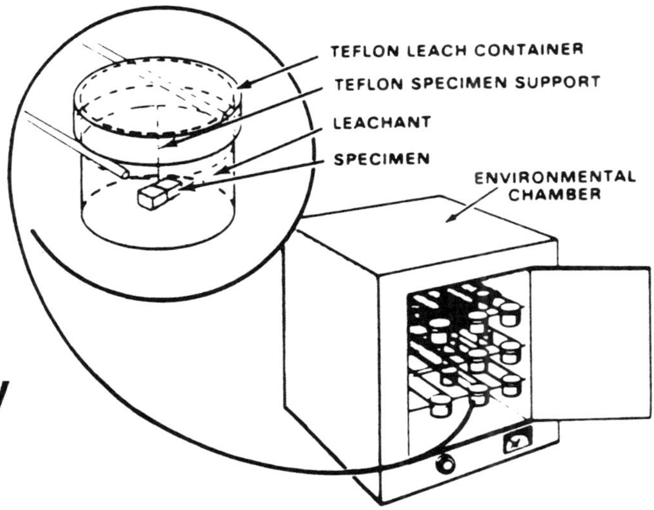

28 day

90 C

No filtration

monolith

mL soln/g solid = 27

static

analyze pH and concentration

no standard glass control

Figure 6-9. Schematic representation of the monolithic MCC-1 test.

hence long term bias corrections cannot be eliminated. The concentrations are reported as normalized elemental mass losses, i.e. NL_i, released from the glass in grams of glass per square meter of glass surface area. This has the advantage that the release concentrations in parts per million are normalized by the weight fraction of that element present in the glass. The normalized mass loss is a function of (1) the mass fraction of that element in the glass, (2) the exposed surface area of the glass, and (3) the leachate volume, as given below:

$$NL_i = \frac{C_i}{F_i \bullet (SA/V)} \tag{6-16}$$

where:

NL_i = normalized elemental mass loss (g_{glass}/m^2)
C_i = mass of element "i" in the solution (g_i/m^3)
F_i = fraction of element "i" in the glass (g_i/g_{glass})
SA = specimen surface area (m^2)
V = solution volume (m^3)

The SA/V ratio is held constant when comparing relative glass durability, e.g. glass monoliths with a geometrically measured surface area of 4 cm^2 are immersed in 40 cm^3 of leachant. Tests can be run in Teflon® or stainless steel vessels. The former representing tests which are considered to be open systems (open to interaction with air) and the latter representing closed systems.

MCC-3 Testing. The Materials Characterization Center MCC-3 test procedure (76) is an agitated crushed glass test for the determination of maximum elemental release rates from glasses or glass-ceramics. Crushed glass is used so that the leachate is dominated by the glass chemistry so that the maximum upper concentration limits released by a glass under closed system conditions can be determined (Figure 6-10). The minimum test duration is 28 days (76). Variations in the surface area of the glass, the volume of the solution, and the test duration can be used to accelerate laboratory leach rates to long times. The recommended ratio of the ml of solution to the grams of solid is 10. Recommended test temperatures are 40°, 90°, 110°, 150°, and 190°C so that acceleration in the leaching kinetics due to temperature can be determined provided that no change in leaching mechanism occurs over the temperature interval tested. The leachant can be deionized water or simulated groundwater. Leachates are filtered to remove colloids and/or particulates and are analyzed for both pH and elemental concentrations of the glass species of interest. A standard glass is not used and, hence long term bias corrections cannot be eliminated. The concentrations are reported as normalized elemental mass losses, i.e. NL_i, released from the glass in grams of glass per square meter of glass surface area as given in Equation 6-16. The use of both Teflon® vessels and stainless steel vessels is allowed as in the MCC-1 procedure.

28 days

90 C

Filter

0.11 mm

mL soln/g solid = 10

constant agitation by rolling

analyze pH and concentration

no standard glass control

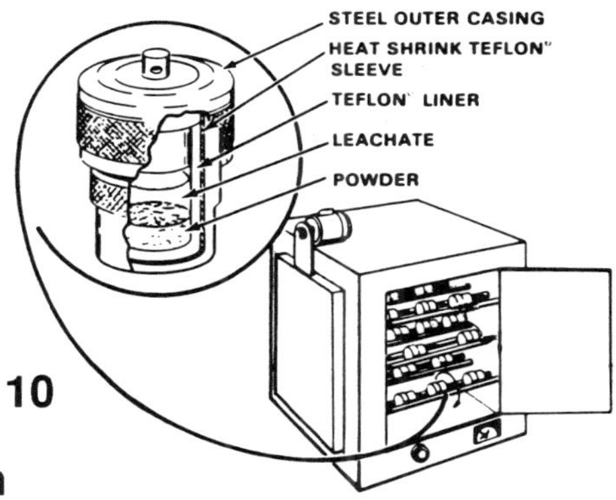

Figure 6-10. Schematic representation of the agitated crushed glass MCC-3 test.

Product Consistency Test (PCT). The Product Consistency Test was developed to provide confirmation of the consistency of nuclear waste glasses during production (78). The test is sensitive to the glass composition and homogeneity and can be run in a remote environment with radioactive or hazardous glasses because of the simple sample preparation and test procedure. The test was developed (78-82) to aid in the characterization of glasses for the Defense Waste Processing Facility (DWPF) at the Savannah River Site (SRS). The test can be performed rapidly compared to the MCC-1 and MCC-3 test protocols and has a higher precision and accuracy than other existing durability tests including ASTM and DIN standard tests for glass (80,82).

Crushed glass, preferably of 100-200 mesh, at the recommended ratio of 10 mL of solution to 1 gram of solid, is immersed in ASTM Type I water for a minimum of 7 days at 90°C (Figure 6-11). Groundwaters could be used but are not preferred since this is a durability test to determine the effects of glass composition and homogeneity. Leachates are filtered to remove colloids and/or particulates and analyzed for pH and the elemental concentration of the glass species of interest. A standard glass is used as a control to eliminate long term

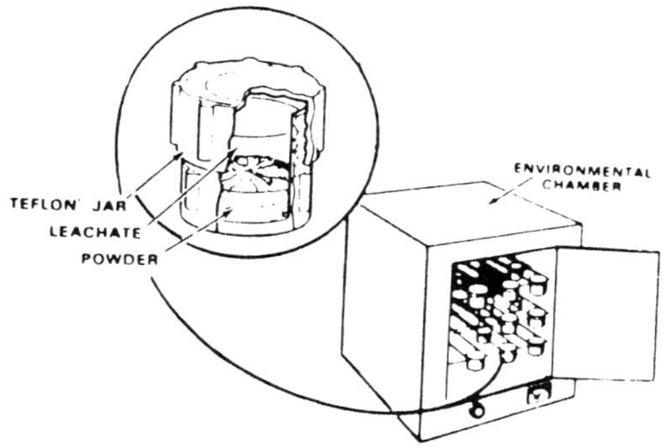

7 day

90 C

Filter

0.11 mm

mL soln/g solid = 10

static

analyze pH and concentration

standard glass as control

Figure 6-11. Schematic representation of the crushed glass PCT test.

bias in the experimental procedure and in the analytic procedures. The concentrations can be reported as normalized elemental mass losses, NL_i, released from the glass in grams of glass per square meter of glass surface area (Equation 6-16) or as ppm. The use of both Teflon® vessels and stainless steel vessels is allowed as in the MCC-1 and MCC-3 test procedures.

Controlled Environmental Testing. Many glasses, including radioactive waste glasses, contain elements which are redox active, i.e. can occur in more than one redox state. The valence state of an ion can be important in determining its equilibrium solubility and its rate of release from glass. Moreover, the adsorption and migration of multivalent elements in natural environments are highly dependent upon the redox state of the groundwater. This is particularly important for some natural environments, e.g repository groundwaters and some seawaters, which are expected to be anoxic (low or negative oxidation potential, Eh).

The MCC-1, MCC-3, and PCT test protocols described above test both redox containing and non-redox containing glasses in oxidizing leachants (deionized water or oxidized groundwaters). In order to determine redox active glass solubilities as a function of temperature, pH, Eh, and water chemistry any of the test protocols described above can be run in the absence of air (closed system conditions) with anoxic leachants. Simulated groundwaters can be equilibrated with the anoxic species present in the rock or soil from which the original groundwater composition was derived (83). Dearation of the leachant is achieved with sparging 99.999% Ar (83-86). Leachants can be heated on a hot plate with ground rock or soil present to equilibrate the groundwater with the redox containing species in the solid (Figure 6-12). To prevent oxidization of the leachant, this should be done in a glovebox continuously purged with 99.999% Ar.

The Eh and pH of the solution can be measured in the equilibration vessel or on an aliquot of the leachant. The latter is recommended since the Eh and pH probes leak calomel and KCl into the equilibration vessel solution at elevated temperatures. In order for the Eh measurements to be significant, 1) the redox couple being measured must be known, 2) the system must be poised, and 3) the Pt surface of the Eh measuring probe must be cleaned in the Ar atmosphere to get a stable reading (83-86). Knowing the redox couple being measured avoids erroneous measurements of mixed couples. In poorly poised systems the concentration of the redox couple is low and Eh measurements drift. In well poised systems, the redox capacity is large, e.g. greater than 10^{-5}M of the redox species is present, and Eh measurements are reproducible (83,87). The Pt probe must be cleaned to remove the Pt-O which forms in air and/or in oxygenated leachants. If the Pt-O is not removed, it causes erroneous Eh readings (83,88).

Once an anoxic leachant is prepared and the Eh and pH are measured, then Teflon® or stainless steel vessels containing the glass of interest can be filled (Figure 6-12). The presence of the host rock does not have a significant effect on the results of anoxic leaching in groundwaters (86). If Teflon® vessels are used, the entire experiment has to be performed in the glovebox or environmental chamber continuously purged with 99.999% Ar. If stainless steel Parr bombs are used (representing closed system conditions), the vessels can be filled and sealed in the environmental chamber under controlled atmosphere. Once sealed, the vessels can be removed from the chamber and run in test ovens in air. The vessels must be returned to the glovebox before they are opened so that the final Eh and pH are measured inside the glove box. Any testing for redox species, e.g. the amount of Fe^{2+} vs Fe^{3+} in solution must be done in the inert environment of the glovebox (85).

GLASS DISSOLUTION: A FUNCTION OF THERMODYNAMIC EQUILIBRIA

Variations in Glass Composition.

Statistical Determinations. All test parameters except the glass composition were kept constant. The durability of 50 nuclear waste glasses were compared

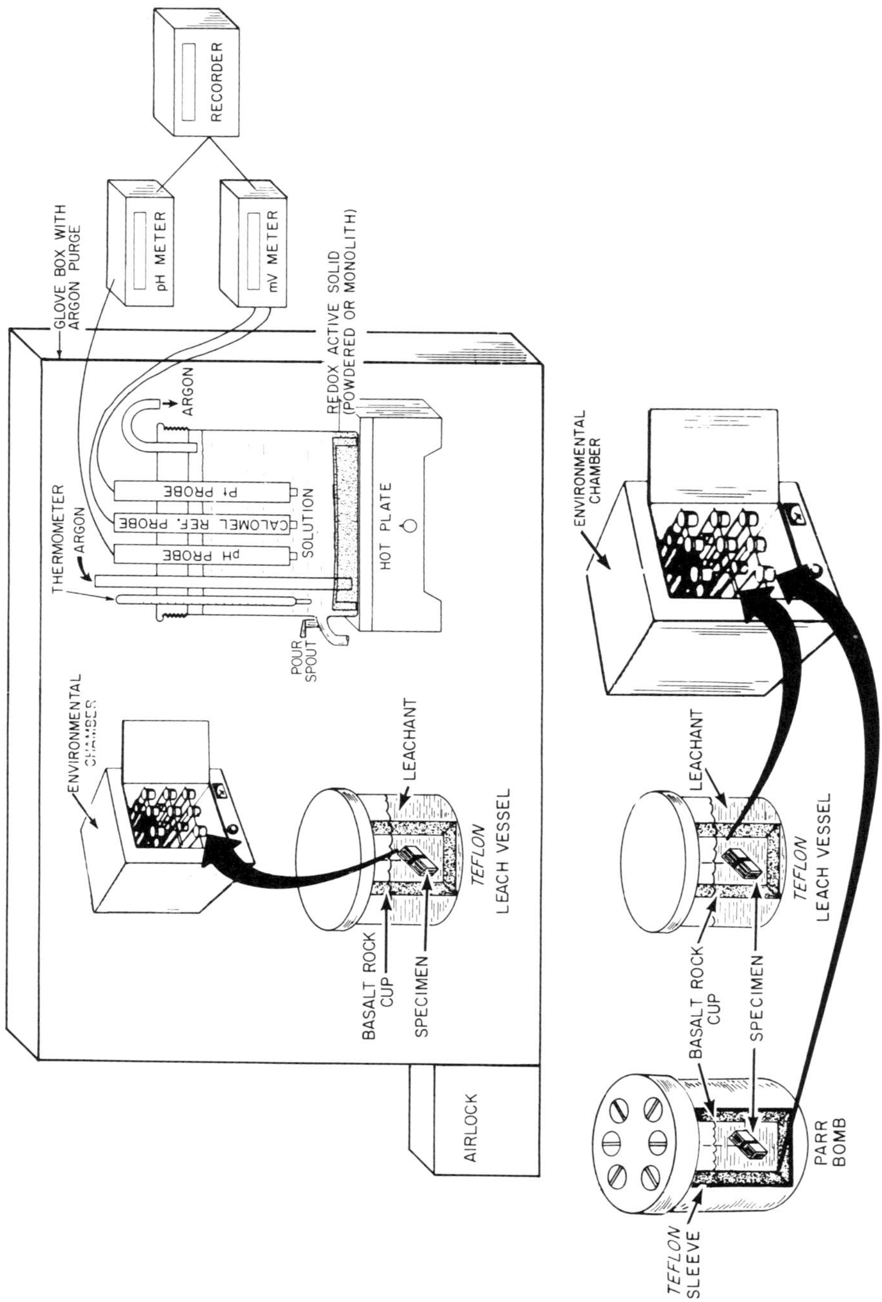

Figure 6-12. Schematic of an MCC-1 test being performed on glass in an anoxic leachant (basalt groundwater) in the absence of air.

to 100 natural and man-made glasses (8,9,11). The 150 different glasses included natural obsidians, tektites, basalts, pure SiO_2, pyrex, modern window glass, and simulated medieval window glasses from the European Science Foundation (ESF). Four actual ancient Roman and Islamic glasses (Jalame ca. 350 A.D., Nishapur 10-11th century A.D., and Gorgon 9-11th century A.D.) obtained from the Corning Museum of Glass were also included. The MCC-1 leach test protocol described above was performed on the 150 glasses. All tests were run in duplicate. The leachant was ASTM Type I water, the constant SA/V ratio was $0.1 cm^{-1}$. The temperature of all the tests was 90°C and the test duration was 28 days.

The ΔG_{hyd} was calculated from the analyzed glass compositions alone and from a combination of the glass compositions and the final solution pH. Statistically determined regression equations and the 95% upper and lower confidence limits were determined (9,11) for binary combinations of the following variables: ΔG_{hyd}, $\log(NL_{Si})$, $\log(NL_B)$, and $-\log(H^+)$ or pH. Logarithmic concentrations were used throughout the statistical analysis because the ion concentrations are assumed to be proportional to the ideal ion activities in the thermodynamic treatment (11).

A simple linear regression of the ΔG_{hyd} and $\log(NL_{Si})$ data demonstrated that glass durability could be determined from glass composition for the 304 experimental data points (Figure 6-13a). The computer generated equation of best fit relating glass composition to silicon released from the glass is plotted in Figure 6-13a and the mathematically determined slope is -0.1638 when ΔG_{hyd} is the independent variable. This correlation has an R^2 of 0.58. A similar regression of the pH corrected ΔG_{hyd} and $\log(NL_{Si})$ data gave an almost identical slope (-0.1636) with an R^2 of 0.73 (Figure 6-13b). The computer-calculated 95% confidence intervals are shown by the dotted lines in Figure 6-13. The statistical analysis indicated that the primary contribution to the 95% confidence interval was from the ΔG_{hyd} term: errors in glass analysis are more significant than errors in leachate analyses (9).

Figure 6-13 illustrates that the more negative the ΔG_{hyd} term, the less durable the glass and the more Si is released to solution. Using this approach, the durability of the most durable nuclear waste glasses is $\approx 10^6$ years by comparison with the durability of basalts of $\approx 10^6$ years (18). The waste glasses are slightly less durable than the high-temperature tektite and obsidian glasses. The least durable waste glasses are comparable to the most durable simulated medieval window glasses of $\approx 10^3$ years. In this manner, the durability of nuclear glasses can be interpolated to be $>10^3$ years and $\leq 10^6$ years.

Since many natural and man-made glasses do not contain boron, only 140 experimental leachate analyses (70 glasses) could be statistically fit (Figure 6-14a). Boron release follows a pattern similar to that of silicon: the more negative the pH corrected ΔG_{hyd} term, the less durable the glass, and the more boron is released to solution. The relative positions of the obsidian, tektite,

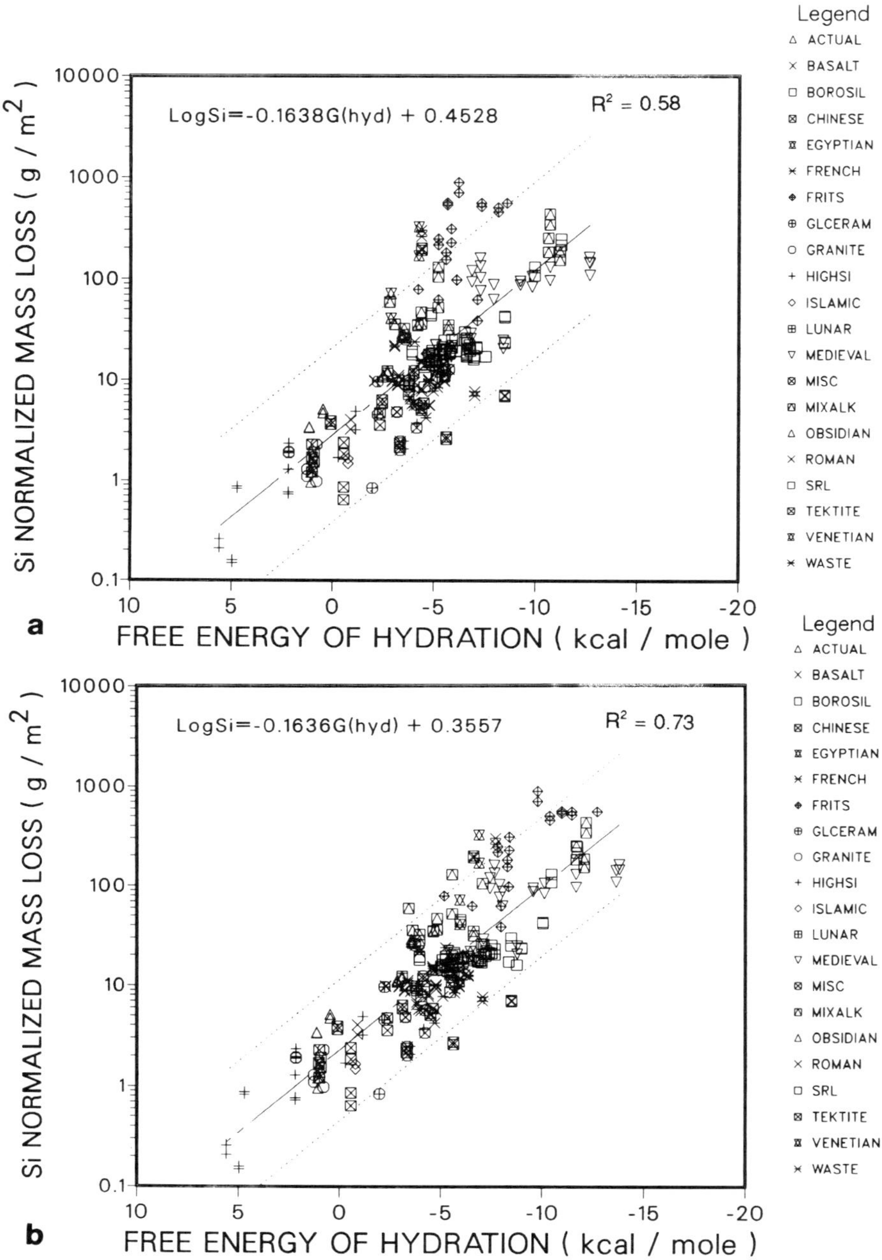

Figure 6-13. (a) Linear regression plot of over 300 leach tests relating ΔG_{hyd} calculated from glass composition alone to log silicon release for the 28 day static MCC-1 leach test, (b) linear regression plot of over 300 leach tests relating ΔG_{hyd} calculated from glass composition and pH to log silicon release for the 28 day static MCC-1 leach test.

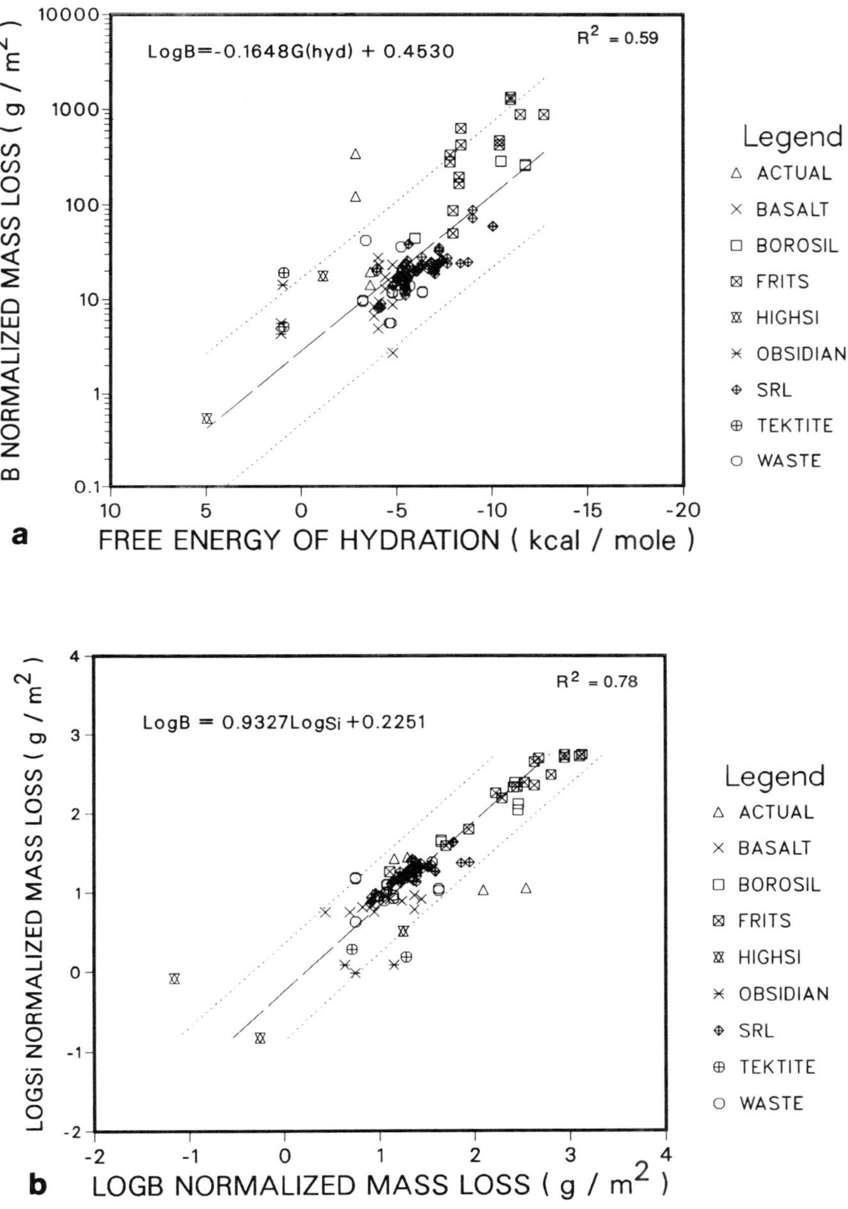

Figure 6-14. (a) Linear regression of the ΔG_{hyd} to log boron release for 140 leach tests (70 glasses which contained boron), (b) linear regression of the log silicon release versus the log boron release in the leachates.

nuclear waste glasses, and the medieval window glasses remain the same as shown in Figure 6-14a and b.

The slope relating the pH corrected ΔG_{hyd} and $\log(NL_B)$ is -0.1648. This is similar to the -0.1636 slope calculated for the relation between the pH corrected ΔG_{hyd} and $\log(NL_{Si})$. The high silica glasses, including the tektites and obsidians, contain very little boron and, hence, the positive free energy portion of the curve has a poorer statistical fit and the overall R^2 is only 0.59. Since the slopes relating pH corrected ΔG_{hyd} to $\log(NL_{Si})$ and to $\log(NL_B)$ are similar, it is not surprising that a plot of $\log(NL_{Si})$ and $\log(NL_B)$ has a slope of ≈ 1 (Figure 6-14b) (9,11). This is anticipated for glasses which undergo congruent dissolution and can be approximately calculated by dividing the slope of Figure 6-13b by that of Figure 6-14b. Although, the regression equations have similar slopes, the absolute value of $\log(NL_B)$ is greater than $\log(NL_{Si})$ for a given glass as predicted from the relative ion activities of these solution species (see Figure 6-3).

The slopes of the ΔG_{hyd} vs $\log(NL_{Si})$ and the $\log(NL_{Si})$ vs ΔG_{hyd} do not invert, indicating bias in the data set. This bias may be from the measurement of the glass composition as discussed above or due to the poor response of leachant-dominated MCC-1 test. Likewise, the slope of the $\log(NL_{Si})$ vs $\log(NL_{Si})$ does not invert, indicating that Si may precipitate in the leachate since the leachates are not diluted nor filtered in the MCC-1 test procedure. Alternatively, the measurement of B near the detection limit, e.g. in leachates of low B containing glasses, may contribute significant error.

Multiple regression analysis indicated colinearity among the regressors (independent variables). The ΔG_{hyd} term unadjusted for solution pH excursions correlated with $\log(NL_{Si})$, $\log(NL_B)$, and solution pH (11). The variables $\log(NL_{Si})$ and $\log(NL_B)$ were highly correlated with each other and with the solution pH. The colinearity was statistically examined for the pH-adjusted and the pH-unadjusted ΔG_{hyd} term and the trends were similar. The $\log(NL_{Si})$, ΔG_{hyd}, and the pH are highly colinear (Figure 6-15a) because the pH and ΔG_{hyd} are related through the Nernst equation. In addition, $\log(NL_{Si})$ and $\log(NL_B)$ are related to the pH through the activity-pH relationship and are highly co-linear (Figure 6-15b). The $\log(NL_{Si})$ and $\log(NL_B)$ are also related to ΔG_{hyd} through the equilibrium constants for the hydration of SiO_2 and B_2O_3 and the equilibrium constants for silicic acid and boric acid dissociation (11). The dominance of the ΔG_{hyd}-$\log(NL_{Si})$ relationship between the ΔG_{hyd}-pH-$\log(NL_{Si})$ variables supports the concept of Grambow (44,45) that dissolution of all silicate-based glasses can be described by the activity diagrams for the dissolution of amorphous silica.

The over 300 glasses analyzed by the MCC-1 test were also divided into two subsets. One subset contained 149 glasses without ferric or ferrous iron and the other subset contained 151 glasses containing iron (11). The iron-free glasses gave a regression fit of:

$$\log(NL_{Si}) = -0.2307\ \Delta G_{hyd} - 0.1004 \qquad R^2 = 0.75 \qquad (6\text{-}17)$$

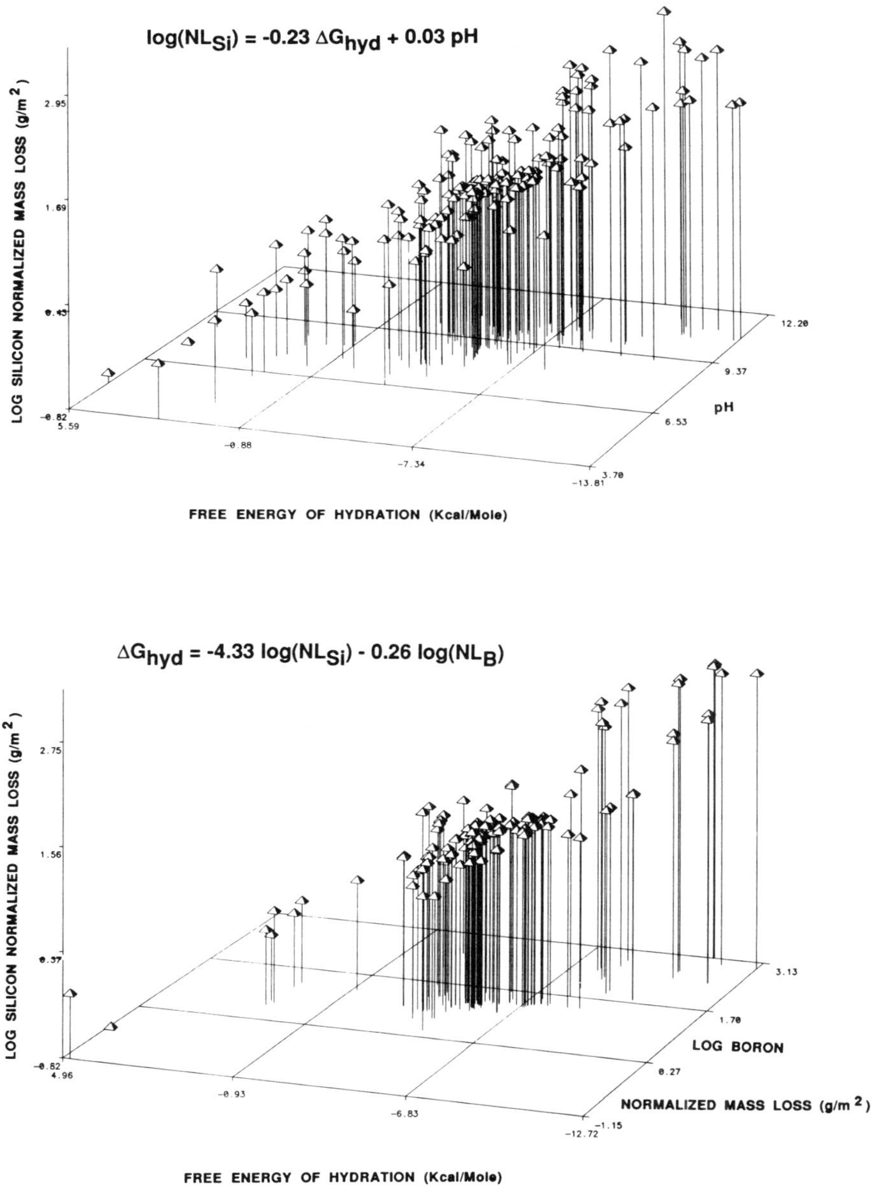

Figure 6-15. Statistically calculated colinearity between the variables ΔG_{hyd}, $\log(NL_{Si})$, and $\log(NL_B)$, and leachate pH.

while the iron-containing glasses gave an equation of best fit of:

$$\log(NL_{Si}) = -0.1620\ \Delta G_{hyd} - 0.2312 \qquad R^2 = 0.75 \qquad (6\text{-}18)$$

When the high alkali and boron-containing glass frits and other glass ceramics were removed from the iron-free glass subset, 76 commercial and ancient glasses remained. The least squares fit for these iron-free glasses was:

$$\log(NL_{Si}) = -0.1815\ \Delta G_{hyd} - 0.3565 \qquad R^2 = 0.84 \qquad (6\text{-}19)$$

The similarity of the slopes of Equations 6-18 and 6-19 demonstrates that the relation between glass durability and glass composition holds for compositionally different glass subsets when the poorly durable high alkali-boron containing frit glasses are eliminated. This may be a function of the poor response of the MCC-1 test for poorly durable glasses due to saturation effects.

Laboratory and In-Situ Examples. Newton and Paul (13) were able to leach poorly durable ground ESF medieval window glasses in water at 25°C and achieve results which were linear with ΔG_{hyd}. The ESF glasses contacted the solution for only 24 hours and the amount of K_2O extracted was measured. A plot of the calculated ΔG_{hyd} and log K_2O in ppm gave a slope of -0.212 (18). The same ESF medieval window glasses were leached by the MCC-1 procedure. A ΔG_{hyd}-$\log(NL_{Si})$ plot for the nine ESF glasses studied yielded a slope of -0.208 (Figure 6-16a), similar to the slope obtained for K_2O release. It is to be remembered that the theoretical slope should be 1/2.303 RT (89) or - 0.733 at 25°C (-0.607 at 90°C) if the leachants are far from saturation with respect to the element being examined. Saturation effects lower the slope from the theoretical slope and depend on the experimental design parameters and whether or not the silica concentrations are normalized.

Newton and Paul (13) also demonstrated that the free energy of hydration appeared to correlate with the logarithm of the loss of thickness of various glasses measured in mm/century. The effects of long-term weathering were simulated by immersing the glasses in water of pH 7 at 25°C. This correlation gave a slope of about -0.289, in agreement with the data shown for the response of ΔG_{hyd} with K_2O in the leachate solutions of the crushed glass tests (17) and with $\log(NL_{Si})$. Analysis of ten of these simulated ESF glasses which were buried in a limestone mound in Ballidon, UK (pH≈9.5) for 5 years (90,91) also correlated with the ΔG_{hyd} (92). Since the pH of the environment was kept constant, the ΔG_{hyd} could be calculated from the glass composition alone or from composition and groundwater pH. The depths of attack were not measured very accurately (91) and a slope of -0.421 (Figure 6-16b) was obtained. When the laboratory release rate, NL_{Si}, was plotted against the in-situ depth of attack for these same glasses, a correlation with a slope of ≈1 was obtained (Figure 6-16c).

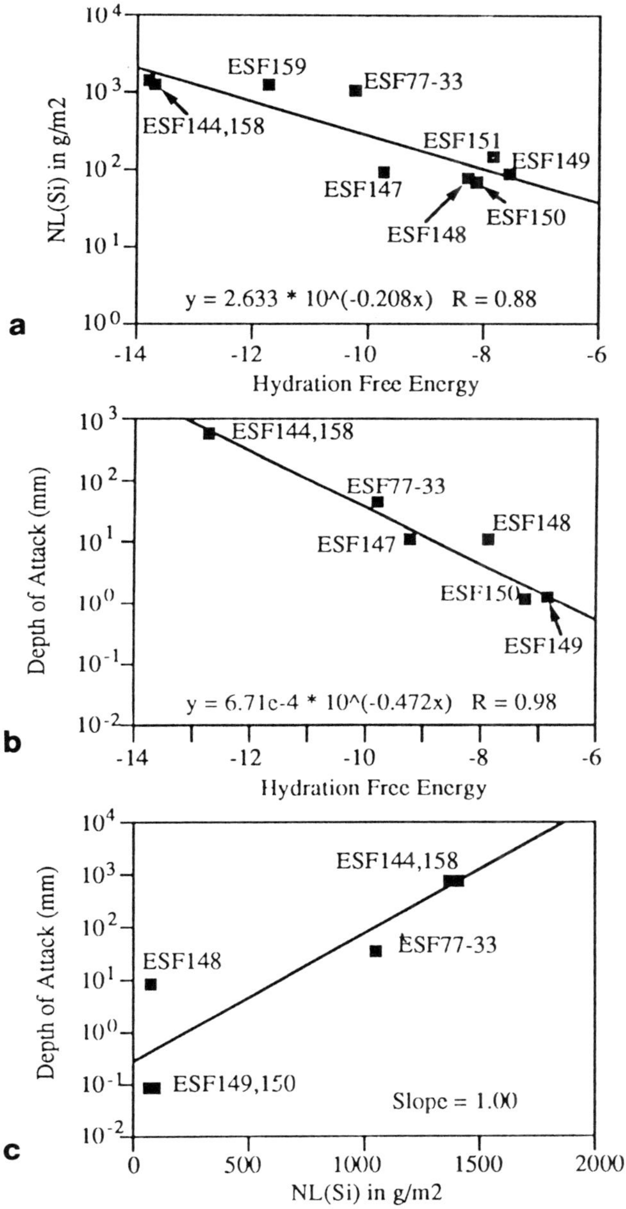

Figure 6-16. (a) Application of hydration thermodynamics to MCC-1 laboratory test data for medieval window glasses, (b) application of hydration thermodynamics to "depth of attack" of the glass surfaces of medieval window glasses after burial in the United Kingdom for 5 years, (c) comparison of the laboratory and in-situ data from Figure 6-16a and b (after Jantzen, Ref. 12)

The relative durabilities of the ESF medieval window glasses predicted from ΔG_{hyd} are the same when plotted against:

(1) K_2O released to solution in a 24 hour laboratory experiment
(2) Si released to solution in a 28 day laboratory experiment
(3) loss of thickness observed in "long-term" laboratory experiments
(3) depth of attack observed in 5 year in-situ burial experiments (12).

More importantly, the relative durabilities of the medieval window glasses predicted by hydration thermodynamics are the ones observed to occur in nature during weathering (13).

The hydration thermodynamic model was also applied to in-situ measured depth of attack for waste glasses which had been exposed to silica saturated groundwaters for two years (93). The waste glasses had been buried in a granitic mine in Stripa, Sweden to simulate burial of waste glass in a granitic repository. The glass monoliths were emplaced in boreholes in the mine which were allowed to fill with the natural silicate groundwater. The glass-groundwater environment in the borehole was kept at 90°C to simulate the heat of radioactive decay. After 2 years burial, the depth of attack was measured accurately by secondary ion mass spectroscopy (SIMS) analysis. The slope of in-situ depth of attack and ΔG_{hyd} for the waste glasses buried in Stripa was -0.270 (Figure 6-17), similar to the slopes given for the elemental release concentrations in the high-purity water experiments and the ESF glass depths of attack.

Variations in Surface Layer Formation

Species may be leached from a glass surface by the two mechanisms represented by Equations 6-1 and 6-2 which are described as dissolution or corrosion. At high pH values, the corrosion mechanism changes from one of primarily ion exchange (Equation 6-1) to one of primarily matrix dissolution (Equation 6-2) as attack on the Si-O bonds is accelerated by increase in the OH- concentration (22).

Glass surfaces have been categorized by Hench and Clark (94) into five groupings defined from surface composition profiles (Figure 6-18). The leached surface categorization is related to the number of Si-O-Si and the Si-O-M (where M is a modifying cation) bonds by IR studies (95,96). It is, therefore, not surprising that these surface categories, labeled I to V, were found (8) to follow a natural progression along the linear relation between the ΔG_{hyd} vs. $\log(NL_{Si})$ plot of Figure 6-13a.

The relation of surface layer type with ΔG_{hyd} is shown in the inset of Figure 6-18. Hench and Clark (94) define a Type I glass surface as having a thin, less than a 50Å, surface hydration layer. Type I glasses are exemplified by pure silica glass. A Type II surface has a silica-rich layer depleted in alkali (typical

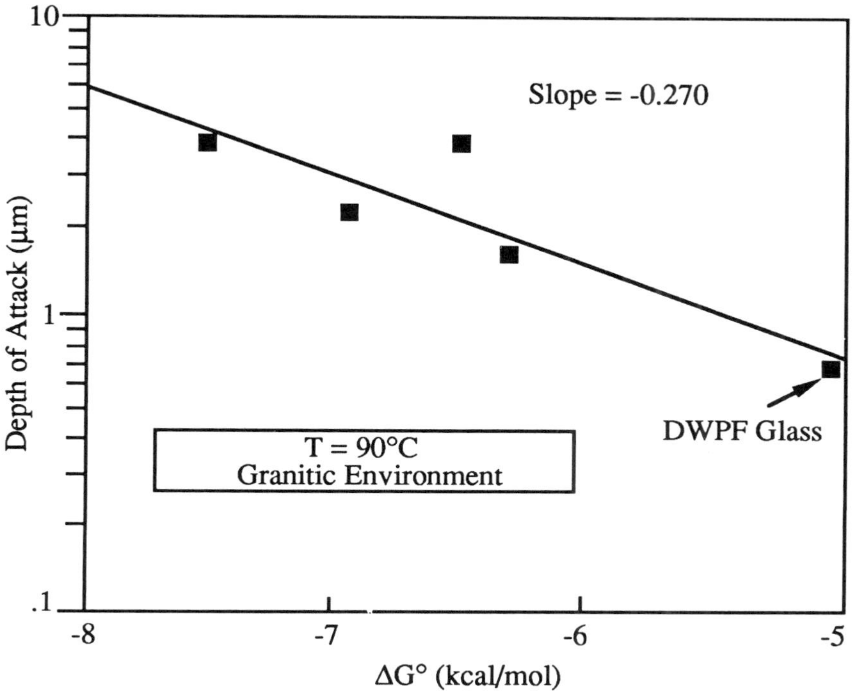

Figure 6-17. Depth of attack of simulated nuclear waste glasses buried in Sweden vs ΔG_{hyd}. The depth of attack was measured by SIMS analysis (after Plodinec, Ref.93).

of commercial container glass). A Type III glass has two layers, one silica-rich and one a surface layer composed of elements such as aluminum or iron leached from the glass. A Type III glass is represented by the aluminosilicates. Type IV glasses also have a silica-rich film but the silica concentration is insufficient to protect the glass from rapid attack. Type V glasses have no layer formation and alkali and silica are lost from the surface at the same rate. Because of the uniform attack, the surface composition of a Type V glass is equivalent to the bulk composition. There is, therefore, little difference in the surface SiO_2 compositional profile between a Type V and a Type I glass (94). However, a glass exhibiting Type V behavior loses considerable amounts of ions into solution relative to a durable Type I glass. Thus, chemical analysis of the corrosion solution is needed to distinguish between Type I and Type V glass surface behavior.

Durable glasses, such as Type I glasses, almost always develop a stable surface film which has a higher concentration of network formers than the bulk composition (94). The Type II glass surfaces form a silica-rich film due to dealkalization and are very durable in solutions of <9 pH. The dual layer formation of Type III glasses can be produced by dealkalization, surface

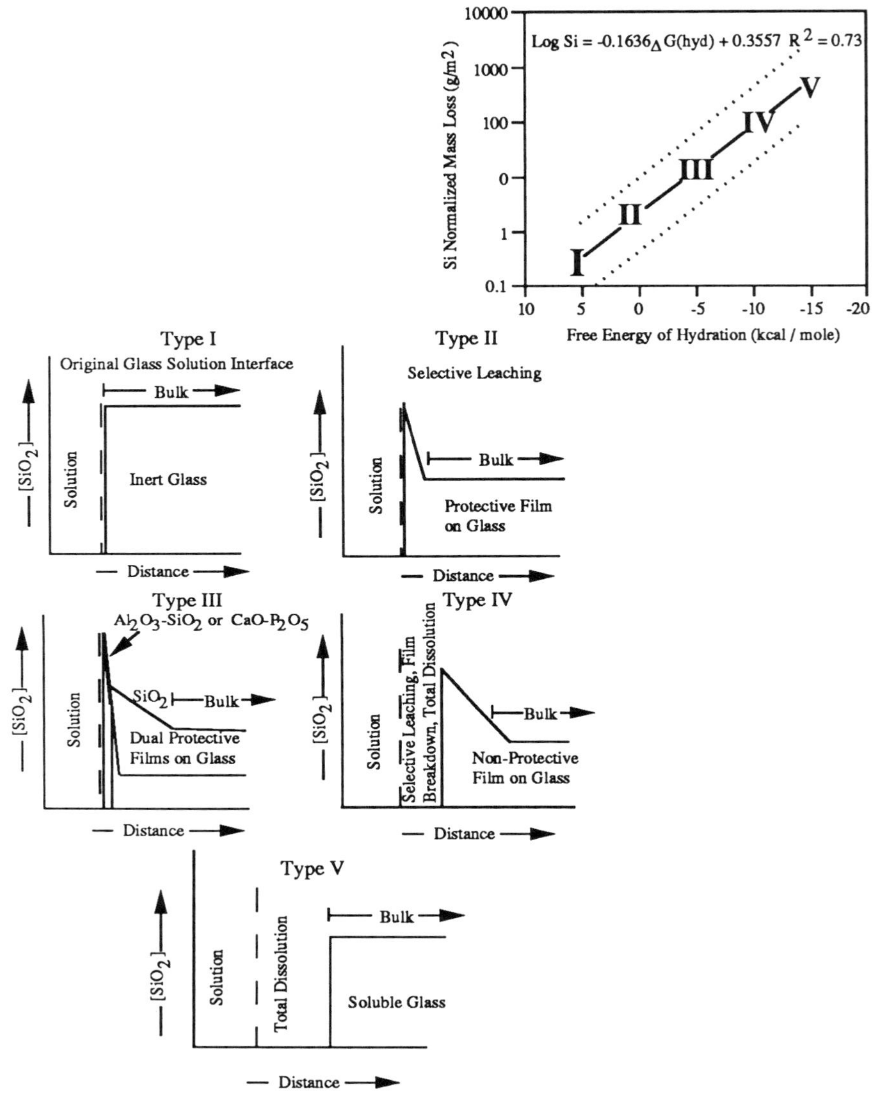

Figure 6-18. Surface layer categorizations of Hench and Clark (after Ref. 94) based on composition profiles and SIMS analysis. Inset shows the relationship of the surface layer types, in numerical sequence, with the calculated ΔG_{hyd}.

structural modifications, and/or precipitation from solution. Nuclear waste glasses have been found to form Type II, III, and IV surfaces depending on the concentration of network formers. Medieval window glasses form Type IV surfaces as well.

The relationship of the free energy to the activities of all the species in a glass conceptually allows the calculation of the ΔG_{hyd} to model the relative roles of amorphous silica dissolution as modified by surface layer formation (11). In this manner, it is conceptually similar to the glass durability modeling of Grambow (23,50,51,89).

Variations in Leaching Solutions/Environments

Many groundwaters are silica-saturated from continuous interaction with silica-rich rocks and/or silica-rich soils. Leaching of nuclear waste glasses in silicate-saturated groundwaters has been found to minimize the dissolution of the glass (97).

In the initial studies (6-11,29) of modeling glass durability (log NL_{Si}) as a function of glass composition (ΔG_{hyd}), leachants of high-purity water were used (see Figures 6-13--6-16). In order to quantify the effects of solution chemistry on glass durability, a "reference set" of $\approx$10 glasses were chosen for study. The glass compositions included pure SiO_2, tektites, basalts, waste glasses, two ESF glasses, and a poorly durable frit glass. The test temperature, test duration (28 days), and SA/V ratio were kept the same as the MCC-1 tests in high-purity water.

The silica-saturated groundwaters were found to minimize the amount of silica released to solution as found in previous experiments (97). For the MCC-1 solution dominated tests the chemistry of the groundwater buffered the final leachate pH; even the poorest glass, the frit glass, did not cause leachate pH excursions.

Since the ΔG_{hyd} defined in the high-purity water experiments is calculated from the glass composition and the solution pH, the linear equation determined is an expression of the response of both glass and solution pH (Figure 6-13b). In the buffered groundwater experiments, the constant pH altered the calculated ΔG_{hyd} term. However, the "buffering capacity" of the groundwater, dominated by the silica solubility, concomitantly lowered the Si released from the glass. The data from the groundwater leachants, therefore, superimposes on the linear equation defined by the high-purity water experiments (Figure 6-19) (12,14).

The successful application of the ΔG_{hyd} model to predictions of glass durability in silica-saturated groundwaters are supported by the data of Grambow (50,51), which suggests that dissolution of silicate and borosilicate glasses can be described by the activity diagrams for the dissolution of amorphous silica. Conditionally, this implies that under static conditions, the performance of a given glass in nature can be predicted by hydration thermodynamic theory if the

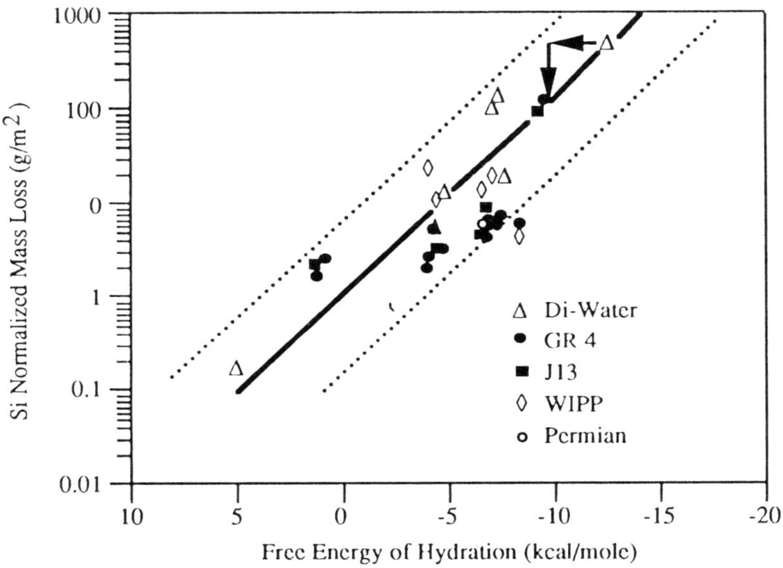

Figure 6-19. Comparison of the hydration thermodynamic model for a reference set of glasses tested by the MCC-1 protocol in varying leachants (after Refs. 12,14).

groundwater pH and the glass composition are known. If the groundwater pH is constant, then only the glass composition need be known in order to determine "relative" glass durabilities in a given environment.

GLASS DISSOLUTION: A FUNCTION OF KINETIC TEST PARAMETERS

The Kinetic (SA/V)•t Parameter

The hydration thermodynamic model furnishes a quantitative frame of reference to understand how the various test parameters given below affect glass durability. By varying glass composition and holding all other parameters constant, the thermodynamic role of glass composition was evaluated (see previous sections). It was also shown that the thermodynamic model was applicable in various types of leachants. Once the relative effects of glass composition and leachant composition are known, the same glass compositions can be used to evaluate the effects of the other parameters: the test variables.

The glass composition was kept constant by choosing a representative "reference set" of about 10 glasses. The test temperature was kept constant at 90°C, but the test duration was varied. Glass monoliths were tested at various test durations by the MCC-1 leach test protocol. Crushed glasses weighing 1.5

grams (same weight of glass used in the MCC-1 monolithic tests) were immersed in 40 cm^3 of high-purity water (ASTM Type I) in a modified MCC-3 leach test protocol. The SA/V ratio was varied by testing glasses crushed to various mesh sizes [200 (74μm diameter), 230 (62μm), 270 (53μm), and 400 (37μm)]. The test durations (t) were varied from 7 to 90 days. The samples were agitated every 48 hours rather than continuously. The solutions were not filtered and the fines were not removed from the crushed glass by an alcohol wash as suggested in the MCC-3 leach test protocol. This is thought to contribute to some of the variability in the data. These glasses are currently being reexamined with the PCT leach procedure to model ΔG_{hyd} with greater precision than possible with the MCC-1 or MCC-3 leach test protocols.

The glass durabilities at the high (SA/V)•t conditions (MCC-3 procedure) were found to follow the same linear trend as the MCC-1 glass monolith tests [low (SA/V)•t] when plotted as a function of ΔG_{hyd} (Figure 6-20a). The amount of silicon released to the solution is a function of the mass fraction of that element in the glass and of the SA/V parameter (Equation 6-16). The SA/V term in the denominator of Equation 6-16 dominates the calculation of NL_i and hence the results of the crushed glass leach tests cluster regardless of the length of time of the test. This also causes the crushed glasses to appear to be more durable than the monolithic samples. However, when the silicon content is plotted in ppm, the increased dissolution of the crushed glasses is readily observed (Figure 6-20b).

The results of the crushed glass tests can be more easily interpreted if the silicon release is plotted against the (SA/V)•t parameter as previously suggested (98,99). The sequence of parabolic curves is plotted against ppm silicon released to solution in Figure 6-21. The silicon released to solution increases in the relative order predicted by the calculated ΔG_{hyd} for these glasses. The same relative durability sequence of the glasses, at any (SA/V)•t can also be observed in Figure 6-20.

The range of saturation values known (37) for amorphous silica at 90°C are overlain on Figure 6-20. Glasses with $\Delta G_{hyd} > -9$ kcal/mole require longer residence times or smaller particles sizes (larger SA) in order to approach steady state saturation with respect to silica, e.g. for the solution concentrations to be close to the saturation values for amorphous silica. For glasses which are poorly durable, the alkali is quickly released to solution and drives the pH to extreme alkaline conditions. These extreme conditions accelerate glass matrix dissolution and hence, the silicon release to solution surpasses the reference saturation level even at small (SA/V)•t. The silica saturation for a glass is, therefore, dependent on glass composition.

The slopes of $\log(NL_{Si})$-ΔG_{hyd} for monolithic tests [low (SA/V)•t] and crushed glass tests [high (SA/V)•t] are the same (9). This occurs because the test geometry alters only the kinetic reactivity parameters, e.g. surface area (SA), leachant volume (V), and the length of time that the glass has contacted the leachant (t). These kinetic reactivity parameters alter the rate at which the saturation ion concentrations are approached.

Measures of reaction progress such as $\log(NL_{Si})$ in g/m^2 or log(Si) in ppm have been shown (Figure 6-21) to be a function of (SA/V)•t for the sequence of "reference glasses" discussed above. The log(Si) versus (SA/V)•t curves for the various glasses increase in the order predicted by their relative ΔG_{hyd} values. Since log(Si) is a function of both (SA/V)•t and ΔG_{hyd}, a plot of over 120 data points collected on the "reference set" of glasses was plotted in three dimensional log(Si)-(SA/V)•t-ΔG_{hyd} space[2]. A scatter plot of the data is shown in Figure 6-22a). The (SA/V)•t axis increases from left to right so that the parabolic curves represented by each glass appear reversed as shown in the inset. A statistically interpolated spline function was fit to the data and the data was found to form a plane in three-dimensional log(Si)-(SA/V)•t-ΔG_{hyd} space (Figure 6-22b). The surface shown in Figure 6-22b was derived from the data in Figure 6-22a with the same axial representations. The plane is fairly flat at the high (SA/V)•t values under steady state conditions (see inset) with the most curvature at the low (SA/V)•t values where the influence of the kinetic parameters dominate. This curvature represents the steep portions of the parabolic curves for the individual glasses. The flatter portion of the curve represents the silica saturation plane as a function of glass composition.

The slopes of the ΔG_{hyd}-$\log(NL_{Si})$ plots for monolithic tests [low (SA/V)•t] and for crushed glass tests [high (SA/V)•t] are similar because they intersect the three-dimensional plane at constant (SA/V)•t (9).

The Kinetic Temperature Parameter

The test temperature can also affect leaching of glass. The effects of temperature are not examined in this study because it is well documented that leaching data can be plotted linearly verses temperature or 1/temperature (31, 100-102, see also Wicks in this book). An apparent rate constant can be determined by fitting the data to an Arrhenius equation (see Equation 6-22). The kinetic temperature parameter can be used to accelerate glass leaching as long as the temperature chosen does not change the leaching mechanism (31, 103,104). In all the studies presented in this chapter, a constant temperature of 90°C was chosen as representative of the heat of radioactive decay that nuclear waste glasses will be subjected to. It is also representative of the higher temperatures used in the ASTM C225 leaching procedure (105) for pharmaceutical and commercial glasses.

---

[2] In this case log (Si) is plotted preferentially to $\log(NL_{Si})$ because $\log(NL_{Si})$ is a function of SA/V•t according to equation 6-16.

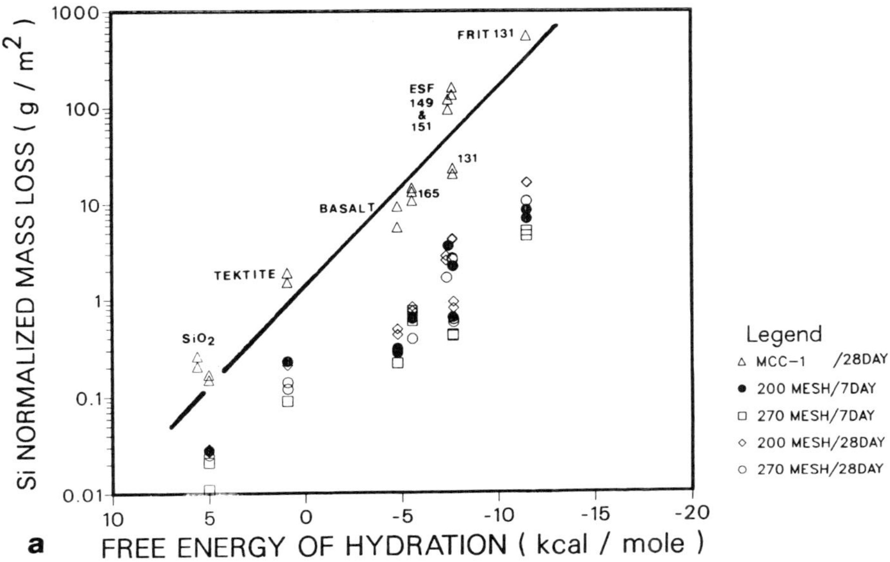

a

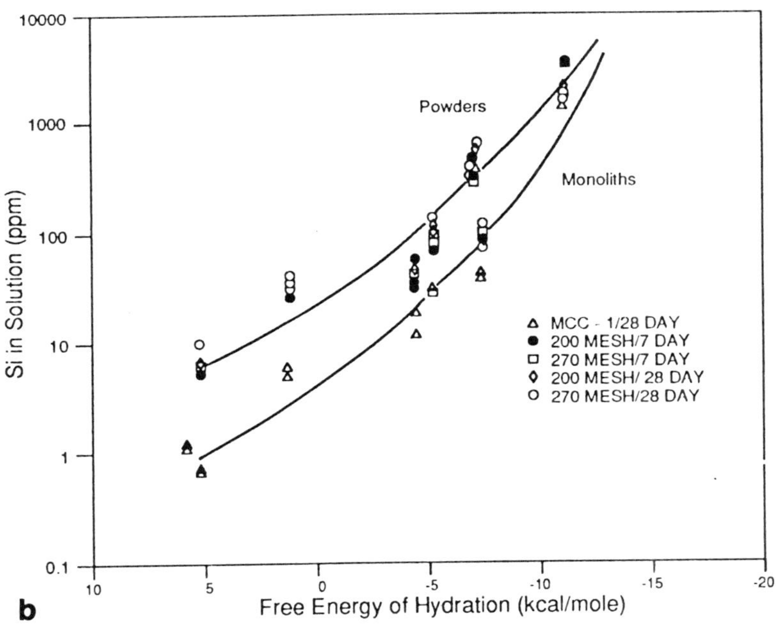

b

Figure 6-20. (a) Free energy of hydration versus NLi_{Si} for a "reference set" of glasses leached as monoliths and also crushed, (b) free energy of hydration versus Si released in ppm for the same data shown in (a).

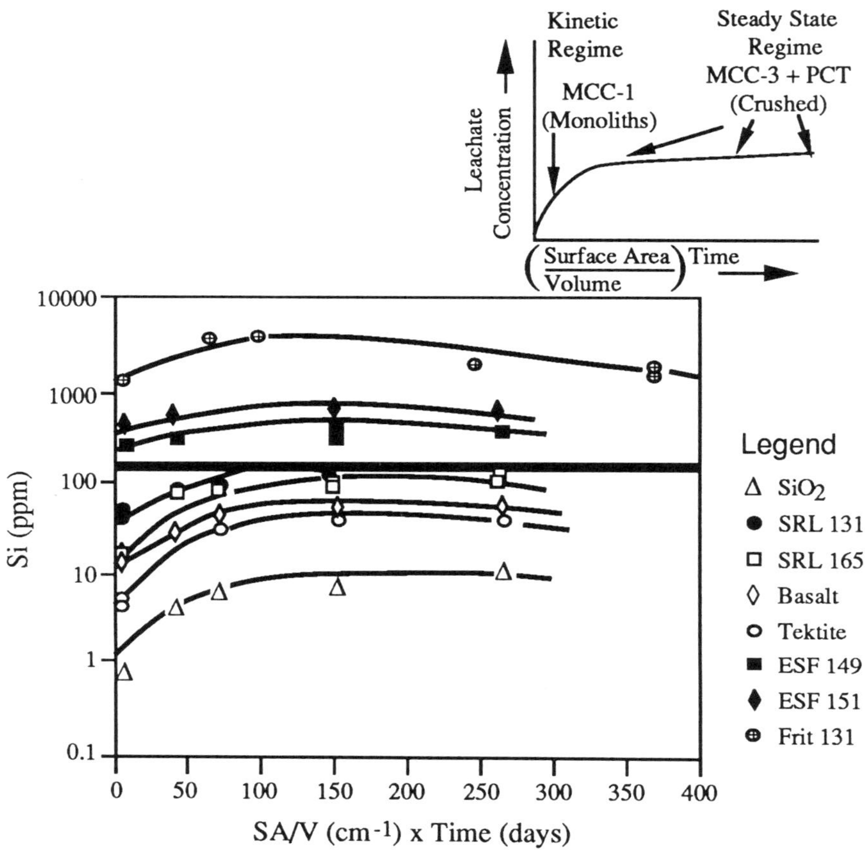

Figure 6-21. (SA/V)•t parameter plotted versus silicon release to solution in ppm for the "reference set" of glasses shown in Figure 6-20. The solid horizontal line is the silica saturation for amorphous SiO_2 at 90°C (after Ref. 37).

Kinetic vs Thermodynamic Contributions

The combined kinetic and thermodynamic contributions to glass durability have been verified experimentally by defining the plane in three dimensional ΔG_{hyd}-concentration-(SA/V)•t space at constant temperature (Figure 6-22b). The flatter portion of the curve represents the silica saturation plane as a function of glass composition. At constant kinetic conditions, e.g. test geometry (SA/V), temperature, and test duration, the three dimensional plane is intersected at constant (SA/V)•t and the ΔG_{hyd}-concentration plots for glass as a function of glass composition have similar slopes. A summary of the mathematical treatment of the combined kinetic and thermodynamic aspects of glass durability which define these interrelationships is given below.

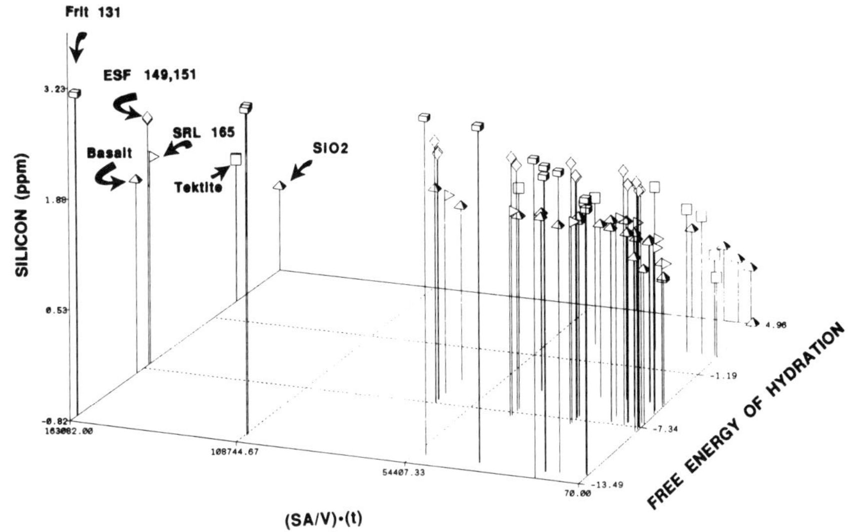

Figure 6-22. (a) Scatter plot and (b) spline function of the three-dimensional relationships between silicon released to solution in ppm, the kinetic (SA/V)•t test parameter, and the thermodynamic ΔG_{hyd} calculated from the glass composition and the solution pH. Inset shows how the plot represents reverse parabolic curves.

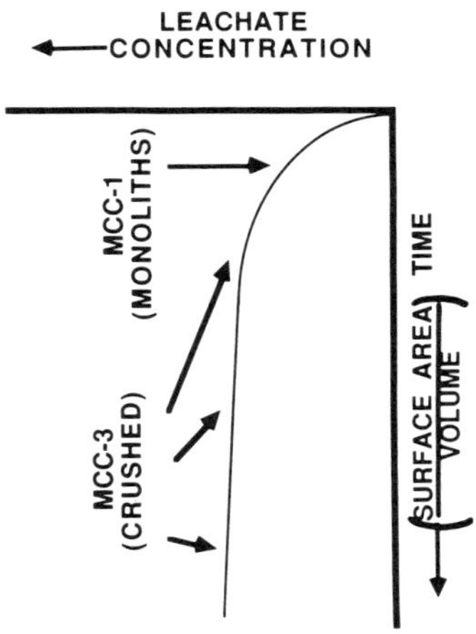

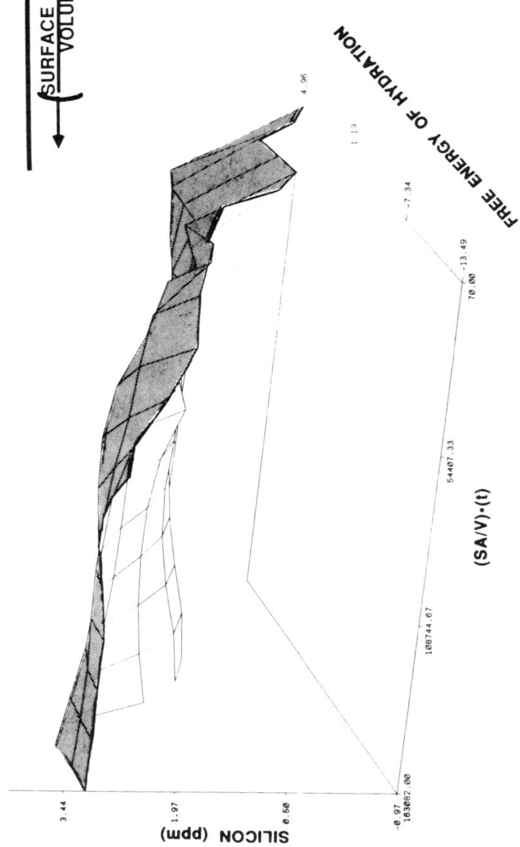

Figure 6-22 continued. (b) spline function

The kinetic dissolution rate of species i in a glass follows a parabolic curve as a function of leaching time (Figure 6-21). The overall dissolution of the glass, expressed as the concentration-dependent release of species i as a function of time, can be expressed (25) as:

$$\frac{dc_i}{dt} = \frac{SA}{V} k_i \tag{6-20}$$

where c_i = concentration of amount leached
t = time
SA = exposed surface area of the solid
V = volume of the leaching solution
k_i = rate of dissolution of species i in solution

The value of k_i can depend on solution pH, temperature, c_i, and the concentration of other species in solution. If the dissolution of glass is treated as congruent, then:

$$\frac{dc_i}{dt} = SA(f_i)(k) \tag{6-21}$$

where f_i = stoichiometric content of i in the glass
k = overall dissolution rate of the glass

In turn k, the dissolution rate of the glass, is affected by the test temperature. The temperature dependence of k follows an Arrhenius equation up to temperatures of ≈120°C (39) and can be expressed (25) as:

$$k = A\, e^{-Eo/RT} \tag{6-22}$$

where E_o = activation energy
T = test temperature (°K)
R = gas constant
A = constant.

If the solutions are undersaturated, and k is dependent primarily on the pH and the test temperature rather than on the concentrations of other components in solution, then the forward rate of dissolution can be expressed (25) as:

$$\frac{dc_i}{dt} = \frac{SA}{V} f_i k^+ (a_{H^+})^n \tag{6-23}$$

where k^+ = forward rate of dissolution $k^+ = A^+ e^{(-E+/RT)}$
n = effect of pH on the rate constant k (a value of ~ 1 in acid regimes and a value of 0.2-0.4 in alkaline regimes)

Static or slow flow leach tests at large SA/V ratios and long time durations permit the accumulation of glass dissolution products in the leachant to become significant, e.g. steady-state and/or equilibrium saturation. As the solution concentration, c_i, approaches equilibrium with the concentration species i in the glass (f_i), precipitation of c_i from solution can occur. In mineral dissolution, a second term is added to Equation 6-23 to represent the precipitation or "back reaction" rate law. This term is negative since it is assumed to represent the reverse mechanism of mineral dissolution (25). This leads to the equation:

$$\frac{dc_i}{dt} = \underbrace{\frac{SA}{V} f_i k^+ (a_{H^+})^n}_{\text{Forward Rate}} - \underbrace{\frac{SA}{V} f_i \frac{Q^m}{K_{eq}^m} k^+ (a_{H^+})^n}_{\text{Precipitation Rate}} \qquad (6\text{-}24)$$

where Q = reaction activity quotient for the precipitating species
m = deviation from equilibrium determined experimentally from rate data near equilibrium

Because

$$\Delta G_r = \Delta G^\circ_r + RT\ln Q = RT \ln Q/K_{eq} \qquad (6\text{-}25)$$

Equation 6-24 can be rewritten as:

$$\frac{dc_i}{t} = \frac{SA}{V} f_i\, k^+ (a_{H+})^n \; (1\text{-}e^{\,m\Delta Gr/RT}) \qquad (6\text{-}26)$$

The general rate law expressed as equations 6-24 and 6-26 enables the extension of laboratory dissolution data and thermodynamic data to a general rate law which is applicable up to and including equilibrium. Equation 6-26 is analogous to expressions derived for mineral dissolution utilizing the progress variable approach and transition state theory (106). In this latter approach, the ΔG_r is replaced by the affinity of the reaction, A, which is a measure of the steady state equilibrium achieved. The affinity of the reaction is equal to the ΔG_r divided by the reaction progress (89). This is because the precipitation rate law depends on the solution composition (25) and in particular on the ion activity product (IAP) and the solubility product (K_L) of the precipitating species.

A detailed mathematical treatment where the effects of the solution composition are considered is given by Grambow (89). The saturation index of the solution (IAP/K_L) is used to approximate the deviation of the solution from equilibrium, e.g. the steady state equilibrium achieved for a given experimental situation. The affinity of the rate limiting reaction (the measure of deviation from equilibrium) is proportional to the IAP/K_{eq} where K_{eq} is the equilibrium constant for the rate limiting reaction.

In Grambow's approach (89), the affinity of the rate limiting reaction, A*, describes the effects of solution saturation on glass dissolution, while the activation energy term describes the influence of temperature. By using simplifying assumptions and integrating the general rate equation (of the type given in Equation 6-26), a first order rate equation of the type shown below can be derived:

$$c_i = c_s + (c_o - c_s)\ e^{-k^+(SA/V)} \tag{6-27}$$

where c_s = saturation concentration of c_i
c_o = initial concentration of ci
k_+ = forward rate of dissolution $k_+ = A^+\ e^{-E^+/RT}$

The overall rate of dissolution of the glass, k, is proportional to (SA/V)(t)(Σ a_i)(k^+/K_{eq}) where a_i is the activity of the species taking place in the reaction. Experimentally, the time-dependent release of an element from a given glass to the leachate was found to be a smooth function of the (SA/V)•t parameter (98,99).

In Grambow's mathematical approach (89, see also Grambow in this book), the influence of glass composition and solution pH are considered to be expressed in the Gibb's free energy term which is related to the affinity and to the saturation concentrations through the activities of the species in solution. In Lasaga's mathematical approach (25), the pH is considered to be a constant because of the buffering capacity of most natural environments (groundwaters). Alternatively, Lasaga considers the activities of the species to be constant at steady state. Since the activity-pH diagrams provide the fundamental correlation between minimum solubility of a species in solution and the free energy (11,16,42), these approaches are reasonable. All the rate equations simplify to the type given in Equation 6-27.

The combined kinetic and thermodynamic contributions to glass durability have been verified experimentally by defining the plane in three dimensional ΔG_{hyd}-concentration-(SA/V)•t space at constant temperature. At constant kinetic conditions, e.g. test geometry (SA/V), temperature, and test duration, the three dimensional plane is intersected at constant (SA/V)•t and the ΔG_{hyd}-concentration plots for glass as a function of glass composition have similar slopes.

GLASS DISSOLUTION: A FUNCTION OF ENVIRONMENTAL CONDITIONS

The effects of aqueous environments on metallurgical corrosion have been described by Pourbaix (15) in an atlas of equilibrium pH-electrochemical potential (E) diagrams. Pourbaix diagrams are also used to assess mineral stability during exposure to the environment, e.g. weathering. Construction of thermodynamically derived Pourbaix diagrams for glass will unify the current understanding of the dissolution behavior of glass as a function of environmental conditions, e.g. chemical environments, natural environments, weathering, and/or groundwater exposure.

Pourbaix diagrams are used in geochemistry and referred to as pH-Eh diagrams. These pH-potential diagrams are used to depict the stability relations among solid phases and to assess mineral (rock) stability during weathering (16) and soil diagenesis. The processes of rock-weathering and soil formation are derived from the effects of dilute, acidic, and oxidizing aqueous environments of the earth's surface upon the mineral species which formed under high-temperature reducing conditions below the earth's surface. These mineral species and volcanic glasses have been in contact with concentrated, alkaline, oxygen-free sub-surface aqueous environments for millions of years without undergoing alteration (15). This is analogous to the corrosion of metals formed under high-temperature reducing conditions and then exposed to the varied environments of the earth's surface. Extending the analogy to the geologic performance of radioactive waste glasses or the environmental performance of commercial glasses, the glasses fabricated under high-temperature oxidizing or reducing conditions interact with dilute acidic oxidizing media (rain, relative humidity) or with dilute alkaline oxidizing media (groundwater).

Quantification of a Pourbaix Diagram for Glass

Construction and compilation of Pourbaix diagrams is based on thermodynamic principles and equilibrium equations which involve changes in pH and/or changes in oxidation potential. Because the thermodynamic approach of Pourbaix is dependent on thermodynamically calculated hydration equilibria, it can be combined with the hydration thermodynamic approach to derive Pourbaix diagrams for glass dissolution.

The statistically-determined relationships between pH-ΔG_{hyd}, pH-log(NL_{Si}), and pH-log(NL_B) based on response of the 300 MCC-1 durability tests discussed above were used to quantify a generic Pourbaix diagram for glass dissolution (10,11). Several regressors were found to be nearly linear combinations of other regressors in the data base, e.g. colinear. The colinearity between the regressors is caused by the fundamental thermodynamic relationships among ΔG_{hyd}, the solution pH, and the concentration of species in solution discussed above. The ΔG_{hyd}, and the pH are highly colinear because the pH and ΔG_{hyd} are related through the Nernst equation (11). In addition, log(NL_{Si}) is related to the pH through the activity-pH relationship. The log(NL_{Si}), log(NL_B), and the pH are highly colinear because of the restriction imposed by congruent dissolution. Each of these solution species is colinear with pH because of the pH dependence of the saturation ion activities. The log(NL_{Si}) and log(NL_B) are related to ΔG_{hyd} through the equilibrium constants for the hydration of SiO_2 and B_2O_3, and the equilibrium constants for silicic acid and boric acid dissociation.

Over the limited pH range (≈5-13 pH) of the deionized water experiments, the statistically determined slopes between pH and the remaining three variables gave (11) the following:

$$pH = -0.489\ \Delta G_{hyd} + 6.12 \tag{6-28}$$

$$\log(NL_{Si}) = +0.429\ pH - 2.73 \tag{6-29}$$

$$\log(NL_{B}) = +0.416\ pH - 2.31 \tag{6-30}$$

Because of the colinearity relationships between pH-log(NL_{Si}) and pH-log(NL_B), a three dimensional Eh-pH-concentration diagram similar to the schematic Eh-pH-activity diagram in Figure 6-2a can be constructed. The concentrations of Si and B in solution have been shown experimentally (85,86) to be almost completely independent of the solution Eh. The Eh independent equilibria will, therefore, parallel the Eh axis as shown schematically in Figure 6-2a. Isopleths of constant solution concentration can be contoured on the Eh-pH plane from the statistically determined slopes given in Equations 6-29 and 6-30.

A quantitative Pourbaix diagram for glass dissolution can, therefore, be developed (Figure 6-23a and 6-23b). Because of the relation of ΔG_{hyd} to Eh and pH through the Nernst equation and the colinearity of ΔG_{hyd} (here ΔG_{hyd} is calculated only from the glass composition) to pH, a similar three dimensional Eh-pH-ΔG_{hyd} diagram can be derived. Since ΔG_{hyd} was theoretically shown (11) to be relatively insensitive to solution Eh, it can also be contoured parallel to the Eh axis on a Pourbaix diagram for glass (Figure 6-23c). The statistically determined ΔG_{hyd}-pH slope given in Equation 6-28 is used to determine the contour interval.

Determination of the Pourbaix Diagram Stability Fields

The data of Adams (107), Wicks (39) and others (108) (Table 6-2) indicates that high silica commercial glasses, and low silica glasses including borosilicate glasses undergo primary network dissolution at pH ≈10. This pH corresponds to the calculated Eh-independent boundaries between $H_2SiO_3/HSiO_3^-$ and $H_3BO_3/H_2BO_3^-$. The pH of the $H_2SiO_3/HSiO_3^-$ boundary is chosen as representative of the alkaline network dissolution boundary for the regime of active corrosion on a Pourbaix diagram for glass (Figure 6-24a). The boundary plotted corresponds to log C = 10^{-4}M H_2SiO_4.

The data of Adams (107) indicates that high-silica commercial glasses do not undergo active corrosion in acidic media as do lower silica content glasses (Table 6-2), including nuclear waste glasses (107,108). The low-silica-containing glasses follow parabolic relationships with solution pH (39,107,108). In order to describe a generalized Pourbaix diagram for these glasses, the Eh-independent

Table 6-2. Regions of Active Corrosion, Passivation, and Immunity for Glass

Solution	Glass	pH	Eh	Observation	Reference
DI Water	Waste	<6	Oxic	Active Corrosion	39
DI Water	Waste	6-8.5	Oxic	Minimal Corrosion Surface Layer Formation	39
DI Water	Waste	8.5-11	Oxic	Active Corrosion Layer Formation	39
pH Buffers	Waste	5-9	Oxic	Minimal Corrosion	40
pH Buffers	Waste	<3	Oxic	No Layer Formation	40
pH 3 Buffer	Waste	3	Oxic	Active Corrosion	40
pH 7 Buffer	Waste	7	Oxic	Some Layer Formation	40
pH 11 Buffer	Waste	11	Oxic	Poor Layer Formation	40
DI Water	SLS*	7-9	Oxic	Minimal Corrosion	108
DI Water	SLS	9-11	Oxic	Active Corrosion	108
DI Water	Waste	7-9	Oxic	Corrosion Controlled by $Fe(OH)_3$	109
DI Water	Waste	7-9	Oxic	Silica Solubility at a Minimum	109
pH Buffers	↑ SiO_2	1-9	Oxic	Minimal Corrosion	107
pH Buffers	↑ SiO_2	9-13	Oxic	Active Corrosion	107
pH Buffers	↓ SiO_2	1-3	Oxic	Active Corrosion	107
pH Buffers	↓ SiO_2	3-9	Oxic	Minimal Corrosion	107
pH Buffers	↓ SiO_2	9-13	Oxic	Active Corrosion	107
DI Water	Waste	?	Slight Anoxic	Surface Layer Enriched in Fe and Si	110
Basaltic Groundwater	Waste	9.75	Anoxic	Layer Formation Retarded Minimal Corrosion, Immunity (?)	85
Basaltic Groundwater	Waste	9.75	Anoxic	Corrosion < Oxic-Immunity (?)	86
Granitic Groundwater	Waste	~8.5	Anoxic	Corrosion < Oxic-Immunity (?)	86
DI Water+Fe^{O}	Waste	8	Anoxic	Layer Formation Retarded Immunity (?)	84

* SLS = Soda-Lime-Silica glass.
↑ = High
↓ = Low

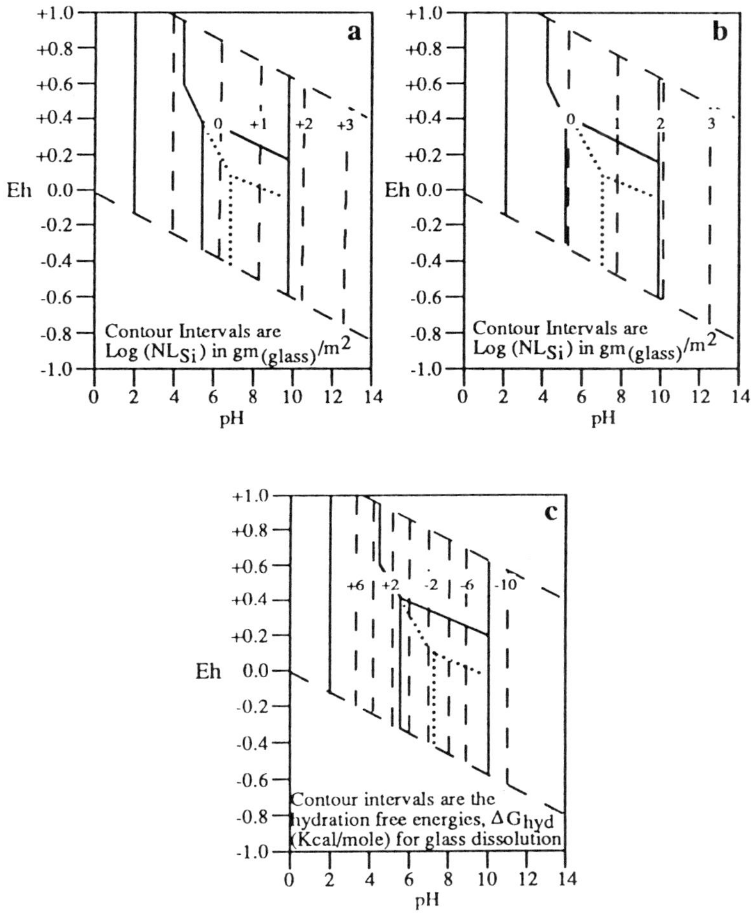

Figure 6-23. Glass dissolution contours based on release of silicon and boron to solution and on the ΔG_{hyd} overlain on the Pourbaix diagrams from Figure 6-24. Contours represent the Eh independent-pH dependent variables for the over 300 glasses studied (see Equations 6-28 to 6-30).

acidic stability field for active glass dissolution is chosen (Figure 6-24a) as the pH of the zero point of charge (ZPC). The zeta potential-pH response for vitreous silica, borosilicate glass, soda-lime-silica glass and soda aluminosilicate glass are all similar (111) and the ZPC for all these glasses is ≈2 (Table 6-2). Horn and Onoda (111) attribute the similarity of the zeta potential-pH response of these compositionally diverse glasses to the dominant role of the silica network and the minimal role of the network modifying cations. Indeed, the

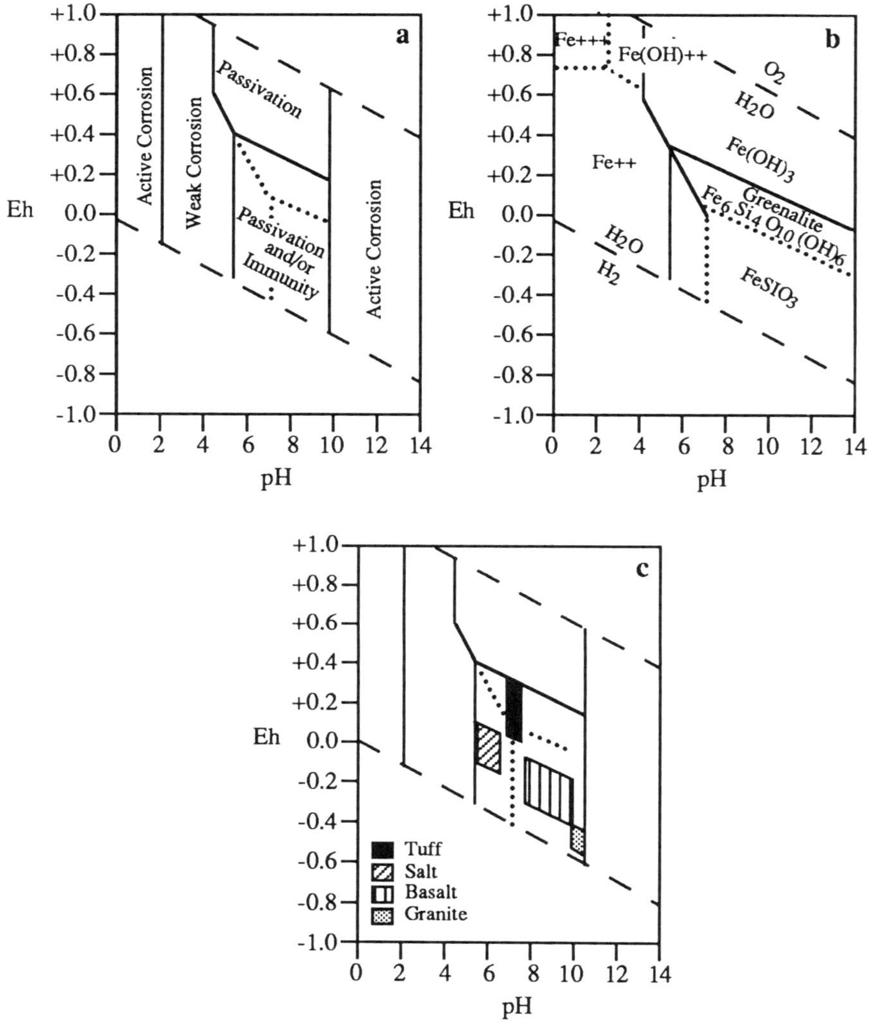

Figure 6-24. (a) The Pourbaix diagram for low silica and borosilicate glasses containing iron, (b) phase stability fields used in determining the Pourbaix diagram, and (c) potential groundwater conditions that nuclear waste glasses would be exposed to (after Refs. 10 and 11).

ZPC of SRP borosilicate nuclear waste glass was measured as exactly 2 (10). Due to the wide range of glass compositions that have a ZPC of ≈2, this pH value is chosen as the approximate boundary for the stability field of active corrosion on the Pourbaix diagram.

At pH values between 2 and 10 in oxidizing environments, hydrolysis of glass occurs by nucleophilic attack. The glass surface charges are negative and surface layer formation occurs by reprecipitation or chemisorption (112) of metal hydroxides (which are positively charged) from solution. For example, $Al(OH)_3$ and $Fe(OH)_3$ are the predominant metal hydroxide species between pH values of ≈3-11 and ≈3-13, respectively (113). Both $Al(OH)_3$ and $Fe(OH)_3$ have been reported (8,39,94,109) as primary components of surface layers observed on leached glasses containing Al_2O_3 and Fe_2O_3. The presence of Al_2O_3 and Fe_2O_3 in glasses is known (113) to increase its durability in the pH range of 4 to 9, indicating that the surface layer formation observed on such glasses is passivating as indicated mechanistically by Wicks (39) and others (94,114,115).

Borosilicate nuclear waste glasses contain aluminum, iron, and other transition metal cations. Iron is the predominant transition metal cation and can be present in both the oxidized and the reduced state. The Eh-independent boundary between Fe^{+++} (as $Fe(OH)^{++}aq$) (16) and $Fe(OH)_3$ at log C=10^{-4}M is proposed as the region of surface layer passivation for glasses containing iron (Figure 6-24a,b). This is based on the predominant role of $Fe(OH)_3$ in the surface layers of nuclear waste glasses (39,109). This boundary occurs at pH values of 3-4 depending on log C and is in agreement with the observations of Wicks (39) that surface layers do not form on nuclear waste glasses at pH values <3. The activity-pH plot derived by Grambow (109) for commercial waste glasses demonstrates that $Fe(OH)_3$ is the dominant species precipitating between pH values of 7 and 9. Similar stability fields can be calculated and plotted on glass-specific Pourbaix diagrams for aluminum (8,32) or aluminosilicate complexes (116), zirconium (117), zinc (109) or any other metal hydroxide species mechanistically found to participate in the formation of surface layers. Glasses such as borosilicate nuclear waste glasses, tektites, basalt, and obsidians contain multivalent metal oxide species such as Fe^{+2}/Fe^{+3} and Mn^{+2}/Mn^{+3} which behave differently in oxidizing and reducing aqueous environments. Metal hydroxides such as $Mn(OH)_3$ and $Fe(OH)_3$ are expected to form from glasses containing Mn_2O_3 or Fe_2O_3 in oxidizing environments or by oxidative dissolution of MnO or FeO in oxidizing environments. The thermodynamic boundary between Fe^{+++} (as $Fe(OH)^{++}$) and $Fe(OH)_3$ has already been defined as representing the Eh-independent portion of the surface passivation boundary. The thermodynamic boundary between Fe^{++} and $Fe(OH)_3$ is, therefore, defined as the Eh-pH dependent extension of the surface passivation boundary for reduced glasses which undergo oxidative dissolution in oxidizing aqueous environments (Figure 6-24b).

Metal hydroxides such as $Mn(OH)_2$ and $Fe(OH)_2$ are stable in reducing aqueous environments (16) and would be expected to form from glasses containing MnO or FeO which are dissolved in anoxic (reducing) solutions. In the presence of silica, Mn^{+2} or Fe^{+2} silicate species are also stable in reducing environments. During dissolution of silicate glass in the presence of Fe^0 and aqueous Fe^{+2}, formation of a hydrated ferrous iron silicate complex, greenalite, was identified coexisting with $Fe(OH)_3$ (84). The identification of this phase was consistent with the measured Eh-pH conditions and coincided with the thermodynamically calculated boundary between a ferrous iron silicate "$FeSiO_3$"

and $Fe(OH)_3$. The formation of an anhydrous ferrous iron silicate complex is predicted by Garrels and Christ (16) in the presence of Fe^{+2} and amorphous silica. However, greenalite is the hydrated analogue, $Fe_6Si_4O_{10}(OH)_8$. This same phase assemblage has been identified as the primary species responsible for the formation of iron ore deposits which form from iron-enriched silica solutions (117). The iron ore silicate solutions are assumed (118) to be about 10^{-4} M H_2SiO_3 because that value represents the concentration of dissolved silica in equilibrium with quartz and clay mineral species (119).

Measured thermodynamic data is not available for greenalite. Klein and Bricker (118) calculated the free energy of formation for this phase to calculate an Eh-pH diagram for iron-ore formation. Using the data of Klein and Bricker (118) for greenalite in equilibrium with H_2SiO_3 at 10^{-4}M, the Eh-dependent thermodynamic boundary between greenalite/Fe(OH) is calculated and shown in Figure 6-24b. ($H_2SiO_3 = 10^{-4}$M is the same concentration used to define the 2SiO3/HSiO3- boundary). The "$FeSiO_3$"/$Fe(OH)_3$ boundary previously calculated by Jantzen (84) in the presence of amorphous SiO_2 is shown for reference.

Because of the estimation of the thermodynamic data for greenalite, the exact boundary can vary between the two calculated boundaries shown on Figure 6-24. The greenalite/$Fe(OH)_3$ boundary is chosen to represent a change in the type of surface layer passivation, e.g. from hydroxide to silicate (Table 6-2) (85,110). However, this may also represent a change in mechanism from surface passivation in oxidizing environments to immunity in reducing environments. If Fe^{+2} in a reduced glass is structurally associated in the glass as an "$FeSiO_3$" component, then in the absence of oxygen there is no driving force for the dissolution of the iron silicate component. Likewise, in the presence of silica-saturated groundwaters, there is no driving force for the dissolution of silica and the ferrous iron silicates would be thermodynamically stable. This has been observed in nature for basalts (composed of ferrous iron silicate minerals and up to 60 volume percent ferrous iron intergranular glass) that have existed for millions of years under reducing groundwater conditions. Laboratory experiments (85) with partially reduced borosilicate glasses in reducing groundwaters have demonstrated the immunity of these glasses under these environmental conditions (Table 6-2).

The example given for the determination of the Pourbaix diagram stability fields (Figure 6-24) was for a borosilicate nuclear waste glass high in iron content. Figure 6-24c defines the anticipated groundwater pH and Eh ranges of the potential geologic repositories in which this waste glass may be buried. By overlapping Figure 6-24c on Figure 6-24a, it is clear that nuclear waste glass dissolution under the anticipated groundwater conditions will be controlled by glass surface layer passivation and/or thermodynamic immunity.

CONCLUSIONS

Hydration thermodynamics has wide applicability to predict the durability of natural, ancient, modern, and nuclear waste glass. The predicted durabilities correlate with those observed in nature and give a means for interpolation of the long-term durability of nuclear waste glasses. Hydration thermodynamics can also be used to determine the relative durability of commercial glasses and for archeological applications. The hydration thermodynamic approach was shown to apply to such measures of reaction progress as releases of Si, B, and K to solution in both short term and long term laboratory tests. The approach was also shown to apply to glass surface "depths of attack" from 2 year and 5 year in-situ burial tests in Sweden and in the United Kingdom.

Hydration thermodynamics furnishes a quantitative frame of reference to explain how the following parameters affect glass durability:

(1) exposed surface area (SA) of the solid
(2) volume (V) of the leaching solution
(3) frequency of replenishing/changing the solution, e.g. time duration (t) of the experiment
(4) temperature (T) of leaching in °C
(5) glass composition

Combined with the thermodynamic concepts of Pourbaix diagram construction, the hydration thermodynamic approach has been shown to predict the response of glass to environmental conditions, e.g. solution pH and oxidation potential.

ACKNOWLEDGEMENTS

I would like to dedicate this chapter to the memory of Amal Paul of the Indian Institute of Technology. His initiatives in the thermodynamic modeling of glass durability inspired this work and the work of several students at various universities in the United States. This application of Paul and Newton's work to the long-term modeling of nuclear waste glass durability is a significant contribution to humanity as well as to science.

Special thanks are due to Robert Brill of the Corning Museum of Glass for providing the synthetic ancient, Chinese, Islamic, medieval, mixed alkali, microtektite, and lunar glasses, as well as actual ancient museum specimens. Robert Weeks of Vanderbilt University donated the Libyan Desert Glass while John Wosinski of Corning Glass Works and John O'Keefe of NASA supplied several of the natural tektites. Theodore Fredericks of A&T Mineral Corporation also generously donated natural tektites, obsidian and nepheline. Roy Newton of the University of Sheffield supplied the synthesized medieval window glasses of the European Science Foundation (ESF).

The clarity of the manuscript was greatly enhanced by James E. Beach of the Savannah River Site. R.L. Postles of the Savannah River Laboratory assisted with the statistical analysis of the data.

The information contained in this article was developed during the course of work under Contract No. DE-AC09-89SR18035 with the U.S. Department of Energy.

REFERENCES

1. U.S. Dept. of Energy, Waste Form Selection for SRP High-Level Waste, Report No. DOE/EA-0179 (1982).
2. Walters, H.V. and Adams, P.B., *J. Non-Cryst. Solids*, 19:183 (1975).
3. Ewing, R.C., *Scientific Basis for Nuclear Waste Management,I*, (G.J. McCarthy, ed.), Plenum Press, NY, p. 57 (1979).
4. Malow, G., and Ewing, R.C., *Scientific Basis for Nuclear Waste Management, III*, (J.G. Moore, ed.), Plenum Press, NY, p. 315 (1981).
5. Malow, G., Lutze, W., and Ewing, R.C., *J. Non-Cryst. Solids*, 67: 305 (1984).
6. Plodinec, M.J., Jantzen, C.M., and Wicks, G.G., *Advances in Ceramics, V.* (G.G. Wicks and W.A. Ross eds.), Am. Ceram. Soc., Columbus, OH, p. 491 (1984).
7. Plodinec, M.J., Jantzen, C.M., and Wicks, G.G., *Scientific Basis for Nuclear Waste Management, VII*, (G.L. McVay, ed.), Elsevier North-Holland, New York, p. 755 (1984).
8. Jantzen, C.M., and Plodinec, M.J., *J. Non-Cryst. Solids*, 67: 207 (1984).
9. Jantzen, C.M., *Advances in Ceramics, V.* 20, (D.E. Clark et al., eds.) Am. Ceram. Soc., Columbus, OH, p. 703 (1986).
10. Jantzen, C.M., *Scientific Basis for Nuclear Waste Management, XI*, M.J. Apted and R.E. Westerman, eds.), Materials Research Society, Pittsburgh, PA, 519 (1988).
11. Jantzen, C.M., Nuclear Waste Glass Durability: I. Predicting Environmental Response from Thermodynamic (Pourbaix) Diagrams, DP-MS-87-2, *J. Am. Ceram. Soc.* (in press).
12. Jantzen, C.M., *Materials Stability and Environmental Degradation*, (I.A. Barkatt, et al., eds), Materials Research Society, Pittsburgh, PA, p. 143
13. Newton, R.G., and Paul, A., *Glass Technology*, 21: 307 (1980).
14. Jantzen, C.M. and Ramsey, W.G., *Scientific Basis for Nuclear Waste Management, XIII*, (V.M. Oversby and P.W. Brown, eds.), Materials Research Society, Pittsburgh, PA (1990). (1988).
15. Pourbaix, M., "Atlas of Electrochemical Equilibria in Aqueous Solutions", Eng. Trans. by J.A. Franklin, NACE, Houston, TX, p. 644 (1974).
16. Garrels, R.M., and Christ, C.L., *Solutions, Minerals, and Equilibria,*

Harper and Row, NY, p. 435 (1965).

17. Paul, A., *Chemistry of Glasses*, Chapman and Hall, NY (1982).
18. Paul, A., *J. Mat. Sci.* 4:12, p. 2246 (1977).
19. Charles, R.J., *J. Appl.Phys.*, 29: 1549 (1958).
20. Iler, R. K., *Colloid Chemistry of Silica and Silicates*, Cornell Univ. Press, Ithaca, NY (1955).
21. Rimstidt, J.D., and Barnes, H.Z., *Geochim. Cosmochim. Acta*, 44: 1683 (1980).
22. Douglas, R.W., and El-Shamy, T.M.M., *J. Am. Ceram. Soc.*, 50: 1 (1967).
23. Grambow, B., *Glastechnische Berichte*, 56: 566 (1983).
24. Wallace, R.M., and Wicks, G.G., *Scientific Basis for Nuclear Waste Management, VI*, (D.G. Brookins ed.), Elsevier North Holland, New York, 23 (1983).
25. Lasaga, A.C., *J. Geophys. Res.*, 89: 4009 (1984).
26. Kittrick, J.A., *ACS Symposium Series*, 93: 401 (1979).
27. Helgeson, H.C., and Mackenzie, F.T., *Deep Sea Res.*, 17:877 (1970).
28. Rimstidt, J.D., *The Kinetics of Silica-Water Reactions*, Unpublished PhD Thesis, Pennsylvania State University, p. 135 (1979).
29. Jantzen, C.M., *Proceedings of the First International Conference on Advances in the Fusion of Glass*, Am. Ceramic Society, Westerville, OH, 24.1 (1988).
30. Hench, L.L., and Clark, D.E., *Surface Properties and Performance Prediction of Alternative Waste Forms*, NUREG/CR-3472, Vol. 2 (1986).
31. Mendel, J.E. (Compiler), *Final Report of the Defense High-Level Waste Leaching Mechanisms Program*, PNL-5157, Battalle Pacific Northwest Laboratories, Richland WA (1984).
32. Jantzen, C.M., Clarke, D.R., Morgan, P.E.D., and Harker, A. *J. Am. Ceram. Soc.*, 65:6, 292 (1982).
33. Bottinga, Y. and Weill, D.F., *Am. J. Sci.*, 272: 438 (1972).
34. Richert, P. and Bottinga, Y., *C.R. Acad. Sci.*, Paris, 295: 1121 (1982).
35. Hormadaly, J., *J. Non-Cryst. Solids*, 79: 311 (1986).
36. Tardy, Y. and Garrels, R.M., *Geochim. Cosmochim. Acta*, 38: 1101 (1974).
37. Morey, G.W., Fournier, R.O. and Rowe, J., *J. Geophys. Res.* 69:1995 (1964).
38. Dilmore, M.F., University of Florida thesis (1977).
39. Wicks, G.G., Mosley, W.C., Whitkop, R.G. and Saturday, K.A., *J. Non-Cryst. Solids*, 49: 413 (1982).
40. Wicks, G.G., O'Rouke, P.E. and Whitkop, P.G., U.S. DOE Report #DP-MS-81-104, E.I. duPont deNemours & Co., Savannah River Laboratory, Aiken, SC (1981).
41. Jantzen, C.M., Schwahn, D., Schelten, J., and Herman, H., *Phys. Chem. Glasses* 22:5, 138 (1981).

42. Cooke, D. and Paul, A., *J. Br. Ceram. Soc.* 77:104 (1978).
43. Krumbein, W.C. and Garrels, R.M, *J. Geology*, 60:1 (1952).
44. Grambow, B. *Adv. in Ceramics, V. 8*, (G.G. Wicks and W.A. Ross, eds.) Amer. Ceram. Soc, Columbus, OH 474 (1984).
45. Grambow, B. and Strachan, D.M. *Scientific Basis for Nuclear Waste Management, VII*, (G.L. McVay, ed.), North Holland, New York, p. 63 (1984).
46. Reimus, P.W., Kuhn, W.L., Peters, R.D., Pulsipher, B.A., PNL-5919 (1986).
47. Macedo, P.B., Barkatt, A., Gibson, B.C. and Montrose, C.J., *Nuclear Technology*, 73, pp 199-209 (1986).
48. Stevels, J.M., *Progress in the Theory of the Physical Properties of Glass*, Elsevier, NY (1948).
49. Stevels, J.M., *Phillips Tech. Rev.*, 22:300 (1960/61).
50. Pettijohn, F.J., *J. Geology*, 49:610 (1941).
51. Loughnan, F.C., *Chemical Weathering of the Silicate Minerals*, Elsevier, NY p. 141 (1969).
52. Newton, R.G., *J. Glass Studies*, 17:161 (1975).
53. Greaves, G.N., *Phil. Mag. B*, 60:6, 793 (1989).
54. Howitt, D.G., *Phys. Chem. Glasses*, 22:29 (1981).
55. White, W.B. and Minser, D.G., *J. Non-Cryst. Solids*, 67:45 (1984).
56. Waff, H.S., *Can. Miner.*, 15:198 (1977).
57. McMillan, P.W., *Glass Ceramics*, Academic Press, New York, p. 229 (1964).
58. Rawson, H., *Inorganic Glass Forming Systems*, Academic Press, New York, p. 317 (1967).
59. Wood, M.I. and Hess, P.C., *Contrib. Miner. Petrol.*, 72:319 (1980).
60. Mysen, B.O., Virgo, D. and Seifert, F. *Carnegie Institute of Washington Year Book*, 80:308 (1981).
61. Goodman, C.H.L., *Glass Technology*, 28:1, 19 (1987).
62. Nelson, C., Furukawa, T. and White, W.B., *Mat. Res. Bull.*, 18, 959-966 (1983).
63. Toop, G.W. and Samis, C.S., *Trans. AIME*, 224:878 (1962).
64. Whittaker, E.J.W. and Muntus, R., *Geochem. Cosmochim. Acta*, 34:945 (1970).
65. Mason, B., *Principles of Geochemistry*, John Wiley & Sons, New York, p 329 (1966).
66. Scholze, H. and Kreidl, N.J. in: *Glass Science and Technology*, Vol. 3, (D.R. Uhlmann and N.J. Kreidl, eds.) Academic Press, Orlando, FL (1986).
67. Eisenman, G., *Biophysical Journal*, 2, 259 (1962).
68. Baes, C.V. Jr. and Mesmer, R.E., *The Hydrolysis of Cations*, John Wiley and Sons (1976).
69. Parks, G.A., *Chem. Rev.*, 65: 177 (1965).

70. Das, C.R., *Glass Industry*, 50:9, 483 (1969).
71. Feng, X., *Composition Effects on Chemical Durability and Viscosity of Nuclear Waste Glasses-Systematic Studies and Structural Thermodynamic Models*, Catholic University of America, unpublished PhD Thesis (1988).
72. Addison, W.E., *Structural Principles in Inorganic Compounds*, John Wiley & Sons, London, p. 183 (1961).
73. Kapustinskii, A.F., *Quart. Reviews* (London), 10:283 (1956).
74. Whittaker, E.J.W., *Geochem Cosmochim Acta*, 31:2275 (1967).
75. Liebau, F., *Structure and Bonding in Crystals, II*, Academic Press, New York, p. 197 (1981).
76. Mendel, J.E. (compiler), *Nuclear Waste Materials Handbook-Waste Form Test Methods*, Prepared by the Materials Characterization Center, Pacific Northwest Laboratories, Richland, WA, U.S. DOE Report DOE/TIC-11400 (1981).
77. Strachan, D.M., Barnes, B.O., and Turcotte, R.P., *Scientific Basis for Nuclear Waste Management, III*, (J.G. Moore, ed.), Plenum Press, New York, p. 347 (1981).
78. Jantzen, C.M. and Bibler, N.E., *Product Consistency Test for DWPF Glass: Part I.* Test Development and Protocol, U.S. DOE Report DPST-87-575, E.I. duPont deNemours & Co., Savannah River Laboratory, Aiken, SC (1987).
79. Bibler, N.E. and Jantzen, C.M., *Waste Management '89*, Vol.1, (R.G. Post,ed.), University of Arizona, Tucson, AZ, 743 (1989).
80. Piepel, G.F., Jones, T.E., Eggett, D.L., and Mellinger, G.B., *Product Consistency Test Round Robin Conducted by the Materials Characterization Center-Summary Report*, U.S. DOE Report PNL-6967, Materials Characterization Center, Battelle Pacific Northwest Laboratories, Richland, WA (1989).
81. Bibler, N.E. and Bates, J.K., *Scientific Basis for Nuclear Waste Management XIII*, (V.M. Oversby and P.W. Brown, eds.), Materials Research Society, Pittsburgh, PA, p. 327 (1990).
82. Jantzen, C.M. and Bibler, N.E., *The Product Consistency Test for the DWPF Wasteform*, U.S. DOE Report WSRC-MS-90-149 for publication in the Proceedings of the 2nd International Seminar on Radioactive Waste Products, Julich, FRG (1990).
83. Jantzen, C.M., *Adv. in Ceramics, V8*, (G.G. Wicks and W.A. Ross, eds.) Amer. Ceram. Soc, Columbus, OH 385 (1984).
84. Jantzen, C.M., *Scientific Basis for Nuclear Waste Management VII*, (G.L. McVay, ed.), Materials Research Society, Pittsburgh, PA, p. 613 (1984).
85. Jantzen, C.M. and Wicks, G.G., *Scientific Basis for Nuclear Waste Management VIII*, (C.M. Jantzen, J.A. Stone, and R.C. Ewing, eds.), Materials Research Society, Pittsburgh, PA, p. 29 (1985).
86. Jantzen, C.M. and Bibler, N.E., *Scientific Basis for Nuclear Waste*

Management IX, (L.O. Werme, ed.), Materials Research Society, Pittsburgh, PA 219 (1985).

87. Drever, J.I., *The Geochemistry of Natural Waters*, Prentice Hall, Englewood Cliffs, NJ (1982).
88. Whitfield, M., *Limnology and Oceanography*, 19: 857 (1974).
89. Grambow, B., *Scientific Basis for Nuclear Waste Management, VIII*, (C.M. Jantzen, J.A. Stone, and R.C. Ewing, eds.), Materials Research Society, Pittsburgh, PA, p. 15 (1985).
90. Newton, R.G., *Glass Technology*, 26:6, 293 (1985).
91. Newton, R.G., *More Results from the Ballidon Glass Burial Experiment, Glass Technology* (in press).
92. Jantzen, C.M. and Newton, R.G., *Application of Hydration Thermodynamics to the Durability of Medieval Window Glasses* (in preparation).
93. Plodinec, M.J., *MRS Bulletin*, XII[5], 61 (1987).
94. Hench, L.L., Clark, D.E., *J. Non-Cryst. Solids*, 28:83 (1978).
95. Hench, L.L., Clark, D.E. and Yen-Bower, E.L., Proceedings, Conference on High-Level Radioactive Solid Waste Forms, NUREG-CP-0005, p. 199 (1970).
96. Hench, L.L., Newton, R.G. and Bernstein, S., *Glass Technology*, 20:144 (1979).
97. Wicks, G.G., Robnett, B.M. and Rankin, W.D., *Scientific Basis for Nuclear Waste Management, V*, (V.W. Lutze, ed.), Elsevier, New York, p. 15 (1982).
98. Oversby, V.M., *Workshop on Leaching Mechanisms of Nuclear Waste Forms*, U.S. DOE Report PNL-4382, Battelle Pacific Northwest Laboratories, Richland, WA, pp 97-129 (1982).
99. Pederson, L.R., Buckwalter, C.Q., McVay, G.L. and Riddle, B.L., *Scientific Basis for Nuclear Waste Management, VI*, (D.G. Brookins, ed.), North Holland, New York, p. 47 (1983).
100. Clark, D.E., Maurer, C.A., Jurgensen, A.R., and Urwongse, L., *Scientific Basis for Nuclear Waste Management, V*, (W. Lutze, ed.), Elsevier, New York, p. 1 (1982).
101. Westsik, J.H. Jr. and Peters R.D., *Scientific Basis for Nuclear Waste Management, III*, (J.G. Moore, ed.), Plenum Press, New York, p. 355 (1981).
102. Barkatt, Aa., Gibson, B.C., Macedo, P.B., Montrose, C.J., Sousanpour, W., Barkatt, A., Boroomand, M., Rogers, V. and Penafiel, M., *Nuclear Technology*, 73:140 (1986).
103. Jantzen, C.M., *Comment on Hydration Aging of Nuclear Waste Glass*, U.S. DOE Report DP-MS-82-102, E.I. duPont deNemours & Co., Savannah River Laboratory, Aiken, SC (1982).
104. Maschmeyer, R.O., *J. Non-Cryst. Solids*, 38, 39:655 (1980).
105. ASTM, *1990 Annual Book of ASTM Standards*, Vol. 15.02, Procedure

C225-85, 65 (1990).

106. Aagaard, P. and Helgeson, H.C., *Am. J. Sci.*, 281:237 (1982).
107. Adams, P.B., *J. Non-Cryst. Solids*, 67:193 (1984).
108. Clark, D.E. and Hench, L.L., *Nucl. Chem. Waste Management*, 2:93 (1981).
109. Grambow, B., *Scientific Basis for Nuclear Waste Management, V*, (W. Lutze, ed.) North-Holland, New York, p. 93 (1982).
110. Manara, A., Lanza, F., Ceccone, G., DellaMea, G., Salvagno, G., *Scientific Basis for Nuclear Waste Management, VIII*, (C. M. Jantzen, J.A. Stone, and R.C. Ewing, eds.), Materials Research Society, Pittsburgh, PA, p. 63 (1985).
111. Horn, J.M. Jr. and Onoda, G. Y. Jr., *J. Am. Ceram. Soc.*, 61:523 (1978).
112. Buckwalter, C.Q. and Pederson, L.R., *J. Am. Ceram. Soc.*, 65:9, 431 (1982).
113. Paul, A. and Zaman, M.S. *J. Mat. Science*, 13:1499 (1978).
114. Macedo, P.B., Barkatt, Aa., Gibson, B.C. and Montrose, C., *J. Nuclear Technology*, 73:199 (1986).
115. Sales, B.C., White, C. W., Begun, G.M., and Boatner, L.A., *J. Non-Cryst Solids*, 67:245 (1984).
116. Barkatt, Aa., Macedo, P.B., Sousanpour, W., Boroomand, M.A., Szoke, P., and Rogers, V.L., U.S. DOE Report PNL-4382, Battelle Pacific Northwest Laboratories, Richland, WA (1982).
117. Das, C.R., *J. Am. Ceram. Soc.*, 64:4, 188 (1981).
118. Klein, C. and Bricker, O.P., *Economic Geology*, 73:1457 (1977).
119. Sillen, L.G. *Advances in Chemistry Series*, No. 67, 57-69 (1967).
120. Geldart, R.W. and Kindle, C.H., *The Effects of Composition on Glass Dissolution Rates: The Application of Four Models to a Data Base*, U.S. DOE Report PNL-6333, Battelle Pacific Northwest Laboratories, Richland, WA (1988).

Appendix A

In order to apply the hydration free energy model, the glasses studied must be well characterized. If the glasses are crystallized, the model may not apply unless the composition of the residual glass matrix is used. The model may not apply to phase separated glasses. The glass composition must be known accurately including the Fe^{2+}/Fe^{3+} ratio and the redox states of any other redox sensitive ions.

The overall free energy of hydration of a glass is calculated as an additive function of the hydration free energies of the silicate and oxide component end members. The formalism is:

$$\Delta G_{hyd} = \Sigma x_i \bullet (\Delta G_{hyd})_i \tag{6A-1}$$

where $(\Delta G_{hyd})_i$ is the free energy change of the thermodynamically most stable hydration reaction of component i at mole fraction x_i. If a stable hydration product is observed on the leached glass surface, and the hydration reaction for this product has a higher free energy of formation then the theoretical $(\Delta G_{hyd})_i$ value given in Table 6A-1, then the free energy of formation for the observed hydration product, $(\Delta G_{hyd})_{obs}$, should preferentially be used.

For example:

Table 6A-1.

Glass Composition		Glass Components	
Oxide Species	(mol%)	Silicate and Oxide Species	(mole fraction of glass = x_i)
Na_2O	14.00	$Na_2O \bullet SiO_2$	0.14
FeO	3.00	$FeO \bullet SiO_2$	0.03
Fe_2O_3	3.00	Fe_2O_3	0.03
CaO	5.00	$CaO \bullet SiO_2$	0.05
SiO_2	75.00	SiO_2	0.50
	100.00		0.75

where the pure SiO_2 contribution is depleted by the amounts of SiO_2 associated with the inosilicate groups. Note that the mole fraction of glass in terms of orthosilicate and residual oxide species is 0.75. This allows for the calculation of ΔG_{hyd} per mole of glass (13). If the mole fraction of glass is renormalized (120), the calculation incorrectly gives the ΔG_{hyd} per mole of components. To complete the calculation the mole fractions of the inosilicate and residual oxide species are multiplied by the values given in Table 6A-1. For example:

Table 6A-2.

Silicate and Oxide Species	(mole fraction of glass = x_i)	$(\Delta G_{hyd})_i$ (from Table 6-1)	$\Sigma(\Delta G_{hyd})_i$ (kcal/mol of glass)
$Na_2O \bullet SiO_2$	0.14	-28.815	-4.034
$FeO \bullet SiO_2$	0.03	-14.609	-0.438
Fe_2O_3	0.03	+15.50	+0.465
$CaO \bullet SiO_2$	0.05	-16.116	-0.806
SiO_2	0.53	+ 5.59	+2.963
		ΔG_{hyd} =	-1.850

The thermodynamic approach assumes that the silicate and borate components of a glass hydrate to silicic and boric acid. At pH values of greater than 9.5, the solubility of silica and borate increases rapidly due to dissociation. An additional contribution to the hydration free energy must be calculated from the dissociation constants of silicic and boric acid at high pH (8,9). The calculated ΔG_{hyd} should be adjusted for the leachate pH by the following:

for Si

$$\Delta(\Delta G_{hyd}) = 1.364[(-\log(1 + \frac{10^{-10}}{10^{-pH}} + \frac{10^{-21.994}}{10^{-2pH}})] \tag{6A-2a}$$

and for B

$$\Delta(\Delta G_{hyd}) = 1.364[-\log(1 + \frac{10^{-9.18}}{10^{-pH}} + \frac{10^{-21.89}}{10^{-2pH}} + \frac{10^{-35.69}}{10^{-3pH}})] \tag{6A-2b}$$

Equation 6A-2a applies for all silicate-based glasses including borosilicate glasses while Equation 6A-2b applies only to boron-containing glasses. At pH values <7 the contribution to the ΔG_{hyd} term from equation 6A-2a is minimal. At pH values <9 the contribution to the ΔG_{hyd} term from equation 6A-2b is minimal. The adjusted free energy term was statistically shown to be more highly correlated with glass durability than the uncorrected hydration free energy for over 300 glasses tested (9). The adjusted ΔG_{hyd} term should, therefore, be used to relate glass composition and leachate pH to glass durability.

7

Nuclear Waste Glasses: Corrosion Behavior and Field Tests

G.G. Wicks

Waste and Environmental Remediation Programs
Westinghouse Savannah River Company
Aiken, South Carolina

INTRODUCTION

Every major country involved with long-term management of high-level radioactive (HLW) waste has either considered or selected glass as the matrix of choice for immobilizing and ultimately, disposing of potentially hazardous, high-level radioactive material. There are many reasons why glass is preferred. Among the most important considerations is the ability of glass to accommodate and immobilize the many different types of radionuclides present in waste, and make them into a highly durable, integrated, solid waste form. The waste glass product produced not only has outstanding technical properties but also possesses excellent processing features, which allow the glass to be produced with relative ease even under difficult remote-handling conditions, necessary for processing highly radioactive material. The single most important property of the waste glass is its outstanding chemical durability and its corrosion resistance to a wide range of environmental conditions. The following treatise will provide an introduction into the field of HLW management and emphasize the behavior and performance of nuclear waste glass systems studied under a variety of conditions.

BACKGROUND-HIGH LEVEL WASTE (HLW)

High Level Waste Inventory

Located in the United States today, are approximately 100 million gallons of high-level radioactive waste (HLW) containing over 1 billion curies of radioactivity (1). Most of this waste has been generated from defense programs and results from the reprocessing of spent fuels used in the production of tritium and plutonium, both of which are used for military applications. Most of the defense waste is stored as a liquid or semi-liquid in large underground tanks at the Hanford Site in Richland, Washington, and at the Savannah River Site (SRS) in Aiken, South Carolina. While the largest volume of this waste is located at Hanford, the largest amount of radioactivity within this waste is contained at Savannah River. In addition to these inventories, a smaller quantity of defense waste has also been generated from the reprocessing of fuels used in naval reactors, and is stored in stainless steel bins as calcine at the Idaho Chemical Processing Plant in Idaho Falls (1,2).

In addition to the large amounts of defense waste currently on hand, another potential source of HLW is from reprocessing associated with nuclear power plants used to produce electricity. While reprocessing is in progress in other countries, commercial reprocessing was discontinued in the United States in 1971. Up to that time, 612,000 gallons of HLW were generated, which is currently being stored at West Valley, New York. Existing inventories of defense and commercial HLW in the United States and their locations are summarized schematically in Figure 7-1 (1,2). These HLW inventories are in addition to a growing inventory of spent fuel.

High Level Waste Characteristics

High level waste (HLW) can consist of three major components; *sludge, supernate, and salt cake.* The sludge is the most important component of the waste in that it contains most of the radioactivity including fission products and long lived actinides. Sludge comprises about 10% by volume of the entire waste inventory and settles to the bottom of the underground storage tanks as a thick, gelatinous precipitate. It consists mainly of aluminum, iron, manganese oxides and hydroxides and contains actinides, fission products and Sr-90. The supernate makes up the remaining 90% of the waste and consists mostly of sodium nitrate and sodium nitrite along with Cs-137. In order to reduce the total amount of waste to be stored, the supernate is concentrated by evaporation and returned to the cooled tanks where salt cake crystallizes out of solution. Waste tanks along with salt and sludge components are shown in Figure 7-2 (3).

High Level Waste Disposal Strategy

Since the Manhattan Project of World War II, HLW in the U.S. has been stored mainly in underground storage tanks in a liquid slurry form. Many of these original tanks are now reaching the ends of their projected lifetimes and as a result, some leaks have developed. Even though no injuries from radiation have occurred as a result of these practices, it should be clear, that a new means to more effectively and more permanently isolate the waste is needed. The strategy for long-term management of HLW will change from a program

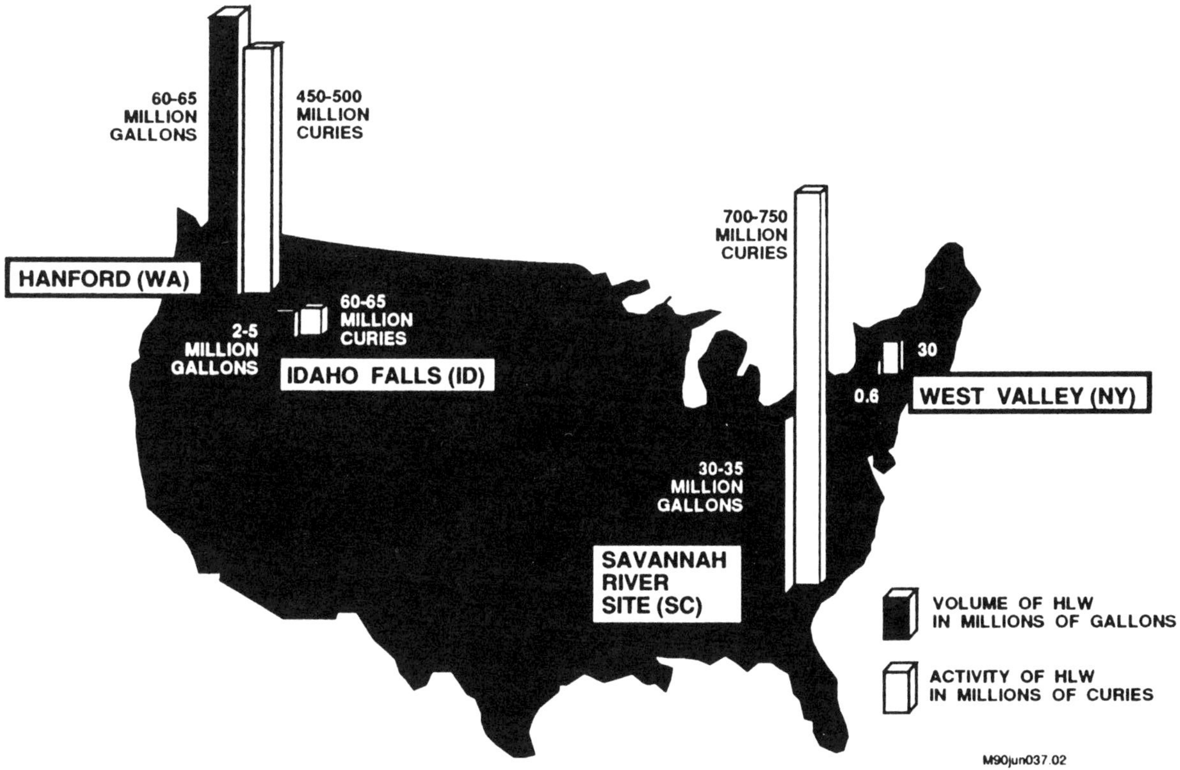

Figure 7-1. Volumes and activities of high level radioactive waste in the United States.

Waste Tanks Under Construction at SRS

Sludge

Supernate/Saltcake

Figure 7-2. High level radioactive waste.

developed in the 1940's of temporarily storing the waste in a relatively mobile liquid form, to a policy of the 1990's of immobilizing the waste into a solid glass product and permanently disposing of the waste glass forms in deep geologic burial, as a part of a multibarrier waste isolation system.

In Figure 7-3 the waste disposal strategy is summarized and in Figure 7-4 a multibarrier isolation system is shown schematically. At the center of this disposal system is the durable, high integrity waste glass product, followed by the casting canister (304L stainless steel for the first HLW to be vitrified in the U.S.), followed by an engineered barrier tailored for the geologic conditions of the repository (may include an overpack and possible backfill), and finally, the geologic barrier. Geologic formations studied include salt, basalt, shale, clay, granite, and tuff, with the tuff site at Yucca Mountain in Nevada being selected for a more detailed study. Each of the elements of the multibarrier isolation system is designed to inhibit the release of radionuclides and prevent them from harmfully entering the accessible environment and consequently, allow the waste products to meet federal release rate criteria. Among the most important regulations relating to waste form and waste package performances are 10 CFR 60 (Nuclear Regulatory Commission (4) and 40 CFR 191 (Environmental Protection Agency (5)). A time schedule for permanently disposing of HLW in the United States has been defined by the Nuclear Waste Policy Act of 1982 (6) and subsequent legislation. This provides the basis for both meeting and solving the problem of long-term management of HLW in our generation.

Unlike other types of hazardous waste, high level radioactive waste actually becomes less hazardous with time, due to decay of radionuclides. For example, about 99% of the radionuclides and daughter products in SRS HLW decay with half-lives ranging from microseconds to less than 30 years (Table 7-1). After a time period of about 1,000 years, the activity of the waste decreases 4 to 5 orders of magnitude and for longer time periods, the activity actually approaches and eventually becomes less, than natural uranium, from which the waste was originally generated. These effects are shown in Figure 7-5 (7).

ADVANTAGES OF GLASS

Many different options and potential waste forms were evaluated to determine the best system and strategy for safety and effectively disposing of HLW (8-14). Borosilicate glass was selected over other waste form alternatives which included calcines and supercalcines, cements, titanates, cermets, Synroc as well as other types of glass systems for immobilization of the first HLW to be treated in the United States, the waste at SRS. This decision was made based upon data and information supplied by more than three decades of national and international research and by reviews supplied by independent committees. Among the peer review committees used in this process, were groups sponsored by the National Academy of Sciences (15), the American Physical Society (16),

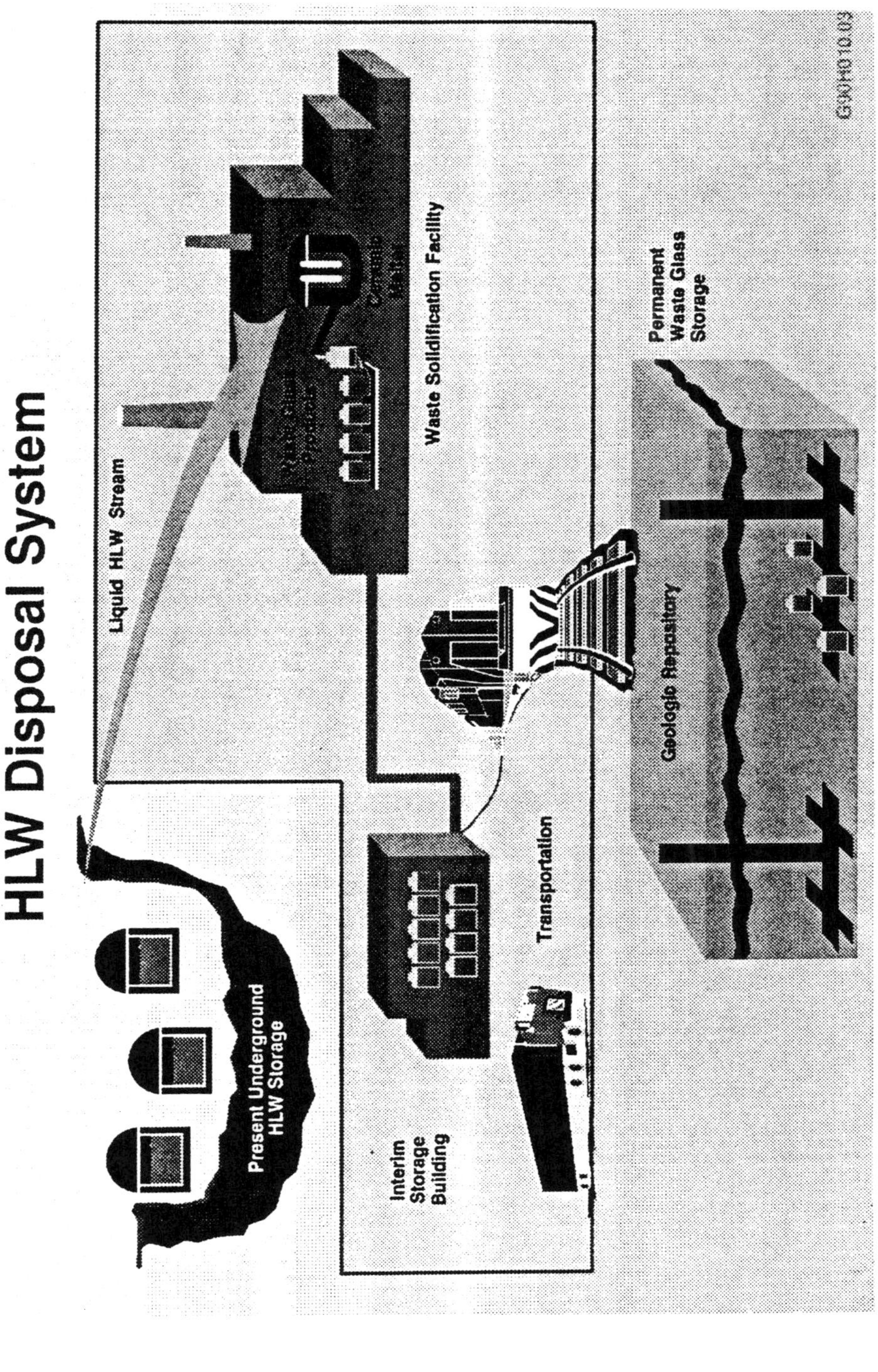

Figure 7-3. HLW disposal system.

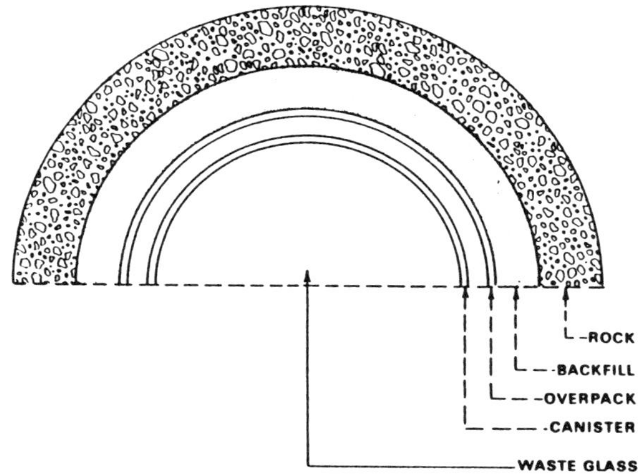

Figure 7-4. Schematic of multibarrier package.

and the U.S. Department of Energy (17-19), who all concluded that borosilicate glass would be a good material for immobilizing the SRS HLW. The selection of borosilicate glass was based on characteristics in two categories; a) *processing considerations* and b) *technical performance*. There were five areas of technical performance evaluated; 1) flexibility/ability to immobilize all waste components 2) mechanical integrity 3) thermal stability 4) radiation stability and 5) chemical durability. Chemical durability was considered the most important property and weighted most heavily during the assessment. Following is a brief description of waste glass performance in each of these categories (3,20).

Processing Considerations

Vitrification Facilities & Practical Operating Experience. The only HLW immobilization facilities in the world are glass-making operations. These include successful production facilities at Marcoule and LaHague in France, as well as a plant located at Mol in Belgium. The first French plant began operations in 1978 and the combined production of both of the French facilities has been more than 630 tons of product containing approximately 150 million curies of radioactivity. There are also additional plants either in production, under construction or being planned in other countries such as Japan, Germany, India, the United Kingdom, and the United States. Following is a brief description of the vitrification process and facility at SRS, which will immobilize the first HLW in the United States.

SRS Vitrification Process. An important feature of the long-term management strategy of SRS HLW is to first separate the waste into radioactive

Table 7-1. Half Lives of Waste Radionuclides & Daughter Products

Nuclide	Half Life	Nuclide	Half Life	Nuclide	Half Life
Ac-225	10 d	Nb-95M	3.61 d	Rb-87	4.89E10 a
Ac-227	21.773 a	Nd-144	2.1E15 a	Rh-103M	56.12 m
Ac-228	6.13 h	Nd-147	10.99 d	Rh-106	29.8 s
Ag-110	24.6 s	Ni-59	7.6E4 a	Rn-219	3.96 s
Ag-110M	249.8 d	Ni-63	100 a	Rn-220	55.6 s
Am-241	432 a	Np-236	1.2E5 a	Rn-222	3.823 d
Am-242	16.01 h	Np-237	2.14E6 a	Ru-103	39.24 d
Am-242M	141 a	Np-238	2.117 d	Ru-106	372.6 d
Am-243	7.37E3 a	Np-239	2.35 d	Sb-124	60.2 d
At-217	32.3 ms	Np-240	61.9 m	Sb-125	2.76 a
Ba-136M	0.306 s	Np-240M	7.22 m	Sb-126	12.4 d
Ba-137M	2.552 m	Po-210	138.38 d	Sb-126M	11 s
Ba-140	12.76 d	Po-211	5.16E-1 s	Se-79	6.5E4 a
Bi-210	3.0E6 a	Po-212	2.98E-7 s	Sm-147	1.06E11 a
Bi-211	2.14 m	Po-213	4.2E-6 s	Sm-148	7.0E15 a
Bi-212	60.6 m	Po-214	163.7 μs	Sm-149	1.0E16 a
Bi-213	45.6 m	Po-215	1.78 ms	Sm-151	90 a
Bi-214	19.8 m	Po-216	0.15 s	Sn-121m	55 a
Co-60	5.272 a	Po-218	3.11 m	Sn-123	129.2 d
Cd-113	9.0E15 a	Pa-231	3.28E4 a	Sn-126	1.0E5 a
Cd-115M	44.6 d	Pa-233	27 d	Sr-89	50.52 d
Ce-141	32.5 d	Pa-234	6.7 h	Sr-90	29 a
Ce-142	5.0E16 a	Pa-234M	1.17 m	Tb-160	72.4 d
Ce-144	284.4 d	Pb-209	3.25 h	Tc-99	2.13E5 a
Cm-242	163 d	Pb-210	22.3 a	Te-125M	58 d
Cm-243	28.5 a	Pb-211	36.1 m	Te-127	9.5 h
Cm-244	18.11 a	Pb-212	10.64 h	Te-127M	109 d
Cm-245	8.5E3 a	Pb-214	26.8 m	Te-129	69.5 m
Cm-246	4.78E3 a	Pd-107	6.5E6 a	Te-129M	33.4 d
Cm-247	1.58E7 a	Pm-147	2.623 a	Th-227	1.87E1 d
Cm-248	3.4E5 a	Pm-148	5.37 d	Th-228	1.913 a
Cr-51	27.7 d	Pm-148M	41.3 d	Th-229	7.3E3 a
Cs-134	2.065 a	Pr-143	13.58 d	Th-230	7.54E4 a
Cs-135	3.0E6 a	Pr-144	17.28 m	Th-231	25.52 h
Cs-136	13.1 d	Pr-144M	7.2 m	Th-232	1.4E10 a
Cs-137	30.17 a	Pu-236	2.85 a	Th-234	24.1 d
Eu-152	13.4 a	Pu-237	45.12 d	Tl-207	4.77 m
Eu-154	8.5 a	Pu-238	87.74 a	Tl-208	3.053 m
Eu-155	4.73 a	Pu-239	2.41E4 a	Tl-209	2.2 m
Eu-156	15.2 d	Pu-240	6.56E3 a	U-232	70 a
Fr-221	4.8 m	Pu-241	14.35 a	U-233	1.59E5 a
Fr-223	22 m	Pu-242	3.76E5 a	U-234	2.45E5 a
Gd-152	1.1E14 a	Pu-243	4.956 h	U-235	7.04E8 a
I-129	1.6E7 a	Pu-244	8.2E7 a	U-236	2.34E7 a
In-115	4.4E14 a	Ra-223	11.434 d	U-238	4.47E9 a
La-140	40.28 h	Ra-224	3.66 d	U-240	14.1 h
Nb-93M	13.6 a	Ra-225	14.8 d	Y-90	64 h
Nb-94	2.0E4 a	Ra-226	1600 a	Y-91	58.5 d
Nb-95	34.98 d	Ra-228	5.76 a	Zr-93	1.5E6 a
				Zr-95	64.03 d

μs: microseconds
ms: milliseconds
s: seconds
m: minutes
h: hours
d: days
a: years

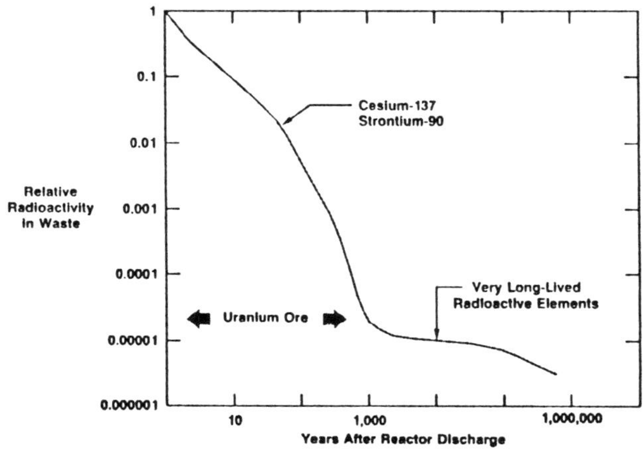

Figure 7-5. Decay of radioactivity in waste.

and non-radioactive parts. Since most of the waste is non-radioactive, this allows the large bulk of the waste to be handled by less expensive and easier processing and disposal techniques. After a series of processing steps, including an in-tank precipitation process which removes radionuclides from the supernate, the decontaminated portion of the waste is segregated and ultimately mixed with a cement-based material to produce a waste product called saltstone, that will be buried in engineered vaults on site (21,22). This disposal strategy is designed to produce a durable waste form and subsequent system that allows groundwater at the disposal location to even meet EPA drinking water standards.

Only the highly radioactive part of the waste will be sent to a vitrification facility for remote processing into borosilicate glass. After processing operations which concentrate radionuclides and produce a waste stream more amenable to subsequent operations (23,24), glass frit is then added to the HLW stream. The mixture will then be liquid-fed into a joule-heated ceramic melter in the vitrification facility. Here the waste mixture will be melted at a temperature of 1150°C using a unique off-gas system that is able to remove 99.999999% of the cesium in the original waste. The molten glass will then be discharged from the melter and cast into stainless steel canisters, 2-ft. in diameter and approximately 10-ft. in height. The units will then be decontaminated by a frit blasting technique (which then recycles the used frit back to the front end of the process to produce subsequent waste glass so no additional waste is generated) and finally, welded shut using resistance upset welding. This process and the products produced are described in more detail elsewhere (20,25-28).

DWPF. The Defense Waste Processing Facility (DWPF) represents the first vitrification plant for immobilizing HLW to be constructed in the U.S. (26-36).

The facility which contains 5 million cubic feet of volume and represents a $920 million dollar investment, is currently under construction at the Savannah River Site and is more than 99% complete. The DWPF, shown in Figure 7-6 will be the largest plant of its type in the world. The design of the facility is based on the 35 years of successful operating experience of reprocessing plants on site. This unique glass making facility is scheduled to begin "cold" operations to "check out" equipment and operating procedures late in 1990 and will run for several years before processing actual radioactive waste, scheduled for the year 1992. Additional facilities are also planned at other HLW sites in the U.S.

Technical Performance

Flexibility. Glass has demonstrated the ability to accommodate not only the 40 or more different elements that are found in waste streams, but also large variations in waste composition (37,28). The reason for this feature is a result of the relatively open random network structure that characterizes glass systems and its ability to accommodate elements or radionuclides of different sizes, charge and characteristics, as well as differing amounts of these constituents (39,40). In Table 7-2, the many different elements found in SRS waste and frit are shown and includes most of the periodic table, and in Table 7-3, glass frits used for immobilizing the waste and component characteristics are summarized (3,41).

Thermal Stability. Waste glass products possess good thermal stability. Upon cooling from the melt or from self-heating due to radionuclide decay, waste glasses can phase separate or crystallize (42). Many different studies have been performed to assess the effect of these processes on performance of the glass (43-45). In one series of studies, Savannah River waste glass was purposely devitrified, even though extensive devitrification is not expected for this system. In Figure 7-7, the resulting time-temperature-transformation (TTT) curve is shown (45). The phases formed were identified and leaching tests showed that even in this "worst-case" scenario, the effects of the crystalline phases on chemical durability were not significant.

Mechanical Integrity. Waste glass products also possess good mechanical integrity. Cracking can occur due to stresses induced during fabrication or from accidents during handling, transportation or storage operations. Various mechanical tests have been performed, ranging from laboratory-scale studies to drop tests of full-size canisters containing simulated waste glass product (46-50). In Figure 7-8, a drop test performed by Battelle Pacific Northwest Laboratory on SRS waste canisters filled with simulated waste glass is shown (46). Based on these data, several important observations can be made. First, glass forms fracture into relatively large chunks inside canisters and fracture is generally localized to the area of impact. Second, the amount of increased surface area

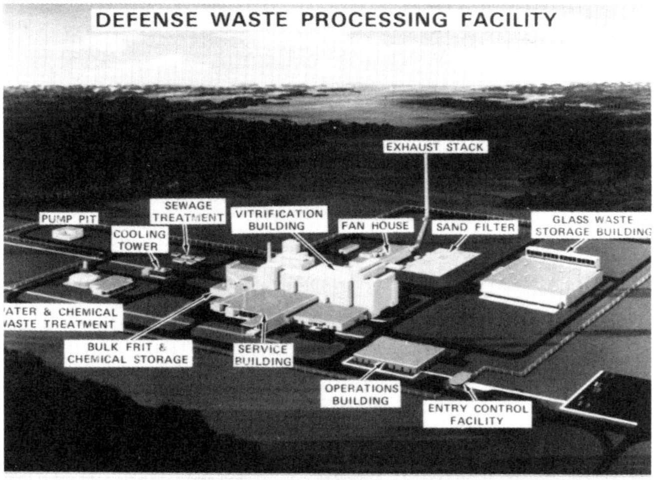

Artist's view of the DWPF complex

DWPF construction nearing completion

Figure 7-6. Defense Waste Processing Facility

Table 7-2. Waste Elements in SRS Waste. Elements in SRS waste shown in shaded boxes.

Period	Group 1-A	II-A	III-B	IV-B	V-B	VI-B	VII-B	VIII	VIII	VIII	I-B	II-B	III-A	IV-A	V-A	VI-A	VII-A	Inert Gases
1	**H**																	He
2	Li	Be											**B**	**C**	**N**	**O**	**F**	Ne
3	**Na**	**Mg**											**Al**	**Si**	**P**	**S**	**Cl**	Ar
4	**K**	**Ca**	Sc	**Ti**	**V**	**Cr**	**Mn**	**Fe**	**Co**	**Ni**	**Cu**	**Zn**	Ga	Ge	As	**Se**	Br	Kr
5	**Rb**	**Sr**	**Y**	**Zr**	**Nb**	**Mo**	**Tc**	**Ru**	**Rh**	**Pd**	**Ag**	**Cd**	In	**Sn**	**Sb**	**Te**	**I**	Xe
6	**Cs**	**Ba**	**La**	Hf	Ta	W	Re	Os	Ir	Pt	Au	**Hg**	**Tl**	**Pb**	Bi	Po	At	Rn
7	Fr	Ra	Ac															
					Ce	**Pr**	**Nd**	**Pm**	**Sm**	**Eu**	Gd	**Tb**	Dy	Ho	Er	Tm	Yb	Lu
					Th	Pa	**U**	**Np**	**Pu**	**Am**	**Cm**	Bk	Cf	Es	Fm	Md	No	Lr

Table 7-3. Silicate Glass Frit Compositions* For Immobilization of HLW

COMPONENT	COMPOSITIONAL RANGE	FRENCH AVH [R7T7] Commercial	GERMAN PAMELA [SM 513] Commercial	US SRS [165] Defense
SiO_2	45-77 wt. %	54.9	58.6	68.0
B_2O_3	5-20	16.9	14.7	10.0
Na_2O	1-20	11.9	6.5	13.0
Li_2O	0-7	2.4	4.7	7.0
MgO	0-4	-	2.3	1.0
ZrO_2	0-1	-	-	1.0
Al_2O_3	0-3	5.9	3.0	-
CaO	0-7	4.9	5.1	-
ZnO	0-10	3.0	-	-
TiO_2	0-5	-	5.1	-

* All compositions given in weight percent. The compositional range denotes a general range of major components that have been investigated for immobilization of a variety of US and international waste compositions, and includes US defense as well as commercial wastes.

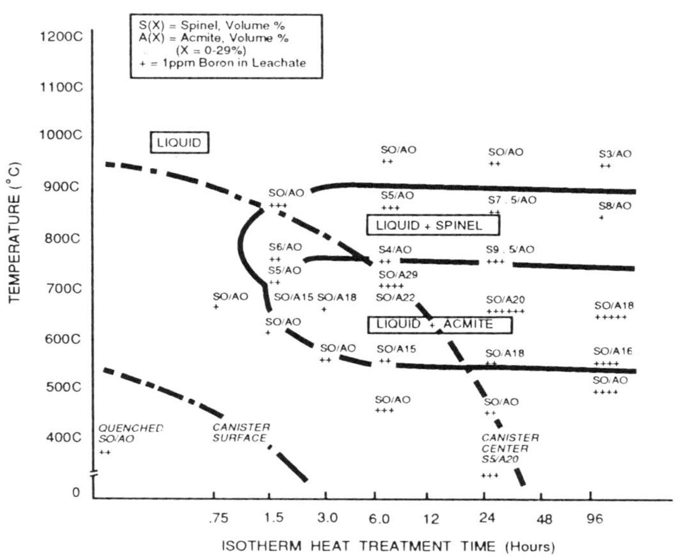

Figure 7-7. TTT for SRL 165/TDS waste glass (ref. 45).

Figure 7-8. Drop tests of canisters containing simulated SRS waste glass.

produced is low along with the amount of resulting fines or small particles. The waste glass product possesses more than adequate mechanical stability under anticipated as well as accident conditions.

Radiation Stability. As radionuclides are incorporated into glass structures, a significant radiation field can be produced. The glass can be irradiated by alpha and beta particles, gamma rays, and neutrons that result from decaying radionuclides. The effects that these components have on important properties of waste forms have been assessed for a variety of parameters, including chemical and mechanical integrity, stored energy, helium accumulation, density changes, and radiolysis (51-55,59). Based on all existing data, waste glass forms perform very well under all of the radiation conditions expected during all stages of solidification and isolation of the HLW. Figure 7-9 is a photograph of a Savannah River waste glass sample which was doped internally with Curium-244. This is an intense alpha emitter, which is considered by many to be the most potentially detrimental radiation effect. The amount of radiation received by the glass sample is equivalent to the radiation that would be accumulated after more than 1 million years of storage. The glass sample remains intact even under these extreme conditions and its leaching rate is within a factor of two of the undoped glass.

Figure 7-9. Radioactive waste glass sample.

Chemical Durability. The most important and most studied property of the solidified waste glass forms is its chemical durability. This provides a measure of how well the waste glass will retain radionuclides under anticipated as well as accident scenarios. Nowhere is this more important than in the final resting place of the waste, the geologic repository. The two most important observations made after evaluating the chemical durability of a variety of waste glass systems under many different repository conditions is that: 1) *leaching of glass is very* *low when subjected to realistic scenarios and conditions;* and 2) *not only is the chemical durability of the waste glass systems good, but durability actually improves with time.* These findings will be discussed in more detail below.

CHEMICAL DURABILITY OF NUCLEAR WASTE GLASS-TESTING & EVALUATION

Leaching Tests

There have been many different types of leaching tests used over the years to help assess the chemical durability of waste glass systems (60), many of which will be discussed in detail by Zoitos and Clark in this book. Leaching tests have been conducted both statically as well as dynamically and can generally be grouped into three main categories: a) *Laboratory Intercomparison Leach Tests;* b) *Basic Glass Corrosion Studies,* and c) *Repository Relevant and Field Testing of Waste Glass Systems.*

Laboratory Intercomparison Tests. Standardized laboratory leaching studies have been an important part of assessing corrosion of nuclear waste glass systems. These tests are characterized by conducting leaching experiments under exact and reproducible conditions which can be controlled in the laboratory. This is necessary in order to: a) compare one waste form to another under identical conditions, b) to provide a means to systematically optimize a specific composition and c) to help assure that reproducible and highly durable products can be made.

One of the most used leaching tests in the nuclear community is the MCC-1 static leach experiment (61). While results produced from these tests are not necessarily relevant to performance in a repository setting, the tests are useful in comparing performance of alternative waste forms and for optimization of compositions.

Other standardized tests have been based on combining the best features of other tests. The Product Consistency Test or PCT was developed at SRS to provide a quick and sensitive means to assure that a good, homogeneous and reproducible glass composition is produced from future DWPF production operations (62). The test involves simple sample preparation, can be performed under remote handling conditions, and produces results in a relatively short time. The PCT is based on standardized tests involving crushed glasses, which include the MCC-3 Agitated Powder Leach Tests, the ASTM Standard Test Method for Shake Extraction of Solid Waste with Water, and the Corning Glass Works Chemical Durability Test.

Basic Glass Corrosion Studies. Standard as well as non-standard leaching experiments have been used to better understand the corrosion mechanisms of waste glass systems. The MCC tests have been used extensively to control test conditions and allow key parameters to be varied systematically (61). By combining solution analyses with detailed studies of leached glass surfaces, the effects of important processing and repository related conditions on waste glass leaching can be assessed, along with providing important information on leaching mechanisms. More insight into mechanisms of dissolution can also be obtained by combining static leaching studies with dynamic leaching studies, such as use of the slow-flow tests developed by Macedo and co-workers at The Catholic University of America (83,84).

Repository Relevant and Field Testing of Waste Glass Systems. Unlike laboratory experiments, field tests are not conducted under controlled conditions but are designed to provide as realistic a repository environment as possible to assess glass leaching. In these tests, simulated (nonradioactive) waste glass forms are buried underground and exposed to actual variations and conditions of geology and hydrology found in nature. Waste glass performance is then evaluated in this realistic setting. The performance of waste glass forms is also assessed as a function of proposed package components and engineered barriers within this environment. In order to obtain as full an understanding as practically possible of the waste glass behavior, these data and interpretations are often combined with laboratory leaching and mechanisms studies. Additional important information can also be determined on performance of species of interest, including migration, solubility and sorption effects, and subsequent mass transport. Ultimately, in order to make predictions of waste glass behavior out to very long time periods, it is desirable to understand and model corrosion behavior and/or to correlate this behavior with natural systems. Natural glass analogues, such as those studied by Ewing and others, are known to exist in stable forms for millions of years in the earth's surface, and can be used to better understand long-time behavior of waste glass systems (63-69).

CHEMICAL DURABILITY OF NUCLEAR WASTE GLASS-PERFORMANCE

Integrated Study Approach

Improved analytical capabilities have allowed investigators to obtain more insight into the corrosion processes of glass. Prior to the late 1960's, the most often used methods for assessing the durability of glasses was to immerse the glass in a corrosive medium under defined conditions and then to either measure weight loss or analyze the solution, to measure elements leached from the glass. Weight loss measurements were often the easiest to perform while solution analyses provided additional information concerning the leaching behavior of specific elements. These techniques continue to be used in many studies today. However, neither method addresses back-reactions or precipitated phases that could form on glass surfaces, especially for more complex glass systems. These events could significantly effect measurements and subsequent interpretation of the data.

In the late 1960's and into the 1970's, investigators such as Clark and Hench of the University of Florida began combining solution analyses with bulk and surface studies of leached waste glasses, to obtain a more complete picture of the leaching process (70). In subsequent years, new and more sophisticated surface analytical tools appeared along with more accurate solution analyses. Since all of the analytical tools used to study nuclear glass systems have limitations along with advantages, an integrated study approach was often used which combined and overlapped each of these techniques. This provides important information on the chemistry or structure of species of interest on and within leached surface layers and correlates these data with that found in solution. This, therefore, provides as complete a picture as possible of the leaching performance and corrosion mechanisms of complex nuclear waste glass systems. Analytical tools used to study leached nuclear glasses are depicted schematically in Figure 7-10, along with their sampling depths (3,70).

Parameters of Importance

In order to assess, understand and ultimately, be able to predict the long term reliability of waste glass systems, chemical durability has been assessed as a function of important parameters that would be encountered during each stage of the solidification, transportation and interim and permanent disposal scenario. These important variables affecting the chemical durability of waste glass include time (71-73), temperature (74-76), solution pH (77-79), Eh (80,81), composition (waste, glass, leachate and homogeneity) (82-84), devitrification (42-45), waste loading (85,86), surface area of sample to leachant ratio [SA/V] (87,88), flowrate (89), pressure (90), surface finish, glass cracking and fines (46-50), radiation effects (51-55), geology, hydrology, and package components [canister metal, possible overpack, potential backfills] (56-58). Based on all data currently available, the chemical durability of waste glass forms should be extremely good when subjected to realistic values of these parameters. Following are examples and a more detailed discussion of the effects of three of the most important parameters on glass leaching:

Time. The most important observation concerning time-dependent leaching behavior of waste glass systems is that glass leaching generally decreases with increasing time. This is important in modeling long-term leaching behavior of waste glass systems since many models use short-time leaching data obtained in laboratory tests. Therefore, some models can overestimate glass corrosion over long time periods.

In Figure 7-11, the normalized elemental mass losses (see Zoitos and Clark chapter) for leaching of Savannah River waste glass at 90°C are shown out to 3.5

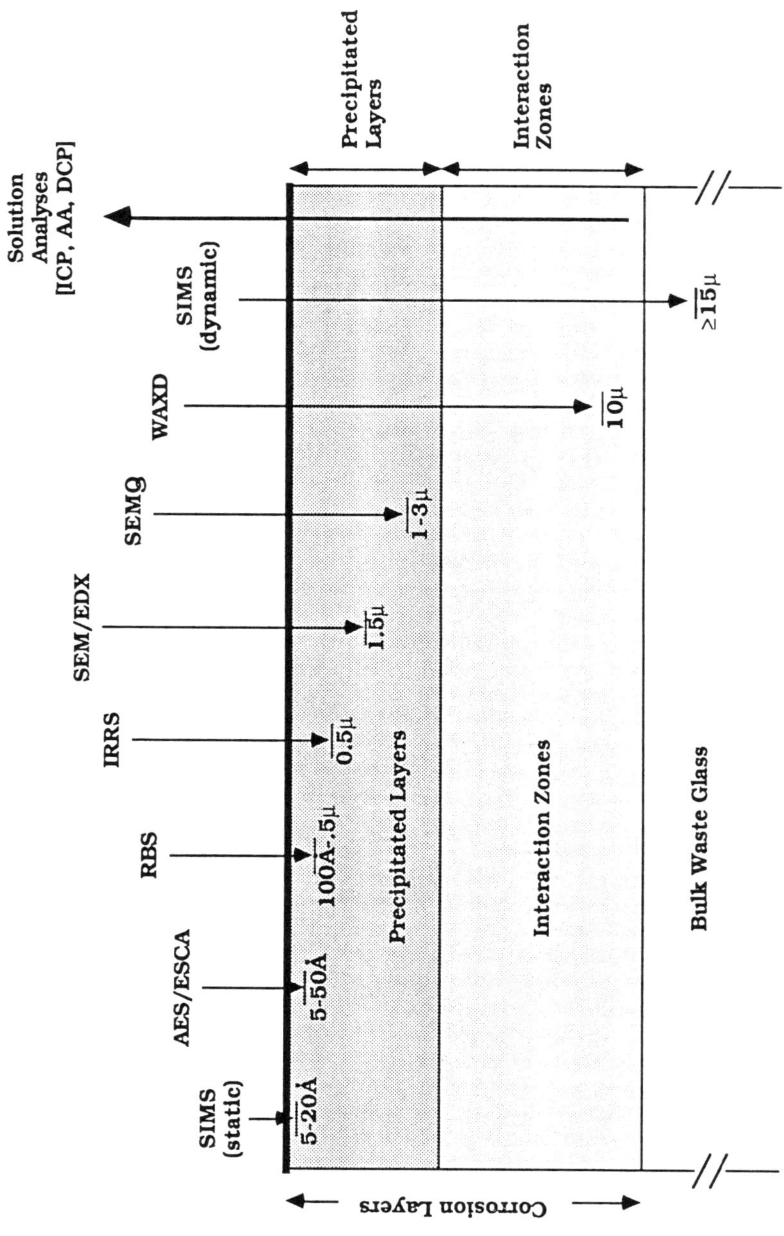

Figure 7-10. Analytical tools used to investigate MIIT waste glasses.

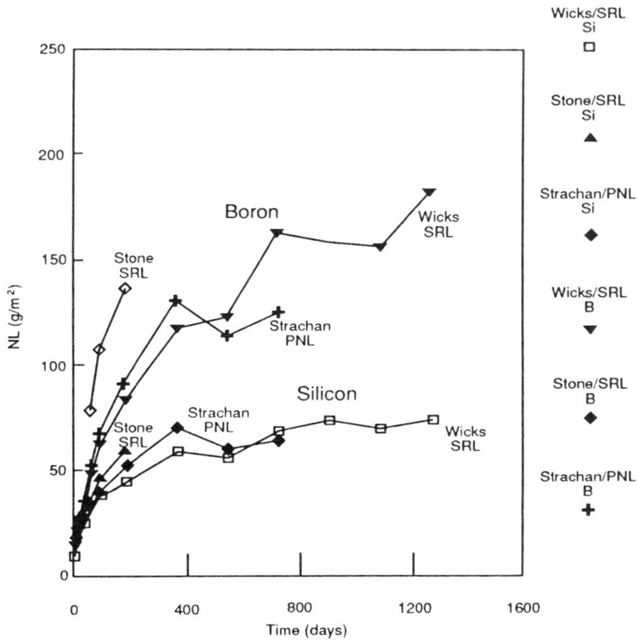

Figure 7-11. Long-term leaching studies. NL for 90°C data. (Based on silicon and boron extraction).

years of leaching, based on silicon and boron released to solution (73). Data from three investigators are superimposed on this plot for leaching of the similar waste glass compositions. These data show the trend of less species of interest being released to solution with increasing time and from changes in slope of these curves, leach rates decreasing with increasing time (72,73).

The time dependent leaching behavior of waste glass systems is best explained by a 3-stage corrosion process. The glass will start leaching by ion-exchange processes and exhibit square root time dependence. The leaching behavior initially will be dependent on conditions of the sample and test, such as surface roughness of the glass and chemistry of the leachant. As corrosion of the glass continues, linear time kinetics are often noted as matrix dissolution occurs. In the case of leaching of waste glass systems and the many different elements present, leaching can be further reduced by formation of precipitates and subsequent leached layers of glass surfaces and by species building up in solution [silica saturation effects]. This can result in a third stage of the corrosion process, which will be discussed in more detail later.

The observed time dependence of waste glass leaching is important from a practical point of view. In an actual repository environment, static or near-static conditions are expected in most of the environments under consideration, so glass leaching would be expected to improve with increasing time, as observed in laboratory tests simulating these conditions. Constituents released to solution, like silica, can eventually approach saturation levels and in many geology/hydrology/glass interactions (see Jercinovic and Ewing chapter), surface layers have been observed, which can also be potentially protective to the glass underneath. Consequently, waste glass performance based on chemical durability would be expected to improve with time in a repository environment.

Temperature. The most important observations of the effects of temperature on waste glass performance is that glass corrosion increases with increasing temperature and that at relatively high temperatures, corrosion mechanisms can change. In Figure 7-12, normalized elemental releases from a leached commercial nuclear waste glass system were examined by Westsik and Peters and are shown at temperatures ranging from 25°C to 250°C (74).

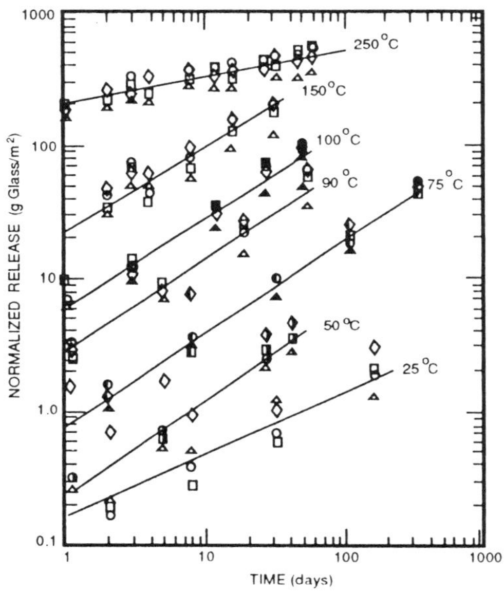

Figure 7-12. Temperature differenc on leaching of waste glass (ref. 74).

There have been many investigations on hydrothermal leaching of waste glass systems (74-76). For leaching of waste glass at or near ambient conditions, interdiffusional processes are important while for temperatures near or greater than 100°C, network dissolution can dominate. In higher temperature tests, 200-600°C, various crystalline phases have been observed to form for a variety of nuclear glass systems. These phases can increase the leachability of the systems and also alter corrosion mechanisms. It is undesirable to store waste glasses in very high temperature environments under hydrothermal conditions in which leachants are present for prolonged periods of time.

In proposed disposal scenarios, peak repository temperatures will be relatively low, about 100°C initially, depending on the geology and storage conditions. There will also be a significant time delay before groundwater is able to contact a contained waste glass product. For a time delay of only 200 to 300 years, the temperature of the waste forms will be significantly reduced to temperatures much lower than 100°C, due to radionuclide decay.

pH. Leachant pH is considered by many to be the most important parameter affecting glass corrosion (77-79). For many nuclear waste glass compositions, a parabolic relationship of pH with glass dissolution has been observed. In the pH range of approximately 5 to 9, glass dissolution is relatively constant and a minimum is noted. In Figure 7-13, the pH dependence for a SRS waste glass composition is shown and superimposed are the pH values of groundwaters in geologic formations considered over the years as possible repositories (3). Fortuitously, groundwater pH values associated with the geologic formations under study, initially fall within this range of minimum glass dissolution. The parabolic relationship and pH curve becomes tighter and shifts to higher values with increasing temperature.

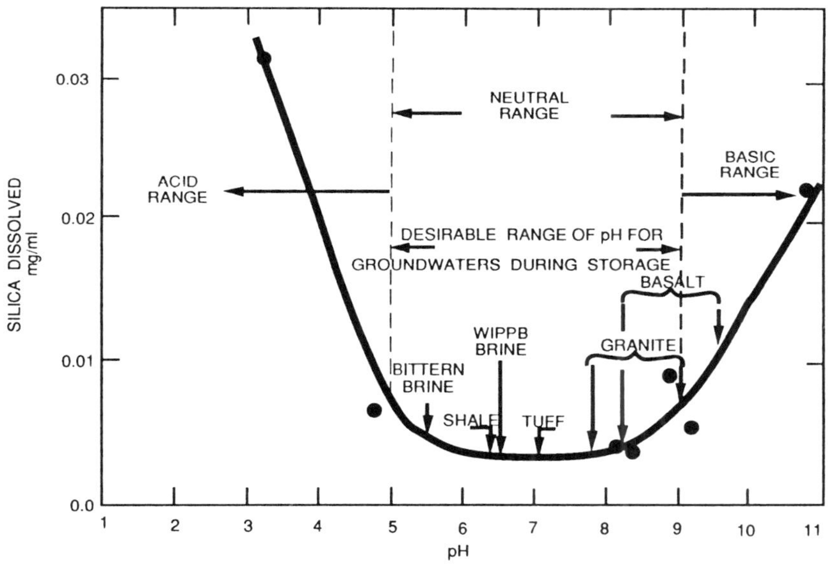

Figure 7-13. Leachability of SRS waste glass as a function of pH.

For pH values less than 5 and greater than 9, glass dissolution can increase significantly. Acidic solutions are particularly corrosive to waste glasses and dissolution has been seen to increase very dramatically at low values of pH. This is a result of the exchange of hydrogen ions from solution with cations that are in the glass, resulting in glass dissolution which can increase linearly with time. In slightly acidic solutions, leached surface layers are noted but for very low pH leachates, surface layers did not form in many of the SRS waste glass compositions.

In basic solutions, glass dissolution increases with increasing pH. For water/glass interactions, this is a direct result of attack of hydroxide ions on Si-O bonds of the random network of the silica-based glass and on increased solubility of silica in higher pH solutions. Alkaline solutions increase the rate of glass dissolution and since silica solubility increases at higher pH, more silica is able to go into solution before approaching saturation.

As alkali ions are released to solution, hydroxides form and increase the pH in unbuffered solutions. However, the pH does not continue to rise to very high values since the system becomes self-buffering, due to extraction of glass species such as silicon and boron, and formation of their salts and acids in solution. The leachant can also be buffered by elements present in the solution as existing in actual groundwaters. After several months of leaching, the solution pH can level off and stabilize as shown in Figure 7-14, for the long term leaching studies of the SRS waste glass systems described in the previous section (73). Groundwater chemistry, geologic conditions and the presence of package components can all contribute to further buffering of the leachate present. In slightly basic solutions, precipitated surface layers of well defined morphology were observed, but for very basic leachates, these layers were very porous and friable.

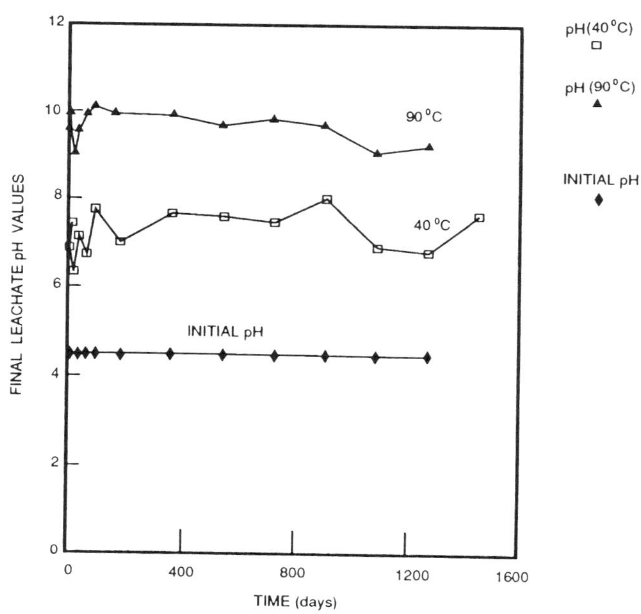

Figure 7-14. Long-term leaching study.

NUCLEAR WASTE GLASS STRUCTURE AND CORROSION PROCESSES

There are many very good reviews on the chemical durability of waste glasses and their ability to resist aqueous corrosion (90-98). Most of the current information is based on original or pioneering studies involving the leaching of simpler, more familiar glass systems (usually binary compositions) exposed to water (99-104). These early efforts led to the identification of two main stages of corrosion, which include interdiffusional processes followed by matrix dissolution. Corrosion of waste glasses is an extension of this concept and recognizes, that for the 40 or more components in nuclear glass systems, corrosion processes may be more complex. For nuclear glasses, the surface interactions and the variety of precipitates that can form on and within leached layers, as well as alterations to the surrounding solution, are believed to be especially important to the corrosion processes and long-term performance of the glass. One approach that has been proposed to describe this process adds a third stage to the two stages already proposed for simpler systems. This model and its relationship to the structure of glass, is summarized schematically in Figure 7-15 and described below (3,105,106).

Waste Glass Structure

Waste glass forms are composed of approximately 70% glass forming chemicals added to about 30% waste constituents. The glass components are generally added to the liquid waste stream in the form of a multicomponent premelted glass or frit prior to waste glass melting operations, which produce solidified radioactive products. Although there are many individual elements that comprise a nuclear waste glass system, these components can play only one of three basic roles in the glass structure: network formers, intermediates or modifiers (39,40).

Constituents such as silica and boric oxide, are generally added to the waste as major components of the frit. The silicon and boron atoms are *network formers* and become located in the center of oxygen polyhedra in the configuration of tetrahedra or tetrahedra and triangles, respectively. These polyhedra are then tied together by sharing corners, generally in accordance with Zachariasen's rules, which then makes up the 'framework' or 'skeleton' of the of the random network structure of the solidified waste glass form. Another structural role that both glass frit and waste elements can play is that of *intermediates*, which is exemplified by major components found in the waste such as alumina. These components can replace the network formers and still retain the framework structure of the glass. Other cations can move to the singly bonded oxygen ions that are created, for charge neutrality. The final role that components can play is the most prevalent, that of *modifiers*. In this case, important waste components such as cesium and strontium, along with alkali and

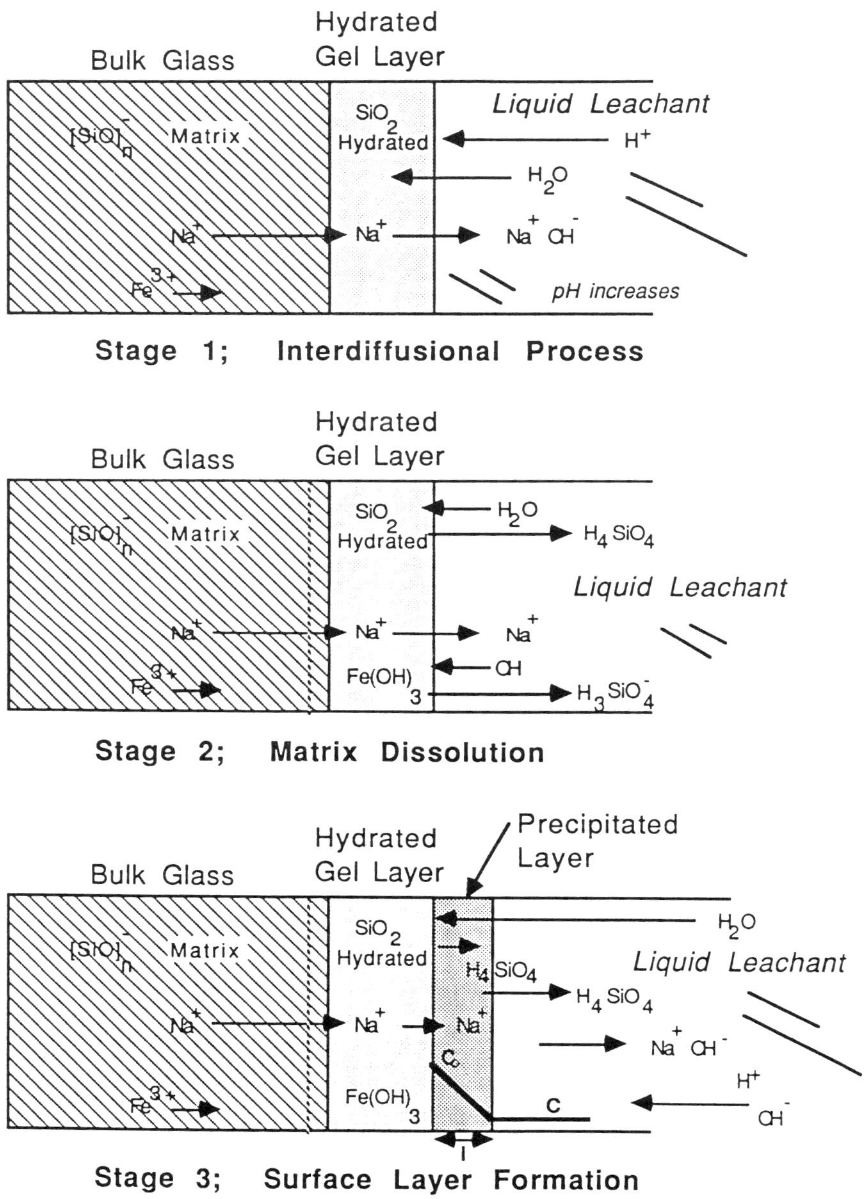

Figure 7-15. Proposed 3-stage corrosion process for waste glasses.

alkali earth constituents, are located within the holes of the random network structure, and can also be associated with nearby singly bonded oxygen ions. In Figure 7-16, a model depicting a simplified waste glass random network structure of the glass is shown (3,40). An important point to note is that both glass and waste glass components become an integral part of the random network structure of the glass. Components are incorporated by primary and/or secondary bonding, which helps explain why the glass is able to retain radionuclides so well during leaching and why different elements can leach at different rates. Based on this picture of waste immobilized in glass, the 3-stage corrosion process for nuclear waste glass systems is described below (105,106).

3 Stage Corrosion Process for Nuclear Waste Glasses

Stage I: Interdiffusion. In the early stages of glass corrosion, interdiffusional processes dominate. Water diffuses into the glass and ion-exchanges with modifiers in the glass, producing a modifier-deficient surface layer. This is ideally diffusion controlled and square root time dependent. As the modifiers are released to solution, the pH rises due to formation of components such as alkali hydroxides.

Stage II: Matrix Dissolution. At intermediate times, matrix dissolution becomes the dominant mechanism and the primary surface structure of the glass begins to break down. Glass dissolution becomes controlled by the leachate pH which determines the solubility of amorphous silica. In an open or dilute system, in which silica concentration is considerably below saturation levels, the dissolution rate of the glass matrix will be constant and the concentration of network formers such as silicon in solution, will increase linearly with time. In a closed system in which leaching products can accumulate, the dissolution rate increases with increasing pH but then slows down as the solution becomes saturated with silicates. For waste glass systems, Stage II is generally reached very rapidly.

Stage III: Surface Layer Formation. During intermediate and long times, surface layers can form during the corrosion process. Examples of surface layers formed on SRS waste glass after leaching in ultra pure water as well as tuff, granite, shale, salt and basalt groundwaters are shown in Figure 7-17. These layers are produced mainly from precipitation of insoluble compounds formed near or within the leached glass surface. The layers can contain a variety of precipitated and/or absorbed species which are found in waste or frit compositions, and can consist of highly crystalline phases as well as amorphous components. These layers and associated precipitation kinetics can be more complex than simpler glass systems (see Hench chapter). The surface layers produced can act as a protective barrier to subsequent glass dissolution, of very high effectiveness or of little or no effectiveness, depending on the chemistry and properties of the layers and on the corrosion environment.

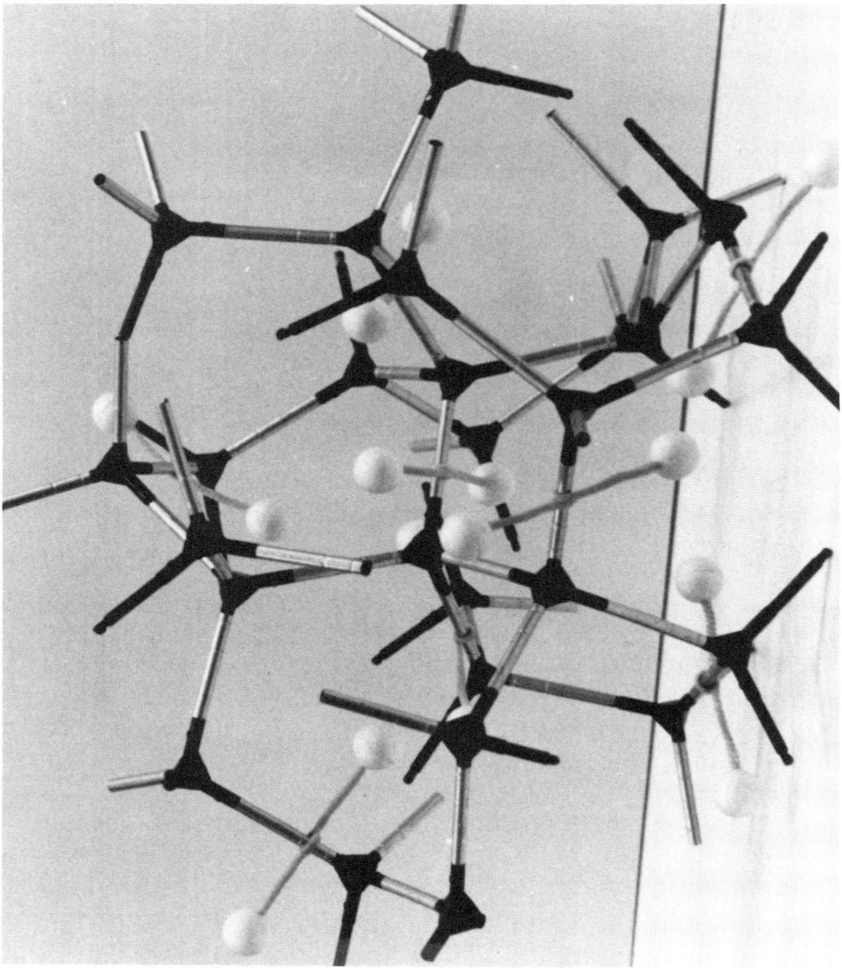

Figure 7-16. Simplified structure of SRS waste glass.

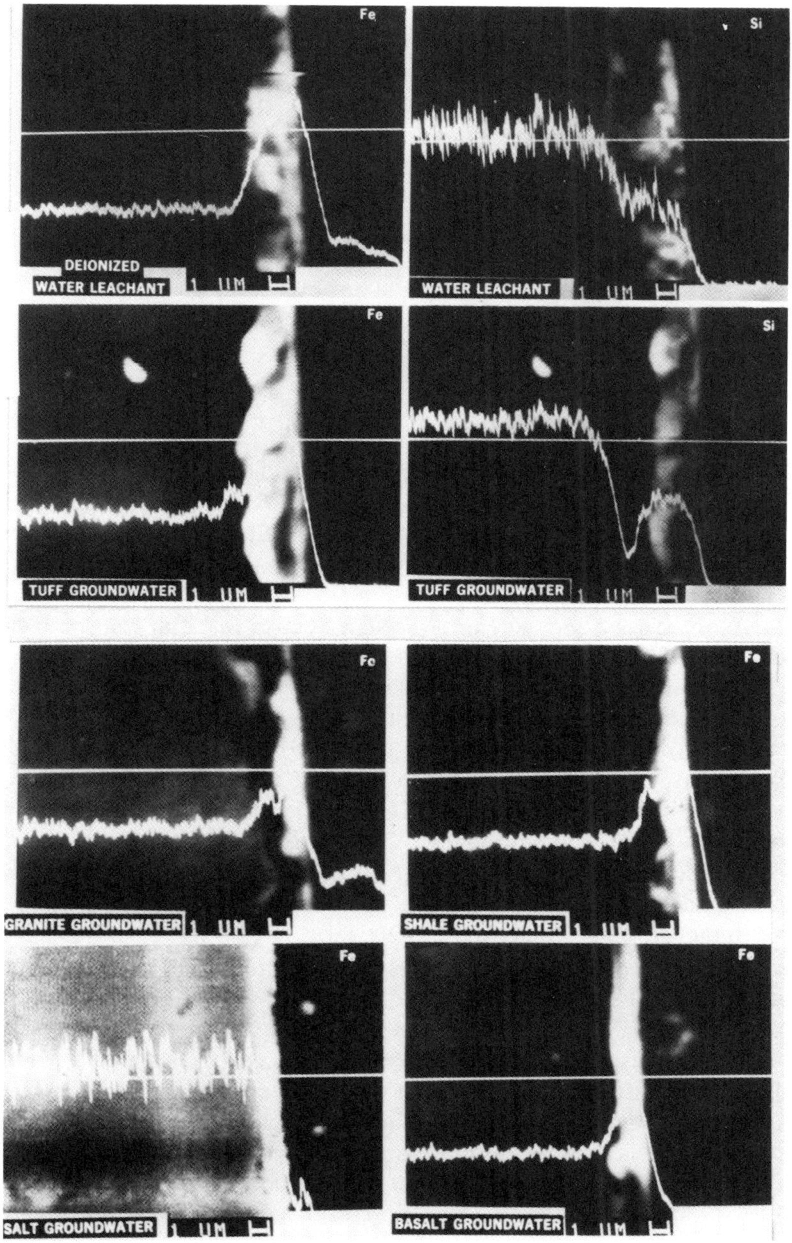

Figure 7-17. Leached layers formed on SRS waste glass.

Modeling of Corrosion in Nuclear Waste Glasses

Leaching of many simple as well as complex glass systems can be described by the qualitative leaching mechanisms given above. In the late 1960's, Douglas and El-Shamy (100,101) described the leaching of simple glass systems as the first two stages of the process just described and by the following empirical representations:

$$Q = at^{1/2} + bt \tag{7-1}$$

or

$$d\log Q \,/\, d\log t = (0.5a + bt^{1/2})/(a + bt^{1/2}) \tag{7-2}$$

where Q is the total amount of alkali released, t the time of leaching at a given temperature, and a and b are constants. In the limits, as t approaches zero, the gradient approaches 1/2 and for t approaching infinity, the gradient approaches 1. Plots of logQ versus logt would be close to linear over extended times, with slopes varying between 1/2 and 1 with increasing time and temperature.

As the glass begins to leach, the first term in equation 7-1 dominates and leaching behavior is proportional to square-root time kinetics. This is expected to be most important for glasses that have good chemical durability, low leaching temperatures and short leaching times, and for glasses being leached in neutral pH solutions. As leaching continues, the second term dominates and linear time kinetics ensue. This is expected to be most important for glasses of lower durability, higher temperatures and longer leaching times, and for glasses leaching in highly acidic or highly alkaline solutions. The general observation of parabolic to linear release of alkali has been observed in many glass systems and is predicted from other modeling approaches.

In the late 1970's and early 1980's, there were a variety of qualitative and quantitative modeling approaches used to describe and predict waste glass behavior (105-119). Some of these approaches involved the derivation of a source term used to describe release of radionuclides from a waste form and subsequent transport while other modeling approaches attempted to describe release and transport from a complete waste package in a geologic environment. A common feature of many of these and subsequent modeling efforts is the use of conservative assumptions and "worst case" scenarios.

One of the earlier efforts to model the behavior of SRS waste glass occurred in the 1980's and was built upon early work which modeled the behavior of simpler glasses. The SRL Leachability Model was developed by Wallace and Wicks and is based on the 3-stage corrosion process described quantitatively above. The main difference between this model and those of simpler glass systems is recognition of Stage III, the discovery of surface layers and the effects they can have on subsequent glass leaching.

As the waste glass system leaches, interdiffusion and matrix dissolution processes (Stage I and II) occur very rapidly. This results in an alkali deficient surface and the release of the various glass components. Some of these soluble elements are released to solution while other elements form insoluble compounds and precipitate within or on the leached glass surface (Stage III). This precipitated layer has the potential of affecting further glass leaching, in that glass components that are subsequently leached from the glass must then pass through this outermost precipitated layer before being released to solution. Depending on the characteristics of the precipitated layer -- its porosity, composition, adherence to the glass underneath -- and depending on the leaching environment and conditions, the layer can have little or no effect on glass leaching or can provide an important protective barrier, that can reduce the rate of glass leaching with increasing time. Data from static leaching tests performed on four different systems and data from very slow flow experiments performed by Macedo and co-workers (83,84), indicate that these layers can contribute to improving the long term leaching performance of the waste glass forms.

The SRL leachability model is a steady state approach and is built upon experimental observations. It assumes that the rate of silica dissolution, R, per unit area at the surface of the gel layer is the same as the rate of diffusion of silicate through the precipitated layer, with the rates given as follows (105,106):

$$\text{dissolution:} \quad R = K(C_s - C_o) \tag{7-3}$$

$$\text{diffusion:} \quad R = D/l(C_o - C) \tag{7-4}$$

where K is a rate constant, D is the diffusion coefficient of dissolved silicate in the layer, l is the thickness of the precipitated layer, and C, C_o, and C_s are the silicate concentration of the leachate, the solution next to the gel layer and a saturated solution (with respect to H_4SiO_4) in the leachate, respectively (see Figure 7-15).

Since $l = qL$, where L is the amount of silica leached up to time t and q is a constant, and since $R = dL/dt$, combining the above equations yields:

$$dL/dt = \frac{KC_s(1 - C/C_s)}{1 + \beta L} \tag{7-5}$$

Here, where $\beta = Kq/D$, the equation is a general representation of silica dissolution from the waste glass system, while C/C_s represents fractional saturation of silica in the leachate.

For the case of a closed system, in which the corrosion products are allowed to build up in solution, $C/C_s = \alpha L$, where $\alpha = SA/V\ (1/C_s)$. Here SA is the surface area of the glass and V is the volume of leachant. Integrating equation 7-5 and using the conditions that $t = 0$, $L = 0$ yields:

$$(1 + \beta/\alpha)(\ln(1-\alpha L)) + \beta L = -\alpha K C_s t \qquad (7\text{-}6)$$

For the case of an open system, where corrosion products are not allowed to accumulate in the leachate, $\alpha = 0$ and:

$$L + (\beta/2)L^2 = KC_s t \qquad (7\text{-}7)$$

A useful means for interpreting these results is to represent leaching by a log-log plot of the amount leached versus time and assess the slopes of the plots produced. These data are summarized in Figure 7-18. As shown in Curve A, the model predicts that initially the slope will be 1 as corrosion begins. The process then becomes dominated by reaction of silica with water, and approaches 1/2 as the layer grows and leaching becomes dominated by diffusion of species through the layer. The slope can then continue at 1/2 for leaching of the glass in a dilute solution (α approaching zero) in which corrosion products do not accumulate, or be significantly reduced as the solution saturates in silica. Ideally, this slope would approach 0 if total saturation can ever be achieved and equilibrium fully realized. Curve B depicts the case of solution volume being small (α large) and saturation being reached before layers can affect the leaching process and cause a diffusion-controlled reaction. The slope again can theoretically approach zero.

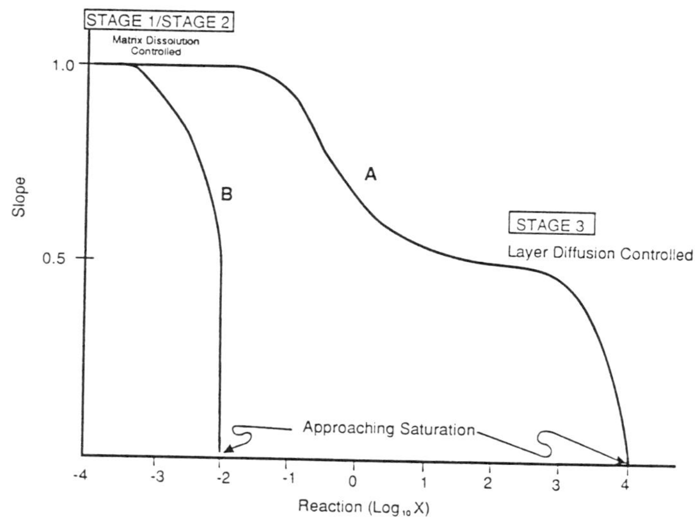

Figure 7-18. SRL kinetic model predictions.

Existing data indicate that leaching of waste glass systems can decrease significantly, but still continue even after many years, although at very low rates, so equilibrium for the complete system is not fully achieved in this time interval.

Other kinetic modeling approaches have emphasized different aspects of the corrosion process. Some have also focussed on the presence of reaction products but have considered glass dissolution dominating over diffusional effects. Other have emphasized surface area of glass to volume of leachant ratio along with leachant renewal frequency. Still other approaches focus on performance after long time periods under conditions where solubility considerations are most important and dominate. One of the more interesting models is that developed by Grambow, which emphasizes saturation effects and considers an equilibrium between solution and glass surface (112-116). With the use of a computer code for geochemical calculations [PHREEQE], it can describe equilibria conditions for reaction products that can form and provide input to another code [GLASSOL], which determines reaction rates (117). This modeling effort is described in more detail by Grambow in this book. Many of the models derived for nuclear waste glass systems successfully describe glass leaching for specific systems investigated and for the specific leaching conditions studied.

Another approach for describing glass behavior was developed by Plodinec and Jantzen and is based on hydration thermodynamics and the original work of Paul (118,119). In this approach, the glass is treated as a mixture of oxides with known free energies of hydration. The resistance of a waste glass system to leaching by water is then calculated to be the sum of the individual component hydration free energies, weighted by the mole fraction of the component in the glass. Therefore, by relating the stability of a system to the free energies of hydration, nuclear waste glasses can be directly compared to other systems, such as natural and ancient glasses, that have been stable for millions of years. One result from this analysis concluded that SRS waste glass should be as stable to aqueous attack as natural basalt. The thermodynamic approach has also been used to predict compounds that can form in the precipitated surface layers described earlier (Stage III), which can subsequently affect further glass corrosion. This approach is described in more detail by Jantzen in this book.

IN-SITU TESTING AND PERFORMANCE OF SRS WASTE GLASS

The ability of geologic formations to retain radionuclides is not without precedent. Two billion years ago, a chain reaction was started in a natural uranium ore deposit located at Oklo in Africa, and continued for hundreds of thousands of years before burning out. The waste that was generated by this natural event was trapped within this site. Hence, nature not only produced the first fission reactor, but also the first geologic repository to contain the waste generated (120).

The first field tests assessing the ability of waste forms to retain radionuclides in geologic formations was begun at Chalk River in the early

1960's by Canada. Following this pioneering effort, subsequent in-situ programs were characterized by cooperative, international undertakings, such as the burial of glasses, waste forms and package components in granite in Sweden; in clay in Mol, Belgium; in a limestone formation in the United Kingdom; and in salt, at the Waste Isolation Pilot Plant [WIPP] in Carlsbad, New Mexico. This latter effort, the WIPP/SRL in-situ testing program, involves the first field tests involving burial of simulated nuclear waste forms to be conducted in the United States. It also represents the single largest, most cooperative venture of this type yet undertaken in the international waste management community.

Materials Interface Interaction Tests [MIIT]

The WIPP/SRL Materials Interface Interaction Tests [MIIT] is a joint effort managed by Sandia National Laboratories in Albuquerque, New Mexico and the Savannah River Laboratory in Aiken, South Carolina and sponsored by the U.S. Department of Energy. MIIT involves the field testing of simulated or non-radioactive waste forms and waste package components supplied by seven different countries. Included in MIIT are over 900 waste form samples comprising 15 different systems, almost 300 individual metal samples of 11 types of potential canister and overpack materials, and over 500 geologic and backfill specimens. In total, there are 1,926 relevant interactions that characterize this effort (121-125).

The MIIT program is part of a larger effort at Savannah River aimed at understanding and being able to predict long term performance of DWPF waste glass in realistic repository environments. All samples, including waste forms, canister and overpack metals and geologic specimens, were fabricated in the shape of pineapple slices. They were then stacked onto heater rods in various stacking sequences in order to produce interfaces and interactions of interest.

There are fifty assemblies in MIIT in seven different stacking sequences, which comprise a 7-part MIIT program, each part with its own set of specific objectives. The assemblies were inserted into brine-filled boreholes about 655 meters below the surface at WIPP in bedded salt deposits of the Salado Formation and heated to a temperature of 90°C (Figure 7-19). Samples and aliquots of brine were obtained after 0.5, 1 and 2 years as part of this 5 year burial study (121,125).

Compositional Correlations. The compositions of the waste forms supplied from seven countries in MIIT have been tied together by Ramsey and Wicks in compositional ternaries (126-128). These correlations were made based on structural considerations, bonding energies, and surface layer characteristics. These efforts suggest that for the variety of waste glass systems studied world-wide, the behavior of many of these systems, such as the borosilicate waste forms, should be very similar, including leaching mechanisms. In Figure 7-20, a compositional ternary relating MIIT systems based on the structural role of components as network formers, intermediates and modifiers is shown.

Installation of the 50 MIIT assemblies and almost 2000 samples in J-room at WIPP

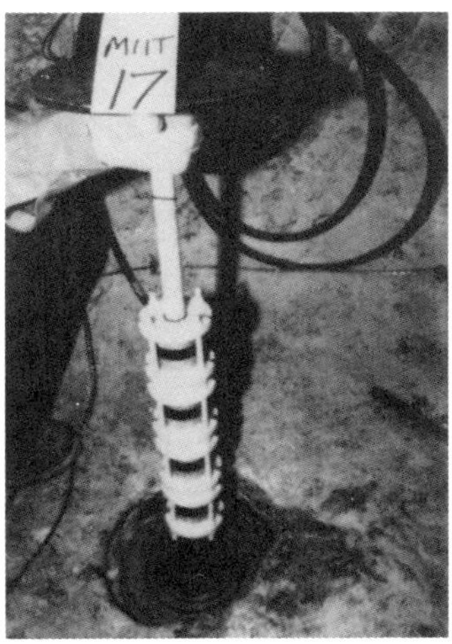

Type IIA assembly being installed. This configuration produced solution analyses that could be used with surface studies and provided time dependent leaching data.

Figure 7-19. WIPP/SRL in-situ testing program.

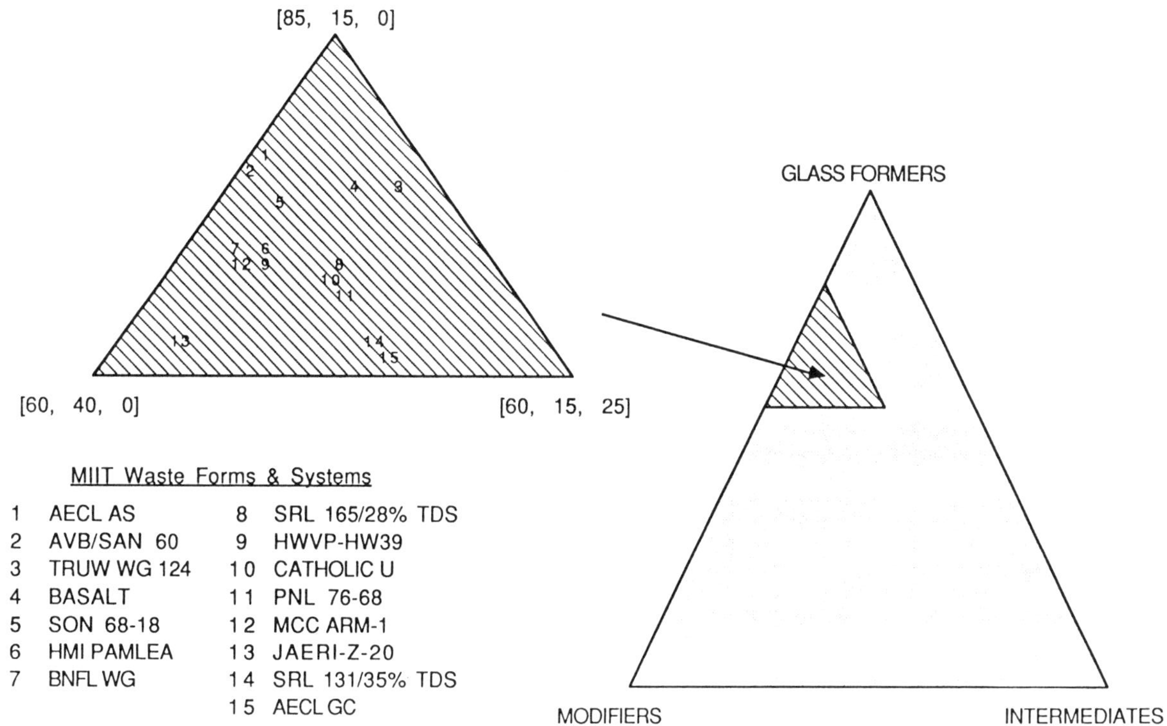

Figure 7-20. Compositional correlations of MIIT waste forms (ref. 127).

Post-Test Analyses of SRL 165/TDS Waste Glass. Based on both solution analyses and surface studies, the performance of SRL 165/TDS waste glass has been very good. In Figure 7-21, a cross-section of SRL waste glass after being buried for two years in WIPP is viewed, using scanning electron microscopy [SEM] along with accompanying energy dispersive x-rays [EDX]. There are two distinct regions noted using these techniques: an outermost precipitated layer, and an inner glass reaction zone (129).

Precipitated and Glass Reaction Zone. Precipitated and glass reaction zones were analyzed and measured as a function of time and other relevant conditions by Taaca and Wicks as shown in Table 7-4 (130). The outermost layer consists mainly of precipitated salt phases and includes crystalline as well as amorphous regions as determined by SEM, EDX and wide angle x-ray diffraction [WAXD]. These phases were attributed to the brine and salt precipitates that form on the glass surface during the MIIT leaching tests. There are two important points to note concerning the glass reaction region underneath the precipitated layers: 1) the amount of interaction of the glass with the surrounding brine environment, as measured by the intrusion of Mg and Cl from the brine into the glass, is very low; and 2) the rate of interaction decreases with increasing time. These observations indicated that the chemical durability of the glass is very good and actually becomes better with increasing time.

The reacted glass surfaces were investigated in more detail by other analytical techniques, including Fourier Transform Reflection Spectroscopy [FTIRRS], by Clark, Zoitos and co-workers at the University of Florida (131), and by Secondary Ion Mass Spectroscopy [SIMS], by Lodding, et al. of Chalmers University in Sweden (132,133). FTIRRS showed that the disruption to the glass structure was very small, even after leaching for two years in WIPP, while SIMS provided one of the most detailed tools for mapping individual elements throughout the leached layers of the burial glasses. SIMS showed that the precipitated layer actually consists of two individual layers and the reaction zone of the glass consists of at least three individual layers (Figure 7-22) (133). The SIMS technique is discussed in more detail by Lodding in this book. By combining studies of SEM, EDX, FTIRRS, WAXD, TEM, EMP and most importantly, the SIMS studies of Lodding, a composite picture of leaching of the SRL waste glasses in WIPP was developed and is shown in Figure 7-23 (134). Each of the various precipitated and reacted glass layers are discussed below.

α_0-Outermost Precipitated Salt Layer. The outermost precipitated layer was studied and characterized by Vernez of the CEA in France (135), Harker of Rockwell International (136), Ewing of the University of New Mexico (137) and SRL. Both amorphous as well as highly crystalline phases were observed, including $MgCl_2$, KCl, $CaSO_4$, NaCl, and a variety of mixed silicates, along with additional minor phases. The layer was generally very heterogeneous and varied considerably in thickness. The chemistry of the layer was dominated by large amounts of Mg and Cl derived from the surrounding brine, along with other components such as Ca, Na, S, and Si. The salt layer, which was formed pri-

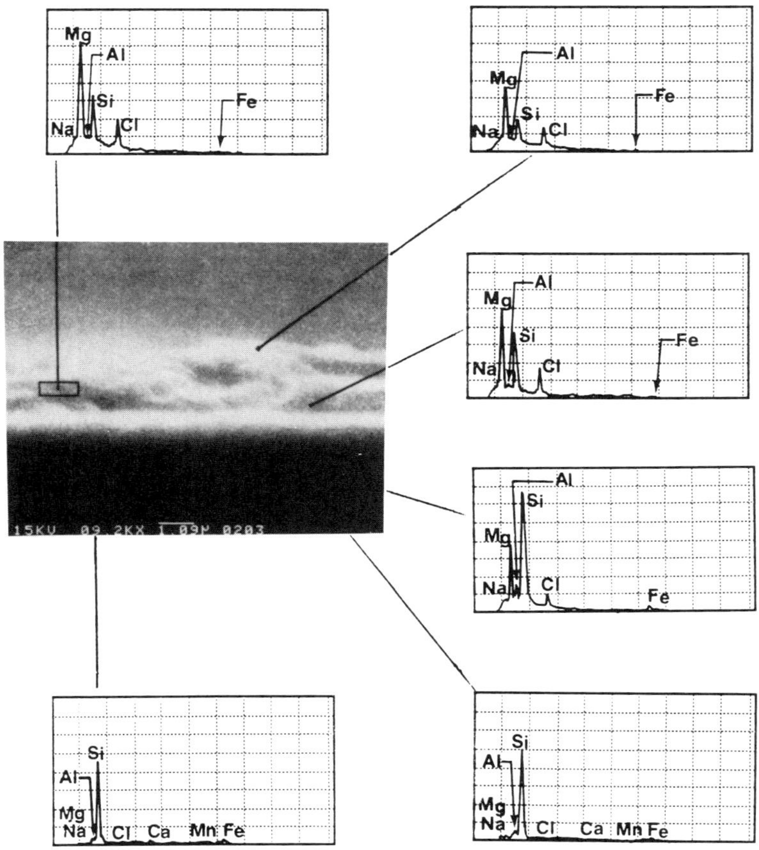

Figure 7-21. SEM-EDX cross-sections of precipitated and leached surfaces (ref. 129).

marily as a result of the geologic environment and not from leaching the glass, is expected to effect subsequent glass leaching.

Table 7-4. Observed Thicknesses of SRS 165/TDS Waste Glass Interaction Regions (ß). *MIIT In-Situ Tests (ref. 130).*

TIME	ORIENTATION	INTERFACE	INTERACTION ZONE (β)
6 mo.	Up	Glass/Salt	< 1 micron
1 yr.	Up	Glass/Salt	2-3 microns
2 yr.	Up	Glass/Salt	2-3 microns
2 yr.	Down	Glass/Salt	2-3 microns
2 yr.	Up	Glass/Glass	1-2 microns

α_1-Precipitated Glass Layer. Immediately adjacent to the glass surface and under the outermost precipitated salt layer, is a thin, precipitated glass layer. This layer formed when elements from the glass were leached and precipitated in this region. The layer is more uniform than the salt precipitated layer and is characterized by large amounts of elements from the brine, including Mg and Cl, along with Si. This layer is also relatively depleted in elements such as Al, Zr, and Fe, which are generally the least leachable species within the glass. This observation -- along with ratios of other components present, morphology and subsequent brine analysis of leachate -- shows that this is a precipitated region and not a selectively leached part of the original glass.

β_o-Major Depletion Zone. Directly under the precipitated layer is where glass begins and represents a major depletion zone in the outermost glass surface. Major components of the glass are depleted in this area. The region is further characterized by the intrusion of major brine components such as Mg. The original glass surface, the α/β interface, is located in this region.

β_1-Gradient Zone. Below the major depletion zone is the gradient zone, which is characterized by depletion of alkali and alkali earth components of the glass and enrichment in the main brine component, Mg. One of the most interesting features of this zone is the presence of a potassium peak. The distance from the α/β interface or glass surface to the potassium peak in the β_1 zone represents the main reaction front of the glass after interacting with the surrounding environment.

β_2-Diffusion Zone. This innermost glass leached layer is believed to be similar to the gel layer which initially forms on simple glasses during leaching. The zone is characterized by depletion of Li from the glass and enrichment in

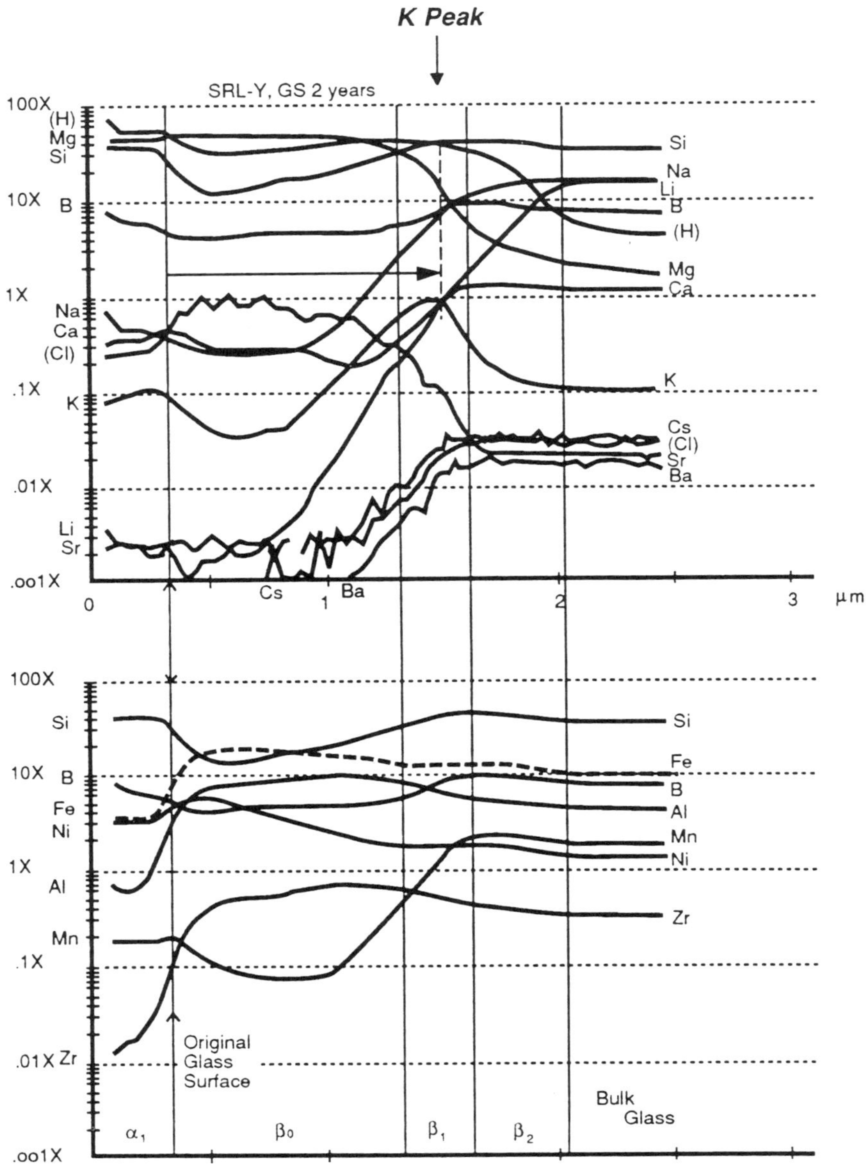

Figure 7-22. SIMS profiles (SRL Y glass, GS/2yrs) (ref. 133).

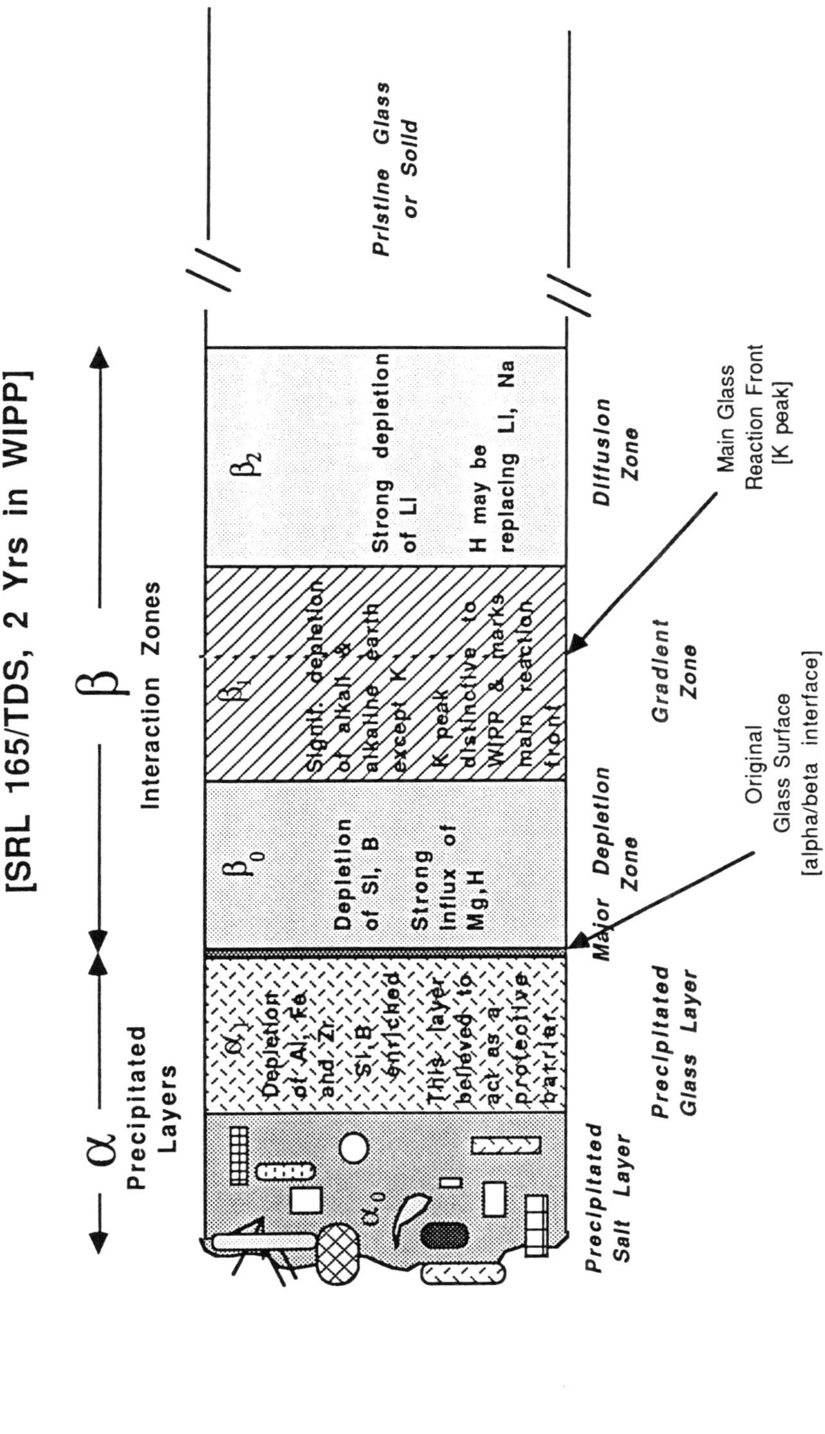

Figure 7-23. Schematic representation of precipitated/leached layers in MIIT glasses.

H from the solution. The thickness is consistent with diffusion calculations of these components in the bulk glass.

Brine Analyses. The MIIT program is the only in-situ testing effort of this type which allows solution analyses to be obtained and subsequently correlated with surface studies. Brine analyses were performed on leachates from selected boreholes containing SRL 165/TDS waste glass undoped, and glass doped with Eu and Yb as chemical tracers. After measuring concentrations found in solution, leach rates were calculated based on sample characteristics and testing conditions. In addition to assessing glass performance in the field, the solution analyses were used to define the original position of the glass surface in more detail. Analyses were performed by Macedo and co-workers at the Catholic University of America and are summarized in Table 7-5 (138,139).

There are several important observations that can be made from these data. First, the leaching behavior of SRL glass is very similar to other important international waste glass compositions used in this study. This includes the Japan-Switzerland-Sweden [JSS] composition. This observation was predicted by the compositional ternaries discussed earlier. Next, the actual leach rates are very low, generally less that $1 g/m^2$-day. This is noted based both on brine analyses and on calculations of SIMS profiles, even for very mobile and non-radioactive species such as Li. Finally, if one takes into account the geometry of the stored waste, leaching depths can also be calculated and related to the Nuclear Regulatory Commission release rate criteria [10 CFR Part 60]. For the MIIT tests, these data showed a release rate of species of interest of less than one part in 100,000 for all elements investigated thus far (134,138,139).

SUMMARY

Based on all data currently available, the performance of nuclear waste glass systems is excellent, when tested under realistic conditions, as determined by many studies performed by many different investigators. In addition to possessing outstanding chemical durability, the durability also improves with increasing time. This behavior has been observed not only in laboratory tests, but in actual field experiments as well. The observed time-dependent leaching behavior can be described and better understood by a variety of existing kinetic models, geochemical models and by the use of thermodynamic analyses. The behavior of the waste glass systems can be shown to be similar to natural glasses such as basalt, which have been stable for millions of years.

While glass systems have demonstrated the ability to immobilize and retain radionuclides very well, it must be emphasized that waste glass forms are only one part of the multibarrier isolation system. Each element of this system is designed to provide additional barriers to prevent radionculides from harmfully entering the accessible environment. This disposal scenario provides an

Table 7-5. **Average Upper Limits of Leach Rates of MIIT Waste Glasses (refs 138, 139).** (ICP-MS brine analysis)

	LR x 10^{-4} [g/m2d]	LD [μm/yr]
SRL 165/TDS		
Zr	1.6	0.02
Li		0.36
La	3.5	0.042
Eu	3.0	0.036
FR JSSA		
Y	7.4	0.090
Ce	1.2	0.014
Pr	3.1	0.037
Nd	1.2	0.015

Regulation **[10CFR60]**

NRC Annual Fractional Release: 10(-5)/yr <u>**From Entire Package**</u>

For a DWPF canister, this required release rate corresponds to an LD of ~1 μm/yr for the glass. CONCLUSION: SRL waste glass meets the NRC criterion in the WIPP tests by itself, even for very mobile and non-radioactive species!

extremely safe and effective means for disposing of and isolating potentially harmful radioactive elements for not only our generation, but for generations yet to come.

REFERENCES

1. U.S. Department of Energy, *Integrated Data Base for 1988: Spent Fuel and Radioactive Waste Inventories, Projections and Characteristics*, Report DOE/RW-0006, Rev. 4 (1988).
2. Crandall, J.L., and Clark, H.J., *Integrated High-Level Waste Immobilization Plan*, Savannah River Laboratory (1977).
3. Wicks, G.G., *Nuclear Waste Glasses*, in: *Treatise on Materials Science and Technology - Glass IV*, M. Tomozawa and R.H. Doremus ed.). Vol. 26, p 57 (1985).
4. U.S. Nuclear Regulatory Commission, *Final Rule for the Disposal of High-Level Waste in Geologic Repositories*, Code of Federal Regulations, 10 CFR 60, June (1983).
5. U.S. Environmental Protection Agency, *Environmental Standards for the Management and Disposal of Spent Nuclear Fuel, High-Level and Transuranic Radioactive Waste*, Code of Federal Regulations, 40 CFR 191, September (1985).
6. Federal Register, *Nuclear Waste Policy Act of 1982; Proposed General Guidelines for Recommendation of Sites for Nuclear Waste Repositories*, 10 CFR Part 960, Vol. 48, No. 26, U.S. Department of Energy (1983).
7. Tacca, J.A., and Wicks, G.G., *Radioactivity Analyses of SRS Waste Glass Canister and Comparison to Natural Uranium Ore*, to be published (1991).
8. Wallace, R.M., Hull, H.L., and Bradley, R.F., *Solid Forms for Savannah River Plant High-Level Waste*, DP-1335, Savannah River Laboratory (1973).
9. U.S. Energy Research and Development Administration, *Alternatives for Long-Term Management of Defense High-Level Waste at the Savannah River Plant*, Report ERDA-77-42 (1977).
10. Stone, J.A., Allender, J.S., and Gould, T.H., *Comparison of Properties of Borosilicate Glass and Crystalline Ceramic Forms for Immobilization of SRP Waste*, DP-1627, Savannah River Laboratory (1982).
11. Gordon, D.E., and Bernadzikowski, T.A., *National Program for Immobilization of High-Level Radioactive Wastes*, presented at the Second Joint ASME/ANS Nuclear Engineering Conference, American Society of Mechanical Engineers, Paper 82-NE-34 (1982).
12. U.S. Department of Energy, *The Evaluation and Selection of Candidate High-Level Waste Forms*, U.S. DOE Report DOE/TIC-11611 (1982).
13. Dunson, J.B., Eisenberg, A.M., Schuyler, R.L., Haight, H.G., Mello, V.E.,

Gould, T.H., Butler, J.L., and Pickett, J.B., *Assessment of Processes, Facilities, and Costs for Alternative Solid Forms for Immobilization of SRL Defense Waste*, DP-1625, Savannah River Laboratory (1982).

14. U.S. Department of Energy, *Environmental Assessment - Waste Form Selection for SRP High-Level Waste*, U.S. DOE Report DOE/EA-0179 (1982).
15. Report of Panel on Waste Solidification, National Academy of Sciences (1978).
16. Report to the American Physical Society by Waste Management, *Rev. Mod. Phys.* 50 (1978).
17. U.S. Department of Energy, *The Evaluation and Review of Alternative Waste Forms for Immobilization of High-Level Radioactive Wastes*, Alternative Waste Form Peer Review Panel, U.S. DOE Report DOE/TIC-10228 (1979).
18. U.S. Department of Energy, *The Evaluation and Review of Alternative Waste Forms for Immobilization of High-Level Radioactive Wastes, Report No. 2*, Alternative Waste Form Peer Review Panel, U.S. DOE Report DOE/TIC-11219 (1980).
19. U.S. Department of Energy, *The Evaluation and Review of Alternative Waste Forms for Immobilization of High-Level Radioactive Wastes, Report No. 3*, Alternative Waste Form Peer Review Panel, U.S. DOE Report DOE/TIC-11472 (1981).
20. Wicks, G.G., and Bickford, D.F., *High Level Radioactive Waste - Doing Something About It*, DP-1777, Savannah River Laboratory (1989).
21. Wilhite, E.L., *Spectrum '86 - Nuclear Management, Decontamination, and Decommissioning*, Vol. 1, p 157, American Nuclear Society (1986).
22. Wilhite, E.L., Langton, C.A., Sturm, G.F., Hooker, R.L., and Occhipinti, E.S., *Spectrum '86 - Nuclear and Hazardous Waste Management International Topical Meeting*, p 99, American Nuclear Society (1986).
23. d'Entremont, P.D., and Walker, D.D., *Waste Management '87*, Vol. 2, p 69, Univ. of Arizona (1987).
24. King, S.M., *Spectrum '88 - Nuclear and Hazardous Waste Management International Topical Meeting*, p 22, American Nuclear Society (1988).
25. Boersma, M.D., *Fuel Processing and Waste Management*, Vol. 1, p 131, American Nuclear Society (1984).
26. Haywood, J.E., and Killian, T.H., *Waste Management '87*, Vol. 2, p 51, Univ. of Arizona (1987).
27. Weisman, A.F., Papouchado, L.M., Knight, J.R., and McIntosh, D.L., *Waste Management '88*, Vol. 2, p 203, Univ. of Arizona (1988).
28. Elder, H.H., McIntosh, D.L., and Papouchado, L.M., *Spectrum '88 - Nuclear and Hazardous Waste Management International Topical Meeting*, American Nuclear Society (1986).
29. Baxter, R.G., *Waste Management '86*, Vol. 2, p 449, Univ. of Arizona (1986).

30. Cowan, S.P., and Fulmer, D.C., *Waste Management '87*, Vol. 2, p 75, Univ. of Arizona (1987).
31. Brumley, W.J., *Advances in Ceramics - Nuclear Waste Management* (D.E. Clark, et al. ed.), Vol. 20, p 37 (1986).
32. Ng, C.K., Hsiu, F.J., and Almuti, A.M., *Spectrum '86, Proceedings of the American Nuclear Society's Topical Meeting on Waste Management and Decontamination and Decommissioning*, p 593 (1987).
33. Kemp, J.B., *Spectrum '86, Proceedings of the American Nuclear Society's Topical Meeting on Waste Management and Decontamination and Decommissioning*, p 2054 (1987).
34. Athey, R.E., *Spectrum '86, Proceedings of the American Nuclear Society's Topical Meeting on Nuclear and Hazardous Waste Management*, p 167 (1987).
35. Gosh, I.K., Alexandrou, C.D., and Tom, H.W., *Spectrum '86, Proceedings of the American Nuclear Society's Topical Meeting on Nuclear and Hazardous Waste Management*, p 365 (1987).
36. DaSilva, *Spectrum '86, Proceedings of the American Nuclear Society's Topical Meeting on Nuclear and Hazardous Waste Management*, p 369 (1987).
37. Soper, P.D., Walker, D.D., Plodinec, M.J., Roberts, G.J., and Lightner, L.F., *Bull. Am. Cer. Soc.* 62: 1013 (1983).
38. Plodinec, M.J., *J. Non-Cryst. Solids* 84: 206 (1986).
39. Kingery, W.D., Bowen, H.K., and Uhlmann, D.R., *Introduction to Ceramics*, 2nd ed., Wiley (1976).
40. Wicks, G.G., *Structure of Glasses*, in: *Encyclopedia of Materials Science and Engineering*, Vol. 3, p 2020, Pergamon (1986).
41. Lutze, W., and Ewing, R.C. ed., *Radioactive Waste Forms for the Future*, Elsevier Science Publishing Co. (1988).
42. Tomozawa, M., Singer, G.M., Oka, Y., and Warden, J.T., in: *Ceramics in Nuclear Waste Management* (T.D. Chikalla and J.E. Mendel ed.), CONF-790420, p 193, Technical Information Center, U.S. Department of Energy (1979).
43. Turcotte, R.P., and Wald, J.W., in: *Scientific Basis for Nuclear Waste Management* (C. Northrup ed.), Vol. 2, p 141 (1980).
44. Robnett, B.M., and Wicks, G.G., *Effects of Devitrification on the Leachability of High-Level Radioactive Waste Glass*, DP-MS-81-60, Savannah River Laboratory (1981).
45. Bickford, D.F., and Jantzen, C.M., in: *Scientific Basis for Nuclear Waste Management VII* (G.L. McVay ed.), Vol. 26, p 557 (1984).
46. Smith, T.H., and Ross, W.A., *Impact Testing of Simulated High-Level Waste in Canisters*, BNWL-1903, Pacific Northwest Laboratory (1975).
47. Smith, P.K., and Baxter, C.A., *Fracture During Cooling of Cast Borosilicate Glass Containing Nuclear Wastes*, DP-1602, Savannah River

Laboratory (1981).

48. Perez, J.M., and Westsik, J.H., *Effects of Cracks on Glass Leaching*, in: *Abstracts from ORNL Conference on the Leachability of Radioactive Solids*, p 35 (1980).
49. Jardine, L.J., Reedy, G.T., and Mecham, W.J., in: *Scientific Basis for Nuclear Waste Management* (S.V. Topp ed.), Vol. 6, p 115 (1982).
50. Mechem, W.R., Jardine, L.J., and Steindler, M.J., in: *Scientific Basis for Nuclear Waste Management* (S.V. Topp ed.), Vol. 6, p. 125 (1982).
51. Weber, W.J., and Roberts, F.P., *Nucl. Tech.* 60: 178 (1983).
52. Turcotte, R.W., and Roberts, F.P., in: *Ceramics and Glass Radioactive Waste Forms* (D.W. Readey and C.R. Cooley ed.), CONF-770102, p 65 (1977).
53. Bibler, N.E., and Kelley, J.A., *Effect of Internal Alpha Radiation on Borosilicate Glass Containing Simulated Radioactive Waste*, DP-MS-75-94, Savannah River Laboratory (1975).
54. Turcotte, R.P., *Radiation Effects in Solidified High-Level Wastes, Part 2, Helium Behavior*, BNWL-2051, Pacific Northwest Laboratory (1976).
55. Roberts, F.P., Turcotte, R.P., and Weber, W.J., *Materials Characterization Center Workshop on the Irradiation Effects in Nuclear Waste Forms, Summary Reports*, PNL-358, Pacific Northwest Laboratory (1981).
56. Plodinec, M.J., Wicks, G.G., and Bibler, N.E., *An Assessment of Savannah River Borosilicate Glass in the Repository Environment*, DP-1629, Savannah River Laboratory (1982).
57. Wicks, G.G., Rankin, W.D., and Gore, S.L., in: *Scientific Basis for Nuclear Waste Management VIII* (C.M. Jantzen, J.A. Stone, and R.C. Ewing ed.), Vol. 44, p 171 (1985).
58. Lutze, W., Muller, R., and Montserrat, W., in: *Scientific Basis for Nuclear Waste Management XI* (M.J. Apted and R.E. Westerman ed.), Vol. 112, p 575, Materials Research Society (1987).
59. Bibler, N.E., in: *Scientific Basis for Nuclear Waste Management* (S.V. Topp ed.), Vol. 6, p 681 (1982).
60. Mendel, J.E., *A Review of Leaching Test Methods and the Leachability of Various Solid Media Containing Radioactive Wastes*, BNWL-1765, Pacific Northwest Laboratory (1973).
61. Mendel, J.E., *Nuclear Waste Materials Handbook - Waste Form Test Methods*, U.S. DOE Report DOE/TIC-11400, Materials Characterization Center, Pacific Northwest Laboratories (1981).
62. Jantzen, C.M., and Bibler, N.E., *Product Consistency Test (PCT) for DWPF Glass: Part I. Test Development and Protocol*, DPST-87-575, Savannah River Laboratory (1987).
63. Ewing, R.C., in: *Ceramic and Glass Radioactive Waste Forms* (D.W. Ready and C.R. Cooley ed.), CONF-770102, p 139, U.S. Energy Research and Development Administration (1977).

64. Adams, P.B., in: *Ceramics in Nuclear Waste Management* (T.D. Chikalla and J.E. Mendel ed.), CONF-790420, p 233, Technical Information Center, U.S. Department of Energy (1979).
65. Brookins, D.G., Abashian, M.S., Cohen, L.H., and Wollenberg, H.A., in: *Scientific Basis for Nuclear Waste Management* (S.V. Topp ed.), Vol. 6, p 231, Elsevier (1982).
66. Murakami, R., Ewing, R.C., and Bunker, B.C., in: *Scientific Basis for Nuclear Waste Management XI* (M.J. Apted and R.E. Westerman ed.), Vol. 112, p. 737, Materials Research Society (1987).
67. Ewing, R.C., and Jercinovic, J., in: *Scientific Basis for Nuclear Waste Management X* (J.K. Bates and W.B. Seefeldt ed.), Vol. 84, p. 67, Materials Research Society (1987).
68. Cowan, R., and Ewing, R.C., in: *Scientific Basis for Nuclear Waste Management XII* (W. Lutze and R.C. Ewing ed.), Vol. 127, p 49, Materials Research Society (1989).
69. Murakami, R., Banba, R., Jercinovic, M.J., and Ewing, R.C., in: *Scientific Basis for Nuclear Waste Management XII* (W. Lutze and R.C. Ewing ed.), Vol. 127, p 65, Materials Research Society (1989).
70. Clark, D.E., Pantano, C.G., and Hench, L.L., *Corrosion of Glass*, Books for Industry (1979).
71. Walker, D.D., Wiley, J.R., Dukes, M.D., and Leroy, J.H., *Leach Rate Studies on Glass Containing Actual Radioactive Waste*, DP-MS-80-96, Savannah River Laboratory (1980).
72. Strachan, D.M., *Scientific Basis for Nuclear Waste Management V*, (W. Lutze ed.), Vol. 11, p 181 (1982).
73. Wicks, G.G., Stone, J.A., Chandler, G.T., and Williams, S., *Long Term Behavior of Simulated Savannah River Plant (SRP) Waste Glass - Part I: MCC-1 Leachability Results, 4 Year Leaching Data*, DP-1728, Savannah River Laboratory (1986).
74. Westsik, J.H., and Peters, R.D., in: *Scientific Basis for Nuclear Waste Management*, (J.G. Moore, ed.), Vol. 3, p. 357 (1981).
75. Flynn, D.F., Jardine, L.J., and Steindler, M.J., in: *Scientific Basis for Nuclear Waste Management* (C. Northrup ed.), Vol. 2, p 103 (1980).
76. Hermansson, H.P., Christensen, H., Werme, L., Ollila, K., and Lundquist, R., in: *Scientific Basis for Nuclear Waste Management* (S.V. Topp ed.), Vol. 6, p 107 (1982).
77. McVay, G.L., and Peterson, L.R., *J. Am. Cer. Soc.* 64: 154 (1981).
78. Wicks, G.G., O'Rourke, P.E., and Whitkop, P.G., *The Chemical Durability of SRP Waste Glass as a Function of Groundwater pH*, DP-MS-81-104, Savannah River Laboratory (1981).
79. Werme, L.O., Hench, L.L., Nogues, J.L., *J. Nucl. Mater.* 116: 69 (1983).
80. Jantzen, C.M., in: *Advances in Ceramics* (G.G. Wicks and W.A. Ross ed.), Vol. 8, p. 385, American Ceramic Society (1984).
81. Jantzen, C.M., and Wicks, G.G., in: *Scientific Basis for Nuclear Waste*

Management VIII (C.M. Jantzen, J.A. Stone and R.C. Ewing ed.), Vol. 44, p 29 (1985).

82. Strachan, D.M., Barnes, B.O., and Turcotte, R.P., in: *Scientific Basis for Nuclear Waste Management* (J.G. Moore ed.), Vol. 3, p. 347 (1981).
83. Macedo, P.B., Barkatt, A., and Simmons, J.H., *Nucl. Chem. Waste Management* 3: 13 (1982).
84. Macedo, P.B., Barkatt, A., and Simmons, J.H., in: *Scientific Basis for Nuclear Waste Management V* (W. Lutze ed.), Vol. 11, p 57 (1982).
85. Clark, D.E., Maurer, C.A., Jurgensen, A.R., and Urwonge, L., in: *Scientific Basis for Nuclear Waste Management V* (W. Lutze ed.), Vol. 11, p 1 (1982).
86. Rankin, W.D., and Wicks, G.G., *J. Am. Cer. Soc.* 66: 417 (1983).
87. Buckwalter, C.Q., Pederson, L.R., and McVay, G.L., *J. Non-Cryst. Solids* 49: 397 (1982).
88. Chandler, G.T. Wicks, G.G., and Wallace, R.M., *Nuclear Waste Management II* (D.E. Clark, W.B. White, and A.J. Machiels ed.), Vol. 20, p 455 (1986).
89. Wicks, G.G., Mosley, W.D., Whitkop, P.G. and Saturday, K.A., *J. Non-Cryst. Solids* 49: 413 (1982).
90. Mendel, J.E., *Summary Report, Workshop on the Leaching Mechanisms of Nuclear Waste Forms*, PNL-4382, Materials Characterization Center, Pacific Northwest Laboratory (1982).
91. Ross, W.A., in: *Ceramic and Glass Radioactive Waste Forms* (D.W. Readey and C.R. Cooley ed.), CONF-770102, p 49 (1977).
92. Hench, L.L., in: *Ceramics and Glass Radioactive Waste Forms* (D.W. Readey and C.R. Cooley, eds.), CONF-770102, p 189 (1977).
93. Mendel, J.E., Ross, W.A., Turcotte, R.P., and McElroy, J.L., *Nucl. and Chem. Waste Management* 1: 17 (1980).
94. McVay, G.L., and Buckwalter, C.O., *Nucl. Technol.* 5: 123 (1980).
95. Chick, L.A., and Turcotte, R.P., *Glass Leaching Performance*, PNL-4576, Pacific Northwest Laboratory (1981).
96. Wicks, G.G., in: *Waste Management '81* (M.E. Wacks and R.G. Post ed.), Vol. 1, p 321, ANS Topical Meeting, Univ. of Arizona (1981).
97. Plodinec, M.J., in: *Waste Management '85* (R.G. Post ed.), p 441, Univ. of Arizona (1985).
98. Wicks, G.G., in: *Spectrum '86*, Proc. of the American Nuclear Society Topical Meeting on Waste Management and Decontamination and Decommissioning, p. 896 (1987).
99. Morey, G.W., *The Properties of Glass*, Chap. 4, 2nd ed., Van Nostrand-Rheinhold (1938,1954).
100. El-Shamy, T.M., and Douglas, R.W., *Glass Technology*, Vol. 13, p 77 (1972).
101. El-Shamy, T.M., Lewins, J., and Douglas, R.W., *Glass Technology*, Vol.

13, p 81 (1972).

102. Doremus, R.H., *Glass Science*, Wiley (1973).
103. Doremus, R.H., in: *Treatise on Materials Science and Technology, Glass II* (M. Tomozawa and R.H. Doremus ed.), Vol. 13, p 25, Academic Press (1979).
104. Bartholomew, R.F., in *Treatise on Materials Science and Technology, II* (M. Tomozawa and R.H. Doremus ed.), Vol. 13, p 25, Academic Press (1982).
105. Wicks, G.G., and Wallace, R.M., *Leachability of Waste Glass Systems - Physical and Mathematical Models*, DP-MS-82-18, Savannah River Laboratory (1982).
106. Wallace, R.M., and Wicks, G.G., in: *Scientific Basis for Nuclear Waste Management VI* (D.C. Brookins ed.), Vol. 15, p 23, Elsevier (1983).
107. Kuhn, W.L., Peters, R.D., and Simonson, S.A., *Nucl. Technol.*, Vol. 63, p 82 (1983).
108. Pescatore, C., and Machiels, A.J., in: *Advances in Ceramics* (G.G. Wicks and W.A. Ross ed.), Vol. 8, p. 508, American Ceramic Society (1984).
109. Chambre, P.L., and Pigford, T.H., in: *Scientific Basis for Nuclear Waste Management VII* (G.L. McVay ed.), Vol. 26, p 985, Elsevier (1984).
110. Chambre, P.L., Lee, W.W.L., Dim, C.L., and Pigford, T.H., in: *Scientific Basis for Nuclear Waste Management X* (J.K. Bates and W.B. Seefeldt ed.), Vol. 84, p 131, Materials Research Society (1987).
111. Chambre, P.L., Kang, C.H., Dim, C.L., and Pigford, T.H., in: *Scientific Basis for Nuclear Waste Management XI* (M.J. Apted and R.E. Westerman ed.), Vol. 112, p 285, Materials Research Society (1987).
112. Grambow, B., and Strachan, D.M., *Leach Testing of Waste Glasses Under Near-Saturated Conditions*, PNL-SA-11554, Pacific Northwest Laboratory (1983).
113. Grambow, B., in: *Scientific Basis for Nuclear Waste Management V* (W. Lutze ed.), Vol. 11, p 93, Elsevier (1982).
114. Grambow, B., in: *Advances in Ceramics* (G.G. Wicks and W.A. Ross ed.), Vol. 8, p 474, American Ceramic Society (1984).
115. Grambow, B., Lutze, W., Ewing, R.C., and Werme, L.O., in: *Scientific Basis for Nuclear Waste Management XI* (M.J. Apted and R.E. Westerman ed.), Vol. 112, p 531, Materials Research Society (1987).
116. Grambow, B., and Strachan, D.M., in: *Scientific Basis for Nuclear Waste Management XI* (M.J. Apted and R.E. Westerman ed.), Vol. 112, p. 713, Materials Research Society (1987).
117. Parkhurst, D.I., *PHREEQE - A Computer Code for Geochemical Calculations*, USGS/wri-80-96, Government Printing Office (1980).
118. Plodinec, M.J., Jantzen, C.M., and Wicks, G.G., in *Advances in Ceramics* (G.G. Wicks and W.A. Ross ed.), Vol. 8, p 491, American Ceramic Society (1984).
119. Plodinec, M.J., Jantzen, C.M., and Wicks, G.G., in: *Scientific Basis for*

Nuclear Waste Management VII (G.L. McVay ed.), Vol. 26, p 755, Elsevier (1984).

120. *Proc. of the Symp. on the Oklo Phenomenon*, Report IAEA-SM-204, IAEA (1975).
121. Wicks, G.G., *WIPP/SRL In-Situ and Laboratory Testing Programs - Part I: MIIT Overview, Nonradioactive Waste Glass Studies*, DP-1706, Savannah River Laboratory (1985).
122. Wicks, G.G., and Molecke, M.A., in: *Advances in Ceramics, Nuclear Waste Management II* (D.E. Clark, W.B. White and A.J. Machiels ed.), Vol. 20, p 657 (1986).
123. Molecke, M.A. and Wicks, G.G., *Test Plan: WIPP Materials Interface Interactions Tests (MIIT)*, Vol. 9, No. 1, Sandia National Laboratories (1986).
124. Wicks, G.G., and Molecke, M.A., in: *Waste Management '88* (R.G. Post and M.E. Wacks ed.), Vol. 2, p 383 (1988).
125. Wicks, G.G., *WIPP/SRL In-Situ Tests - Part III: Pictorial History of MIIT and Final MIIT Matrices, Assemblies, and Samples Listings*, DP-1733, Savannah River Laboratory (1987).
126. Ramsey, W.G., and Wicks, G.G., *WIPP/SRL In-Situ Tests - Part III: Compositional Correlations of MIIT Waste Glasses*, DP-1769, Savannah River Laboratory (1988).
127. Wicks, G.G., Ramsey, W.G., and Molen, K.A., *Testing of High-Level Waste Forms Under Repository Conditions* (T. McMenamin ed.), p 158, Commission of European Communities (CEC), EUR 12 017 EN (1989).
128. Ramsey, W.G., and Wicks, G.G., *WIPP/SRL In-Situ Tests: Compositional Correlations of MIIT Waste Glasses*, presented at the 4th International Symp. on Ceramics in Nucl. Waste Management, to be published in proceedings (1990).
129. *Proc. of the Workshop on Testing of High-Level Waste Forms Under Repository Conditions* (T. McMenamin ed.), Commission of European Communities (CEC), EUR 12 017 EN (1989).
130. Tacca, J.A., and Wicks, G.G., *WIPP/SRL In-Situ Tests: MIIT Program - Surface Studies of SRP Waste Glasses*, presented at the 4th Int. Symp. on Ceramics in Nucl. Waste Management, to be published in proceedings (1990).
131. Clark, D.E., Zoitos, B.K., Wicks, G.G., and Molen, K.A., *Testing of High-Level Waste Forms Under Repository Conditions* (T. McMenamin, eds.) p.140, Commission of European Communities (CEC), EUR 12 017 EN, (1989).
132. Lodding, A.R., Engstrom, E.U., and Odelius, H., *Testing of High-Level Waste Forms Under Repository Conditions* (T. McMenamin ed.), p 127, Commission of European Communities (CEC), EUR 12 017 EN (1989).
133. Lodding, A.R., Engstrom, E.U., Clark, D.E., and Wicks, G.G., *Quantitative Concentration Profiling and Element Balance in SRL Glass*

After Two Years in WIPP, presented at the 4th Int. Symp. on Ceramics in Nucl. Waste Management, to be published in proceedings (1990).

134. Wicks, G.G., Lodding, A.R., Macedo, P.B., Clark, D.E., and Molecke, M.A., *High Level Radioactive Waste Management*, Vol. 1, p 443, American Nuclear Society (1990).
135. Vernaz, E., and Godon, N., *Testing of High-Level Waste Forms Under Repository Conditions* (T. McMenamin ed.), p 81, Commission of European Communities (CEC), EUR 12 017 EN (1989).
136. Harker, A.B., Flintoff, J.F., and Howitt, D.G., *Testing of High-Level Waste Forms Under Repository Conditions* (T. McMenamin ed.), p 152, Commission of European Communities (CEC), EUR 12 017 EN (1989).
137. Jercinovic, M.S., Kaser, S., and Ewing, R.C., *Testing of High-Level Waste Forms Under Repository Conditions* (T. McMenamin ed.), p 183 Commission of European Communities (CEC), EUR 12 017 EN (1989).
138. Sassoon, R.E., Gong, M., Brandys, M., Adel-Hadadi, M., Barkatt, A., and Macedo, P.B., *Analysis of Brine Leachates from MIIT; 1. Leaching of Nuclear Waste Glass Doped with Chemical Tracers*, presented at the 4th Int. Symp. on Ceramics in Nucl. Waste Management, to be published in proceedings (1990).
139. Brandys, M., Gong, M. Sassoon, R.E., Barkatt, A., and Macedo, P.B., *Analysis of Brine Leachates from MIIT; 1. Leaching of Nuclear Waste Glass Doped with Chemical Tracers*, presented at the 4th Int. Symp. on Ceramics in Nucl. Waste Management, to be publ. in proc. (1990).

8

The Chemical and Environmental Stability of Optical Glasses

R. J. Parry

Senior Research Scientist
Compositions Department
Pilkington Technology Centre
Lathom, Lancashire, England

INTRODUCTION

The ability of glass to resist the chemical agencies in its environment is of prime importance and yet glass is commonly regarded as a relatively inert substance which can last for centuries. The power of laboratory glassware to withstand chemical attack is almost taken for granted and indeed many specimens of ancient Egyptian glass still exist in a remarkable state of preservation.

No glass is immune from chemical attack. The resistance to attack will depend on factors such as the concentration, temperature and nature of the attacking reagent, and on the chemical composition, physical condition and structure of the glass. For example Pyrex® borosilicate glass is almost unaffected by boiling concentrated hydrochloric acid, but is severely corroded by dilute hydrofluoric acid at room temperature. Phosphate glasses on the other hand are attacked readily by dilute hydrochloric acid, but can be quite resistant to hydrofluoric acid.

OPTICAL GLASSES

Glasses are used for many applications in science, technology and everyday life. One large, very special group encompasses the glasses used in the manufacture of spectacles, magnifying glasses and filters, and the construction of microscopes, telescopes, binoculars, cameras, enlargers and projection equipment. These are widely known as optical glasses. In fact any glass which is used in order to exploit its optical properties can be regarded as an optical glass. The chemcial durability of optical glasses is the focus of this chapter.

Refractive Index and Dispersion

Optical glasses are characterized by two numbers. These are the refractive index, Nd, measured at the yellow helium d line (wavelength 587.6 nm) and the Abbe Number, Vd. The Abbe Number is a measure of the dispersion of the glass. It is defined as:

$$Vd = Nd\text{-}1/(N_F - N_C) \qquad (8\text{-}1)$$

where N_F is the refractive index for the blue hydrogen F line (486.1 nm) and N_C is that for the red hydrogen C line (656.3 nm).

It is usual to represent optical glasses as points on an Nd versus Vd diagram as shown in Figure 8-1. A generally accepted, but arbitrary, distinction is made between crown and flint glasses, represented by the line Vd = 55 for glasses of refractive index less than 1.6 and Vd = 50 for refractive index greater than 1.6. Each glass type is identified by a six figure reference number obtained from the first three significant figures of the term Nd-1 and the first three significant figures of Vd. The range of optical glasses is divided further into a number of groups, as shown in Table 8-1 together with the major constituents in each

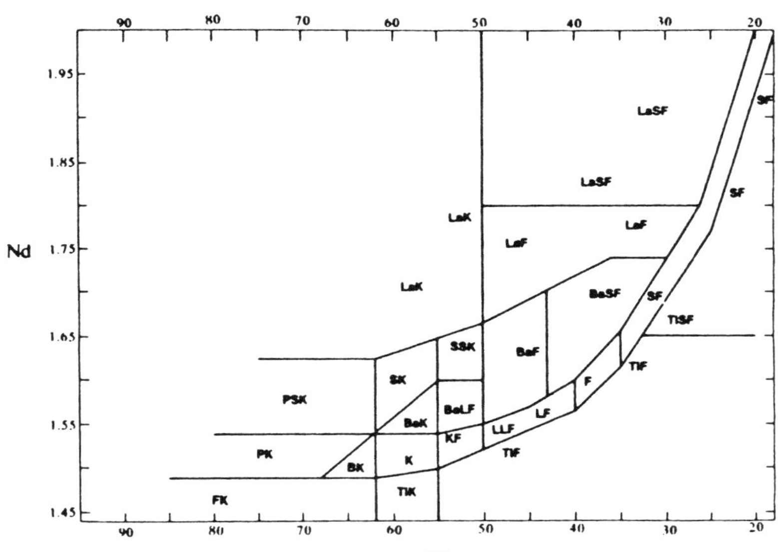

Figure 8-1. Optical glasses represented as points on an Nd versus Vd diagram.

Table 8-1.

Group	Class	Schott Code	Major Constituents
Borosilicate	BSC	BK, PK	R_2O-B_2O_3-SiO_2
Hard Crown	BC	K, BaLK	R_2O-RO-SiO_2
Zinc Crown	ZC	ZK	R_2O-ZnO-B_2O_3-SiO_2
Med. Barium Crown	MBC	BaK	R_2O-BaO-B_2O_3-SiO_2
Dense Barium Crown	DBC	SK, SSK	BaO-B_2O_3-SiO_2
Lanthana Crown	LaC	LaK	La_2O_3-RO-ZrO_2-B_2O_3-SiO_2
Fluor Crown	PC	PK	B_2O_3-SiO_2-R_2O-RF
Titanium Crown	TiC	TiK	SiO_2-TiO_2-RF
Phosphate Crown	PC	PK	P_2O_5-R_2O-RO-B_2O_3-Al_2O_3
Phosphate Heavy Crown	PC	PK	P_2O_5-R_2O-RO-B_2O_3-Al_2O_3
Extra Light Flint	ELF	LLF	SiO_2-PbO-R_2O
Telescope Flint	TF	KZF	SiO_2-R_2O-Sb_2O_3
Light Flint	LF	LF	SiO_2-R_2O-PbO
Dense Flint	DF	F	SiO_2-R_2O-PbO
Extra Dense Flint	EDF	SF	SiO_2-R_2O-PbO
Double Extra Dense Flint	DEDF	SF	SiO_2-R_2O-PbO
Barium Flint	BF	BaF, BaSF	SiO_2-BaO-PbO-R_2O
Borate Flint	Bor F	KZFS	B_2O_3-PbO-Al_2O_3
Lanthana Flint	LaF	LaF, LaSF	La_2O_3-B_2O_3-RO-ZrO_2
Titanium Flint	TiF	TiF, TiSF	SiO_2-TiO_2-KF

R_2O represents alkali oxides and RO divalent metal oxides group. The Nd, Vd region covered by each group is shown also in Figure 8-1. The outstanding contribution to the effect of composition on optical properties is that of Otto Schott in the latter part of the last century. This led to the establishment of the famous German firm Schott u. Gen. which has been foremost in the field of optical glass ever since. Many optical glass users are familiar with the traditional Schott designations, and these have been included in Table 8-1. To illustrate the nomenclature, a borosilicate crown glass with a refractive index Nd of 1.517 and Vd of 64.17 is identified as BSC 517642, or an alkali lead silicate flint glass of Nd 1.805 and Vd 25.43 is identified as DEDF 805254. The equivalent Schott types are BK7 and SF6 respectively.

The optical properties of a glass depend on its chemical composition. In formulating a glass composition to give particular optical properties the glass chemist also has to consider the effect of the composition on the other glass properties. To the user, transmission, density, thermal expansion, hardness and chemical resistance are just as important as the refractive index and dispersion. For manufacture, high and low temperature viscosity, liquidus, rate of crystal growth and electrical resistivity are also important. The cost of raw materials and their toxicity must also be taken into account. The composition selected for production is always a compromise with, perhaps, improvement of some properties at the expense of others.

The Cold Working of Optical Glasses

In the course of its preparation, an optical component, such as a lens or prism, undergoes a number of stages in which the glass is exposed to differing chemical conditions. The behavior of the glass under these conditions can greatly affect the production yields because of the occurrence of defects in the product which are linked to its chemical resistance. Prisms or lenses required in large quantity are usually worked from directly pressed or remolded glass blanks, but medium quantities are invariably made from annealed strip or slabs. The slabs are slit with a diamond saw into sizes convenient for grinding.

The glass is ground firstly with a water slurry of an abrasive such as carborundum (silicon carbide) or corundum (aluminum oxide). It is smoothed by continuing the grinding with finer and finer grades of abrasive. More recently, diamond grinding using diamond pellets bonded in alloy is being increasingly used and, depending on appropriate conditions, is capable of producing a better surface than can be achieved with loose abrasives.

After grinding, the glass is polished using a water slurry of rouge (ferric oxide), cerirouge (cerium oxide) or zirconia. In the case of cerirouge, the slurry is maintained under mildly acidic conditions at between pH 5.5 and 6.5. Polished surfaces are easily scratched. Thus immediately after polishing the first face, the lens must be rinsed in water to remove the polishing compound, carefully dried and the surface protected by brushing with shellac or spraying with cellulose paint. After the second face has been polished the protective layer is removed by washing in a solvent, e.g. methylated spirit or acetone.

At this stage the surfaces will be contaminated with traces of the pitch used for fixing the glass in the polishing tool, residues from the protective layer, finger prints, dust, oil, the remains of the polishing compounds and debris from the glass itself. It is thus necessary to clean the component. The organic contaminants are removed by using an organic solvent such as trichlorethylene or acetone in an ultrasonic tank. This is followed by immersion in a tank containing 50% water and 50% percent isopropyl alcohol with a detergent. The lenses are rinsed finally with water and then with isopropyl alcohol. A harsher method of cleaning is to use synthetic detergents in an ultrasonic tank. The synthetic detergents consist of surfactants and inorganic builders. The surfactants emulsify and dissolve the oily matter and the builders dissolve the glass surface itself to which the contaminants are attached. Commonly used builders are alkali phosphates or silicates which are alkaline in solution.

In most cases lenses and prisms are coated after cleaning in order to reduce reflection losses. The optical components are usually recleaned by suspending in solvent vapor, and dust particles are removed by mounting in a vacuum chamber and using a static eliminator. When the chamber is evacuated the glass surfaces are discharge-cleaned by exposure to the ionic bombardment of a high tension glow discharge. The material to be deposited, already placed in a tungsten or molybdenum filament, is heated until it evaporates. The vapor condenses on the glass surfaces and a film is formed. The most common material used for single layer coating is magnesium fluoride but for multilayer coatings other substances such as silica, titania, alumina, bismuth oxide or zinc sulphide can be used.

For a comprehensive account of optical glass working, the reader is referred to the books by Horne (1) and Izumitani (2).

Durability of Optical Glasses

The general term "durability" denotes the resistance of glass towards the corroding action of water, atmosphere and aqueous solutions of acids, bases and salts. Compared to window glass, laboratory ware and container glass, optical glasses are not exposed normally to severe conditions. However, this is not always the case. For example, the outer lenses of periscopes and underwater cameras come in contact with sea water, and optical instruments are used in the hot, humid, tropical conditions which promote attack even on window glass.

Durability Related Defects in Polished Optical Glass

There are three types of defect in polished optical glass components which are linked to the durability of the glass. These are known in the optical industry as dimming, staining and latent-scratch.

Dimming refers to the white opaque or crystalline deposits formed on the glass surface. In Japan, these areas of surface differing in scattering power are described by the term shiroyake (shiro = white) (3). These deposits occur due to the interaction of the glass surface with water vapor in the air, often in conjunction with acid vapors such as carbon dioxide, sulphur dioxide etc. Dimming tends to occur during storage of optically worked components. It can also arise if water is allowed to dry in contact with the glass during the polishing process.

Staining

Staining, or tarnish as is it often called, refers to the variously colored stains sometimes observed on the surface of polished optical glass. It is known to the Japanese optical industry as aoyake (ao = blue). This type of surface deterioration tends to occur during lens fabrication. Soluble glass components can be removed from the glass surface by water leaving a thin film on the surface. The refractive index of the film is lower than that of the bulk glass. This gives rise to colors due to interference of light, and the observed color of the stain will depend on the thickness and refractive index of the film. Interference colors are seen due to the interference of the light reflected from the upper and lower surfaces of the film, the latter acting as a plane parallel plate. The rays diverging from the front and back surfaces come originally from the same source and hence are in a condition to interfere. This occurs when:

$$2\mu t \cos\theta = (n + 1/2)\lambda \qquad (8\text{-}2)$$

where μ is the refractive index, t the thickness, θ the angle of refraction and n is an integer (n = 0, 1, 2, etc.).

Latent Scratch

Latent scratch refers to the tendency of some glasses to develop scratches, and lenses which appear to be perfect after final polishing become rejectable for scratches which only show up after cleaning.

The Need for Durability Testing

The capacity of an optical glass to withstand the chemical influences encountered during the preparation of optical components and their subsequent storage and use is clearly very important. As Hovestadt (4) comments: "The glass must not be tarnishable; that is, it must not be attacked by the air. It must be strong enough to bear the manipulation necessary in grinding and polishing."

It is desirable for the optical instrument manufacturer to have some indication of the likely behavior of a glass under conditions of use. This will enable him to take suitable precautions when using less-durable glasses. For example, the lens designer can adjust a lens system so that the use of "problem" glasses is avoided. If it is necessary to use such a glass then it may be possible to place it in an internal position in the lens system so that its polished surfaces are not directly exposed to the atmosphere. The user can ensure that an otherwise desirable glass, but having a poor alkali resistance, is not cleaned in an alkaline detergent solution.

The user needs appropriate information on the chemical durability of the product, based on suitable tests. Such tests must determine the durability of different types and varieties of optical glasses without awaiting the results of actual use over an extended period. The tests must rank the glasses in the order which matches the experience gained in use with types which are already familiar. The tests will also help the glassmaker to develop more resistant glasses.

The Durability Testing of Optical Glasses

The deterioration of the surface of optical elements is often extremely slow, and some considerable time may elapse before the defects become visible. It is questionable whether results from accelerated laboratory tests are in accord with those obtained under normal conditions, and it could be misleading to classify optical glasses as to their stability from any single series of tests. There are many durability test methods for optical glasses. Some of them are of a general nature, but others were developed to deal with special problems. Methods of testing have been discussed by Gliemeroth and Peters (5), Haigh (6), Adachi, Miyade and Izumitani (7), Walters and Adams (8), and Houston (9).

Most of the durability testing methods devised to determine the long term behavior of glass in use rely on similar principles. The corroding action is accelerated by using concentrated reagents, increasing the surface area by use of powders, or by using higher temperatures. Sensitive methods are employed to detect and measure the results of the tests.

Durability tests can be conveniently divided into three types, namely weathering, acid resistance and alkali resistance, and further discussion will be carried on under these three headings.

Weathering

Several methods have been devised for testing the weathering properties of optical glasses. Some of these are open to criticism because the conditions of test bear little relation to conditions of natural exposure. Tests suitable for container or flat glass are, in the main, too severe for use with optical glasses.

Autoclave methods, in which samples are heated with water under pressure have been used. Temperature, time, pressure, and whether the glass was exposed to steam or immersed in the water, have profound effects on the results. If the temperature is too high, or the duration of heating too long, it becomes too difficult to discriminate between samples. During the action of the superheated water on the glass, a considerable amount of hydrolysis and solution takes place. After autoclaving, the test samples can show considerable differences in surface appearance. Some of the more durable glasses can be unaffected, others show heavy, white incrustations and others a dimmed surface. Some glasses can be completely decomposed as reported by Morey & Bowen (10). The glasses can be ranked qualitatively as to their appearance, say in the manner described by Haigh (6) after autoclaving for 4 hours at 151°C at 4 atmospheres.

Powdered glass has been used as a means of accelerating attack in a number of hydrolytic resistance tests. These involve titrating the alkali extracted from the glass by water with a mineral acid, or determining the change in weight. The tests may be carried out at an elevated temperature to increase further the rate of corrosion. A powder test which is more appropriate for optical glasses is that adopted by the Japanese optical glass manufacturers Hoya (11) and Ohara (12). Water durability is rated into 6 classes according to the percentage weight loss when glass particles of specified grain size are boiled for 1 hour in water.

Powder methods can be useful, but great care has to be taken to achieve reproducible results, and these have to be treated with caution particularly in the case of optical glass compositions. In conducting International Organization for Standardization (ISO) tests, for example, the specified procedure has to be followed exactly to obtain consistency.

In the preparation of test samples by ball-milling, Reid et al. (13) found that the particle size distribution varied considerably with the period of ball-milling. The average particle size after the same period of ball-milling appeared to depend on the composition of the glass. It is interesting to note that Heywood (14) observed a similar phenomenon when grinding different types of coal. He suggested the preferred size of particle produced was determined by the inner structure of the material and not by the internal forces applied. Sykes (15) has discussed the precautions necessary to obtain a reproducible surface area per unit volume. The presence of aggregates of small particles, which are difficult to remove, can greatly affect the amount of alkali extracted. Sewell (16) attributed differences in surface area to three main causes: the presence of fine particles adhering to the normal grains, the presence of agglomerates of fine particles, and possible variations in the basic surface areas of the grains. Pyare et al. (17) have pointed out that the surface area of a fixed weight of glass powder may vary considerably according to the density of the glass.

Optical glasses vary widely in composition and many do not contain any alkali. Thus tests based on measuring the amount of alkali extracted are inappropriate. To take an extreme example, phosphate glasses produce an acidic solution when attacked by water. The differences in structure and density leading to differences in total surface area of powdered samples could lead to erroneous conclusions when comparing the weathering resistance of optical glasses which are unrelated in composition.

Practically all glasses used for optical purposes have their surfaces ground and polished and it is generally accepted that the chemical resistance properties of a polished surface can differ considerably from that of the bulk glass. For this reason, and the problems associated with powder methods outlined above, tests involving polished glass samples have gained support.

Jones (18), for example, developed a dimming test in which polished pieces are exposed to a water-saturated atmosphere at 50°C. The samples were examined after a month and classified according to the haze produced. In this test, however, droplets of water tended to form which could wash off the corrosion products. Simpson (19) devised a test in which samples were cycled between 72% and 100% relative humidity which overcame the droplet problem. The surface deterioration was measured using a haze meter and he recorded the percentage haze against time over a period of 32 days for several different glasses. Simpson found that there was a good correlation between his laboratory tests and the appearance of the glasses stored in an unheated warehouse for twelve months in Puerto Rico where the climatic conditions were known to be severe. In order to give a closer simulation of natural conditions the optical glass manufacturer Chance (now Pilkington) introduced the Thermodyne Test (20). This made use of the most important features of atmospheric exposure, namely, variation in temperature and humidity.

Both Pilkington and Schott currently use the same climatic resistance test which is as follows:

> Polished uncoated glass plates are exposed to a water vapor saturated atmosphere with the temperature being alternated between 40°C and 50°C in a cycle of approximately 45 minutes. During each cycle water can condense on the glass in the heating phase and dry up during the cooling phase. After having been exposed for 30, 100 and 180 hours, the glasses are removed from the test chamber and classified according to the percentage light scatter caused by surface changes as shown in Table 8-2.

Table 8-2.

Climatic Resistance Group (CR)	Time in hours to exhibit 2-5% light scatter
1	> 180
2	from 180 - 100
3	from 100 - 30
4	< 30

No surface deterioration of glasses in Class CR1 is expected under normal conditions of processing and storage, but particular care should be taken with glasses in class CR4. Classes CR2 and CR3 contain glasses having climatic resistances lying between the two extremes.

Acid Resistance

Modern lens systems often have many components and consequently a large number of reflecting surfaces. For maximum light transmission it is necessary to reduce the reflection losses by "blooming" the lenses, ie. by applying antireflection coatings. The application of blooming techniques has overcome to a great extent the problem of deterioration in use due to atmospheric attack because of the protective qualities of the coating materials. However, the occurrence of stains in the polished lenses which were not detected prior to blooming can become obvious when the lenses are coated. This can cause a high rejection rate of finished components at the end of a costly production process.

The susceptibility of a glass to staining can be assessed by measurement of its acid resistance. The soluble glass components are leached out by acid, but if the glass contains insoluble components, especially silica in sufficient quantity, these remain on the surface in the form of a protective layer. Such layers are exceptionally cohesive and optically homogenous and tend to inhibit further attack. If the thickness of these layers reaches 1/4 wavelength colored spots appear due to interference. The optical thickness of a layer, or its refractive index and physical thickness separately, can be determined by optical means such as the observation of the interference color, measurement of reflectivity, ellipsometry etc. This has been discussed in some detail by Kinosita (3).

Berger (21) suggested that a relation of the type

$$100t = ax + bx^2 \tag{8-3}$$

can represent the relationship between acid attack and time with a sufficient degree of accuracy, where t denotes the time in hours and x the layer thickness in microns. Schott (22) formerly quoted in their catalogue the logarithms of the factors a and b for the attack of 0.5 N HNO_3. The glasses were classified according to the time taken to destroy a glass film of 0.1 micron. Less resistant glasses were further tested but using 4.6 pH standard acetate solution as the attacking agent.

In the case of some glasses the value for b is negligible and this denotes a uniform attack which always occurs when the attacked glass layer is removed completely. The protective action of the formed surface layer is indicated by the increasing value of b. This protective action can become so effective that the value of a can be ignored.

A Pilkington modification of the Schott method was devised as follows:

Dividing equation (8-3) by x

$$100\ t/x = a + bx \tag{8-4}$$

This shows that a plot of t/x against x should give a straight line with intercept a and slope b. In the case when b is zero 100 t/x should be a constant, and when a is zero the graph will be a straight line through the origin. The difference in level between the attacked and unattacked surface of polished pieces placed for varying lengths of time in 4.6 pH standard acetate solution at 25°C was measured. This was done by counting the number of fringes displaced when a sample was observed under a microscope fitted with a Nomarski interferometer attachment (23).

It was found in general that there are three types of attack.

(a) The surface becomes stained

(b) A layer is completely removed leaving a clean etched surface

(c) An opaque layer is formed.

Photographs of the interference patterns obtained for the three types of attack are reproduced in Figures 8-2, 8-3 and 8-4, respectively.

It is often difficult to measure the fringe displacement as in case (c) but by reducing the period of attack it is sometimes possible to obtain a reasonable fringe pattern as shown in Figure 8-5. The results using this method for DBC615562 are given in Table 8-3.

Figure 8-2. Interference pattern of a stained surface.

Table 8-3.

Time (hours)	Thickness of layer (microns)	100 t/x	log 100 t/x
2.5	0.075	3330	3.52
5.0	0.150	3330	3.52
7.25	0.262	2770	3.44
10.00	0.300	3330	3.52

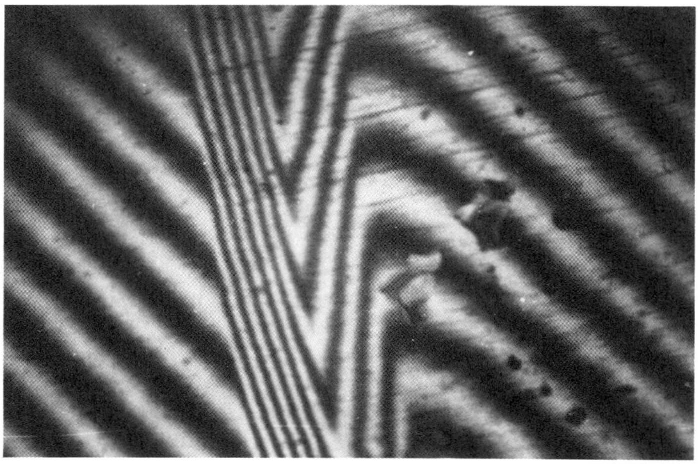

Figure 8-3. Interference pattern of second type of attack. (Layer removed, clean etched surface remains).

Figure 8-4. Inteference pattern of opaque layer.

Figure 8-5. Fringe pattern.

It was found that equation (8-4) does not apply to all glasses, as shown by the results for DBC 614599 in standard acetate solution (Table 8-4). In this case the glass was obviously attacked as shown by the color of the stains produced in the initial stages. From the direction of displacement of the fringes it is possible to decide whether a swelling or etching has occurred. Although it was not possible to measure the displacement, it appeared that up to 2 hours immersion time the glass surface swelled slightly. After this time the samples had a layer of crystalline material adhering to them which could be removed by rubbing with a cloth to give a fairly clean etched surface.

Table 8-4.

Time hours	Thickness of layer (microns)	Color of stain	100t/x
0.5	0.0	blue-black	-
0.75	0.0	dark blue	-
1.0	0.0	blue	-
2.0	0.09	green	2200
3.75	0.90	-	417
5.0	1.05	-	475
6.0	1.20	-	500
7.0	1.41	-	495
22.0	2.10	-	1050

The method was not very satisfactory because some glasses were clearly attacked, as shown by staining, but did not give a measurable difference in level. In other cases the build up of a crystalline layer on the attacked surface prevented measurement of the fringe pattern.

Walters and Adams of Corning Glass Works (8) modified the American Optical test used for ophthalmic glasses. Polished test pieces are exposed to 10% HCl at 25°C. One specimen is partially immersed in the acid for 20 seconds and examined for signs of discoloration. A second sample is totally immersed for 10 minutes and the loss in weight determined. More stable glasses are exposed for a two hour period.

Reid et al. (24) devised the test used by Pilkington until recently. This consisted of placing polished pieces in 0.5N nitric acid at 25°C and determining the time for the first signs of attack to become apparent. The nature and extent of surface corrosion is expressed by a 3 term code. The first indicates the time, the second the severity of attack and the third the manner in which the surface is affected. Two British instrument manufacturers, Rank Precision Industries and Wray Optical, were asked to grade the optical glasses used by them in the order of difficulty of processing. With two exceptions the acid resistance measurements placed them in the same order. The two exceptions were BF700412 which was better than expected and BF614439 which was worse.

Powder methods have been used to assess acid resistance. For example, the Japanese manufacturers Hoya and Ohara rank their glasses according to the percentage weight loss when glass grains of size 420-590 microns are boiled for 60 minutes in 0.01N HNO_3. Another powder method devised by Bredow (25) was used by SOVIREL (26) (now part of the US company Corning) to grade their glasses.

The multiplicity of tests makes it difficult for the optical glass user to compare glasses from different sources and this has been recognized by the glass manufacturers. In 1980 an ISO sub-committee was set up comprising of expert representatives of each of the major optical glass manufacturers. The purpose of the committee was to devise a method of testing the resistance of optical glasses to aqueous acid solutions at 25°C, and a classification of the glasses according to the results of the test. This fine example of international co-operation has led to the publication of ISO 8424 which is outlined below.

The acid resistance of a glass is defined by the time taken to remove a layer of 0.1 micron thickness. Two strengths of acid are employed depending on the resistance to this form of attack. For highly resistant glasses, 0.5N nitric acid is used to locate them in one of four groups. The samples are polished pieces of size 30 x 30 x 2mm and the layer removed is determined by weighing. If a 0.1 micron layer is removed in less than 6 minutes, 4.6 pH standard acetate solution is used as the attacking medium. The glasses are graded as shown in Table 8-5.

Table 8-5

Acid Resistance Group	pH of attacking solution	Time in hours to remove 0.1 micron layer
1	0.3	> 100
2	0.3	from 10 - 100
3	0.3	from 1 - 10
4	0.3	from 0.1 - 1
5	0.3	< 0.1
51	4.6	from 1 - 10
52	4.6	from 0.1 - 1
53	4.6	< 0.1

Changes in the surface of the sample which are visible after the group has been established are assessed qualitatively with the naked eye. These are indicated in addition to the acid group as follows:

- 0.0 no visible change
- 0.1 clear but irregular surface
- 0.2 staining and/or interference shows
- 0.3 thin whitish layer (cloudy surface)
- 0.4 loosely adhering thick layers

Alkaline Resistance

In former lens production methods it was unusual for optical glass to meet alkaline conditions. However, modern methods of bulk manufacture of lenses have changed the situation. In order to raise polishing rates, increased pressure and harder polishing agents, such as cerium oxide fired at higher temperatures, are used. This increases the tendency for scratching to occur. Immediately after polishing, these scratches are too small to be observable and if the lenses were cleaned with an organic solvent they probably would pass inspection.

Organic liquids are expensive, toxic and inflammable. Thus lens manufacturers have resorted to cleaning with solutions of inorganic detergents to avoid their use. These detergent solutions are generally alkaline and, when the polished lenses are washed, the previously invisible scratches stand revealed. Kinosita (3) says the latent scratches are those regions which have been rendered sensitive by local severe stress during the polishing or grinding, and become visible when the depth of corrosion reaches some tens or hundreds of angstroms. He observes that latent scratches can also be developed by acid treatment, though to a lesser extent. Izumitani (2) maintains that the ease with which scratching occurs depends on the hardness of the glass, whereas the tendency of the scratches to grow depends on the chemical durability of the glass with respect to the inorganic builders involved. Fletcher et al. (27) have examined 27 commercially available inorganic detergents and most of them are strongly alkaline.

Higher speed grinding and polishing increases the temperature generated due to friction, and as the glass is abraded, the water-based media tend to become more alkaline especially when the solutions are recycled. In order to indicate the susceptibility of optical glasses to alkaline attack Schott (28) introduced a new test which has also been adopted by Pilkington (29). The test is designed to simulate the effect of a hot alkaline wash on a polished glass surface. A solution of sodium hydroxide (pH 10) at 90°C. is used, and the time in minutes required to remove a layer of 0.1 micron thickness is determined. A two digit figure is used to express the resistance. The digit before the decimal point indicates the AR class, and that after the decimal point indicates the surface change observed with the naked eye. The classes are shown in Table 8-6.

Table 8-6.

Alkaline Group (AR)	Time in minutes to remove a 0.1 micron layer
1	> 120
2	from 120 to 30
3	from 30 to 7.5
4	< 7.5

The surface change is characterized by

.0	no change
.1	clear but irregular surface
.2	interference colors
.3	whitish staining
.4	white coating (thick layers)

Hoya (11) use a test in which the loss in weight of a polished sample in 0.01M NaOH is determined after a period of 15 hours. They also use another test with 0.01M sodium tri-polyphosphate solution because this is more corrosive towards lanthanum and lead-containing glasses than caustic soda alone.

Andrianova and Molchanov (30) described a test which they applied to non-silicate glasses. It consisted of finding the weight loss of rectangular samples, polished on all sides, when exposed to 0.5N NaOH at 90°C. for 4 hours. They concluded that alkali washing of optical components is permissible only with those glasses whose surfaces are etched by no more than 3-4 microns.

Summary of Test Methods

Tests for corrosion resistance such as the ones outlined above vary in merit, but most of them will give an indication of glass durability. Ideally, a suitable test should be quick, simple and reproducible. It should be capable of comparing all glasses no matter what their compositions may be, and rank them in order gained by experience. Because of the difference in response to weathering, and the attack of acids, alkalies and detergents, more than one test is necessary to predict the resistance of optical glasses to the chemical and environmental conditions they may encounter.

The optical glass manufacturers are aware of the need for uniformity in test procedure so that users can compare glasses obtained from different suppliers. Under the chairmanship of Dr. A. Peters of Schott, the ISO sub-committee mentioned previously is engaged in devising suitable tests for climatic, alkaline and detergent resistance. It is expected that these tests will be introduced within the next two years.

The Effect of Composition on Durability

Because of the complexity of optical glass compositions and the widely different response to solutions of different pH, it is difficult to discuss the effect of composition on durability in a general way. However, there are a number of guiding principles. The most important component governing durability is the glass former itself, and the nature of the glass former is reflected in the glasses derived from it. The main glass formers are silica, boron trioxide and phosphorus pentoxide, and the glasses based on each of them are dealt with separately.

Silicate Glasses

The majority of optical glasses have silica as the glass former, and most of the published work on durability has been carried out on silicate glasses. Fused silica is not attacked by water at normal temperatures and it is resistant to acids, except hydrofluoric and phosphoric acids. It is susceptible to attack by alkaline solutions.

Many silicate glasses contain alkali and it appears that the leaching rate in water and acid solution increases with the alkali content. Zhdanov (31) compared the leaching rate in 0.1N HCl of binary lithium, sodium and potassium silicate glasses in which the molar percentage of alkali was the same. The leaching rate was greatest for the potassium glass and least for the lithium.

Pyare et al. (17) examined the effect of divalent elements on water solubility when substituted in a base glass of molar percentage composition 16 Na_2O.10 RO.74SiO_2 where RO was MgO, CaO, SrO, BaO, PbO and ZnO. They showed that the higher the coulombic force between the divalent cations and a singly charged oxygen ion the lower is the solubility. The coulombic force is given by:

$$ZZ' / (r + r_o)^2 \tag{8-5}$$

where Z and Z′ are the electrostatic charges on the divalent cation and the oxygen ion, and r and r_o are the respective ionic radii. Results were included also for Al_2O_3, TiO_2 and ZrO_2 substitutions in the base glass. These show that the solubility decreases further as the coulombic force increases although zirconia appears better than might have been expected on this basis. Earlier work by Dimbleby and Turner (32) on the effect on water durability of replacing sodium oxide by other oxides in a sodium silicate base glass placed the oxides in the same order.

Brewster, Kreidl and Pett (33) working for the US company Bausch & Lomb made molar substitutions of various oxides for alumina in a simple barium borosilicate composition. The compositions were tested against 1% sodium metaphosphate solution, 1% nitric acid solution and 50% hydrochloric acid solution. They concluded "The chemical resistances of the glasses containing molecular equivalents of La_2O_3, ThO_2, ZrO_2, ZnO and SiO_2 are, in this order better than those containing equivalent amounts of the alkaline earths, but are nevertheless surpassed by those containing TiO_2, Nb_2O_5 and Ta_2O_5."

Aposhian and Kreidl (34) made molar replacements of TiO_2 ZrO_2 Ta_2O_5 and Nb_2O_5 in a potassium lead silicate base glass. These were tested against 1% nitric acid solution at 25°C or 50% hydrochloric acid at 50°C or 90°C and the time to develop a 1/4 wavelength thick film was determined. Big improvements in resistance were seen when lead was replaced. In nitric acid, the order was $TiO_2 = Nb_2O_5 > ZrO_2 > Ta_2O_5$ whereas in HCl the order was $Nb_2O_5 > ZrO_2 > TiO_2 = Ta_2O$.

With regard to water and acid corrosion resistance it is possible to make a general statement that the higher the coulombic force the better the durability. The ionic radii and coulombic forces for a number of cations are shown in Table 8-7.

There can be exceptions to the general rule however. For example, in the case of alkali-containing glasses it has been observed that a mixture of sodium and potassium oxides gives a better water resistance than the same amount of alkali in the form of sodium oxide or potassium oxide alone. If one alkali oxide is replaced by another in a series of glasses, some physical properties such as expansion or electrical resistivity vary in a non-linear manner, and, depending on the property, reveal maxima or minima at a particular alkali ratio. This phenomenon is known as the mixed alkali effect. Peddle (35) stated that to obtain maximum durability in an alkali-lead-silicate glass, the optimum ratio should be 7 parts of potassium oxide to 3 parts of sodium oxide. This ratio holds good for all percentages of alkali below 20, and is independent of the amount of silica or lead oxide in the glass. Sun Yuzhen et al. (36) found when

Table 8-7

Cation	Radius (Å)	Coulombic Force $(ZZ'/(r + r_o)^2)$
Cs^+	1.67	0.106
K^+	1.33	0.134
Na^+	0.97	0.178
Li^+	0.68	0.231
Ba^{2+}	1.34	0.266
Pb^{2+}	1.20	0.296
Sr^{2+}	1.12	0.315
Ca^{2+}	0.99	0.350
Cd^{2+}	0.97	0.356
Zn^{2+}	0.74	0.437
Mg^{2+}	0.66	0.471
La^{3+}	1.016	0.514
Th^{4+}	1.02	0.683
Al^{3+}	0.51	0.822
Zr^{4+}	0.79	0.834
Ti^{4+}	0.68	0.925
Ge^{4+}	0.53	1.1074
Nb^{5+}	0.69	1.145
Ta^{5+}	0.68	1.156
B^{3+}	0.23	1.129
Si^{4+}	0.42	1.206
P^{5+}	0.35	1.633

working on alkali-titanium-silicate glasses that minimum water solubility occurred when the K_2O/Na_2O ratio was approximately 2.5.

Another departure from the general rule was noted in the work by Aposhian and Kreidl mentioned above. They showed that a combination of titania and zirconia gave a higher acid resistance than the same molar concentration of either oxide alone.

With regard to alkaline resistance, Molchanov and Prikhid'Ko (37) examined the effect on alkaline corrosion of molar substitutions of divalent metal oxides for silica in a base glass having the composition 13% Na_2O 87% SiO_2. The attacking agent was 0.5N NaOH and the thickness of layer destroyed in 4 hours at 90°C was measured. They found that for up to 20% molar replacement of silica there was practically no change in alkali resistance. With further silica replacement there was a gradual decrease in stability and differences between the various oxides began to appear.

The oxides can be ranked according to the extent to which they increase the corrosion resistance. The order is BeO > CaO > CdO > MgO > SrO > BaO > ZnO > PbO. Using the same test conditions and base glass composition,

Molchanov and Makarova (38) examined the effect of oxides of multivalent elements. They found the order of alkaline resistance to be:

$La_2O_3 > ZrO_2 > ThO_2 > Nb_2O_5 > Ta_2O_5 > CeO_2 > TiO_2 > CrO_3 > MnO > B_2O_3 > Al_2O_3 > NiO > CoO$

They suggested that the higher the solubility of the metal hydroxide the greater is the attack on the glass. It is significant that the only distinction between lanthanum and the other trivalent elements is the insolubility of its hydroxide in alkalies and water. This also applies to the hydroxides of zirconium and thorium. Similarly, the higher resistance of cadmium glasses in comparison with zinc glasses is attributed to the fact that zinc easily forms zincates in alkaline solution whereas there is no corresponding effect with cadmium. The ability of niobium to form niobates, and tantalum to form tantalates can be used to explain why these higher valency elements are not as effective as lanthanum in lowering the solubility in alkaline solutions.

Borate Glasses

Vitreous boron trioxide is hygroscopic and soluble in acids and alkalies. Although boron trioxide is mainly used as an additive to glasses in which silica is the main glass former, there are numerous optical glasses which are truly borate glasses. As a family, borate glasses have poor durability, but they are used extensively for their valuable optical properties. Boron trioxide in combination with lanthanum oxide is the basis of the high index, low dispersion glasses discovered by the eminent American George Morey (39). Boron trioxide glasses are peculiar in that the rate of change of refractive index with wavelength in the blue region of the spectrum is less than that of silicate glasses having the same Abbe Number. This enables lens designers to correct for the so called "secondary spectrum" in achromatic combinations. Unfortunately, these borate flint, or KZFS, glasses are amongst the worst for climatic, acid and alkaline durability in the glass manufacturers' catalogues.

In vitreous boron trioxide the glassy matrix is formed by the linkage of boron and oxygen atoms in which each boron atom is surrounded by 3 oxygen atoms, i.e. it has a coordination number of 3. When another oxide, such as an alkali metal oxide, is added to the boron trioxide, some of the boron atoms can surround themselves with four oxygen atoms, i.e. the coordination number changes to 4, with electrical neutrality being maintained by adjacent alkali ions. The durability is closely connected with the ratio of BO_4 to BO_3, which in turn is dependent on the amount of alkali oxide added. As the ratio of alkali oxide to boron trioxide is increased, the solubility in both acid and alkali falls. This has been illustrated by work carried out by El-hadi et al. (40)

Mazelev (41) conducted experiments on the water solubility of a series of lithium borate glasses to which was added divalent elements. He concluded that the solubility was minimal when the whole of the boron was bound in borates. The formation of any borates, even lithium borate, sharply decreased the solubility. Glasses with a high content of divalent metal borates were similar in solubility to normal industrial silicate and borosilicate glasses. The degree of solubility was not simply due to changes in the amount of oxides, but mainly structural factors: the ratio of the lithium and divalent oxides to boron trioxide, the degree of binding of B_2O_3 in borates, the character of the borates formed, and the corresponding change in the coordination state of the boron in the glass. 35 mole% replacements of cadmium, zinc, lead strontium and barium oxides for lithium oxide in a base glass of molar composition 40% Li_20 60% B_20_3 increased durability in the order: Cd0 > Zn0 > Pb0 > Sr0 > Ba0.

This indicates zinc, strontium and barium are in the same order as might have been expected from the coloumbic forces shown in Table 10, but cadmium and lead are better than expected.

Brewster et al. (33) examined glass in the ternary BaO - La_2O_3 - B_20_3 system. They found all the glasses have very poor chemical resistances except the binary lanthanum borate. A big improvement in resistance to sodium metaphosphate solution was achieved when part of the lanthana was replaced by tantalum pentoxide. Chakraborty et al. (42) in work on binary lanthanum borate glasses stated that glasses containing more than 20 mole% lanthana are very durable under ambient conditions. The response of one glass, 25 mole% La_2O_3 75 mole% B_2O_3, to the effect of solutions of varying pH showed it to be much more soluble at low pH than at higher pH.

Phosphate Glasses

Vitreous phosphorous pentoxide has a great affinity for water and phosphate glasses are generally rather soluble. Phosphate glasses have good ultraviolet light transmission and their rate of change of refractive index with wavelength in the blue region of the spectrum is greater than that of silicate glass having the same Abbe Number. By using a phosphate crown element together with a borate flint element in an achromatic doublet it is almost possible to eliminate the secondary spectrum. Because of these special properties phosphate optical glasses are in demand in spite of problems with durability.

Bunker et al. (43) studied the dissolution of a series of phosphate glasses in aqueous solutions. The glasses having the composition (50 - X) M_2O.X Ca0.50P_2O_3, where M_2O represents alkali metal oxide, appeared to dissolve uniformly in contrast to simple silicates which are selectively leached. The trends observed in durability as a function of composition mirrored, in general, those seen for silicate glasses. The more alkali the glass contained, the lower the durability. The durability increased in the order Li > Na > K. When calcium oxide was substituted for barium oxide the solubility decreased.

Phosphate glasses of good water corrosion resistance can be obtained by melting the metaphosphates of divalent elements. Magnesium metaphosphate and zinc metaphosphate glasses are particularly stable. The most durable phosphate glasses, however, are obtained by the addition of aluminum oxide; even quite small additions can lead to an appreciable decrease in solubility.

Minami and Mackenzie (44) examined the water solubility of potassium phosphate glasses to which was added a third oxide component. These included B_20_3, Ba0, Bi_20_3, Ca0, $Ge0_2$, La_20_3, Nb_20_5, $Mo0_3$, $Si0_2$, Ta_20_5, V_20_5, Y_20_3, Zn0, Al_20_3 and $W0_3$. The last two oxides were the most effective in increasing durability. In the systems R_20-Al_20_3-P_20_5 and R_20-$W0_3$-P_20_5, where R_20 represents alkali metal oxide, they found in both systems that the dissolution rate in water at 100°C increased in the order Cs > Rb > K > Na > Li. Smooth curves were obtained when the dissolution rate was plotted against the radius of the alkali ion. When combinations of Li - Cs, Na - K and K - Cs were tested there was evidence of a mixed alkali effect.

Parker (45) examined the effect of 5 mole% substitutions of various oxides for P_20_5 in a glass of molar composition P_20_5-50% Ca0-15% Mg0-15% Na_20-20%: He determined the loss in weight of polished samples of size 25 x 25 x 5 mm after boiling in 1 liter of water for 6 hours. He showed that the solubility fell when any oxide replaced P_20_5. When alkali oxides were substituted the order was Li > Na > K > Cs. The divalent element oxides gave the order: Zn > Mg > Ca > Sr > Ba > Pb.

Aluminum oxide gave the same solubility value as zinc oxide. Apart from zinc and lead, these elements decreased the solubility in the order anticipated from their coulombic forces. When the glass formers $Si0_2$, B_20_3 and $Ge0_2$ were substituted, the solubility decreased in the order Ge > B > Si. This is the reverse of the order which might have been expected from their coulombic forces, and possibly indicates the onset of microphase separation. It is known that combinations of the glass formers tend more or less towards phase separation (46) and it is also known that micro phase separation can influence durability. If the more durable phase is dispersed as droplets in a matrix of the less durable phase, the durability is reduced. If the less durable phase is dispersed within a connected structure of the more durable phase, the durability can be increased (47,48).

With regard to alkali resistance, Andrianova and Moldanov (30) observed that in tests on a series of optical glasses, the four best phosphate glasses were decomposed about five times faster than silicate glasses. When the test results were compared with the known glass compositions, they concluded that the least resistant glasses were those incorporating zinc. Barium and calcium phosphates were the major constituents of the most resistant glasses and magnesium phosphate was intermediate between zinc and barium in terms of imparting alkaline resistance.

Glasses in General

Although silicate, borate and phosphate glasses differ in their familial durability characteristics, there are some common compositional features.

The effect of the various elements appears to be governed by the valency of the element and the ionic radius of its cation. In the case of water and acid resistance, generally the higher the coulombic force between the cation of an element and a singly charged oxygen ion, the greater will be the effect of its oxide on increasing resistance. Ions of elements, such as lead or zinc, with 18 electrons in the penultimate shell do not always obey this general rule. For example, Tindyala and Ott (49) found that molar substitutions of lead oxide for zinc oxide in a lithium-zinc-alumino-phosphate base glass decreased the solubility in water. Dimbleby and Turner (32) showed that zinc oxide in relatively high concentration was more effective than titania, and identical to alumina, in reducing water solubility.

The effect of a particular oxide on durability can vary with its concentration in a glass. For example, the acid resistance of lead containing flint glasses becomes progressively worse as the lead content is increased. Ge et al. (50) found that the alkaline resistance of glasses in the three-component $Ca0$-Al_20_3-$Si0_2$ system fell with increasing alumina content. This was not simply a concentration effect, but depended on the structure of the glasses. The resistance fell as the coordination number of the aluminum ions changed from 4 to 6.

Ions of elements with high coulombic forces can compete for oxygen ions to satisfy their coordination requirements and this can lead to microphase separation. Microphase separation can either increase or decrease durability.

The mixed alkali effect can produce improvements in durability and it appears that this is only one particular example of a general phenomenon, the mixed oxide effect. The mixed oxide effect has been discussed by Dietzel (51) on the basis of differences in the coulombic forces, or field strengths, of the cations.

The effect of a particular oxide on the properties of a glass cannot be regarded in isolation from the other glass components. All the glass ingredients influence its properties to a greater or lesser extent.

The Effect of Micro-organisms

Under certain environmental conditions optical glass components can deteriorate due to the growth of molds or bacteria on the surfaces. The presence of such micro-organisms can obscure lenses and prisms, and lead to considerable loss in light transmission. In addition, the surfaces of some glasses can be damaged by etching.

Fungi

As early as 1830 Faraday (52) commented "In plate glass the change is of another kind, and is shown by the appearance of minute vegetations or crystallizations, which spread, obstructing the light whenever they occur. Mr. Dolland, who has shown me cases of both kinds of injury in flint and plate glass, is inclined to believe that the latter has, during his long experience, proved most injurious."

According to Turner et al. (53), fungal spores germinate on the moist surfaces of lenses or prisms, or more frequently on the organic debris on the surfaces. The mycelium spreads thence over the whole surface, and the fine hyphal threads are often surrounded by minute condensed water droplets, or droplets of alkali-soluble substances from the glass itself. Richards (54) stated that not all branching forms are due to molds. One such "growth" on a glass surface was shown by microchemical methods to be crystallized potassium carbonate, which had been formed from potassium extracted from the glass.

The growth of micro-organisms is enhanced by a hot, humid atmosphere as found in tropical regions. If organic matter is present as a source of food, simple forms of plant and animal life may grow quite vigorously on a glass surface. Foreign material that collects on a glass surface is often organic in nature; for example, dust, grease, fingerprints and dead microscopic insects. The glass does not act as a food for molds, but provides a surface on which they can grow and water can condense. The mold growth that takes place is initiated by the sorption of water. Teitell and Berk (55) suggested the reserve food present in the mold spores, along with the sorbed water allows sufficient mold growth to interfere with the operation of optical instruments.

Turner et al. (53) said that the fungi which grow in optical instruments belong to the groups Phycomycetes, Aseomycetes and Fungi imperfecti. Kerner-Gang (56) reported that 66 fungal species had been isolated from microscopes from all parts of the world and the majority were known to cause glass to deteriorate. The question of how specific to the tropics are the fungi which grow on glass surfaces is unsettled. Mold spores from one location can be transported on the instrument to another where conditions are such that rapid growth can occur. The spores may be introduced during the construction of the instrument, or enter from outside during packing, storage or use in the field.

Nagamuttu (57) was of the opinion that the frequency of fungal growth depended on the material of the lens. Razumovskya and Mitushova (58) examined fungal growth on a range of optical glasses. They found that the fungi tested grew best on the natural surface of quartz. Fungi spores can germinate over quite a wide range of pH values, but for the majority of fungi an acidic medium favors spore germination. Substantial differences in the growth on the glasses were not observed except in the case of one glass with a high alkali content. As far as this glass was concerned either no development of fungi occurred, or only the growth of a few spores and a little development of mycelia was noted. This was attributed to the alkali liberated from the glass by moisture which raised the pH of the surface.

Jones (59) mentioned that organic acids can be produced by the growth processes of fungi which can cause etching of glasses susceptible to acid attack.

He also noted that if the fungi are cleaned from a glass surface before they can take part in chemical reactions, the glass is not damaged. Richards (54) found that fungal growth will not take place if the glass surfaces are perfectly clean.

Baker (60) has discussed the preventive measures to inhibit fungal attack. These include hermetical sealing of instruments and use of a fungicide. He found that ethyl mercury chloride was completely effective, but should only be regarded as a secondary measure to that of the production of an instrument capable of being properly sealed and desiccated.

Figure 8-6 (X275) shows the growth of aspergillus penicillioides on a spectacle lens after 4 months incubation at 90% relative humidity. No nutrient was used. Cleaning removed the fungus completely leaving a good surface.

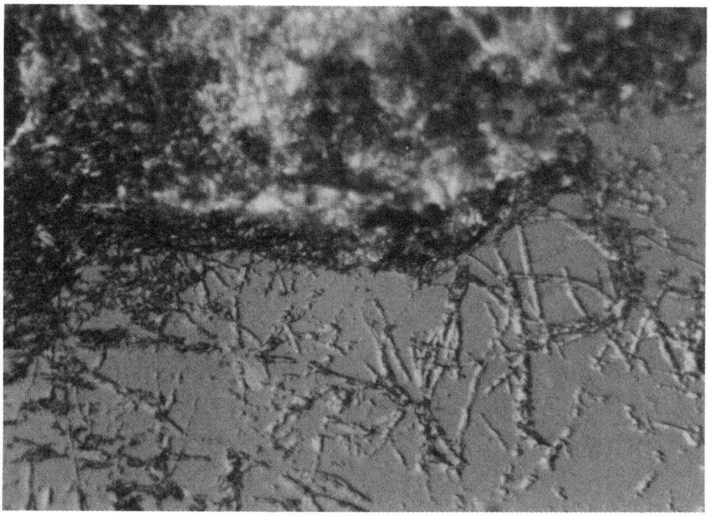

Figure 8-6. Aspergillus penicilloides on spectacle lens.

Figure 8-7 (X275) shows the etched surface of a spectacle lens after removal of fungus. The fungus was aspergillus pencillioides incubated for 6 months at 90% relative humidity. The nutrient was yeast.

Bacteria

Razumovskya and Mitushova (58) said that the main agents responsible for overgrowth of glass surfaces are lower fungi, although sometimes bacteria are encountered. These bacteria were judged to be of lesser significance.

Prod'homme (61) conducted laboratory experiments on three optical glass types, namely BSC518642, DBC590605 and EDF693310. After preliminary tests on different sorts of bacteria she found that Escherichia Coli, which is well known and very widespread, developed best in contact with glass. Two nutritive

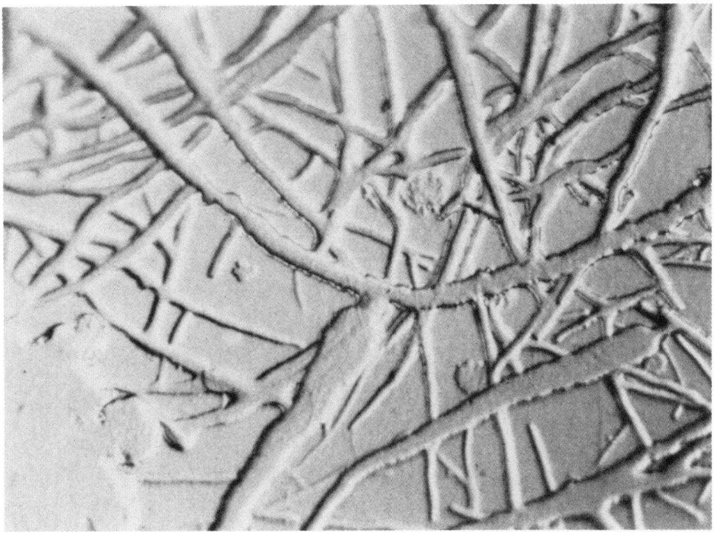

Figure 8-7. Etched surface of spectacle lens after removal of fungus.

media were used; a solid medium made up of nutritive gelose, and a liquid medium made up of 95% Stoll and 5% glycerol. The BSC glass showed no deterioration in one month with the nutritive media alone, but the medium seeded with E. Coli showed deterioration after 15 days. The samples exhibited lines and points which became increasingly numerous and pronounced. They appeared most frequently in the form of single lines, or lines in a herring-bone pattern, and resembled faults obtained by chemical attack. The DBC and EDF glasses were both attacked by the nutritive media within 1 month, but the attack was greater when seeded with E. Coli. The action of the micro-organisms appears to be that of increasing and promoting the attack at certain preferential sites, but a nutritive medium is essential for the development of bacteria.

Khan and Khan (62) isolated fourteen organisms obtained from a range of glass items showing microbial growth on their surfaces. These were tested for their preference towards glass for growth and sporulation. It was found that the organisms utilized inorganic ingredients for their growth and sporulation. They were alkalophilic, independent of light and did not require an organic carbon supply. They were also tolerant of drought conditions, and capable of growing on apparently dry glass surfaces.

SUMMARY

The chemical and environmental stability of optical glasses is of great technical and commercial importance.

In the course of their conversion from raw glass to ground and polished lenses and prisms, optical glasses encounter a variety of corrosive media. Such media can give rise to defects which are linked to the durability of the glass. These defects are known as dimming, staining and latent scratch. During storage and in operation optical glass surfaces can deteriorate due to weathering, or attack by micro-organisms.

In order to make reasoned estimates of the likely behavior of optical glasses under conditions of service, use is made of accelerated laboratory tests. Such tests must determine the durability of different types and varieties of optical glasses without awaiting the results of actual use over an extended period. The test results should rank the glasses in accord with the order gained by experience in use. There are many tests available and there is a need for standardization. This problem is being addressed by the International Organization for Standardization with the active co-operation of the major optical glass manufacturers. Suitable tests are being devised to determine climatic, alkaline and detergent resistance. A test for acid resistance has already been published. Because of the difference in response of glass to weathering, and the attack of acids, alkalies and detergents, more than one test is required in order to pre-determine its overall durability.

The durability of a glass depends on its chemical composition and structure. In the case of optical glasses, the number and variety of ingredients necessary to provide particular optical properties make it very difficult to define how each component functions in the complex mixtures of which even the most simple optical glasses are composed. The durability is determined mainly by the glass former, but it can be modified by the other glass components. In the case of water and acid attack it appears, generally, that the higher the coulombic force between the cation of an element and a singly charged oxygen anion, the greater will be its effect in increasing chemical resistance. In the case of alkaline attack, it seems that the greater the insolubility of the hydroxide of an element the greater will be its effect in increasing alkaline resistance. Durability can be dependent on other factors such as microphase separation, mixed oxide effects and concentration effects.

This paper is published with the permission of The Directors of Pilkington and of Dr. A. Ledwith, Director of Group Research.

REFERENCES

1. Horne, D.F., *Optical Production Technology*, Adam Hilger Ltd. (1972).
2. Izumitani, T.S., *Optical Glass*, American Institute of Physics (1986).
3. Kinosita, K., *Progress in Optics* (E. Wolf ed.), Vol. 4, Wiley Interscience, (1965).
4. Hovestadt, H., *Jena Glass*, Macmillan & Co. (1902).

5. Gliemeroth, G., and Peters, A., *J. Non-Cryst. Solids* 38 & 39: 625-630 (1980).
6. Haigh, T., *Durability of Optical Glasses*, British Scientific Instrument Research Association (1921).
7. Adachi, S., Miyade, E., and Izumitani, T., *J. Non-Cryst. Solids* 42: 569-578 (1980).
8. Walters, H.V., and Adams, P.B., *Applied Optics* 7: 845-848 (1968).
9. Houston, J.B., *Optical Engineering* 14: 158-160 (1975).
10. Morey, G.W., and Bowen, N.L., *J. Soc. Glass Tech.* 11: 97 (1927).
11. *Hoya Catalogue*, Hoya Corporation, Tokyo (1984).
12. *Ohara Catalogue*, Ohara Optical Glass Mfg. Co. Ltd., Tokyo (1982).
13. Reid, A.M., Parry, R.J., and Blackburn, J., *Conference on Physics of Optical Glass Proc.*, Lathom, England (1963).
14. Heywood, H., *J.I.C. Chem. Eng. Soc.* 6: 26-38 (1950-52).
15. Sykes, R.F.R., *Glass Tech.* 6: 178-183 (1965).
16. Sewell, P.A., *Glass Tech.* 10: 9-14 (1969).
17. Pyare, R., Srivastava, M.R., and Nath, P., *J. Mater. Sci.* 17: 2932-2938 (1982).
18. Jones, F.L., *J. Am. Ceram. Soc.* 24: 119 (1941).
19. Simpson, H.E., *Glass Ind.* 36: 515-517, 534, 536, 540 (1955).
20. Hampton, W.M., *Proc. Phys. Soc.* 54: 400 (1942).
21. Berger, E., *J. Soc. Glass Tech.* 2O: 257-278 (1936).
22. *Schott Catalogue*, Schott & Gen., Mainz (1957).
23. Nomarski, G., and Weill, A.R., *Rev. de Metall.* 52: 121-134 (1955).
24. Reid, A.M., Parry, R.J., and Blackburn, J., *Jap. J. App. Physics* 4: 300 (1965).
25. Bredow, H., *Glas Email Keramo Technik* 10: 297-299 (1959).
26. *Sovirel Catalogue*, Sovirel, Paris (1966).
27. Fletcher, W., Keir, E.S., Johnson, P.G., and Slingsby, B., *Glass Tech.* 3: 195-200 (1962).
28. *Schott Catalogue*, Schott & Gen., Mainz (1982).
29. *Pilkington Catalogue*, Pilkington Special Glass Ltd., St. Asaph, U.K. (1989).
30. Andrianova, L.V., and Molchanov, V.S., *Sov. J. Opt. Tech.* 38: 681-683 (1971).
31. Zhdanov, S.P., *Fiz. i Khimistekla* 4: 505-514 (1978).
32. Dimbleby, V., and Turner, W.E.S., *J. Soc. Glass Tech.* 10: 304 (1926).
33. Brewster, G.F., Kreidl, N.J., and Pett, T.G., *J. Soc. Glass Tech.* 31: 153-169 (1947).
34. Aposhian, M.M., and Kreidl, N.J., *J. Am. Cer. Soc.* 34: 103-107 (1951).
35. Peddle, C.J., *J. Soc. Glass Tech.* 4: 120 (1920).
36. Sun, Y., Su, Y., and He, B., *J. Non-Cryst. Solids* 8O: 335-340 (1986).
37. Molchanov, V.S., and Prikhid'ko, *IZV.AN SSSR OkhN* 1:3 (1958).

38. Molchanov, V.S., and Makarova, T.M., *ZH. PR. KH.* 34: 100-107 (1961).
39. Morey, G.W., British Patent 462,304, March 3, 1937. Assigned to Eastman Kodak, New York.
40. El-Hadi, Z.A., Gammal, M., Ezz-El-Din, F.M., and Moustaffa, F.A., *Glass & Ceramic Bull.* 32: 15-19 (1985).
41. Mazelev, L.Y., *Borate Glasses*, Consultants Bureau, New York (1960).
42. Chakraborty, I.N., Shelby, J.E., and Condrate, R.A., *J. Am. Ceram. Soc.* 67: 782-785 (1984).
43. Bunker, B.C., Arnold, G.W., and Wilder J.A., *J. Non-Cryst. Solids* 64: 291-316 (1984).
44. Minami, T., and Mackenzie, J. D., *J. Am. Ceram. Soc.* 60: 232-235 (1977).
45. Parker, B., unpublished internal Pilkington Report (1990).
46. Vogel, W., *Chemistry of Glass*, American Ceramic Soc. Inc., 72 (1985).
47. Wang, C., and Zhou, L., *J. Non-Cryst. Solids* 80: 360-370 (1986).
48. Taylor, P., Ashmore, S.D., and Owen, D.G., *J. Am. Ceram. Soc.* 70: 333-338 (1987).
49. Tindyala, M.A., and Ott, W.R., *Ceram. Bull.* 57: 432-437 (1978).
50. Ge, D., Han, Z., Yan, Y., Chen, H., Lou, Z., Xu, X., Han, R.. and Yang, L., *J. Non-Cryst.Solids* 80: 341-350 (1986).
51. Dietzel, A.H., *Phys. & Chem. of Glasses*, 24: 172-180 (1983).
52. Faraday, M., *Trans. Roy. Soc.*, Part 1, 1 (1830).
53. Turner, J.S., McLennan, E.I., Rogers, J.S., and Matthaei, E., *Nature* 158: 469-472 (1946).
54. Richards, O.W., *J. Bacteriol.* 58: 453-455 (1949).
55. Teitell, L., and Berk, S., *Ind. & Eng. Chem.* 44: 1088-1095 (1952).
56. Kerner-Gang, W., in: *Biodeterioration Investigation Techniques* (A. H. Walters ed.), Applied Science Publishers (1977).
57. Nagamuttu, S., *Int. Biodetn. Bull* 3: 25-27 (1967).
58. Razamovskya, Z.G., and Mitushova, *Optiko-Mekan Promysh* 3: 69-72 (1957).
59. Jones, F.L., *J. Am. Cer. Soc.* 28: 32 (1945).
60. Baker, P.W., *Int. Biodetn. Bull* 3: 59-64 (1967).
61. Prod'homme, M., *Proc. 7th Int. Congress On Glass,* Paper 17 (1965).
62. Khan, M.R., and Khan Z.M., *Bangladesh J. Bot* 6: 89-98 (1977).

9

Surface Modification of Bioactive Glasses and Ceramics

Larry L. Hench

Advanced Materials Research Center
University of Florida
Alachua, Florida

INTRODUCTION

Certain compositions of glasses, ceramics, glass-ceramics, and composites have been shown to bond to bone (1-7). These materials have become known as "bioactive ceramics". Some even more specialized compositions of bioactive glasses will bond to soft tissues as well as bone (7). A common characteristic of bioactive glasses and bioactive ceramics is a time dependent, kinetic modification of the surface that occurs upon implantation. The objective of this paper is to review the concept of bioactivity, the controlled surface reactions that lead to bioactivity, the compositional limits of bioactive glasses and ceramics, the biological factors that influence bioactivity, and the use of bioactive materials in the clinic.

BIOACTIVITY

Bioactivity is the characteristic of an implant material which allows it to form a bond with living tissues. Materials that are not bioactive form a non-adherent layer of fibrous tissue at the implant interface. Materials that are bioactive develop an adherent interface with tissues that resists substantial mechanical forces. In many cases the interfacial strength of adhesion is equivalent or greater than the cohesive strength of the implant material or the tissue bonded to the bioactive implant.

Bonding to bone was first demonstrated for a certain compositional range of glasses, later termed *"Bioglasses®"*[*] (1-5). These bioactive glasses contained SiO_2, Na_2O, CaO, and P_2O_5 in specific proportions. There were three key compositional features to these glasses that distinguished them from traditional soda-lime-silica glasses: 1) less than 60 mole percent SiO_2, 2) high Na_2O and CaO content, and 3) high CaO/P_2O_5 ratio. These compositional features make the surface highly reactive when exposed to an aqueous medium.

FIVE TYPES OF GLASS SURFACES

The five types of glass surfaces and six surface conditions resulting from glass-environment interactions are illustrated in Figure 9-1 (8,9). The ordinate in Figure 9-1 represents the relative concentration of SiO_2 (or other oxides in Type III surfaces) in the glass and the abscissa corresponds to the depth into the glass surface. If species are selectively dissolved from the glass surface, the relative SiO_2 concentration will increase producing an SiO_2-rich surface layer. If all species in the glass are dissolved simultaneously (congruent dissolution) the relative concentration of SiO_2 will remain the same as in the original glass. When combinations of selective dissolution, congruent dissolution and precipitation from solution occur then heterogeneous mixtures of the 6 surface conditions shown in Figure 9-1 are possible.

Type I glasses undergo a surface hydration reaction that is only a few monolayers thick, and no compositional profile is measurable. High purity vitreous silica exposed to neutral solutions is an example of a Type I surface. Exchange of alkali and alkaline earth ions with hydrogen and/or hydronium ions (i.e., selective dissolution) results in Type II glass surfaces if there is sufficient concentration of network formers in the surface film to stabilize it. Type III surfaces are described later. If network formers are not sufficient, or if the environment is rich in OH^- or other species which can break Si-O-Si network bonds, the surface layer is unstable and a Type IV surface is produced. A glass that is undergoing total network attack (also referred to as congruent dissolution) is described as having a Type V surface. Often from the perspective of average surface composition there is little distinction between Type I and Type V surfaces. However, large quantities of ions are lost from a Type V surface during corrosion and consequently extensive surface pitting can result due to localized heterogeneous attack. Additionally, large dimensional changes often accompany corrosion of glasses with Type V surfaces (8,9).

---

[*]Registered trademark, University of Florida, Gainesville, FL 32611.

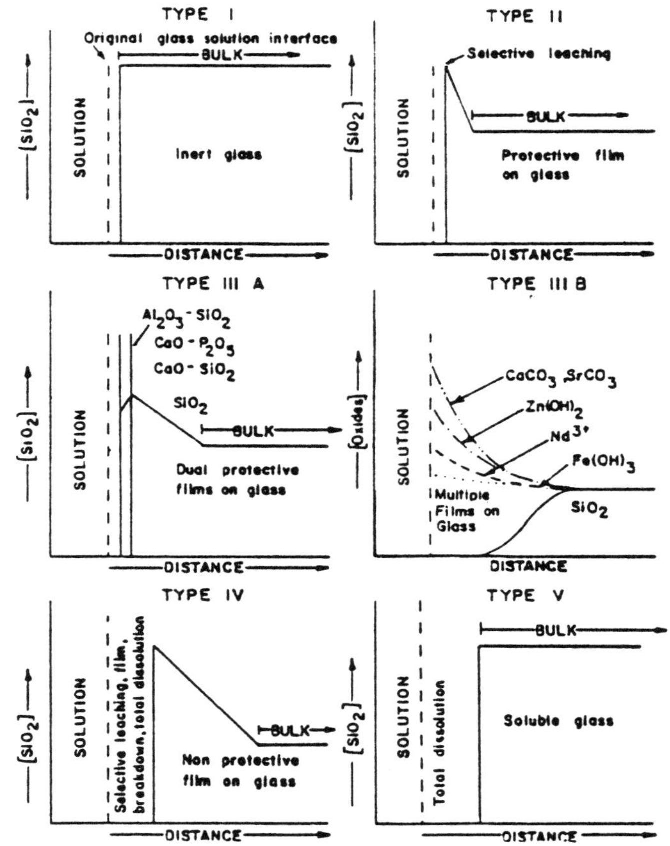

Figure 9-1. Five types of silicate glass surface reactions.

The transition from a Type I to Type V surface that results from static corrosion and a localized increase in solution pH is illustrated in Figure 9-2. The initial condition of the glass, pt (a), is one of very little surface reaction, only surface silanol (Si-OH) formation. However, as the glass is exposed to water or a humid environment selective leaching occurs resulting in a Type II surface, pt (b). At longer times the interfacial pH becomes greater than 9, and attack of the Si-O-Si network increases. Consequently, the thickness of the ion-exchange layer reaches a maximum, pt (c), because at longer times (c $\rightarrow$ d) the surface

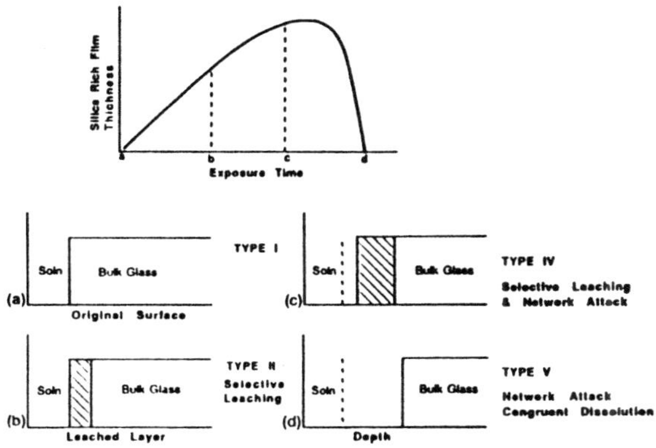

Figure 9-2. Time dependent changes of glass surfaces with static leaching or flow.

SiO_2-rich layer is dissolved by network dissolution faster than it is formed by ion exchange. Eventually, a congruently dissolving Type V glass exists, pt (d).

KINETICS OF SILICATE GLASS SURFACE REACTIONS

The general equation which describes the overall rate of glass surface reactions (R) has at least four terms:

$$R = \underbrace{-k_1t^{0.5}}_{\text{Stage 1}} \underbrace{- k_2t^{1.0}}_{\text{Stage 2}} + \underbrace{k_3t^x}_{\text{Stage 3}} + \underbrace{k_4t^y}_{\text{Stage 4}} + k_nt^z \quad (9\text{-}1)$$

The first term, developed by Douglas and El-Shamy (10), describes the rate of alkali extraction from the glass and corresponds to what is often termed a Stage 1 reaction (see Figure 9-3). A Type II glass surface is primarily undergoing Stage 1 attack.

<u>Stage 1</u>: The initial or primary stage of attack is a process which involves ion exchange between alkali ions from the glass and hydrogen ions from the solution, during which the remaining constituents of the glass are not altered. Reaction (1) in Figure 9-3 tends to dominate during Stage 1. During Stage 1, the rate of alkali extraction from the glass is parabolic in character (Lyle (11); Zagar and Schillomoeller (12); Rana and Douglas (13)).

The second term describes the rate of network dissolution which is associated with a Stage 2 reaction (Figure 9-3). A Type IV surface is experiencing a combination of Stage 1 and Stage 2 attacks. A Type V surface is dominated by a Stage 2 attack.

Stage 2: The second stage of attack is a process whereby breakdown of the silica structure occurs and total glass dissolution ensues. Reaction (2) in Figure 9-2 is the dominant reaction in Stage 2 processes.

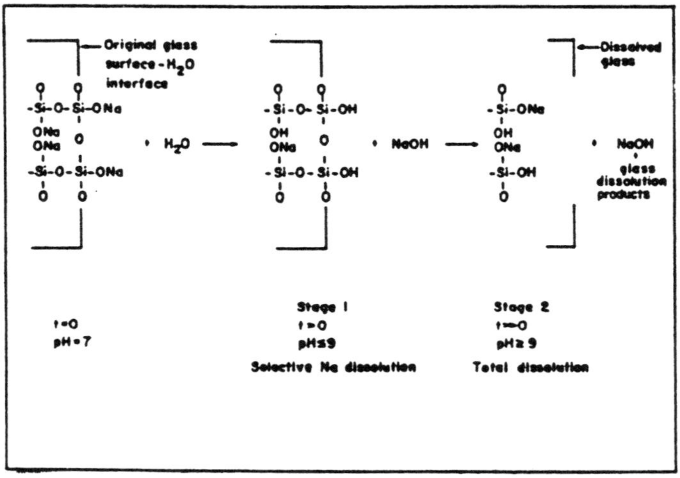

Figure 9-3. Mechanisms of glass corrosion for an alkali-silicate glass.

Studies by Wang and Tooley (14,15), Lyle (11), Rana and Douglas (13), Das (16) and Das and Douglas (17), Bacon and Calcamuggio (18), Douglas and Isard (19), Tsuchihashi and Sekido (20), Weyl (21) in addition to the classic paper of Douglas and El-Shamy (10) provide the chemical basis for understanding Stage 1 and 2 processes.

The experimental evidence of the formation of surface compositional profiles that arise as a consequence of Stage 1 and Stage 2 reactions was accumulated in the 1970's as new surface sensitive analytical tools became available.

A number of studies of glasses of various compositions have shown depth compositional profiles for Type II, IV, and V surfaces. The techniques used include Auger electron spectroscopy (AES) (22-25), secondary ion mass spectroscopy (SIMS) (26), resonant nuclear reactions (RNR) (27), electron spectroscopy for chemical analysis (ESCA) (28,29), secondary ion photoemission spectroscopy (SIPS) (30,31), infrared reflection spectroscopy (IRRS) (32-34).

Film depths in the range of 0.01-1.0 µm are generally observed for Type II glasses. Type IV glasses typically exhibit films of 1.0-100 µm depth. Addition of network modifiers of high electric field strength to the glass changes Type IV surfaces to Type II, greatly decreases the thickness of the silica-rich films, and increases the film density and glass durability.

TYPE III GLASS SURFACES

Previous surface studies of biologically active silicate glasses containing CaO and P_2O_5 (Bioglasses) (5,35,36) and certain compositional ranges of Li_2O-Al_2O_3-SiO_2 glasses (37) showed that dual films developed on the glass. The secondary films were either hydrated CaO-P_2O_5 or Al_2O_3-SiO_2 layers. Chemical depth profiles using AES showed that the dual films formed on top of a silica rich film which resulted from rapid ion exchange of alkali for protons (or hydronium ions). Other work showed that such a dual film could be formed by addition of either phosphate (36) or aluminum ions in solution (38) as well as through release of phosphate or aluminum ions from the glass.

Analysis of anion concentrations of the dual apatite film formed on bioactive glasses showed some exchange of CO_3^{2-} ions for OH^- ions similar to that found for bone (39). Incorporation of CaF_2 in the glass results in a fluorapatite film forming on the glass (40).

Studies also showed that the effectiveness of a number of alkaline corrosion inhibitors (32), such as soluble calcium or beryllium salts, are due to formation of dual films composed of insoluble alkaline earth-silicate compounds (41,42).

This type of glass surface with a dual protective film, is designated Type IIIA in Figure 9-1. The thickness of the secondary films can very considerably, from as little as 0.01 μm for Al_2O_3-SiO_2-rich layers to as much as 30 μm for CaO-P_2O_5-rich layers.

The formation of a Type III surface is due to a combination of the repolymerization of SiO_2 on the glass surface by the condensation of the silanols formed from the Stage 1 reaction, e.g.,

$$\text{Si-OH} + \text{OH-Si} \rightarrow \text{Si-O-Si} + H_2O \qquad (9\text{-}2)$$

This Stage 3 reaction contributes to the enrichment of surface SiO_2 characteristic of Types II, III, and IV surface profiles (see Figure 9-1). It is described by the third term in eq. 9-1. This reaction is probably interface controlled with a time dependence of $+k_3t^{1.0}$ i.e., x = 1.0 in eq. 9-1, however the quantitative measurements to establish values for k_3 and x have not been made**.

The fourth term in eq. 9-1, $+k_4t^y$, (Stage 4) describes the precipitation reactions which result in the multiple films characteristic of Type III glasses. When only one secondary film forms, such as hydroxyapatite, the surface is Type

**For a discussion of this reaction see refs. 57,58.

IIIA. When multiple films form due to several partially soluble cationic species reaching their solubility limits a Type IIIB surface is formed. This behavior is characteristic of nuclear waste glasses (9,43-45).

TYPE IIIA BIOACTIVE GLASS AND CERAMIC SURFACES

One of the most important aspects of the CaO-P_2O_5 rich Type IIIA surface is that it is bioactive, i.e., such compositions form a chemical bond to living bone (1). When the bioactivity is sufficiently high the material forms a bond to living soft tissues as well as bone (7). The bonding is due to the formation of a hydroxyl-carbonate apatite (HCA) layer on top of the SiO_2-rich layer at the same rate as living cells are forming interfacial metabolites. Thus, the kinetics of the Stage 4 precipitation reaction closely match biomineralization reaction rates in repairing connective tissues. The biochemical reaction rates are controlled by a combination of genetics and local chemical and biomineralization factors. The reaction rates of the bioactive material in Stage 4 are controlled by composition and local chemical environments. When the biomineralization and physical chemical rates are matched, bonding occurs.

The HCA crystallites nucleate and bond to the interfacial metabolites, such as mucopolysaccharides, glycoproteins, and collagen (1-7). This mechanism of incorporation of organic biological constituents occurs very rapidly within the dynamic HCA surface and is the initial step in establishing bioactivity and bonding of tissues to the implant surface. Type IIIA compositions lie within the bioactive bonding region (A) in Figure 9-4 (3).

Increasing the SiO_2 content of the glass towards 60 mole % reduces the rates of ion exchange, network dissolution, silica repolymerization, and HCA crystallization. Bioactivity is reduced proportionally; i.e., as SiO_2 content of the

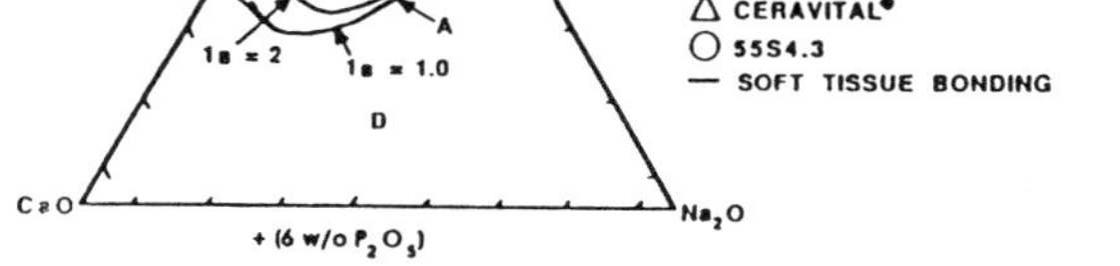

Figure 9-4. Ternary diagrams showing index of Bioactivity (I_a) as a function of composition. Note: All compositions in Region A contain 6 wt% P_2O_5.

glass increases the values of k_1, k_2, k_3 and k_4 decrease and bioactivity decreases. At glass SiO_2 contents of 60 mole % and higher, the HCA layer does not form within 2-4 weeks, and the glasses are not bioactive. This is region (B) in Figure 9-4. The glasses in region B have Type I and II surfaces. The biomineralization reaction rates are much faster than Stage 4 kinetics for Type I and II surfaces. Consequently, unmineralized fibrous tissue fills the space adjacent to the implant, and since growing HCA crystallites are not present, there is no interfacial bonding. Thus, the Type I and II glasses are bioinert.

The closer the composition is to the center of region A, the more rapid is the rate of formation of the apatite layer. When the bonding layer forms sufficiently fast it is able to bond even to soft tissues, region E, as well as bone. Thus, the overall surface reaction rate equation (1) has a direct effect on the biological reaction rates at the interface. This is because of varying types of attachment complexes on cell membranes which are specific to the ions in the glass surface (46).

BONDING MECHANISMS OF BIOACTIVE CERAMICS

We now know that the tissue-bonding mechanisms of bioactive silicates involve a complex series of physiochemical and ultrastructural phenomena. These include:

1) Rapid exchange of Na^+ or K^+ with H^+ or H_3O^+ from solution,

2) Loss of soluble silica to the solution resulting from breaking of Si-O-Si bonds and formation of Si-OH and $Si(OH)_4$ groups at the glass interface,

3) Condensation and repolymerization of an SiO_2-rich layer on the surface,

4) Migration of Ca^{2+} and PO_4^{3-} groups to the surface through the SiO_2-rich layer,

5) Formation of $CaO-P_2O_5$-rich film on top of the SiO_2-rich layer,

6) Growth of the SiO_2-rich layer by diffusion-controlled alkali ion exchange,

7) Growth of the amorphous $CaO-P_2O_5$-rich film by incorporation of soluble calcium phosphates from solution,

8) Crystallization of the amorphous $CaO\text{-}P_2O_5$ film by incorporation of OH^-, CO_3^{2-}, or F^- anions from solution to form a mixed hydroxyl, carbonate, fluorapatite layer, and

9) Agglomeration and chemical bonding of the apatite crystallites around collagen fibrils and within adsorbed mucopolysaccharides and other proteins produced by osteoblasts or fibroblasts.

The consequence of this sequence of reactions is the creation of an interfacial bonding zone between tissue and the implant that is a 30-μm layer of hydrated $CaO\text{-}P_2O_5$ crystals on top of a 100 to 120 μm layer rich in SiO_2. Electron and optical micrographs in Refs. 1-7 show the direct bonded interface between rat tibial bone and several bioactive glass and glass-ceramic implants from 7 to 365 days after implantation.

Steps 1 through 5 occur within minutes of implantation, thereby disguising the silicate implant to the physiological environment as if it were mineralizing apatite. Steps 6 through 9 require 7 to 10 days to complete and are controlled by the type and concentration of progenitor stem cells at the implant interface. The ratio of silica in the glass to alkali and alkaline earth ions ($SiO_2/R_2O + RO$) determines the rate of formation and thickness of the bonding zone. As the bone-bonding boundary (Figure 9-4, region A) is approached, the rate of formation and thickness of the bonding zone decreases.

In the presence of osteogenic precursors, the bioactive silicates favor formation of osteoblasts which govern the further steps of bone development (3). Gross and Strunz (6) summarize their ultrastructural findings of the bioactive silicate bone bonding process, "*Within the extracellular matrix and between small bundles of fibrils, matrix vesicles appear and display small, electron-dense, needle-like crystallites assumed to be apatite. After rupture of the vesicle membrane, calcifying fronts are formed. Often this process begins and is therefore more pronounced in the surroundings of the implant and the adjacent osteoblast, but may also start in the area around the osteoblast and the involve the surroundings of the already mineralized seam of amorphous cementing substance at the interface. Later on, the whole area is mineralized, the osteocytes being rather evenly distributed and often arranged with their long axes parallel to the surface of the implant. This feature is found in different species--rats, dogs, pigs, chickens, and humans--and proves the morphologic basis for the biomechanical quality of the bond bonding.*" Gross and Strunz (6) have also shown that addition of the multivalent cations Al^{3+}, Ta^{4+}, Sb^{3+}, or Zr^{4+} inhibits bonding. Unmineralized osteoid and chondroid persist at the implant interface of glasses with these elements, indicating the presence of substances that impede mineralization or the absence of substances that promote it. The inhibited cells at these interfaces do not switch from the production of metachromatic ground substance and Type II collagen to the production of Type I collagen and organelles for mineralization. In contrast, bioactive glasses and glass-ceramics with less than critical concentrations of such multivalent cations do show bonding with osteoblasts and produce matrix vesicles and normal mineralization at the interface. Type IIIA bioactive silicate implants release monophosphates at their interface, whereas nonbonding compositions release tri-, tetra-, or polyphosphates. Local phosphate concentrations and the formation or function of matrix vesicles can be altered by the local monophosphate/polyphosphate ratios, and mineralization is thereby inhibited or enhanced.

Interfacial concentrations of soluble Ca and Si are also strongly influenced by silicate implant compositions. Multivalent cations, which convert Type IIIA bioactive bonding surfaces to Type II bioinert non-bonding surfaces, inhibit release of soluble Ca and Si. The concentration and function of osteocalcin, Gla proteins, osteonectin, and other extracellular glycoproteins at the implant interface may well be controlled by soluble Ca or Si concentrations and this type of molecular complexation.

COMPOSITIONAL EFFECTS ON BIOACTIVITY

Compositions near the middle of Figure 9-4 have the highest levels of bioactivity. As SiO_2 content increases, the bioactivity decreases due to the change from an amorphous structure comprised primarily of ionically bonded 2-dimensional sheets or chains to a 3-D structure.

By changing the compositionally controlled reaction kinetics (eq. 9-1), the rates of formation of hard tissue at a bioactive implant interface can be altered, as shown in Figure 9-5. Thus, the level of bioactivity of material can be related (47) to the time for more than 50% of the interface to be bonded, ($t_{0.5bb}$) e.g.: (Index of Bioactivity) $I_B = (100/t_{0.5bb})$. It is necessary to impose a 50% bonding criterion for an index of bioactivity since the interface between an implant and bone is irregular. The initial concentration of osteoblasts, chondroblasts, and fibroblasts varies as a function of the fit of the implant and the condition of the bony defect (6). Consequently, all bioactive implants require an incubation period before bone proliferates and bonds (Figure 9-5). The length of the incubation period at which this process occurs varies over a wide range depending on composition.

The compositional dependence of I_B suggests that there are isoI_B contours within the bioactivity boundary. Comparative data from the literature and extrapolation based upon relative reaction kinetics suggest the isoI_B contours shown in Figure 9-4. The change of I_B with $SiO_2/(Na_2O + CaO)$ ratio is very large as the bioactivity boundary is approached, as indicated by Auger electron spectroscopy analyses of the compositional dependence of rates of HCA formation, measured by Ogino et al. (36,48).

Addition of multivalent ions to a bioactive glass or glass-ceramic will serve to shrink the isoI_B contours. Greenspan and the author showed that only 3%

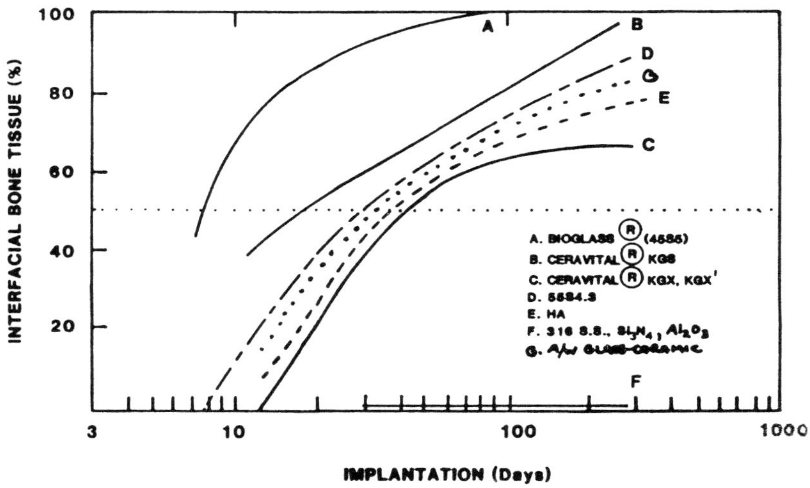

Figure 9-5. Time dependence of bone bonding for various bioimplants (A-E) compared with bioinert implants (F).

Al_2O_3 added to a 45S5 Bioglass® destroys bone bonding (49). Thus, the isoI_B contours shrink to nothing as the percentage of Al_2O_3 increases. The studies of Gross and Strunz have also shown that only a small percentage of multivalent cations eliminate bioactivity (6). Thus, the isoI_B contours shown in Figure 9-4 indicate the contamination limits for bioactive glasses and glass-ceramics. If a starting implant composition is near the I_B boundary, it may take only a few ppm of multivalent cations to shrink the I_B boundary to zero and eliminate bioactivity. Also, the sensitivity of fit of a bioactive implant and length of time of immobilization post-op depends upon I_B value and the closeness to the I_B = zero boundary. Implants near the I_B boundary require more precise surgical fit and longer fixation times before loading.

Because of the sensitivity of I_B to multivalent cations, the research to develop bioactive glasses that bond well to surgical alloys has been complicated (3,6). The open 2-D network structure of a bioactive glass that enables rapid proton-alkali exchange (Stage 1, eq. 1) also yields rapid diffusion of metallic transition elements through the glass at elevated temperatures. Such elements are required to produce a durable chemical bond between metal and glass. However, adding the metallic elements, such as Co, Ni, Cr, or Fe, to the glass in sufficient concentration to achieve glass-metal bonding shrinks the I_B boundary to zero. Effective solutions to this problem have included use of dual coating techniques, a rapid immersion method, and flame spraying (3). However, reliability of the glass-metal interface of these coatings has always been difficult to achieve.

SOFT TISSUE BONDING

Although bioactivity is defined as the bonding of any tissue to an implant, most of the above discussion has been concentrated on the bonding of hard tissue. Most of the data and evidence regarding bioactivity have come from bone implants. In fact, very early work reported by the author and associates cited evidence that bioactive glasses did not bond to soft tissues (2). However, a later study by Wilson, et al. (7) showed quite conclusively that when non-adherent fibrous capsules formed around bioactive glass implants in muscle it was an artifact of the experiment caused by continuous micromotion at the implant-tissue interface. When the implant-tissue interface was immobilized, Wilson, et al. (7) proved that collagen fibers of the soft tissue became embedded and bonded within the growing silica-rich and HCA layer on the 45S5 bioactive glass.

Soft tissue bonding was also observed in subcutaneous implants, when they were sufficiently immobile, and the collagen fibers were attached to remnants of decalcified 45S5 Bioglass® after 8 weeks subcutis in rats (7). The nature of the collagen-bioactive glass interface *in vivo* appears identical to that observed when the same glass is exposed to collagen fibers *in vitro*. When the amorphous calcium phosphate phase crystallizes in the presence of collagen fibers, *in vitro* or *in vivo*, the collagen becomes structurally integrated within the apatite agglomerates and vice versa. This is the case whether the collagen is generated by fibroblasts or osteoblasts. However, the presence of mucopolysaccharides, such as chondroitin sulphate D, significantly enhances the physiochemical interaction between the crystallizing apatite layer and collagen (3,5). Between collagen fibers and surface-active glass *in vivo* there is an amorphous zone 80 to 100 nm thick, which can be mineralized. This zone or seam apparently consists of extracellular ground substance which may contain mucopolysaccharides, glycoproteins, and various nectins, and provides a basis for attachment of collagen (6). Such a seam is present only at the interface of bioactive materials bonded to tissues. Non-bonded interfaces do not show this extracellular bonding zone coincident with collagen embedded within the apatite layer on the implant. Likewise, bioactive glass (45S5 Bioglass®) implants in the mouse middle ear bond to the tympanic membrane, and a regenerated layer of respiratory epithelium covers the implant, as shown by Merwin, et al. (50,51). This tissue response to the bioactive glass reproduces the normal tissues in the middle ear.

Maxillofacial implants are used in situations where, by its very nature, surgery has produced sites where variable interfaces with both bone and soft tissues occur. Bioactive glass implants with high I_B values show attachment to both bone and soft tissues as needed when used in maxillofacial applications (52). These compositions of high I_B are in the middle of Figure 9-4. Implants without bioactivity, $I_B = 0$, bond to neither bone nor soft tissue in maxillofacial applications (51,52).

Bioactive implants with intermediate I_B values do not develop a stable soft tissue bond; instead the fibrous interface progressively mineralizes to form bone (6). Consequently, there appears to be a critical isoI_B boundary beyond which bioactivity is restricted to stable bone bonding and inside which bioactivity includes both stable bone and soft tissue bonding depending on the progenitor stem cells that are in contact with the implant. This soft tissues isoI_B limit is shown by the dark dashed contours in Figure 9-4 and is discussed by Wilson and Nolletti (53).

An important manifestation of the combined hard tissue - soft tissue response to bioactive implants occurs in their use to treat periodontal defects (54,55). Use of the most bioactive powders (45S5 Bioglass®) led to a more rapid and complete repair of surgically created periodontal defects in Patus monkeys than was observed for the lower bioactivity HA or tricalcium phosphate (TCP) powders. Powders with high bioactivity also resulted in reattachment of the periodontal ligament and little or no epithelial downgrowth. Wilson, et al. report (55) *"The bioactive glass particles are associated with development of new osteocementum which, by 9 months, lines the entire defect and which permits epithelial growth only to its upper limit, which is generally close to the original, normal level of the gingiva".* The highly bioactive glass powders (I_B = 8 to 10) showed both osteoconduction and osteoproduction of bone, whereas the powders with lower I_B values (I_B = 3 to 4) exhibited only osteoconduction. Osteoproduction has been defined by Wilson (55) as, *"The process whereby a bioactive surface is colonized by osteogenic stem cells free in the defect environment as a result of surgical intervention."* This differs from osteoconduction where the implant simply provides a biocompatible interface along which bone migrates. Recall that osteoinduction is considered to be the production of bone in an ectopic site, i.e., one in which bone is not normally found, such as muscle or skin.

The critical isoI_B contour for osteoproduction is likely to be similar to that for soft tissue bonding (Figure 9-4). The isoI_B contour for osteoconduction is likely to be close to the I_B = 0 limit of bone-bonding.

EFFECT OF COMPOSITION ON INTERFACIAL STRENGTH

Comparison of the research papers in a recent volume that summarizes data on a large number of bioactive materials (56) indicates that the thickness (d) of the bonding zone between a bioactive implant and bone is proportional to its index of bioactivity (I_B). The failure strength (S_f) of a bioactively fixed bond appears to be inversely dependent on the thickness of the bonding zone (d). For example, 45S5 Bioglass® with a very high I_B develops a gel bonding layer of 200 µm which has a relatively low shear strength. In contrast A/W glass-ceramic, with an intermediate I_B value has a bonding interface in the range of 20-50 µm and a very high resistance to shear. Consequently, the interfacial bonding strength appears to be optimum for I_B values of ~4. However, it is important to recognize that the interfacial area bonding is time dependent, as shown in Figure 9-5. Therefore, interfacial strength involves time dependent variables that include morphological factors such as the change in bonding interfacial area with time, progressive mineralization of the interfacial tissues, and increases in the elastic modulus of the bonded interface with time.

CONCLUSIONS

This analysis shows that both the biomineralization and the biomechanical factors of bioactive fixation are influenced by the values of the surface chemical rate constants of bioactive implants, k_1-k_4. Therefore, knowledge of these rate constants and their time dependence should make it possible to design compositions of bioactive implants for specific combinations of bonding rate, bond thickness, and bond strength. Different combinations of these features are needed for load-bearing vs. non load-bearing prostheses for example.

Specific requirements for bioactive fixation of prostheses for osteoporotic, arthritic, or geriatric patients may perhaps be achieved by varying the rate constants. Quantitative values for k_1 to k_4 and t^y and t^z need to be measured as a function of implant composition and tissue pathology. Eventually such studies should yield a quantitative theory for bioactive fixation. This should produce the design criteria for implant compositions that can be molecularly designed for specific clinical applications. This is a worthy goal to achieve for bioactive implants and could potentially improve life for millions of persons..

REFERENCES

1. Hench, L.L., Splinter, R.J., Greenlee, T.K. and Allen, W.C., *J. Biomedical Maters. Res.* 2: 117 (1971).
2. Hench, L.L., and Paschall, H.C., *J. Biomed. Maters. Res. Symp.* 4: 25 (1973).
3. Hench, L.L., and Ethridge, E.C., *Biomaterials--An Interfacial Approach,* Academic Press (1982).
4. Hench, L.L., and Clark, A.E., *Biocompatibility of Orthopedic Implants,* (D.F. Williams ed.) Vol. 2, Chap. 6, CRC Press (1982).
5. Hench, L.L., *Fundamental Aspects of Biocompatibility,* (D. F. Williams ed.) Vol. 1, Chap. 4, CRC Press (1981).
6. Gross, U.M., and Strunz, V., *Clinical Applications of Biomaterials,* (A.J.C. Lewe, T. Albrektsson, and P. I. Branemark ed.) p. 237, Wiley (1989).
7. Wilson, J., Pigott, G.H., Schoen, F.J., and Hench, L.L., *J. Biomed. Maters. Res.* 15: 805 (1981).

8. Hench, L.L., and Clark, D.E., *J. Non-Cryst. Solids* 28: 83 (1978).
9. Hench, L.L., *J. De Physique*, Colloque C9, supplement au no. 12, Tome 43, C9-625 (1982).
10. Douglas, R.W., and El-Shamy, T.M., *J. Amer. Ceram. Soc.* 50(1): 1-8 (1967).
11. Lyle, A.K., *J. Amer. Ceram. Soc.* 26(6): 201-204 (1943).
12. Zagar, L., and Schillmoeller, L., *Glastechn, Ber.* 33(4): 409-116 (1960).
13. Rana, M.A., and Douglas, R.W., *Phys. Chem. Glasses* 2(6): 179-195 (1961).
14. Wang, F.F., and Tooley, F.V., *J. Amer. Ceram. Soc.* 41(11): 467-469 (1958).
15. Wang, F.F, and Tooley, F.V., *J. Amer. Ceram. Soc.* 41(12): 521-524 (1958).
16. Das, C.R., *Glass Ind.* 50: 422-427, 483-485 (1969).
17. Das, C.R., and Douglas, R.W., *Phys. Chem. Glasses* 8(5): 178-184 (1967).
18. Bacon, F.R., and Calcamuggio, G.L., *Am. Ceram. Soc. Bull.* 46(9): 850-855 (1967).
19. Douglas, R.W., and Isard, J.O., *J. Soc. Glass Tech.* 33: 289 (1949).
20. Tsuchihashi, S., and Sedido, E., *Bull. Chem. Soc. Japan* 32(8): 868-872 (1959).
21. Weyl, W.A., *Glass Ind.* 28(8): 408-412, 428-432 (1947).
22. Rynd, J., and Rastogi, A.K., *Am. Ceram. Soc. Bull.* 53: 631 (1974).
23. Chappell, R.A., and Stoddart, C.T.H., *Phys. Chem. Glasses* 15: 130 (1974).
24. Pantano, C.G., Jr., Dove, D.B. and Onoda, G.Y., Jr., *J. Non-Crystal. Solids* 19: 41 (1975).
25. Clark, D.E., Pantano, C.G., Jr., and Hench, L.L., *Corrosion of Glass*, Books for Industry (1979).
26. Gossink, G., de Grefth, H.A.M., and Werner, J.W., *J. Amer. Ceram. Soc.* 52: 4 (1979).
27. Lanford, W.A., Davis, K., Lamarche, P., Laursen, T., Groleau, R., and Doremus, R.H., *J. Non-Crystal. Solids* 33: 249 (1979).
28. Chick, L.A., McVay, G.L., Millinger, G.B., and Roberts, R.F., *Battelle Memorial Inst., PNL Report* 3465 (1980).
29. Budd, S.M. in *Glass Surfaces*, (D.E. Day, ed.), p. 55, North Holland Publishing Co. (1975).
30. Bach, H., and Bauke, F.G.K., *Phys. & Chem. of Glasses* 15: 123 (1974).
31. Baucke, F.K., *J. Non-Crystal. Solids* 14: 13 (1974).
32. Sanders, D.M., Person, W.B., and Hench, L.L., *Appl. Spectroscopy* 26: 530 (1972).
33. Sanders, D.M., Person, W.B., and Hench, L.L., *Appl. Spectroscopy* 28: 247 (1974).

34. Clark, D.E., Ethridge, E.C., Dilmore, M.F., and Hench, L.L., *Glass Technol.* 18: 121 (1977).
35. Pantano, C.G., Jr., Clark, A.E., Jr., and Hench, L.L., *J. Amer. Ceram. Soc.* 57: 412 (1974).
36. Ogino, M., Ohuchi, F., and Hench, L.L., *J. Biomedical Maters. Res.* 14: 55 (1980).
37. Dilmore, M.F., Clark, D.E., and Hench, L.L., *Amer. Ceram. Soc. Bull.* 57: 1040 (1978).
38. Dilmore, M.F., Clark, D.E., and Hench, L.L., *Amer. Ceram. Soc. Bull.* 58: 1111 (1979).
39. Fujiu, T., Ogino, M., Kariya, M., and Schimura, T., private communication.
40. Hench, L.L., Spilman, D.B., and Nolletti, D.R., in *Biological and Biomechanical Performance of Biomaterials*, (P. Christel, A. Meunier and A.J.C. Lee ed.), pp 99-104, Elsevier (1986).
41. Oka, Y., Ricker, K.S., and Tomozawa, M., *J. Amer. Ceram. Soc.* 62(11-12): 631-632 (1977).
42. Oka, Y., and Tomozawa, M., *J. Non-Crystal. Solids* 42: 532 (1980).
43. McVay, G.L., and Buckwalter, C.Q., *Nuclear Technologies*, Vol. 51 (1980).
44. Hench, L.L., Clark, D.E., and Yen-Bower, E.L., *Nuclear Chemical Waste Management* 1: 50 (1980).
45. Wicks, C.G., in *Proceedings of Waste Management 1981 Conference*, Tucson, AZ, Feb. 23-27, 1981.
46. Seitz, T.L., Noonan, K.D., Hench, L.L., and Noonan, N.E., *J. Biomed. Maters. Res.* 16(3): 195-207 (1982).
47. Hench, L.L., in *Bioceramics: Materials Characteristics Versus In Vivo Behavior*, (P. Ducheyne and J. Lemmons ed.) pp 54-71, Annals of the New York Academy of Sciences, Vol. 523 (1988).
48. Fujui, T., and Ogino, M., *J. Biomed. Maters. Res.* 18: 845-859 (1984).
49. Greenspan, D.C., and Hench, L. L., *J. Biomed. Maters. Res.* 10(4): 503-509 (1976).
50. Merwin, G.E., Atkins, J.S., Wilson, J., and Hench, L.L., *Otolaryngol Head Neck Surg.* 90: 461-469 (1982).
51. Merwin, G.E., Rogers, L., Wilson, J., and Martin, R., *Arch. of Otolaryngol Head and Neck Surg.* 122: 280-284 (1986).
52. Wilson, J., and Merwin, G.E., *J. Biomed. Maters. Res.* 22A2: 159-177 (1988).
53. Wilson, J. and Nolletti, D., in *CRC Handbook Bioactive Ceramics* (T. Yamamuro, J. Wilson, and L. L. Hench, ed.) Vol. 1, Chap. 28, CRC Press (1990).
54. Fetner, A.E., Low, S.B., Wilson, J., and Hench, L.L., *J. Dent. Res.* 65: 347 (1986).
55. Wilson, J., Low, S.B., Fetner, A., and Hench, L.L., in *Biomaterials and*

Clinical Applications (A. Pizzoferrato, P.G. Marchetti, A. Ravaglioli, and A.J.C. Lee ed.) Elsevier (1987).

56. Yamamuro, T., Wilson, J., and Hench, L.L., ed., *CRC Handbook Bioactive Ceramics*, Vols. I & II, CRC Press, Boca Raton, Florida (1990).
57. Bunker, B.C., Tallant, D.R., Headley, T.J., Turner, G.L., and Kirkpatrick, R.J., *Phys. Chem. of Glasses*, 29(3): 106 (1988).
58. Hench, L.L. and West, J.K., *Chem. Reviews* 90: 33-72 (1990).

10

Corrosion of Heavy-Metal Fluoride Glasses

Catherine Simmons

Department of Materials Science & Engineering
Advanced Materials Research Center
University of Florida
Alachua, Florida

INTRODUCTION

Heavy-metal fluoride (HMF) glasses are being investigated and developed in laboratories throughout the world. Predictions of ultra-low intrinsic optical losses coupled with their broad optical transparency, from the near UV (0.25 μm) to the mid-infrared (8-10 μm), make them potentially valuable for several applications including fiber waveguides, laser windows and IR lenses.

The study of HMF glasses is relatively young in terms of the history of glass science. However, since their discovery in 1975, (1) a large body of data has been accumulated leading to a surprisingly good understanding of the physical, chemical, and optical properties of a broad range of HMF compositions. These glasses differ markedly from silicates in several respects. Fabrication of HMF glasses is frequently hampered by their strong tendency for crystallization, most systems having a relatively limited region for stable glass formation. The structure of HMF glasses is quite different from that of silicates in that fluorine, being univalent, does not form a true bridging bond. Rather, glass formation appears to occur through intermolecular bonding due to excess charges from high coordination near-neighbor shells surrounding the glass formers (2). A combination of the unstable nature of HMF glasses and their weakly bonded structure leads, in part, to another major difference between HMF glasses and silicates, their behavior in corrosive environments. This chapter deals with the mechanisms for corrosion of HMF glasses and, where appropriate, makes comparisons to silicates.

GLASS PREPARATION

A variety of preparation techniques have been employed to produce HMF glasses. Typically metal-oxide and/or metal-fluoride compounds are mixed with a fluoridating compound (e.g. $NH_4F.HF$) and melted at 800 - 1000°C. Frequently reactive atmospheres containing fluorine are introduced during melting to avoid contamination by atmospheric OH^-. It has been shown that most compositions melted with sufficient care to avoid this contamination exhibit reproducible chemical durability results independent of starting material (3). Table 10-1 gives a partial list of typical compositions studied.

Table 10-1. Typical HMF Glass Compositions

Table 10-1 - Typical Compositions (mol%)								
Mnemonic	ZrF_4	BaF_4	LaF_3	AlF_3	LiF	NaF	PbF_2	GdF_3
ZBL	62.0	33.0	5.0					
ZBLA	57.5	34.5	4.0	4.0				
ZBLAL	51.8	20.0	5.3	3.3	19.6			
ZBLAN	54.0	15.0	6.0	4.0		21.0		
ZBLALPb	50.4	15.5	4.9	3.1	20.2		4.9	
ZBGA	60.0	32.0		4.0				4.0
	BaF_2	ZnF_2	LuF_3	YbF_3	ThF_4	NaF		
BZLT	19.0	27.0	27.0		27.0			
BZYbT	19.0	27.0		27.0	27.0			
BZYbTN	10.0	27.0		27.0	27.0	9.0		
BTYbZ	15.0	28.3		28.3	28.3			
	BaF_2	UF_4	YF_3	AlF_3	FeF_3	MnF_2	ZnF_2	CaF_2
BUYFe	20.0	40.0	20.0		20.0			
BUYAMn	30.4	30.4	4.8	2.0		32.4		
BUYAZn	30.4	30.4	4.8	2.0			32.4	
BCYA	22.0		16.0	40.0				22.0

TEST METHODS

In chemical durability testing there are many variables to be considered (e.g. surface preparation, container selection, solution composition and pH, temperature, surface area to solution volume (SA/V) ratio, etc.). Large differences in the leach rates of individual glass components have been seen for samples placed in stagnant solutions, while those exposed to highly agitated (well mixed) acidic solutions were seen to dissolve more rapidly and more congruently (4). Since test conditions can exert a major influence on the results obtained, it is critically important to select carefully and to attempt to identify the role that each of these variables plays in the leaching process.

Dissolution rates are determined either by following the mass loss of the glass with time (5-10) or by following the change in concentration of the dissolution products in solution with time (4, 8-19). The latter method yields more useful information about leaching mechanisms since glasses seldom dissolve in a totally congruent manner and corrosion products, as well as water, are usually adsorbed onto the leached surface, making weight loss measurements difficult to interpret accurately.

The studies discussed in this chapter, unless otherwise stated, involve samples which were soaked in deionized water in polymethylpentene (PMP) containers at temperatures below 100°C. For comparison between various glasses, leach rates are reported as composition normalized leached rates (NLR_i) for each element i, given by the equation:

$$NLR_i\ (g/cm^2d) = \frac{X(10^6)V}{SA \cdot t \cdot f_i} \qquad (10\text{-}1)$$

where X is the concentration of each cation in solution (μg/ml), V is the solution volume (ml), SA is the sample surface area (cm^2), t is the soak time interval (days) and f_i is the mass fraction of the cation in the unleached glass (see Clark/Zoitos chapter for similar equation).

Static tests yield very different results from flow tests (18) and, therefore, are described separately.

AQUEOUS CORROSION PROCESSES

Reactions of HMF Glasses with Water

In general, when HMF glasses first come into contact with pure water it sets off a chain of reactions which will be referred to as Stage I corrosion, as follows: 1) selective leaching of soluble metal-fluoride species, 2) an ion-exchange, or hydrolysis, reaction $F^-(gl)/OH^-(sol)$ at the glass surface, 3) diffusion of water into the glass surface, and 4) hydrolysis of dissolved species in solution leading to a pH drift into the acidic range. Reactions 1 - 3 produce a porous, hydrated surface transform layer which grows with time. The duration of Stage I depends strongly on the glass surface area to solution volume (SA/V) ratio and, to some extent, on glass composition. For the purposes of this discussion we will assume stagnant conditions and moderate SA/V ratios, where corrosion products are allowed to build up in solution. In this case, the most damaging reaction of Stage I corrosion is reaction 4 which can lead to extensive pH drift and greatly accelerated chemical attack. Once the build-up of hydrolyzed corrosion products has begun to occur in solution, an acidic environment is quickly established and Stage II corrosion processes dominate. These can be described as 1) solubility controlled dissolution of additional glass components (including some matrix elements), 2) further diffusion of water into the transform layer, 3) hydrolysis of some glass components within the transform layer, 4) hydrolysis of dissolved species in solution, leading to a further drop in solution pH, and 5) precipitation of saturated corrosion products.

Studies of aqueous corrosion in HMF glasses have shown that, under similar test conditions, their behavior can be classified according to composition; glasses with the same primary glass former exhibit the same general leaching characteristics with only slight differences being induced by modifiers. Now that we have identified the basic corrosion processes, further discussion will be divided according to primary glass former, into 1) fluorozirconate glasses, 2) thorium glasses, 3) uranium glasses, and 4) other HMF glasses, as listed in Table 10-1.

The Effect of pH

The effect of solution pH on the corrosion of HMF glasses is dramatic. Differences in leach rates of as much as five orders of magnitude have been measured for a single glass (ZBLAL) immersed in solutions buffered at pH 2 through 10. Figure 10-1 demonstrates this effect for a variety of glass compositions. It is interesting to note that while the thorium glasses (BZYbT, BZYbTN) exhibit better durability than the fluorozirconate (ZBLAL) in acidic environments, the reverse is true in neutral and basic solutions.

The relationship between corrosion rate and solution pH is particularly important to understanding the aqueous leaching behavior of fluoride glasses. Hydrolysis occurring both in solution and on the glass surface creates a locally acidic environment. This translates into an accelerated corrosion rate, as demonstrated in Figure 10-1. While in low to moderate solution volumes corrosion products are carried away from the surface by normal diffusion through the solution, there is evidence that a largely static, highly concentrated liquid envelope forms at the glass-solution interface (4). This region, being acidic, promotes a higher leach rate than the bulk solution pH would predict and corro-

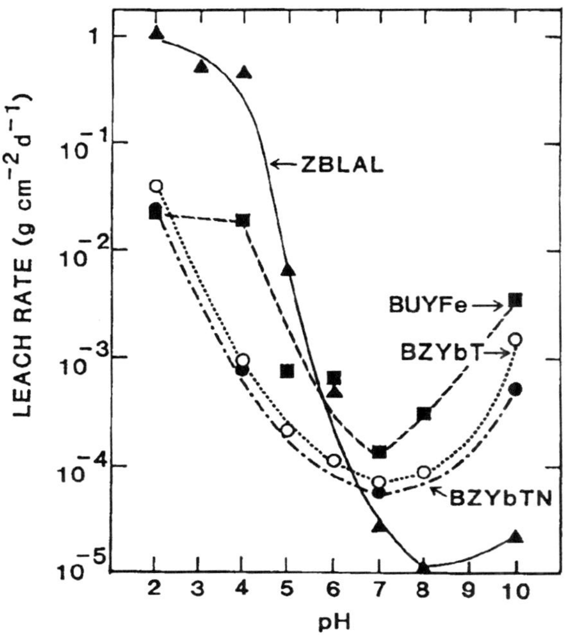

Figure 10-1. Comparison of leach rates as a function of solution pH for fluoride glasses based on ZrF_4, BaF_2-ThF_4 and UF_4 (Guery et al., 1988).

sion products accumulated in this region begin to form crystalline deposits on the glass surface within hours, long before saturation occurs in the solution as a whole. The effects of pH drift and local corrosion product build-up can be largely eliminated through vigorous stirring and large solution volumes (e.g. high flow rates).

Fluorozirconate glasses undergo the most severe hydrolysis during leaching. This is attributed primarily to the reactions of zirconium fluoride with water. An F-/OH- anion exchange reaction at the surface produces zirconium-hydroxyfluoride which remains in the transform layer as:

$$ZrF_7\ (gl) + x\ H_2O \rightarrow (ZrF_{7-x}\ (OH)_x)\ (gl) + x\ HF \tag{10-2}$$

while some ZrF_4 goes into solution without a prior hydroxylation step as:

$$ZrF_4\ (gl) \rightarrow ZrF_x^{4-x}\ (aq) + (4-x)F^-\ (aq) \tag{10-3}$$

Titration tests (15) have confirmed that subsequent hydrolysis of the $ZrF_x^{4-x}(aq)$ leads to the formation of species containing, on average, 2 OH^- ions as:

$$ZrF_4 + 2(H_2O) \rightarrow ZrF_2(OH)_2\ (s) + 2H^+\ (aq) + 2F^-(aq) \qquad (10\text{-}4)$$

and

$$ZrF_3^+(aq) + 2(H_2O) \rightarrow [ZrF(OH)_2]^+\ (aq) + 2H^+(aq) + 2F^-(aq) \qquad (10\text{-}5)$$

with

$$H^+ + F^- \Leftrightarrow HF \qquad (pH = 3.45) \qquad (10\text{-}6)$$

In stagnant, low volume solutions a pH drift from the initial value of 5.6 (deionized water) to an equilibrium value of 2.5 was measured (15). In the case of uranium based and thorium based glasses, where U and Th were found to hydrolyze, the equilibrium pH was found to be 3.5 and 4.8, respectively. Therefore, it is not surprising that these three glass families exhibit aqueous leach rates in the order Zr > U > Th.

This knowledge also leads to the prediction that HMF glasses immersed in unbuffered water will exhibit leach rates in excess of those corresponding to the initial solution pH. This prediction is born out in Figure 10-2, where the range of leach rates in deionized water is shown for the three glass families discussed above. In all cases the short-term leach rates exceed those corresponding to the initial pH of 5.6. At longer times, surface effects can complicate the leaching process.

SURFACE CONDITION

On exposure to water, HMF glasses rapidly develop opaque surface films that severely degrade their optical transmission characteristics, both in the visible and near IR.

There are a variety of techniques available for studying the reactions of HMF glass surfaces with water. Among the most useful are infrared transmission spectroscopy (IR, FTIR), scanning electron microscopy (SEM), and electron probe micro-analysis (EPMA), and x-ray photoelectron spectroscopy (XPS).

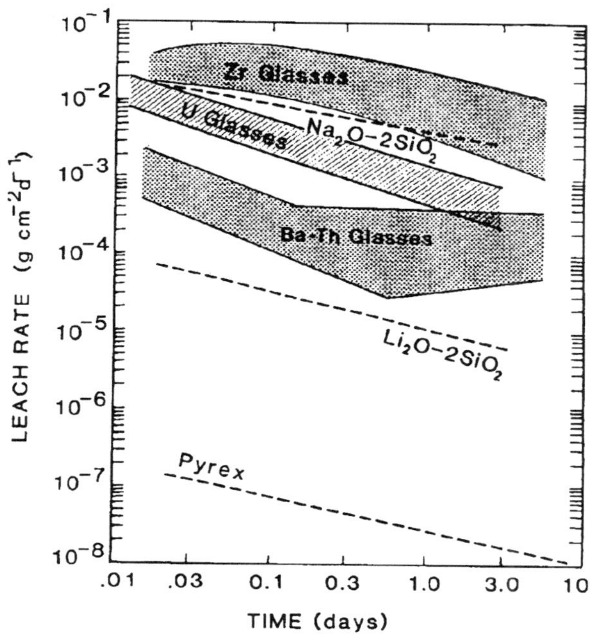

Figure 10-2. Comparison of leach rates in deionized water of fluoride glasses based on ZrF_4, BaF_2-ThF_4, and UF_4 (Guery et al. 1988).

Transform and Crystal Layer Formation

During Stage I corrosion the composition of the glass surface in contact with water is altered by selective leaching of the more soluble species (15,20) and by an $F^-(gl)/OH^-(sol)$ anion exchange in the glass (4,9,11,15,21,22). These processes produce a porous, hydrated, weakly adherent transform layer at the surface which thickens with time. Several studies using IR Spectroscopy (3), leachate analysis (6,13), pH titration (15), and direct layer thickness measurements (17,20) have shown a linear dependence of layer thickness on the square root of time. This $t^{1/2}$ behavior suggests that the kinetics of leaching and layer formation are largely diffusion controlled.

During Stage II corrosion, as water diffuses through the porous transform layer to the bulk glass interface some dissolved species hydrolyze, creating a more acidic environment (4,15,22). This increases the solubility of all components, leading to further dissolution of the glass. Materials transported out of the transform region encounter a more neutral environment upon entering the solution. This lowers corrosion product solubility, causing supersaturation of less soluble species and precipitation of several crystal forms on the sample surface.

The appearance of the leached surface varies markedly with solution pH and solution agitation conditions. For example, rapidly flowing, stirred solutions carry away corrosion products as they form, while unstirred stagnant solutions enhance the formation of precipitated surface crystals. In the case of fluorozirconate glasses, these are primarily of hydrated zirconium fluoride (needle-like) crystals and zirconium-barium fluorides (spherical crystallites) SEM as shown in Figure 10-3. The porous transform layer beneath the crystal is largely amorphous in nature and exhibits a mud-cracked appearance. This cracking has been observed in all samples, independent of layer thickness, and

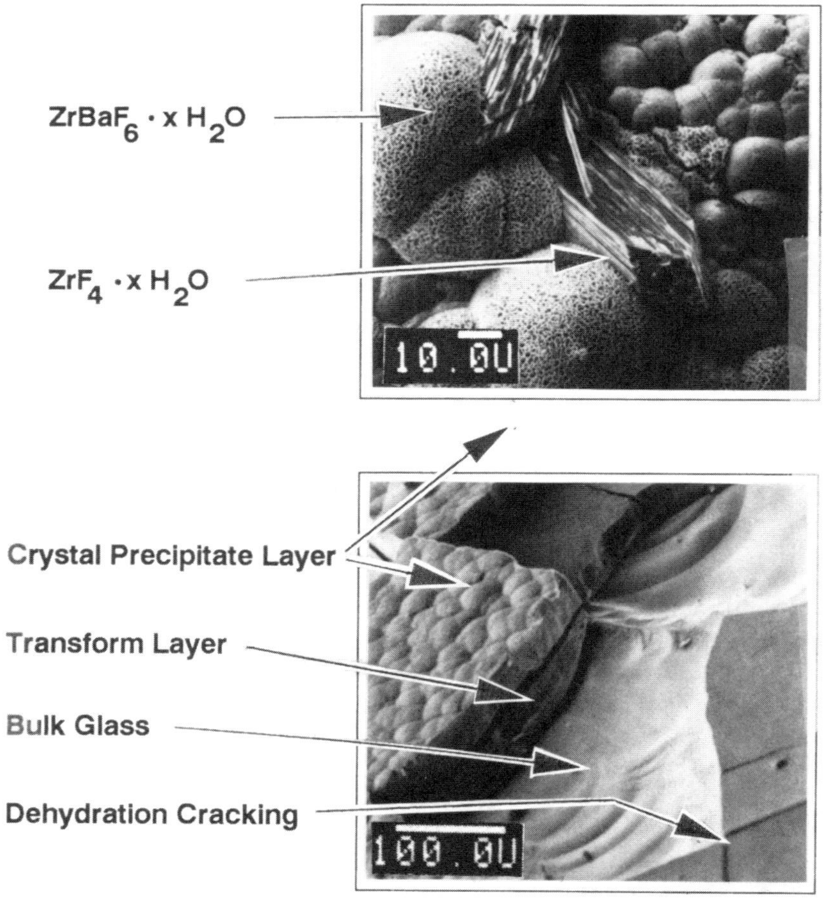

Figure 10-3. Corrosion layers formed on ZBLA glass, staticly corroded at SA/V =1cm^{-1} for 7 days. Top shows magnification of crystal precipitates; Bottom shows the crystal precipitate and transform layers as well as unleached glass. (Chen, 1987; Chen et al., 1987).

is likely the result of shrinkage stresses developed during drying prior to observation under the microscope. The bulk glass, exposed through spalling of the transform layer, appears to be uncorroded.

The rate of growth of the transform layer with exposure time also is strongly influenced by test conditions (17,21). Figure 10-4 shows SEM micrographs of typical cross-sections for leached ZBLA glass exposed to a stagnant SA/V=1cm^{-1} solution (17). Crystal precipitates are seen to form at the original glass/water interface while the boundary of the porous transform layer advances into the bulk. A plot of the measured thickness with time of the transform layer only is shown in Figure 10-5. In contrast, identical samples exposed to a stirred, high volume solution (flow rate = 7.5 liter/hr) showed an absence of surface crystal deposits and a much slower rate of growth for the transform region, as plotted in Figure 10-6.

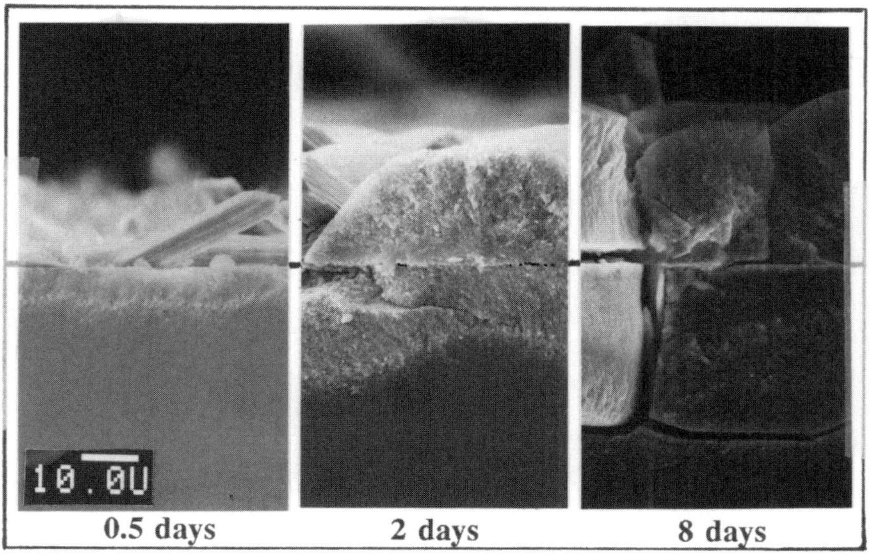

Figure 10-4. Formation of crystalline and transform layers on ZBLA glass during stagnant corrosion test in D.I. water, SA/V = 1. (Chen et al., 1987).

FORMATION OF HYDROXYL SPECIES

It is difficult, if not impossible, to accurately identify the processes occurring within the transform region if crystal precipitates are permitted to form on the glass surface. Therefore, test conditions must be selected to prevent, or

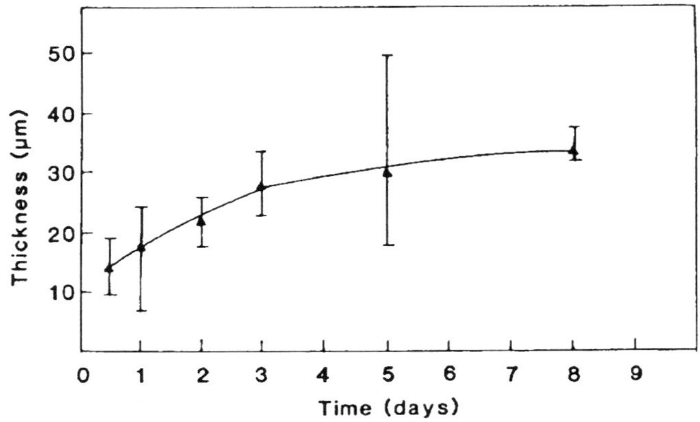

Figure 10-5. Transform layer thickness vs. corrosion time for ZBLA glass during stagnant test in D.I. water, SA/V = 1. (Chen et al., 1987)

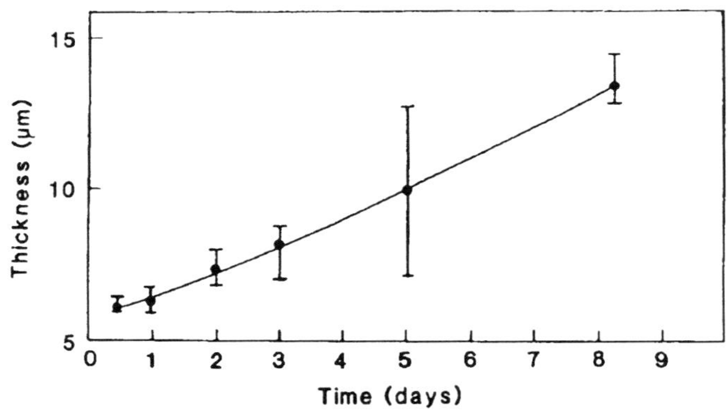

Figure 10-6. Transform layer thickness vs. corrosion time for ZBLA glass during stirred flow test in D.I. water, flow rate = 7.5l/hr. (Chen et al., 1987).

minimize, these deposits. LeToullec et al. (22) conducted tests in stirred, rapid flow conditions, as described above, while Pantano and Brow (23) and Lohr et al. (3) employed large solution volumes with vigorous agitation. Under these conditions the solution pH was maintained near its initial value (5.2 - 5.6) throughout the soak period, and the sample surfaces were relatively free of crystal deposits. After exposure the samples were examined by FTIR, IR or XPS analysis.

FTIR Analysis

LeToullec et al. studied the formation of hydroxides in corroded ZBL,

ZBLA and BZYbT glasses by FTIR. The investigators first established the peak positions in the IR absorption spectra of $Zr(OH)_4$, ZrO_2, $Th(OH)_4$, ThO_2, $Al(OH)_3$, $Ba(OH)_2$, $La(OH)_3$, and $Zn(OH)_2$ with individual measurements on commercially prepared, high purity powders incorporated in KBr pellets. The ratio of the integrated peak areas of the OH stretching ($3430cm^{-1}$) and HOH bending ($1620cm^{-1}$) absorbance bands for pure molecular water was determined using a ZnSe liquid cell. Based on this ratio, it was possible to separate from the total OH stretching band the portion due solely to the penetration of molecular water into the glass from the portion associated with the formation of metal-hydroxyl species resulting from F^-/OH^- ion exchange (see Figure 10-7).

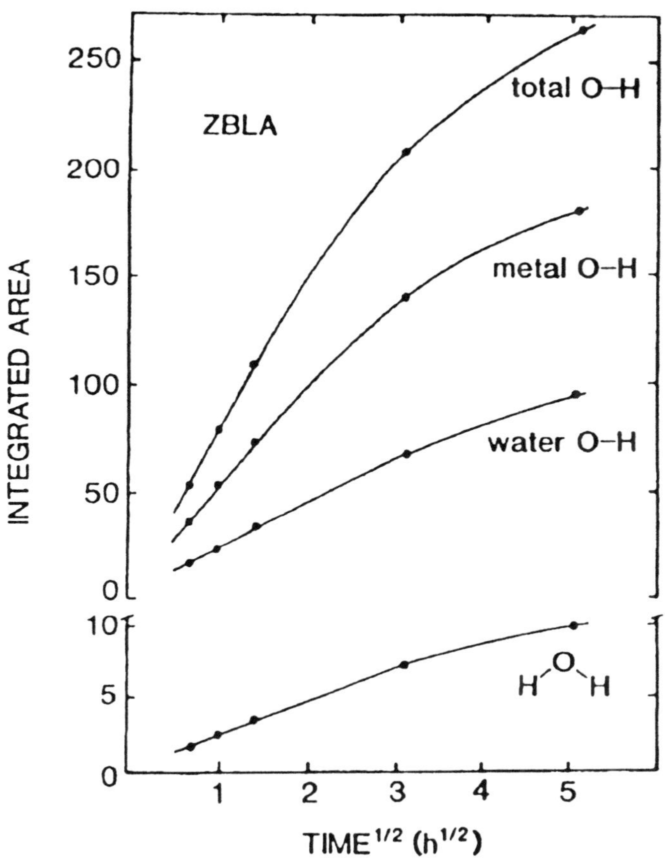

Figure 10-7. Evolution with immersion time in water of the total OH integrated absorption peak area at $340cm^{-1}$, the individual contributions of metal OH and water OH to the toal peak area at $3430cm^{-1}$ and the HOH peak area at $1630cm^{-1}$. (LeToullec et al., 1988).

Subsequent comparison between the spectra of leached glasses with those of the individual compounds enabled the investigators to identify several metal-hydroxyl species present in the transform layers. Such a comparison (Figure 10-8) of two fluorozirconate glasses with $Zr(OH)_4$ shows excellent agreement for the peak positions at 1550 and 1340 cm^{-1} and demonstrates the presence of Zr-OH bonds. The same analysis was conducted on leached BZYbT and showed a much lower penetration of water into the glass, with the majority of the OH vibration (90%) resulting from metal-hydroxyl compounds, primarily Th-OH species.

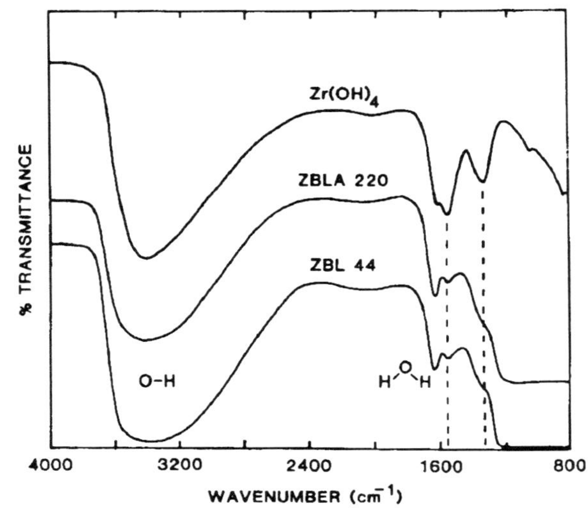

Figure 10-8. FTIR spectra of ZBLA glass leached for 220hr, ZBLA glass leached for 44 hr, and $Zr(OH)_4$ in a KBr pellet. (LeToullec et al., 1988).

Lohr et al. (3) observed that the ratios of the peak **intensities** for the stretching (3430cm^{-1}) and bending (1620cm^{-1}) vibrations of water remained roughly constant with soak time but were unable to detect the presence of hydroxyl species in the leached layer. Based on extinction coefficients for water in fluorozirconate glasses, measured by Fonteneau et al. (24), they estimated that the hydrated layer contains as much as 1 mole H_2O per ZrF_4 unit. They also noted that the $t^{1/2}$ behavior they observed suggests diffusion-limited kinetics for both layer formation and leaching.

XPS Analysis

Pantano and Brow (23) conducted an XPS analysis of surface reaction products resulting from hydrolysis of ZBLA. By analyzing the chemical shifts of the Zr-3d photoelectron from surfaces exposed to water, they estimated that the dominant surface species were ZrO_2F_5 or $Zr(OH)_2F_5$, supporting the conclusions of Simmons et al. (11), Simmons and Simmons (15), Ravaine and Perera (9) and LeToullec et al. (22) that a strong F^-_{glass}/OH^-_{sol} ion exchange process accompanies the corrosion of these materials. Pantano and Brow also found that, while the composition of the (hydr)oxy-fluoride surface species did not change with solution pH, the concentration was much lower in acidic solutions and much higher in basic solutions where dissociated $(OH)^-$ radicals were abundant.

COATINGS

Despite the poor chemical durability of fluorozirconates and their great technical importance, there have been few coatings studies. A variety of authors have looked briefly at the effect of several dielectric and metallic coatings, including diamond-like carbon (DLC) coatings, MgO, MgF_2, phosphates, chalcogenides, Al, Sn and Ti, on the chemical corrosion behavior of fluorozirconate glasses.

Simmons and Simmons (15) found that DLC coatings exhibited promise in reducing the corrosion of the ZBLA glass. By comparing the before and after infrared transmission spectra for short-term (4 hr) aqueous corrosion tests on identical ZBLA samples, coated and uncoated, they show that very little change has occurred in the coated sample, while fairly extensive surface hydration is observed in the uncoated ZBLA. This observation was confirmed by SEM; the surface of the uncoated sample was covered with the typical dehydration cracking discussed above, while the coated surface showed only isolated pit corrosion and no cracking or spalling of the surface. The pit corrosion is thought to have resulted from pinholes in the experimental coating and it is believed that improved deposition techniques can overcome this problem.

MgO coatings on ZBLAN have protected the glass with no visible sign of attack for 100 hrs (25). Phosphate (26) and chalcogenide (27) overcoatings have been somewhat effective in protecting glass fibers from attack in a humid atmosphere.

SUMMARY

The mechanisms controlling aqueous corrosion of heavy-metal fluoride glasses are complex and vary widely depending on the test environment. In stagnant unbuffered water, particularly at high SA/V ratios, accumulation of corrosion products can lead to severe modifications in the leaching process. Preferential leaching of the more soluble species leads to the formation of an altered surface layer. Hydrolysis of some fluoride species (eg. $ZrF_x^{(4-x)+}$), consuming solution -OH, causes a rapid decrease in pH of the liquid in contact with the glass surface which, in turn, results in increased solubility and an accelerate dissolution rate. Ultimately, saturation of the less soluble species occurs, causing precipitation of surface crystals. Both pH drift and crystal precipitation can be largely eliminated through the use of low SA/V ratios in a well-mixed or flowing bath at constant pH.

Dissolution rates for several HMF glass families have been studied as a function of controlled pH and glass composition. The results demonstrate a dramatic change in corrosion rate when going from neutral to acidic solutions: a factor of 10^5 for ZBLAL and a factor of 250 for BZYbT. When these results were compared with results from identical tests on glasses in deionized water, it was found that solution pH drift acts as a serious liability, especially in the case of the Zr^- based glasses.

It is expected that further studies in this area will produce glasses with improved chemical durability without substantial losses to IR optical transparency.

REFERENCES

1. Poulain, M. Poulain, M. and Lucas, J., *Mat. Res. Bull.* 10: 243 (1975).
2. Simmons, J.H., Simmons, C.J., Ochoa, R. and Wright, A.C. in: Fluoride Glass Fiber Optics (I. Aggarwal and G. Lu ed.), *Physical Acoustics*, Vol. XX, pp 37-84, Academic Press (1990).
3. Loehr, S.R., Bruce, A.J., Mossadegh, R., Doremus, R.H. and Moynihan, C.T., *Mater. Sci. Forum* 5: 311-322 (1985).
4. Simmons, C.J., *J. Am. Ceram. Soc.* 70: 654-661 (1987).
5. Mitachi, S., *Phys. Chem. Glasses* 24: 146 (1983).
6. Frischat, G.H. and Overbeck, I., *J. Am. Ceram. Soc.* 67: C-238 (1984).
7. Loehr, S.R., Bruce, A.J., Mossedegh, R., Doremus, R.H. and Moynihan, C.T., *Mater. Sci. Forum* 19-20: 327 (1987).
8. Tregoat, D., Liepmann, M.J., Fonteneau, G., Lucas, J. and Mackenzie, J.D., *J. Non-Cryst. Solids* 83: 282 (1986).
9. Ravaine, D. and Perera, G., *J. Am. Ceram. Soc.* 69: 852 (1986).
10. Seddon, A.B., Shah, W.A., Clare, A.G. and Parker, J.M., *Mater. Sci. Forum* 19-20: 465 (1987).
11. Simmons, C.J., Sutter, H., Simmons, J.H. and Tran, D.C., *Mat. Res. Bull* 17: 1203-1210 (1982).
12. Barkatt, A. and Boehm, L., *Mater. Lett.* 3: 43 (1984).
13. Doremus, R.H., Murphy, D., Bansal, N.P., Langford, W.A. and Burman, C., *J. Mater. Sci.* 20: 4445-4453 (1985).
14. Simmons, C.J., Guery, J., Chen, D.G. and Jacoboni, C., *Mater. Sci. Forum* 5: 329-334 (1985).
15. Simmons, C.J. and Simmons, J.H., *J. Am. Ceram. Soc.* 69: 661-669

(1986).

16. Simmons, C.J., *J. Amer. Ceram. Soc.* 70: 295-300 (1987).
17. Chen, D.G., Simmons, C.J. and Simmons, J.H., *Mater. Sci. Forum* 19-20: 315-320 (1987).
18. Guery, J., Chen, D.G., Simmons, C.J., Simmons, J.H. and Jacoboni, C., *Phys. Chem. Glasses* 29[1]: 30-36 (1988).
19. Seddon, A.B. and Shaw, W.A., *Mater. Sci. Forum* 32-33: 255-260 (1988).
20. Tick, P. and Mitachi, S., *Bull. Am. Ceram. Soc.* 69: 543 (paper #5-JXIX-90) (1990).
21. Chen, D.G., *University of Florida Graduate Thesis*: 101 (1987).
22. LeToullec, M., Simmons, C.J. and Simmons, J.H., *J. Am. Ceram. Soc.* 71: 219-224 (1988).
23. Pantano, C.G. and Brow, R.K., *J. Am. Ceram. Soc.* 71: 577-581 (1988).
24. Fonteneau, G., Tregoat, D. and Lucas, J., *Mat. Res. Bull.* 20: 1047 (1985).
25. Schultz, P.C., Vacha, L.J.B., Moynihan, C.T., Harbison, B.B., Cadian, K. and Mossedegh, R., *Mater. Sci Forum* 19-20: 343 (1987).
26. Vacha, L.J.B., Shultz, P.C., Moynihan, C.T. and Crichton, S.N., *Mater. Sci. Forum* 19-20: 419 (1987).
27. Nakata, A., Lau, J. and Mackenzie, J.D., *Mater. Sci. Forum* 6: 717 (1985).

11

Corrosion of Geological and Archaeological Glasses

Michael J. Jercinovic and Rodney C. Ewing

Department of Geology
The University of New Mexico
Albuquerque, New Mexico

INTRODUCTION

The corrosion of glass in the geologic environment is of considerable significance in geology and archaeology. The chemistry of seawater is partly controlled by alteration processes occurring at the surface of, or within, the oceanic crust (1). The elemental exchange occurring during alteration results, in part, from glass/seawater interactions (e.g., hydrolysis of the glass). Additionally, release of elements in these hydrolytic reactions in large-scale hydrothermal systems within the oceanic crust near seafloor spreading centers may be significant in the formation of economically-important mineral deposits. Glass alteration is also significant in determining the hydrochemistry of some fresh water systems, such as those in Iceland (2) and in the Columbia River Plateau region of the northwestern United States (3).

Low temperature (0 to 20°C) corrosion processes tend to be dominated by chemical kinetics. Therefore, glass alteration can be used as an age dating technique (applicable to both geologic and archaeologic materials) if the corrosion mechanisms and geologic histories are well understood. Glass alteration processes may also be important in understanding the genesis of planetary regolith materials, particularly on terrestrial planets which have atmospheric or surface volatiles (4). In such cases, glasses may be derived from both volcanic and impact processes, subsequently altered, with alteration products then being physically fractionated and dispersed by atmospheric processes.

The primary high level nuclear waste repository in the United States will be located in silicic tuffs, material consisting primarily of fragmental, devitrified volcanogenic glass. Thus, understanding the geochemistry of the repository environment involves the consideration of glass alteration. The assessment of the long-term corrosion behavior of synthetic nuclear waste form glass (e.g., borosilicate glass) when exposed to the geologic environment requires the development of predictive models and a means of verification. The existence of an appropriate natural analogue may provide possible constraints for determining the long-term corrosion rates of waste form glasses and may also provide a means of model verification (5-7).

THE OCCURRENCE OF NATURAL GLASSES

Glasses occur in nature as a result of the rapid cooling (quenching) of silicate melts (magmas) as they come into contact with a quenching medium (water, ice, atmosphere). Silicate melts, consisting of molten liquid, crystals (except in superheated melts), and gases (dissolved or as bubbles), are derived from pre-existing rocks or condensed solids by heating via internal sources, resulting in magmatic activity, or due to meteoritic impact. Magmatism is expressed in the geologic record as volcanic rock (melt erupted or extruded, and crystallized at the planetary surface) and plutonic rock (melt crystallized at depth). Volcanism is most prevalent along the margins of crustal plates, primarily at constructive boundaries (mid-ocean ridges) and destructive margins, where plates collide and oceanic crust is "consumed" as it is subducted beneath less dense continental crust (Figure 11-1).

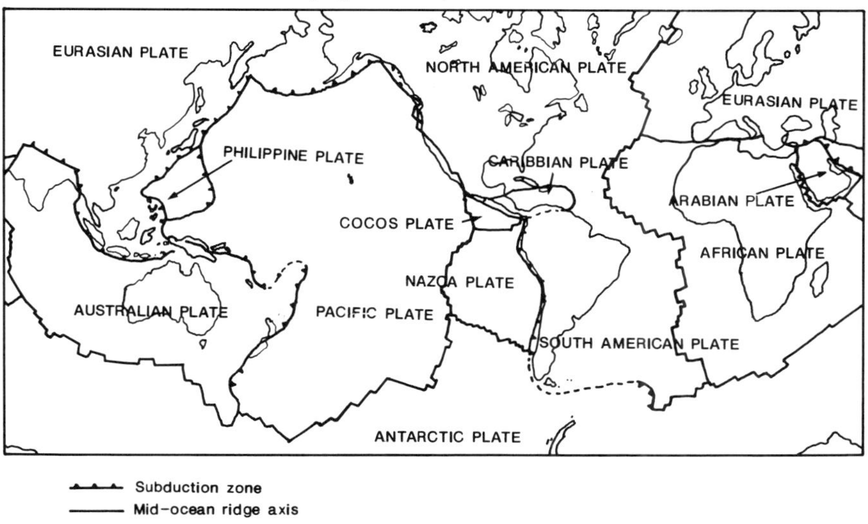

Figure 11-1. Major crustal plate boundaries of the Earth, indicating regions of active volcanism and glass formation (adapted from (68)).

The igneous rocks which crystallize from silicate magmas are broadly classified based on their SiO_2 content: acid (SiO_2 > 66 wt%); intermediate (SiO_2 52 to 66 wt.%), basic (SiO_2 45 to 52 wt.%), and ultrabasic (SiO_2 < 45 wt.%). Ultrabasic rocks are generally derived by accumulation of low-silica minerals which settle during crystallization within a magma chamber. There are examples of rocks derived from ultrabasic liquids on Earth, mostly older than 2.5 billion years.

Glasses created by the quenching of natural silicate melts occur in rocks representing each of the major groups listed above. Glasses found in association with ultrabasic rocks are not themselves ultrabasic, having formed as a "residual liquid" following crystallization of minerals which have left the remaining melt much higher in silica and in incompatible elements than the bulk rock. Glasses form on the surface of the earth anywhere magma reaches the surface and is rapidly cooled. The quenching media necessary to produce glasses depend on the cooling rate required of the particular melt. Magmas rich in silica may only require exposure to the atmosphere to produce significant quantities of glass. Melts lower in silica (e.g., basic melts) generally require quenching by liquid water during extrusion or direct contact with cold rock during intrusion to obtain the cooling rates necessary for glass formation. Eruptions occurring at mid-ocean ridges produce significant quantities of glass from the interaction of low silica melts and seawater. Such subaqueous eruptions generate pillow basalts (Figure 11-2) and hyaloclastites (glassy, clastic debris) in the deep sea environment (Figure 11-3). Eruptions beneath glacial ice also result in glassy

Figure 11-2. Pillow basalts from the central rift zone of Iceland, along the Mid-Atlantic Ridge.

Figure 11-3. Hyaloclastite (bedded fragmental glass) at Tuya Butte, British Columbia, a volcano which erupted beneath glacial ice approximately 10,000 years ago. The exposed escarpment is about 100 m high.

subaqueous deposits, mostly involving low silica melts, but there are examples of high silica subglacial hyaloclastites and pillows as well (8). Rock textures produced during magmatic crystallization depend, to a large extent, on cooling rate. Commonly, igneous rocks contain relatively large (macroscopic) crystals in a fine-grained groundmass. In many cases, this groundmass contains significant quantities of glass which, as in the case of glasses in ultrabasic rocks, contains high concentrations of elements which are incompatible to the structures of the early-formed, higher temperature crystals. This glassy "mesostasis" is relatively susceptible to corrosion when the rock is exposed to aqueous solutions. Thus, the weathering of the rock may proceed predominantly by groundmass destruction. The solution chemistry resulting from this corrosion process will, therefore, more reflect glass-water, compared to crystal-water, interactions. Eruptions of basic magma on the lunar surface have produced glass beads, probably the result of quenching during eruption fountaining. The interaction of basic magma produced at depth beneath subduction zones and the overlying silicic crust produces magmas with intermediate compositions. Glasses are easily produced in silicic volcanism as rapid cooling, even in air, will result in glass at high silica content. Thus, ash flow tuffs, produced in massive eruptions of differentiated magmas rich in silica and volatiles can consist predominantly of glass. High silica glasses are also found in the deep sea as ash layers from eruptions on continental margins. Rapid heating of quartz-rich (SiO_2) sand by lightning strikes produces limited quantities of glass, forming "fulgerites". Glasses also can be of meteoritic origin, occurring in both projectile and target. Glass produced in the target (impacted body) occurs as impact melt sheets or tektites (melt droplets ejected by impact into the upper atmosphere). Meteoritic glass (ancient primary glass within the meteoritic projectile) originated in the parent body regolith and examples are found which approach 4.5 billion years in age.

GEOLOGIC SYSTEMS

The study of corrosion processes in natural environments is complicated by the nature of the systems in question. In comparison to experimental corrosion studies, natural systems occur on a much larger spacial scale, are never completely isolated, involve many components, and exist over long time periods. Geologic systems can be evaluated on a size range from micrometers to thousands of kilometers, and classifying them as open or closed usually depends on the scale of the system chosen. Geologic systems thermodynamically range from relatively simple, involving few components at high temperatures with phases approaching equilibrium, to highly complex systems, involving metastable solids, an aqueous phase, and possibly gases. Phases themselves range from simple, one-component metals or oxides, to complex, multi-element materials, typically silicates, some of which are actually "phase regions" in solid solutions. The number of phases, components and variables can be quite high.

Because reaction rates are strongly temperature dependent, kinetically-dominated, low-temperature alteration systems may have many phases existing metastably. In fact, nearly all rocks exposed at the surface of the Earth are metastable, having been generated under pressure and temperature conditions quite different than those of the Earth's surface. Only the extremely slow kinetics of gas-solid reactions preserve the assemblages with little alteration for geologic time periods. Rocks exposed to aqueous media, or to elevated temperatures may alter rapidly, but only rarely to an assemblage which approaches thermodynamic equilibrium at the surface of the Earth. Therefore, a major question in evaluating any assemblage of phases in geologic systems is "To what extent do the constituent phases represent ancient conditions, and what record has been left of changes in the system with time?"

Variables such as water flow rate, temperature, and water chemistry generally change with time during the history of the system in question. Systems initially open may become closed with time. The temperatures which govern reaction kinetics depend on the climatic and thermal histories of the deposits. In terms of glass alteration processes, time is obviously quite important when considering reaction rates. Sample age only provides a boundary for determining alteration rates as the true exposure time of the rock (or glass) to the aqueous solution is unknown, but somewhat less than the sample age. The complexity of geologic systems must be kept in mind when attempting to extend the results of controlled experiments to natural systems or to extend the results of geologic history to predicting behavior in the future. The larger the system chosen and the greater the length of time under consideration, the more difficult quantitative assessment of the system becomes.

In order to evaluate what a naturally-altered sample has been exposed to from the time it was deposited to the present, that is, the alteration environment and how that may have changed, requires the ability to assess geologic systems based on geochemistry and hydrology. Even the simplest geologic system requires generalizations and assumptions when geochemical and hydrologic modelling are attempted. For instance, modelling fluid flow in a natural deposit can be exceedingly complex. Simply extending the relationships of the microscopic realm (pore volume, grain dimensions and shapes) to model a macroscopic deposit is unrealistic as all deposits are inhomogeneous at some level, and macroscopic features such as bedding and grain size differences or fractures can radically alter fluid flow. The one effect which is perhaps the most difficult to assess is time and how the system evolved, particularly considering the influences of "external" factors such as climate and tectonics. The composition and flow rate of aqueous solutions affecting a deposit may drastically change with time in response to both internal and external factors.

GENERAL ALTERATION AND CORROSION CONCEPTS AS APPLIED TO NATURAL GLASSES

Alteration is the process by which a thermodynamically metastable or unstable material converts chemically or physically into another, more stable material (or assemblage of materials). Corrosion implies breakdown of the material, not necessarily including the formation of more stable materials.

The process of conversion of one material to a more thermodynamically stable configuration (lowering the total free energy of the system) is generally done in steps, with intermediate, metastable phases forming because nucleation or growth rates favor them as they may lower the system free energy rapidly (Ostwald step sequence). Each step in the sequence requires overcoming the appropriate activation energy to initiate reaction. Thus, in most natural systems, we find them in some metastable state intermediate between the fresh starting phase assemblage and the thermodynamically most stable assemblage.

Glass corrosion in nature involves two general mechanisms: diffusion-hydration and dissolution-hydrolysis. Hydration is the process by which H_2O (probably as H_3O^+) penetrates the glass from the exposed outer surface of the material inward. This process is usually accompanied by interdiffusion of ions from the glass outward into the solution (ion-exchange). Hydrolysis is the process of breakdown of the glass network (dissolution) when exposed to the aqueous environment, releasing some elements as aqueous species and leaving behind the insoluble residue. As a certain amount of hydration and interdiffusion must take place for hydrolysis to proceed, both processes may occur simultaneously when glass is exposed to water. The relative influence of each process must be assessed in order to establish overall reaction mechanisms and rates. White (9) describes the one-dimensional interdiffusion of Na^+ and H^+, including a moving boundary (due to dissolution), as following a modified version of Fick's second law:

$$(\delta c/\delta t)_y = \delta/\delta y[D(\delta c/\delta y)_t]+R(\delta c/\delta y) \quad (11\text{-}1)$$

where c is the concentration of a glass component at distance y (y is estimated from: $y = x\text{-}Rt$, where x is the initial distance of the component from the glass surface), D is the diffusion coefficient, t is time, and R is the dissolution rate of the glass. Diffusion coefficients can be calculated from the comparison of experimentally-produced diffusion profiles and solving equation 11-1 by substituting the binary diffusion coefficient for D and solving for concentration (9). Diffusion coefficients are strongly dependent upon pH and temperature. The overall glass corrosion reaction is described in terms of contributions from diffusion-hydration, releasing alkalis into solution in accordance with D (m^2/sec) above, and on the matrix dissolution rate a (m/sec) (10). For situations where time $t < a^2/D$, diffusion dominates and the quantity of alkalis released (Q, in mol/m^2) is given by:

$$Q = C_o * 2\ [(D^*t)/\pi]^{1/2} \quad (11\text{-}2)$$

where C_o is the alkali concentration in the glass. Where $t > a^2/D$, dissolution dominates and

$$Q = C_o(a^*t+D/a) \quad (11\text{-}3)$$

where D/a gives the depth of water penetration and a^*t is the depth of dissolution (10,11). Grambow (7) models the glass dissolution process as occurring with two rates: 1) an initial, or forward, rate (silica undersaturated); and 2) a final rate (silica saturated). The rate of glass hydrolysis is controlled by the solution activity of silicic acid:

$$r_m = k+ * (1\text{-}a_{H4SiO4}/K^*) \quad (11\text{-}4)$$

where r_m is the rate of matrix dissolution, $k+$ is the forward rate constant, a_{H4SiO4} is the activity of silicic acid and K^* is the equilibrium rate constant for the rate limiting reaction, $SiO_2 + 2H_2O \rightarrow H_4SiO_4$ (10). As hydrolysis proceeds, the increasing silicic acid concentration in solution will eventually lead to surface silica saturation and a lower dissolution rate (7). Considering the effect of a barrier to mass transport and a residual (final) rate of matrix dissolution at silica saturation, the rate of matrix dissolution becomes:

$$r_m = k+ [D/L * (K^* - a_{H4SiO4}) + r_{fin}]/[D/L * K^* + k+] \quad (11\text{-}5)$$

where D is the silica diffusion coefficient, L is the thickness of the transport barrier, and r_{fin} is the final rate of reaction which exists because glass corrosion proceeds even at silica saturation, as indicated by experiments (10).

As is the case for any thermally-activated process, the relation between the rate of reaction and temperature can be described by the Arrhenius relationship:

$$k=Ae^{-E/RT} \quad (11\text{-}6)$$

where k is the hydration rate, A is a constant (the frequency factor), E is the activation energy of hydration, R is the gas constant, and T is the absolute temperature. Activation energies for various glass corrosion processes and glasses are summarized in Table 11-1. Because reaction rates increase exponentially with temperature, higher temperatures due to hydrothermal activity or, to a lesser extent, climate, can rapidly (over a period of years to hundreds of years) decompose or devitrify glass which may otherwise persist metastably for millions of years.

Table 11-1. Activation Energies for Glass Alteration Processes.

Process	Activation Energy (kj/mol)
Low Silica Glass	
Dissolution in seawater	65(29)
Dissolution in fresh water	32(48)
High Silica Glass	
Dissolution (tektites)	80(121)
Na diffusion (obsidian)	96(9)
Hydration	84(122), 80-92(81)

ALTERATION OF LOW SILICA GLASS

Introduction

Low silica glass exposed to surface water or groundwater alters to a material known as palagonite. This term was introduced in reference to the fine-grained matrix of a glassy, granular-volcanogenic deposit (tuff) from Palagonia, Sicily (12). The term palagonite has, historically, been rather loosely applied, referring in some cases to entire deposits (13), and in other cases to microscopic materials associated with altered low silica glass (14,15). Because of this ambiguity in terminology, Honnorez (16) argues against the use of the term "palagonite" as a reference to any specific phase. As microanalytical capabilities have improved, the term has become more restricted, and is now generally used in reference to a vitreous, red, yellow, or brown, alteration product of clay-like chemical composition which replaces low silica glass on exposed surfaces (17-20). The term will be used here in this restricted sense.

Palagonite: General Properties and Composition

Surface layers form on all glass surfaces exposed to aqueous solutions, including pillow rim surfaces, hyaloclastite grain surfaces, vesicle walls, and along fractures (Figure 11-4). The boundary between fresh glass and the material that replaces it (palagonite) is sharp, with the glass surface being scalloped in cross section, due to dissolution (4,21). In some samples, the glass is penetrated by thin tubes which extend tens of micrometers into the glass

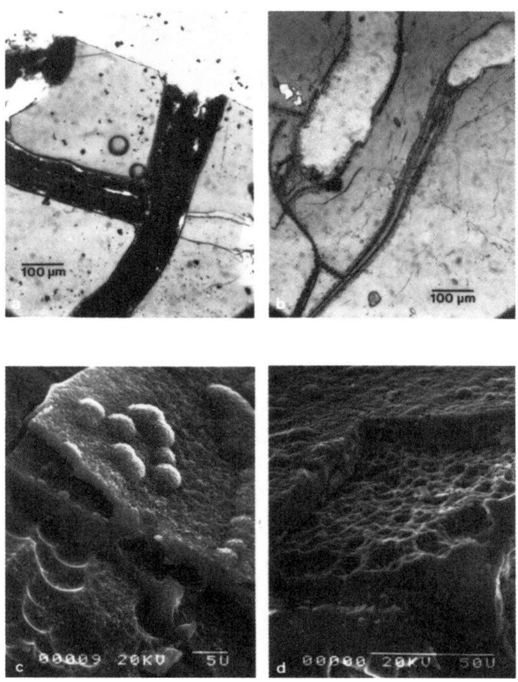

Figure 11-4. Thin section micrographs and scanning electron images of glasses and corresponding palagonite rinds in deep sea dredge samples. a) and c) USNM 113521-69; b) and d) USNM 113715 (adapted from (20)).

beneath the palagonite rind (1,22). The palagonite rind is, in all cases, more red than the parent glass, a manifestation of the Fe^{2+}, present in the parent glass, being oxidized to Fe^{3+} during palagonitization. The altered glass at Palagonia has had approximately 80% of the original Fe^{2+} oxidized to Fe^{3+} (23). Palagonite rinds are optically isotropic or weakly anisotropic, indicating an amorphous or poorly crystalline nature. The refractive index of palagonite is in the range 1.46 to 1.70 (18,25,26), and has a specific gravity which ranges from 1.82 to 2.85, with higher values related to Fe-Ti rich glass (18,24,26-28).

Palagonitization is a non-isochemical alteration process, not simple hydration of the glass; that is, elements are generally depleted, but to varying degrees, in palagonite relative to parent glass, with an increase in H_2O. The composition of palagonite differs from one occurrence to another in relative concentrations of elements but generally consists of Si, Al, Fe, and Mg in proportions suggesting ferromagnesian clay minerals (smectite clays). Altered low silica glasses from locations representing many different alteration environments and ages have been chemically analyzed (Table 11-2). Palagonite rinds of different chemical composition can form on glasses of essentially identical composition, indicating the important influence of the alteration environment (solution composition, temperature, flow rate of water through the sample).

Table 11-2. Average Microprobe Analyses (wt%) of Glass and Corresponding Palagonite Rinds.

	Tantalus, Hawaii(24)		USNM 113715 Deep sea dredge glass(20)		A5A Ash Mountain, B.C.(39)		TB2 Tuya Butte, B.C.(39)	
	Glass	Palagonite	Glass	Palagonite	Glass	Palagonite	Glass	Palagonite
SiO_2	39.4	41.4	50.4	41.6	51.7	30.0	50.9	38.9
TiO_2	2.8	4.1	1.86	1.83	1.81	3.24	2.78	4.63
Al_2O_3	11.9	7.2	15.3	18.2	15.4	23.3	13.6	8.8
FeO^*	11.8	11.8	10.0	13.0	11.1	19.3	13.0	20.7
MnO			.19	.02**	.17	.04**	.20	.12**
MgO	4.8	6.0	7.7	2.78	6.4	.59	4.39	1.62
CaO	12.5	3.5	10.8	1.57	9.2	1.17	9.7	3.15
Na_2O	7.8	.2	3.02	1.91	3.30	.03	3.83	.44
K_2O	3.0	.7	.14	2.29	0.69	.09	1.40	.46
P_2O_5			.19	.26	.30	.71	.78	1.32
total	94.0	76.1	99.6	83.5	100.1	78.5	100.6	80.1

*All Fe reported as FeO.
**Below detection limit.

As a result of palagonitization, most major elements are lost from the glass to the local aqueous solution, with the general exception of Ti and Fe. Al can be retained or lost during palagonitization, depending on solubility relationships which are solution dependent. The alteration of glass high in Fe and Ti results in palagonite rinds which have Fe and Ti concentrations considerably higher than palagonite rinds on glasses lower in Fe and Ti (28). Thus, there is some dependence of palagonite composition on the composition of parent glass, at least with respect to immobile elements.

A difficulty in the interpretation of analytical data of glasses and corresponding palagonite rinds is that elemental concentrations are expressed as weight percents so that true gains or losses cannot be evaluated. Ti has been suggested to be insoluble over a wide range of conditions, thus being conserved during palagonitization, and may be used to monitor the mobility of other elements (1,29). Another approach to estimating true elemental gains and losses is to consider palagonitization to be a constant volume process, such that the volume of glass reacted is equivalent to the present volume of palagonite (18,20). This assumption appears reasonable based on the pseudomorphic relationships seen in thin section (18). Knowing that the relative volumes are equal, elemental gains and losses can be obtained by multiplying the weight fraction of an element (or oxide) in glass and palagonite by the density of the material to obtain mass per unit volume of that element in glass and palagonite (Figure 11-5). The mass/volume difference for an element between glass and palagonite thus represents true gain or loss if the reaction is truly isovolumetric, and if the densities of glass and palagonite are accurate.

All palagonites show losses of Si, whereas Fe and Ti are retained, or even enriched, relative to associated glass. For major elements, the greatest variability is in the concentrations of Al, alkaline-earths, and alkalis. Al solubility is pH-dependent, and is relatively insoluble under neutral pH conditions. The behavior of Al is of particular importance in the formation of aluminosilicate cements at high reaction progress (see section on secondary mineralization below). Fresh water alteration generates palagonite rinds which are depleted in alkalis and alkaline-earths relative to parent glass. Seawater alteration results in palagonite rinds in which K is generally highly enriched and Na is retained. The K concentration of palagonite may be useful as a "maturity index", relevant to the submarine alteration of low silica glass (30). Palagonites formed on glasses dredged from the sea floor can be classified into two distinct compositional groups, one high-Fe, low-Si, and the other low-Fe, high-Si (31). As all deep sea dredge samples are exposed to the same major environment throughout the majority of their residence time on the sea floor (i.e., seawater, open ocean, 0 to 10°C), and the glasses are of nearly identical composition, the difference must arise from microenvironmental variations on the sea floor, perhaps similar to those responsible for the formation of metal-rich deposits near mid-ocean ridge crests (32-35). Palagonitization of low silica glasses obtained from drill cores of the Ocean Drilling Program generally resemble the low-Fe, high-Si dredge sample palagonites in terms of major element chemistry.

Compositional zoning has been documented from electron microprobe studies of some palagonite rinds (20,30,36,37). As shown in Figure 11-6, most commonly the outermost (first-formed) portions of the rind are lower in Al and Ca, and correspondingly higher in Fe and Mg, compared to the inner (later-

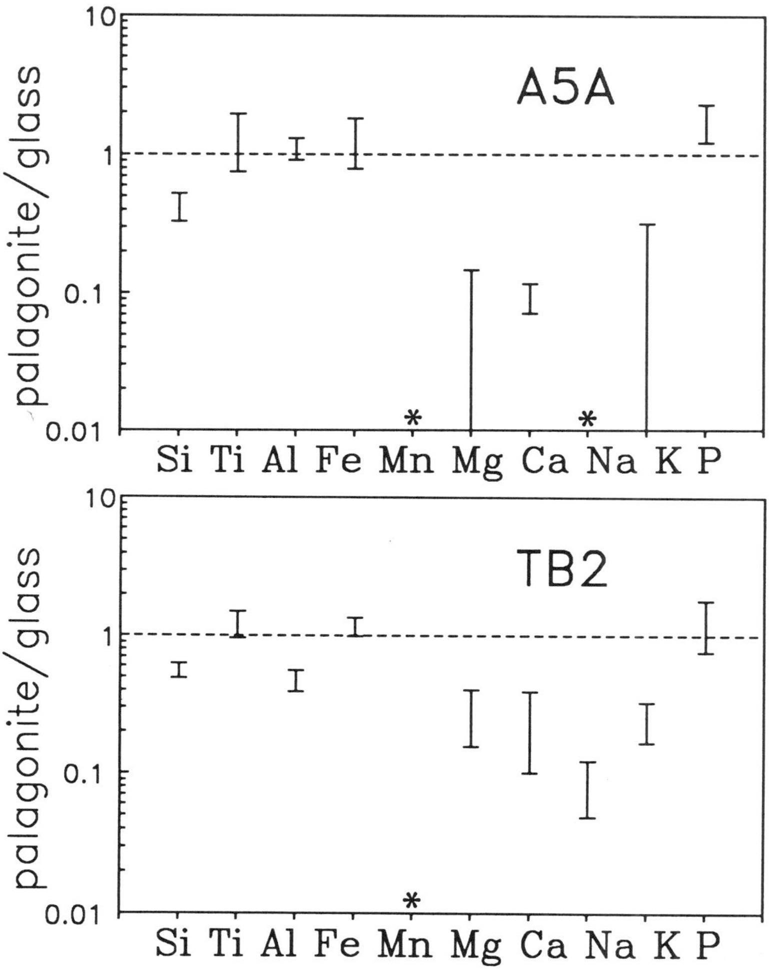

Figure 11-5. Major element palagonite composition normalized to parent glass in altered low silica glasses from British Columbia. Weight percents have been converted to mass/volume using the density values obtained for glass (d_g = 2.75 to 2.85 g/cm^3) and palagonite (d_g = 1.93 to 2.14 g/cm^3) from Koko Craters, Hawaii (18,24). The weight per volume of each oxide is then calculated from:

$d_g \cdot wt\%_{i,g}$ = wt./volume for elements in glass; and

$d_p \cdot wt\%_{i,p}$ = wt./volume for elements in palagonite,

(where $wt\%_{i,g}$ is the weight percent of oxide component i in glass and $wt\%_{i,p}$ is the weight percent of component i in palagonite. If palagonization is isovolumetric, then values plotting below 1 indicate loss to the solution and values greater than 1 indicate accumulation in the rind above the amount available in the glass progenitor (adapted from (39)).

formed) portions of the rinds. Compositional gradations across the rinds suggest changing solution composition with time. The strongest compositional gradients correlate with thicker rinds and secondary aluminosilicate cements, particularly zeolites, an indication of high reaction progress under closed system conditions (20,38,39).

The microstructure of palagonite rinds formed on naturally and experimentally altered low silica glasses have been investigated by transmission electron microscopy (TEM) and analytical electron microscopy (AEM). Experimental alteration of low silica glass in seawater at 50°C produces an amorphous outer zone, an intermediate zone of Mg, Al-hydroxycarbonate, and a clay-like inner zone (40). The outer zone evolves to Mg-smectite and the hydroxycarbonate composition changes with time, becoming higher in Al and Si, and correspondingly lower in Mg and Ca. Experiments at pressures of 1 bar and 350 bars at 60°C produce microcrystalline Fe-hydroxide after 4 days, which is overgrown by hydrotalcite ($Mg_6Al_2CO_3(OH)_{16} \cdot 4H_2O$) after 5 days (29). Crovisier *et al.* (29), conclude that the dissolution mechanism and rate are not pressure dependent in this range. Gibbsite ($Al(OH_3)$), in an amorphous Fe-Al-Ti matrix, forms as a major phase in alteration rinds on low silica glass altered experimentally in deionized water or NaCl solutions at 70°C for 301 days (41). Similar experiments in solutions containing $MgCl_2$ produce layers consisting of an amorphous Fe-Al-Ti-Mg silicate matrix and smectites (41).

TEM/AEM observations of palagonite rinds on naturally-altered low silica glasses generally suggest an amorphous or poorly crystalline nature, with the degree of crystallinity being variable within a particular rind (29,31,42,43). More crystalline areas exhibit "fibers" of material which compositionally and microstructurally suggest precursors to smectite clay (Figure 11-7). High-Fe, low-Si palagonites in deep sea dredge samples are more crystalline than their low-Fe, high-Si counterparts, with the more coarsely crystalline domains possibly a result of reconstruction of amorphous material (31). Microanalysis suggests that glass dissolution is congruent on a nanometer scale and no hydrated glass is observed, supporting the suggestion that dissolution/reprecipitation reactions dominate for at least 10^5 years (29,43). Palagonite rinds from seawater-altered samples, now subaerially exposed, from Palagonia, Sicily consist predominantly of poorly crystalline smectite clays (29), similar in microstructure to the fibrous high-Fe, low-Si deep sea dredge sample palagonites (31). Fresh-water-altered low silica glass from subglacial volcanics in north-central British Columbia exhibit less evidence for crystallinity, having domains which are entirely amorphous and others which have fibers with lattice fringes consistent with smectite clay and chlorite. Larger crystallite sizes in some palagonite rinds, such as the high-Fe, low-Si dredge sample palagonites (31), may be indicative of either lower supersaturation levels, allowing more crystal growth, or may indicate a rind aging phenomenon (recrystallization).

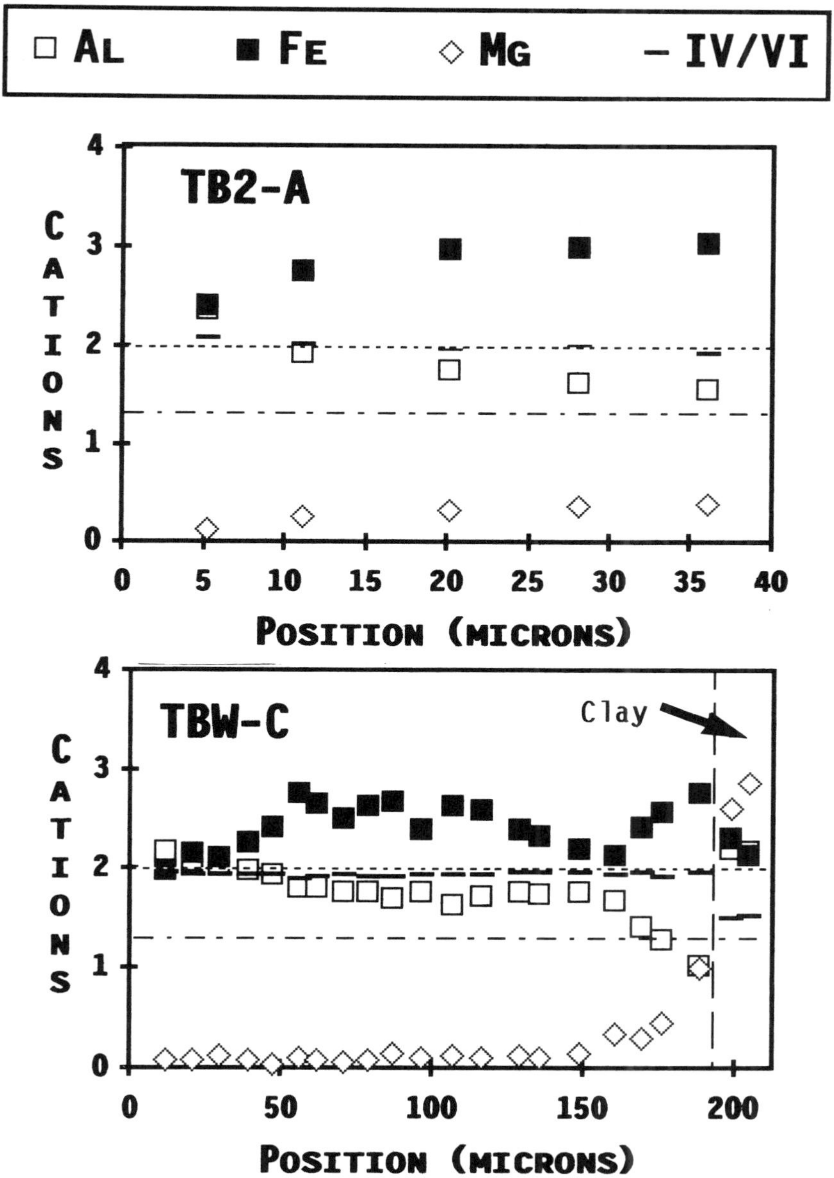

Figure 11-6. Elemental zoning profiles across palagonite rinds on subglacially produced glasses from British Columbia. The value IV/VI represents the ratio of the sum of tetrahedral cations to the sum of octahedral cations based on a 22 oxygen S (clay mineral) normalization. Analyses were done by electron microprobe analysis (adapted from (39)).

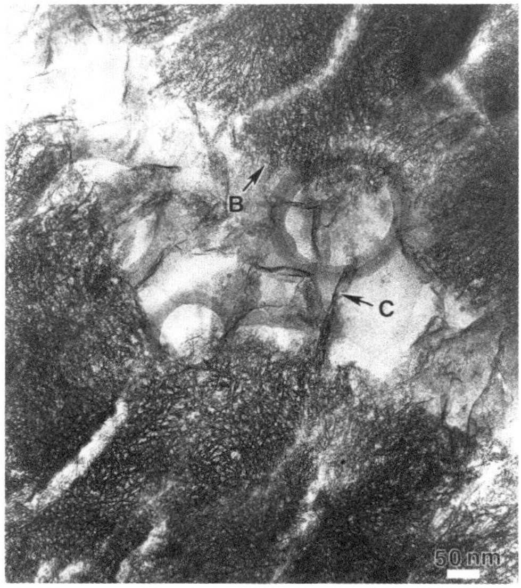

Figure 11-7. Analytical electron micrograph of a palagonite rind formed on a pillow basalt rim dredged from the Mid-Atlantic Ridge. Areas marked B and C refer to microstructure types which appear to represent different degress of formation of crystalline "fibers" (adapted from (20)).

Mechanism and Rate

The chemical composition and microstructure of palagonite indicates that the original glass has undergone network dissolution, and that palagonite forms by concomitant reprecipitation of insoluble material at the interface (21,24). The generally amorphous to poorly crystalline nature of palagonite implies that 1) it is thermodynamically metastable; and 2) that it forms under conditions of high supersaturation, precipitating mainly as critical nuclei with little crystal growth. As amorphous palagonite rinds are found on all natural low silica glasses exposed to aqueous solutions, the condition of high supersaturation exists at the glass surface independent of solution composition or glass surface area to solution volume considerations. If palagonite precipitation required attainment of saturation in the bulk solution, flow rates or initial solution volume constraints would prevent the formation of rinds in some systems and only glass dissolution would occur. Therefore, a high chemical gradient must be established between the dissolving glass and the bulk solution, and the palagonite must precipitate before mass transport begins to level that gradient. Rapid, kinetically-dominated precipitation accounts for the small grain size of palagonite rinds and the initiation of rind precipitation at low values of glass surface area/solution volume. Dibble and Tiller (44) suggest that the formation of alteration products follows an Ostwald step sequence in which a series of irreversible reactions occur to form a more stable assemblage. They state:

> *". . . metastable reactions occur because formation of less stable phases can lower the total free energy of the system faster than growth of the stable phase assemblage."*

In aluminosilicate systems, fast nucleating phases which incorporate large masses of Si and Al per unit growth time relative to more thermodynamically stable materials will form because they lower the total system free energy more rapidly. Thus, metastable materials, such as palagonite, will often form as intermediate phases.

Bonatti (45) proposed that palagonitization may occur rapidly with eruption at magmatic temperatures. Because of the relative lack of alteration of igneous minerals and absence of high temperature alteration phases, it is generally inferred that palagonitization occurs at low temperatures over thousands of years (18,24,46). Furnes (46) has determined from experimental palagonitization studies that:

> *". . . the process can be initiated as a post-eruptional (or post-cooling) phenomenon at any time when sideromelane encounters an aqueous environment, and that the initiation of the process does not require high temperatures."*

Activation energy gives an important clue to the rate controlling process. Surface analysis of experimentally-reacted low silica glass indicates that H penetrates into the glass to a depth of as much as 0.5 µm at 300°C, accompanied by interdiffusion of ions outward to the solution (47). Berger *et al.* (47) suggest that dissolution of low silica glass is diffusion controlled. However, the relatively high activation energy of the process (32 kJ/mol in "fresh water" experiments (48), and 65 kJ/mol in seawater experiments (49) when compared to diffusion-controlled reactions, 21 kJ/mol (50) indicates surface reaction control (48).

Moore (51), Moore *et al.* (52), and Hekinian and Hoffert (53) compared palagonite rinds formed in the deep sea (on dredge samples) with manganese coatings on the same samples. Well-constrained rates of Mn precipitation allowed the derivation of age/rind thickness relationships. Hekinian and Hoffert confirmed their Mn thickness ages using fission track dating techniques on samples from the Mid Atlantic Ridge. Estimated palagonitization rates are summarized in Table 11-3.

Experimental studies (46) indicate that palagonitization rates increase rapidly with temperature and are non-linear with time. The rate of palagonization of alkaline-low silica glass is .30 µm/month during the initial 10 months of exposure from 20 to 70°C. After 10 months at temperatures above 70°C, the rate

Table 11-3. Summary of Palagonitization Rates (from Jercinovic and Ewing(20)).

Rate	Conditions	Reference
Natural palagonitization		
3 μm/1000 yr	seawater	Morgenstein and Riley(120)
2.6 to 4.3 μm/1000 yr	seawater 5°C	Hekinian and Hoffert(53)
3 to 30 μm/1000 yr	seawater 5 to 50°C	Moore(51) Moore *et al.*(52)
15 μm/1000 yr	seawater 4°C	Jakobsson and Moore(55)
3 to 20 μm/1000 yr*	fresh water and seawater	Grambow *et al.*(38)
.1 μm/1000 yr**	fresh water and seawater	Grambow *et al.*(38)
Experimental palagonitization		
.3 μm/month	fresh water and seawater 20 to 70°C	Furnes(46)
3.3 μm/month	fresh water and seawater above 70°C after 10 months	Furnes(46)
100 μm/1000 yr	seawater	Crovisier *et al.*(49)
180 μm/1000 yr	seawater 3°C and 350 bars	Crovisier *et al.*(56)
12 to 17 μm/month	NaCl solution at 200°C	Malow *et al.*(93)

*Forward rate.
**Final rate.

increases to 3.3 µm/month. The eruptive emergence of Surtsey off the south coast of Iceland from November 1963 through June 1967 provides the opportunity to observe the palagonitization process as it occurs. Palagonitization of the Surtsey glass is documented at temperatures as low as 40°C and up to 150°C within a few years after the eruption. Jakobsson and Moore (54,55) have estimated palagonitization rates as a function of temperature based on detailed drill core studies of the Surtsey glass. The predicted rate from Surtsey is 15 µm/1000 yr for 4°C considering that the rate doubles with a rise in temperature of 12°C. Moore *et al.* (52) found the rate to increase 1.8X for a temperature rise of 10°C in seawater near Hawaii. Crovisier *et al.* (56) have experimentally noted an increase in the release of dissolved silica with pressure, implying dissolution rates for 60°C of 10 µm/yr at 1 bar and 14 µm/yr at 350 bars for glass in seawater. However, the actual dissolution rates are probably much closer as the higher implied rate at higher pressure is likely due to more effective penetration of the solution into cracks.

The palagonitization rate may not be constant with time, although rinds apparently do not act as a diffusive barrier to mass transfer between glass and aqueous solution (29). The rate may slow due to the attainment of solution silica saturation relative to the glass surface (38). Naturally-altered samples from British Columbia and the deep sea suggest that the forward (silica undersaturated) rate is 3 to 20 µm/1000 yr and the final (silica saturated) rate is 0.1 µm/1000 yr (38). If a deposit can maintain high glass surface area relative to solution volume, final rate conditions could be rapidly attained, and glass could be preserved for a considerable length of time. A centimeter thick pillow rim would be completely palagonitized in less than five million years at a rate of 2 µm/1000 yr, and 2mm sized glass shards in fragmental deposits would be completely palagonitized in one million years at this rate. However, fresh glass is found in deep sea deposits that are considerably older, such as DSDP Holes 335 (13 m.y. (57)), 408 (20 m.y. (58)), 407 (35 to 36 m.y. (58)), 382 (70 to 80 m.y. (59)), and 417 and 418 (105 to 110 m.y. (60)). Unaltered glass is also found in fresh-water-altered deposits of considerable age, such as the Miocene Frenchman Springs Flow of the Wanapum Basalts (12 to 14 million years old) in the Columbia River Basalt Group (61). Very old glasses exposed to mixed environments are also found, such as in the Upper Triassic (200 to 210 million years) Hound Island Volcanics (36) and the Carboniferous (340 million years) Lahn Dill deposits (62). In order to preserve glass in deposits of great age, a number of authors have suggested that the deposits may have been effectively sealed by secondary mineralization. Sealing may have preserved glasses in DSDP leg 46 deposits for the last 13 million years (63). Calcite precipitation apparently sealed the Hound Island volcanics (36). The 35 to 55 million year old Siletz River volcanics also have some fresh glass remaining, probably an effect of sealing (64). The utility of palagonite rind formation as an aging tool is clearly limited by the difficulty in determining rates for well documented samples. This is because actual exposure ages are unknown due to hydrologic changes (e.g. sealing) despite, in some cases, well known sample ages, and because of the possibility of rates changing with time.

Secondary Mineralization

The precipitation of secondary cements can affect the porosity of deposits, such as hyaloclastites, and may inhibit the chemical communication between the glass and the aqueous solution. The concentration of dissolved components must remain below solubility limits (by maintaining small glass surface area/solution volumes) if extensive palagonitization is to occur (46,65). The occurrence of secondary minerals with palagonitization may provide insight into the system conditions (temperature, solution chemistry, flow rate) during and/or after palagonitization.

Zeolites, clay minerals (particularly smectites), and calcite are the most common secondary minerals associated with altered low silica glasses (Table 11-4). Analcime, phillipsite, and chabazite are the most commonly reported zeolites (Figure 11-8). There is some evidence that zeolite type may be influenced by initial glass composition (66,67), and mineral compositions are environmentally dependent (68). Zeolite species are temperature dependent. Mordenite, heulandite, stilbite, epistilbite, chabazite, thomsonite, and mesolite-scapolite form in hydrothermal deposits below 100°C and are replaced by laumontite at 100 to

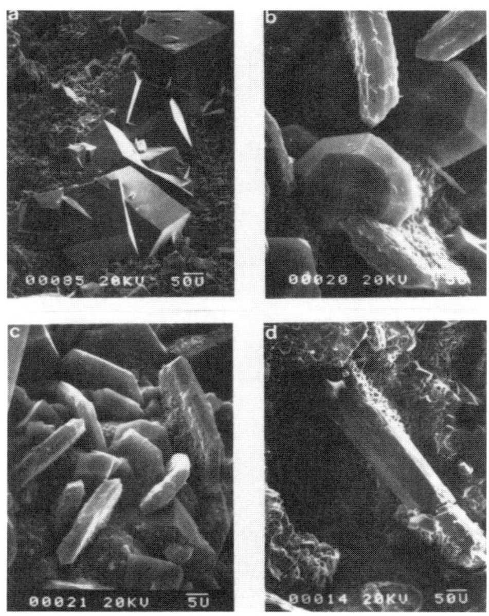

Figure 11-8. Scanning electron micrographs of zeolites in an altered low silica hyaloclastite from Iceland. a) chabazite; b) analcime; c) heulandite (?); d) zeolites overgrowing calcite (adapted from (20)).

Table 11-4. Secondary Minerals Commonly Associated with Altered Glasses.

Mineral	Formula
Clay Minerals	
Smectites	
Nontronite	$(.5Ca,Na)_{.66}Fe_4Al_{.66}Si_{7.34}O_{20}(OH)_4$
Montmorillonite	$(.5Ca,Na)_{.66}Mg_{.66}Al_{3.34}Si_8O_{20}(OH)_4$
Saponite	$(.5Ca,Na)_{.66}Mg_6Al_{.66}Si_{7.34}O_{20}(OH)_4$
Fe-saponite	$(.5Ca,Na)_{.66}(Fe,Mg)_6Al_{.66}Si_{7.34}O_{20}(OH)_4$
Kandites	
Kaolinite	$Al_4(Si_4O_{10})(OH)_8$
Illite Group	
Illite	$KAl_4SI_7AlO_{20}(OH)_4$
Celadonite	$(K,Ca,NA)(Al,Fe,Mg)_2(Al_{.11}Si_{3.89}O_{10})(OH)_2$
Chlorite Group	
Chlorite	$(Mg,Al,Fe)_{12}(Si,Al)_8O_{20}(OH)_{16}$
Prehnite	$Ca_2Al_2(OH)_2Si_3O_{10}$
Apophyllite	$KFCa_4(Si_4O_{10})_2 \cdot 8H_2O$
Mixed Layer	
Chlorite/montmorillonite	
Saponite/nontronite	
Smectite/chlorite	
Zeolites	
Analcime Group	
Analcime	$Na_{16}[(AlO_2)_{16}(SiO_2)_{32}] \cdot 16H_2O$
Wairakite	$Ca_8[(AlO_2)_{16}(SiO_2)_{32}] \cdot 16H_2O$

Table 11-4. Secondary minerals commonly associated with altered glasses (continued).

Mineral	Formula
Natrolite Group	
Natrolite	$Na_{16}[AlO_2)_{16}(SiO_2)_{24}]\cdot 16H_2O$
Thomsonite	$Na_4Ca_8[(AlO_2)_{20}(SiO_2)_{20}]\cdot 24H_2O$
Mesolite	$Na_{16}Ca_{16}[(AlO_2)_{48}(SiO_2)_{72}]\cdot 64H_2O$
Scolecite	$Ca_8[(AlO_2)_{16}(SiO_2)_{24}]\cdot 24H_2O$
Gonnardite	$Na_4Ca_2[(AlO_2)_8(SiO_2)_{12}]\cdot 14H_2O$
Chabazite Group	
Chabazite	$Ca_2[(AlO_2)_4(SiO_2)_8]\cdot 13H_2O$
Gmelinite	$Na_8[(AlO_2)_8(SiO_2)_{16}]\cdot 24H_2O$
Erionite	$(Ca,Mg,K_2,Na_2)_{4.5}[(AlO_2)_9(SiO_2)_{27}\cdot 27H_2O$
Phillipsite Group	
Phillipsite	$(K,Na)_{10}[(AlO_2)_{10}(SiO_2)_{22}]\cdot 20H_2O$
Laumontite	$Ca_4[(AlO_2)_8(SiO_2)_{16}]\cdot 16H_2O$
Heulandite Group	
Heulandite	$Ca_4[(AlO_2)_8(SiO_2)_{28}]\cdot 24H_2O$
Stilbite	$Ca_4[(AlO_2)_8(SiO_2)_{28}]\cdot 28H_2O$
Clinoptilolite	$Na_6[(AlO_2)_6(SiO_2)_{30}]\cdot 24H_2O$
Mordenite Group	
Mordenite	$Na_8[(AlO_2)_8(SiO_2)_{40}]\cdot 24H_2O$
Epistilbite	$Ca_3[(AlO_2)_6(SiO_2)_{18}]\cdot 18H_2O$
Carbonates	
Calcite	$CaCO_3$
Aragonite	$CaCO_3$
Dolomite	$CaMg(CO_3)_2$
Calc-silicates	
Gyrolite	$Ca_4(OH)_2Si_6O_{15}\cdot 3H_2O$

Table 11-4. Secondary minerals commonly associated with altered glasses (continued).

Mineral	Formula
Calc-silicates	
Reyerite	$(Na,K)Ca_7Si_{11}AlO_{29}(OH)_4 \cdot H_2O$
Tobermorite	$Ca_5H_2(Si_3O_9)_2 \cdot 4H_2O$
Sulfates	
Gypsum	$CaSO_4 \cdot 2H_2O$
Anhydrite	$CaSO_4$
Barite	$BaSO_4$
Silica	
Quartz	SiO_2
Opal	SiO_2
Chalcedony	SiO_2
Epidote Group	
Epidote	$Ca_2FeAl_2O \cdot OH \cdot Si_2O_7 \cdot SiO_4$
Clinozoisite	$Ca_2Al \cdot Al_2O \cdot OH \cdot Si_2O_7 \cdot SiO_4$
Pumpellyite	$Ca_4(Mg,Fe,Mn)(Al,Fe,Ti)_5(OH)_3Si_6O_{23} \cdot 2H_2O$
Feldspar	
K-feldspar	$KAlSi_3O_8$
Albite	$NaAlSi_3O_8$
Sulfides	
Pyrite	FeS_2
Pyrrhotite	Fe_7S_8-FeS
Oxides and Hydroxides	
Hematite	Fe_2O_3
Goethite	$FeO \cdot OH$
Mn-hydroxide	
Fe-hydroxide	
Fe-Mn hydroxide	

120°C (67). Tomasson and Kristmannsdottir (69) report the complete disappearance of zeolites above 230°C.

Nontronite, Fe-saponite, montmorillonite, and chlorite are the most common clay minerals associated with palagonitization. Fe-saponite is the most abundant clay in the submarine environment (70). Nontronite and celadonite form in the deep sea at low temperatures (0 to 20°C) under oxidizing conditions (71). Saponite forms from 30 to 200°C under "non-oxidative" conditions (72). Chlorite generally occurs with glasses altered in the deep sea or to glasses that have been hydrothermally altered. Chlorite appears in hydrothermally altered deposits in Iceland between 230 and 280°C, with interlayered chlorite/montmorillonite from 200 to 230°C (67,69,73). Calc-silicates and epidote occur in hydrothermally altered basic (low silica) rocks, at temperatures below 200°C for calc-silicates and above 200°C for epidote (67).

The sequence of cementation minerals precipitated (the paragenetic sequence) can provide a record of the chemical evolution of the system. The zeolite sequence, phillipsite (K-rich), chabazite (Ca-rich), analcime (Na-rich) is common in fresh-water-altered low silica hyaloclastites (18,24,39). In deep sea drill core samples the sequence saponite-gyrolite-phillipsite-chabazite-gmelinite-analcime-apophyllite-opal+calcite has been reported (74). Reaction path modelling (Figure 11-9) suggests that phillipsite should be the first zeolite formed, followed by chabazite, under conditions of relatively high glass surface area to solution volume (38). Chabazite is restricted to relatively high pH values, and will only form in seawater when Mg has been scavenged by smectite precipitation reactions. Closed or restricted systems may, therefore record increasing solution pH as glass hydrolysis proceeds. The appearance of zeolites is associated with Al depletion in palagonite relative to parent glass. Solubility limits are reached in solution when the Al, Ca, and alkali concentrations become high enough. Thus, zeolite precipitation depends critically on initial solution pH (to mobilize Al), initial glass surface area/solution volume, solution flow rate, and the amount of glass hydrolyzed. In palagonitized deposits from the Canary Islands, chabazite crystallized first, followed by phillipsite, then analcime and clay (75). One of the more detailed studies of authigenic mineral paragenesis in glassy low silica rocks is that of the Siletz River volcanics of Oregon (64). The following sequence is indicated: smectite-pyrite-calcite-thomsonite-natrolite-analcime-scolecite-mesolite-stilbite-heulandite-apophyllite-chabazite-mordenite-calcite-laumontite-quartz. These deposits may have altered under hydrothermal conditions.

Effects of Higher Temperatures or Long Exposure Times

Recrystallization of palagonite to a coarser grained, thermodynamically more stable assemblage (generally zeolites and smectite clays) of minerals may take place over extended time periods (16) or under the influence of increased temperature (47). At relatively high temperatures, the palagonite itself may

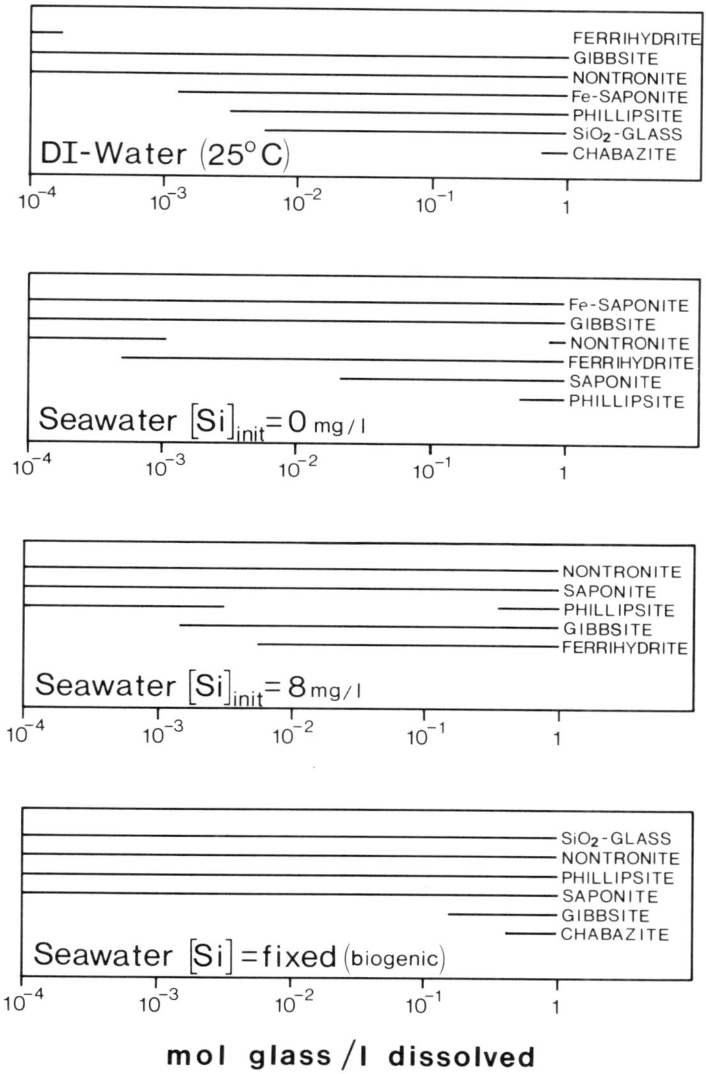

Figure 11-9. Sequence of phase formation calculated by reaction path modelling as a function of reaction progress assuming closed system conditions. Calculations were done for low silica glass in deionized water, in seawater with an intial Si concentration of 8mg/l, and in seawater with an initial Si concentration of 0 mg/l, and in seawater with Si fixed biogenically (adapted from (38)).

precipitate in a more coarsely crystalline form, resulting in the rind becoming optically anisotropic, consisting almost entirely of chlorite (76).

ALTERATION OF HIGH SILICA GLASS

Introduction

Unlike palagonitization, which is predominantly a hydrolytic reconstruction reaction, the alteration of high silica glasses is dominated by hydration of the glass, with concurrent ion exchange. In the hydration of high silica glass, water apparently diffuses into the silicate matrix of the glass. As this happens, volume and index of refraction increase (77). In general, the hydration rate for high silica glass follows a $t^{1/2}$ rate law (Figure 11-10), but age dates obtained using

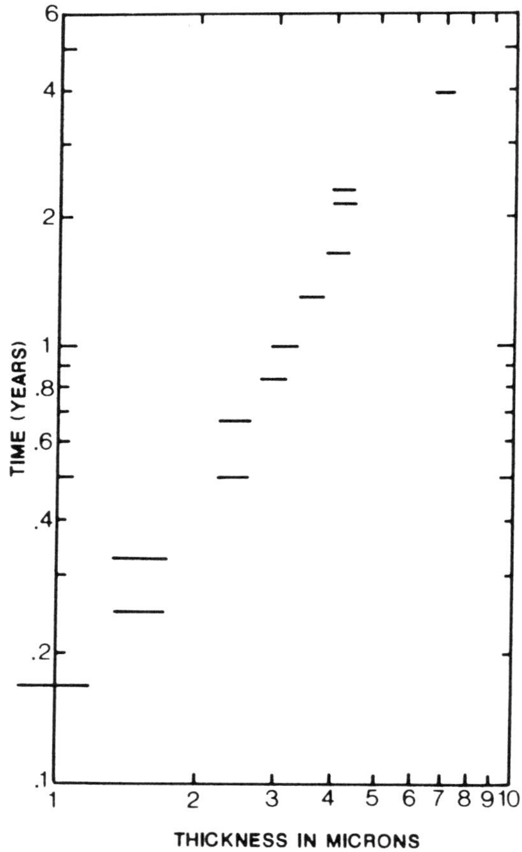

Figure 11-10. Time-hydration rind thickness relationships experimentally determined for high silica glass at 100°C (adapted from (77)).

hydration rinds on obsidian do not always correspond to ages obtained by other methods (78). If the mineralogic and geochemical nature of the "hydration layer" is sufficiently well understood, then the mechanism can be better evaluated. That is, is the process involved only one of hydration and ion exchange or is some form of hydrolysis and reprecipitation also included in forming the alteration layer?

Mechanism and Rate

High silica glass is hydrated by a process of ionic interdiffusion. H_3O^+ ions enter the glass network and exchange with alkali cations (79,80). Ion exchange is strongly influenced by the composition of the glass (81,82). Alkali leaching kinetics are also complicated by the formation of surface layers, and depend critically on pH (9,83). Low pH favors ion exchange, while alkaline solutions enhance glass dissolution (83). Hydration experiments (25°C, pH 1.0 to 6.2 for up to 3 months) on obsidian indicate that Na and K release rates are parabolic while release rates for Si and Al are linear (9). The weathering mechanism inferred to explain the resulting leach profiles and solution compositional changes involves both glass surface dissolution and diffusion of Na, K, and Al. The calculated interdiffusion coefficients (equation 11-1) are 5×10^{-19} cm^2/sec for Na and 1.2×10^{-20} cm^2/sec for H at 25°C and pH = 1.0 (9). The temperature dependence, derived from an Arrhenius type relationship (equation 11-6) is given by:

$$D_{Na} = 0.0503 \exp(-23{,}130/RT) \tag{11-7}$$

and the pH dependence given by:

$$\log D_{Na} = -18.5 + 0.493 \log a_{H+} \tag{11-8}$$

where a_{H+} is the H^+ activity (9).

The dissolution of high silica glass is observed to occur by a two stage process under hydrothermal conditions (130 - 150°C) in alkali carbonate solutions (84). The first stage, or "induction" period, is characterized by a relatively slow dissolution rate. The duration of this stage decreases with temperature and solution carbonate concentration. The second stage is characterized by a much higher dissolution rate, expressed by:

$$dC/dt = k(C_s - C) \tag{11-9}$$

where k is the first order rate constant, C_s is the SiO_2 concentration at saturation, and C is the SiO_2 concentration at some time t. Silica concentrations increase in solution until zeolite precipitation begins. Phillipsite is the first zeolite formed, and is followed by clinoptilolite and then mordenite. The rate of glass dissolution increases with temperature and solution carbonate concentration, which lowers the time necessary for the appearance of zeolites.

The nature of the hydrated surface layers clearly influences the interdiffusion kinetics as diffusion through bulk glass will not be the same as diffusion through either a hydrated glass or surface precipitate. Precipitation of secondary phases onto hydrated synthetic nuclear waste glass surfaces has been noted in some experiments (85,86). Transmission electron microscopy of alkali silicate glasses leached in aqueous solutions from 20 to 80°C and pH 1 to 11, indicates that a gel layer forms and then repolymerizes, resulting in a phase separated structure which is not an effective diffusion barrier (87).

The rate of hydration of high silica glass is dependent on the chemical composition of the glass and temperature (77). Glasses higher in Ca, Mg, and H_2O relative to Si hydrate more slowly (Figure 11-11). Jezek and Noble (82) suggest that hydration will proceed without recrystallization of the glass until a bulk H_2O content of 3 wt.% is reached. Further hydration allows the formation

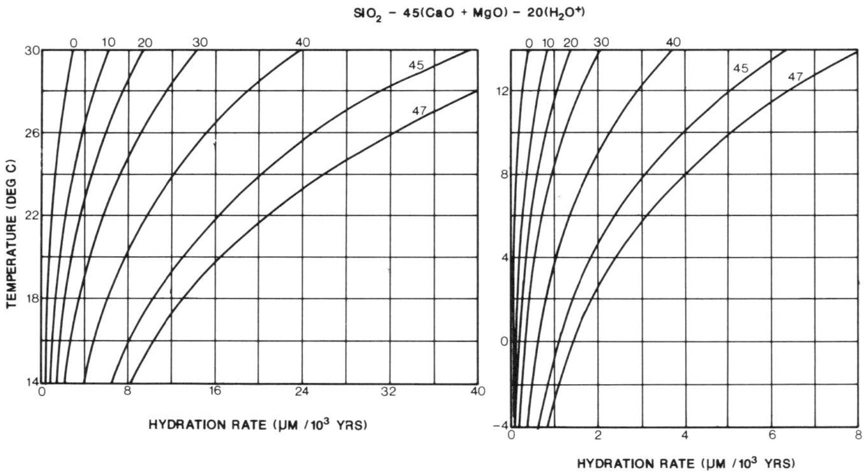

Figure 11-11. Hydration rate-temperature relationships as a function of glass composition (adapted from (81)). Curves represent different values of SiO_2 -45(CaO +MgO) - 20(H_2O^+).

of zeolites and clay minerals. The hydration rate for a given glass composition increases with temperature by a factor given by the Arrhenius relationship (81) (Figure 11-11).

Influence of Elevated Temperatures

Hydrothermal alteration of high silica glass is dominated by devitrification of the glass and precipitation of cementation-authigenic secondary phases. Experimental hydrothermal alteration (150°C to 250°C, pH 2.0 to 10.5, up to 84 days) shows that the paragenesis and mineralogy of secondary phases is a function of temperature, initial solution chemistry, glass composition, and closed or open system conditions (88). High silica glass reacts more slowly than lower silica glasses (88). Secondary minerals include zeolites (phillipsite, mordenite, and analcime, in that sequence), clay minerals (montmorillonite and kaolinite), and alkali feldspar (200°C after 20 days in NaOH solution).

Metamorphism of high silica glass leads to the formation of a variety of authigenic minerals, mostly clay minerals and zeolites. High silica glass initially alters to montmorillonite, opal, and alkali zeolites (89). As burial depth and temperature increase, the mineral assemblage becomes progressively more dehydrated, finally ending with an anhydrous, alkali aluminosilicate (albite), mica minerals (chlorite, illite, and prehnite) and pumpellyite, a complex Ca-Fe orthosilicate (Figure 11-12).

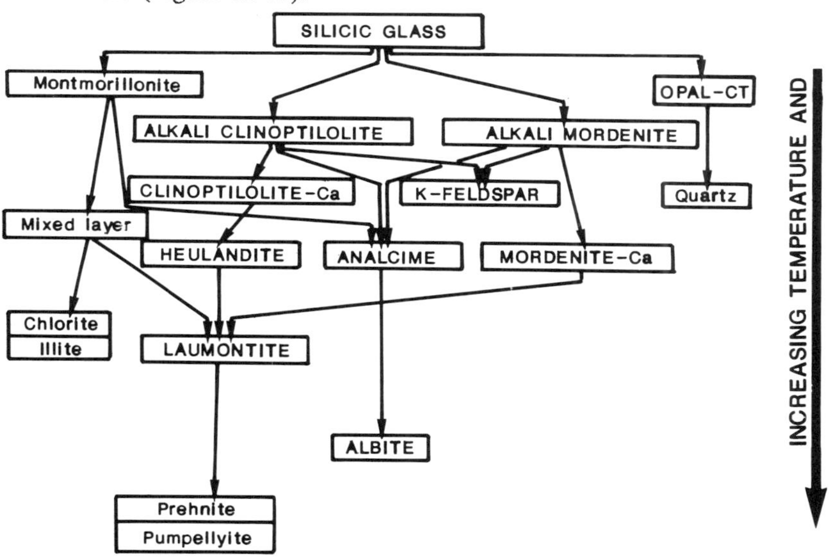

Figure 11-12. Development of authgenic minerals from high silica glass as a function of increasing temperature (adapted from (89)).

IMPLICATIONS

Nuclear Waste Disposal

Naturally-altered glasses provide a means of verifying predictive models (7,90-92) which have been developed to assess the long term corrosion of nuclear waste form glasses (5). Laboratory tests can only demonstrate short term alteration behavior. Extrapolating experimental results to time periods relevant to that which would be necessary for a nuclear waste repository (10^3 to 10^5 years) requires detailed knowledge of long term processes. If corrosion mechanisms and/or rates change over long time periods, then predictions based solely on short term laboratory tests will be misleading. Grambow (5) suggests that glass alteration proceeds at two rates, a forward, or initial rate occurring when the solution is below silica saturation relative to the glass surface, and a final, much slower rate, occurring when the solution has reached saturation. Grambow *et al.* (38), in a study of naturally-altered low silica glasses, have shown that there may, in fact, be two predominant reaction rates in nature, implying a forward rate of 3 to 20 µm/1000 year, and a final rate of 0.1 µm/1000 year. If theoretical modelling fails to "predict" the results of natural alteration of glass in geologic environments over time periods of at least 10^5 years, then the chosen model should be rejected if the natural glass and waste form glass are analogous. Theoretical calculations predict the following sequence of phase formation during the alteration of low silica glass in seawater: nontronite-saponite-phillipsite-chabazite; which compares well with alteration-paragenetic sequences in nature: palagonite-smectite-phillipsite-gyrolite-chabazite-analcime (38). Laboratory tests indicate that natural low silica glass and waste form borosilicate glass are similar in terms of alteration processes and rates (93,94), and also have been shown to have similar hydration energies (95). Natural analogues provide information regarding the types, amounts, and chemical compositions of alteration phases, and may be useful in estimating long term corrosion rates.

Glass corrosion processes are also significant in repository evaluation. Glass forms a significant portion of deposits which have been considered for the storage of high-level nuclear waste, such as the low silica volcanic rocks forming the Columbia River Plateau of the northwestern U.S., and the high silica tuffs forming Yucca Mountain at the Nevada test site.

Archaeology

Alteration rates of glasses have been used as an age dating method, useful in archaeology, geology, and geomorphology. The most widely applied method utilizes the hydration rate of high silica glass, generally to obsidian artifacts in archaeology (77,78,81). If diffusion of water into the glass controls hydration, then Fick's second law, applied to a semi-infinite medium applies, and the hydration rind thickness can be expressed as:

$$l = 2(Dt/\pi)^{1/2} \qquad (11\text{-}10)$$

where l represents the hydration rind thickness, D is the diffusion coefficient, and t is the age of the glass (more precisely, the time of exposure to water or water vapor). The increase in refractive index of the glass as a result of hydration provides a means of quickly measuring rind thickness optically. Sputter-induced optical emission spectroscopy can also be used to measure hydration rind thickness (96). As is pointed out by Friedman and Smith (78), glass ages obtained by hydration rind thickness do not always correspond to ages obtained for the same material by other methods, such as at the Rowe Excavation in New Mexico (Figure 11-13). This may be due to uncertainties in the temperature history of the materials in question, variation in exposure of the material to water, or reaction progress effects such as surface layer formation or silica saturation (97). Solution chemistry may, therefore, be significant in determining the overall hydration rates (97).

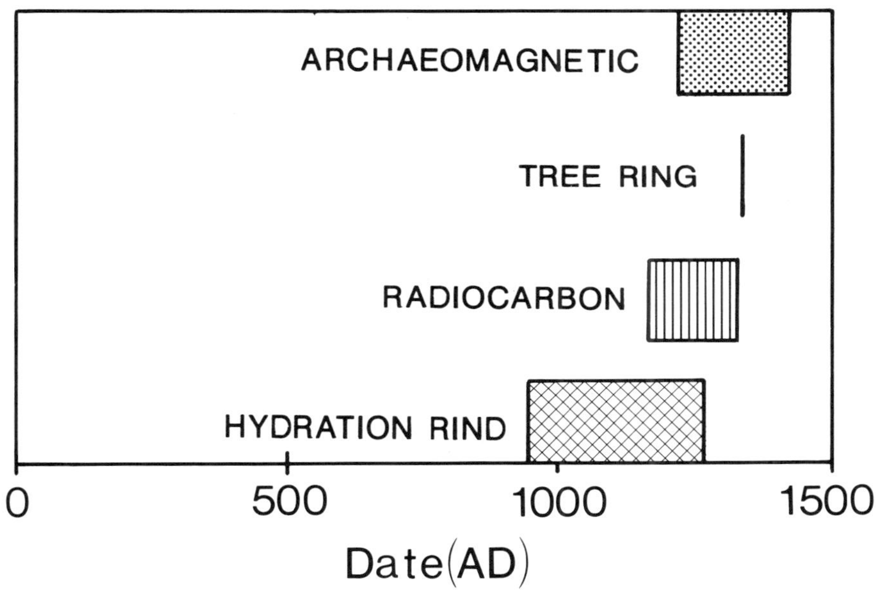

Figure 11-13. Comparison of site ages obtained by various methods for the Rowe Excavation, New Mexico (adapted from (97)).

Average temperature cannot be used to estimate the hydration rate as k does not vary linearly with temperature (equation 11-6). Thus the temperature history of a sample must be well established if the applicable rates are to be calculated and used to estimate the age based on the hydration rind thickness (81).

Planetary Geology

Mineralogical modelling of the x-ray florescence results of the Mars Viking landers, based on known terrestrial assemblages, suggests that the best match for the chemistry of the Martian surface fines is a mixture of smectite clay, sulfates and, perhaps, ferric oxides and carbonates, materials which may have been derived primarily from the weathering of basic (low silica) igneous rocks (98,99). The abundance of such products may be explained by the generation of large amounts of glass via magma/ground-ice interactions, analogous to terrestrial subglacial volcanism, and subsequent palagonitization (99-102). Glass is a more likely weathering progenitor than crystalline silicates due to the propensity of glass for rapid decomposition (103). Clay minerals on the Martian surface would exist metastably, and would most likely have originated from reactions involving liquid water (104). Hydrothermal solutions, heated by impact response or volcanism, may also have been involved (4,105,106). Some proportion of the regolith material may have been generated by hydrothermal alteration of glassy impact melt sheets (105); however, the low abundance of glass and associated alteration products in terrestrial impact melt sheets suggests that such materials may be relatively minor constituents of the Martian regolith (107). In all likelihood, both impact events and volcanism played a role in melt production and formation of the fine material of Martian "bright regions" as present Martian geomorphology is dominated by volcanic and impact-generated landforms. Spectral reflectance studies (108,109) indicate that crystalline silicates are unlikely to be dominant constituents of the Martian bright regions (which include the two Viking landing sites) and argue instead that amorphous gels, similar to certain terrestrial palagonites, are much better analogues. Allen *et al.* (4) have shown that such products are also compatible with the magnetic and grain size characteristics of the Martian fines.

There is strong evidence for the presence of volatiles, particularly water, in the Martian regolith (110). The regolith would thus act as a volatile reservoir, containing the water (as ice or liquid water) which is initially necessary to quench the melt to glass and subsequently is available for alteration reactions.

Palagonite compositions cover a range which can resemble the composition of the Martian regolith fines (4) (Figure 11-14). The following scenario is suggested:

1) Basic (low silica) magma erupts into a volatile-rich regolith or surface ice forming deposits laden with glass;

2) Heat from volcanism maintains liquid water in the vicinity of the new deposits;

3) Palagonitization of the glass takes place as long as liquid water is in contact with the glass;
4) The palagonitized deposits are abraded and deposited on a planetary scale by eolian processes.

Terrestrial palagonite compositions can vary considerably in even the apparently simplest of environments, such as the deep sea, open ocean. Low-Al palagonites are generated in both fresh water and seawater environments. Al

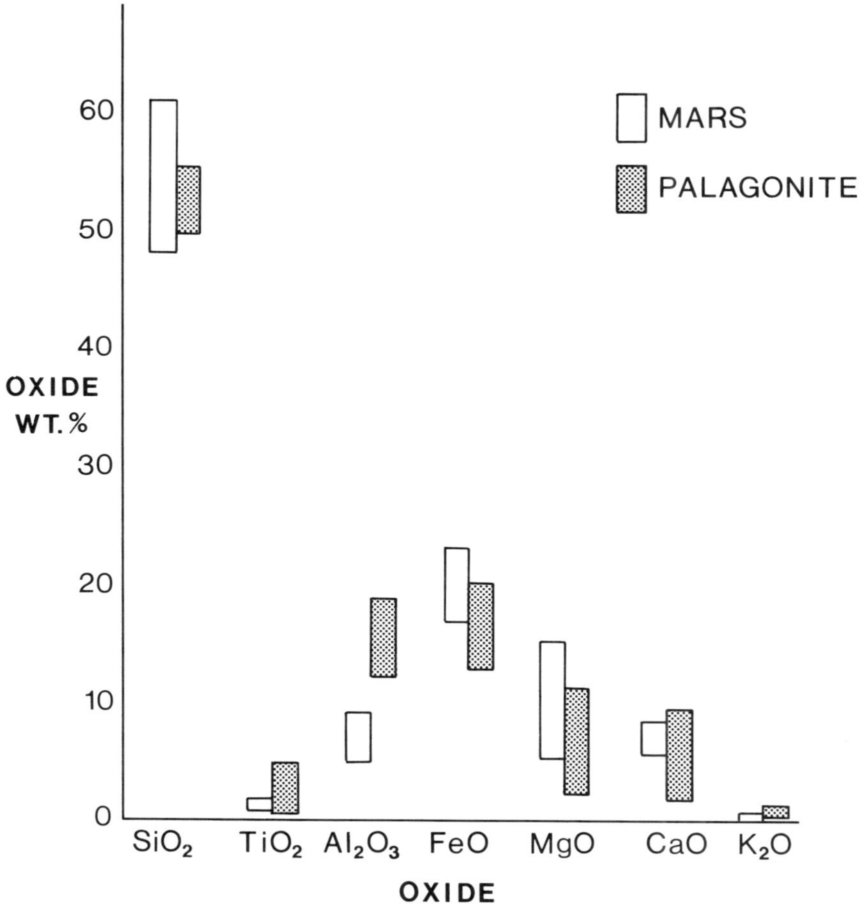

Figure 11-14. Comparison of Icelandic palagonite composition to Martian regolith fines (adapted from (4)).

depletion in palagonite rinds relative to parent glass has been attributed to acid alteration environments (27,111).

Some terrestrial palagonites have compositions which are nearly identical to the composition of the Martian regolith fines. However, these low-Al terrestrial palagonites are almost always associated with cementation authigenesis of aluminous zeolites and/or clays, preserving the system mass balance with respect to the major mineral-forming cations. Loss of aluminum from the system is rare in terrestrial samples. This implies that closed-system conditions may be necessary for initial Al mobilization. A different scenario must therefore be considered for Mars if palagonitization is chosen as the analogue regolith-forming process. For Mars, regolith generating processes must account for fractionation of Al from the system, or alteration products must be generated from low-Al starting materials.

The Geochemistry of Seawater, Groundwater, and Mineral Deposits

Geochemistry of seawater. The alteration of the basic igneous rocks (basalts) which comprise the upper portion of the sea floor plays a significant role in the determination of the composition of seawater. Hall and Robinson (112) indicate that over one-half of the volume of oceanic basement rocks consists of pillow lavas. Pillow lavas commonly have centimeter-thick rims of low silica glass, and there may also be a considerable amount of fragmentary interpillow glass (hyaloclastite). Pillow interiors also contain considerable amounts of glass as interstitial groundmass material. Staudigel and Hart(1) estimate that pillow basalts contain at least 20% glass, not including interstitial glass, and fragmental deposits contain at least 60% glass. Analysis of sea floor pillow basalts indicates that the glass shows a greater extent of alteration compared to the more crystalline pillow interiors (113). The low temperature alteration of crystalline basalt and glass do not show the same trends for all elements (Figure 11-15). Ca and Fe tend to be depleted during palagonitization of the glass, but are generally enriched in altered crystalline cores (113). Na and Al are generally enriched in palagonitized glass and depleted from altered crystalline rock. Both materials act as sinks for K and sources for Mg during alteration.

Staudigel and Hart (1) have estimated the effects of glass alteration on the chemical budget of the sea, concluding the following:

1) Elements released during alteration of low silica glass in seawater (all major elements except for Fe and Ti) participate in the local formation of clays, zeolites, and carbonates, but are also potential sources of elements for hydrothermal solutions (see section below on mineral deposits);
2) Other processes, in addition to glass alteration, must occur to explain mass balance discrepancies.

Influence on groundwater composition. Just as palagonitization is an important factor in seawater chemistry, the alteration of glass is important in determining the groundwater chemistry in areas where glass forms a significant portion of the near-surface geology. Gislason and Eugster (2) have shown that cold groundwaters in northeastern Iceland derive their SiO_2, Na, K, Ca, Mg, SO_4, and Cl contents from the alteration of glassy basalts, with silica concentrations being compatible with the solubility of low silica glass.

White *et al.* (114) have shown that the chemistry of groundwater at Rainier Mesa, Nevada is controlled by the incongruent dissolution of the fragmental deposits of high silica glass which compose much of the regional terrain. Mg, Ca, and Na are preferentially released into the groundwater, and K is retained during dissolution of the glass. Mg and Ca are selectively removed from the aquifer in areas where zeolite (clinoptilolite) and montmorillonite precipitate.

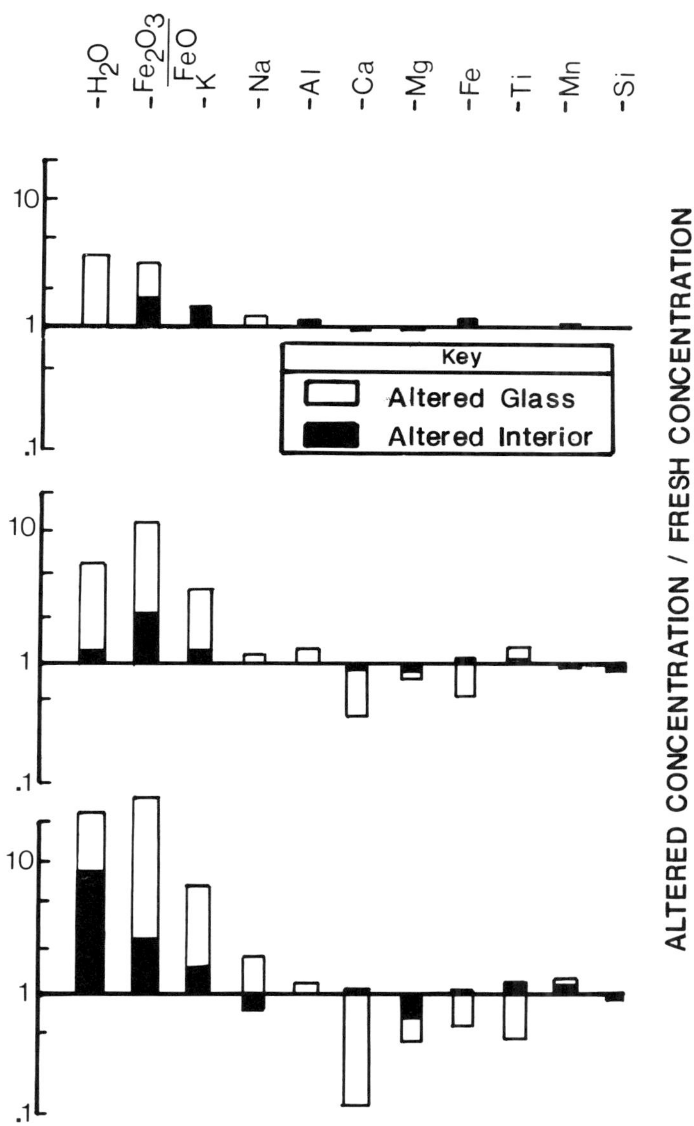

Figure 11-15. Comparison of alteration effects in altered margins and interiors of pillow basalt relative to unaltered basalt (adapted from (113)). Each plot represent analyses from separate pillows.

The fragmental nature of the deposits provides a high surface area relative to solution volume and also increases the solution retention time. Thus, glass corrosion can dominate groundwater chemistry even in areas where glassy deposits are volumetrically less significant than crystalline deposits.

Mineral deposits. Deposition of metallic minerals, generally as massive sulfides, from hydrothermal solutions in median valleys of the mid-ocean ridge system involves the mixing and circulation of seawater and magmatic solutions through the upper oceanic crust (33,115,116) (Figure 11-16). Such sulfide de-

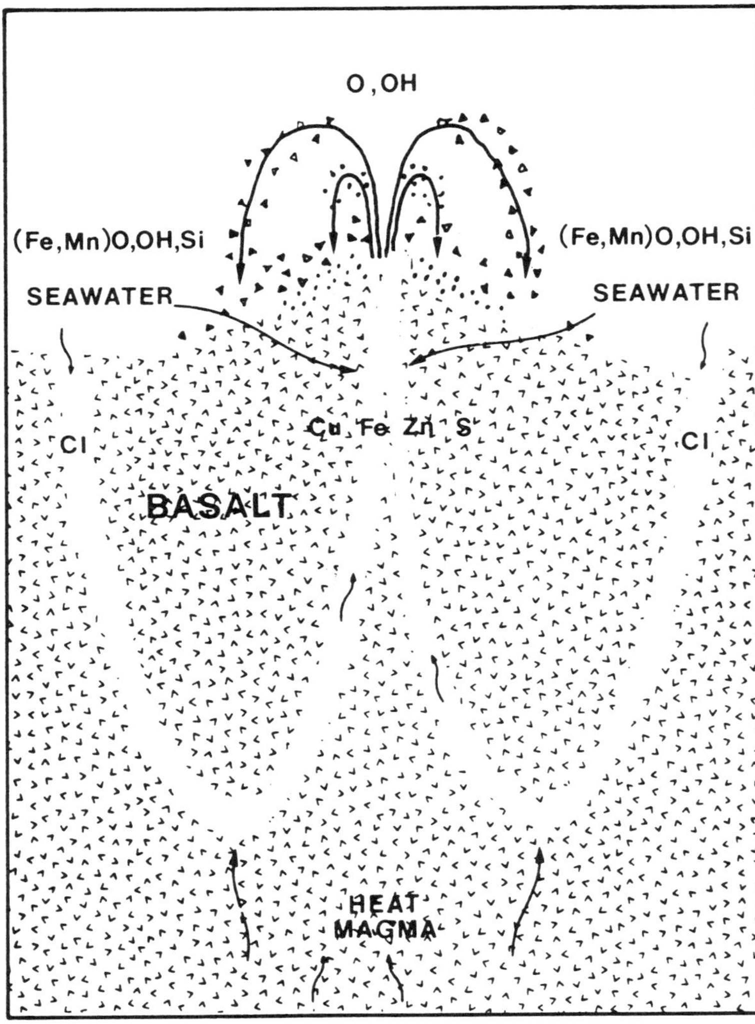

Figure 11-16. Schematic representation of hydrothermal circulation at mid-ocean ridges resulting in the production of massive sulfides (adapted from (15)).

posits are later exposed in areas where plate tectonics have resulted in the attachment of portions of the oceanic crust to continental margins. The chemistry of metal-rich solutions is influenced, to a large degree, by the interactions of the circulating seawater with low silica glass. The glass is readily accessible to the circulating seawater and, because of its abundance and metastable nature, alteration of the glass will account for much of the elemental redistribution affecting hydrothermal solutions.

The alteration of high silica glass may play a significant role in elemental mobilization and redistribution forming some uranium deposits (117). Economically important deposits of zeolites are formed in saline, highly alkaline brines which develop in some closed system lake environments and involve considerable reaction of high silica glass (118,119). As the rate of glass dissolution increases at high pH, such an environment is suitable for rapid corrosion of the glass and precipitation of zeolites (118).

SUMMARY

1) Glasses form a significant portion of the geologic record, on Earth as well as other planetary objects. Alteration of these glasses is an important mechanism of geochemical fractionation in the upper crust of the Earth and possibly also on other planetary surfaces. Glass alteration is a significant process in the chemical budget of the sea and in the formation of some economically important mineral deposits. The evaluation of both the long term corrosion of nuclear waste form glass and repository environments requires the understanding of the alteration of natural glasses in the geologic environment.

2) Glasses are metastable and corrode when exposed to aqueous solutions in the geologic environment. Corrosion takes place by an initial hydration and ion exchange step followed by breakdown of the glass network (dissolution) with release of silica as silicic acid and metal hydroxides into the local aqueous solution. The relative significance of hydration and dissolution is dependent upon the silica content of the glass. Dissolution slows down but does not stop as the silica concentration in solution reaches saturation relative to the surface of the glass.

3) Low silica glasses alter predominantly by hydrolytic reconstruction of the glass, with hydration and ion exchange occurring on a scale of tens of nanometers. The glass is replaced on surfaces exposed to aqueous solutions by palagonite, a hydrous, clay-like aluminosilicate gel, the chemical composition of which is dependent on microenvironment and glass composition. The formation of palagonite does not protect the glass from further dissolution, and is itself metastable. High degrees of

reaction progress (glass dissolution) are accompanied by the formation of cementation-authigenic precipitates, generally smectite clays, zeolites, calcite or opaline silica. High degrees of reaction progress tend to be in closed, or restricted systems. Elevated temperatures lead to the formation of low-water zeolites and Mg-micas, particularly chlorite. The uncertainties in the details of the alteration environment throughout the geologic history of deposits makes the use of palagonitization as an age dating technique almost impossible.

4) High silica glasses alter predominantly by hydration and devitrification. Hydration and ion exchange may take place on a scale of millimeters. Obsidians may be hydrated with little evidence for loss of cations to solution. High silica glasses devitrify to mixtures of silica polymorphs, feldspars, micas, and zeolites at slightly elevated temperatures in the presence of volatiles.

REFERENCES

1. Staudigel, H., and Hart, S.R., *Geochim. Cosmochim. Acta* 47: 337-350 (1983).
2. Gislason, S.R., and Eugster, H.P., *Geochim. Cosmochim. Acta* 51: 2841-2856 (1987).
3. Deutsch, W.A., Jenne, E.A., and Krupka, K.M., *Chem. Geol.* 36: 15-34 (1982).
4. Allen, C.C., Gooding, J.L., Jercinovic, M.J., and Keil, K., *Icarus* 45: 347-369 (1981).
5. Ewing, R.C., and Jercinovic, M.J., in: *Scientific Basis for Nuclear Waste Management X*, pp 67-83, Materials Research Society (1987).
6. Plodinec, M.J., Jantzen, C.M., and Wicks, G.G., in *Scientific Basis for Nuclear Waste Management VII* (G.L. McVay ed.), pp 755-762, Materials Research Society (1984).
7. Grambow, B., in: *Scientific Basis for Nuclear Waste Management VIII* (C.M. Jantzen, J.A. Stone, and R.C. Ewing ed.), pp 15-27, Materials Research Society (1985).
8. Furnes, H., Friedleifsson, I.B., and Atkins, F.B., *J. Volcanol. Geotherm. Res.* 8: 95-110 (1980).
9. White, A.F., *Geochim. Cosmochim. Acta* 47: 805-816 (1983).
10. Grambow, B., *Nuclear Waste Glass Dissolution: Mechanism, Model and Application*, SKB Technical Report 87-02 (JSS Project), 68p., SKB (1987).
11. Doremus, R.H., in: *Treatise on Materials Science and Technology 17* (M. Tomozawa and R.H. Doremus ed.), pp 41-69, Academic Press (1979).
12. Von Waltershausen, W.S., *Gott. Sud.* 1: 371-431 (1845).

13. Pjetursson, H., *Scottish Geographic Mag.* 16: 265-293 (1900).
14. Peacock, M.A., *Geol. Mag.* 63: 385-399 (1926).
15. Peacock, M.A., and Fuller, R.E., *Am. Min.* 13: 260-382 (1928).
16. Honnorez, J., in: *The Sea, Vol. 7: The Oceanic Lithosphere* (C. Emiliani ed.), pp 525-587, John Wiley and Sons (1981).
17. Nayudu, Y.R., *Int. Symp. of Volcanology, Japan* (abstract) (1962).
18. Hay, R.L., and Iijima, A., *Geol. Soc. Amer. Memoir* 116: 338-376 (1968).
19. Jakobsson, S.P., and Moore, J.G., *Geol. Soc. Amer. Bull.* 97: 648-659 (1986).
20. Jercinovic, M.J., and Ewing, R.C., *Basaltic glasses from Iceland and the deep sea: Natural analogues to borosilicate nuclear waste-form glass*, SKB Technical Report 88-01 (JSS Project), 221p., SKB (1988).
21. Hay, R.L., and Jones, B.F., *Geol. Soc. Amer. Bull.* 83: 317-332 (1972).
22. Geptner, A.R., *Litoligiya i Poliznye Iskopaemye* 5: 113-130 (1978).
23. Honnorez, J., *La palagonitization: l'alteration sous-marine du verre volcanique basique de Palagonia (Sicile)*, Ph.D. Thesis, University of Brussels (1967).
24. Hay, R.L., and Iijima, A., Contrib. Min. Pet. 17: 141-154 (1968).
25. Stokes, K.R., *Min. Mag.* 38: 205-214 (1971).
26. Noak, Y., and Crovisier, J.L., *Bull. Mineral.* 103: 523-527 (1980).
27. Furnes, H., *Chem. Geol.* 22: 249-264 (1978).
28. Furnes, H., and El Anbaawy, I.H., *N. Jb. Min. Abh.* 139: 279-302 (1980).
29. Crovisier, J.L., Honnorez, J., and Eberhart, J.P., *Geochim. Cosmochim. Acta* 51: 2977-2990 (1987).
30. Zhou, Z., and Fyfe, W.S., *Am. Min.* 74: 1045-1053 (1989).
31. Jercinovic, M.J., Murakami, T., and Ewing, R.C., in *Proceedings of the Sixth International Symposium on Water/Rock Interaction*, pp 337-340, Balkema (1989).
32. Bischoff, J.L., and Dickson, F.W., *Earth Planet. Sci. Lett.* 25: 385-397 (1975).
33. Edmond, J.M., Measures, C., McDuff, R.E., Chan, L.H., Collier, R., Grant, B., Gordon, L.I., and Corliss, J.B., *Earth Planet. Sci. Lett.* 46: 1-18 (1979).
34. Edmond, J.M., Measures, C., Mangum, B., Grant, B., Sclater, F.R., Collier, R., Hudson, A., Gordon, L.I., and Corliss, J.B., *Earth Planet. Sci. Lett.* 46: 19-30 (1979).
35. Von Damm, K.L., Edmond, J.M., Grant, B., Measures, C.I., Walden, B., and Weiss, R.F., *Geochim. Cosmochim. Acta* 49: 2197-2220 (1985).
36. Muffler, L.J.P., Short, J.M., Keith, T.E.C., and Smith, V.C., *Amer. J. Sci.* 267: 196-209 (1969).
37. Furnes, H., *Chem. Geol.* 43: 271-285 (1984).
38. Grambow, B., Jercinovic, M.J., Ewing, R.C., and Byers, C.D., in *Scientific Basis for Nuclear Waste Management IX* (L. Werme ed.), pp 263-272,

Materials Research Society (1985).
39. Jercinovic, M.J., Keil, K., Smith, M.R., and Schmitt, R.A., *Alteration of basaltic glasses from north-central British Columbia*, to be published in *Geochim. Cosmochim. Acta.*
40. Crovisier, J.L., Thomassin, J.H., Juteau, T., Ebrerhart, J.P., Touray, J.C., and Baillif, P., *Geochim. Cosmochim. Acta* 47: 377-387 (1983).
41. Murakami, T., Ewing, R.C., and Bunker, B.C., in *Scientific Basis for Nuclear Waste Management XI* (M.J. Apted and R.E. Westerman ed.), pp 737-748, Materials Research Society (1988).
42. Eggleton, R.A., and Keller, J., *N. Jb. Miner. Mh.* 1982: 289-311 (1982).
43. Murakami, T., Banba, T., Jercinovic, M.J., and Ewing, R.C., in *Scientific Basis for Nuclear Waste Management XII* (W. Lutze and R.C. Ewing ed.), pp 65-72, Materials Research Society (1989).
44. Dibble, W.E., and Tiller, W.A., *Clays Clay Mins.* 29: 323-330 (1981).
45. Bonatti, E., *Bull. Volcanol.* 28: 257-269 (1965).
46. Furnes, H., *Contrib. Min. Pet.* 50: 105-113 (1975).
47. Berger, G., Schott, J., and Loubet, M., *Earth Planet. Sci. Lett.* 84: 431-445 (1987).
48. Gislason, S.R., and Eugster, H.P., *Geochim. Cosmochim. Acta* 51: 2827-2840 (1987).
49. Crovisier, J.L., Fritz, B., Grambow, B., and Eberhart, J.P., in *Scientific Basis for Nuclear Waste Management IX* (L. Werme ed), pp 273-280, Materials Research Society (1985).
50. Lasaga, A.C., *Kinetics of Geochemical Processes, Reviews in Mineralogy 8*, Mineralogical Soc. America (1981).
51. Moore, J.G., *U.S. Geol. Surv. Prof. Paper 550-D*, pp 163-171 (1966).
52. Moore, J.G., Fornari, D.J., and Clague, D.A., *U.S. Geol. Surv. Bull. 1663*, 11 (1985).
53. Hekinian, R., and Hoffert, M., *Marine Geol.* 19: 91-109 (1975).
54. Jakobsson, S.P., and Moore, J.G., *Surtsey Research Progress Report 9*: 76-93 (1982).
55. Jakobsson, S.P., and Moore, J.G., *Geol. Soc. Amer. Bull.* 97: 648-659 (1986).
56. Crovisier, J.L., Eberhart, J.P., and Honnorez, J., in *Extended Abstracts of the Fifth International Symposium on Water-Rock Interaction*, pp 142-145, International Association of Geochemistry and Cosmochemistry (1986).
57. Aumento, F., Melson, W.G., et al., *Initial Reports of the Deep Sea Drilling Project 37*, U.S. Government Printing Office, 1008p (1977).
58. Luyendyk, B.P., Cann, J.R., et al., *Initial Reports of the Deep Sea Drilling Project 49*, U.S. Government Printing Office, 1020p (1978).
59. Tucholke, B.E., Vogt, P.R., et al., *Initial Reports of the Deep Sea Drilling Project 43*, U.S. Government Printing Office, 1115p (1979).
60. Donnelley, T., Francheteau, J., et al., *Initial Reports of the Deep Sea*

Drilling Project 51, 52, 53, U.S. Government Printing Office, 1613p (1979).

61. Jercinovic, M.J., Ewing, R.C., and Byers, C.D., in *Advances in Ceramics 20: Nuclear Waste Management II*, pp 671-697, American Ceramic Society (1987).
62. Schmincke, H.-U., and Pritchard, G., *Naturwissenschaften* 67: 615 (1981).
63. Schmincke, H.-U., Robinson, P.T., Ohnmacht, W. and Flower, M.F.J., in: *Initial Reports of the Deep Sea Drilling Project 46* (L. Dmitriev, J. Heirtzler et al. ed.), pp 341-355, U.S. Government Printing Office (1978).
64. Kieth, T.E.C., and Staples, L.W., *Clays Clay Mins.* 33: 135-144 (1985).
65. Furnes, H., *Bull. Volcanol.* 38: 173-186 (1974).
66. Iijima, A., and Harada, K., *Am. Min.* 54: 182-197 (1968).
67. Kristmannsdottir, H., in: *International Clay Conference, Developments in Sedimentology 27* (M.M. Mortland and V.C. Farmer ed.), pp 359-367, Elsevier (1978).
68. Fischer, R.V., and Schmincke, H.-U., *Pyroclastic Rocks*, 472p, Springer-Verlag (1984).
69. Tomasson, J., and Kristmannsdottir, H., *Contrib. Min. Pet.* 36: 123-134 (1972).
70. Cole, T.G., and Shaw, H.F., *Clay Minerals* 18: 239-252 (1983).
71. Kurnusov, V.B., Kholodokevich, I.V., Kokorina, L.P., Kotov, N.V., and Chudaev, O.V., in: *Proc. Int. Clay Conf.*, pp 547-556 (1982).
72. Stakes, D.S., and O'Neil, J.R., *Earth Planet. Sci. Lett.* 57: 285-304 (1982).
73. Kristmannsdottir, H., *The Transactions of the Geol. Soc. Sweden* 97: 289-292 (1975).
74. Honnorez, J., in: *Natural Zeolites: Occurrence, Properties, Use* (L.B. Sand and F.A. Mumpton ed.), pp 245-258, Pergamon (1978).
75. Brey, G., and Schmincke, H.-U., *Bull. Volcanol.* 43: 15-33 (1980).
76. Viereck, L.G., Griffin, B.J., Schmincke, H.-U., and Pritchard, R.G., *J. Geophys. Res.* 87: 6459-6476 (1982).
77. Friedman, I., and Trembour, F.W., *Amer. Scientist* 66: 44-51 (1978).
78. Friedman, I., and Smith, R.L., *Amer. Antiquity* 25: 476-522 (1960).
79. Doremus, R.H., *J. Non-Cryst. Solids* 19: 137-144 (1975).
80. Laursen, T., and Lanford, A., *Nature* 276: 153-156 (1978).
81. Friedman, I., and Long, W., *Science* 191: 347-352 (1976).
82. Jezek, P.A., and Noble,D.C., *Am. Min.* 63:266-273 (1978).
83. Bunker, B.C., Arnold, G.W., and Beauchamp, E.K., *J. Non-Cryst. Solids* 58: 295-322 (1983).
84. Hawkins, D.B., *Clays Clay Mins.* 29: 331-340 (1981).
85. Strachan, D.M., Pederson, L.R., and Lokken, R.O., *Results from the Long-Term Interaction and Modeling of SRL-131 Glass with Aqueous Solutions*, PNL-5654, Pacific Northwest Laboratory (1985).

86. Diebold, F.E., and Bates, J.K., in: *Advances in Ceramics 20: Nuclear Waste Management II* (D.E. Clark, W.B. White, and A.J. Machiels ed.), American Ceramic Society (1986).
87. Bunker, B.C., Headly, T.J., and Douglas, D.C., in *Mat. Res. Soc. Symp. 32*, Elsevier (1984).
88. Holler, H., and Wirsching, U., in: *Natural Zeolites: Occurrence, Properties, Use* (L.B. Sand and F.A. Mumpton ed.), pp 329-336 Pergamon (1978).
89. Iijima, A., in: *Natural Zeolites: Occurrence, Properties, Use* (L.B. Sand and F.A. Mumpton ed.), pp 175-198, Pergamon (1978).
90. Pescatore, C., and Machiels, A.J., in: *Advances in Ceramics VIII, Nuclear Waste Management* (G.G. Wicks and W.A. Ross ed.), pp 509-518, American Ceramics Society (1983).
91. Harvey, K.B., and Litke, C.D., *J. Amer. Cer. Soc.* 67: 533 (1984).
92. Barkatt, A., Macedo, P.B., Gibson, B.C., and Montrose, C.J., in *Scientific Basis for Nuclear Waste Management VIII* (C.M. Jantzen, J.A. Stone, and R.C. Ewing ed.), pp 3-14, Materials Research Society (1985).
93. Malow, G., Lutze, W., and Ewing, R.C., *J. Non-Cryst. Solids* 67: 305-322 (1984).
94. Lutze, W., Malow, G., Ewing, R.C., Jercinovic, M.J., and Keil, K., *Nature* 314: 252-255 (1985).
95. Jantzen, C.M., and Plodinec, M.J., *J. Non-Cryst. Solids* 67: 207-223 (1984).
96. Tsong, T.S.T., Houser, C.A., Yusef, N.A., Messier, R.F., White, W.B., and Michels, J.W., *Science* 201: 334-339 (1978).
97. McGrail, B.P., Pederson, L.R., Strachan, D.M., Ewing, R.C., and Cordell, L.S., in *Proc. Symp. on Materials Issues in Art and Archaeology* (P.B Vandiver, J. Druzik, and C. Stevenson ed.), pp 263-269, Materials Research Society (1988).
98. Baird, A.K., Toulmin, P., III, Clark, B.C., Rose, H.J., Jr., Keil, K., Christian, R.P., and Gooding, J.L., *Science* 194: 1288-1293 (1976).
99. Toulmin, P., III, Baird, A.K., Clark, B.C., Keil, K., Rose, H.J., Jr., Christian, R.P., Evans, P.H., and Kelliher, W.C., *J. Geophys. Res.* 82: 4625-4634 (1977).
100. Soderblom, L.A., and Wenner, D.B., *Icarus* 34: 622-637 (1978).
101. Allen, C.C., *J. Geophys. Res.* 84: 8048-8059 (1979).
102. Allen, C.C., *Advances in Planetary Geology* (NASA TM-81979), pp 161-264 (1980).
103. Gooding, J.L., and Keil, K., *Geophys. Res. Lett.* 5: 727-730 (1978).
104. Gooding, J.L., *Icarus* 33: 483-513 (1978).
105. Newsom, H.E., *Icarus* 44: 207-216 (1980).
106. Berkley, J.L., and Drake, M.J., *Icarus* 45: 231-249 (1981).
107. Allen, C.C., Gooding, J.L., and Keil, K., *J. Geophys. Res.* 87: 10083-10101 (1982).
108. Evans, D.L., and Adams, J.B., *Proc. Lunar Planet. Sci. Conf. 11*, pp 757-76 (1980).

109. Singer, R.B., *J. Geophys. Res.* 87: 10159-10168 (1982).
110. Arvidson, R.E., Goettel, K.A., and Hohenberg, C.M., *Rev. Geophys. Space Phys.* 18: 565-603 (1980).
111. Ailin-Pyzik, I.B., and Sommer, S.E., *J. Geophys. Res.* 86: 9503-9510 (1981).
112. Hall, J.M., and Robinson, P.T., *Science* 204: 573-586 (1979).
113. Thompson, G., *Trans. Amer. Geophys. Union* 54: 1015-1019 (1973).
114. White, A.F., Claassen, H.C., and Benson, L.V., *U.S. Geol. Surv. Water-Supply paper 1535-Q*, 34p, U.S. Government Printing Office (1980).
115. Rona, P.A., *Earth-Sci. Rev.* 20: 1-104 (1984).
116. Janecky, D.R., and Seyfried, W.E., Jr., *Geochim. Cosmochim. Acta* 48: 2723-2738 (1984).
117. Zielinski, R.A., *U.S. Geol. Surv. Open File Report 77-744*, U.S. Government Printing Office (1979).
118. Hay, R.L., *Geol. Soc. Amer. Special Paper 85*, 130p (1966).
119. Surdam, R.C., and Sheppard, R.A., in: *Natural Zeolites: Occurrence, Properties, Use* (L.B. Sand and F.A. Mumpton ed.), pp 145-174 Pergamon (1978).
120. Morganstein, M., and Riley, T.J., *Asian Perspectives* 17: 145-159 (1975).
121. Barkatt, Aa., Boulos, M.S., Barkatt, A., Sousanpour, W., Boroomand, M.A., Macedo, P.B., and O'Keefe, J.A., *Geochim. Cosmochim. Acta* 48: 361-372 (1984).
122. Friedman, I., Smith, R.L., and Long, W.D., *Geol. Soc. America Bull.* 77: 323-327 (1966).

12

Corrosion of Glazes and Enamels

Richard A. Eppler

Eppler Associates
Cheshire, Connecticut

INTRODUCTION

A major reason for using vitreous materials for coatings is their potential for chemical resistance in service. Among the many products which have a vitreous coating for reasons of chemical resistance in service are chemical reactors, dental crowns, dinnerware and hotel china, food and beverage containers, stove tops, wall and floor tile, washing machine baskets, and water heater linings.

These products are expected to withstand prolonged exposure to various corrosive liquids, especially water. They must not lose their gloss or change color, pit, or permit penetration of the liquid. The superiority of vitreous coatings for such service is indicated by examples of glazed pottery made several thousand years ago, which still exhibit a glossy surface. Water heater linings are expected to last 10 to 20 years, while constantly exposed to 140°F water. Hotel china is expected to survive several daily washings in hot alkaline water without loss of gloss, or other damage to the appearance of the surface.

Nevertheless, all vitreous surfaces are affected to some extent by contact with water, acid or alkaline liquids. At the other extreme from the durable examples just cited are the sodium silicates, or water glass, which dissolve readily in water. In addition, even otherwise durable glasses are readily attacked by hydrofluoric acid.

Chemical resistance is an important parameter in the usefulness of most glass coatings. Chemical attack can alter the appearance of glass surfaces in several ways. A common effect is loss of gloss. Occasionally, pitting of the surface occurs. Commonly, development of layers of reaction product is observable to the naked eye. Liquids stored in glazed vessels can become contaminated with elements extracted from the ceramic coating. Hence, it is important to inquire into the nature of the interaction between liquids and silicate glass surfaces and to develop coatings which minimize corrosion effects. That is the objective of this chapter.

CORROSION PROCESSES

When a glass coating is brought into contact with an aqueous solution, alkali ions are usually extracted into the solution in preference to silica and the other constituents of the glass (1). An alkali-deficient leached layer is formed on the glass surface. The thickness of these silica-rich films, and probably their compactness, vary with the composition of the glass and, for the same glass, with the test conditions of time, temperature and solution pH. Generally, a poorly durable material will develop a thicker film. The film thickness will be limited by the rate at which it is, in turn, corroded by the solution.

Ion Exchange

The first of these reactions can be viewed as an ion exchange, wherein an entity from the solution, which will be called a "proton" penetrates the glassy network, replacing an alkali ion, which escapes to the solution (2):

$$H^{+}_{sol'n} + Me^{+}O^{-}\text{-}Si_{glass} \rightarrow H^{+}O^{-}\text{-}Si_{glass} + Me^{+}_{sol'n} \quad (12\text{-}1)$$

Understand that what has been called the attacking "proton" is probably a hydrate thereof, as H^+ is not energetically favored in solution (3). Fortunately, this complication is basically irrelevant to the critical phenomena in durability.

As H^+ ions replace Me^+ ions there is produced a surface film, resembling vitreous silica or silica gel, and having different properties than the parent glass (4). This film swells, acts as a barrier to further reaction, and decreases diffusion rates into and out of the surfaces, thereby inhibiting further attack (1,5-7). Studies have shown this to be the principal mode of attack in acidic and neutral media.

Hydroxyl Attack

The presence of alkali ions in the solution tends to raise the pH to the point at which the silica-rich layer can itself be attacked by hydroxyl ions (2):

$$Si\text{-}O\text{-}Si_{glass} + Me^{+}OH^{-}_{sol'n} \rightarrow Si\text{-}OH_{glass} + Si\text{-}O^{-}Me^{+}_{glass} \quad (12\text{-}2)$$

This latter reaction is autocatalytic, as the open oxygen can interact with another water molecule to regenerate the hydroxyl ion, which is then available for a repeat reaction:

$$Si\text{-}O^{-}_{glass} + HOH \rightarrow Si\text{-}OH_{glass} + OH^{-}_{sol'n} \quad (12\text{-}3)$$

Hydroxyl attack is strongly dependent on the concentration of alkali ions available in the solution, and hence on the pH. In alkali silicates the transition between the region where hydroxyl attack predominates and that where the ion exchange process predominates is rather sharp, between pH values 9 and 10 (8). In more durable glasses the transition between the regions of ion exchange and hydroxyl attack is more gradual. In some of the very durable hot water tank enamels the transition pH is 4. Hence, the rate of ion exchange is more sensitive to glass composition than is the rate of hydroxyl attack (9).

Kinetics of Corrosion

For the ion exchange process, studies such as that shown in Figure 12-1 demonstrate that the rate of alkali extraction varies linearly with the square root

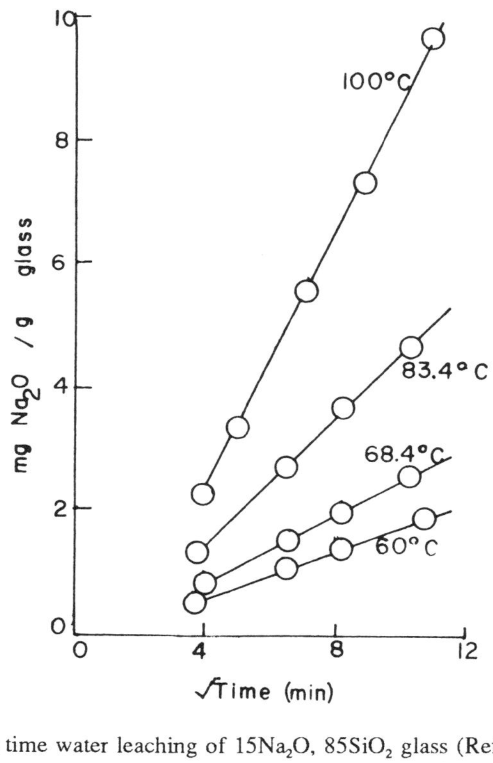

Figure 12-1. Short time water leaching of $15Na_2O$, $85SiO_2$ glass (Ref. 1).

of time at short times and at low temperatures (1,9). This graph plots the weight of Na_2O extracted under dynamic conditions and very low surface area/volume of solution (SA/V) against the square root of time for the early stages of extraction. By contrast, Figure 12-2, which shows data for longer periods of time, demonstrates that the rate of alkali extraction approaches a linear time dependence under conditions of longer time and higher temperature.

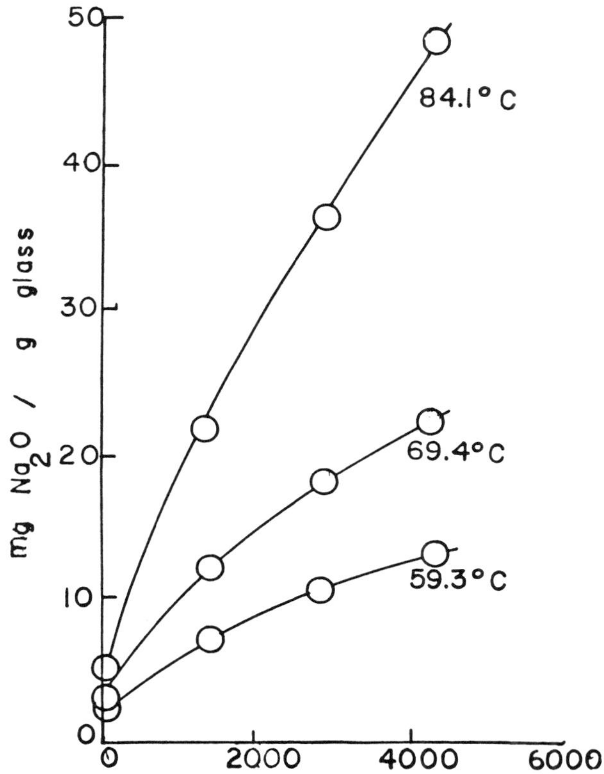

Figure 12-2. Long time water leaching of $15Na_2O$, $85SiO_2$ glass (Ref. 1).

This sort of extraction behavior can be represented by a rate equation of the form:

$$Q = at^{1/2} + bt \tag{12-4}$$

where t is the time and, a and b are constants. This equation has a limiting slope of 0.5 as t approaches 0 and 1 as t approaches infinity. Over limited times approximately linear plots of log Q vs. log t would be expected (2) with slopes varying between 0.5 and 1.

In mineral acids the increased concentration of "protons" increases the rate of the diffusion-controlled ion exchange process. In many cases only this process is observed. In alkaline media the rate of the linear dissolution process increases until it is the only process observed.

These effects are illustrated (10) by the data shown in Figure 12-3, which

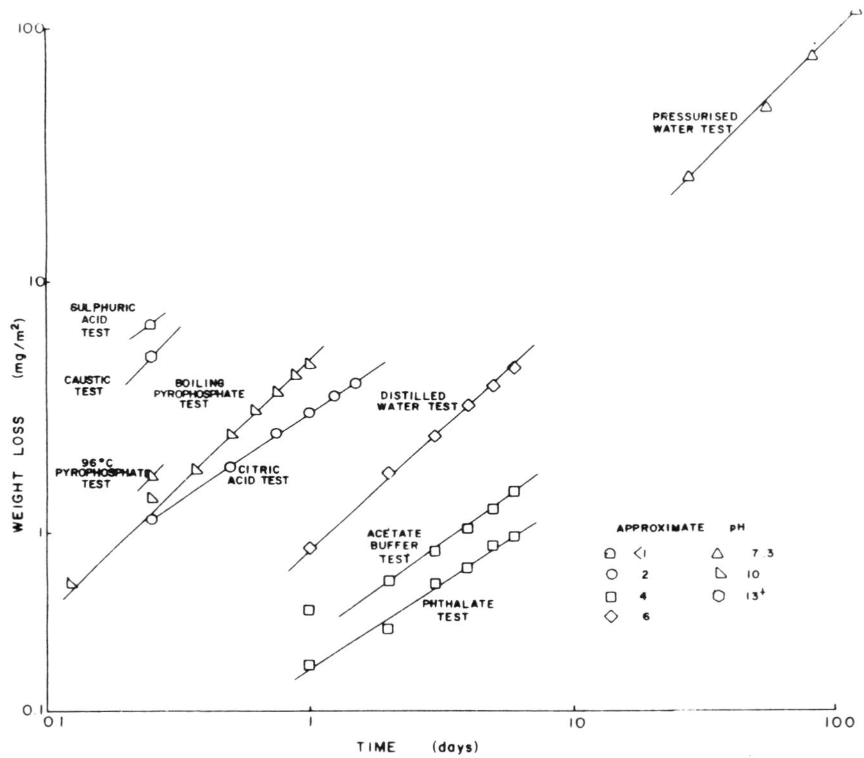

Figure 12-3. Weight loss data at 93°-104°C for enamel H1 (Ref. 6).

plots the weight loss due to leaching of a hot water tank porcelain enamel at 93-100°C. A systematic reduction in corrosion rate is noted as the pH is increased from less than 1 for the 5% sulfuric acid test to 2 for the 5% citric acid test, to 4 for the phthalate test. The slopes of these curves are about 0.6, as one would expect for an ion exchange-dominated corrosion.

As the pH is further increased, the corrosion rate is increased. From the phthalate test, one proceeds to the distilled water tests at pH 7 to the pyrophosphate tests at pH 10 to the caustic tests at pH values greater than 13. Here the slopes of the curves are about 0.9, as would be expected for a process dominated by bulk dissolution.

Figure 12-4 shows some data on the same hot water tank enamel discussed previously, corroded at 65-66°C. Again, the slope is near 0.5 for the citric acid test, while it is near 1 for tests in alkaline media. Of course, rates are much lower at this lower temperature.

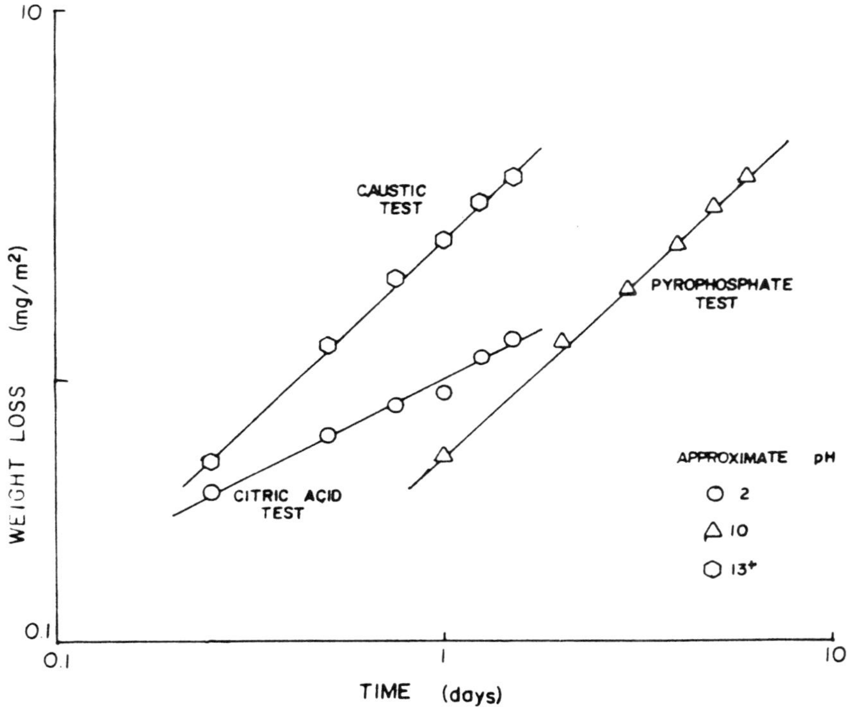

Figure 12-4. Weight loss data at 65°-66°C for enamel H1 (Ref. 6).

The curves drawn through the data points on these two figures were obtained by fitting the data to:

$$\log N_i = H\log t - E(1000/T) + F - GpH \qquad (12\text{-}5)$$

for the ion exchange process, and:

$$\log N_d = A\log t - B(1000/T) + C + D(100/pH^2) \qquad (12\text{-}6)$$

for the dissolution process. A through H are constants determined by the least squares fitting procedure. The total measured weight loss is:

$$N = N_i + N_d \qquad (12\text{-}7)$$

These equations demonstrate that the principal factors influencing the corrosion rate of a glass coating are time, temperature and pH. Generally, these three parameters alone can account for better than 90% of the observed rates.

One consequence of the pH sensitivity of glass corrosion is that exposure to water vapor produces a different corrosion pattern than exposure to liquid water (11). Water vapor deposited on a vitreous surface rapidly increases in pH, due to the high SA/V ratio, leading rapidly to a zone where hydroxyl attack predominates. This rise in pH occurs much more slowly when the volume of liquid is larger.

While t, T and pH are the principal factors governing the attack of liquids on glass, there are some secondary effects (12). The ions in the attacking solution may form complexes, which reduce the activity of the solution. Other ions may interact chemically with hydrous silica ions on the surface of the glass coating. Figure 12-5 compares the weight loss of a hot water tank enamel due to attack by "pure" water with attack due to solutions buffered with citrate and acetate

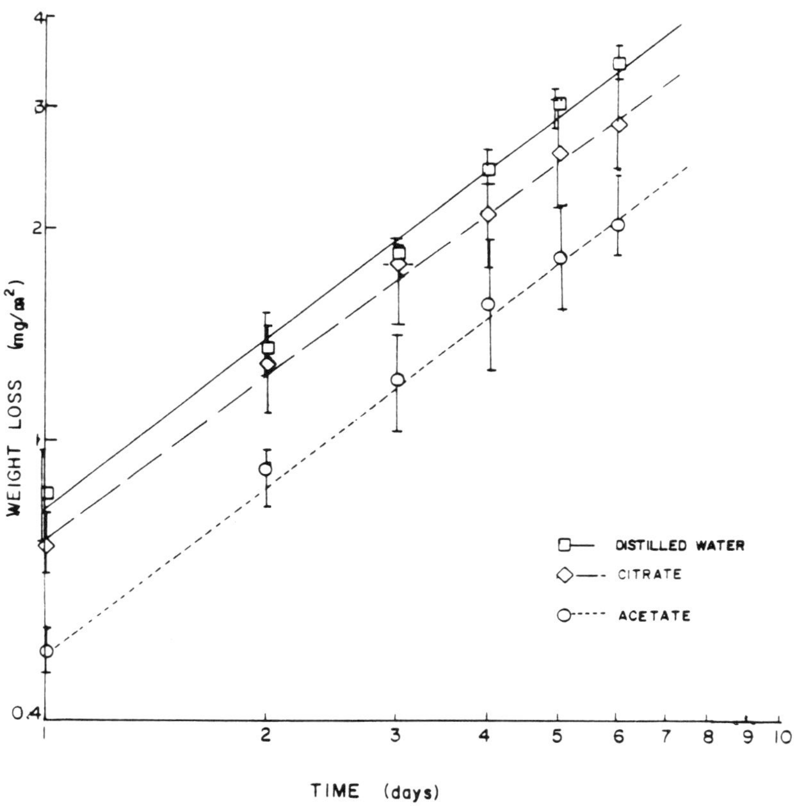

Figure 12-5. Weight loss from corrosion of enamel H4 in water and in two buffered media (Ref. 7).

anions. Figure 12-6 shows similar data for maleate, phosphate and succinate anions. A variation of as much as a factor of two is observed in the six solutions between the least aggressive (acetate) and the most aggressive (maleate). These anions have formed complex ions in solution which increase or decrease the solubility of a glass coating.

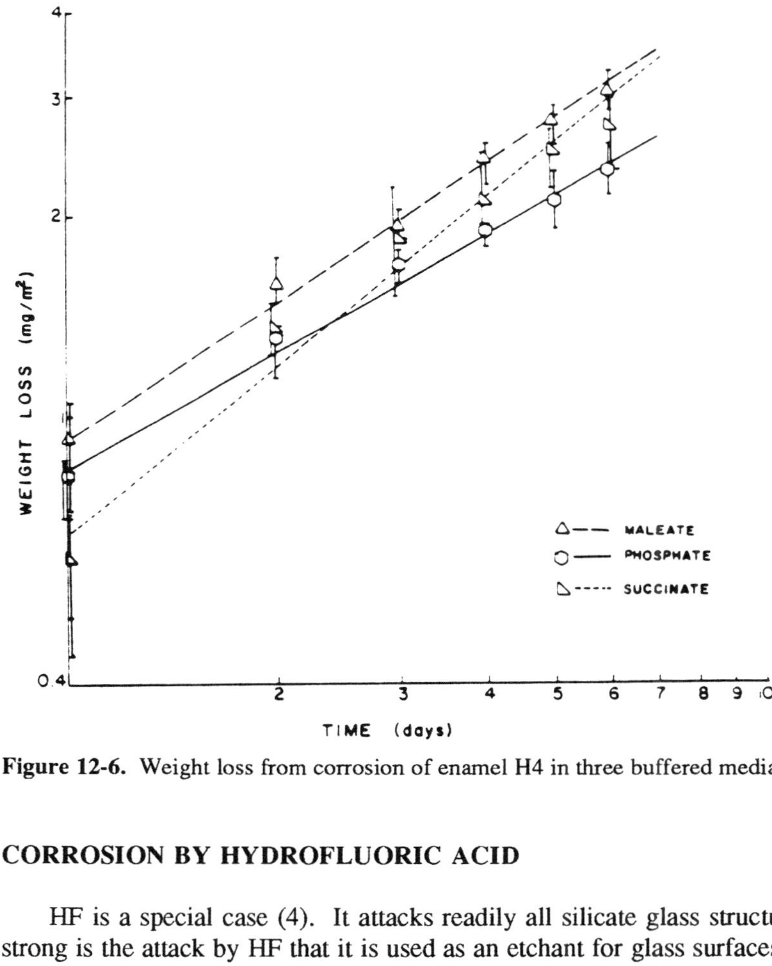

Figure 12-6. Weight loss from corrosion of enamel H4 in three buffered media (Ref. 7).

CORROSION BY HYDROFLUORIC ACID

HF is a special case (4). It attacks readily all silicate glass structures. So strong is the attack by HF that it is used as an etchant for glass surfaces. In the presence of hydrofluoric acid or acidic fluorides silicofluoride ions are formed according to the following equation:

$$SiO_2 + 6HF \rightarrow SiF_6^{-2} + 2H_2O + 2H^+ \quad (12\text{-}8)$$

These silicofluoride ions are slightly soluble, so that mineral acids are used in etching processes to hasten the removal of the silicofluorides. Hydrofluoric acid also attacks the glass structure directly:

$$\text{Si-O-Si}_{\text{glass}} + \text{H}^{+}\text{F}^{-}_{\text{sol'n}} \rightarrow \text{Si-OH}_{\text{glass}} + \text{Si}^{+}\text{F}^{-}_{\text{sol'n}} \quad (12\text{-}9)$$

Even acid-resistant glass coatings are attacked, and a mixture of two parts HCl and one part HF gives rapid surface removal of glass coatings (13).

EFFECT OF COATING COMPOSITION

The rate of extraction of ions from any glass coating is determined by the coating's overall composition, and by the extraction process which is operative (14,15).

Resistance to Ion Exchange

As one might expect, the rate of release of alkali ions from the coating by the ion exchange process increases with increasing alkali content (16). But this is only the first effect. Usually, K_2O is more soluble than Na_2O, which is more soluble than Li_2O (16-17). Mixtures of alkalis are usually less soluble than a single alkali (18-19).

Additions of alkaline earth ions decrease coating durability, but much less so than alkalis (16). Hence, substituting alkaline earth for alkali on a mole-for-mole basis will appear to improve durability. The effect of ZnO on acid resistance is somewhat ambiguous. Its solubility is less than that of alkalis or alkaline earths, but more than that of SiO_2 or similar oxides. Additions of CdO lower the durability.

The effect of PbO additions on acid resistance varies with the presence of other ions. For example, in lead-alkali silicates the total modifier content (PbO + alkalis) is directly proportional to the amount of modifier extracted (20).

Additions of alumina are particularly beneficial in increasing acid resistance (16-18,21). The only circumstances when alumina is not beneficial occur when large amounts are used, sufficient to force some Al^{+3} from 4 to 6 coordination. Additions of SiO_2, TiO_2 and ZrO_2 also improve acid durability.

Up to 12% B_2O_3 improves the durability of alkali silicate glasses (16). In most other circumstances, however, addition of B_2O_3 drastically lowers the acid resistance. Only the alkalis lower durability more than B_2O_3 (22). Additions of P_2O_5 and F, which interrupt the silicate network of the coating also drastically lower the acid durability.

A parameter which correlates all these findings regarding ion exchange durability is (14,15):

$$FM = \text{Good}/(\text{Bad})^{1/2} \quad (12\text{-}10)$$

where FM is a figure of merit for a glass, expressed in terms of the molar concentrations of its constituents:

$$\text{Good} = 2(Al_2O_3) + (SiO_2) + (TiO_2) + (ZrO_2) + (SnO_2) \tag{12-11}$$

$$\begin{aligned}\text{Bad} = {} & 2[(Li_2O) + (Na_2O) + (K_2O) + (B_2O_3) + (P_2O_5)] \\ & + (MgO) + (CaO) + (SrO) + (BaO) + (F) \\ & + (ZnO) + (PbO)\end{aligned} \tag{12-12}$$

The development of this parameter is discussed below in the treatment of lead release. For the moment, let us consider what it tells us about ion exchange durability. Most important is the observation that all components of a glass contribute to its durability, not just the constituents which are extracted. The oxides in this relationship are expressed as single ion concentrations, reflecting the ionic nature of this process. Note that the only desirable constituents are the refractory components of the glass matrix. All of the modifier ions, as well as B_2O_3 and P_2O_5 are deleterious. Hence, the problem in coating formulation is to balance the durability requirements against the ease of forming the coating in the firing process.

The one limitation on this correlation is when the coating is subject to phase separation (14). Many glass coatings, particularly those containing B_2O_3 and alkaline earths, exhibit phase separation, either on a macro scale or on a micro scale. In such a case, one of the phases will usually be high in SiO_2, the other an alkali or alkaline earth borate (23). If the borate phase is interconnected, it will determine the coating durability, not the overall glass. Even when it is not interconnected, the presence in the coating surface of a phase of low chemical durability may result in pitting.

Resistance to Hydroxyl Attack and Stability Diagrams

Alkaline durability is a very different phenomenon from ion exchange durability. It has already been noted that it is corrosion of the silica structure itself.

Figure 12-7 is a stability diagram for silica in aqueous solution. It is a plot of the activity (which is related to the concentration, and approaches concentration at low values) of silicate species in water as a function of pH. The data are derived from thermodynamic calculations, and are an indication of the driving force for removal of silica from a glass exposed to water. The lower the thermodynamic activity of any soluble species, the better the coating durability. The activity of soluble silicate species such as $HSiO_3^-$ and SiO_3^{-2} is negligibly

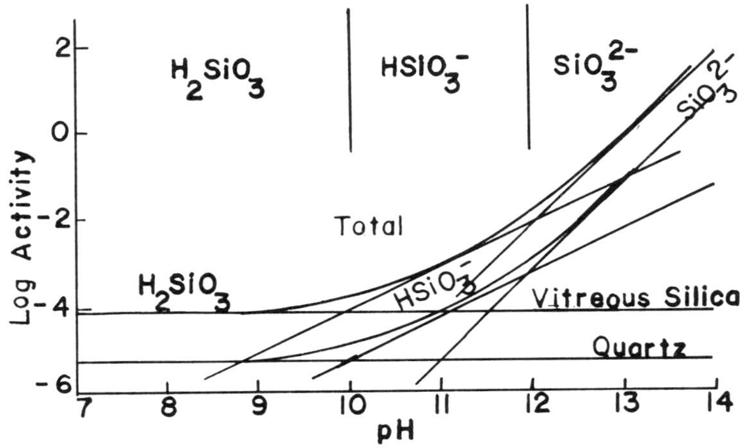

Figure 12-7. Stability diagram of quartz and vitreous silica in squeous solution at 25°C and various pH (Ref. 1).

small below pH 9, and only becomes appreciable at pH 12. Hence, it is not surprising that there are very few materials which will improve the alkaline durability of a silicate-based glass coating.

Figure 12-8 shows a corresponding stability diagram for ZnO. Below pH 7, zinc oxide is a deleterious material, as the activity of Zn^{+2} exceeds that of the glass. In the alkaline region, however, it is an interesting material. The isoactivity points for $HZnO_2$ and ZnO_2 are near pH 13. Hence, ZnO can be a desirable addition in the pH 9 to 12 range.

Figure 12-9 shows a stability diagram for PbO. Below pH 7, lead oxide is soluble as Pb^{+2} and to a much lesser extent, as $PbOH^+$. In the alkaline region, however, the solubility of $HPbO_2^-$ becomes significant only above pH 14. PbO thus improves alkaline durability, at the same time it decreases ion exchange durability.

By contrast, Figure 12-10 shows the stability diagram for Al_2O_3. The soluble alumina species found here are Al^{+3} below pH 3.2 and AlO_2^- above pH 11. Hence, while alumina is desirable for improving ion exchange durability, it is only helpful for alkaline durability in the pH range from 7 to 10.

Figure 12-11 shows why zirconia is widely known to improve the durability of silicate glasses more than any other element. Even small concentrations (2%) yield significant improvements in both ion exchange and alkaline durability. Figure 12-11 shows that, although the hydration of ZrO_2 is energetically very favorable, soluble ionic species like ZrO^{+2}, Zr^{+4} and $HZrO_3^-$ occur only below pH 2 and above pH 17. Hence, zirconia is a very desirable addition for alkaline

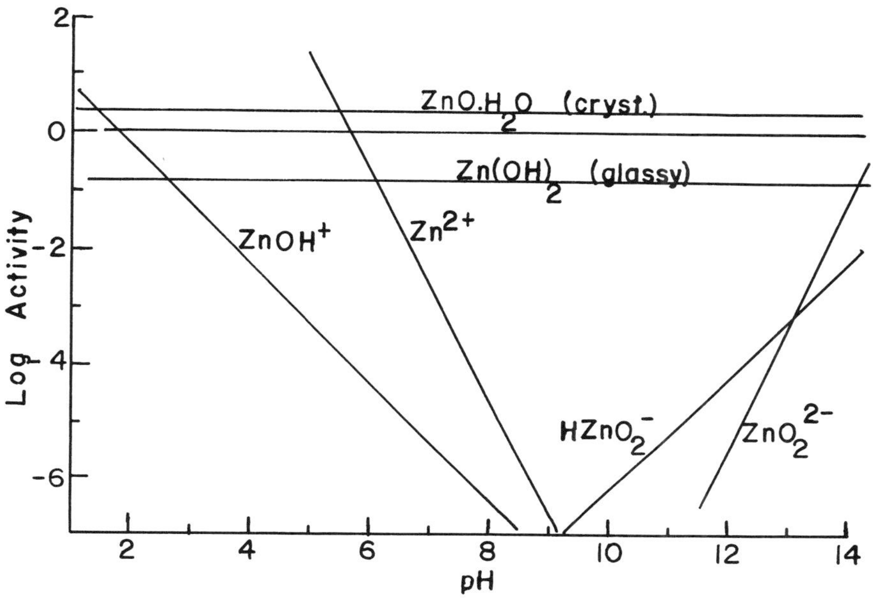

Figure 12-8. Stability of zinc oxide in aqeous solutions at 25°C and various pH (Ref. 1).

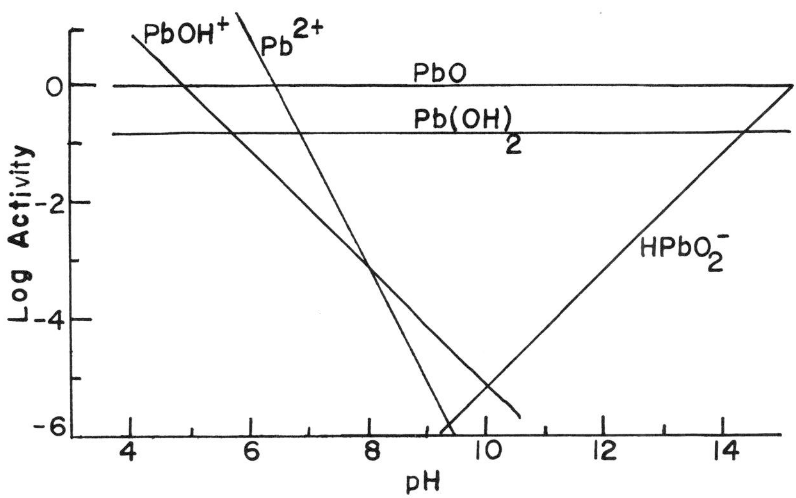

Figure 12-9. Stability of lead oxide in aqeous solutions at 25°C and various pH (Ref 1).

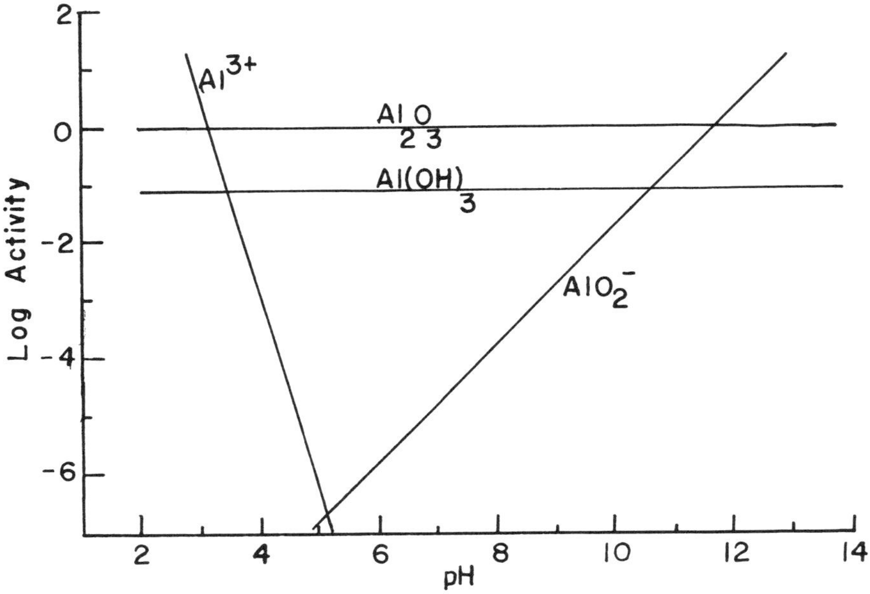

Figure 12-10. Stability of alumina in aqeous solutions at 25°C and various pH (Ref. 1).

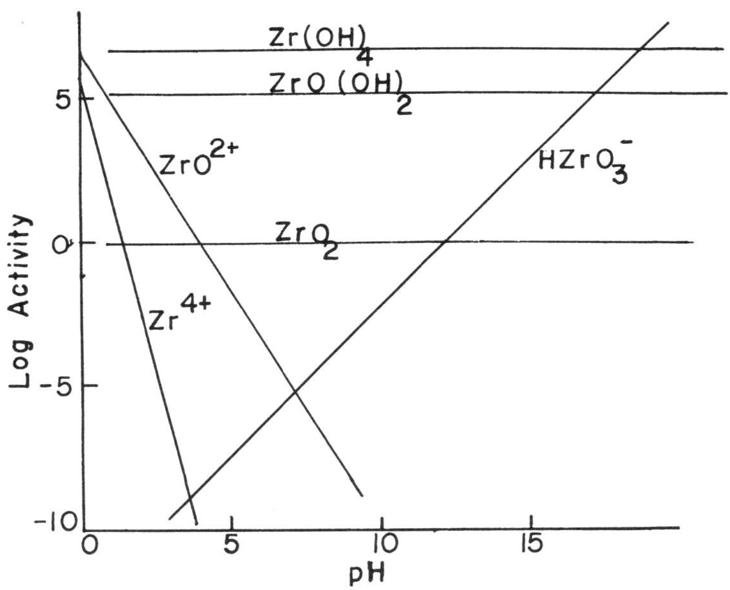

Figure 12-11. Stability of zirconia in aqeous solutions at 25°C and various pH (Ref. 1).

durability over the whole range from neutral (pH 7) to strongly alkaline (pH 12 to 14).

TESTS FOR CORROSION RESISTANCE

There are a number of tests which are used in the industry to quantify the durability of a glass. Usually they involve exposing the sample to a controlled temperature, time and pH. Depending upon the use envisioned, the attacking medium might be an acid, an alkali or water. For example, home laundry enamels are evaluated in alkaline phosphates at pH 10. Hot water tanks are evaluated in superheated water. Dinnerware are evaluated in mild acids (usually acetic) to simulate acidic foods, and in alkaline phosphates to simulate dishwashing.

The details of the various tests will be found in volumes 2.05 (tests for porcelain enamels) and 15.02 (tests for ceramic glazes) of the ASTM standards. Table 12-1 lists the principal methods currently (1989) used for the testing of ceramic coatings. They are divided into three categories: tests of acid resistance, tests of alkaline resistance, and other tests. The ASTM books of standards are reissued annually, as all the methods are under active management of a committee of experts on the relevant method.

Table 12-1. Corrosion Resistance Tests.

Tests for Acid Resistance	
C282	Acid Resistance of Porcelain Enamels (Citric Acid Spot Tests) (24).
C283	Boiling Acid, Resistance of Porcelain Enameled Utensils To (24).
C1034	Lead and Cadmium Extracted from Glazed Ceramic Cookware (25).
C738	Lead and Cadmium Extracted from Glazed Ceramic Surfaces (25).
C895	Lead and Cadmium Extracted from Glazed Ceramic Tile (25).
C872	Lead and Cadmium Release from Porcelain Enamel Surfaces (24).
Tests for Alkaline Resistance	
C614	Alkali Resistance of Porcelain Enamels (24).
C556	Detergents, Resistance of Overglaze Decorations to Attack by (25).
Other Corrosion Tests	
C650	Chemical Substances, Resistance of Ceramic Tile to (25)

LEAD AND CADMIUM RELEASE FROM CERAMIC COATINGS

Lead and cadmium release from ceramic coatings is an example of the use of one of these testing methods, as well as an important subject in itself. Particular emphasis will be placed on those surfaces used for consumption of food and drink.

Lead oxide is widely used in the formulation of vitreous coatings (15). There are a number of reasons for this. The strong fluxing action of lead oxide permits greater flexibility in glaze formulation. PbO reduces viscosity and allows for satisfactory processing over a wider firing range and under the varying conditions that may occur in a production-scale piece of equipment. PbO imparts low surface tension, high index of refraction and resistance to devitrification. This combination of properties cannot entirely be reproduced on a production scale in leadless systems. It is for these reasons PbO continues in use today.

Cadmium is much less widely used than lead, but is considerably more toxic (26). Its use is confined to the CdS-Se pigments which are used to impart red color to glass coatings. These pigments contain large amounts of Cd, and they are fairly soluble in acidic media. Moreover, when CdS-Se is used in a glass coating it is usually necessary to add 3 - 5% CdO to the coating formulation to improve the color stability.

Over the years, occasional episodes of lead toxication have resulted from the use of improperly formulated and fired lead-containing coatings on ceramic ware (27). Usually, these episodes of lead toxication have resulted from the use of an improperly made vessel for acid beverages consumed regularly in large quantities for a prolonged period of time. The pieces of ware implicated in these cases have come from a variety of sources, but in most cases have been manufactured by hobbyists or some artware manufacturers ignorant of the proper methods to assure production of safe coatings (28).

In one such case, a brown earthenware pitcher was purchased from a studio potter and used for storing fruit juice (29). An entire family suffered various effects of lead poisoning resulting from lead oxide leached into the fruit juice from the earthenware pitcher.

In another case, a two-year-old boy died, and his four-year-old brother suffered severe illness from lead poisoning after consuming large quantities of apple juice which had been stored in a ceramic pot (30). Subsequent analysis showed that the apple juice contained a large concentration of lead leached from the vessel. The pot was purchased from a roadside stand and was made by an amateur potter. In a third case, a doctor was afflicted with lead poisoning after consuming large quantities of cola in a mug made by his son in a pottery art class (27). These cases illustrate the similarity of instances in which glazed pottery has been implicated in lead poisoning.

Cases of cadmium poisoning from ceramic ware have not been reported, to my knowledge. However, a potentially dangerous situation was uncovered. This situation involved the lid of a porcelain enameled cooking pot. The manufacturer had placed a red coating on the inside of the lid. Under use conditions, steam condensation occurred, as would be expected, and with it leaching of Cd from the red coating.

Kinetics of Lead Release

The kinetics of the lead release process have received particular attention (32-34). Three stages have been identified (31). The initial logarithmic mechanism is observable only in the most durable glazes, because it occurs almost instantaneously in less durable coatings. It is associated with the release of loosely held Pb^{+2} ions near the surface of the glaze (33). The intermediate mechanism is superficially linear in time (31), but recent studies (35) have identified the mechanism as a combination of chemical reaction between lead in the glass and a "proton" from the solution, with the diffusion of the Pb^{+2} ions to the reaction site. The long time logarithmic mechanism has been identified (31,33) as a diffusion of Pb^{+2} ions through a barrier layer.

Procedures for Controlling Lead and Cadmium Release Under Production Conditions

How is this problem controlled under production conditions? A standard test has been developed for determining the lead and cadmium release from glass surfaces (C738 - see Table 12-1), so that various surfaces can be examined and rated (25). This test is specific for Pb and Cd, accurate to low release levels in the ppm range, readily reproducible in different laboratories and under field conditions.

In this test, a sample is first washed with detergent and rinsed in distilled water. It is then exposed to a 4% acetic acid solution for 24 hours at room temperature while covered with a watch glass or other suitable cover. The leachate solution is then placed in an atomic absorption spectrometer and the ppm of Pb and Cd are determined.

Current United States Food and Drug Administration standards (20) are 7 ppm average of six samples for lead released from flatware, 5 ppm maximum of six samples for small hollowware and 2.5 ppm maximum of six samples for large hollowware. For cadmium, the standards are 0.5 ppm average of six samples for flatware, 0.5 ppm maximum of six samples for small hollowware and 0.25 ppm maximum of six samples for large hollowware. Revisions to these levels are currently under consideration (37). Moreover, because of statistical fluctuations during production, operating standards must be less than half these values (38).

Typical Results

The results obtained in testing of dinnerware for lead release are quite dramatic (15). Commercial dinnerware glazes almost always give readings of 0.5 to 2 ppm. Properly designed commercial artware will be somewhat higher but still considerably under the standard of 7 ppm. On the other hand, the hobbyware materials involved in the health cases discussed above have all given tests in excess of 50 ppm.

The results with Cd are more specialized (26). Most glazes have no added cadmium and will not give a measurable reading. Those that do contain cadmium will show 0.1 - 0.2 ppm even when properly formulated. For this reason it is best if CdS-Se red colors are not used for glass surfaces which will come in contact with food or drink.

Formulating a Coating for Low Lead Release

There are a number of factors which must be considered in formulating and processing a glass to achieve low lead and cadmium release (14). These include total glass composition, including opacifiers and colorants when used, the thermal history of the glass during processing, effectiveness and uniformity of glaze application when used as a coating, glaze-body solution at the interface in coating applications, and atmospheric conditions which exist during firing, in particular the flow of air over the ware during high temperature processing.

The formulation of glass for acceptable lead release is not a simple matter (14). Several factors must be considered and balanced against one another. On the other hand, certain guidelines and relationships have been shown to be valid. Of greatest importance, it has been shown that lead release is directly related to the acid resistance of the glaze. Materials which are acceptable are those which are impervious to the action of food acids such as citric acid. Materials which are not acceptable are subject to attack by food acids.

Prediction of the Lead Release of a Ceramic Coating

Studies have shown that the lead release of a given formula can be predicted from experimental knowledge concerning the acid resistance of various glazes

(14). It has been shown that silica, alumina, zirconia and similar oxides such as titania and tin oxide are effective in lowering the lead release of a glaze.

Referring to the Equation 12-11, the concentrations are expressed in molar ratios, and reflect the ionic nature of the ion exchange process. The 2 arises from the fact that there are 2 equivalents of aluminum ions in each equivalent of Al_2O_3.

It has also been shown that alkalis, alkaline earths, B_2O_3, fluoride, phosphate, ZnO, CdO, and PbO are all more or less detrimental to the lead release in a glaze. This is reflected in Equation 12-12. This term consists of the soluble glass former B_2O_3, those ions such as F which disrupt the silicate network, together with all of those ions which do not enter the basic glass structure.

These factors have been related to a data base of 77 glazes (Figure 12-12), and examined by a regression technique. As a result the figure of merit mentioned earlier was developed (Eq. 12-10).

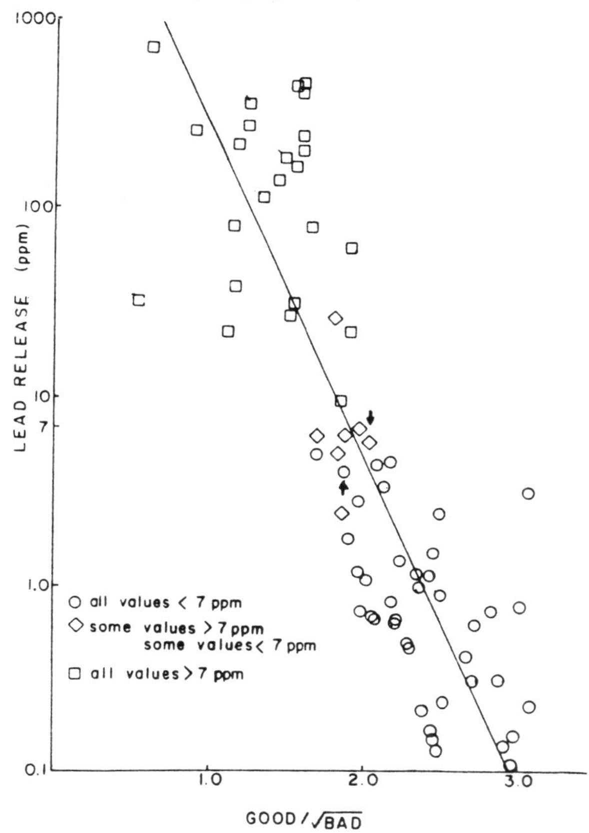

Figure 12-12. Comparison of lead release data with the figure of merit (Ref 8).

It is observed (14) that when the figure of merit exceeds 2.05 the lead release is below the FDA standard (36). When it is less than 1.80, some measurements are always greater than the standard. The figure of merit is not able to predict acceptability in the small interval 1.80 to 2.05. Of the glazes studied, 38% fall in the interval less than 1.80, 17% fall in the intermediate range, and 45% fall in the interval greater than 2.05.

Thus, in most cases, the figure of merit provides a simple, straightforward calculation to predict the acceptability of the lead release of a glaze. It is, therefore, very useful in guiding development work. In terms of industrial practice, the figure of merit almost always gives a definite prediction of the lead release, since glazes of marginal lead release have been emphasized in this study, but occur infrequently in the real world. Note that this calculation assumes proper glazing and firing procedures (14). Also, while the figure of merit is useful for formulating an acceptable lead-containing glaze, the lead release should always be verified by testing, using the standard test procedures (ASTM C-738).

Only a few qualifications have been noted. As we mentioned before, this calculation does not apply to a multiphase system, as the formulation of the least durable component is the durability to be used. Second is the effect of CuO. CuO in the glass has the unique effect of turning an otherwise safe glass into an unsafe one (18). Thus, glasses containing CuO should never be used in contact with food or drink.

SUMMARY

Chemical resistance in service is a major consideration leading to the selection of a vitreous material as a coating material. Ceramic coatings can be designed to be highly resistant to practically all liquid media except hydrofluoric acid and related fluorides.

There are two processes operative in the aqueous corrosion of a vitreous surface. In acidic media an ion exchange process dominates. In alkaline media the process is one of bulk dissolution of the glass surface.

The ion exchange process is a function of the total composition of the coating, and not merely those constituents which are extracted. A figure of merit has been developed to predict the acid resistance of various formulations.

There are only a very few elements which improve the alkaline resistance of a silicate-based glass coating. The principal additive for alkaline resistance is zirconia.

Over 90% of the aqueous corrosion of ceramic coatings can be accounted for by a consideration of only three parameters - the time and temperature of corrosion, and the pH of the attacking medium.

While the basic phenomena involved in the corrosion of ceramic coatings are well understood, many of the details of the processes are not. The properties of the corrosion layers which form in response to ion exchange have been examined only superficially. The exact nature of the species in the attacking solution sometimes make a difference in the corrosion rate.

The role of glaze composition has been worked out to the extent of the first-order correlation discussed above. Beyond that, understanding of compositional effects is fragmentary, largely limited to the effects of alkali adjustment.

REFERENCES

1. Paul, A., *Chemistry of Glasses*, pp 108-147, Chapman and Hall (1982).
2. Eppler, R.A., *Am. Ceram. Soc. Bull.* 56: 1068-70 (1977).
3. Ernsberger, F.M., *Phys. Chem. Glasses* 21: 146-49 (1980).
4. Taylor, J.R., and Bull, A.C., *Ceramics Glaze Technology*, pp 167-183, Pergamon (1986).
5. Tsuchihashi, S., *Bull. Chem. Soc. Jap.* 24: 161-64 (1951).
6. Rana, M.A., and Douglas, R.W., *Phys. Chem. Glasses* 2:179-95 (1961).
7. Das, C.R., and Douglas, R.W., *Phys. Chem. Glasses* 8:178-84 (1967).
8. Boksay, Z., Bouquet, G., and Dobos, S., *Phys. Chem. Glasses* 9: 69-71 (1968).
9. El-Shamy, T.M., PhD Thesis, University of Sheffield (1966).
10. Eppler, R.A., *Am. Ceram. Soc. Bull.* 60: 618-22 (1981).
11. Clark, D.E., and Ethridge, E.C., *Am. Ceram. Soc. Bull.* 60: 646-49 (1982).
12. Eppler, R.A., *Am. Ceram. Soc. Bull.* 61: 989-95 (1982).
13. Barkatt, A., Barkatt, A., Boroomand, M.A., et al., *Adv. Ceram.* 8: 482-90 (1985).
14. Eppler, R.A., *Am. Ceram. Soc. Bull.* 54: 496-99 (1975).
15. Eppler, R.A., *Proceedings International Conference on Ceramic Foodware Safety* (J.F. Smith and M.G. McLaren ed.), pp 74-96 (1974).
16. Mellor, J.W., *Trans. J. Brit. Ceram. Soc.* 34: 113-90 (1935-35).
17. Koenig, J.H., *Ohio State Univ. Eng. Expt. Sta. Bull.* #95, pp 116 (1937).
18. International Lead Zinc Research Organization, *Lead Glazes for Dinnerware, ILZRO Manual Ceramics*, No.1 (1970).
19. Dilmore, M.F., Clark, D.E., and Hench, L.L., *J. Am. Ceram. Soc.* 61: 439-43 (1978).
20. Yoon, S.C., PhD Thesis, Rutgers (1973).

21. Singer, F. and German, W.L., Ceramic Glazes, Borax Consolidated (1964).
22. Moore, H., *Trans. J. Brit. Ceram. Soc.* 55: 589-600 (1956).
23. McMillan, P.W., *Glass-Ceramics*, 2nd ed., Academic Press (1979).
24. *ASTM Annual Book of Standards*, part 2.05, "Metallic and Inorganic Coatings."
25. *ASTM Annual Book of Standards*, part 15.02, "Glass, Ceramic Whitewares."
26. Eppler, R.A., and Carr, D.M., *Cadmium 81, Proceedings, 3rd Intern. Cadmium Conf.*, pp 31-33, International Lead Zinc Research Organization (1982).
27. Harris. R.W. and W.R. Elsea, *J. Am. Med. Assoc.*, 202:208-210 (1967).
28. Eppler, R.A., *Glass Science and Technology*, Vol. 1, "Glass-Forming Systems" (D.R. Uhlmann and N.J. Kreidl ed.), pp 301-38, Academic Press (1983).
29. Block, J.L., *Good Housekeeping*, p 63, Nov. 1969.
30. Beiler, Z., *Montreal Star*, p 41, Dec. 1969.
31. Eppler, R.A., and Schweikert, W.F., *Am. Ceram. Soc. Bull.* 55: 277-81 (1976).
32. Norris, A.W., and Bennett, H., *Trans. Brit. Ceram. Soc.* 50: 225-39 (1951).
33. Yoon, S.C., PhD Thesis, Rutgers (1973).
34. Escardino, A., De La Torre, J., and Blasco, A., *J. Brit. Ceram. Soc.* 86: 47-51 (1987).
35. Escardino, A., De La Torre, J., and Blasco, A., *J. Brit. Ceram. Soc.* 86:118-23 (1987).
36. U.S. Food and Drug Administration, *Compliance Policy Guides*, "Cadmium" 7117.06, and "Lead" 7117.07.
37. Calderwood, J.A., *Soc. Glass Ceram. Decorators 1988/89 Seminar Proceedings*, pp 12-15 (1989).
38. Moore, C.F., *Trans. J. Brit. Ceram. Soc.* 76: 52-57 (1977).

13

Corrosion and Conservation of Ancient Glass and Ceramics

Pamela B. Vandiver

Conservation Analytical Lab
Smithsonian Institution
Washington, D.C.

INTRODUCTION

The field of conservation in the West has developed rapidly during the last three decades, defining its function as follows: artifacts are to be stabilized and consolidated with the goal that any treatment will be reversible, will allow close-up observation, but will be unobtrusive from a distance (1). Western conservation has disavowed the practice of restoration, as the altering of objects by applying current judgments retrospectively to determine what the object should have or might have looked like. Since artifacts have histories of weathering, restoration and sometimes embellishment or symbolic destruction, it is often difficult to decide to what state an object should be returned, and at what point added embellishments are historically significant and represent the cultural views of people who appreciated a particular object at a certain point in time.

Corrosion of ancient glasses and ceramics is a more ubiquitous problem than most people, even corrosion scientists, believe. Low-fired ceramics often decompose upon excavation because the clay, fired just above dehydroxylation temperature, rehydrates, or because unstable glass compositions formed during sintering have weathered during burial. In addition, secondary salts may have precipitated and swelled in pores, exceeding fracture strength. Ancient glasses and glazes may solarize, pit, crizzle, craze, form weathered layers, or even be discovered in the ground in recognizable shapes but as unrecoverable white powders. Some very ancient glasses have decomposed to gels, and they crack and powder on drying. Glazes and enamels frequently have different expansion coefficients than inclusions, or substrates of ceramic or metal, and, even if they do not have incipient damage from cooling during manufacture, changes in temperature, humidity and basicity during burial, excavation, conservation, display, shipment or storage may lead to differential moisture expansion, delayed crazing, chipping, flaking, peeling and eventual delamination. Some ceramics, such as plasters, may alter in microstructure during local surface diffusion or dissolution and reprecipitation processes, such as Ostwald ripening, and may lose sufficient strength that they no longer support their own weight. Many ceramics are composites or have been assembled from many materials or parts which may differentially corrode, causing conflicts in conservation treatments, because what may be acceptable for one material or part may be detrimental to the longevity of another. An example is the washing of a glazed ceramic to remove salts in the body, a process which corrodes the glaze and sometimes delaminates it.

What we know about the past is based upon selected, collected and excavated documents and artifacts which have been filtered through vagaries and mechanisms of preservation--in other words, preservation is based on random, chaotic and situational processes as well as physico-chemical principles. Conservation science is a recently-developed discipline which studies empirically the mechanisms of corrosion and which isolates the conditions and treatments which promote longevity through modeling and accelerated aging experiments. In the past, composition and environment were taken as the most important variables, and other potentially significant factors such as variability in raw materials, manufacturing technology and use-wear were ignored or have been judged too difficult to measure. "Use-wear" is a term archaeologists apply to the effects of use, and reuse, on an artifact within the cultural context in which it was made prior to discard. Because of the difficulty of differentiating the wear caused only by use from that caused by postdepositional wear and site formation processes, the two are linked as one field of inquiry. Some examples are functional studies of bone, stone tools, and ceramics (2-5).

Corrosion and weathering of ancient ceramics, plasters, glasses and glazes are described in this chapter through case studies. These studies include the rehydroxylation of low-fired Paleolithic ceramics, liquid-phase sintering and Ostwald ripening of Neolithic plasters, spherulite and layer formation, hydration and gelation of Near Eastern Bronze Age glasses and Parthian period glazes of the soda-lime-silica family and a Roman glass from Samaria. Examples of crazing and delayed crazing are given, glaze fit is calculated and the relationship of crazing to corrosion is shown. Degradation of high-fired stoneware and porcelain glazes is also demonstrated.

CONTRIBUTING FACTORS

- Particle Packing
- Sintering Aids
- Use-Wear
- Rehydroxylation
- Hydration
- Montmorillonite Impurities
- Ostwald Ripening
- Liquid-Phase Sintering in Aqueous Solution
- Gelation
- Differential Shrinkage
- Fracture
- Weathering
    - Condensation-Evaporation
    - Condensation-Runoff
- Crazing
- Delayed Crazing
- Surface Heterogeneities

CASE STUDIES IN CORROSION

Low-Fired Paleolithic Ceramics

The earliest known clay-based and fired ceramics are from the site of Dolni Vestonice, Czechoslovakia, which are dated to the Upper Paleolithic period about 26,000 years ago (6a). More than 10,000 figurine fragments have been excavated that were made of local soil, a loess consisting of quartz and micas, with minor amounts of illite, chlorite, dolomite, feldspar and calcite. The loess was mixed with a liquid, formed, and fired to temperatures of 500 to 700°C in open pits and low-walled kilns. The loess also contains minor impurities of salts, such as phosphate, which act as fluxes during firing. Even though a low earthenware-temperature range was used during firing, replication and autoclaving experiments have shown that the fragments are quite durable (6b).

At other Upper Paleolithic period Eurasian sites, very few figurines have been found: one horsehead excavated at Zazaragi, near Sendai, Japan (ca. 18,000 B.P.); one human figurine at Maina, Siberia, (ca. 15,000); and three fired fragments from a hearth at Kostienki (ca. 22,000 B.P.)(7-10). I have not seen the figurine from Maina, but I have examined the fragments from Kostienki and Zazaragi and have concluded based on preliminary studies using an optical microscope, that they probably have been fired. They have a mohs hardness of about 4 which makes them unlikely to be unfired clays, and color variation ranging from tan to red to gray and black, in some cases layered from exterior to interior, and in other cases juxtaposed along the surface (although such color variation might be caused by staining from iron and manganese oxides). Modeling studies of fired local soils from Kostienki and Sendai show evidence of deterioration during autoclaving which may explain why more Paleolithic ceramics have not been preserved. In the case of Kostienki, samples of the three fired fragments have been compared with similar local soils. A very coarse soil (2 microns - 3 mm maximum diameter) was used for the fired fragments, which did not have enough particle-to-particle contacts to sinter and form a nonfriable, durable ceramic. In the Zazaragi example, microscopic examination revealed that sufficient fine clay to form a durable ceramic was present in the horsehead. Local clays have been identified by x-ray diffraction as illites and, by scanning electron microscopy (SEM), as having a relatively coarse average clay particle size (ca. 5 microns). The particles have good crystallinity and a relatively high clay dehydroxylation temperature, established by differential thermal analysis as a minimum decomposition temperature of 720°C with onset at 690°C (11,12). If fired to 770°C with a 30 min. ramp and 30 min. soak at maximum temperature, the sintered microstructure in Figure 13-1 results. When autoclaved for 10 hrs. at 150 psi, based on changes in the phases observed by x-ray diffraction and scanning electron microscopy (13), the clay rehydroxylates and the glassy bond hydrates and is not durable (Figure 13-2). The ceramic is reduced from a mohs hardness of 4 to 2 in areas where it remains intact. No salts were found in the acidic volcanic soil samples, which would have acted as a sintering aid to promote glass formation, as at Dolni Vestonice.

These preliminary tests establish that weathering of low-fired ceramics occurs because of poor particle-packing of the clay and grit in the clay body during manufacture, weathering processes of clay rehydroxylation and glass hydration, and compositional variables such as the presence or absence of salts in the original clay body. Use-wear is another significant contributor to the degradation of Paleolithic ceramics, making them less durable than expected--the internal cracks accumulated during use, discard, trampling, final burial, earth movement and freeze-thaw cycles. Thus, even though there is evidence of Upper Paleolithic ceramics, it is likely that we will have only a limited opportunity to glimpse at a few places and points in time what was a more common and widespread ceramic figurine technology than we thought.

One further aspect of usage has contributed to the deterioration of the Dolni Vestonice figurines. They have fracture surfaces characteristic of thermal shock, even though the loess soil is an excellent refractory with low thermal expansion and would not be expected to fracture this way. Replication tests have indicated that it is difficult to thermal shock this clay body. We have, therefore, suggested that these figurines were used, not primarily as visual images, but in performance which involved the figurines as material indicators to reinforce ritual activity--activity which involved the use of thermal shock. In this case, we suspect that social function led directly to deterioration.

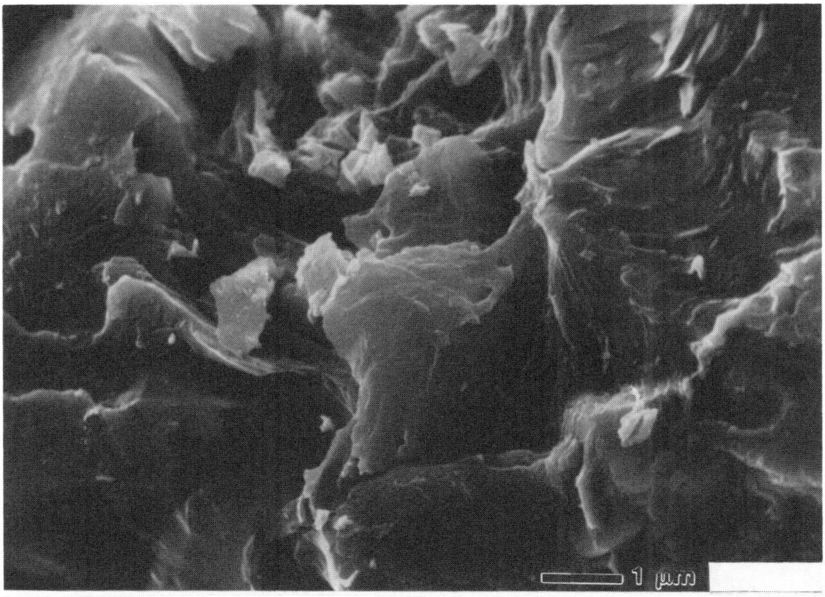

Figure 13-1. Fired soil of illitic clay and quartz, Sendai, Japan.

Figure 13-2. The same clay rehydroxylates and the glassy bond degrades when autoclaved.

Many aspects of low-fired ceramics remain to be studied. For instance, this preliminary discussion of mechanism falls far short of an understanding of decomposition rate, critical factor dependence, or rate limiting steps in the corrosion process. Even though historical precedent has involved autoclave trials to model durability, no experiments have been conducted which test their appropriateness to model long-term deterioration processes or to quantify their use. Even the corrosion mechanisms require further study because of a lack of detail and the necessity to elucidate some of the assumptions which have been made. For instance, are the model soils the same as the ones used for the ancient ceramics, and are they really representative samples? For example, the similarity of the particular Japanese clay used as a standard could not be tested, although it came from an appropriate level at the site, because the horse, being a unique national treasure, cannot be sampled, nor can the original firing temperature be established.

In summary, low density particle packing, too high a coarse fraction, poorly durable or lack of sintering aids, use-wear, rehydroxylation of the clay fraction and hydration of the glassy fraction of low-fired ceramics, usually in combination, lead to deterioration.

Neolithic Plasters

The extensive use of lime and gypsum plasters for a wide variety of uses such as floor and wall plasters, bowls and storage vessels, gaming pieces, objects of personal adornment and sculpture in the Near East dates to the Pre-Pottery Neolithic (ca. 7200-6000 B.C.) about 1000 years before the widespread development of clay-based pottery, about 6400 B.C. (14,15). Production of lime and gypsum plasters is a multi-step process requiring skill in the selection and collection of raw materials, heating of limestone at 800-900°C (gypsum at 150-200°C) followed by slaking the quicklime in water to form the hydroxide, mixing with various additives, shaping or applying as a paste, and often coating with a slip coat, sometimes colored with pigments, and burnishing. This early craft activity was complex, requiring considerable organization of resources, particularly fuel, to produce the quantities of plaster found on many sites. Other materials were also used as plasters, for instance, "mud" plasters of clay or marl (a calcareous silt and clay mixture or a calcareous sediment, respectively), often with such inclusions as straw or grit temper. These, however, are not water resistant or durable, but they do not require the effort and energy of pyrotechnological processing.

Many ancient lime and gypsum plasters currently are fragile and friable--both unable to support their own weight and subject to further degradation in conditions of changing temperature and relative humidity. Some plasters were formed with vegetable fiber temper or around reed bundle armatures. The decomposition of this fiber has weakened the integrity of the objects. In addition, many plasters contain about 10-15 vol% unfired montmorillonite clay which, when subject to changes in humidity, will absorb moisture and swell, in some cases to many times its dry volume (Figure 13-3). Such additions, which would have ameliorated the low plasticity and poor formability which is characteristic of plasters, can lead to cracking and deterioration in unsuitable conditions (Figure 13-4).

Changes in microstructure and properties occurring over long time periods in lime and gypsum plasters may lead to deterioration. Gypsum plaster has a solubility in water of about 0.2 wt%; the solubility of lime plaster depends on acidity, but is about 0.0015 wt%. Because of their solubility, both gypsum and lime plasters undergo metamorphic changes when exposed to high relative humidities (over 70%) over periods of years (15). The fine particles undergo grain coarsening by Ostwald ripening and densification by liquid-phase sintering in aqueous solution (16,17). Autoclaving experiments have shown that the needlelike particles of gypsum plaster (Figure 13-5a) shorten, thicken and grow in size and the fine submicron spherical particles of lime plasters coarsen and sinter (Figure 13-5b). If conditions for growth are maintained constantly, the plasters may be strengthened (Figure 13-6). If these conditions are cyclical and/or acidic pollutants are present, the plasters, being quite porous, will have surfaces periodically etched and, particles will become detached at joints. Porous agglomerates form, and the structure is weakened (Figure 13-5b). Preservation requires storage in stable conditions of low relative humidity (15) and, when necessary, consolidation (18). Although the corrosion of marble and other calcareous stones is being studied (19-22), no long term investigations of plaster stabilization are known to me.

The deterioration of Neolithic and other plasters is caused by unfired montmorillonite impurities, Ostwald ripening of the partially-soluble gypsum and, to a lesser extent, lime particles, and their liquid-phase sintering in aqueous environment over long time periods. Cycling of environmental conditions which are deleterious with those which give stability is particularly harmful.

Near Eastern Bronze Age Aqueous Glass Corrosion

The corrosion of glasses has been investigated more thoroughly than that of any other ceramic material. In the middle of the last century the importance of composition to durability was established in studies of glasses potentially suitable for use in chemical experiments and in lenses (23). At the same time, the mechanism of hydration was isolated as critical to glass corrosion, and the iridescence on Roman and Pre-Roman glass was shown to be the effect of weathering rather than the intentional product of glassmakers' efforts (23,24). The visual features associated with corrosion and technology as well as the usual criteria of shape and decoration have long been catalogued in art historical and archaeological studies of ancient glasses (25,26), making technical studies important contributors to the interpretation of ancient glasses.

In most accounts of the origins of manmade glasses, glassmaking is said to

Figure 13-3. Weathered and corroded mixture of gypsum and montmorillonite clay plaster (clay at arrow).

Figure 13-4. Secondary gypsum precipitated in an expanding crack in gypsum plaster.

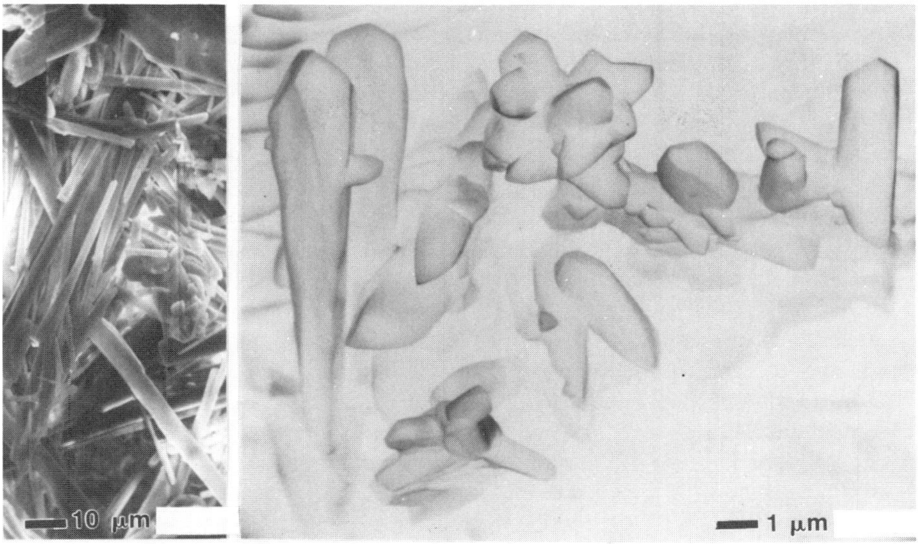

Figure 13-5. Neolithic period gypsum plaster from Abu Hureyra before (left) and after (right) cyclic autoclaving for five one hour periods which causes decrease in strength due to formation of porous agglomerates.

Figure 13-6. Fractured surface of gypsum plaster autoclaved continuously for 5 days resulting in sintering, grain growth and densification.

begin in the Near East with the manufacture of vessels in the mid-second millennium B.C. in Syria and Iraq, along the Levantine coast and in northern Egypt. Textural accounts of glassmaking on unbaked clay tablets date to the ninth century B.C.; however, many of the words mentioned in the texts are related philologically to Sumero-Akkadian words of the middle of the second millennium--words which were developed in the preceding half millennium (27). The textual evidence, thus, links glassmaking to as early as 2300 B.C., and the archaeological evidence which consists of a few glass beads and some small lumps of glass reinforces this early date (28,29). Unfortunately, most of these small and fragile glass finds have been lost or destroyed before they could be studied by modern techniques; only two beads from Nippur have been thoroughly studied, and one lump from Eridu is currently being investigated (30).

The widespread occurrence of a technology is easily documented in the archaeological record, but the invention of a technology is virtually impossible to pinpoint because of the large spatio-temporal intervals in archaeological sampling and because of the problems of preservation. Often a type of object, known to come from a later time period, will be classed as intrusive, if found in an earlier-than-anticipated context. In order to establish an earlier period of production, a context with dated artifacts in association with the object is required, or one which is sealed by a layer of established date. Cyril Smith has suggested that the sequence of development of ancient materials and technologies follows the same type of S-shaped curve found in nucleation and growth of crystals from a matrix, where the number or volume fraction is plotted on the y-axis and time is plotted on the x-axis (31). For a technology to be recognized in the archaeological record, it is likely that the maximum of the curve is being approached, although, with care and luck, earlier instances in the sequence of development can be tracked. However, also necessary is that the preservation of materials extend farther back in time than the rise in this curve. For many ceramics, glasses, metals and plasters, the earliest occurrences are in objects with considerable deterioration, thus coinciding often with the longest time-depth of their preservation (32).

In the 1989 field season of the University of Chicago at Nippur, a major site in southern Iraq, two glass beads were excavated in an Akkadian context (ca. 2350 B.C.) in association with tablets and architecture. One was a spherical bead with a meniscus at each end of a circular hole, evidence of vitreous forming. The other was a cylindrical bead with a hexagonal cross section. Both beads were found in moist, almost neutral (6.8 pH) soil, consisting of quartz sand, calcareous mixed-layer illite-montmorillonite clay, and a small amount of gypsum and other salts. Both beads had some loss of material, but over half of each surface was glassy-looking and reflective. The colors were bright and consisted of translucent yellow and turquoise blue on one, and olive green and opaque white on the other. Both surfaces, however, could be scratched with a fingernail. The beads were brought to the dighouse lab encased in the moist soil they were found in and were sealed in plastic. The beads were kept in a moist environment, cleaned of dirt using a dental pick under a microscope, examined and photographed. Within an hour, a change in glossiness occurred as drying progressed, and within 24 hours microcracks developed at the exterior which were observed under a microscope to propagate into the interior as shrinkage progressed. The appearance of the glasses changed from translucent to opaque, but most of the surface remained smooth and glossy. Once this initial, rapid transformation occurred, the rate of change slowed. One year later, the major cracks in the glass beads had opened, but otherwise the external appearance of the beads was unaltered.

This same process of rapid deterioration of glasses upon excavation has been recognized by archaeologists at other Southwest Asian sites (33). At Nippur some glazed ceramics and glasses from later periods were observed which demonstrated similar behavior. Drying of these other artifacts, for instance Seleucid glazes, resulted in an opaque, white surface layer of considerable green strength and durability, as can be observed in many similar objects in museums. In one case, probably a very porous gel, the glass artifact powdered during drying. Swabbing with water or acetone to clean the surface increased differential shrinkage between the core and exterior and led to partial dissolution and subsequent spalling during drying. Application of very slight heat, for instance as microscope illumination, led to spalling and delamination of the glassy surface, adding another example to Bimson's warning about the danger of heating ancient glasses (34).

Essentially, a relatively static corrosion process is found, which is closer to aqueous corrosion than to atmospheric weathering as described by Clark et al. (35). The time is long, being over 4000 years, so that an equilibrium may be expected to have been achieved. The surface area to volume ratio is intermediate between the static weathering of stored window panes and the aqueous corrosion inside a closed beverage bottle--the models used by Clark et al. The moisture content of the soil is quite low, perhaps 20 vol%, and aqueous flow through the soil is negligible. The find spot was above the level of ground water, as well as the present plain surface, by 2 to 5 meters and below the tell surface by 10 meters. Both reactants and products appear to have remained in situ, the soil layer nearest the glass beads being alkaline to the taste. Both hydration and network breakdown have occurred with considerable local variability, and the residuum, which is visibly identifiable as a glass, may have effectively formed a gel or partially gelled state (36). The critical part of this problem is isolation of the mechanism, either condensation-evaporation (atmospheric weathering) or continuous ion exchange (aqueous corrosion), which occurs in the near surface layer of soil surrounding the glass, or perhaps both. Experiments are required in environmental monitoring of the soil parameters at the bead-soil interface and their variability with distance from the bead, in controlled drying rates on model glasses/gels with corresponding variations in relative humidity and other parameters, as well as characterization of the compositional and structural variability in these glasses.

A considerable number of important studies have investigated bulk and trace compositions of ancient glasses (37-41), but it is important to recognize that what has been studied is only the surviving material, and very likely not the range of composition melted in antiquity, nor the local range of variability within a single glass. Because the weak link in composition and structure corrodes or weathers first, it is important to determine and locate the rate-limiting (with respect to preservation), or rate-accelerating (with respect to deterioration), phenomenon (42-50). Some of the variation in structure of a single glass bead is shown in Figure 13-7. The bulk composition of the blue glass is 69 wt% SiO_2, 0.8% Al_2O_3, 14.9% Na_2O, 2.6% K_2O, 8.2% CaO, 3.0 MgO, 0.2% Fe_2O, 1.2% CuO, 0.1% As_2O_3 (51,52). This is a reasonable composition for optimal workability and durability, a model for which is 70-75% SiO_2, 10-15% alkaline earths, 15-20% alkalis. This glass, although corroded, is sufficiently hard to achieve a good polish. The variation in arrangement of gelled structures belies a complex prior history and calls into question simplistic concepts of equilibrium.

A Roman Glass Vessel from Samaria

On an example of a near-surface find of weathered glass from Samaria, West Bank, Israel (ca. 250-400 A.D.), the exterior surface can be differentiated into

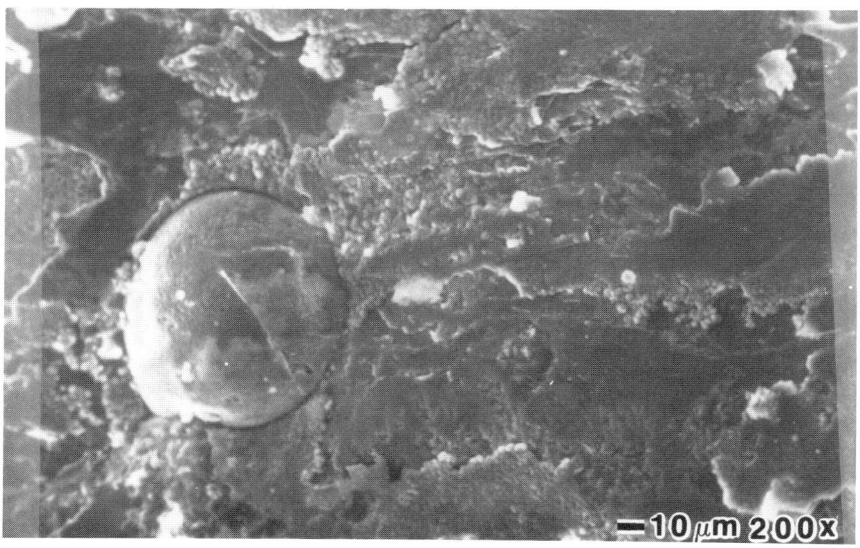

Figure 13-7. Section fractured parallel to the surface of a mid-second millennium B.C. turquoise blue and white glass bead from Nuzi, Iraq, showing fine gel-like particles corroded in layered and globular structures. (Figure 13-7 is continued on the following pages).

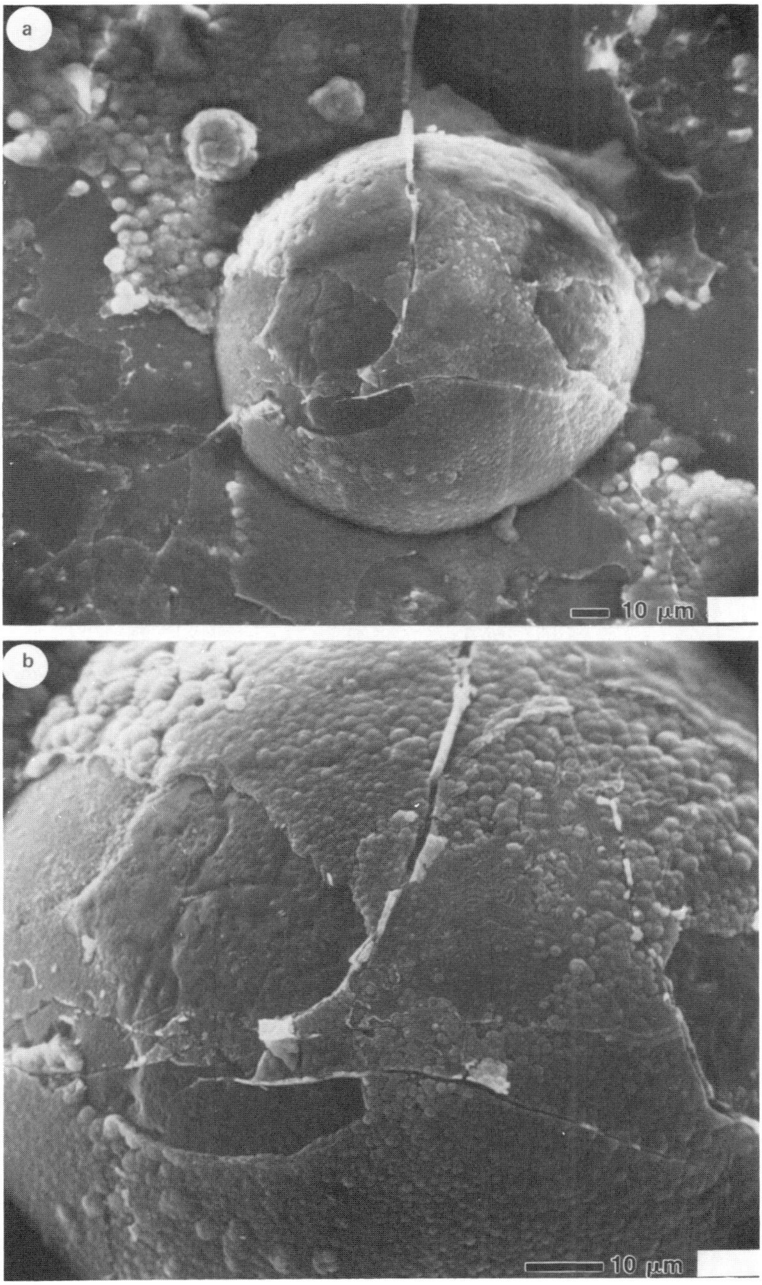

Figure 13-7 continued: (a,b) interior surface of a pore.

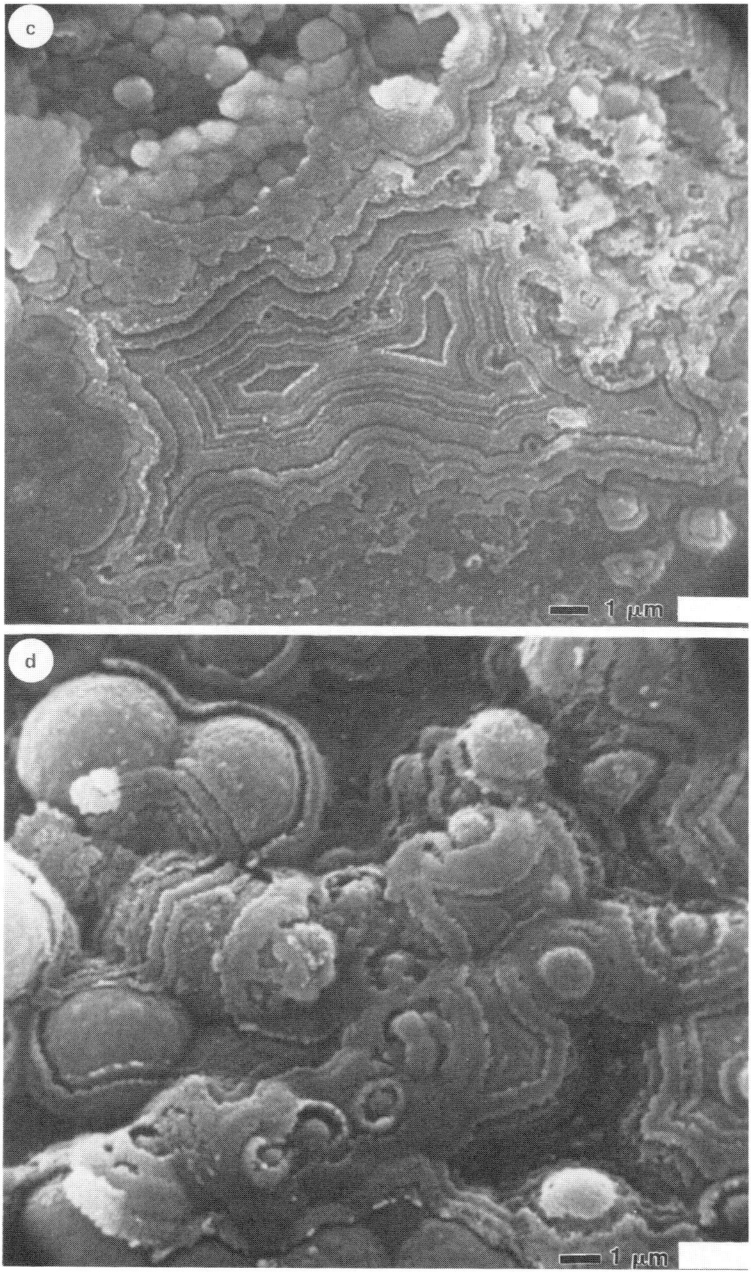

Figure 13-7 continued: (c,d) surface of weathering layer.

Figure 13-7 continued: (e) surface of weathering layer; (f) submicron gel particles.

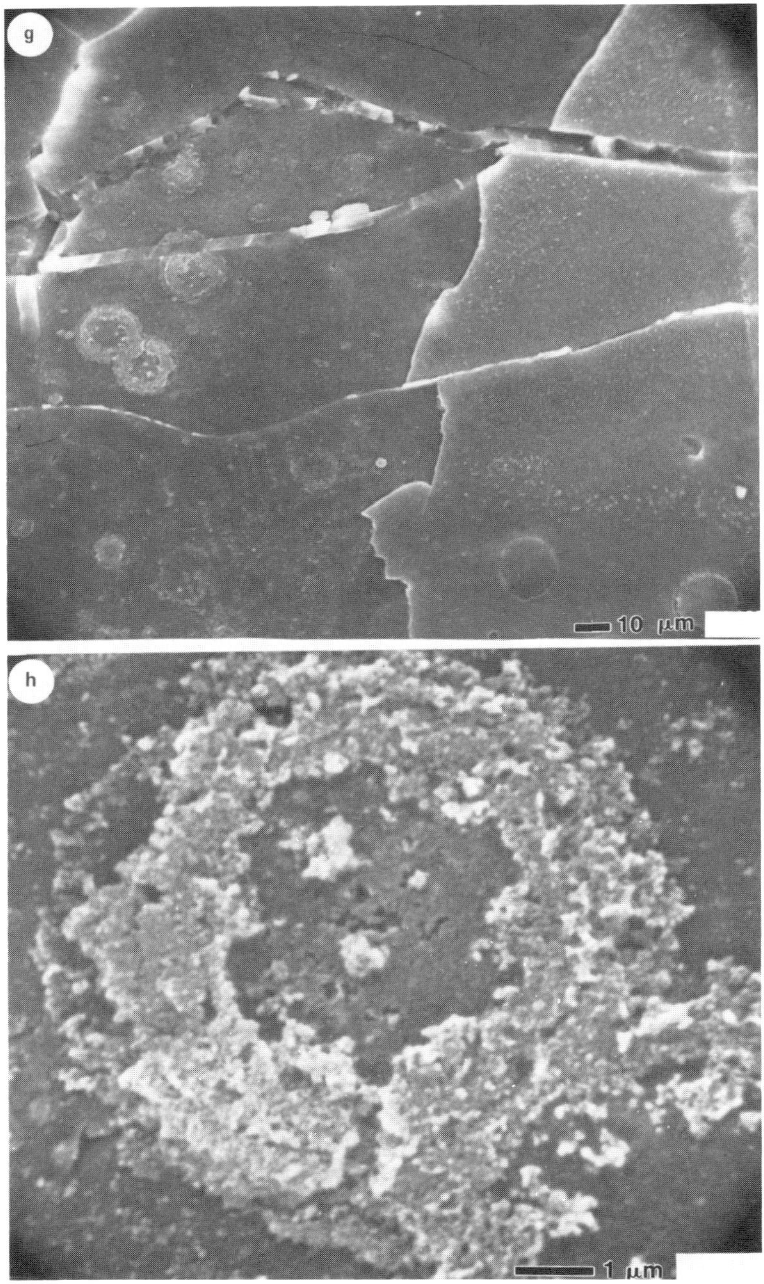

Figure 13-7 continued: (g,h) pits in corroded layer probably caused by re-solution of gel.

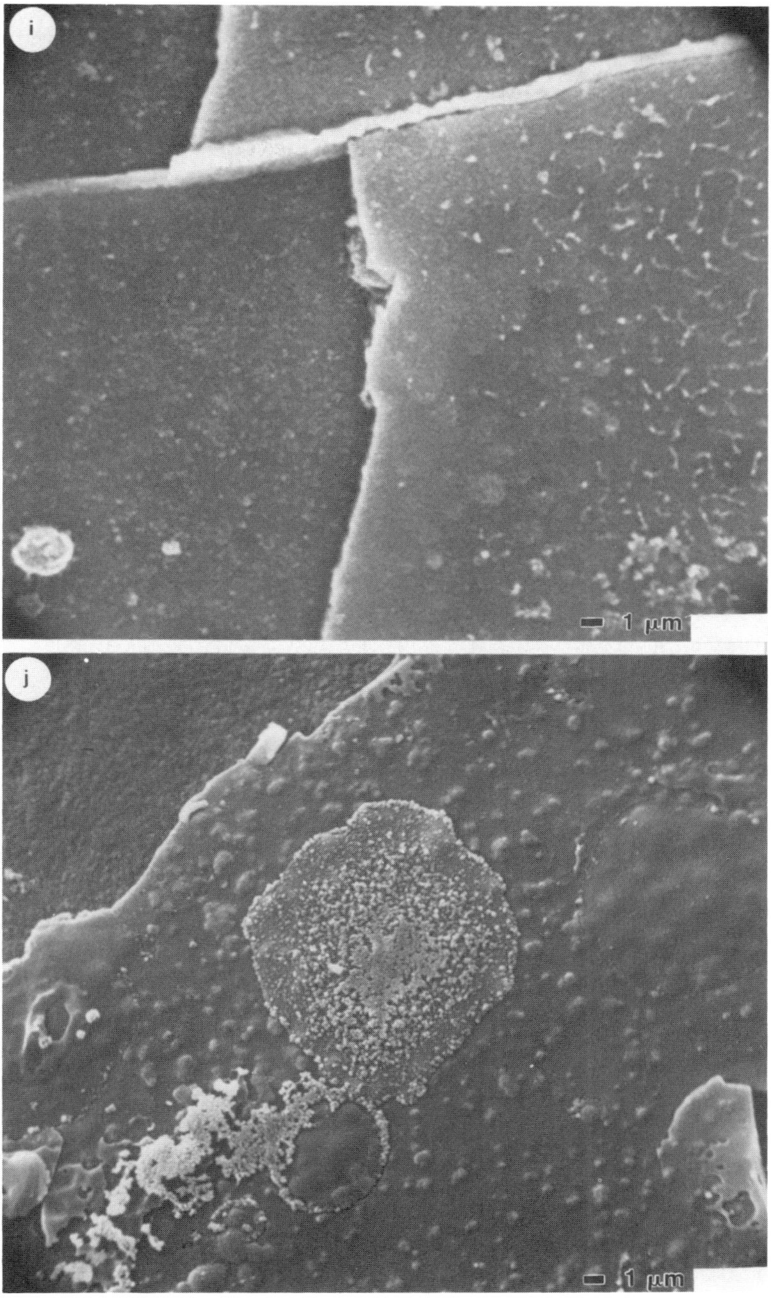

Figure 13-7 continued: (i,j) variation in structure of layers.

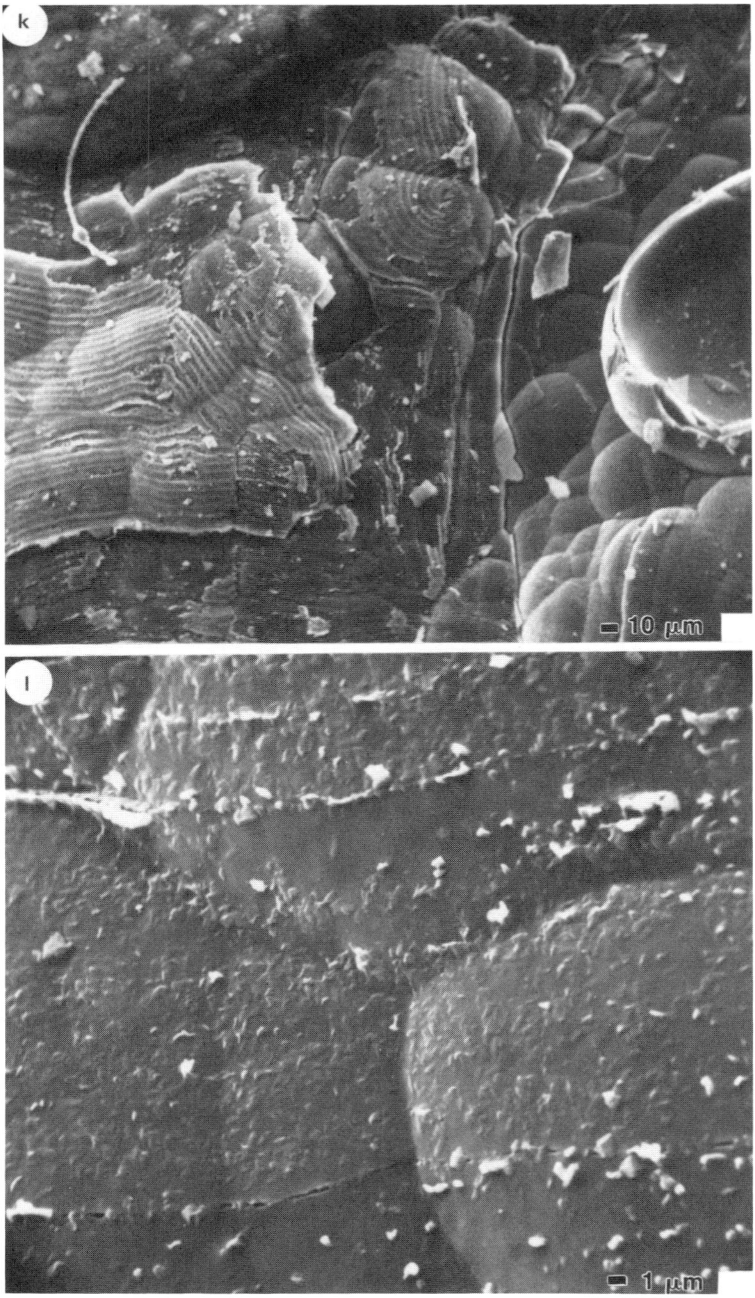

Figure 13-7 continued: (k) zoned layer structures extend through cracks in globules; (l) horizontal striae of original crystals of calcium antimonate opacifier in globular gel structure.

three regions by the type of weathering product (53). One region has hemispherically-shaped pits or craters which are not uniform in size and which vary in depth (Figure 13-8a). They usually overlap one another. The surface has more than fifty iridescent layers (Figure 13-8b) which are composed of submicron, rounded particles of gel. Another area is comparatively smooth and intact, with only small scratches and relatively little iridescence. However, various salts, dirt and dust particles are present, some of which are organic (Figure 13-9a, arrow) and give rise to only a background signal of continuum radiation in the EDS signal (Figure 13-9b). Some of the salt precipitates are disrupted (Figure 13-9a), and others show a structure consistent with interrupted growth (Figure 13-9b).

Near the boundary between these two areas and occurring on the smoother surface are small brown spots, or spherulites, which have a ringed structure containing a layered cross-section (Figure 13-10) and often a shrinkage crack at the boundary with the unweathered glass (Figure 13-10a, arrow). By applying a small force to the edge of one of these spherulites, the hemispherically-shaped cleavage develops, and the spherulite easily detaches from the surface, leaving a pit. In general, the largest spherulites tend to be smaller than the smallest single craters. Some pits are surrounded by a brown ring, indicative of an incompletely detached spherulite. Some spherulites are oval in shape and have elongated along the axis of a surface scratch.

The composition of spherulites shows increased concentrations of the oxides of iron, manganese and aluminum, decreased concentrations of the oxides of calcium and magnesium, and intermediate concentrations of silica, soda and potassia (Table 13-1). In another test using Auger electron spectroscopy on the dark brown rings of a spherulite, the local MnO concentration was about 20% and the FeO concentration 5%.

Table 13-1. Compositions of Samarian Glass.

	Scanned Fresh Fractured Surface (n=30)	Weathered Surface (n=30)	Spherulite (n=8)
SiO_2	68.48 (0.84)	86.30 (0.65)	80.97 (3.80)
Al_2O_3	2.61 (0.11)	3.96 (0.17)	5.76 (0.44)
CaO	6.74 (0.44)	5.48 (0.10)	3.97 (0.85)
MgO	0.68 (0.34)	0.55 (0.03)	0.34 (0.03)
Na_2O	19.73 (0.75)	0.30 (0.10)	0.49 (0.23)
K_2O	0.77 (0.16)	1.89 (0.48)	1.23 (0.39)
MnO	0.65 (0.07)	0.94 (0.03)	5.76 (2.08)
FeO	0.29 (0.05)	0.57 (0.06)	1.45 (0.54)
AsO	0.07 (0.05)	0.05 (0.02)	0.04 (0.04)
Total	100.00	100.04	100.04

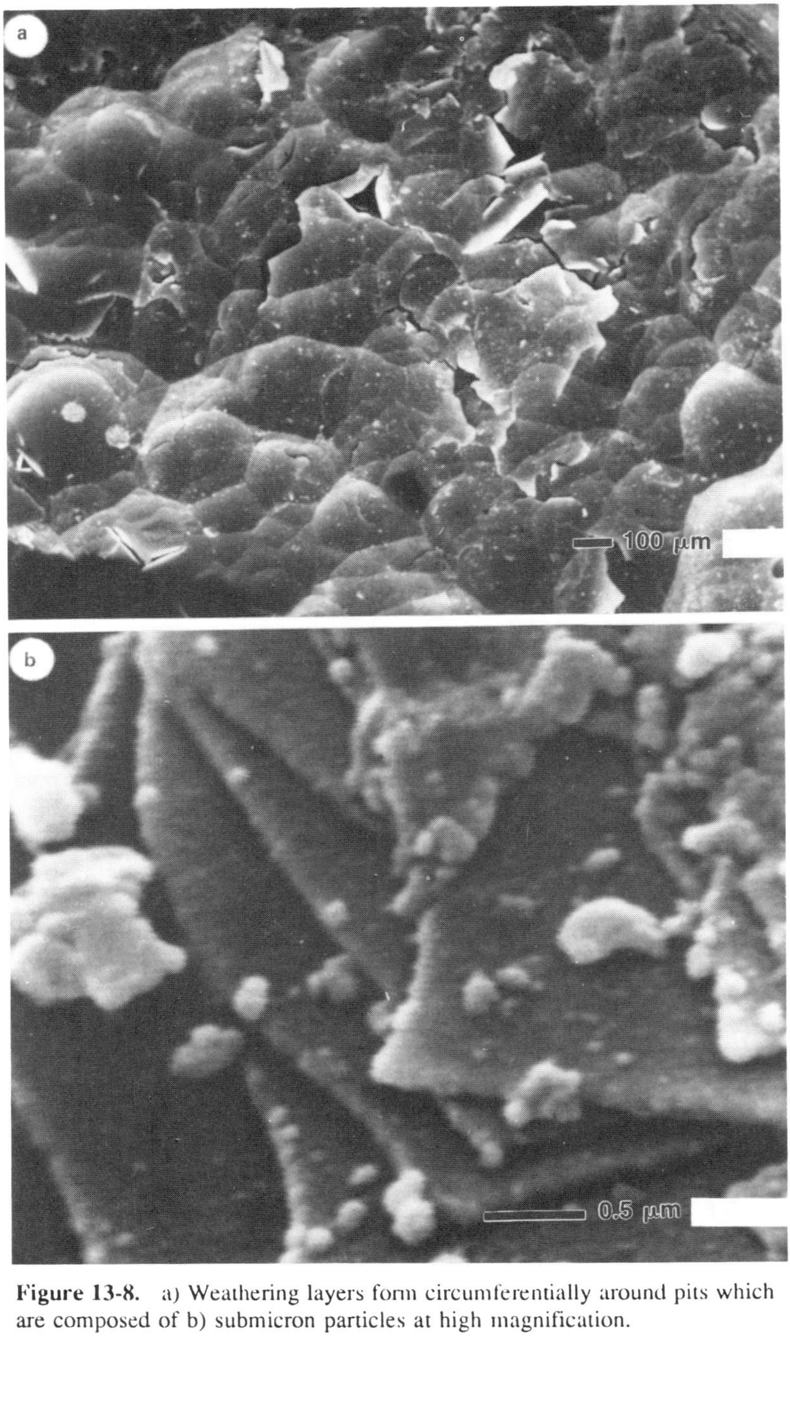

Figure 13-8. a) Weathering layers form circumferentially around pits which are composed of b) submicron particles at high magnification.

Figure 13-9a. Surface weathering processes include accumulation of acicular salts (white), rounded organics (upper left) and fine clay agglomerates (lower center, gray), as well as small pits probably from condensation-runoff (upper left, black). The structure of the salt precipitates is disrupted which would be likely during aqueous runoff or soil movement.

Figure 13-9b. These sodium-containing salts have grown to almost a micron when the conditions for growth were changed and growth slowed, as seen by the rounded, mushroom-shaped tops on the crystals, perhaps due to re-solution and re-precipitation. Horizontal lines are caused by the SEM scan generator and are not in the sample.

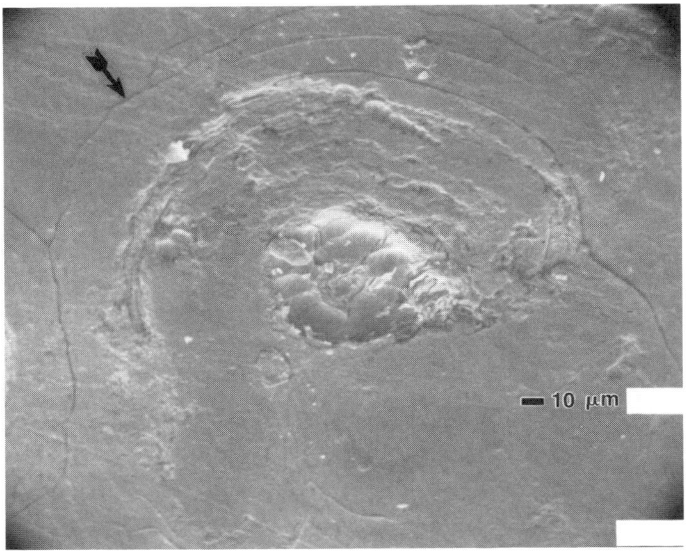

Figure 13-10a. Incipient pit formation with brown, circumferentially layered spherulite in exterior of a Roman glass. A shrinkage crack surrounds the spherulite (arrow). Accumulation of iron- and manganese-containing material occurs in the center. Later, the surface was swabbed with water. Drying shrinkage and syneresis later detached this spherulite, leaving a millimeter-sized pit. Cleaning with acetone produced the same result.

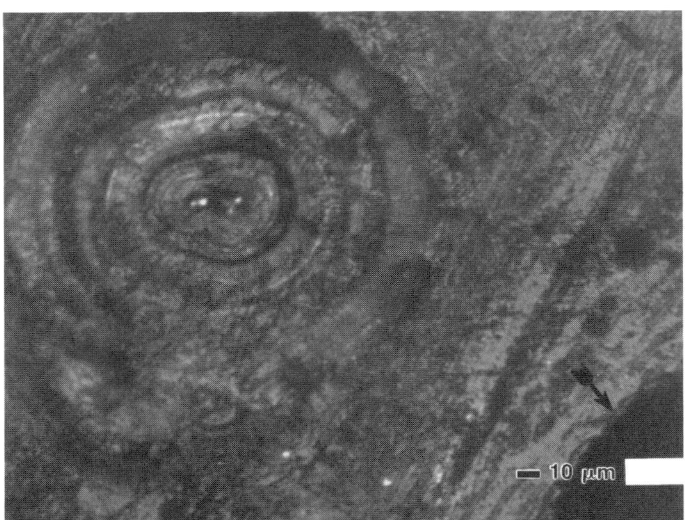

Figure 13-10b. Polarized light micrograph showing ringed structure, flush with scratched surface. Delaminated spherulite left a pit in lower right (arrow).

Based on structure, composition and properties, spherulites provide another mechanism besides surface abrasion that accelerates relatively slow surface weathering; the formation of millimeter-diameter, gelled spherulites leads to delamination of about a half millimeter of surface. Clay, transition metal ions, salts and organics from the soil and salts which have precipitated from the alkalis leached from the glass all aggregate where water droplets have evaporated in the alkali-rich and rough areas of the surface. They accumulate to form irregular islands of surface deposit with an increased surface area which act as microenvironments for corrosion. During successive alternations of rainy and dry seasons, such heterogeneities repeatedly serve as chemically active sites for weathering at an increased rate compared with the rest of the surface, and give rise to the growth of spherulites in the glass.

Glass, buried in the near-surface level of a site with occasional rainfall and farming activity, deteriorates by a weathering process involving condensation-evaporation and condensation-runoff. In the rare sample where both pitted and relatively pristine surfaces are found, the process of spherulite development can be observed often, and its function in initiating pitting can be isolated. The identification of such crystals growing from the surface as shown in Figures 13-9a and b remains to be made, and the role of such crystals in gel formation remains to be studied.

Glazes on Pottery and Tiles

In a soda-lime-silica glaze from a large storage jar with knobbed and incised decoration, made at Nippur, Iraq, during the Parthian period (ca. 200 B.C.-200 A.D.) and fired between 850 and 1000°C, the surface appears a bright, glossy turquoise blue. This vessel was found near the surface, about 1-1/2 meters down, in relatively dry, sandy soil. The glaze exhibits conchoidal fracture patterns when broken and has a mohs hardness of 5, a little softer than window glass, indicating that it is a reasonably durable glass. The surface is crazed with a fine network of cracks about 5mm apart which form a pattern of interconnected "T" cracks characteristic of tensile fracture. These cracks provide a large and confined surface area for sustained chemical reactions.

Even though no corrosion other than a slight dulling and some scratches is visible on part of the surface, considerable corrosion has occurred at the craze lines (Figure 13-11). In SEM backscattered emission a difference greater than two in average atomic number can be seen between the darker, lower "z" glass in the near-crack region and the lighter, higher "z" glass in the bulk. The corroded glass has lost alkali and alkaline earth ions, according to energy dispersive x-ray analysis, and the silicate network has partially broken down and shrunk, as evidenced by a second set of fractures which occur in the corroded glass near the boundary with the less-affected bulk glass (Figure 13-11a). The corrosion front appears to have moved out into the bulk glass from certain points or pits as a concentric and eventually coherent-wavefront, as can be barely seen

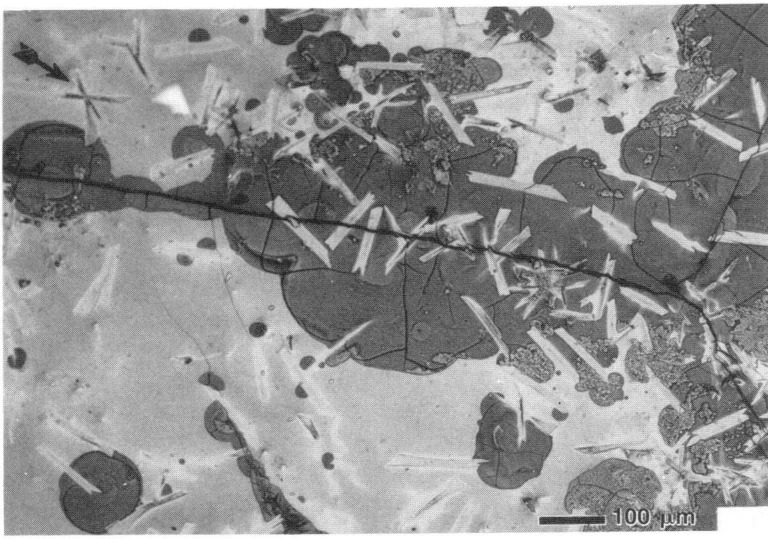

Figure 13-11a. Surface of Parthian turquoise blue, soda-lime-silicate glass from Nippur, Iraq, (ca. 200BC - 200AD). Image in backscattered emission shows deteriorated, low atomic number glaze composition (dark gray) along craze lines with many shrinkage cracks and gypsum crystals which lie on are etched into the surface (arrow).

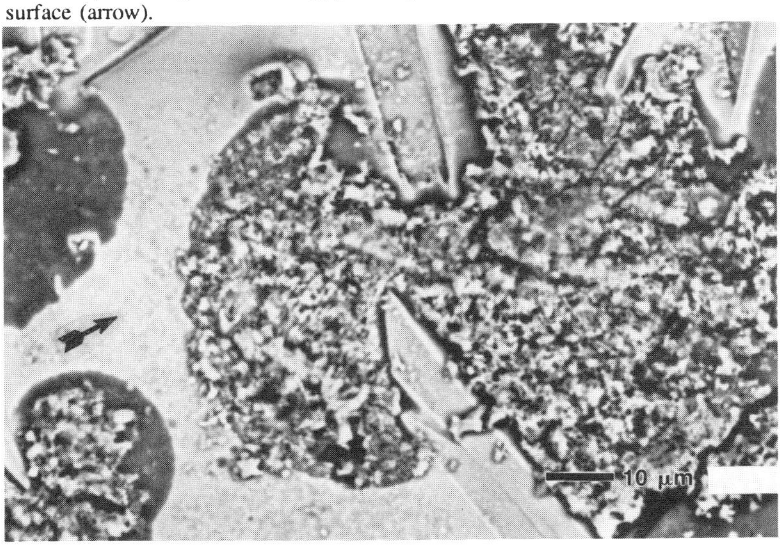

Figure 13-11b. Backscattered image of corrosion products adhering to weathered glaze (mottled accretion, right) and gypsum needles which appear first to inhibit, then to accelerate corrosion (from lower left of Figure 13-8). Note submicron pits of lower z in base glaze (arrow) indicative of weathering. Weathered gel is to left (gray).

in the rings in the corroded glass at the center of Figure 13-11a. At high magnification, submicron particles characteristic of a gel can be seen in the corroded glass.

The surface of the corroded areas of glass is partially coated with a layer of porous, micron-thick particulate corrosion products. The shard was washed in water with a brush, such that much of the layer of corrosion product which originally covered the corroded glass has been removed. Acicular crystals of a calcium-magnesium-silicate have precipitated from the soil and/or glass onto the surface and are etched into the surface. Those crystals which are loosely attached to the surface appear to have slowed the corrosion front; those which are embedded in the surface appear to have surface active corrosion sites at the three-phase boundary interfaces. The surface of the glass shows evidence of pitting corrosion on a submicron scale with associated local compositional variability and loss of alkalis in the pits (Figure 13-11b, arrow). The corrosion structures are similar to those found on the glasses from Samaria and Nuzi, although both examples are simpler than the complex behavior of the Nuzi glasses.

Similar phenomena also occur in a green-glazed, porcelainous celadon made during the twelfth century A.D. in Kangjin, Korea, as a large, closed-mouth Meiping jar with white and black inlaid decoration. No evidence of deterioration can be seen on most of its surface, with the exception of a small area at the base of the foot. This celadon fragment, as is true of most celadon vases, has been washed many times and has been judged by conservators to be stable in ambient climate as well as most storage conditions.

Examination of the surface by SEM, however, shows that corrosion processes are causing deterioration of the glaze in areas with no visible signs of corrosion. Small 10-30 micron quartz particles have floated to the surface during firing and provide a subtle relief texture which adds to the translucent, jadelike appearance (Figure 13-12a). At 573°C during the cooling ramp of the firing, quartz transforms by rotating bonds from the beta to alpha phases with a concurrent 2% volumetric shrinkage. This amount of shrinkage is sufficient to form tensile cracks around and/or through the quartz particles near the quartz-glaze interface. Such cracks provide active sites for localized corrosion (Figure 13-12a and b). The surface also contains small pits which occur in regions of higher calcia and silica and lower potassia and alumina than the base glaze, as shown by EDS spectra. In some regions, needle-like crystals are arrayed radially around small pits or craze lines and more rarely have grown at scratches (Figure 13-12c). Sometimes alkaline salt crystals and transition metal ions, such as iron, are found in the center of a concentration of salt crystals which provide an active corrosion site which reacts with and etches into the glass (Figure 13-12c, arrow).

In the corroded region of the jar near the base of the foot, extensive corrosion has enlarged craze lines, scratches and pinholes (Figure 13-13a and b). Pitting corrosion has occurred but without formation of layered spherulites.

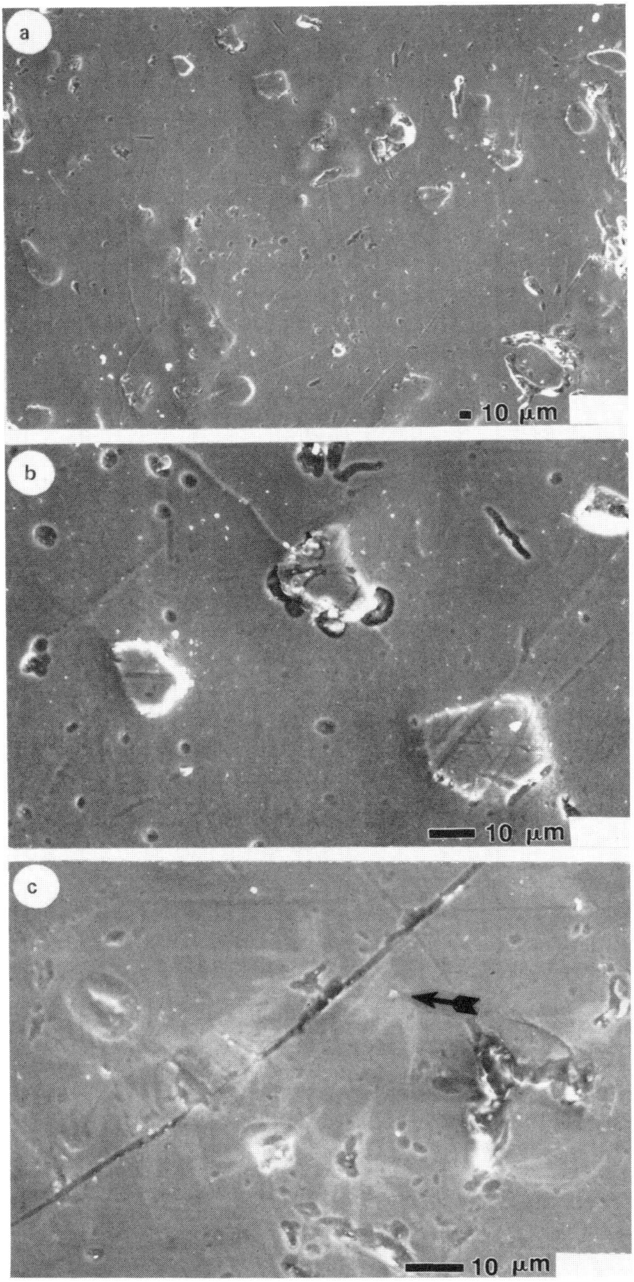

Figure 13-12. Surface of Korean Koryo dynasty celadon glaze from Kangjin, about 12th C. A.D., which shows no visible evidence of weathering. However, microscopic evidence of deterioration includes: a) small pits, scratches and preferential corrosion around quartz particles which have floated to surface during firing where some have been surrounded by circumferential crack formed during quartz inversion on cooling, b) pits in the glaze occur in high CaO regions, c) transition metal ion and alkali-containing salts (light gray) precipitate along scratches and pits. Often a large particle of the mixture will precipitate in the center (arrow) providing an active corrosion site which etches into the glass. Etch pits around quartz particles are present to the upper left and center right of the crack.

Pinholes, which are bubbles in the glaze which have broken open but which could not heal over during firing because the glaze was too viscous, provide active sites for corrosion, as the accumulation of calcium and other salts and presence of Cl and S in pores shows (Figure 13-13a). Pits around quartz particles open further in the more corroded region, sometimes leading to spalling of the quartz (Figure 13-13b, arrow).

Areas on each side of craze lines at the base have discolored to brown. The surface, although particulate, is similar in shape to the pitted surface seen in Figure 13-8a. The surface is composed of agglomerated particulates with some residual structures which still appear vitreous (Figure 13-13c, upper left and lower right). These more compacted gels are presumably more durable compositions and contain more potassium and alumina; whereas the more porous, less durable regions are higher in calcium and silica. This same compositional variation is found in both the early and final stages of corrosion (compare Figure 13-9b with 13-10c). The glaze is inhomogeneous, probably as a result of raw materials and manufacturing technology. The glaze was composed of particles of limestone and China stone - a high potassium aluminum-silicate - from which sericite and mica weather (in the geological sense), which were ground and mixed prior to application. The variations in composition are very minor, because backscattered imaging which detects a variation in average atomic number of 2 does not show the difference as gray contrast. However, the significance of the unexpected higher potassium concentration in the more durable areas of glaze is that the potassium is bonded with alumina and silica in a "tight" and durable glass with a compact crystalline precursor structure; whereas the less durable, high calcia and silica glass derives from a less homogeneous and less dense mixture of ground limestone, quartz and China stone.

Of the relatively durable porcelain and stoneware glazes of East Asia, the Yue and Koryo celadons are the least durable, with some Guan and Longquan celadons being slightly more durable. These glazes contain more lime than other oriental glazes, which gives them a higher thermal expansion and a likelihood of crazing. The Koryo Meiping described above has an average composition of 59.6 wt% SiO_2, 12.3% Al_2O_3, 17.6% CaO, 2.7% MgO, 3.8% K_2O, 0.4% Na_2O, 1.2% FeO, 0.6 MnO, 0.9% P_2O_5, 0.1% TiO_2 and 0.07% SO_3 (totaling 99.17%), which composition does not include the minor amount of quartz and magnetite particles in the glaze (54). The range of calcia values is about 3%, with alumina being about 2.6% and potassia less than 1% in 15 3-5 micron samplings, thus being quite homogeneous. The glaze was fired between 1100 and 1150°C, and, with the exception of crazing and minor pinholing, is virtually free of defects. The high lime content of the glaze relative to the body is the major difference. The expansion coefficient of CaO is high and causes the glaze to shrink on cooling sufficiently more than the body such that the glaze is placed in tension, and its fracture strength is exceeded (55-57). In extreme cases the glaze delaminates from the body in flakes, as is common in American and European

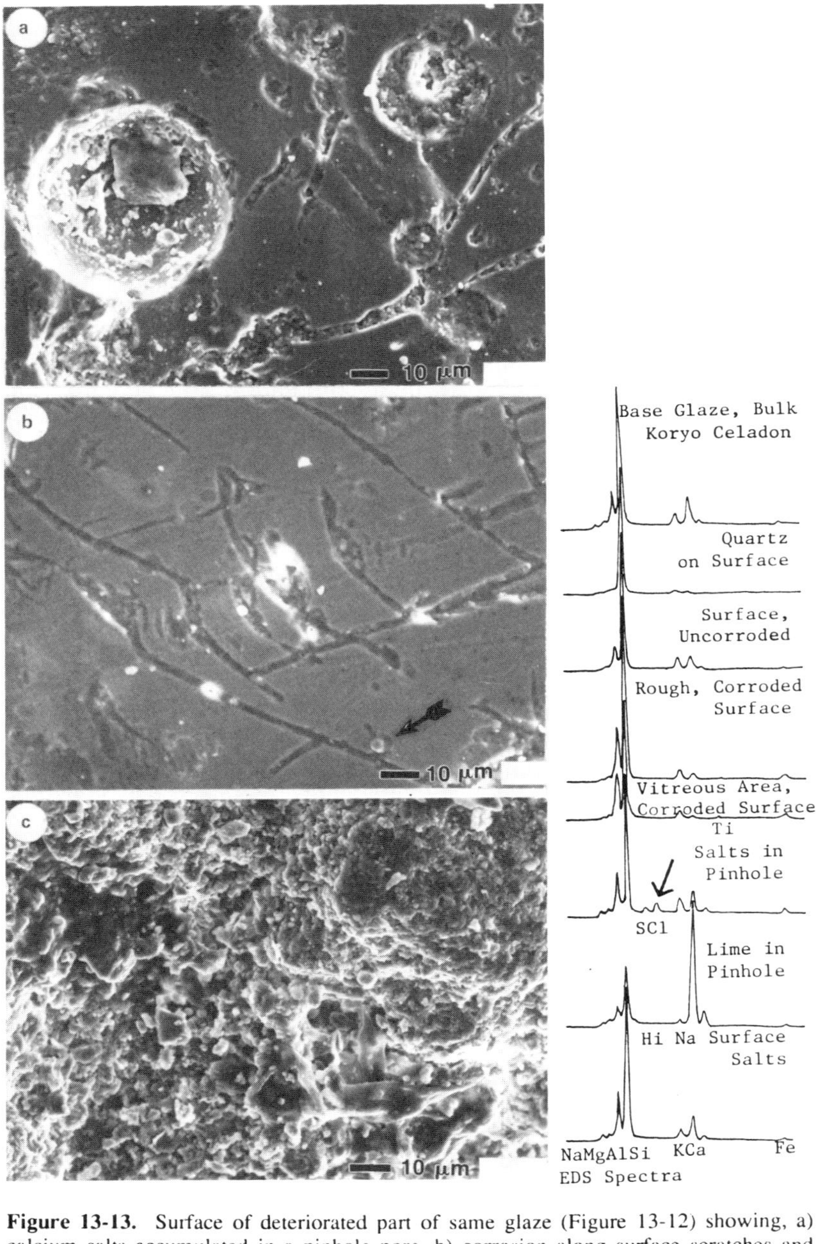

Figure 13-13. Surface of deteriorated part of same glaze (Figure 13-12) showing, a) calcium salts accumulated in a pinhole pore, b) corrosion along surface scratches and around quartz (arrow), c) deteriorated surface with residual vitreous structure in the high alumina and potassia regions of upper left and lower right.

tiles made before the 1930's. At that time solutions were formulated by F.H. Norton and others which consisted of adding low expansion materials, like borax, to the glaze and slowing the cooling rate and adjusting the glaze fit sufficiently to allow for moisture expansion of the earthenware body. The opposite case of glaze compression and body tension causes shivering or wrinkling of the glaze and in extreme cases dunting, or breaking of the ceramic into shards during cooling in the kiln.

Crazing is a well known phenomenon, but it is not the only cause of glaze cracks. A porous ceramic body may expand due to the absorption of moisture. The moisture expansion of the body places the glaze in tension which, if sufficiently large, results in delayed crazing (55-57). Delayed crazing may occur years after the ceramic was made. If there are inclusions or other heterogeneities, such as cord or striae, in the glaze, these concentrate forces locally and provide paths for crack growth. Thus, the presence of glaze cracks indicates either of these two phenomena, but the cause cannot be determined by visual or microscopic inspection. The worst treatment for a crazed glaze is to add more moisture by washing, soaking or swabbing. Thus even though getting the salts out of a porous ceramic body is often recommended for the body, it is deleterious to the glaze.

Crazing is not caused by a simple contraction of the glaze on cooling, but instead only what occurs below the strain point of the glaze. Strain point is defined as a viscosity of $10^{14.5}$ poise cooled at 4°C/min at which point the glaze acts like a rigid solid but internal strain can be relieved after several hours at the corresponding temperature. Above or prior to this point, the glaze flows like a glass to relieve stress. Below this point, glaze fit can be predicted by calculating the stress on the body and glaze and comparing them with the Modulus of Rupture for the glaze, typically about 11,000 psi. The change in length is the product of the coefficient of linear expansion and the temperature gradient. The stress on the glaze, assumed to be a linear elastic material, is an application of Hooke's Law which for glaze stress calculation for a soda-lime glaze from an earthenware body from Nuzi is given as:

$$\sigma_{gl(TOT)} = E_{gl}\Sigma_{overall} = E_{gl}(T_o\text{-}T')(\alpha_{gl}\text{-}\alpha_{bd}) + E_{gl}\Sigma_{ME}, \tag{13-1}$$

where: σ_{gl} = stress on glaze due to thermal expansion
= +13,403 psi > Modulus of Rupture = 11,000 psi, therefore, crazing occurs

$\sigma_{gl(TOT)}$ = stress on glaze due to thermal and moisture expansion = +18,478 >> MOR
therefore, severe crazing develops

σ_{bd} = stress on body due to thermal contraction
= -502.5 psi (compressive)

E_{gl} = modulus of elasticity = 10,150,000 psi

T_o-T' = strain point temperature less ambient
= 500 - 25°C

α_{gl} = coefficient of expansion of the glaze
= 105.8 x 10^{-7}/°C

α_{bd} = coef. of expansion of the body
= 7.8 x 10^{-7}/°C

Σ_{ME} = strain due to moisture expansion
= 0.0005

Some typical values calculated for the thermal expansion are soda-lime silicate glasses, 92-110 x 10^{-7}/°C; alkali-lead glasses, 90; lime-alumino-silicate or porcelain glazes, 45-50; Egyptian faience glazes, 40; and borosilicate glasses, 33. Some typical values for moisture expansion of earthenwares fired to about 900°C are 0.01-0.05% change in length (56). Similar calculations show that high lime porcelainous glazes, soda-lime-silicates and lead alkali-silicates on earthenware will craze due to thermal expansion which will only be exacerbated by moisture expansion, and that low-lime porcelain glazes, Egyptian faience and borosilicate glazes will not craze due to thermal expansion, but if on a porous enough substrate they are predicted to craze due to moisture expansion.

Enamels on metal and enamels on glazes (overglaze enamels) display the problem of crazing, but not delayed crazing because the substrate is not porous. They are composed of lead glasses of highly variable composition which have been separately compounded and fritted to have nearly the same melting temperature. Because of the high lead content, they scratch easily, and many are neither durable nor stable glasses. Overglaze enamels are typically in tension and delaminate in flakes from the higher fired, usually porcelain glaze. The expansion coefficients of enamels on metals are high for glasses and glazes, but not as high as those of metals, such as copper, silver or gold. The higher contraction of the metal places the enamel in compression, but often delamination and spalling occurs, unless enamel is placed on both sides of the metal and special intermediate grip or flux coats are used as intermediate layers. A special problem in the conservation of cloisonne enamels is that the grip coat which is soluble effloresces at metal-enamel joints. This problem is aggravated if the object was not fired high enough. To avoid the problem of spalling, many enamels on metals, particularly metal jewelry from pre-Roman western contexts were preformed pieces of glass which were glued in place with bitumen or another adhesive. Another ancient solution was the use of quartz frits as inlaying materials which were either fused or glued in place. Such mixed material interfaces provide active sites of corrosion, and their stabilization requires close observation and special effort to not allow accumulation of moisture.

TREATMENTS

Historical Perspective

Past solutions to the conservation treatment of ceramics have concentrated on cleaning, joining and stabilizing (59-67). Practice has been to glue broken shards or fragments with such reversible adhesives as acrylics and acetates in varying concentrations of acetone to increase penetration of the adhesive into the ceramics to be joined. Joining follows cleaning during which dirt and salts are removed by soaking, poulticing, swabbing, picking or other mechanical or chemical procedures. If necessary, iron stains are removed. In general, the less durable, more fragile and friable the glass, glaze or ceramic, the less water or other cleaning agent is used. In unfired clay tablets, firing is carried out before soaking to remove salts (66,68). The temperature needs to be determined experimentally. If calcium carbonate is present in the surface accretions or clay body, only a narrow range of firing temperature is possible (Figure 13-14).

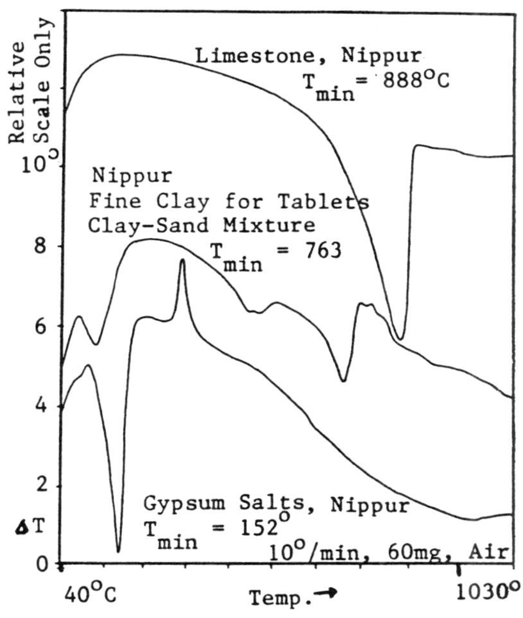

Figure 13-14. Differential thermal analysis (DTA) plots of gypsum, limestone, and a clay-sand mixture suitable as a tablet clay. In firing tablets from Nippur a narrow range separates clay dehydroxylation from calcium carbonate decomposition, which if CaO rehydration and recarbonation occurs, leads to spalling after firing. For these tablets fired at 700°C is too low for the clay to dehydroxylate. 750°C is about right, and 800°C is too high.

Solutions with the least intervention are preferred, the aim being to stabilize the object and to consolidate it with coatings or impregnation only where required for handling. Originally plaster was used to replace missing parts and extensive restoration and in-painting was carried out to fill in the design. At present, such restoration is minimized and lighter density, moldable plastics are being used for reinforcement.

Only pieces of glass which are very thin or have sprung open are glued with a relatively irreversible adhesive, such as epoxy, which has been formulated with a refractive index similar to that of the glass to be repaired. The goal is to employ reversible treatments which are recognizable at close view but which are aesthetically consistent with the object when viewed at a distance (1).

Most polymer coatings do not provide moisture barriers to prevent further weathering of glasses. Some solutions, particularly in the case of Medieval stained and leaded glass windows, have been to replace missing or badly corroded glass, to abraid or etch badly corroded glass, as well as to glue fragments within a leaded frame and to in-paint missing areas of glass enamel or stain. Mica has been tried as a moisture barrier in a polymer coating (69). In addition, solutions of encasing the glass window with protective glazing on the exterior have been employed.

Future Prospects

Another direction which has not been tried is to extend the insoluble quartz-like or feldspar-like part of the glass structure by laser, CVD or sol-gel methods. This is a more long term solution, if it works, but such a suggestion involves a greater risk because of the lack of reversibility and requires a thorough empirical investigation of the features and mechanism, and rates of corrosion of particular glass systems, such as the investigations of Brill into high-potassium "weeping" glasses (65).

Industrial solutions to glass corrosion have been to etch surfaces to blunt crack tips, to slowly cool glasses to increase their density over that obtained during rapid cooling and to allow sufficient time for proper annealing to eliminate internal stresses, and thus make a glass more durable. Another solution has been to place the surface in compression such that the fracture strength of the glass which is usually about 10,000 or 11,000 p.s.i. will increase. Thermal tempering gives a three-fold increase in strength, ion exchange gives greater compressive stress at the surface but is active only over a relatively small diffusion distance of 10-50 microns. Pristine glass which has just been manufactured gives a ten-fold increase in strength. If handled or allowed to sit in air the strength rapidly degrades.

Given the composite nature of most somewhat-degraded ancient materials, and the prohibition against heating such fragile composites, a possible scenario for future conservation treatments of ceramics might be the following:

Stabilize
- Control temp and RH to maintain intermediate values
- Do not wash, pick clean
- Stabilize corrosion products, gypsum and salts as well as glasses and gels

Prevent further reaction
- Reduce surface area by etching or laser treatment
- Compress surface with ion exchange
- Make the glass surface inert: extend the glass surface by building up nonreactive silicate structure, perhaps by CVD

If necessary, add a moisture barrier coating, such as mica flakes

If necessary, consolidate with vapor phase under vacuum to get penetration, and not depend on liquid penetration by relatively weak capillary forces

SUMMARY

Ancient ceramics provide many examples of deterioration which has occurred over long periods of time. Many phase-change paths have been isolated for glasses and glazes, such as vaporization-condensation and condensation-runoff in atmospheric weathering, and ion exchange (hydration and modifier leaching) and silicate network breakdown in aqueous corrosion. Many physical processes accompany structural changes, such as gelation of glasses, rehydroxylation of the clay fraction and deterioration of the glassy fraction of low-fired or porous ceramics, crazing of glazed ceramics, Ostwald ripening and liquid-phase sintering in plasters. Most ceramics retain a memory of initial structure and composition during corrosion, but surface chemistry and surface roughness overcome such initial heterogeneities. Because deterioration is produced by the interaction of many variables, several of which cannot be measured, ancient ceramics are difficult to characterize. The factors which most affect degradation are:

Raw Materials
- Composition, Surface or Bulk, Local or Average
- Homogeneity, Degree of Natural Sorting, Mixing and Refining
- Heterogeneous Phases and Impurities

Manufacturing Technology
- Processing, preparation (fineness of grinding, mixing), forming and finishing, surface decoration or treatment, and thermal history (fritting, annealing),

- Properties and Interactions of each material in a composite and their manufacturing histories
- Impurities and heterogeneities introduced by processing

Use-Wear and History of Deterioration

- Prior History of Use and Reuse: breakage, surface alteration with respect to surface roughness and composition
- Post-deposition Processes, e.g. trampling, displacement, soil formation/erosion
- Excavation and Conservation
- Display and Storage

Environment

- pH of solution
- Relative humidity
- Temperature
- Continuous or cyclical time period
- Composition and structure of environment
- Energy inputs: stress, light, radiation, EM field

Much of the range and scope of our thinking is limited to our lifetimes, and conservation has been no exception. Conservation treatments frequently are judged successful if they will stabilize an object for 100 years; yet, once the decision has been made that an object is sufficiently significant that it should be kept, we should be envisioning longer duration treatments of perhaps 1000 years which will both stabilize and preserve artifacts. Such treatments pose a moral dilemma because they will require greater risk and have less chance of reversibility. Most importantly, insuring artifact stabilization and longevity will require further studies of the mechanisms and rates of corrosion.

Another moral dilemma is posed by archaeological investigations which cause objects to be excavated which cannot or will not be preserved; the decision of what to keep is extremely difficult. The function of preserving some objects is both their appreciation and the learning which understanding them requires, which is not the same as knowledge of their use. It involves a reconstruction and interpretation of material culture and its cultural context. Preservation in the archaeological record depends on the simultaneous coming together of chance preservation and chance discovery, rather than intelligent choice.

For the future, however, many such choices are not only possible but necessary. We are accumulating large quantities of ceramic, glass, plaster and cement waste products, only a small part of which is being reused. We can also decide what modern ceramic and glass objects, and what aspects of their cultural contexts, we would like to keep as teaching tools and as adequate representations of our time. The commonplace criteria for selection, collection and conservation are artistic, cultural, religious and historical significance, and we have the choice of enlarging these to include such other criteria as economic, daily life (folklife), science and technology.

ACKNOWLEDGEMENTS

Grateful acknowledgement is given to many people who inspired or helped with this study, including D.R. Uhlmann who years ago taught me about glass structure, W.D. Kingery who made me think about glaze fit, Lisa Klein and Christine Wang with the Samarian glass, Blythe McCarthy with the glazes from Nippur, and Alan Parks, Microprobe Lab, M.I.T. Department of Geology, Eugene Jarosewich, Mineral Sciences Probe Lab, Museum of Natural History, Smithsonian with the probe analyses. Samples were entrusted to this study by Carney Gavin, Harvard University Semitic Museum (Samarian and Nuzi glasses, Iraq), Andrew Moore, Yale University (plasters from Abu Hureyra, Syria), McGuire Gibson, Oriental Institute (glass and glazes from Nippur, Iraq), and Bohuslav Klima, Archeologicky Ustav (ceramics from Dolni Vestonice, near Brno, Czechslovakia). Thanks to Harold Dougherty for printing the photographs.

REFERENCES

1. *The Murray Pease Report: The Code of Ethics for Art Conservators*, pp 55-68, American Institute for Conservation of Historic and Artistic Works, (1968, rev. 1979).
2. Binford, L., *Bones: Ancient Men and Modern Myths*, Academic Press (1981).
3. Klein, L.H., *Experimental Determination of Stone Tool Uses*, University of Chicago (1980).
4. Hayden, B., *Lithic Use-Wear Analysis*, Academic Press (1979).
5. Vandiver, P.B., and Koehler, C.G., in: *Ceramics and Civilization* (W.D. Kingery ed.), Vol. 2, pp 173-216, A. Cer. S. (1986).
6. (a) Vandiver, P.B., Soffer, O., Klima, B., and Svoboda, J., *Science* 246: 1002-1008: (1989); (b) ibid., to be published in *Ceramics and Civilization* (W.D. Kingery ed.), Vol. 5, A. Cer. S. (1990).
7. *Colloquium for Lithic Cultures, Excavation Report of Zazaragi Site I*, Tohoku, Sendai, Japan (1978).
8. Stolyar, A.D., *Proishozhdeniie Izograzitel'nogo Iskusstva*, pp 211-226, Iskusstvo, Moscow (1985).
9. Vasil'evskii, R.S., and Ermolova, N.M., in: *Paleolit Sibiri* (R.S. Vasil'evskii ed.), p 65, Nauka, Novosibirsk (1983).
10. Abramova, Z.A. et al., in: *Arkheologischeskie Otkrytia 1973 goda* (B.A. Rybakov ed.), pp 38-40, Nauka, Moscow (1974).
11. MacKenzie, R.C., *Differential Thermal Analysis*, Vol. 1, pp 517-8, Academic (1970).

12. Heller, L., et al., *Clay Miner. Bull.* 5: 56-72 (1962).
13. Newman, A.C.D., *Chemistry of Clays and Clay Minerals*, pp 338-344, Wiley and Mineralogical Society 6 (1987).
14. Gourdin, W.H., and Kingery, W.D., *J. Field Archaeol.* 2: 133-150 (1975).
15. Thomson, G., *The Museum Environment*, pp 84-85, Butterworths (1978).
16. Kingery, W.D., Vandiver, P.V., and Prickett, M., *J. Field Archaeol.* 15(2): 219-244 (1988).
17. Kingery, W.D., Bowen, H.K., and Uhlmann, D.R., *Introduction to Ceramics*, pp 414-420, 425-431, 469-513, and 572-573, Wiley (1976).
18. Tubb, K.W., *Mitteilungen der Deutschen Orientgesellschaft* 117: 117-134 (1985).
19. Margolis, S.V., Preusser, F., and Showers, W.J., in: *Materials Issues in Art and Archaeology* (Sayre, E.V., et al. ed.), M.R.S. Proc. Vol. 123, pp 53-58 (1988).
20. Winkler, E.M., and Vos, B.H., in: *Applications of Science in the Examination of Works of Art* (W.J. Young ed.), pp 139-153, Boston Museum of Fine Arts (1970).
21. *The Treatment of Stone, Proc. of the Meeting of the Joint Committee for the Conservation of Stone* (R. Ross-Manaresi, and G. Torraca ed.), 1971, ICOM-ICOMOS (1972).
22. *The Conservation of Stone II, Centro per la Conservazione delle Sculpture all'Aperto* (R. Ross-Manaresi ed.), Bologna (1981).
23. Newton, R., and Davison, S., *Conservation of Glass*, p 135, Butterworths (1989).
24. Douglas, R.W., and Frank, S., *A History of Glassmaking*, Foulis (1972).
25. Harden, D.B., *Roman Glass from Karanis*, Univ. Michigan (1936).
26. Harden, D.B., Painter, K.S., Pinder-Wilson, R.H., and Tait, H., *Masterpieces of Glass*, British Museum (1968).
27. Oppenheim, A.L., *Glass and Glassmaking in Ancient Mesopotamia*, Corning Museum of Glass (1970 and 1989).
28. Moorey, P.R.S., *Materials and Manufacture in Ancient Mesopotamia: The Evidence of Archaeology and Art: Metals and Metalwork, Glazed Materials and Glass*, B.A.R. Int'l Series 237 (1985).
29. Lucas, A., and Harris, J.R., *Ancient Egyptian Materials and Techniques*, E. Arnold (1962).
30. Vandiver, P.B., Gibson, M., and McMahon, A., in prep.; and Freestone, I., in study.
31. Smith, C.S., *A Search for Structure*, pp 325-331, M.I.T. Press (1981).
32. Kaplan, M.F., *Ancient Materials Data as a Basis for Waste Form Integrity Projections*, D.O.E. Contract B-22596-A-E, TASC Report TR-1749-(1979).
33. Starr, R.F.S., *Nuzi: Report on the Excavations at Yorgan Tepe Near Kirkuk, Iraq: 1927-1931*, Vols. 1 and 2, Harvard University Press (1939 and 1937).

34. Bimson, M., and Werner, A.E.A., *J. Glass Studies* 6: 131-132 (1972).
35. Clark, D.E., Pantano, C.G., and Hench, L.L., *Corrosion of Glass*, pp 1-21, Books for Industry (1979).
36. Brinker, C.J., in: *Glass Science and Technology* (D.R. Uhlmann, and N.J. Kreidl ed.), pp 169-230, Academic (1990).
37. Caley, E.R., *The Analyses of Ancient Glasses*, Corning (1962).
38. Turner, W.E.S., *J. Soc. Gl. Tech.* 38: 445T-456T (1954), 40: 162T-186T (1956); and *Trans. Soc. Gl. Techn.* 40: 277-300 (1956).
39. Brill, R.H., in: Oppenheim, pp 105-130, Ref. 27.
40. Sayre, E.V., in: *Applications of Science in the Examination of Works of Art*, pp 145-154, Boston Museum of Fine Arts (1965).
41. Sayre, E.V., and Smith, R.W., in: *Archaeological Chemistry: A Symposium* (M. Levey ed.), pp 279-312, Am. Chem. Soc. (1967).
42. Doremus, R.H., *Glass Science*, pp 213-252, Wiley (1973).
43. Clark-Monks, C., and Parker, J.M., *Stones and Cord in Glass*, Soc. Gl. Tech. (1980).
44. Begley, E.R., *Guide to Refractory and Glass Reactions*, Cahners (1970).
45. Paul, A., *Chemistry of Glasses*, pp 108-147, Chapman and Hall (1982).
46. Bamford, C.R., *Colour Generation and Control in Glass*, Elsevier (1977).
47. Pulker, H.K., *Coatings on Glass*, Thin Films, Vol 6., Elsevier (1984).
48. *Amorphous Materials* (R.W. Douglas, and B. Ellis ed.), Wiley (1970).
49. *Fractography of Glasses and Ceramics* (J.R. Varner, and V.D. Frechette, ed.), A. Cer. S. (1988).
50. Ollier, C.D., *Weathering*, Longman (1975).
51. Vandiver, P.B., in: *Early Pyrotechnology: The Evolution of the First Fire-Using Industries* (T.A. and S.F. Wertime ed.), pp 73-92, Smithsonian Press (1982).
52. Vandiver, P.B., *J. Glass Studies* 25: 239-248 (1982).
53. Vandiver, P.B., Klein, L., and Wang, C., *Weathering Phenomena in a Fragment of Glass from Samaria, 250-400 A.D.*, unpub. ms., M.I.T. (1976).
54. Vandiver, P.B., Cort, L.A., and Handwerker, C.A., in: *Cross-craft and Cross-cultural Interactions* (P. McGovern and M. Notis ed.), Ceramics and Civ., Vol. 4, pp 347-388, A.Cer.S. (1989).
55. Parmelee, C.W., and Harman, C.G., *Ceramic Glazes*, pp 579-600, Cahners (1973).
56. Singer, F. and S.S., *Industrial Ceramics*, pp 88, 549-557, 581, and 796-797, Chapman and Hall (1963).
57. Grimshaw, R.W., *The Chemistry and Physics of Clays and Other Ceramic Materials*, pp 706-707, Wiley (1980).
58. Vandiver, P.B., *The Ancient Problem of Glaze Fit*, unpub. ms., M.I.T. (1980).
59. Gedye, I., in: *The Conservation of Cultural Property*, pp 109-114, UNESCO (1968 and 1975).

60. Plenderleith, H.J., *Siliceous and Related Materials, The Conservation of Antiquities and Works of Art*, pp 299-354, Oxford (1971).
61. Sease, C., *A Conservation Manual for the Field Archaeologist*, Arch. Res. Tools, Vol. 4, pp 69, 71, 92, 108, and 112, Inst. of Archaeol., U.C.L.A. (1987).
62. Larney, J., *Restoring Ceramics*, Watson-Guptill (1975).
63. Wihr, R., in: *Recent Advances in Conservation* (G. Thomson, ed.), Butterworths (1963).
64. Frank, S., *Glass and Archaeology*, Academic (1982).
65. Anon., *Conservation in Archaeology and the Applied Arts: 1975 IIC Congress Preprints*, Stockholm, pp 37-68 and 93-142, I.C.C. (1975).
66. Black, J., *Recent Advances in the Conservation and Analysis of Artifacts*, Jubilee Conservation Conf., pp 59-92, Institute of Archaeology, University of London (1987).
67. Various articles on glass weathering and conservation in the *J. Glass Studies*, e.g., vols. 14,17,20,21,22,24,26,27.
68. Bateman, C.A., Crawford, V.E., Dales, G.F., and Majewski, L.J., *Preservation and Reproduction of Clay Tablets and the Conservation of Wall Paintings*, Quaritch (1966).
69. Schmidt, H., and Fuchs, D., in: *Materials Issues in Art and Archaeology II* (P. Vandiver, J. Druzik, G. Wheeler, eds.), Proc. MRS Spring 1990 Meeting, forthcoming.

Section III.

Corrosion of Ceramics

14

Corrosion of Glass-Ceramics

Walter J. McCracken

GE Aerospace
General Electric Company
Philadelphia, Pennsylvania

INTRODUCTION

Glass-ceramics are defined as polycrystalline solids prepared by controlled crystallization of glasses (1). These materials form a new class of ceramics which is relatively new compared to porcelains, glasses, sintered ceramics and clay bodies. Glass ceramics are used in many high technology and high performance applications due to their unique processibility and physical properties.

The history of research into glass-ceramics starts in the 18th century with M. de Reaumur (2). He studied the devitrification (crystallization) of heat-treated glass bottles that were slowly cooled by being packed in sand. The bottles had a coarse needle-like crystalline structure and lower mechanical strength than the original glass bottle. In the 1820's, "rocklike ceramics" were made by the crystallization of glasses. This was done on an industrial scale in Europe but never reached widespread acceptance due to processing difficulties (3,4). The need for the control of the crystallization process was first recognized by H. H. Blau in 1933 (5). He attempted to control the opacity of glasses via crystallization and his concepts of controlled crystallization were widely studied in Europe in the early 1950's (1). These studies finally led S. D. Stookey at Corning Glass Works in 1956 to develop the first Pyroceram® glass-ceramic objects (6). These were formed as a glass and then heat treated, using controlled nucleation and crystallization, to produce a fine-grained glass-ceramic.

Since the late 1950's, glass-ceramic materials have become established in a wide variety of technical uses due to their unique physical properties associated with controlled crystallization. Many industries have developed specific glass-ceramic materials for use in electronics, military devices, vacuum technology, domestic applications, and other uses requiring high mechanical wear, resistance to chemical attack, and biological compatibility (1,7-10). Examples of such uses are listed in Table 14-1.

The unique material properties of glass-ceramics such as chemical durability, mechanical strength, thermal expansion coefficient, dielectric constant, and electromagnetic radiation transmittance are due to the development of a crystalline microstructure that is usually multiphase. Thus it is necessary to characterize the structure of each phase of the glass-ceramic in order to attempt to understand its behavior. It is especially important to characterize the role of

Table 14-1. Some Technical Uses of Glass-Ceramics

Applications	Compositions
Electronics	
photomachinable circuit boards	Li_2O-Al_2O_2-SiO_2
capacitors	Li_2O-ZnO-PbO-SiO_2
Insulators	Li_2O-2SiO_2
Military	
radomes	MgO-Al_2O_3-TiO_2-SiO_2
ceramic seal for bomb triggers	Li_2O-Al_2O_3-SiO_2
Vacuum Technology	
glass to metal seals for CRT	ZnO-B_2O_3-SiO_2
Domestic	
cooking ware	Na_2O-Al_2O_3-SiO_2
tableware	Li_2O-Al_2O_3-SiO_2
heating surfaces	MgO-Al_2O_3-SiO_2
Industrial	
corrosion resistant tubing	Li_2O-CaO-Al_2O_3-SiO_2
pump impellor	Li_2O-MgO-Al_2O_3-TiO_2
telescope mirror blanks	Li_2O-Al_2O_3-SiO_2
Biological	
implants and implant coatings	Na_2O-CaO_2-P_2O_5-SiO_2
dental materials	Li_2O-CaO-Al_2O_3-SiO_2

each phase with respect to environmental sensitivity in order to predict the performance of the glass-ceramic material.

In considering the durability of glass-ceramics, these materials must be viewed as multiphase systems with each phase having individual corrosion characteristics and possible unique reactions at the phase boundaries. The multiphase microstructure can be subdivided into two categories: 1) the crystalline phase, and 2) the amorphous phases (or the residual glass phase).

Most research into glass corrosion has assumed that the glass surface is a homogeneous single phase structure. This research has been reviewed in a number of articles (11-16), bibliographies (17-18) and in a book by Clark, Pantano, and Hench (19). The mechanisms presented for glass durability fall into five categories: 1) ion leaching--the ion exchange of alkali ions from the glass with hydrogen (or hydronium) ions of the solution, 2) total dissolution--the breakdown of the silicate structure at the surface (generally after leaching occurs, but not necessarily), 3) saturation and precipitation--the concentration of species in solution reach the saturation limit and precipitate on the glass surface, thus altering the surface characteristics of the glass and its subsequent corrosion resistance, 4) pitting--the uneven attack of the glass surface due to the heterogeneities or stress concentrations in the surface, and 5) weathering--intermittent exposure of the surface to water vapor with or without condensation, resulting in ion exchange and formation of salts on the surface.

Some durability studies have been done on multiphase systems such as phase-separated glasses, photochemical machinable glass-ceramics, and commercial glass-ceramics. Phase-separated glasses, such as the borosilicates, can demonstrate large differences in the rate of corrosion between the phases. This fact is used in the manufacturing of Vycor® vitreous silica where the alkali-rich phase dissolves away in a mild acid solution leaving the silica-rich phase intact (20). Phase separation can also result in a difference in thermal expansion coefficients for each phase, causing stress-accelerated attack at the phase boundary (21). Glass-ceramics contain phases that can differ both in composition and structure. The crystalline phases are generally more dense and have lower ion mobilities that the parent glass (22). However, in certain compositional ranges the crystals are most easily leached. The varying relative durability of the glass-ceramic phases have been used to produce photochemical machinable articles. Such materials as Corning's Code 8603 (based on the lithia silicates), are nucleated by UV or visible light in controlled areas of the surface. They are subsequently heat treated for crystal growth and exposed to a corroding media. The crystallized region corrodes away leaving the glass intact with only minor surface leaching (23). In contrast many other glass-ceramic systems show increased durability over the parent glass system (24-28).

A study of glass corrosion mechanisms can give some insight into the durability of a multiphase system by considering each phase individually. However, the overall durability performance of a glass-ceramic cannot be solely determined by separate phase investigations. It is essential to take the phase boundaries into consideration as well.

The chemical durability of glass-ceramics has been of scientific interest since the development of Pyroceram®. Berezhoni (30) presents a review of some studies of the durability of several glass-ceramic systems (including lithia-disilicate) (31,32) in acidic and basic aqueous solutions. In these studies the indicator of the extent of corrosion is weight loss measurements (and a few solution ion concentration determinations) after corrosion. The mechanisms by which corrosion proceeds for multiphase glass-ceramics are not discussed. However, micrographs of dilute HF acid corroded lithia-disilicate do show preferential attack along some grain boundaries and overall dissolution of the glass from around the crystalline material.

The major factors influencing the overall durability of glass-ceramics appear to be: 1) composition of the crystalline phases as compared to the residual glass, 2) structure and density of the crystalline phase, 3) volume fraction of crystalline phase, and 4) grain size of crystalline phase. There has been no previous effort to quantify the relative effect of these factors on the durability of glass-ceramics. A model of corrosion based on several mechanisms of attack is presented in this chapter. It should be useful in understanding the behavior of other multiphase ceramics, with one phase being glass, in an aggressive environment. There are similar applications of this model in liquid-phase-sintered ceramics and the glassy grain boundaries in silicon nitrides.

CORROSION OF A MODEL SYSTEM

Glass-ceramics can be microstructurally considered as numerous small crystals in a glass matrix. These polycrystalline materials are typically prepared from preformed cast glass articles via a specific nucleation-crystallization heat treatment. This process normally produces a very fine crystal (1μm) that is randomly oriented in the matrix. For this study the stoichiometric Li_2O-$2SiO_2$ (33L) system was used, producing a glass-ceramic with the crystalline phase and the glass phase having the same composition (33).

The lithia-silica glass system is the binary basis for a large group of commercial glass-ceramics composed of lithia-alumino-silicates. This binary system has been used in numerous fundamental studies in the areas of: 1) nucleation and crystallization (34), 2) phase separation (35), 3) aqueous durability (29,36), and 4) mechanical strength (37,38). The glass composition chosen for many of the previous studies and this investigation is composed of 33 mole% Li_2O and 67 mole% SiO_2 (Li_2O-$2SiO_2$ and also referred to as 33L). This composition is stoichiometric with the crystalline lithia-disilicate phase that is produced by controlled nucleation and crystallization of this glass (34). Thus, the 33L glass-ceramic provides a two-phase solid with the same composition but with one phase amorphous and the other crystalline.

The corrosion behavior of 33L glass has been studied extensively (35,39,40).

During the early stages of reaction with a neutral aqueous environment, Li^+ is selectively leached from the glass surface via ion exchange with H^+ (or H_3O^+) from the solution. In a closed system in which the ratio of glass surface area to volume of solution (SA/V) is high, and pure water is used as the leachant, the pH of the solution will increase rapidly with exposure time (see Clark and Zoitos' chapter). If the pH is permitted to go above approximately 9, a second mechanism of glass corrosion becomes important in the corrosion behavior (41,42). This mechanism involves dissolution of the glass network and is due to OH^- attack on the silicon-oxygen bonds (see Eppler's chapter). The relative importance of the ion exchange and network dissolution reactions is dependent on numerous factors including solution pH, exposure temperature, time and SA/V. The corrosion mechanisms and kinetics of vitreous 33L have been studied by Sanders and Hench (39) and Ethridge (36) in a wide range of environments. Additionally, the crystallization kinetics for this glass have been well characterized by Hench et al. (43) and Freiman and Hench (34,44).

The objective of this study was to establish the effects of controlled crystallization on the chemical durability of the Li_2O-$2SiO_2$ glass-ceramic systems. The 33L composition has been chosen for two reasons: 1) as already mentioned, the crystallization and the corrosion behavior of the glass are well characterized, and 2) since this is a stoichiometric composition, the crystalline phase has the same composition as the glass phase. This stoichiometry permits the evaluation of corrosion, independent of major compositional variations that may accompany crystallization of nonstoichiometric compositions.

Experimental

The glass was melted in an electric muffle furnace at 1350°C for 24 h. Cylinders 2.5 cm in diameter and 6.0 cm long were cast in a graphite mold and annealed at 350°C for 4 h. These glass cylinders were nucleated at 475°C for 24 h and crystallized at 550°C for various times yielding 0.20, 0.60 and 0.90 volume fraction (V_v) crystallization. The percentage volume fraction crystallization was determined by using optical scanning techniques of applied stereology (45). The cylinders were sliced with a diamond wafering saw into 0.3 cm thick disks, the surfaces of which were polished through 600 grit with SiC paper prior to exposure.

Specimens of 33L glass and partially crystallized 33L glass were exposed to an environment of either demineralized water, 0.1M NaOH, or 0.1M HCl maintained at 100°C for up to 10 h. Duplicate specimens were tested for each exposure condition to evaluate reproducibility. The maximum difference in the solution data between any of the duplications was 46 ppm (265 + 23 ppm) of SiO_2 and 11 ppm (135 + 5.5 ppm) of Li^+. The ratio of surface area of exposed material to solution volume (SA/V) was 2.0 cm^{-1}. Infrared reflection spectra were obtained for all specimens in the spectral region 1300-600 cm^{-1} both prior to and after aqueous exposure. Scanning electron micrographs were also taken of selected samples. The corrosion solutions were analyzed by measuring pH, Li^+ concentration, and SiO_2 concentration at room temperature.

Results and Discussion

Scanning electron micrographs for 33L glass containing four volume fractions of crystallization (V_v = 0, 0.20, 0.60, 0.90) are shown in Figures 14-1 thru 14-5. Micrographs of uncorroded specimens (Figure 14-1) show polished flat surfaces with some polishing scratches. Figures 14-2 thru 14-5 illustrate representative surfaces for each material after exposure to acidic, neutral and basic solutions. Significant differences in the surface microstructural features as a function of percent crystallization can be seen, particularly after exposure to neutral and basic solutions. The crystalline phase is more pronounced on the specimens exposed to the neutral or basic solutions. This is not surprising because the pH of the neutral solution increases with exposure time under static exposure conditions. The increase in solution pH is due to exchange between Li^+ ions from the solid and H_3O^+ ions from the solution, resulting in an increase in OH^- ions in solution (19,46). Thus, long exposure times to a static neutral solution favor network dissolution as do even short exposures to a basic solution.

When V_v = 0 (Figure 14-2), two types of corrosion are observed over a wide range of pH values (pH 1-13). In acidic solutions the primary mode of corrosion is ion exchange. There are two features in the micrographs in Figure 14-2 that support this conclusion: 1) surface cracks, and 2) polishing scratches. The cracks are related to the ion exchange mechanism and usually appear after the specimen has been removed from solution and permitted to dry, or after the specimen has been subjected to a vacuum for SEM analyses. The extent of ion exchange required to produce surface cracks is not known but is probably dependent on the composition of the glass. The fact that the polishing scratches do not exhibit noticeable dimensional changes during exposure to the acidic solution suggests that network dissolution is minimal in the low pH environment. In contrast, the specimens exposed to the basic solutions exhibit broadened polishing scratches and no surface cracks. Thus, the dominant mechanism of corrosion is network dissolution. The small particles on the surface are evidence of precipitation of some compound from the corrosion solution. The specimens exposed to the neutral solution show indications of both ion exchange and network dissolution with the latter being more apparent after long exposure times.

Three types of attack are observed on the glass-ceramic containing 0.20 Vv crystallization (Figure 14-3). In acidic solutions, only ion exchange occurs as was the case with glass (V_v = 0). In both neutral and basic solutions network dissolution of the glassy phase, and preferential attack of the boundary between the glassy and crystalline phases are significant. The preferential phase boundary attack may be due to stresses or minor compositional gradients at the interface. Similar phase boundary attack has been reported by Baylor and Brown (21) for

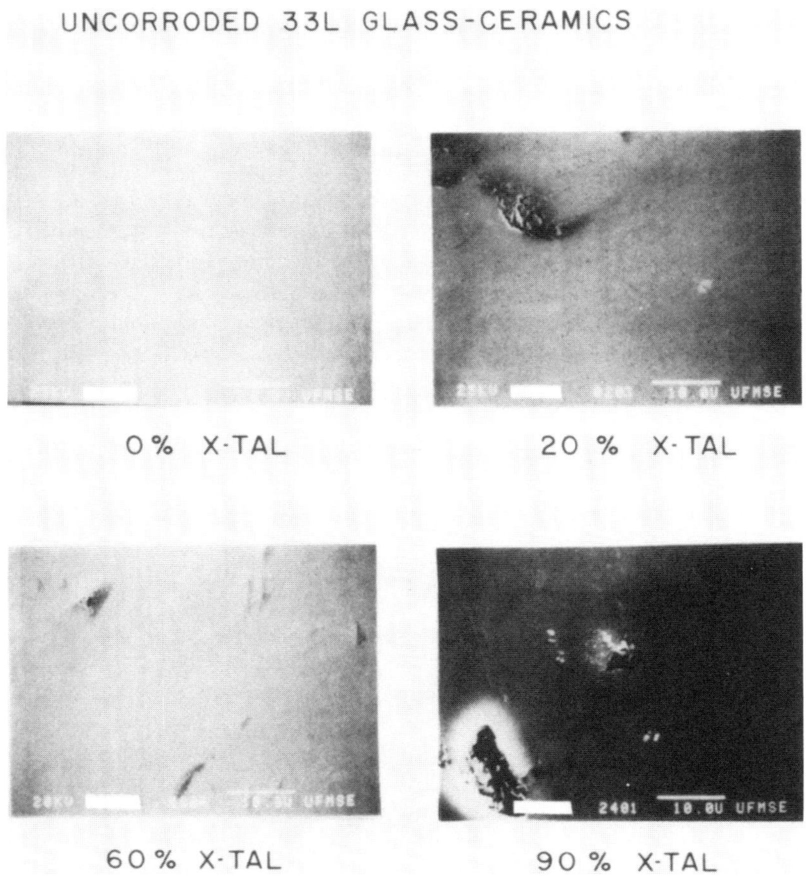

Figure 14-1. Scanning electron micrographs of uncorroded 33L glass-ceramic with 0, 0.20, 0.60 and 0.90 V_v crystallinity. Surfaces polished with 600 grit SiC paper.

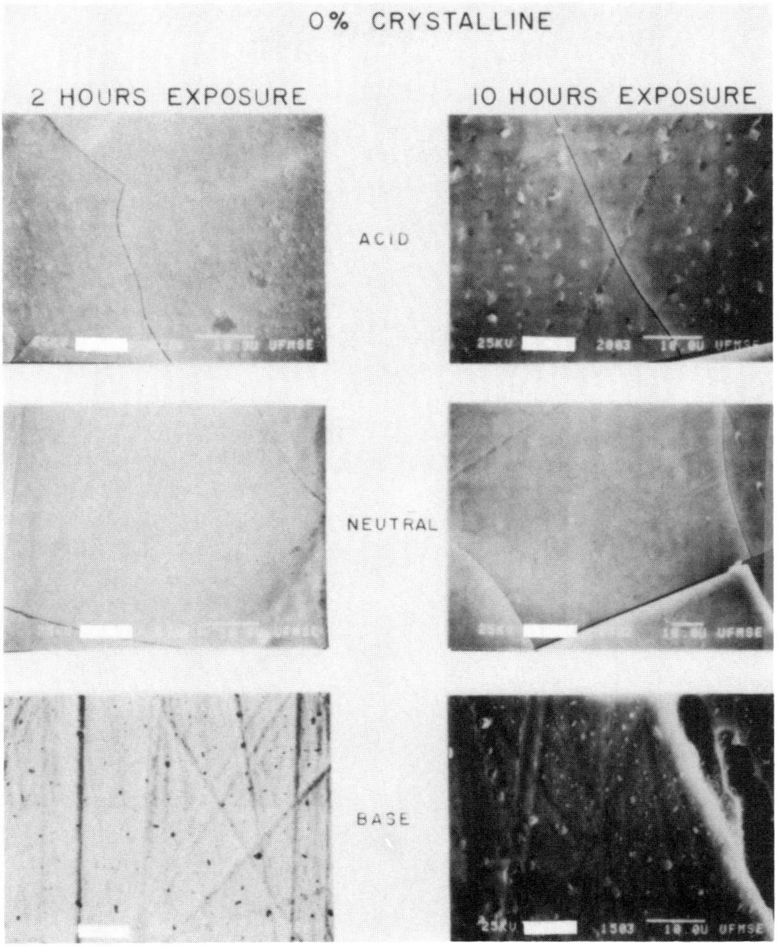

Figure 14-2. Scanning electron micrographs of 33L glass after 2 h and 10 h exposure in acid, neutral and base aqueous solution at 100°C with SA/V = 2.0 cm^{-1}.

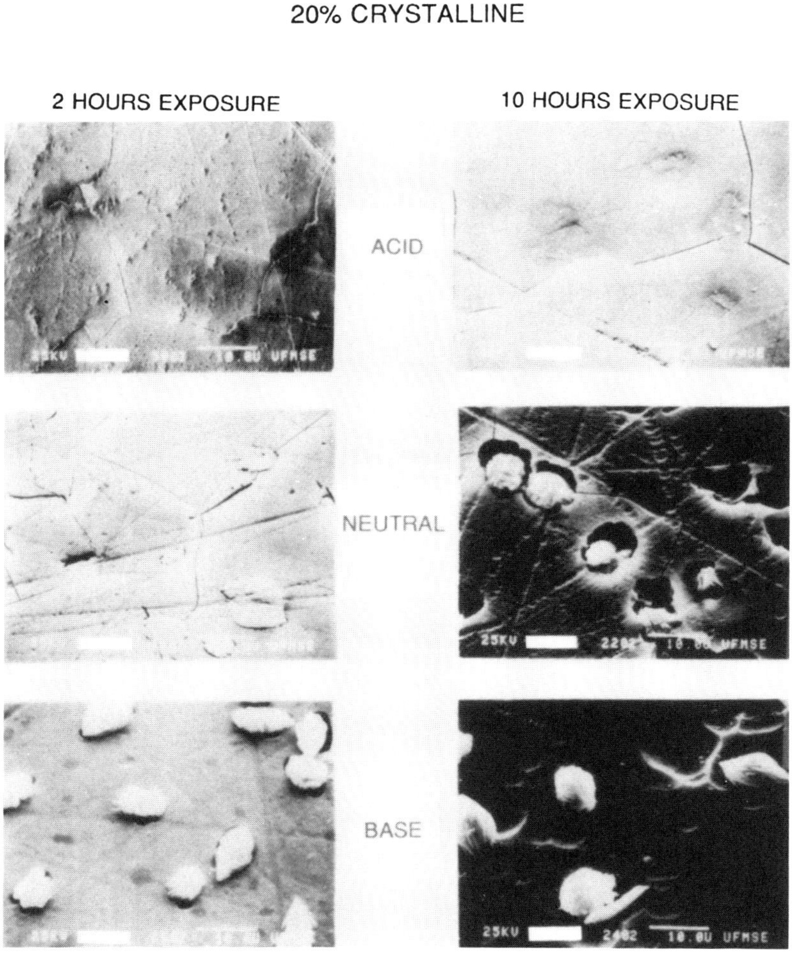

Figure 14-3. Scanning electron micrographs of 33L-0.20 V_v crystalline glass-ceramic after 2 h and 10 h exposure in acid, neutral and base aqueous solution at 100°C with SA/V = 2.0 cm^{-1}.

phase-separated borosilicate glasses. Surface cracks in the glassy phase of the specimen exposed to the neutral solution indicate the presence of an ion exchanged layer even after 10 h of exposure. These cracks do not occur in the crystalline phase. Possibly the most important information provided by Figure 14-3 is the relative durabilities of the glassy and crystalline phases. The fully exposed crystals shown in this figure demonstrate that the crystalline phase is more resistant to network dissolution than the glassy phase.

As the crystalline phase increases form $V_v = 0.20$ to $V_v = 0.60$ and $V_v = 0.90$, the surface morphology changes from that of isolated crystals in a vitreous matrix to one dominated by the crystalline phase (Figures 14-4 and 14-5). This change is most easily seen on the specimens exposed to the neutral and basic solutions. Network dissolution preferentially removes the glass and highlights the crystalline phase. The surfaces of the crystals appear to be degraded by the high pH solutions as illustrated by the removal of material between the layers in the crystal. This intracrystalline attack possibly could be the result of the dissolution of a residual glassy phase trapped between the crystal layers during crystal growth or the preferential attack of the crystals along selected crystallographic planes. Exposure to the acidic solution produces little, if any, network dissolution but does permit ion exchange as evidenced by the surface cracks visible on the 10 h specimens in Figures 14-4 and 14-5.

Infrared reflection spectra shown in Figure 14-6 for 33L glass exposed to a wide range of pH solutions are consistent with those reported in previous work (30,40). In the uncorroded glass spectrum, the peak located at 1030 cm^{-1} (LS) is due to symmetrical Si-O-Si stretching vibrations in a network containing Li^+, and the peak at 930 cm^{-1} (NS) is due to Si-nonbridging oxygen vibrations. The spectrum of vitreous silica is also included in this figure for the purpose of instrument calibration and comparison. The peak at 1100 cm^{-1} (S) for vitreous silica is due to Si-O-Si stretching vibrations in a pure SiO_2 structure. During corrosion of 33L glass in the acidic solution the LS peak shifts to higher wavenumbers and increases in intensity (% reflection) progressively approaching the S peak of vitreous silica as corrosion time increases. These alterations in the spectra are caused by the leaching of Li^+ (ion exchanging with H_3O^+) from the glass, resulting in the development of a SiO_2-rich film on the glass surface (47). Infrared spectra show that the short time corrosion behavior of 33L glass in the static neutral solution is similar to that in the acidic solution. That is, initially there is a development of a SiO_2-rich film on the surface of the glass. However, the peak intensity decreases with long exposure times (i.e., > 4 h). This is usually indicative of surface roughening due to network dissolution of the SiO_2-rich film (19). As discussed earlier, the pH of the static neutral solution increases with exposure time, and the resulting high OH^- concentration enhances network dissolution. The general shape of the infrared spectra for 33L glass exposed to the basic solution does not change. The spectral intensity, however, continuously decreases with exposure time. These spectral variations indicate that no SiO_2-rich film is developed during exposure to the high pH solution.

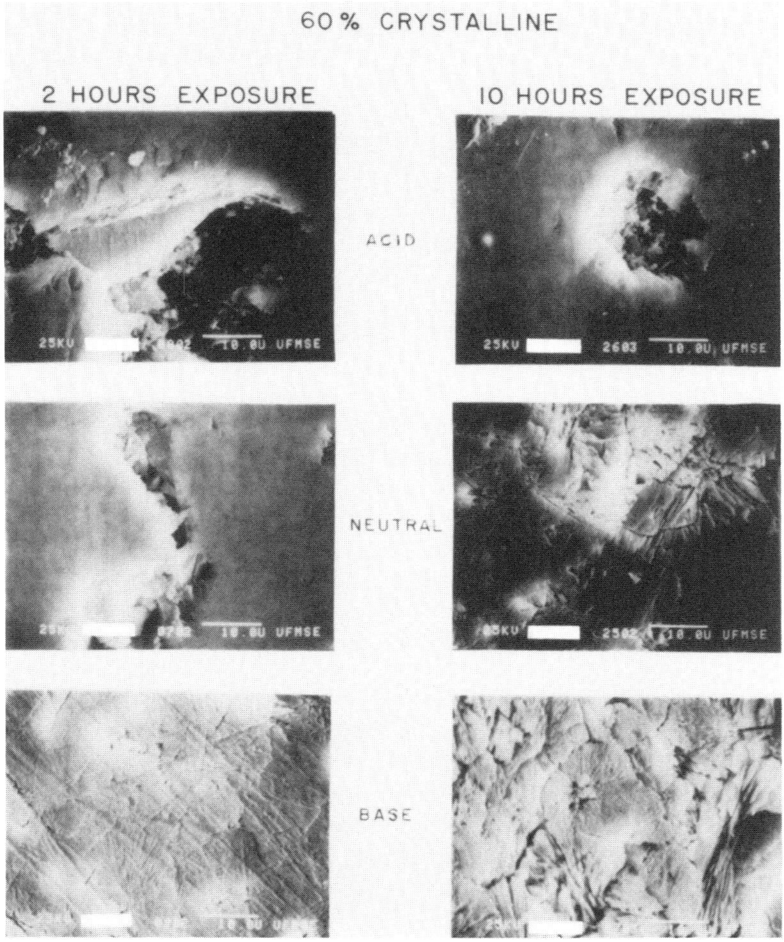

Figure 14-4. Scanning electron micrographs of 33L-0.60 V_v crystalline glass-ceramic after 2 h and 10 h exposure in acid, neutral and base aqueous solution at 100°C with SA/V = 2.0 cm^{-1}.

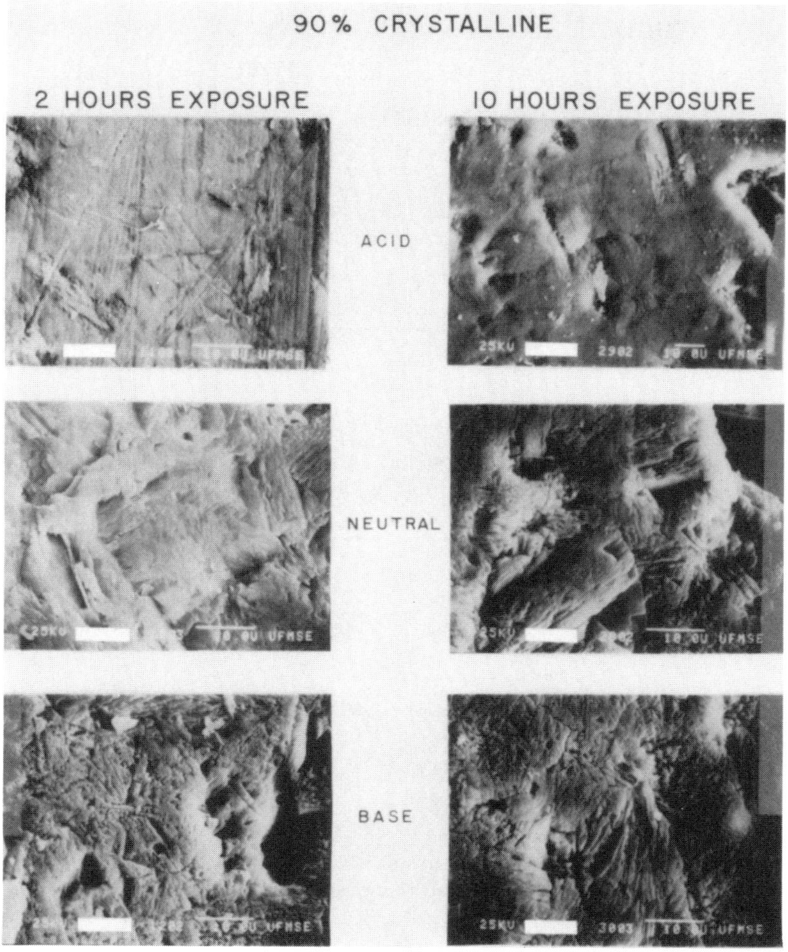

Figure 14-5. Scanning electron micrographs of 33L-0.90 V_v crystalline glass-ceramic after 2 h and 10 h exposure in acid, neutral and base aqueous solution at 100°C with SA/V = 2.0 cm^{-1}.

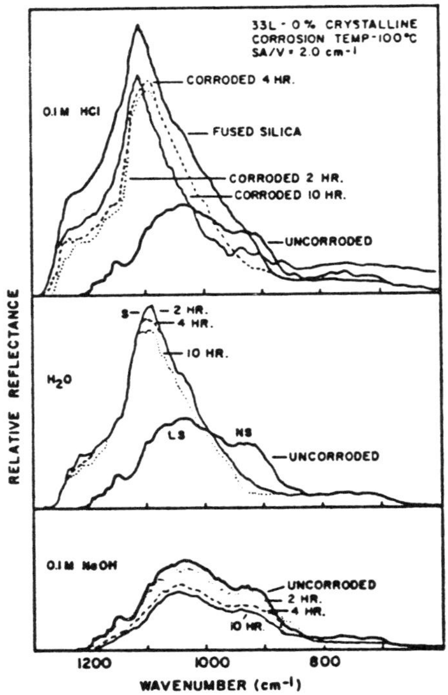

Figure 14-6. Infrared reflection spectra of 33L glass before and after acid, neutral and base aqueous corrosion.

Surface roughening due to network dissolution is responsible for the decline in intensity. Network dissolution enlarges the polishing scratches and causes pitting as shown in the micrographs in Figure 14-2.

Figure 14-7a shows the infrared reflection spectra for 33L glass with 0.20 V_v crystallization exposed to the acidic, neutral and basic solutions. These spectra are nearly the same as those for the glass exposed to similar environments. The major exceptions are the spectra corresponding to the 10 h exposures in neutral and basic solutions. The lower intensities observed on the 33L-0.20 V_v specimen for these exposures indicate increased surface roughening. The increased surface roughening is due to preferential network dissolution of the glassy phase exposing the crystals as shown in Figure 14-3. These data suggest that the glassy phase of 33L-0.20 V_v glass-ceramic is primarily responsible for the observed surface corrosion.

Infrared reflection spectra for 33L glass with 0.60 V_v and 0.90 V_v crystallization exposed to acidic, neutral and base solutions are shown in Figures 14-7b and 14-7c, respectively. The similarity of the spectra for these two materials both prior to and after corrosion facilitates their joint discussion. These

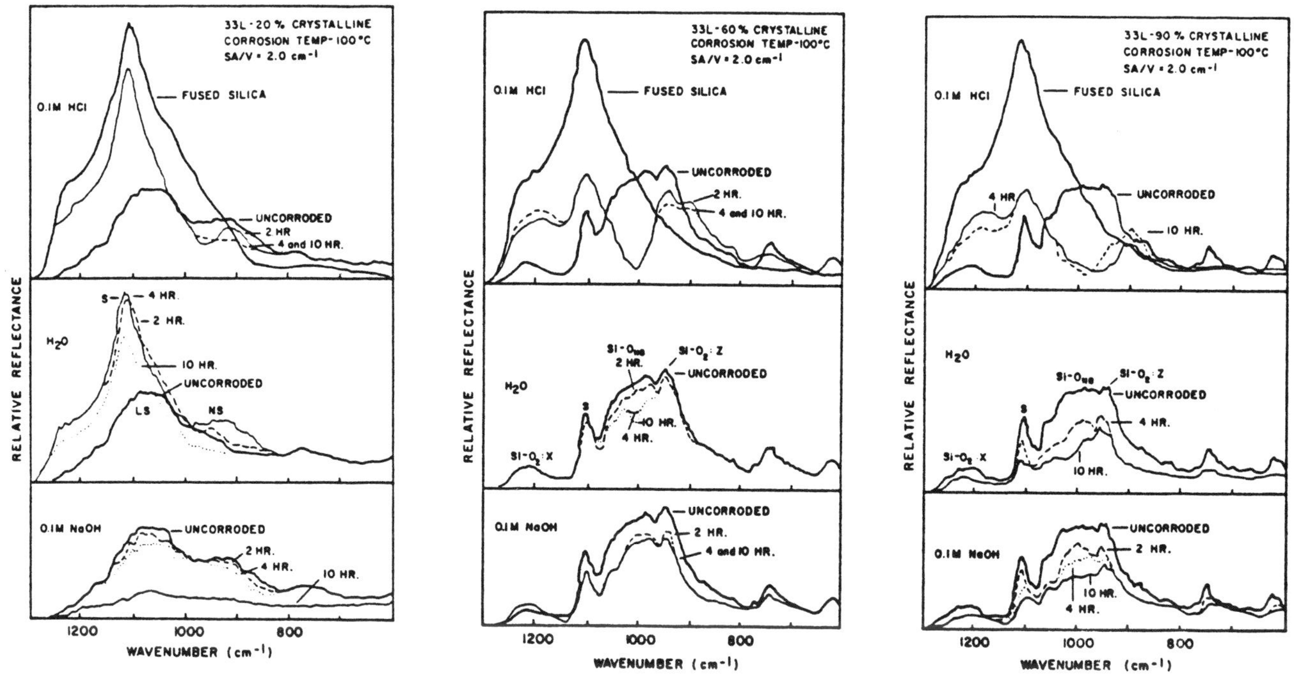

a b c

Figure 14-7. Infrared reflection spectra of 33L crystalline glass-ceramic before and after acid, neutral and base aqueous corrosion. a) 0.20 V_v; b) 0.60 V_v; c) 0.90 V_v.

spectra show some definite changes in surface structure and corrosion behavior compared to the 33L glass and 33L glass with 0.20 V_v crystallization. The IRRS spectra obtained from uncorroded 33L-0.60 V_v and 33L-0.90 V_v consist essentially of vibrations from only the crystalline phase (48). The broad peak at 1000 cm^{-1} is due to the overlapping of the two Si-ONB vibrations found in the lithia-disilicate single crystal. Intracrystalline attack, as seen in the SEM micrographs in Figures 14-4 and 14-5, results in the reduction of the Si-ONB:Y peak at 995 cm^{-1}. This reduction is due to the preferential removal of material along the micaceous layer (010) in the crystals. Therefore, this exposes more X-Z planes for interaction with the infrared radiation. The well defined peaks at 1210 cm^{-1}, 1110 cm^{-1}, and 960 cm^{-1} are assigned to the structural vibration of silicon-nonbridging oxygen stretching, silicon-oxygen-silicon stretching, and another silicon-nonbridging oxygen stretching, respectively.

The positions of these latter three peaks for 33L-0.60 V_v and 33L-0.90 V_v are practically unaltered during exposure to neutral and basic solutions but their intensities generally decrease with time. However, significant peak position alterations are observed when the materials are exposed to the acidic solution. The major changes in the spectra are due to the noticeable loss of the two Si-ONB peaks at 995 cm^{-1} and 1040 cm^{-1}, plus the increase in the height of the peaks at 1110 cm^{-1} and 1210 cm^{-1}. These peak alterations are thought to be caused by the exchange of Li^+ ion from the glass-ceramic surface with the H^+ (or H_3O^+) ions from the solution and the development of a SiO_2-rich film of an apparent amorphous nature on the surface (49). There is also an additional peak at 890 cm^{-1} after acidic attack of 33L-0.60 V_v and 33L-0.90 V_v glass-ceramics. This peak does not correlate with any lithia-disilicate vibration previously observed. During acidic attack, the 33L-0.60 V_v and 33L-0.90 V_v glass-ceramics develop a silica-rich layer on the surface and the lithia-disilicate crystal structure on the surface is leached of Li^+ ions. This is similar to the acidic attack behavior for glasses except that the crystal structure of the lithia-disilicate glass-ceramics remains apparent in the infrared spectra after Li^+ ion leaching.

Solution data are presented in Figure 14-8 for each exposure condition and material. In general the concentrations of Li^+ and SiO_2 increased in solution as exposure time increased for all exposure times and materials. The pH of the neutral solution also increased with time for all materials. This pH increase is due to exchange of Li^+ ions from the material surface with H^+ or H_3O^+ ions from solution. Both the change in the pH and Li^+ ion concentration are less for the glass-ceramics containing large percentages of crystals exposed to the neutral solution (Figure 14-8c). These data suggest that the ion exchange of Li^+ is slower from the crystalline surface than from the glass surface since the glass-ceramics have a smaller area of glass exposed to the solution. The same trend in Li^+ ion concentration is observed in the materials exposed to the basic and acidic solutions (with the exception of the 10 h exposure -- 0.90 V_v crystal).

The concentrations of SiO_2 in solution are an indication of the extent of network dissolution that has occurred in the materials (Figure 14-8b,d,f). As expected, the samples exposed to the acidic solution exhibited the smallest change in SiO_2 concentration with time, and those exposed to the basic solutions exhibited the largest change. In acidic and neutral solution the glass-ceramics with 0.60 V_v crystalline phase was the most durable material; it was the least durable material in the basic solution. As discussed earlier, considerable network dissolution of the glassy phase occurs in all of the materials exposed to the basic solution or to the static neutral solution for long periods of time (see micrographs in Figures 14-2 thru 14-5). The relative high concentrations of SiO_2 in the basic solution for the 0.60 V_v crystalline specimen are thought to be related to two factors: 1) extensive preferential attack on the phase boundaries, and 2) the relative high percentage of glass in the structure (i.e., 0.40 V_v). Preferential boundary attack does not occur in glass with 0 V_v crystallization; only uniform attack occurs. In the glass-ceramic containing 0.90 V_v crystallization, preferential phase boundary attack is significant but the percent of glass affected is small (i.e., 0.10 V_v). However, in the glass-ceramic containing 0.60 V_v crystallization not only is preferential phase boundary attack significant but also the volume of glass affected is large.

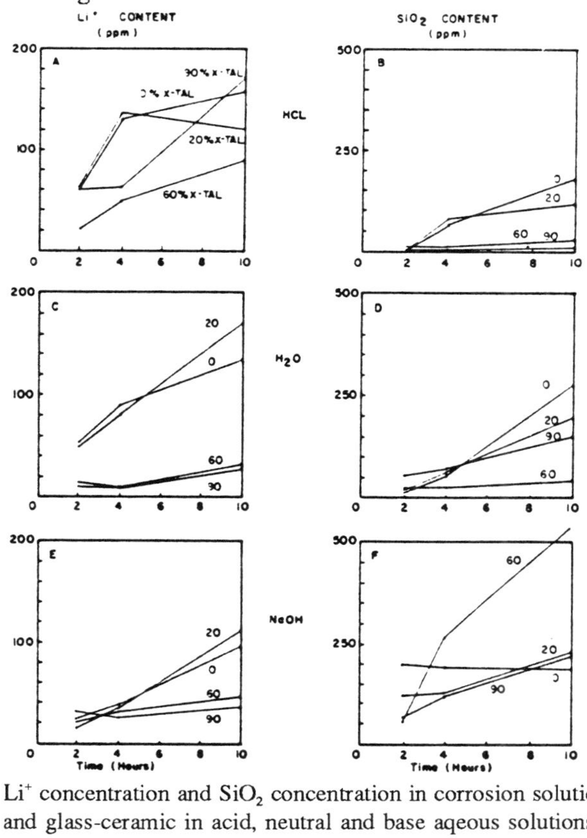

Figure 14-8. Li^+ concentration and SiO_2 concentration in corrosion solution versus time for 33L glass and glass-ceramic in acid, neutral and base aqeous solutions.

The concentration of lithium ions and silica in the corrosion solution can be used to determine the corrosion parameter alpha, which is commonly used to determine the corrosion mechanism for glasses (50). The formula for alpha is as follows (see Clark and Zoitos' chapter):

$$\alpha = [(\text{ppm } SiO_2/\text{m.w. } SiO_2)/(\text{ppm } LI^+/(2 \text{ X m.w. } Li^+))] \text{ X } (1\text{-}PSiO_2/PSiO_2) \qquad (14\text{-}1)$$

$PSiO_2$ = mole fraction of silica in bulk solid
ppm = parts per million of species in solution
m.w. = molecular weight

From this relation, when alpha is near zero the corrosion mechanism is selective ion leaching, creating a silica-rich layer on the surface. When alpha approaches one the corrosion mechanism is congruent dissolution. The original development of the alpha parameter was done for glasses with a uniform surface (i.e., homogeneous and single phase). For glass-ceramics, both phases can be corroding simultaneously by different mechanisms. Therefore, alpha can only be used as an attempt to coordinate and simplify the corrosion solution data.

Figure 14-9 shows alpha vs. time for 0.1M HCl corrosion of the 33L glass and glass-ceramics. Note that the scale for alpha has been expanded to illustrate

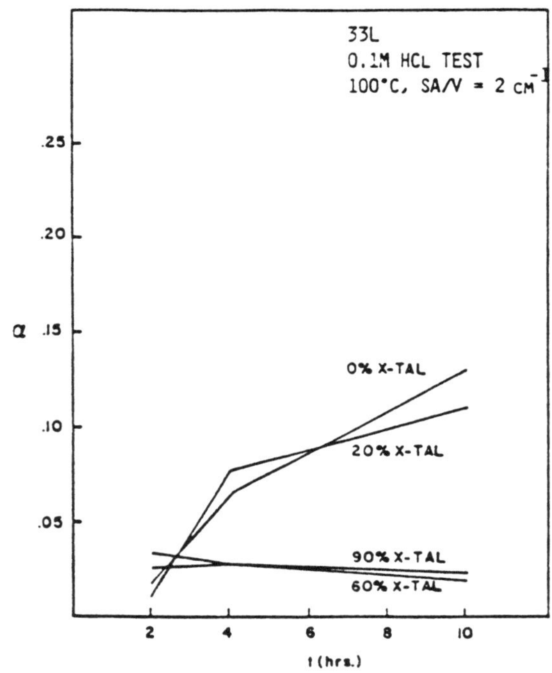

Figure 14-9. Corrosion parameter, alpha versus time of 0.1M HCl corrosion test at 100°C for 33L glass and glass-ceramic.

the slight changes during the test. For acid corrosion all the materials were attacked predominately by the selective ion leaching mechanism. The results for the H_2O corrosion test are shown in Figure 14-10. The 33L-0 V_v and 33L-0.20 V_v, which are predominately glass, illustrate typical binary glass corrosion mechanisms--starting with ion leaching and gradually changing toward network dissolution as the solution pH increases. In contrast the 33L-0.60 V_v and 33L-0.90 V_v show a greater tendency toward network dissolution even in the early stage of corrosion. This behavior is probably due to phase boundary attack in which the glass network between crystals is dissolved congruently.

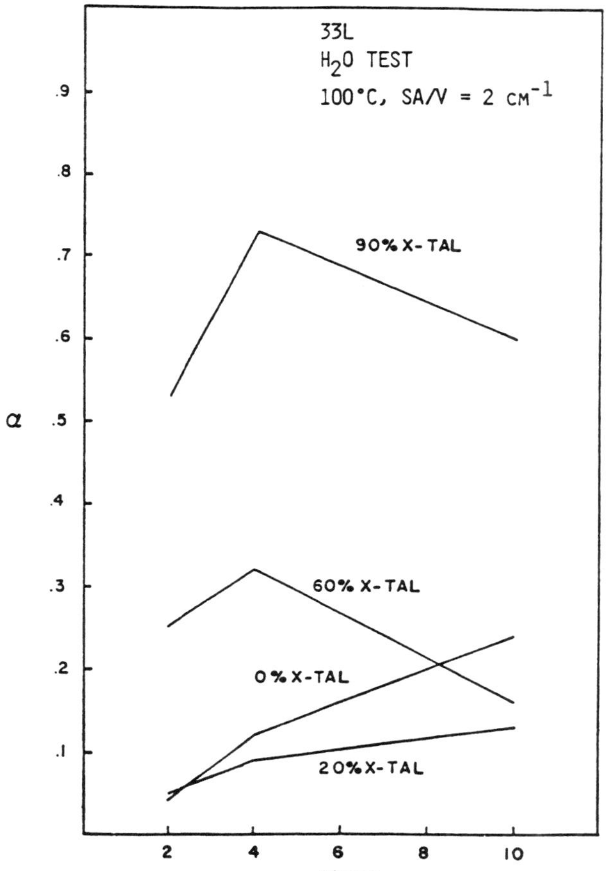

Figure 14-10. Corrosion parameter alpha versus time of H_2O corrosion test at 100°C for 33L glass and glass-ceramic.

In the 0.1M NaOH corrosion test, Figure 14-11, alpha values for 33L-0 V_v and 33L-0.20 V_v indicate initially that the network dissolution dominates the corrosion behavior. Then the mechanism proceeds toward ion leaching being

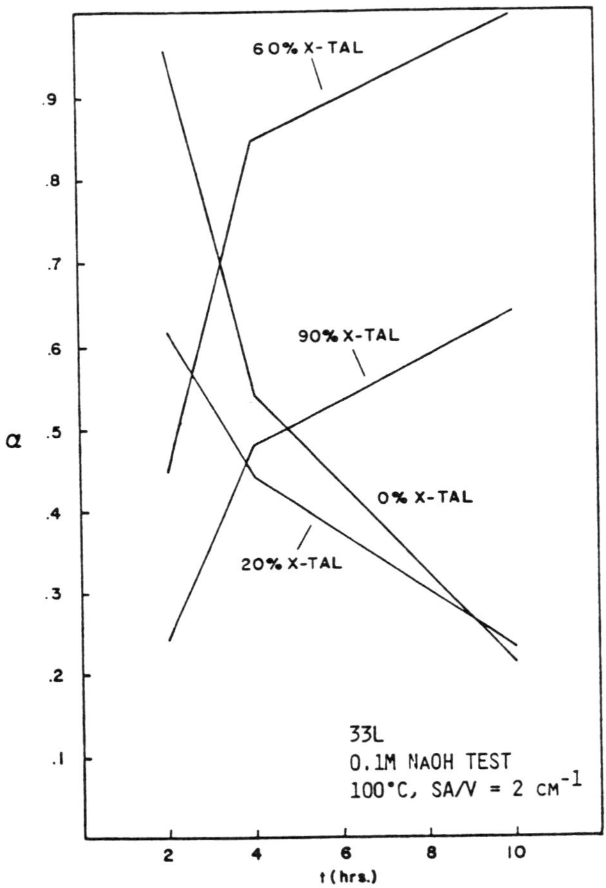

Figure 14-11. Corrosion parameter, alpha versus time of 0.1M NaOH corrosion test at 100°C for 33L glass and glass-ceramic.

more dominant. This is not reasonable according to previous work on basic attack of glass (19). One possibility for this disagreement would be the precipitation of sodium silicates onto the surface of the sample as shown in Figure 14-3. The nature of basic corrosion for 33L-0.60 V_v and 33L-0.90 V_v is opposite from that of the more glassy samples. The corrosion mechanism seems to progress from leaching to network dissolution. The basic solution preferentially attacks the micaceous cleavage plane, in which the lithium ions are located, releasing lithium ions but not by a leaching mechanism. Simultaneously, the remaining glass phase would be dissolving by network dissolution. Since 33L-0.60 V_v contains more glass, the alpha values for 33L-0.60 V_v are overall greater than for the 33L-0.90 V_v.

The glass corrosion parameter, alpha, is not an accurate measure of the corrosion mechanism in glass-ceramics (or any multiphase material). It does not take into account that each phase and the phase boundaries react differently during corrosion. Also there is no correction for precipitation of compounds from the corrosion solution that may occur. However, alpha values are useful to correlate the ion concentrations in the corrosion solution to the bulk composition of the glass-ceramic. It should be pointed out that alpha values do not indicate the extent of corrosion, only the mechanism. The values also may not be representative of the whole surface, but rather of the phase that is dominating the corrosion behavior.

Crystallization significantly affects the corrosion behavior of Li_2O-SiO_2 glass. This chapter has demonstrated that both the mechanism of attack and extent of attack for glass-ceramics are dependent on the percentage of crystallinity, environmental exposure conditions, and duration of exposure. Figure 14-12 illustrates some of the modes of corrosion in lithia-disilicate glass-ceramics.

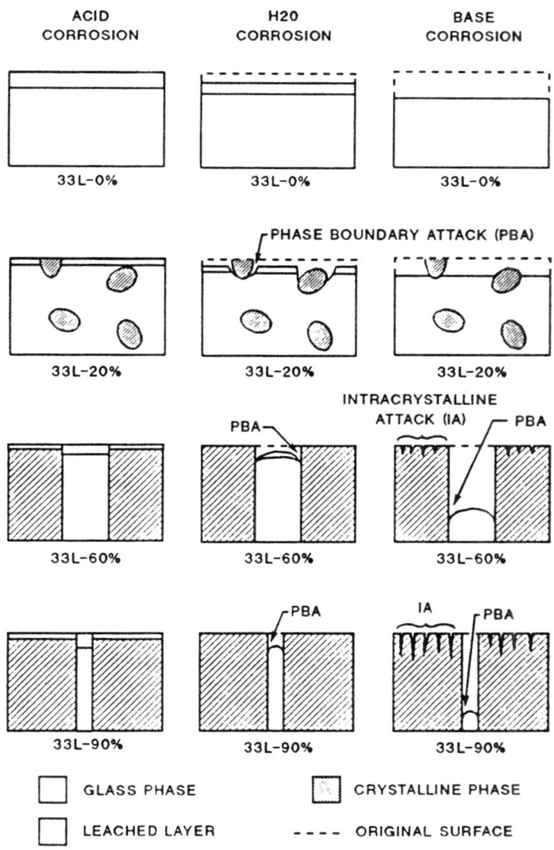

Figure 14-12. Various modes of glass-ceramic corrosion for 33L glass-ceramics.

Certain modes of corrosion such as phase boundary attack and intracrystalline attack are found only in the glass-ceramic materials because of their heterogeneity. The overall durability of the material increases with percent of crystallization. However these non-uniform modes of attack could lead to localized degradation of the lithia-disilicate glass-ceramic material.

SUMMARY

Crystallization significantly affects the corrosion behavior of Li_2O-SiO_2 glass. In general, the least surface damage is observed on all the materials when exposed to acidic solution and the most extensive surface alterations occur on all the materials exposed to the basic solutions. On glass, network dissolution occurs uniformly over the entire exposed surface. On glass-ceramics, network dissolution proceeds at different rates over the surface, with the highest rate occurring on the glassy phase. In addition to ion exchange and network dissolution, phase boundary attack between the glass and crystalline phases and intracrystalline attack are important in glass-ceramic systems exposed to basic solutions, or to static neutral solutions for long times. The intracrystalline attack mode occurs along the micaceous layer (010) of the lithia-disilicate crystal, exposing more X-Z planes, and reducing the magnitude of the Si-ONB:Y peak in the infrared data. Phase boundary attack and intracrystalline attack do not appear to be significant on the glass-ceramics exposed to acidic solutions. However, the crystalline phase maintains its crystal structure after ion exchange with the acid solutions.

REFERENCES

1. McMillian,P.W., *Glass-Ceramics*, Academic Press, Inc., London, 2-15 (1964).
2. de Reamur, M., *Mem. de l' Acad. des Sci.*, 370-388 (1739).
3. Bigot, A., *Chim. 9*, 851 (1923).
4. Portevin, A., *Mem. Soc. Ing. Civils France*, 266 (1923).
5. Blau, H.H., *Ind. Eng. Chem.* 25: 848 (1933).
6. Stookey, S.D., *Brit. Patent No. 752,243*, (1956).
7. Pincus, A.G., *Advances in Nucleation and Crystallization in Glass*, American Ceramic Society, Columbus, Ohio, 210-223 (1971).
8. Doremus, R.H., *Glass Science*, John Wiley & Sons, New York (1973).
9. Kingery,W.D., Bowen, H.K., and Uhlmann, D.R., *Introduction to Ceramics*, John Wiley & Sons, New York (1976).
10. Hench, L.L., Greenlee, T.K., Allen, W.C, and Piotrowski, G., *An Investigation of Bonding Mechanisms at the Interface of a Prosthetic Material*, Report 2, Contract DADA 17-70-C-0001, University of Florida

(1971).

11. Morey, R.W., *The Prospects of Glass*, 2nd edition, Reinhold, New York 101-131 (1954).
12. Das, C.R., *Indian Ceram. Soc.* 24:(1), 12-23 (1965).
13. Weyl, W.A., and Marboe, E.C., *The Constitution of Glass: A Dynamic Interpretation Vol. 2*, John Wiley & Sons, New York, 1010-1278 (1967).
14. Holland, L., *The Properties of Glass Surfaces*, Chapman and Hall, London, 121-192 (1964).
15. Hench, L.L., and Clark, D.E., *J. Non-Cryst. Solid* 28: 83-105 (1978).
16. Bacon, F.R., *Glass Ind.* 49 (8), 438-439, 442-447; (9) 494-499; (10) 554-559 (1968).
17. Beattie, I.R., *J. Soc. Glass Technol.* 36: 34-35 (1952).
18. Ernsberger, F.M., *Annual Review of Material Science, Vol. 2*, p. 529-572, (R. A. Huggins, ed.) Annual Reviews, Inc. Palo Alto, Califronia (1972).
19. Clark, D.E., Pantano, C.G., Jr. and Hench, L.L., *Corrosion of Glass*, Books for Industry, Inc., New York (1979).
20. Doremus, R.H., *Glass Science*, John Wiley & Sons, New York, 8 (1973).
21. Baylor Jr., R., and Brown Jr., J.J., *J. Am. Ceram. Soc.* 59:(3-4) 131-136 (1976).
22. McMillian, P.W., *Glass-Ceramics*, Academic Press, Inc., London, 160 (1964).
23. Berezhnoi, A.E., *Glass-Ceramics and Photo-Sitalls*, Plenum Press, New York, 363-371 (1970).
24. Locesi, B.P., *Symposium on Nucleation and Crystallization in Glasses and Melts*, American Ceramic Society, Columbus, Ohio, 71-76 (1962).
25. McCracken, W.J., Clark, D.E., and Hench, L.L., *Am. Ceram. Soc. Bull.* 61 11: 1218-23 (1982).
26. Chick, L.A., Ryan, R.O., and Thomas, L.E., *Am. Ceram. Soc. Bull.* 62:(4), 505-9, 516 (1983).
27. Shelby, J.E., and Nichols, M.C., *J. Am. Ceram. Soc.* 66:(3), 200-4 (1983).
28. Chao, Y., and Clark, D.E., *Glass Technol.* 25:(3), 152-6 (1984).
29. Dilmore, M.F., Clark, D.E., and Hench, L.L., *Am. Ceram. Soc. Bull.* 58: (11), 1111-1124 (1979).
30. Berezhoni, A.E., *Glass-Ceramics and Photo-Sitall*, Plenum Press, New York, 339-351 (1970).
31. Takizawa, K., Sakaino, T., and Moriya, T., *Bull. Tokyo Inst. Technol.* 53: 1-37 (1963).
32. Totesh, A.S., Averyanov, V.I., Streltsina, M.V., and Raskova, G.P., *The Structure of Glass*, Vol. 5, 150-159, Trans. Russian, Consultants Bureau, New York (1965).
33. Simon, I., *J. Opt. Soc. Am.* 41: 336 (1951).
34. Freiman, S.W., and Hench, L.L., *J. Am. Ceram. Soc.* 51:(7), 382-387 (1968).

35. Borgman, V.A., Leko, V.K., and Markargan, V.K., *The Structure of Glass*, Vol. 8, 66-68. Trans. Russian, Consultants Bureau, New York (1973).
36. Ethridge, E.C., *Ph.D. Dissertation*, University of Florida (1977).
37. Freiman, S.W., and Hench, L.L., *J. Am. Ceram. Soc.* 55:(2), 86-90 (1972).
38. Palmer, R.A., Lindberg, W.R., and Hench, L.L., *J. Am. Ceram. Soc.* 62:5-6, 319-320 (1979).
39. Sanders, D.M., and Hench, L.L., *J. Am. Ceram. Soc.* 56:(7), 373-377 (1973).
40. Sanders, D.M., Person, W.B., and Hench, L.L., *Appl. Spect.* 28: 247-255 (1974).
41. El-Shamy, T.M., Lewins, J., and Douglas, R.W., *Glass Technol.* 13:(3), 81-87 (1972).
42. Douglas, R.W., and El-Shamy, T.M., *J. Am. Ceram. Soc.* 50:(1), 1-8 (1967).
43. Hench, L.L., Freiman, S.W., and Kinser, D.L., *Phys. and Chem. of Glasses* 12:(3), 58-63 (1971).
44. Freiman, S.W., and Hench, L.L., *J. Am. Ceram. Soc.* 52:(2), 111-112 (1969).
45. Freiman, S.W., *Characterization of Ceramics*, (L. L. Hench and R. W. Gould, Marcel Dekker, eds.) New York, 555-579 (1971).
46. Clark, D.E., Dilmore, M.F., Ethridge, E.C., and Hench, L.L., *J. Am. Ceram. Soc.* 59:(1-2), 62-65 (1976).
47. Clark, D.E., Ethridge, E.C., Dilmore, M.F., and Hench, L.L., *J. Glass Technol.* 18:(4), 121-124 (1977).
48. McCracken, W.J., Person, W.B., and Hench, L.L., *J. Mater. Sci.* 20:(11), 3853-64 (1985).
49. Kay, J.F., and Doremus, R.H., *J. Am. Ceram. Soc.* 57:(11), 480-492 (1974).
50. Schmidt, Y.A., and Dubrova, S.K., *The Structure of Glass*, Vol. 1, Tran. Russian, Consultants Bureau, New York, 319 (1955).

15

Corrosion of Ceramic Construction Materials in Acidic Environments

James P. Bennett

U. S. Bureau of Mines
Tuscaloosa Research Center
Tuscaloosa, Alabama

INTRODUCTION

Ceramic construction materials for chemical processes consist of two categories of materials, acidproof brick (red shale and fireclay) and specialty materials (silica, silicon carbide, high alumina, etc.). These materials, examples shown in Figure 15-1, are also known as chemical-resistant masonry. They are used in chemical environments because of their ability to withstand corrosion under severe service conditions. Red shale and fireclay materials are generally considered suitable for use with inorganic acids such as HCl, HNO_3, H_2SO_4, and H_3PO_4; in salt solutions; and in mildly alkaline solutions. Specialty materials such as silica, silicon carbide (SiC), and high alumina are used in applications where high temperature, wear, or chemical compatibility is important. Although these and other specialty materials exist, all have a higher cost than acidproof brick that limits their usage to extremely severe service applications. Depending on the use, ceramic construction materials should be evaluated in an environment simulating actual use conditions.

Chemical-resistant ceramic construction materials are used either as structural materials or as a protective layer over non-chemical-resistant materials. When a chemical resistant vessel is built, a layered structure is typically used, with the inner surface a chemical-resistant material, backed up by an impervious membrane and some kind of support structure (see Figure 15-2). If the use involves high temperature, insulating material may be placed between the impermeable membrane and the chemical resistant brick. The inner structure is composed of ceramic construction material for corrosion, and in some cases,

Figure 15-1. Chemical-resistant brick. A) red shale; B) fireclay; C) silica; D) SiC (silicate bonded); and E) high alumina.

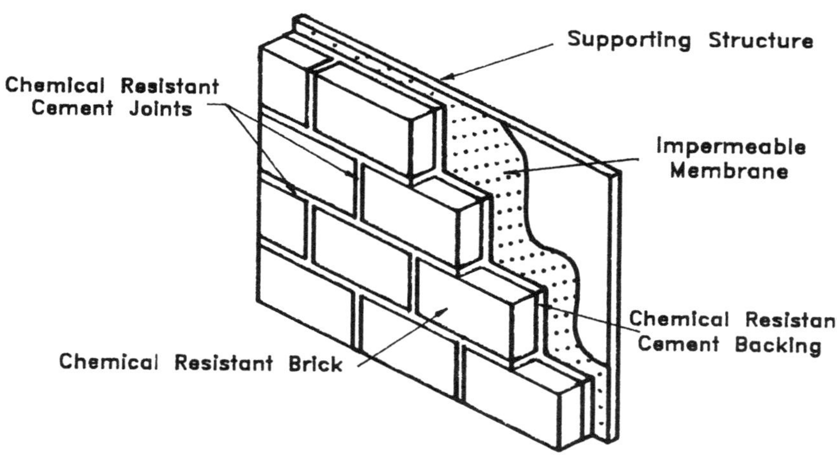

Figure 15-2. Chemical resistant vessel structure.

erosion resistance. Corrosion resistance is imparted through the chemical composition of the brick, and is aided by the low porosity these materials typically have. Low porosity causes corrosion to be a diffusion-controlled phenomenon, slowing the movement of ions in or out of the material.

Applications of chemical-resistant masonry include floors, gutters, chimney liners, pickling tanks, pulp and paper digesters, and chloride vessels. Industries that use the chemical-resistant structures include power, steel and metal working, chemical, pulp and paper, pharmaceutical, food and beverage processing and waste treatment. Corrosion-resistant structures may be formed by bricks held together with mortars or by a monolithic lining. One primary advantage of corrosion-resistant materials is their relatively low cost, which often makes them the only reliable, cost effective lining material for long term usage. Improper selection of a material for a corrosive environment can lead to premature or catastrophic failure of a structure, costing time, money, and creating possible safety hazards.

The purpose of this chapter is to summarize what is currently known about the corrosion resistance of ceramic construction materials based on information in the literature as well as on research conducted at the Bureau of Mines' Tuscaloosa Research Center. This information will aid in the selection of materials for use in acidic environments; however, all materials should be evaluated in an environment simulating actual use conditions. The information contained in this chapter will discuss physical property changes and ion removal rates in different ceramic construction materials, and thus will address only kinetic rather than thermodynamic stability.

Standard Tests

Acceptable standardized tests for evaluating the corrosion resistance of ceramic materials in all corrosive environments do not exist. Industrial equipment designers, fabricators, and material suppliers frequently base recommendations for corrosion resistant ceramic materials upon experience and suppliers advise that materials be tested in an environment simulating the actual conditions of the industrial application. Lay (1) summarizes methods of evaluating corrosion resistance that include visual inspection, use of a dye penetrant, evaluating physical property or phase changes, and monitoring the chemical solution for ion removal. The chemical and physical performance of red shale and fireclay materials are evaluated commercially using published ASTM tests C-279, C-410, or C-980, which evaluate a material's resistance to sulfuric acid. According to C-279, a material's chemical resistance is determined by the amount of weight loss from a crushed brick sample (between 6.75 and 4.75 mm in particle size) after 48 hours exposure in boiling (200°C) concentrated (78 wt.%) sulfuric acid. The weight loss, along with an unexposed material's modulus of rupture and percent water absorption data, determine the general service limitations of the brick. This test and similar ones attempt to classify materials by a rapid, simple test. Caution must be exercised when using these results, as mentioned in ASTM C-279, because no one test indicates all physical property changes that might occur in a variety of chemical environments. Balabanovich, Vinnikov, and Lesokhin (2) exposed crushed corrosion-resistant materials to different concentrations of boiling sulfuric acid and found the greatest weight losses occurred in a 33 wt.% concentration, not the 78 wt.% used in ASTM tests.

Published Data

Information available in the literature on the corrosion resistance of ceramic materials is limited. A general rating of the performance of 15 ceramic materials subjected to different acids, with no reference to acid concentration, temperature, or material properties, was reported by Campbell and Sherwood (3). Materials were evaluated in HNO_3, H_2SO_4, HCl, HF, and H_3PO_4 and received a 5 scale rating from "No reaction, material entirely suitable for service indicated" to "Rapid reaction, material not acceptable." A similar listing based on a key of "complete solution" to "not dissolved" was compiled from the literature by Samsonov (4) and Samsonov and Vinitskii (5). Materials listed included refractory oxides, carbides, borides, and nitrides. Corrosion resistance was based on the amount of insoluble residue remaining after exposures from minutes to thousands of hours.

A ranking of ceramic materials (Table 15-1) by Lay (6) compared acid and base corrosion resistance in general terms based on information from commercial manufacturers and the literature. Although the manufacturer of materials tested was listed, material chemistry or physical properties were not included in his report, making application of the information of limited value.

Table 15-1. Resistance of Ceramic Materials to Acid Attack (6).

	Acids				
Materials	HCl	HNO_3	H_2SO_4	H_3PO_4	HF
Al_2O_3	B	NA	B	A	A
MgO	B	B	B	B	B
$MgAl_2O_4$	A	A	A	A	B
SiC (reaction bonded-free Si present)	A	A	A	A	C
SiC (reaction bonded-SiC)	A	NA	A	NA	A
SiO_2 (vitreous)	A	A	A	B	C
Si_3N_4 (reaction) bonded)	A	A	A	NA	B
ZrO_2	B	B	B	NA	B
$ZrSiO_4$	A	A	A	A	C

A - resistant to attack
B - some attack
C - appreciable attack
NA - not analyzed

More extensive corrosion tables are reported by Lay (1) and Morrell (7) that include research conducted at the National Physical Laboratory as well as information from commercial manufacturers and the literature. In discussing this data, it was noted that most ceramics are not pure compounds. They contain secondary phases added as grain growth inhibitors, sintering aids, or mineralizers. These secondary phases may be more prone to attack than the major components, weakening the structure as they are removed. Again, application of this information is limited because of incomplete chemical or physical property information.

Research was conducted on reaction-bonded SiC by Clinton, Lay, and Morrell (8). Weight changes that occurred in samples exposed to different acid and base solutions for different times and temperatures were monitored; however, detailed chemical and physical properties of the test materials were not reported. SiC that was reaction-bonded with elemental Si was found to have more severe acid attack than alkaline solution attack due to the presence of unreacted elemental Si in the bond phase. Although the SiC material was unaffected by individual acids, combinations such as HF and HNO_3 caused deterioration of the bond phase between SiC grains as free Si was removed. Materials exposed to combinations of acids may exhibit increased corrosion compared to exposure to individual acids. Solution pH was also found to be a factor in the bond phase removal.

A comparison of the corrosion of a sintered SiC material with a reaction sintered SiC and a high purity alumina was made by Lashway (9). Although material properties such as chemistry or physical properties were not listed, a high degree of corrosion resistance in SiC was also found to occur when free elemental Si was absent (Table 15-2).

Table 15-2. Corrosion Resistance of Silicon Carbide and High Alumina in Different Acids and Bases (9).

	Corrosive weight loss, mg/(cm²)(yr)			
Reagent, wt. %	Temp °C	Sintered alpha SiC (No free Si)	Si/SiC composites (12% Si)	Aluminum oxide (99% pure)
98% H_2SO_4	100	1.8	55	65
53% HF	25	< 0.2	7.9	20
85% H_3PO_4	100	< 0.2	8.8	>1000
70% HNO_3	100	< 0.2	0.5	7
25% HCl	70	< 0.2	0.9	72
10% HF + 57% HNO_3	25	< 0.2	>1000	16

Test time: 125 to 300 h of submersive testing, continuously stirred.

Corrosion weight-loss guide:

>1,000 mg/(cm²)(yr)	Completely destroyed within days
100 to 999 mg/(cm²)(yr)	Not recommended for service longer than a month

50 to 100 mg/(cm^2)(yr)	Not recommended for service longer than a year
10 to 49 mg/(cm^2)(yr)	Caution recommended for long-term service
0.3 to 9.9 mg/(cm^2)(yr)	Recommended for long-term service
<0.2 mg/(cm^2)(yr)	Recommended for long-term service; no corrosion other than as a result of surface cleaning was evidenced.

The corrosion resistance of ceramic materials to 96 wt.% H_3PO_4 at temperatures from 22° to 204°C was evaluated by Marcus and Ahrens (10). Although they do not list chemical or physical properties of test materials, they mention the manufacturer and describe their test apparatus. Results (Table 15-3)

Table 15-3. Corrosion Rate of Ceramic Materials in 96 wt% H_3PO_4 (10).

Material name	Chemical formula	Temp., °C	Soak time, h	Corrosion/ dissolution rate, mm/year
Oxides				
Alumina	99 wt pct Al_2O_3 + 1 wt pct SiO_2	22	1,292	0
Alumina	99 wt pct Al_2O_3 + 1 wt pct SiO_2	204	145	451.10
Alumina	94 wt pct Al_2O_3 + 6 wt pct SiO_2	22	1,292	.003
Alumina	94 wt pct Al_2O_3 + 6 wt pct SiO_2	204	145	35.23
Beryllium oxide	BeO	215	112	>22.96
Cordierite	$2MgO \cdot 2Al_2O_3 \cdot 5SiO_2$	204	145	15.37
Mullite	$3Al_2O_3 \cdot 2SiO_2$	204	145	3.43
Silica (fused)	SiO_2	82	1,125	.008
Silica (fused)	SiO_2	204	350	1.02
Spinel	$MgO \cdot Al_2O_3$	204	139	36.27
Zircon	98 wt pct $ZrO_2 \cdot SiO_2$ + 2 wt pct clay ($Al_2O_3 \cdot 2SiO_2$)	204	164	5.61
Zircon	95 wt pct $ZrO_2 \cdot SiO_2$ + 5 wt pct clay ($Al_2O_3 \cdot 2SiO_2$)	204	301	1.25
Zircon	$ZrO_2 \cdot SiO_2$	205	253	.76
Carbides				
Silicon carbide	SiC	208	336	.102
Silicon carbide	SiC	210	120	.147
Silicon carbide	SiC + 30 vol pct Si	204	253	1.70
Nitrides				
Silicon nitride (reaction sintered)	Si_3N_4	204	120	.25
Silicon nitride	Si_3N_4	204	120	.51
Silicon nitride	Si_3N_4	204	119	1.37
Silicon nitride (hot pressed)	Si_3N_4	204	110	4.72

indicated that nominally the same material from different manufacturers can behave differently. For oxide materials, zircon and silica were found to be the least attacked, while Al_2O_3 was attacked the most. Material additives to Al_2O_3, such as SiO_2, greatly increased corrosion resistance. Free elemental Si in SiC was found to cause high rates of material loss. The chemical resistance of Si_3N_4 in hot phosphoric acid varied widely, possibly due to the influence of impurities. Because of the different corrosion behavior of materials such as zircon, silicon carbide, and silicon nitride, the corrosion results in Table 15-3 indicate that nominally the same material can behave differently, depending on how it is produced. At 204°C, oxide materials were not found to be as resistant to phosphoric acid as were some of the carbides and nitrides.

CHEMICAL AND PHYSICAL PROPERTIES OF CERAMIC MATERIALS

The lack of existing definitive data on the corrosion resistance of ceramic materials is in part due to the wide variations in material properties, and to the wide variety of applications where these materials are used. Researchers have evaluated materials in chemical environments using an assortment of sample geometries, often failing to list important material properties such as chemistry or porosity. Even when a specific product is listed by brand name, material properties can change over time as a manufacturer modifies a product or process. The most frequently used chemical resistant brick can be grouped into two categories (11) to describe their chemical and physical properties: acidproof brick (red shale and fireclay) and specialty materials (silica, SiC, and high alumina).

Acidproof Brick

Red shale and fireclay materials are typically prepared by extrusion or dry pressing and fired in a kiln at temperatures from 980° to 1,150°C for red shale and to 1,260°C for fireclay. Chemical composition and crystalline phases present in the red shale and fireclay materials are shown in Table 15-4. Chemical composition varies depending on the local clay deposit used in brick manufacture. The lower firing temperatures, higher glass content, and lower absorption generally associated with red shale brick are due to the presence of more alkali and iron compared to fireclay, as shown in Table 15-4. Materials with low porosity have been found to perform the best in chemical environments.

The crystalline mineral phases present in red shale and fireclay brick are also listed in Table 15-4. The firing time and temperature determine the degree of conversion of the starting materials into glassy and crystalline phases that provide the desired physical properties. Both brick types contain primarily quartz, mullite, and an amorphous (glassy) material. The higher firing temperatures of fireclay can produce a cristobalite phase not observed in red shale. Hematite and

Table 15-4. Range of Chemical Composition and Crystalline Phases Present in Red Shale and Fireclay Brick.

Property	Red shale	Fireclay
Chemical composition, wt pct:		
SiO_2	61.4 - 67	56.8 - 68.6
Al_2O_3	18.6 - 29.4	22.9 - 38.7
Fe_2O_3	4.7 - 6.8	.8 - 3
K_2O	2.5 - 4.6	1 - 3.2
TiO_2	1 - 1.6	1 - 2.8
MgO	.7 - 1.3	.1 - 1.2
Na_2O	.5 - .7	.2 - .5
CaO	.1 - .4	.01 - .8
Phases identified:		
Quartz	Major	Trace-major
Mullite	Trace-major	Minor-major
Cristobalite	None	None-major
Hematite	Minor	None-trace
Rutile	None-trace	None-trace
Amorphous	Major	Minor-major

rutile exist in the red shale after firing, and both can be leached in certain process environments.

The physical properties of red shale and fireclay brick, listed in Table 15-5 along with chemical composition and mineral phases, determine the brick's application. Generally, red shale brick have slightly lower porosity and absorption and higher modulus of rupture, cold crushing strength, and bulk density than fireclay. The red shale brick also have a higher acid resistance (ASTM C-279-H_2SO_4 test), but poorer thermal shock resistance than fireclay brick. Limited information is available on thermal conductivity, thermal expansion, and modulus of elasticity (Table 15-5).

Table 15-5. Physical Property Ranges of Red Shale and Fireclay Brick.

Property	Red shale	Fireclay
Apparent porosity, pct.	3.2 - 12.5	5 - 13.3
Absorption, pct.	.4 - 5	1 - 6
Bulk density, g/cc.	2.27 - 2.50	2.17 - 2.40
H_2SO_4 acid resistance (ASTM C-279), pct wt loss.	.7 - 6	3 - 10
Modulus of rupture, MPa.	14 - 26	8 - 24
Compressive strength, MPa.	69 - 152	34 - 124
Thermal conductivity*, W/m·° C.	1 - 1.3	9 - 1.4
Thermal expansion coefficient*, (25° - 425° C), cm/cm·°C X 10^{-6}	2.2 - 3	1.1 - 1.9
Modulus of elasticity*, GPa.	25 - 92	21 - 69

*Limited information available.

Specialty Materials

The majority of ceramic construction materials used in corrosive enviroments are red shale and fireclay, although specialty materials are used in applications where high temperatures, wear, or chemical compatibility are important. Many specialty ceramic materials can be considered; however, only three of the more common ones: silica, SiC, and high alumina, will be discussed.

Silica. The use of silica materials in chemical environments is limited, because of high cost, to applications requiring a high degree of chemical resistance, such as concentrated phosphoric acid free of fluorine (11). However, silica brick cannot be used in strong alkaline environments or in any concentration of hydrofluoric acid.

Silica bricks of high purity and improved low-temperature thermal shock resistance have been developed for acid resistance usage. A typical brick is produced from high-purity quartz that is fired at a high temperature to a vitreous structure, crushed and graded to a desired particle size, and mixed with a binder and sintering aid. Bricks or other shapes are formed by techniques such as dry pressing or slip casting. Materials are fired at about 1,010°C. Care must be taken in firing and in brick usage to avoid temperatures above 1,100°C, where quartz and cristobalite transformations can occur in the highly vitreous brick. In cyclic temperature service, 1,100°C is the highest use temperature, and with continuous service, the maximum use temperature is 1,650°C.

Table 15-6. Chemical Composition and Crystalline Phases Present in Silica Brick.

Property	Range
Chemical composition, wt pct:	
SiO_2	98.9 - 99.6
Al_2O_3	.2 - .5
Fe_2O_3	.02 - .3
TiO_2	.01 - .02
MgO	.02 - .1
CaO	.02 - .03
Alkali (Na_2O, K_2O, Li_2O)	.01 - .2
Phases identified:	
Quartz	None-minor
Cristobalite	None-minor
Amorphous	Major

Chemical analysis of silica brick is listed in Table 15-6. Silica is the major constituent, with less than 0.5% Al_2O_3 or Fe_2O_3, and less than 0.2% MgO or alkali. Phases in the silica brick identified by x-ray diffraction indicate that the original quartz is predominately converted into an amorphous phase and cristobalite.

The physical properties of silica brick, listed as Type 1 and Type 2 in Table 15-7, depend on silica purity and the manufacturing process. The brick's high silica content makes bond formation between grains during firing difficult, accounting for the low strength and high porosity of fired brick.

Type 1 properties are for a 98% rebonded vitreous silica brick usually manufactured by dry pressing, although casting or air ramming may be used for

Table 15-7. Physical Properties of Silica Brick.

Property	Type 1	Type 2
Apparent porosity, ...pct..	12 - 16	7 - 16
Absorption, ...pct..	5.5 - 7.2	3 - 14
Bulk density, ...g/cc..	1.86 - 1.92	1.79 - 2.05
H_2SO_4 acid resistance (ASTM C-279)*, ...pct wt loss..	1.4	NA
Modulus of rupture, ...MPa..	3.4 - 5.5	3.4 - 13.8
Compressive strength, ...MPa..	31 - 48	14 - 83
Thermal conductivity*, ...W/m•° C..	.7 - 1.2	.6 - 1.2
Thermal expansion coefficient*,		
(25° to 425° C), ...cm/cm•° C X 10^{-6}..	1.2 - 1.6	.2 - .4
(425° to 1,100° C), cm/cm•° C X 10^{-6}..	.1 - .4	.2 - .4
Modulus of elasticity*, ...GPa..	7.6	24 - 35

*Limited information available.
NA - Not analyzed.

nonstandard shapes. This material contains some cristobalite formed during firing that is located at grain boundaries, giving the material its high thermal expansion behavior below 430°C.

The second type, called Type 2, is manufactured from plus 99.5% pure rebonded fused silica. This material does not contain as much cristobalite as found in Type 1, and has lower thermal expansion below 430°C. Manufacturing is by slip casting or by a proprietary method, resulting in different physical properties, but similar chemistry.

Silicon Carbide. Because SiC has outstanding chemical resistance, it has many potential applications. Cost has limited its use, however, to specialized areas where it outperforms other materials because of properties such as resistance to thermal shock, abrasion, impact or high temperature. SiC is made into shapes for chemical applications ranging from refractory brick to pump seals by a variety of techniques such as dry pressing, extrusion, slip casting, injection molding, isostatic pressing, or hot isostatic pressing, and can have a variety of bonds (12) between grains (Table 15-8). Depending on the fabrication technique,

Table 15-8. Categories of SiC Materials (10).

Nomenclature	Bonding	SiC content (wt-pct)
ceramic bonded SiC	clay, silicates, Si_3N_4, Si_2N_2O	up to 95
recrystallized SiC	SiC, Si + SiC, or C + SiC	100
reaction bonded SiC (including infiltrated)	SiC, Si + SiC, or C + SiC	90
pressureless sintered SiC	SiC, Si + SiC, or C + SiC	98
hot pressed SiC (including HIP)	SiC, Si + SiC, or C + SiC	>99.5
HIP-postdensified sintered SiC	SiC, Si + SiC, or C + SiC	98

chemistry and physical properties can vary over a wide range and are listed in Table 15-9. Since SiC is inert to most chemicals, chemical attack typically occurs at the bond phase between grains. Chemical use limitations are thus typically determined by the bond.

Table 15-9. Physical Property Ranges of Chemical Resistant SiC Materials.

Property	Range
SiC content.......................... wt pct..	78 - 100
Porosity,..............................pct..	2 - 18
Bulk density,.........................g/cm^3..	2.55 - 3.10
Modulus of rupture,.....................MPa..	21 - 586
Compressive strength,...................MPa..	138 - >3,900
Thermal conductivity*, (1,200° C)*, W/m•° C...	14.4 - 34.3
Thermal expansion coefficient (20° - 1,400° C),.........cm/cm•° C X 10^{-6}..	4.7 - 5.3
Modulus of elasticity*..................GPa..	117 - 410

*Limited information available

High Alumina. High alumina materials have a wide range of chemical and physical properties. Materials from traditional refractory brick to advanced technical ceramics are used in chemical applications, but because of their expense, they are not commonly used in chemical environments. Alumina content varies from about 79 to 100%, with porosity varying from 26% to near zero. Because of the wide variations in manufacturing techniques and in chemical and physical properties, only a few general properties of alumina materials are listed in Table 15-10.

Table 15-10. Physical Property Ranges of Chemical Resistant Aluminas.

Property	Range
Al_2O_3 content.................... wt pct. .	79 - 99.5
Porosity,.............................pct..	0 - 26
Bulk density,.......................g/cm^3..	2.55 - 3.89
Modulus of rupture,..................MPa..	3 - 25
Compressive strength,................MPa..	38 - 2,600
Thermal conductivity, (400° C)*, W/m•° C..	6.7 - 12.1
Thermal expansion coefficient (25° - 500° C)*,......cm/cm•° C X 10^{-6}..	5 - 7
Modulus of elasticity*,.............GPa..	221 - 372

*Limited information available

ACID EXPOSURE TESTS AND RESULTS

The corrosion resistance of seven ceramic materials (two red shale, two fireclay, a silicate bonded SiC, a silica, and a high alumina) to HCl (13, 14), HNO_3 (15) and H_2SO_4 (16) has been evaluated for 110 days by the Bureau of Mines. Chemical and physical properties of the ceramic materials tested are listed in Table 15-11. Red shale and fireclay materials of high and low porosity were investigated.

Acid Corrosion

The selected ceramic materials were exposed to various acid conditions in the test apparatus shown in Figure 15-3. The system, described in a Bureau publication (13), exposed 30 samples at a time.

Table 15-11. Chemical Composition and Physical Properties of Ceramic Materials (13-16).

Property	Red Shale A	Red Shale B	Fireclay A	Fireclay B
Chemical composition, wt pct:				
SiO_2	64.6	63.3	59.4	68.6
Al_2O_3	20.6	20.7	31.7	22.9
Fe_2O_3	6.4	5.9	2.0	1.9
K_2O	3.6	4.6	3.0	1.5
TiO_2	1.6	1.6	1.7	1.4
MgO	.99	1.3	.35	.56
Na_2O	.58	.57	.27	.27
CaO	.39	.05	.03	.06
BaO	NA	NA	NA	NA
SiC	NA	NA	NA	NA
Apparent porosity, pct	10.9	3.3	5.7	11.1
Bulk density, g/cm^3	2.39	2.56	2.38	2.26
Cold crushing strength, MPa	130	140	68	67

Property	SiC	Silica	High alumina
Chemical composition, wt pct:			
SiO_2	9.6	98.4	8.5
Al_2O_3	.76	.51	85.6
Fe_2O_3	.37	.12	.31
K_2O	.046	.049	.07
TiO_2	.11	<.05	.12
MgO	.035	.056	3.1
Na_2O	.017	.53	0.2
CaO	.22	.12	1.1
BaO	NA	NA	.77
SiC	87.9	NA	NA
Apparent porosity, pct	10.9	11.9	.5
Bulk density, g/cm^3	2.64	1.90	3.70
Cold crushing strength, MPa	99	28	403

NA Not analyzed.

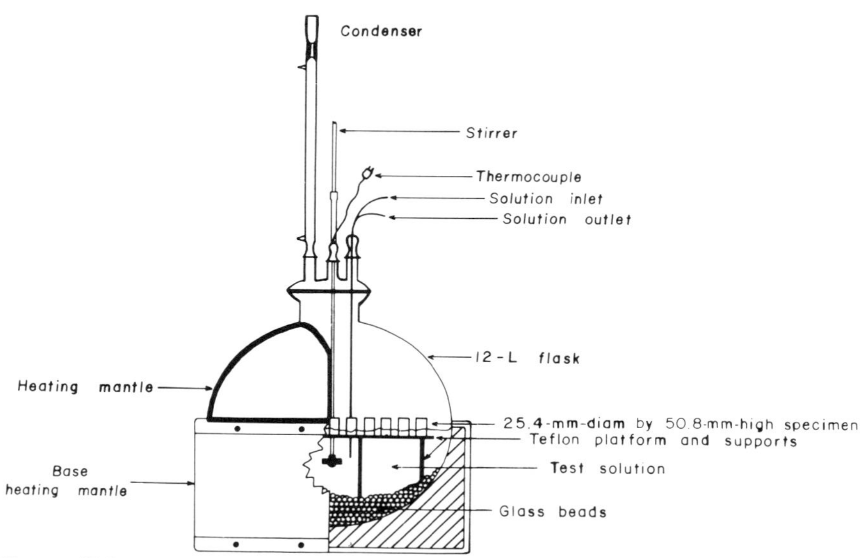

Figure 15-3. Construction detail of corrosion test apparatus.

Weight, cold crushing strength, and microstructural changes were determined on each material. Atomic absorption analysis was used for monitoring Al, Ca, Fe, K, Mg, Na, Si, and Ti ions in the leach solution. Weight changes occurring in the samples closely followed ion removal trends, so only ion removal trends are discussed. An acid-filled vessel containing no test samples was used to provide a control solution for establishing background ion concentrations. Corrosion results from exposing samples to a range of HCl, HNO_3, and H_2SO_4 environments are discussed below.

HCl. The samples listed in Table 15-11 were exposed to 20 wt.% HCl at 50°, 70°, and 90°C and to 30 wt.% HCl at 70°C for 110 days (13, 14). The amount of ions leached from samples was dependant on acid concentration and increased with increasing temperature, exhibiting a maximum ion loss in 20 wt. % HCl at 90°C (Figure 15-4). The total ion weight loss of the red shale and fireclay materials is directly related to their initial apparent porosity. Samples with higher porosity have higher rates of ion removal.

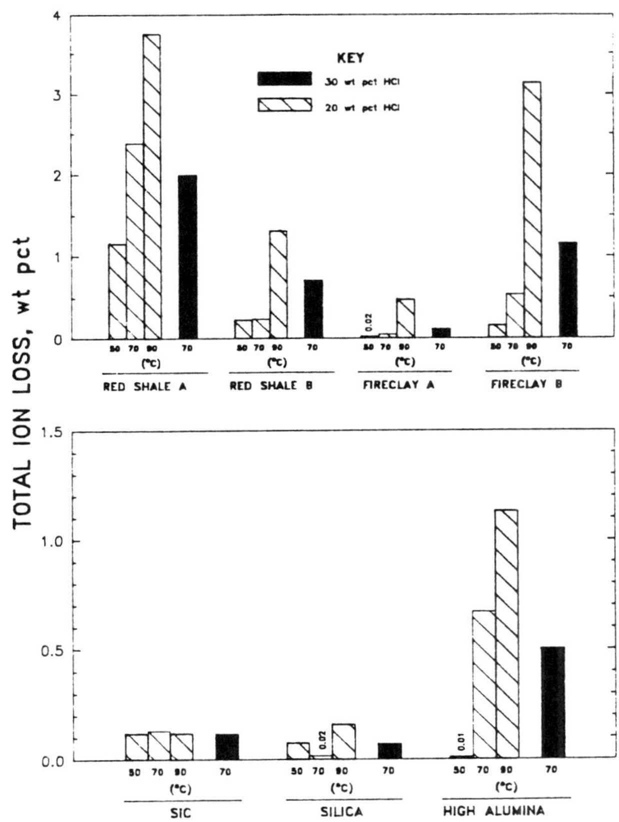

Figure 15-4. Total ions leached from acidproof brick (top) and specialty materials (bottom) in 110 days of HCl exposure.

Specific ions monitored for removal from samples are shown in Table 15-12. In the red shale and fireclay materials, Al and Fe ions had the highest rate of removal, while Si had the lowest, never higher than in the control test. Iron removal from hematite in the microstructure of the red shale and fireclay samples (Figure 15-5) was visible.

Table 15-12. Ions Leached from Ceramic Materials after 110 Days Exposure to HCl, wt%.

Brick and exposure conditions[1]	Ion leached							Total ions leached
	Al	Ca	Fe	K	Mg	Na	Ti	
Red shale A:								
50° C, 20 HCl.....	0.081	0.0450	0.97	0.015	0.026	0.0009	0.013	1.15
70° C, 20 HCl.....	.294	.0700	1.80	.065	.074	.0238	.052	2.38
90° C, 20 HCl.....	.972	.0666	2.14	.248	.161	.0554	.103	3.75
70° C, 30 HCl.....	.186	.0593	1.57	.040	.066	.0138	.059	1.99
Red shale B:								
50° C, 20 HCl.....	.031	.0227	.159	.010	.0090	.0031	.0029	.23
70° C, 20 HCl.....	.039	.0141	.153	.018	.0099	.0042	.0050	.24
90° C, 20 HCl.....	.314	.0719	.675	.142	.0568	.0231	.0221	1.31
70° C, 30 HCl.....	.093	.0145	.528	.030	.0240	.0075	.0071	.71
Fireclay A:								
50° C, 20 HCl.....	.006	.00017	.0118	.0009	ND	.00022	.0013	.020
70° C, 20 HCl.....	.023	.00016	.0219	.0025	.00117	.00053	.0051	.054
90° C, 20 HCl.....	.351	.00607	.0667	.0200	.00660	.00435	.0254	.480
70° C, 30 HCl.....	.055	.00024	.0365	.0029	.00241	.00083	.0137	.112
Fireclay B:								
50° C, 20 HCl.....	.03	.0061	.095	.004	.0026	.0022	.015	.16
70° C, 20 HCl.....	.24	.0145	.186	.018	.0114	.0091	.056	.54
90° C, 20 HCl.....	2.22	.0234	.514	.146	.0415	.0506	.147	3.14
70° C, 30 HCl.....	.52	.0254	.359	.031	.0241	.0145	.175	1.15
SiC:								
50° C, 20 HCl.....	.00432	.00120	.096	.000425	.000705	.000341	.0163	.119
70° C, 20 HCl.....	.00563	.00141	.118	.000409	.000725	.000180	.0055	.132
90° C, 20 HCl.....	.00884	.00121	.109	.000296	.000574	.000354	ND	.120
70° C, 30 HCl.....	.00375	.00080	.095	.000140	.000799	.000639	.0152	.116
Silica:								
50° C, 20 HCl.....	.0178	.0103	.0368	.00326	.0056	.00094	.000207	.075
70° C, 20 HCl.....	.0037	.0017	.0089	.00104	.0015	.00002	.000036	.017
90° C, 20 HCl.....	.0543	.0291	.0578	.00497	.0110	.00272	ND	.160
70° C, 30 HCl.....	.0264	.0117	.0215	.00125	.0056	.00087	ND	.067
High alumina:								
50° C, 20 HCl.....	.006	.003	ND	.0001	ND	.0011	ND	.01
70° C, 20 HCl.....	.410	.206	.00139	.0072	.0060	.0429	.00086	.67
90° C, 20 HCl.....	.693	.327	.00383	.0123	.0245	.0672	.00121	1.13
70° C, 30 HCl.....	.295	.159	.00112	.0053	.0050	.0313	.00067	.50

ND Not detected.
[1]HCl in weight percent.

Silica and high alumina materials show an increase in weight loss with an increase in temperature. The SiC brick had a consistent weight loss of about 0.12 wt.%, regardless of temperature or acid concentration. The ion showing the greatest removal in SiC is Fe, in silica it is Al and Fe, and in the high alumina Al and Ca. Regardless of temperature or acid concentration, the SiC and the silica materials had very low total ion losses, followed by the low porosity fireclay A material.

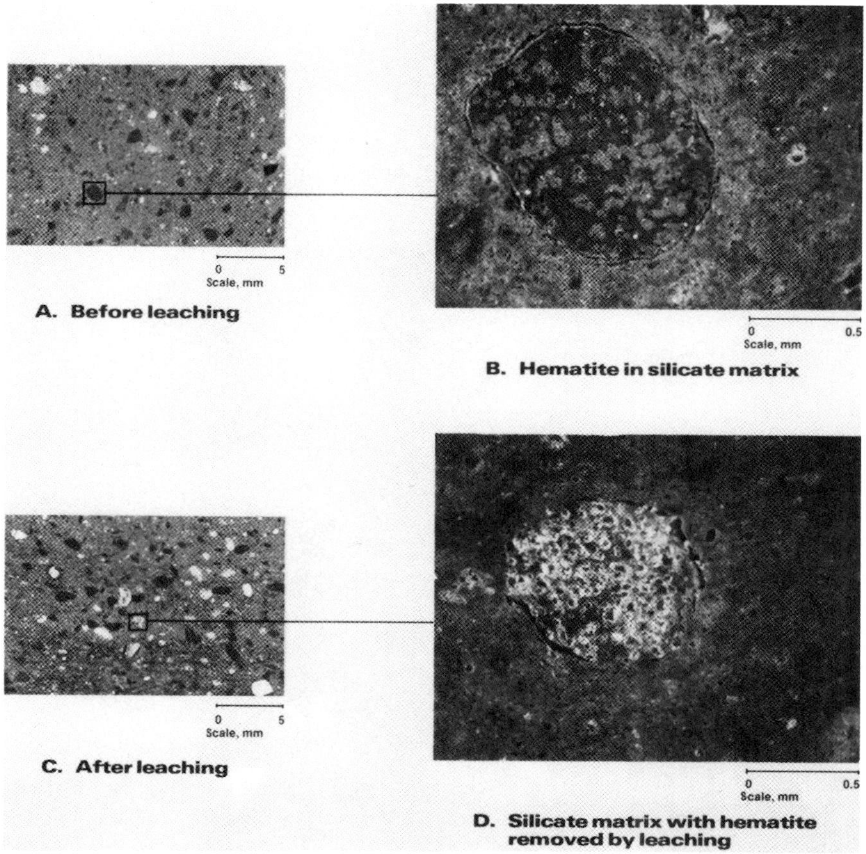

Figure 15-5. Red shale material before and after exposure to 20 wt% HCl at 80°C for 110 days.

Selective leaching was observed in the high alumina brick (Figure 15-6). A lighter colored outer zone had a noticeable increase in small pores compared to the interior. This increase in porosity, due to removal of material from the grains or at the grain boundaries, had no effect on crushing strength over the 110 day test period, but may have an effect over longer exposure times. No significant trends in crushing strength were observed for any samples with reference to changes in acid concentration or temperature during the 110-day test period in HCl. During the 110 days of exposure, the rate at which ions were leached from samples at any given temperature and acid condition tended to follow a second order parabolic curve (see Figure 15-7), in its general form listed as:

$$y = a_1 + a_2x + a_3x^2, \tag{15-1}$$

where:
y = ion concentration in solution,
x = days,
a_1, a_2, a_3 = constants.

As ions were continually leached from a sample, diffusion played a more important role, causing leach rates to decrease and the slope of ion removal to change.

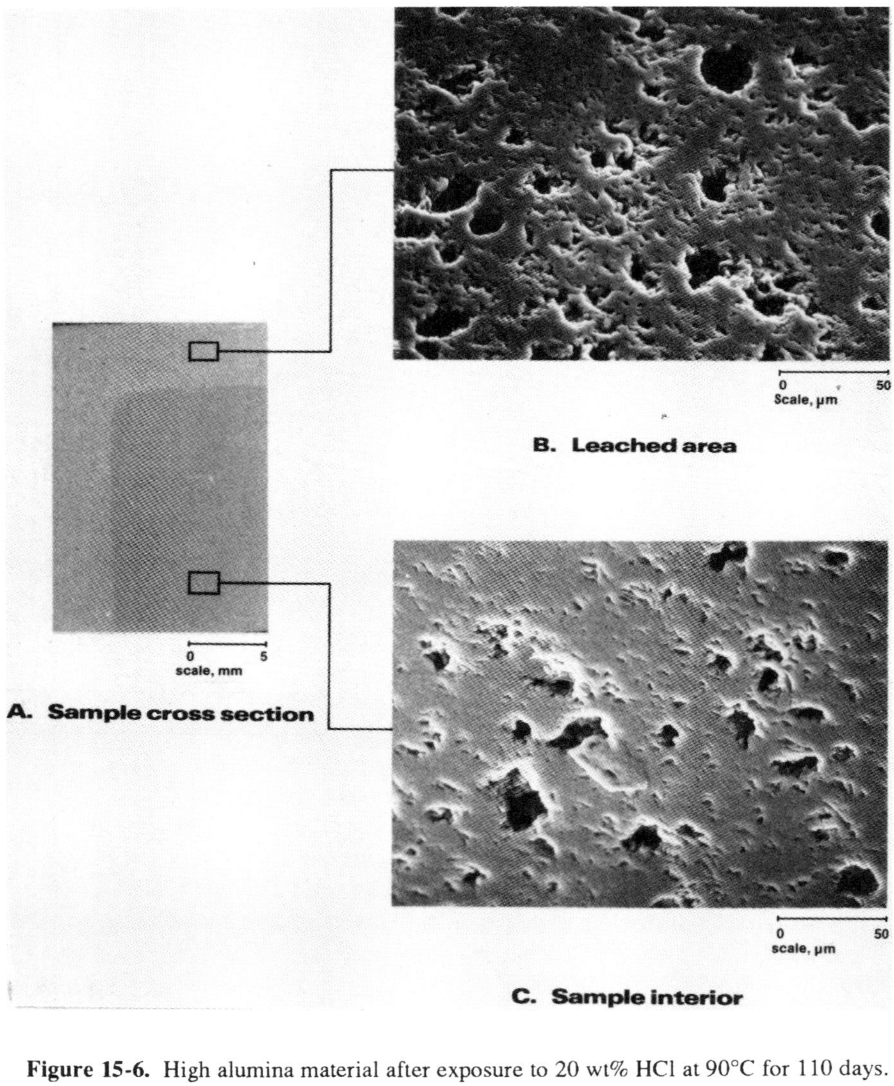

Figure 15-6. High alumina material after exposure to 20 wt% HCl at 90°C for 110 days.

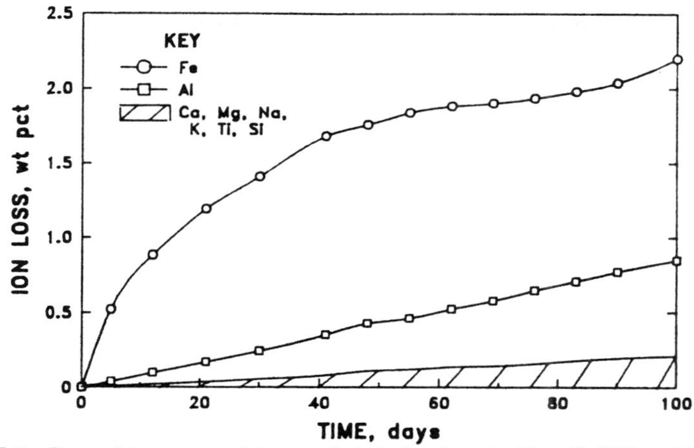

Figure 15-7. Rate of ion removal from red shale brick A in 20 wt% HCl at 90°C.

HNO_3. When HNO_3 (15) corrosion testing of the seven acid-resistant materials listed in Table 15-10 was initiated, NO_x fumes developed in the test vessels with the silicon carbide material, so testing of it was discontinued. The SiC material evaluated is not recommended for use in the high concentrations of HNO_3 evaluated. The remaining materials were evaluated in 40 wt.% HNO_3 at 70° and 90°C and in 60 wt.% HNO_3 at 50°, 70°, and 90°C for 110 days.

Cold crushing strength data did not reveal any significant trends in spite of the weight changes observed in the samples. Ions removed from samples at any given temperature or acid concentration tended to follow the general second-order parabolic equation observed for HCl. Silicon was not found to be removed from any samples in amounts above the control test data.

The more porous bricks (red shale A and fireclay B) exhibited several times more total ions leached (Figure 15-8) at all of the five test conditions than did the denser types (red shale B and fireclay A). In the red shale or fireclay bricks, Fe or Al were the predominant ions removed. Iron removal from hematite occurred in red shale samples, as shown earlier in Figure 15-5. The ions removed from the high alumina samples were predominately Al and Ca. Selective leaching changes in the high alumina samples were similar to those observed after HCl exposure (Figure 15-6). The silica brick had the best acid resistance properties, followed by the low porosity red shale B and fireclay A materials. Ions leached from all samples evaluated were generally higher for the 40 wt.% than for the 60 wt.% HNO_3 solutions.

H_2SO_4. Sulfuric acid corrosion testing (16) of the seven acid-resistance materials listed in Table 15-11 was conducted in 10 wt.% H_2SO_4 at 50°C and 10, 30, 50, 70, and 90 wt.% H_2SO_4 at 90°C for 110 days. Cold crushing strength data did not indicate any significant trends, in spite of the weight changes observed in some samples (Figure 15-9).

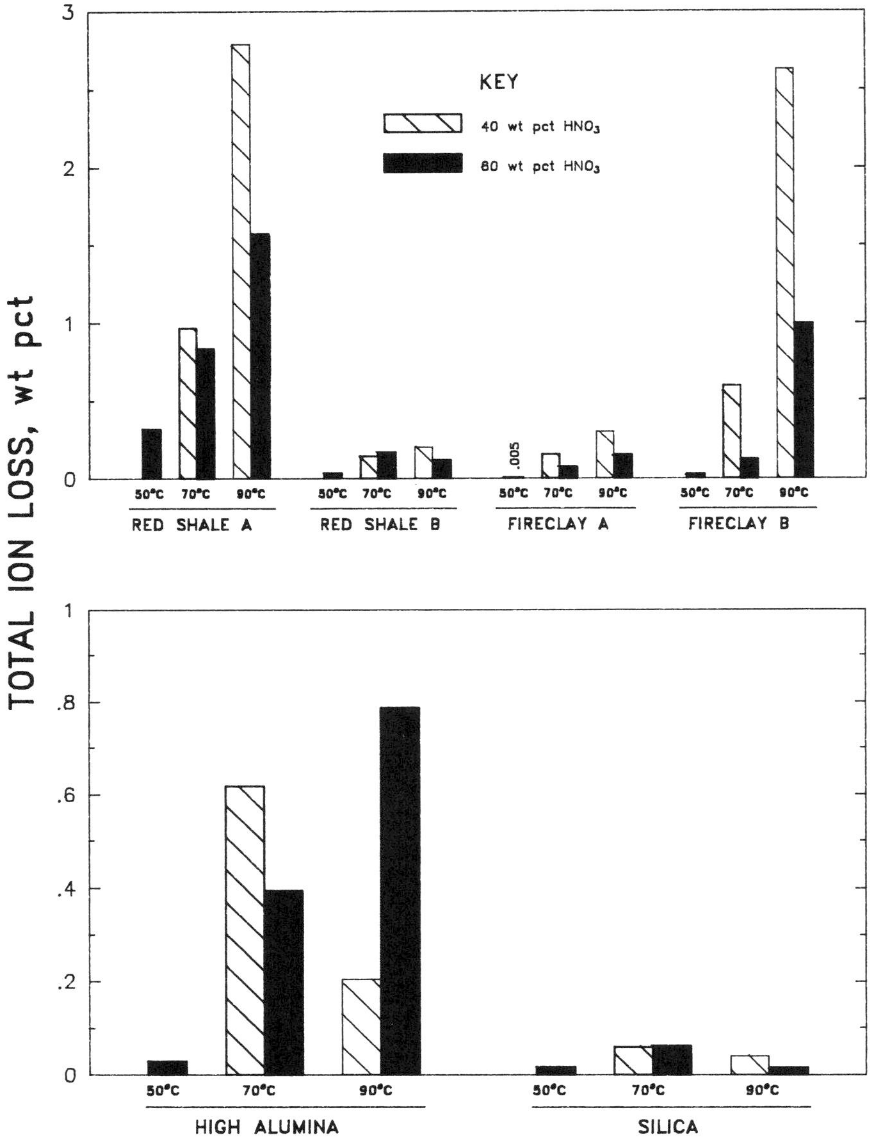

Figure 15-8. Total ions leached from acidproof brick (top) and specialty materials (bottom) in 110 days of HNO_3 exposure.

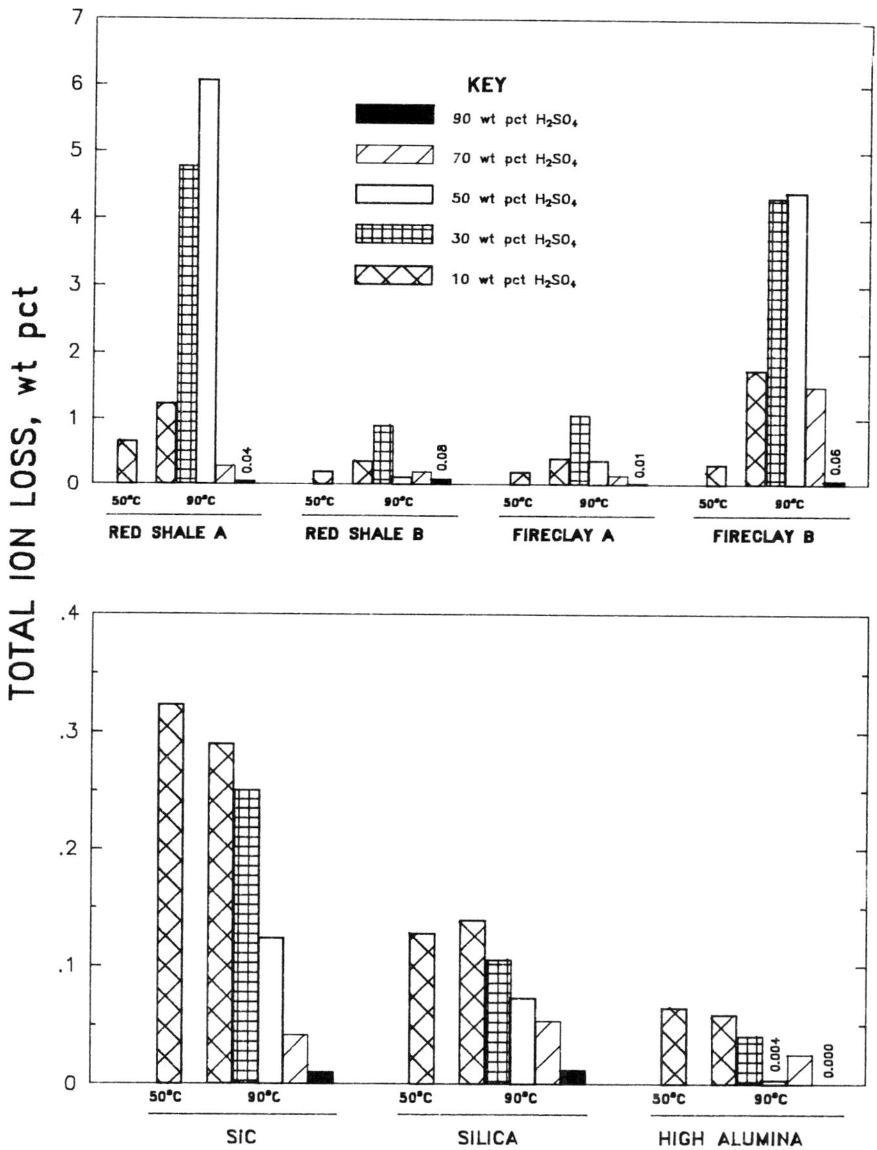

Figure 15-9. Total ions leached from acidproof brick (top) and specialty materials (bottom) in 110 days of H_2SO_4 exposure.

Total ions removed from the red shale and fireclay samples increased with an increase in temperature at the same acid concentration. When acid concentration was increased from 10 to 90% at 90°C, the quantity of ions removed increased to a maximum value at either 30 or 50 wt.% H_2SO_4. Increasing acid concentration to higher values resulted in a decrease in ion removal. The quantity of ions removed from the SiC, silica, and high alumina samples was fairly constant with an increase in temperature and generally decreased when acid concentration was increased from 10 to 90%. In all materials, either the Al or the Fe ion was leached in the largest amounts. Ions removed from samples at any given temperature or time tended to follow the general second-order parabolic equation discussed for HCl. The more porous bricks (red shale and fireclay B) exhibited higher quantities of total ions removed from samples than did any of the denser materials (red shale B and fireclay A).

Microscopic examination of samples indicated that hematite was leached from the microstructure of the red shale and fireclay materials in a way similar to that observed with HCl in Figure 15-5. In addition, XRD analysis indicated a decrease in the mullite phase.

Volume Expansion

Volume expansion, also known as irreversible growth or swelling, occurs in ceramic materials exposed to chemical environments and is generally thought to be similar to moisture expansion observed in structural clay products. Linear expansion of up to 0.35% in red shale or fireclay brick used in chemical applications has been documented (17), but no published technical information or standard test exists to measure expansion or the influence of such factors as temperature, pressure, load, or exposure environment on expansion. Sheppard (18) mentions that a length increase of about 0.16% typically occurs in red shale or fireclay materials used in chemical environments, noting that no one has determined the mechanism of expansion or its relationship to raw material, composition, firing temperature, or firing time. Expansion of structural units in chemical environments is a critical factor in vessel design. Improper design or construction may result in the brick face spalling, the brick lining arching or buckling away from the substrate, the impermeable backup membrane rupturing or tearing, or the transfer of damaging stress loads into the substrate structure.

Different explanations proposed by researchers as to the cause of moisture expansion were summarized by Demediuk and Cole (19) into the following three areas: capillary water, physically-absorbed water on pore surfaces, or chemically-bound water. Volume expansion in clay products was found by Young and Brownell (20) to be related to the noncrystalline phases and chemical composition (sodium and potassium aluminosilicate cause high expansion and lithium aluminosilicates low expansion). Volume expansion was found to be increased by the presence of alkali, and to be decreased by firing the materials to higher temperatures. Robinson (21) found that raw materials sources and brick firing temperatures influenced expansion.

Expansion due to moisture was broken into two components by Lomax and Ford (22), a reversible component (usually less than 0.02%) associated with wetting and drying of a brick, and an irreversible expansion (up to 0.2%) that continues for many years. The volume expansion of a particular brick was thought to be dependent on the firing treatment it received rather than the raw material composition.

Long-term volume expansion studies (up to 20 yr) on floor tile (0 to 8.4% porosity) were conducted by Slyh (23) and compared with autoclave-treated or water-soaked materials. Twenty to twenty-five percent of the expansion occurred in the first month, with 75 to 85% occurring within 10 years. Although the rate of expansion became very low after 10 years and although expansion was measured only through 20 years; extrapolated curves predicted expansion would continue for much longer times, but at a very low rate in the presence of water. Twenty-year expansions ranged from -0.001% for low porosity tiles to 0.057% for the high porosity tiles. Tile expansion after 20 years was similar to a 1-h autoclave treatment at 200 psi and 195°C.

Actual expansion that occurs in structural clay products under industrial use conditions is normally less than predicted by laboratory studies. Ritchie (24) found autoclaved bricks that were restrained (50 psi load) produced 4 to 24% less expansion than unrestrained samples. Brick movement over a 21-year period evaluated by Beard, Dinnie, and Sharples (25) indicated brick in lightly restrained walls experienced expansion values 25% less than unrestrained bricks and heavily restrained brick were 50% less. Numerous researchers have also found that some or all of volume expansion can be eliminated by reheating samples to temperatures of 500°C or higher.

The Bureau of Mines evaluated the volume expansion of the two red shale, two fireclay, silica, and silicate bonded SiC materials (listed in Table 15-11) exposed for times up to 300 days to different temperatures and acid concentrations of HCl, HNO_3, and H_2SO_4; as well as autoclaved for short time periods (26,27). The selected ceramic materials were 25.4mm diameter by 38.1mm high and were prepared for length change monitoring by diamond drilling 2.0mm diameter holes in the center of the 25.4mm diameter surface to a depth of 2.0mm. A bench micrometer with 5.0mm hardened steel balls attached was used to measure length changes to 0.0025mm. The 5.0mm steel ball was centered above the 2.0mm hole and the dimensional change monitored. The ceramic materials were exposed to the different acid conditions in the test apparatus shown in Figure 15-3, which had the teflon platform removed. Samples rested on glass beads and were completely immersed in the test solutions.

Ambient exposure. Linear expansion data of ceramic materials after exposures up to 300 days to different acid environments are listed in Table 15-13. Regardless of the exposure time or condition, the high-porosity red shale A and fireclay B materials had statistically significant larger expansions than the

Table 15-13. Linear Expansion of Ceramic Materials after 100 and 300 Day Exposures, %.

Exposure condition	Red shale A	Red shale B	Fireclay A	Fireclay B	Silica	SiC
100-day exosure:						
90° C, H_2O......	0.050	0.032	0.018	0.049	0.025	0.015
70° C, 20 HCl...	.045	.030	.021	.044	[1].028	.020
90° C, 20 HCl...	.080	.050	.041	.074	.050	.026
70° C, 30 HCl...	.041	.032	.025	.037	.032	[1].009
90° C, 40 HNO_3..	.064	.034	.018	.055	.025	NA
50° C, 60 HNO_3..	.016	.007	.009	.016	[1].009	NA
90° C, 60 HNO_3..	.057	.039	.030	.050	.021	NA
90° C, 10 H_2SO_4.	.061	.045	.028	.065	[1].021	.015
90° C, 50 H_2SO_4.	.048	.034	.028	.058	.027	[1].009
300-day exposure:						
90° C, H_2O......	.123	.055	.022	.099	[1].011	[1].002
70° C, 20 HCl...	NA	NA	.060	.109	[1]-.022	.069
90° C, 20 HCl...	.449	.142	.135	.359	.076	.087
70° C, 30 HCl...	NA	NA	.037	.065	.049	.042
90° C, 40 HNO_3..	.220	.098	.065	.207	.071	NA
50° C, 60 HNO_3..	.039	.027	[1].006	.024	.030	NA
90° C, 60 HNO_3..	.151	.092	.064	.135	[1].039	NA
90° C, 10 H_2SO_4.	NA	NA	.050	.184	[1].045	.052
90° C, 50 H_2SO_4.	NA	NA	.058	.231	[1].018	[1].034

NA Not analyzed.
[1]Not statistically significant; other data are statistically significant changes at 95-pct-confidence level (Student's t-test).

corresponding low-porosity red shale B and fireclay A materials. When the temperature was raised at constant acid concentration (20 wt.% HCl or 60 wt.% HNO_3), the amount of expansion increased. Because the slope of the linear expansion curves in Figure 15-10 indicates that dimensional changes may still be increasing (20 wt.% HCl at 90°C) or begin to level off (60 wt.% HNO_3 at 90°C) by the 300th day, no clear volume expansion behavior is observed.

A general ranking of the volume expansion of the materials tested (largest to smallest) is as follows: high-porosity red shale A and fireclay B >> low-porosity red shale B and fireclay A > SiC and silica. A general ranking (largest to smallest) of the effect of acid environments is as follows: HCl > H_2SO_4, HNO_3 > H_2O. Acid exposures, whether HCl, HNO_3, or H_2SO_4, resulted in greater sample expansion when compared to water exposure at the same temperature. The increased expansion of samples in an acid environment may be due to the leaching of specific ions from samples.

Autoclave Exposure. The results of accelerated autoclave volume expansion tests in different environments is shown in Table 15-14. Regardless of the exposure condition, the high-porosity red shale A samples had greater expansion than the corresponding low-porosity materials. This same trend was noted for 8300-day volume expansion measurements listed in Table 15-13.

Eight hours of autoclave exposure to 50 wt.% H_2SO_4 at 230°C resulted in the most expansion for all the ceramic materials tested. The expansion ranged from 0.094% for the high-porosity red shale A to -0.002% for silica. The expansion observed in the silica and SiC materials was usually small or negative. Samples autoclaved in water for 24-h had expansions ranging from a high of 0.105% for

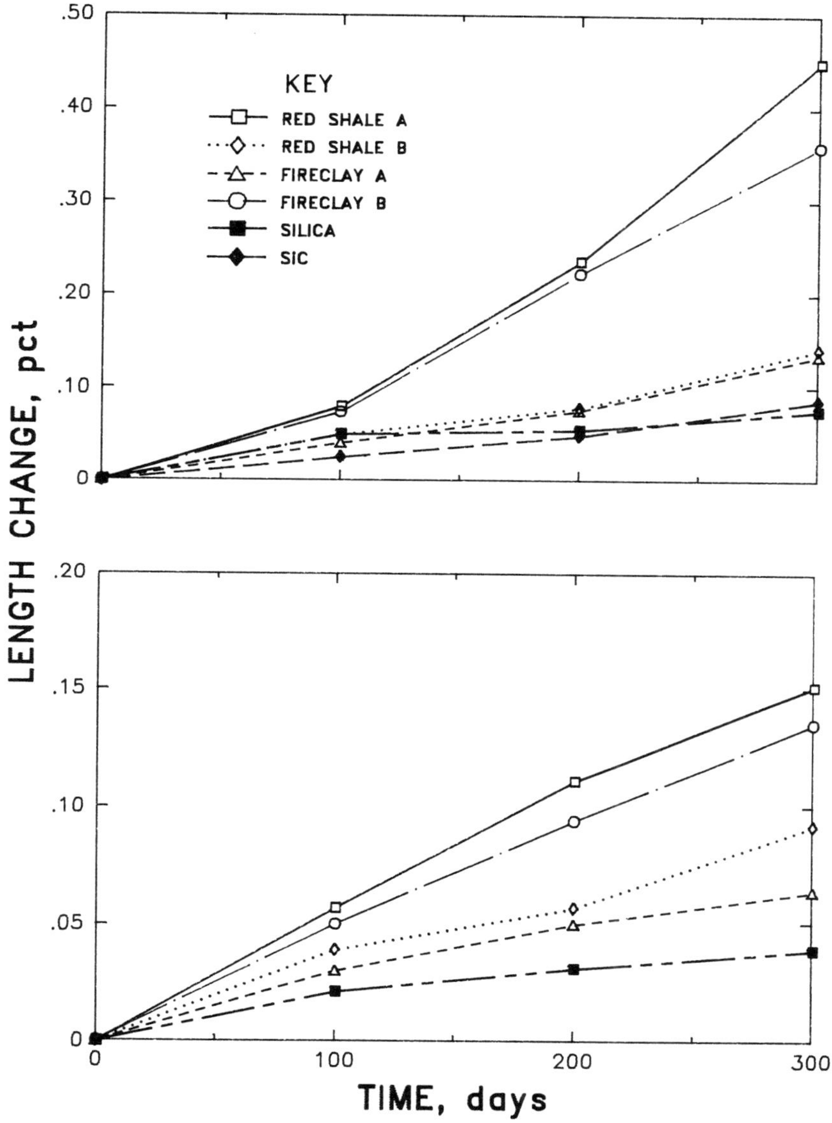

Figure 15-10. Linear expansion of ceramic materials exposed to 20 wt% HCl (top) and 60 wt% HNO_3 (bottom) at 90°C.

Table 15-14. Linear Expansion of Ceramic Material After Autoclaving at 1.5 MPa, Under Four Exposure Conditions, %.

Ceramic material	H_2O, 200° C, 24 h	20 wt pct HCl, 200° C, 8 h	40 wt pct HNO_3, 210° C, 8 h	50 wt pct H_2SO_4, 230° C, 8 h
Red shale A.......	0.105±0.009	0.077±0.023	0.041±0.008	0.094±0.012
Red shale B.......	.049± .004	[1].006± .030	.017± .012	.032± .011
Fireclay A.......	.023± .014	[1].013± .026	[1].008± .017	.061± .017
Fireclay B.......	.103± .010	.054± .022	.024± .012	.088± .013
Silica....	-.029± .021	[1]-.037± .048	-.019± .016	[1]-.002± .007
SiC.......	.059± .040	-.064± .024	NA	-.065± .018

NA Not analyzed.
[1]Not statistically significant; other data are statistically significant changes 95-pct-confidence level (Student's t-test).
NOTE.--Plus-minus (±) values are at 95-pct confidence intervals.

the high-porosity red shale A to a low of -0.029% for the silica brick. The relative expansion trends between 300-day atmospheric exposures and autoclaved samples were comparable, suggesting autoclaving of samples as an accelerated method for comparing volume expansion of different ceramic materials exposed to acid environments.

SUMMARY

1. Ceramic construction materials for chemical processes consist of two categories of materials, acidproof brick (red shale and fireclay) and specialty materials (silica, silicon carbide, high alumina, etc.)

2. Ceramic construction materials for use in chemical environments should be evaluated under conditions that simulate actual use conditions.

3. Autoclave testing of chemical resistant materials accelerates expansion, suggesting autoclaving as an accelerated method for evaluating volume expansion.

4. A study of the effects of HCl, HNO_3, and H_2SO_4 exposures on red shale, fireclay, silica, SiC, and high alumina materials indicated the following:

   a. Ion removal from materials tended to follow a second order parabolic equation.

   b. In general, Fe and Al ions have the highest removal rates, with Ca, Mg, Na, K, and Ti removal being minor. Si was not removed from samples.

   c. The total ion weight loss of acidproof materials was directly related to their initial apparent porosity.

d. Strength changes did not occur in samples.

e. The amount of ions removed from a material depended on the acid concentration and temperature. Increasing temperature generally increased ion leach rates.

f. A general ranking of the volume expansion of materials tested (largest to smallest) is: high porosity red shale and fireclay >> low porosity red shale and fireclay > SiC and silica.

g. A general ranking (largest to smallest) of the effect of acid environment on volume expansion is: HCl > H_2SO_4, HNO_3 > H_2O.

5. The overall performance of ceramic materials evaluated in different acid environments can be rated as follows:

a. in HCl - SiC, silica best, followed by low porosity fireclay.

b. in HNO_3 - silica best, followed by low porosity red shale and fireclay, SiC not recommended.

c. in H_2SO_4 - silica, high alumina, and SiC best, followed by the low porosity red shale and fireclay.

REFERENCES

1. Lay, L., *Corrosion Resistance of Technical Ceramics*, pp 34-44, 50-55, 69-85, 98-107, National Physical Laboratory, Teddington, Middlesex (1983).
2. Balabanovich, G.N., Vinnikov, L.I., and Lesokhin, I.G., *Tr. Leringr. Tekno. Inst. im. Lersoveta* 59: 95-100 (1961).
3. Campbell, I.E., and Sherwood, E.M., *High-Temperature Materials and Technology*, pp 141-147, Wiley (1967).
4. *The Oxide Handbook* (G.V. Samsonov ed.), p 524, IFI/Plenum (1973).
5. Samsonov, G.V., and Vinitskii, I.M., *Handbook of Refractory Compounds*, pp 299-341, IFI/Plenum (1980).
6. Lay, L.A., *National Physical Laboratory Report Chem 96*, "The Resistance of Ceramics to Chemical Attack," pp 1-27, Middlesex, England (1979).
7. Morrell, R., *Handbook of Properties of Technical and Engineering Ceramics - Part 1*, "An Introduction for the Engineer and Designer," pp 183-203, Her Majesty's Stationery Office (1985).
8. Clinton, D.J., Lay, L.A., and Morrell, R., *National Physical Laboratory Report Chem 113*, "An Appraisal of the Resistance of Refel Silicon Carbide to Chemical Attack," pp 1-6, Middlesex, England (1980).

9. Lashway, R.W., *Chemical Engineering* 92: 121-122 (1985).
10. Marcus, L., and Ahrens, R.R., *Amer. Cer. Soc. Bull.* 60: 490-493 (1981).
11. Bennett, J.P., and Eckert W.W., in: *Acid Brick and Silica Brick, Corrosion and Chemical Resistant Masonry Materials Handbook* (W. L. Sheppard ed.), pp 144-154, Noyes (1986).
12. Schwetz, K.A., *Radex-Rundschau* 1: 26-39 (1989).
13. Bennett, J.P., *BuMines RI 8650*, "Corrosion Resistance of Ceramic Materials to Hydrochloric Acid (20 wt.% at 50°C)," p 11 (1982).
14. Bennett, J.P., *BuMines RI 8807*, "Corrosion Resistance of Ceramic Materials to Hydrochloric Acid," p 14 (1983).
15. Bennett, J.P., *BuMines RI 8851*, "Corrosion Resistance of Selected Ceramic Materials to Nitric Acid, p 12 (1984).
16. Bennett, J.P., *BuMines RI 9011*, "Corrosion Resistance of Selected Ceramic Materials to Sulfuric Acid," p 11 (1986).
17. Flynn, R.C., and Stephens, L. C., *Pulp and Paper Industry Corrosion Problems*, pp 46-50, Int. Assoc. Corrosion Eng., (1974).
18. Sheppard, W.L., *Chemical Engineering* Dec. 8, 1986: 149-154.
19. Demediuk, T., and Cole, W.F., *J. Amer. Cer. Soc.* 43: 359-367 (1960).
20. Young, J.E., and Brownell, W.E., *J. Amer. Cer. Soc.* 42: 571-581 (1959).
21. Robinson, G.C., *Am. Cer. Soc. Bull.* 64: 712-715 (1985).
22. Lomax, J., and Ford, R.W., *Trans. and J. Br. Ceram. Soc.* 82:79-82 (1983).
23. Slyh, J.A., *Am. Ceram. Soc. Bull.* 63: 1495-1497 (1984).
24. Ritchie, T., *Build Res. Note 103*, p 5, Natl. Res. Council Can., Div. Build. Res. (1975).
25. Beard, R., Dinnie, A., Sharples, A.B., *Trans. and J. Br. Ceram. Soc.* 82: 82-86 (1983).
26. Bennett, J.P., and Clancy, T.A., *BuMines RI 8962*, "Volume Expansion of Acidproof Brick Exposed to 20 Wt.% HCl at 90°C," p 7 (1985).
27. Bennett, J.P., and Maginnis, M.A., *BuMines RI 8962*, "Dimensional Changes of Select Ceramic Materials Exposed to HCl, HNO_3, and H_2SO_4 Acid Environments," p 10 (1988).

16

Degradation of Ceramic Cutting Tools

P. N. Vaidyanathan

Assistant Director
MICROFABRITECH®
University of Florida
Gainesville, Florida 32611

INTRODUCTION

Cutting tools, although small in size and price compared to the machine tools on which they are used, form the critical link in the man-machine productivity chain. In addition, they pose a very challenging application for materials. The cutting tool in its usage encounters abrasion, high temperature, mechanical and thermal shock, and chemical interaction with the workpiece material. It is important to understand the nature of the degradation that occurs in the tool material. This will permit selection of the proper tool material for a particular application and the design of better cutting tool materials.

The degradation of cutting tools directly impacts manufacturing productivity, which in turn is linked to the cost of every manufactured product. Careful analysis reveals that cutting tools are used directly or indirectly in the manufacturing cycle of almost every product. Everything from aircraft to kitchen appliances to ships have cutting tools used in their manufacturing cycle. Thus the degradation of the cutting tool is an important factor contributing to manufacturing productivity.

Although the study and understanding of the degradation of cutting tools is very important for all manufacturing processes, there is very little information available in the literature (1,2,3). A survey of the literature of the past few years shows a lack of scientific activity in this area. This chapter is an attempt to bring this problem to the attention of the research community and to inform all the interested parties of the need to carry out more research.

Before addressing the degradation of ceramic cutting tool materials, it is necessary to understand the reasons for the use of such tools. It is also necessary to learn about the causes of tool wear related to chemical degradation. Understanding these elements will facilitate discussion of the materials currently in use, the problems associated with them, and the future directions in materials development.

BRIEF HISTORY OF CUTTING TOOL MATERIALS

The history of modern cutting tool materials starts with the development of specialty steels. These include the carbon steels and the high speed steels developed later. These developments occurred from the early part of the nineteenth century to the beginning of twentieth century. Figure 16-1 shows progressive improvement of cutting tool materials as a function of increasing cutting speeds, resulting in productivity improvement. These tool materials contributed significantly to increased productivity and are still in use. But they do pose significant problems when enhanced performance demands are placed on them. The main demand is better performance in terms of offering higher productivity. In the metal cutting manufacturing arena, this translates directly into higher cutting speeds. The high speed steels fail miserably when such demands are placed on them. They fail due to inadequate strengths and chemical instability at the higher temperatures which are encountered at higher cutting speeds.

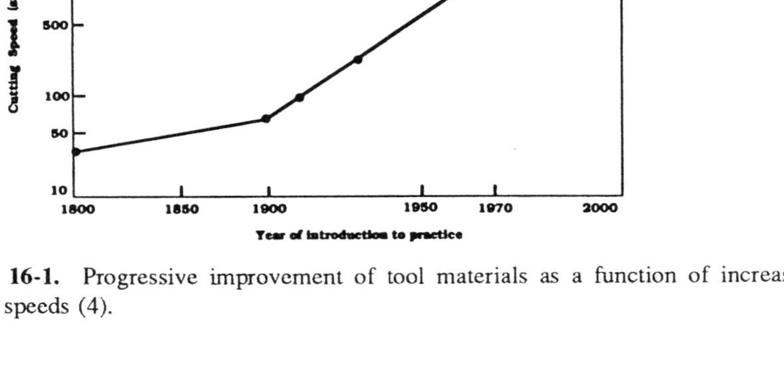

Figure 16-1. Progressive improvement of tool materials as a function of increasing cutting speeds (4).

Generally speaking, ceramic materials are chemically inert and can withstand higher temperatures than the metal-based materials. Hence, they were the obvious choice to be used for metal cutting applications, once it was realized that high temperature stability and chemical inertness would translate into higher productivity. Ceramics are generally hard, which is another necessary property for this application. It was also realized that the raw materials needed for their manufacture were easily obtained. The first ceramic cutting tool materials were introduced in the early to mid 1940's.

Investigations and trials have led to the use of several materials such as the tungsten carbide/cobalt composites, alumina/zirconia composites, alumina/titanium carbide composites, titanium nitride/titanium carbide composites, alumina/silicon carbide whisker reinforced composites, alumina/zirconia phase transformation toughened composites and silicon-nitride-based composites. In addition to these bulk forms, ceramic materials are also used as coatings that are deposited on different substrates by a variety of methods such as chemical vapor deposition and physical vapor deposition. Common ceramics that are currently used as coatings are titanium carbide, titanium nitride, hafnium carbide, hafnium nitride, boron carbide and aluminum oxide. Coating processes used in other industries such as semiconductor fabrication have also been used in the manufacture of cutting tools. This occurred in the early 1970's. The science and technology of fabricating the solids and creating the coatings are discussed in detail elsewhere. Table 16-1 shows the properties of some of the typical cutting tool materials.

DEGRADATION MECHANISMS

Degradation, as used in this chapter, refers to a sufficient change in the chemistry, microstructure or other physical parameters of a material so that reduced tool performance results. Although we are mainly concerned with the chemical wear mechanisms that contribute to the instability of the cutting tool materials, a cursory glance at all the important tool wear mechanisms is in order. This will enable us to understand the significance of "chemical wear".

Abrasive Wear

When a cutting tool is in the process of cutting the work piece, the tool rubs against the material that is being machined. The chip that is formed during the cutting operation passes over the tool. In both these situations, there is severe abrasion on the tool. This leads to the wearing away of the tool material, resulting in a shape change or loss of strength of the tool. This in turn results in the reduction in the performance of the tool, as it is unable to maintain the dimensional accuracy of the part that is being machined. In most cases wear

Table 16-1. Properties of Some Typical Cutting Tool Materials (5).

Property	Carbide	Alumina	Silicon Nitride
Hardness (GPa)	12.3...13.51	15.3....15.9	12.2....15.2
(Ra)	91-92	93.2-93.6	90.8-93.5
Upper use Temp.(°C)	< 800*	<< 2273**	< 1200***
Transverse Rupture (MPa) Strength at 25°C	1000-2400	700-840	830
Fracture Toughness (MPa•m$^{1/2}$)		2.2-2.5	3.6-5.2
Density(g/cm^3)	12-15.1	3.8-3.9	3.35

* Plastic deformation of cobalt at this temperature
** Melting temperature of aluminum oxide
*** Oxidation problems above this temperature

results in fracture of the tool. Figure 16-2 illustrates schematically the wear process that occurs during the use of the tool. Ceramics, being hard, are generally abrasion-resistant and are preferred whenever they can be used.

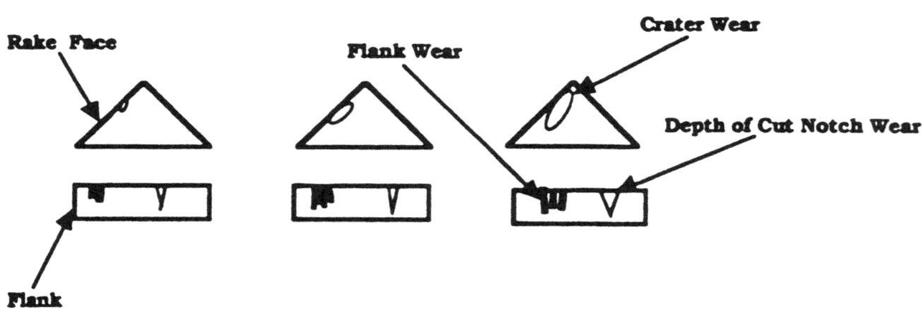

Figure 16-2. Schematic progression of tool wear.

Plastic Deformation

When the tool material is in the cut, the temperature can reach over 1000° C. In materials such as tungsten carbide/cobalt, the hard ceramic particles are held in place by the cobalt metal. At higher temperatures, the metal flows and the cutting tool deforms, resulting in an inability to perform as desired. Other deformation mechanisms such as creep and grain boundary sliding can be important at higher temperatures, even in the more refractory ceramics. The strength, hardness, and tool dimensions have to be maintained at the operating temperatures for acceptable performance.

Fracture

When the tool enters and leaves the cut there is a mechanical shock imparted to the tool. Also, in operations where coolants are used, the coolant is very rarely delivered to the cutting edge and the coolant flow is interrupted by the cutting process itself. Hence, there is an intermittent flow of the coolant reaching the tool which leads to thermal shock. In some cutting operations such as milling, the tool enters and leaves the cut several times a minute, depending on the tool and the speed at which the cutting is carried out. In these cases the tool is subjected to mechanical and thermal shock which could result in premature failure.

Chemical Degradation

As cutting speeds increase, chemical wear limits the performance of the cutting tool material. The temperature increases as the cutting speed increases and, as mentioned already, the tool can experience a temperature of over 1000 C. Figure 16-3 shows the typical temperature profile on a cutting tool during metal cutting. At these temperatures a number of processes occur simultaneously, resulting in the degradation of the tool. Chemical degradation as we presently understand it occurs by two mechanisms: chemical dissolution and diffusion wear.

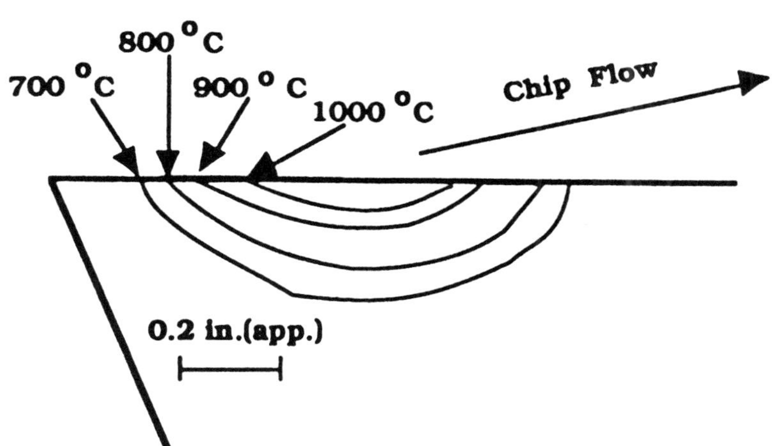

Figure 16-3. Temperature distribution near the tool surface.

Chemical Dissolution. When the temperature of the tool-workpiece interface increases, this form of degradation becomes more noticeable. The tool and the workpiece could be incompatible in the sense that the materials react with one another. If this is the case, the tool could literally dissolve into the work material, thus chemically altering the material. As material is being dissolved away from the cutting tool, it loses its form and also strength. Short of failure, the chemical alteration that occurs in the workpiece can result in problems as the material is altered and likely will not perform as desired.

Diffusion Wear. This type of wear also occurs at high cutting speeds. In this case, the tool material diffuses into the chip that moves past the top (rake face) of the cutting tool. Slowly, enough of the tool material is removed to form a depression on the top of the tool. Figure 16-2 shows the progression of such tool wear. This is termed crater wear, as a "crater" appears on the top face of the cutting tool. This type of wear is mainly due to the fast movement of the freshly-cut material past the tool face, producing a combination of abrasion and high temperature. At the same time, the chemical reaction on the front (flank face) of the tool leads to wear on this face, which is manifested in both the depth of cut notch and as serrations on the cutting tool.

Some examples of chemical interaction between tool materials are as follows:

Aluminum oxide reacts with iron to form a solid solution according to the equation

$$Al_2O_3\ (s)\ (tool) + FeO(s)\ (work\ piece) \rightarrow FeAl_2O_4\ (s) \tag{16-1}$$

(Note: Iron always contains some oxide)

Silicon carbide reacts with iron according to the equation

$$SiC\ (s)\ (tool) + 4Fe\ (work\ piece) \rightarrow FeSi + Fe3C \tag{16-2}$$

$$Si_3N_4 + 2Fe + 3O_2(air) \rightarrow 3SiO_2 + 2Fe_2N + N_2 \tag{16-3}$$

Such reactions are detrimental to tool performance.

Complications Due to Cutting Fluids

To reduce the temperature seen by the cutting tool and the workpiece, coolants are generally used in the cutting operation. In addition to assisting in heat transfer, the coolant may also act as a lubricant, thus facilitating the cutting process. But here again care has to be taken that the cutting fluid does not degrade the tool material or have any detrimental effect on the workpiece during the manufacturing process or when the part is in service. The chemical reactions and associated degradation mechanisms are not well understood (6). An extensive review of the literature revealed that very little work has been done in this area and that additional basic research is warranted.

TAYLOR'S TOOL LIFE EQUATION

F. W. Taylor (7) first proposed and employed the following equation to analyze cutting tool wear. He proposed the following equation:

$$VT^n = C \qquad (16\text{-}4)$$

where V = cutting speed (surface ft/min)
T = tool life (min)
C and exponent n are constants for the material or conditions used.

The tool life in this equation (16-4) is process dependent i.e., it is defined taking into consideration the particular metal cutting application. For example, in close tolerance machining, when the part dimension has to be held very closely, even a very small amount of wear could end the useful life of the tool. Whereas, in rough machining, the tool could be used until it wears to a great extent. However, for relative tool wear studies, it could be defined as the time taken for a finite amount of (measurable) wear of the tool.

This was a first attempt to express in a mathematical form the tool degradation mechanisms and relate them to the performance of a cutting tool. The degradation of the tool is a function of the cutting speed, the tool material and the work piece material. However, it should also be mentioned that the mechanisms of degradation not being well understood, it is difficult to generalize and mathematically express the processes without additional research.

CERAMIC CUTTING TOOL MATERIALS

This section reviews tool materials currently in use and the problems associated with them, along with possible avenues for solving the problems.

Tungsten Carbide/Cobalt Composites

The mainstay of the cutting tool industry is the WC/Co composite and its variations which are very versatile in machining many materials currently in use. At normal cutting speeds, between 150 and 800 surface feet per minute, these tools perform adequately for most operations. As higher productivity demands are placed on the tools, the performance degrades. The chemical reactivity problems associated with machining different materials are discussed individually below.

Steel Machining. In machining steels, at slow to moderate speeds, the WC/Co composite performs adequately. As the cutting speeds increase, the chemical reactivity of the tool material is seen both as crater wear, due to diffusion of the tool material, and as flank wear, due to chemical dissolution into the work piece. As the temperature rises due to higher cutting speeds, these mechanisms accelerate and result in higher wear rates.

Aluminum Machining. In machining aluminum alloys, tool tip temperatures are fairly low. This is mainly due to the high thermal conductivity of aluminum and the ease with which it shears, due to its lower strength. Hence, in most cases, these alloys can be machined without significant problems. Chemical reactivity problems are generally not encountered.

Nickel-Based Alloys. Machining tough nickel-based alloys has been a challenge. These alloys are tough and react chemically. There is also a sticking phenomenon that occurs, due to localized welding, leading to tool failures. In most cases, the WC/Co composites can be used only at slow speeds. These composites are also inadequate in terms of strength to allow machining of such tough alloys.

Titanium alloys. Titanium has low thermal conductivity, a high melting point and is very reactive. Hence it poses special problems. The components in the composite, namely tungsten carbide and cobalt, react with the titanium (8). This tends to form built-up edge and crater wear at almost any speed. Although these composites are used in machining titanium, due to the reactive nature of titanium, the productivity is very low.

Non Metallic Composites. As the use of non-metallic composites becomes more widespread, due to development of new composites and to the development of proper application guidelines, machining of these composites will become more important. Although WC/Co composites are useful in some cases, the chemical compatibility of the composite dictates the use of the proper cutting tool. The abrasive nature of most of these composites renders WC ineffective in most cases.

Aluminum-Oxide-Based Composites

Aluminum oxide composites currently in use as cutting tools are the particulate-strengthened variety which use titanium carbide or zirconium oxide particles, the whisker reinforced variety which use silicon carbide whiskers, and the transformation-toughened variety, which use zirconium oxide and other additives. These materials have been used successfully in machining a variety of materials.

Steel Machining. Aluminum-oxide based materials offer the best solution for machining steels. Although in the initial development phase aluminum oxide lacked the desired toughness for steel machining, the particulate- strengthened systems, especially those using titanium carbide, have been very successful in increasing the productivity of the machining operation. Although there is some reactivity of the ingredients with the steel, as shown by equation 16-1, it is minimal and no major problems are seen. The speed of operation is limited by other parameters and not the tool material. The recent development of the phase-transformation-toughened composite system has contributed to higher productivity by nature of its lower reactivity and higher toughness.

Nickel-Based Alloys. Aluminum-oxide-based materials are best suited in machining the nickel based alloys (9). The toughness of the composites was inadequate until the introduction of the whisker-reinforced composites. The silicon-carbide-whisker-reinforced aluminum oxide composite has dramatically increased productivity in the machining of nickel-based alloys, which are used extensively in the jet engines of aircraft. The chemical reactivity of the composite with the ingredients of the alloys is minimal, thereby eliminating problems that would otherwise be encountered at higher cutting speeds and higher temperatures.

Non-Metallic Composites. As mentioned earlier, in the machining of non metallic composite materials, the temperatures encountered are fairly low. But the abrasive nature of the materials play a very significant role, thus leaving diamond as the only suitable cutting tool material.

Silicon-Nitride-Based Composites

Steel Machining. Silicon-nitride-based materials are unsuitable for machining steels as the chemical reactivity is significant (Eq. 16-3) and renders the tool useless. The relative dissolution of the tools is very high and they cannot be used even at moderate speeds.

Cast Iron Machining. In machining cast iron, silicon-nitride-based materials perform the best. They seem to have the needed mechanical strength and the chemical reactivity with the workpiece is very low, thus enabling very high speeds in machining. This is the tool material of choice in machining cast iron.

Nickel-Based Alloys. The chemical reactivity of silicon nitride with nickel is fairly low. Hence it could be used in the machining of nickel-based alloys. However, at higher speeds reactivity is higher than that of aluminum oxide. Thus tough aluminum-oxide based materials perform better and yield higher productivity.

COMBATING DEGRADATION

Selection of Tool Materials

The best solution for the problem of cutting tool wear due to chemical interaction and degradation during the cutting process is to select the proper cutting tool. The compatibility of the tool with the workpiece should be determined before a particular tool material is chosen. The cutting tool vendors do provide technical assistance for the selection, but the user has to educate himself so that the proper selection can be made. In some cases, such as titanium machining, there is at present, no tool material with all the desired properties. One has to be content with a less than optimal solution.

Coatings - The Interim Solution

One effective way to eliminate the problem is by providing a passive barrier on the substrate tool material. This is at best an interim solution, but if done properly provides some benefits. The chemical vapor deposition and the physical vapor deposition processes are used to apply a thin adherent chemically- inert ceramic coating on the substrate. But care has to be taken to ensure that the coating material is "passive" and provides the needed barrier. Although significant improvements can be achieved with commercially-available coatings, care should be taken to ensure that a proper selection of the substrate material is made and that the coating process does not degrade the substrate (1).

THE FUTURE

Although a number of good cutting tool materials and coatings have been developed and are in use, there are still some significant problems. Moreover there are also ongoing developments in new materials and composites, such as rapidly-solidified alloys and polymers, and metal and ceramic matrix composites, all of which need to be machined to be converted into useable end products. Hence there is a continuous demand for new and better cutting tool materials. These new developments are totally dependent upon our understanding the mutual chemical reactivity of the materials in question. Research in these areas is needed for a thorough understanding of the degradation process and mechanisms. Such work is bound to result in new and better cutting tool materials thus contributing to greater productivity and a better standard of living.

REFERENCES

1. Kramer, B.M., *J. of Eng. for Ind.*, 109:[2] (1987).
2. Yeomans, J. A. and Page, T. F., *Wear*, 131[1]: 163-75 (1989).
3. Stewart, H. A., et al., *Carbide & Tool J.*, 18[1]: 2-7 (1986).
4. Whitney, E. D. and Vaidyanathan, P. N., *Advanced Ceramic Cutting Tools for High Speed Machining, Ceramic Developments*, (C.C. Sorrell, C. C. and B. Ben-Nissan, B., eds.), Materials Science Forum, Vols. 34-36, pp. 85-94, Trans. Tech. Pub. Switzerland (1988).

5. Whitney, E. D. and Vaidyanathan, P. N., *Engineered Ceramics for High Speed Machining*, Tool Materials for High Speed Machining, (J.A. Swartly-Loush, ed.), ASM International (1987).
6. Schey, J. A., *Tribology in Metal Working*, American Society for Metals, June (1984).
7. Shaw, M. C., *Principles of Metalworking*, Oxford Science Publications, (1989).
8. Hartung, P. D. and Kramer, B. M., *Tool Wear in Titanium Machining*, Annals of CIRP, 31[1]: 75-80 (1982).
9. Hartung, P. D. and Kramer, B. M., *Theoretical Considerations in the Machining of Nickel-Based Alloys, Cutting Tool Materials*, Proceedings of an International Conference, Organized by Frank Gorsler, ASM, pp 57-74 (1980).

17

Degradation of TZP Ceramics in Humid Atmospheres

T.T. Lepistö and T.A. Mäntylä

Institute of Materials Science
Tampere University of Technology
Tampere, Finland

INTRODUCTION

Tetragonal zirconia polycrystals (TZP) are a newcomer among engineering ceramics. They possess a favorable combination of mechanical properties at room temperature, e.g. high mechanical strength and moderate fracture toughness. These materials also exhibit good wear properties, and their coefficient of thermal expansion is close to that of iron and iron-based alloys which make them attractive for engineering applications. However, these favorable properties are shadowed by structural instability at low temperatures in humid environments, which may lead to total destruction of useful properties after short service periods.

Although many experimental results have been published recently on the effect of different environmental and structural parameters on the structural instability of TZP ceramics, the actual mechanism is still in dispute. In this chapter, the structure of different zirconia ceramics is briefly described so that their sensitivity to environmental instability can be understood. In the main part of the text, published literature on the effect of different environmental factors and structural parameters on instability and material properties is reviewed. Models for degradation are presented and means to avoid harmful changes in the structure are proposed. Some comments will also be given on the relevance of the data which is obtained with commonly-used characterization and analysis techniques.

ZIRCONIA CERAMICS

At ambient pressures, zirconia has the following polymorphs:

$$\text{melt} \overset{2680°C}{\leftrightarrow} \text{cubic (c)} \overset{2370°C}{\leftrightarrow} \text{tetragonal (t)} \underset{\underset{1170°C}{\leftarrow}}{\overset{950°C}{\rightarrow}} \text{monoclinic (m)} \qquad (17\text{-}1)$$

The t $\rightarrow$ m transformation is martensitic in nature (1-3). It is accompanied by a volume increase of 3% at 950°C to 4.9% at room temperature (4). The volume change can lead to severe microcracking during cooling which destroys the mechanical properties of the material. The transformation can be avoided by lowering the M_s temperature below room temperature. This can be done by decreasing the grain size of the tetragonal phase (GS < 1μm), e.g. decreasing the volume and thus the chemical free energy which drives the transformation, and by using suitable stabilizing oxides, such as yttria. If tetragonal crystals are maintained untransformed in the structure, considerable strengthening and toughening can be obtained when tetragonal crystals transform to monoclinic structure during loading (5-9).

At room temperature this type of small tetragonal zirconia particle is metastable, which means that the tetragonal phase is thermodynamically unstable but kinetically stable. Although it is agreed that the phase transformation is nucleation-controlled, there is disagreement on the mechanism of nucleation. Chen and Chiao (10,11) showed that homogeneous nucleation in zirconia is essentially impossible due to a high activation energy barrier. They proposed that nucleation is a heterogeneous process that occurs on microstructural defects. Defect-induced strains in the structure can reduce activation energy so that nucleation occurs more easily and this kind of interaction may even lead to barrierless transformation. Ruhle and Evans have discussed this nucleation process in their recent review (12) proposing that a nucleus is formed at specific lattice defects, called localized soft modes, adjacent to t-ZrO_2 interfaces in the structure.

The strengthening and toughening mechanisms related to the t$\rightarrow$m transformation are widely utilized in different ceramics. Zirconia-based ceramics are normally divided into three main groups: partially stabilized zirconias (PSZ), tetragonal zirconia polycrystals (TZP) and fully stabilized zirconias (FSZ). Figure 17-1 describes schematically how these structures are formed as a function of the amount of stabilizing oxide.

Metastable tetragonal zirconia particles can also be used in other matrices such as Al_2O_3, SiC, Si_3N_4, TiB_2, mullite, etc. in which case the materials are referred to as DZC ceramics (Dispersed Zirconia Ceramics). The most well-known example of this group is zirconia toughened alumina (ZTA).

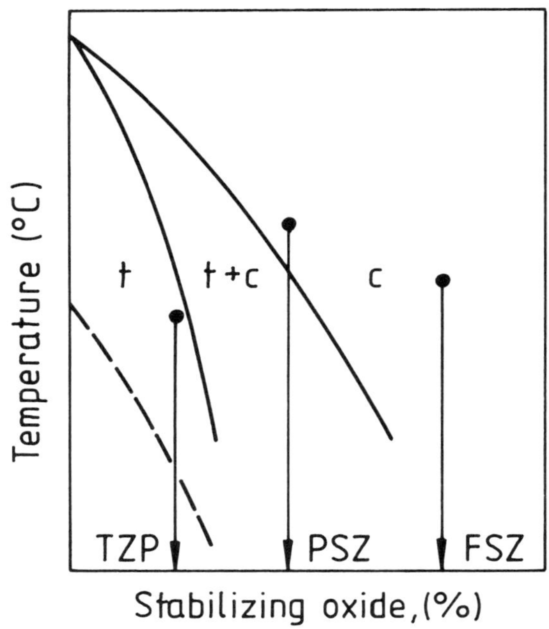

Figure 17-1. Schematic presentation of the formation of different zirconia ceramics. Adapted from (13).

Partially Stabilized Zirconia (PSZ)

PSZ ceramics have a structure in which small metastable tetragonal zirconia precipitates are embedded in a large grained cubic structure. The most common commercial materials are based on MgO stabilization, although CaO and Y_2O_3 are also used for stabilization. When MgO is used, the amount of stabilizer is 8-10 mol%. The solubility of MgO in monoclinic zirconia is very small, and the solubility of MgO in tetragonal zirconia is also low, being only 1 mol% at 1400°C. The solubility in the cubic structure is higher, about 13-14 mol% at the eutectic temperature of 1400°C. During sintering the materials are kept in the single phase region just above the c/c+t phase boundary. The desired tetragonal precipitates are obtained by heat treatment which is done either just above the eutectic temperature or at temperatures which are few hundred degrees below the eutectic temperature - close to 1100°C. The properties of the obtained materials are adjusted by controlling the size of tetragonal particles. The structure can be aged to optimize strength and toughness or over-aged to yield good resistance against thermal shocks.

Tetragonal Zirconia Polycrystals (TZP)

When zirconia ceramics are sintered from small-grained, 10-200nm, co-precipitated powders in $ZrO_2+Y_2O_3$, ZrO_2+CeO_2 or $ZrO_2+Y_2O_3+CeO_2$ systems, totally or almost totally tetragonal microstructures can be obtained. Small particle size and homogeneous distribution of constituents make it possible to sinter these materials to full density at relatively low temperatures, e.g. at the temperature range of 1300-1500°C, resulting in small grain size of less than 1μm. In commercial Y-TZP ceramics the yttria content varies in the range of 1.75-3.5 mol% (3.5-8.7 wt%). Depending on sintering temperature, time and the amount of stabilizing species, the final structure contains 60-100% of tetragonal zirconia. The other constituents in the structure may be c-ZrO_2, small amounts of nontransformable t′-phase and some m-ZrO_2 on the surface, resulting from machining. In grain boundaries there exists a glassy phase containing mainly silica and yttria. The amount of this glassy phase depends on the processing method used. Typical commercial ceramics in this group have 3 mol% of yttria; their strength is > 1000 N/mm^2 and they exhibit fracture toughness on the order of 8-10MPa•$m^{1/2}$.

If ceria is used as a stabilizer element, 12-20 mol% is needed to obtain tetragonal structure at room temperature. These materials are sintered at 1500°C and the final grain size is on the order of 2-3μm, which is somewhat larger than in Y-TZP ceramics. Ce-TZP ceramics are much more stable in humid environments than Y-TZP ceramics. They have high fracture toughness and relatively modest strength.

Fully Stabilized Zirconia (FSZ)

In FSZ ceramics, MgO, CaO and Y_2O_3 are the most common stabilizing oxides. When the amount of stabilizing oxide is large enough, phase transformations can be hindered and the cubic structure can be retained to room temperature. Although such a structure is thermodynamically metastable, the phase transformations are kinetically hindered at these low temperatures. FSZ ceramics have a high coefficient of thermal expansion, higher than that of pure zirconia. This, together with low thermal conductivity, leads to poor resistance against thermal shock. The mechanical strength and fracture toughness are also lower than those of PSZ and TZP ceramics. Due to these drawbacks, FSZ ceramics are not normally used in structural applications.

ENVIRONMENT-INDUCED PHASE TRANSFORMATION

In spite of good mechanical properties which are difficult to exceed with most of the other ceramics, TZP ceramics suffer from structural instability at low temperatures in humid environments. Although many of the structural and environmental parameters which affect this transformation are known, the detailed mechanism is still unclear, which makes it difficult to increase structural stability in these materials without losing favorable mechanical properties.

Environment-induced low temperature phase transformations were first found by Kobayashi et al. (14) when they studied the effect of low temperature annealing in air on the mechanical properties of 4.5-8Y-PSZ ceramics. They noticed that during annealing the amount of monoclinic phase increased in ceramics which had 4.5 mol% of yttria, and the strength of the ceramic decreased drastically. Environment-induced t-ZrO_2→m-ZrO_2 transformation accompanied by degradation of properties is commonly found in different TZP ceramics. The main features of observed phenomena are (15):

1) The degradation occurs significantly in a specific temperature region, at 200 to 300°C in air.
2) The degradation proceeds from the surface to the interior of the specimen.
3) Micro- and macrocracking occur because of the spontaneous phase transformation from t-ZrO_2 to m-ZrO_2.
4) The presence of water accelerates the transformation and the degradation.
5) The degradation can be prevented by increasing the amounts of stabilizing oxides or by reducing grain size.

The factors determining the stability of t-ZrO_2 particles confined in ceramic matrices are considered to be the matrix constraint, chemical composition, and the transformational nucleation barrier which is a function of ZrO_2 particle size (16). In a humid atmosphere, this nucleation barrier is somehow decreased by water so that the transformation can proceed.

PSZ ceramics are not prone to structural instability in humid environments at low temperatures since the cubic structure shields metastable tetragonal particles from direct contact with the environment. This situation is also found in DZC materials. The FSZ ceramics do not contain metastable tetragonal phase and are not prone to this environmental degradation. Although c-ZrO_2 is also metastable at room temperature such instability as in TZP ceramics is not found under corresponding conditions.

EFFECT OF ENVIRONMENTAL FACTORS ON PHASE TRANSFORMATION

Temperature

The dependence of aging-induced transformation on temperature in TZP-ceramics is shown in Figure 17-2, which describes the monoclinic phase content on the surface of 3Y-TZP ceramics after aging in air in the temperature range of 100 to 500°C. The test materials were sintered at different temperatures so that the effect of grain size on the transformation can also be seen. In the material sintered at 1600°C the aging-induced transformation was found in the temperature range of 100 to 400°C, with the maximum transformation in the range of 200 to 250°C. With decreasing sintering temperature and thus with decreasing grain size, the critical temperature below which the degradation

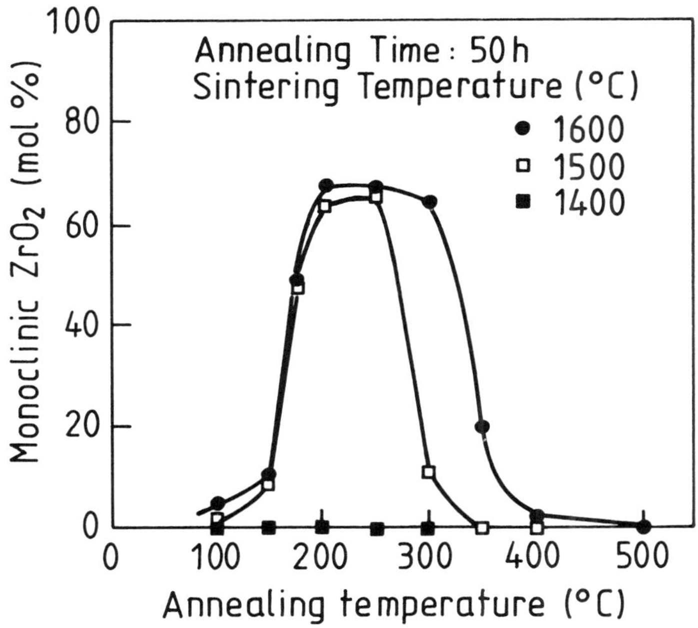

Figure 17-2. The monoclinic phase content in the surface of 3Y-TZP ceramics after aging in air for 50 hours. Adapted from (17).

occurred shifted to lower temperatures. This was suggested to be due to the lowering of the M_s-temperature with decreasing grain size (17). No phase transformation was observed in the material sintered at 1400°C.

Axelsson (18) suggested that the relationship between the phase transformation rate and the aging temperature is due to competition between the water adsorption, dissociation and desorption rates, which determine the available dissociated water content on the surface of the material. It was concluded that the chemical interaction on the solid-gas interface must be incorporated in a model to describe the structural changes which occur in TZP-ceramics under humid conditions.

Environment

It is generally agreed that the aging-induced transformation in TZP-ceramics is enhanced by water or water vapor in the surrounding atmosphere. Even the humidity in air has been found to be sufficient to cause degradation. As the partial pressure of water in the atmosphere increases the phase transformation will increase, as shown in Figure 17-3. Sato et al. (19) found that after aging 50 hours at low temperatures (< 200°C), the transformed monoclinic phase content was much higher in the samples aged in water compared to those samples aged in air. In the temperature range of 200-300°C the phase transformation rate was independent of aging medium, suggesting that the transformation rate is controlled by kinetics of surface reactions and their dependence on material properties.

Adsorption studies by Axelsson (18) indicated that at saturation the absolute adsorbed amounts of H_2O on an oxidized zirconium specimen were less than expected for a dense two-dimensional layer. This may result either from adsorbate-adsorbate interaction or from adsorption at defect sites.

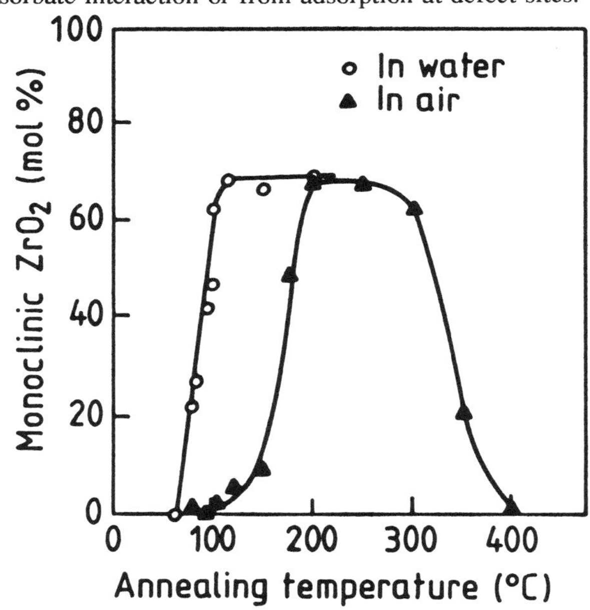

Figure 17-3. The transformed monoclinic phase content on the surface of 3Y-TZP ceramics after aging in water and air for 50 hours. Adapted from (19).

Aging-induced transformation has been found (20) to occur in many water solutions and also in nonaqueous solvents. In aqueous solutions the amount of the transformed monoclinic phase was practically independent of the pH of the solutions, as shown in Table 17-1.

Table 17-1. Transformed Monoclinic Phase Contents in the Surface of the 3Y-TZP Ceramics After Aging in Various Liquids at 95°C for 120 Hours. Adapted from (20).

Environment	Amount of m-ZrO_2 (mol%)	Concentration of water (ppm)
(A) Aqueous solutions		
H_2O	48.8	
5M HCl	53.3	
5M H_3PO_4	52.6	
5M $HClO_4$	56.8	
5M HNO_3	51.1	
5M H_2SO_4	45.7	
1M H_2SO_3	50.9	
5M NaOH	53.8	
30 wt% NH_3	49.7	
(B) Nonaqueous solvents with lone-pair electron orbital opposite proton donor site		
$HOCH_2CH_2OH$	37.0	15 000
$HOCH_2CHOHCH_2OH$	45.2	*
$H_2NCH_2CH_2OH$	38.8	*
C_2H_5OH	25.9	6 850
$(C_2H_5)_2NH$	16.6	1 870
$CH_3CH_2CH_2NH_2$	37.1	1 060
NH_2CHO	36.1	1 730
(C) Nonaqueous solvents with lone-pair electron orbital without opposite proton donor site		
CH_3CN	18.1	1 750
$C_6H_5NO_2$	16.4	1 730
$C_4H_8O_2$	21.0	1 600
(D) Nonaqueous solvents without lone-pair electron orbital		
$C_6H_5CH_3$	0	890
C_6H_{12}	0	190
i-C_8H_{18}	0	180

* Peak position of water is same as that of solvent.

Table 17-1 also shows that other nonaqueous liquids with a molecular structure similar to that of water can accelerate the transformation whereas totally or slightly different molecular structures have little or no effect on the transformation. It was therefore concluded that the degradation in TZP ceramics is caused by the same kind of water induced stress-corrosion mechanism as proposed for vitreous silica ceramics by Michalske and Freiman (21).

THE EFFECT OF MATERIAL PARAMETERS ON PHASE TRANSFORMATION

Stabilizing Agent and Stabilizer Content

The stability of the tetragonal phase in a humid atmosphere has been found to be improved by increasing the stabilizer content. Figure 17-4 describes the effect of yttria content on the phase transformation of Y-TZP ceramics after aging in air at 200°C. No monoclinic phase was found in the surface of 6Y-

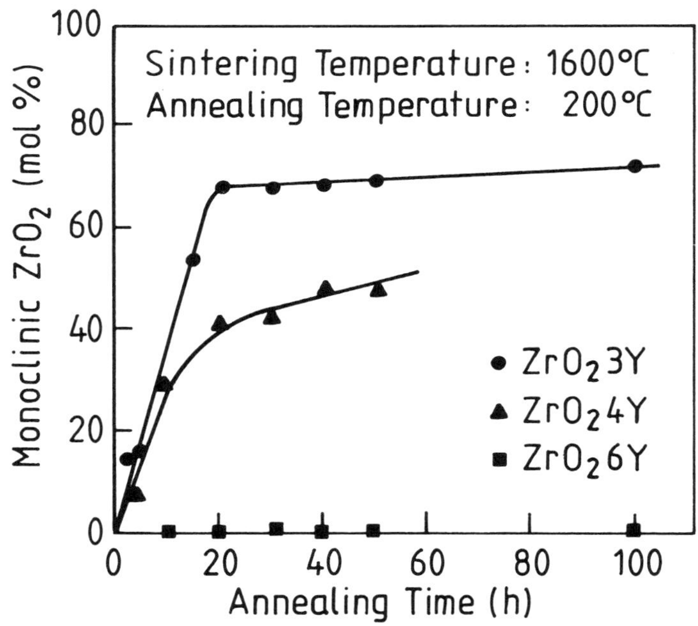

Figure 17-4. Monoclinic phase content on the surface of Y- TZP ceramics after aging in air at 200°C. Adapted from (17).

FSZ samples after the aging (17). However, it must be remembered that the structure of 6Y-FSZ ceramics is fully cubic and therefore stable under humid conditions. It is therefore possible to improve the stability of the structure by increasing stabilizer content, but this will also change the structure from tetragonal to cubic phase with an accompanying decrease in mechanical properties since transformation toughening during loading is prevented.

Ceria has been found to be an effective stabilizing agent for improving the aging properties of TZP ceramics without the loss of fracture toughness (22,23). However, ceria is not a suitable stabilizing agent when used alone because of the poor strength of ZrO_2+CeO_2 ceramics. Therefore, ceria has been used in connection with other stabilizing agents and especially with yttria. Figure 17-5 shows the monoclinic phase content in the surface of ZrO_2+3mol%Y_2O_3+CeO_2

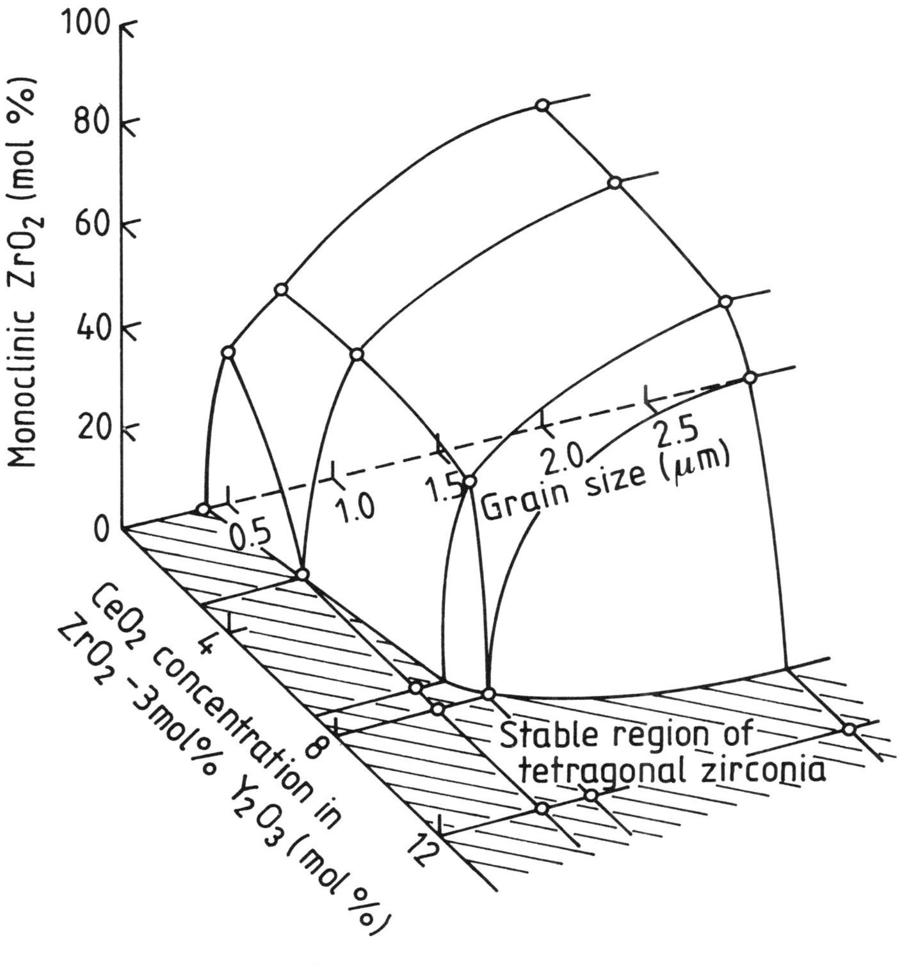

Figure 17-5. Transformed monoclinic phase content in the surface of ZrO_2+3mol%Y_2O_3+CeO_2 ceramics after aging in water at 100°C for 7 days. Adapted from (24).

ceramics after aging in water at 100°C for 7 days. The shaded region corresponds to the grain size and ceria content of the materials in which no phase transformation was found after aging. Thus it is possible to control the phase transformation of Y-TZP ceramics by adding ceria to the material.

Because ceria doping decreases the strength of TZP ceramics, it has been proposed that the aging properties of TZP ceramics can be improved by doping ceria only into a surface layer whose thickness is less than the critical crack length of the material. Promising results have been obtained by using this technique, although the thickness of the doped layer could not be controlled satisfactorily (24).

Grain size

It was shown in Figure 17-2 that by decreasing the grain size of the material the critical temperature for degradation decreases, thus making it possible to stabilize the structure against aging-induced transformation. It has been suggested that this is based on the increase of surface free energy with decreasing grain size (25) which decreases the driving force for transformation.

A number of methods have been presented to determine the critical grain size for transformation. Usually this has been done by plotting the amount of transformed monoclinic phase (measured by XRD) or the rate of transformation against average grain size of the material. Thus the critical grain size coincides with the zero monoclinic content or the zero transformation rate. If the average grain size is used to describe the grain size distribution of the material, this method gives too small a value for the critical grain size because the larger grains which are included in the distribution used to calculate average grain size have retained the tetragonal structure during the aging.

The critical grain size for transformation will depend among other things on the stabilizer content. For example, Watanabe et al. (26) have given the critical grain sizes for ZrO_2+Y_2O_3 ceramics to be in the range of 0.2-0.6μm, corresponding to stabilizer contents of 2-5 mol% of yttria, respectively.

Density

The density of the material has also been shown to affect the aging properties of Y-TZP ceramics (27). The stability of the structure against aging-induced transformation was greater in the hot pressed and the hot isostatically pressed materials than in the sintered materials because of the higher density and smaller grain size of the former materials. Experimentally measured values for critical grain sizes and densities as a function of yttria content are presented in Figure 17-6. Aging-induced transformation was not found in the materials whose densities and grain sizes lie above the curves shown in this figure.

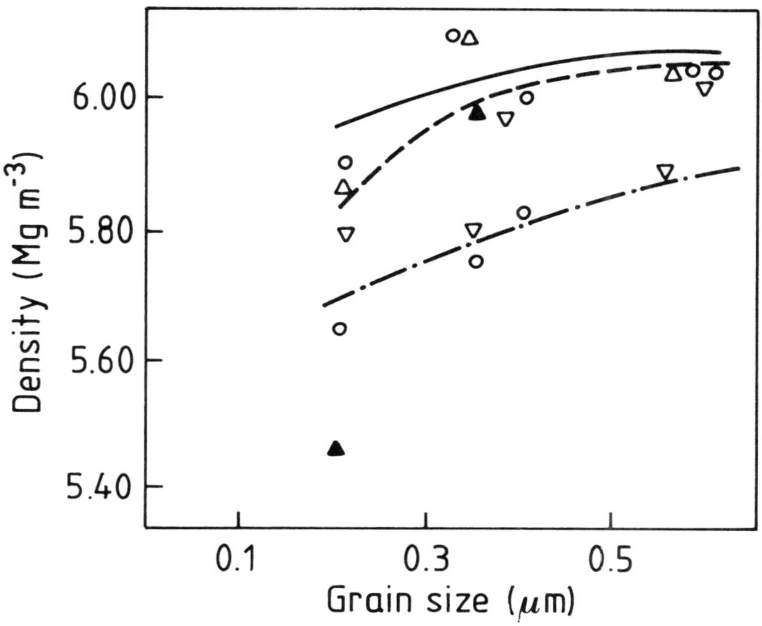

Figure 17-6. Critical grain sizes and densities for aging-induced transformation in Y-TZP ceramics in air at 200 or 250°C. Y_2O_3-contents: (—) 2.5mol%; (---) 3.0mol%; (-··-) 4.0mol%. Adapted from (27).

Grain Boundary Phase and the Shape of the Grains

Conflicting results have been presented on the effects of grain boundary phase on the stability of tetragonal phase in humid atmosphere. Because grain boundary phase also affects the grain morphology, these two factors will be discussed together.

It has been shown that Y-TZP ceramics have SiO_2-rich glassy phases at grain boundaries due to the contamination of the starting powder (28). This glassy phase has been presented to affect the morphology of the grains and also the aging properties of Y-TZP ceramics at low temperatures (29). The materials with lower glassy phase content at grain boundaries had regular hexagonal-shaped grains with sharp edges while the materials with higher glass contents had more rounded grains.

Mecartney (29) exposed 2.5Y-TZP ceramics in an autoclave at 250°C with the partial pressure of water at 2Pa. He varied the glass content of the materials by varying the milling time of the starting powder in an attritor with SiO_2-containing balls. He found that the materials with high glass content degraded least during the aging.

Grain boundary glassy phase can affect the residual stress level of the sintered materials and therefore the sensitivity of the material to phase transformation under an applied stress field. Residual stresses may also affect the degradation phenomenon in TZP-ceramics. Schmauder and Schubert (30) have shown that the residual stresses caused by thermal expansion mismatch are highest at the grain boundaries between two neighboring grains with a and c axes parallel to each other. The magnitudes of the residual stresses from thermal expansion mismatch depend on the glassy phase content of the material. Higher glass contents yield lower residual stress levels in the material after cooling from sintering temperature, because the viscous glassy phase at grain boundaries can accommodate stresses caused by thermal expansion mismatch during cooling until the temperature has decreased below the glass transition temperature. If the structure contains high residual stresses, the activation energy barrier for transformation will be decreased making the material more prone to transformation.

EFFECT OF AGING ON MATERIAL PROPERTIES

Strength

Aging-induced transformation has generally been shown to degrade the mechanical properties of Y-TZP ceramics. A decrease in mechanical properties has been found to coincide with the increase in monoclinic phase content and accompanies micro- and macrocracking during the exposure.

Figure 17-7 compares the strength of sintered and hot isostatically pressed ZrO_2+Y_2O_3 ceramics after 2000 hours exposure in air. The sintered materials with 2-3mol% Y_2O_3 degraded in strength during the aging while the aging treatment was not found to affect the strength of ZrO2+5mol%Y_2O_3 ceramic. In the hot isostatically pressed materials strength decreased only in 2Y-TZP ceramic during the exposure. This is believed to be due to higher density and somewhat smaller grain size. It can be seen that aging-induced transformation and the accompanying decrease in strength cannot be prevented even in the fully dense material if the yttria content is 2mol% or less.

Sato et al. (22) have shown that the strength after aging of 2Y-TZP ceramics depended on the thickness of the transformation layer formed during the exposure in the surface of the samples. After the exposure the bending strength was nearly constant until the thickness of the layer exceeded 50μm. With thicker layers in the surface the strength decreased when the thickness of the layer increased.

Wang et. al. (31) have reported that aging-induced transformation can strengthen and toughen the material by introducing compressive surface stresses and microcracks.

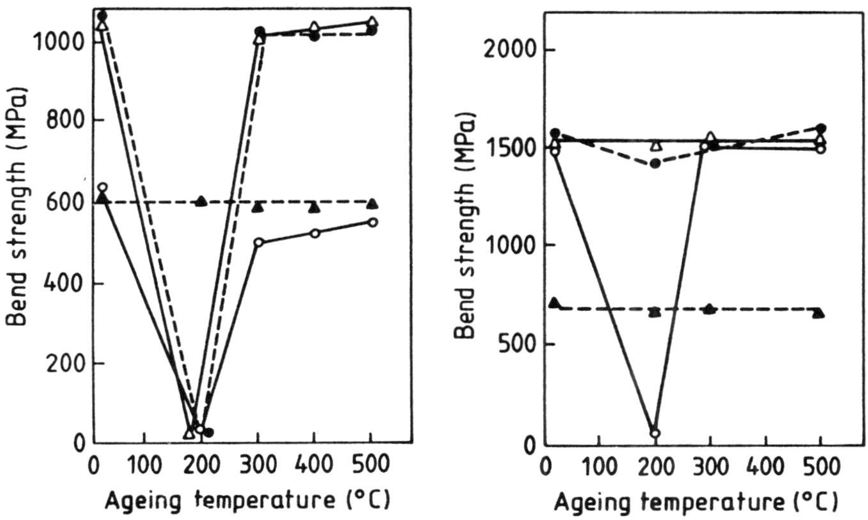

Figure 17-7. Flexural strength of sintered (a) and hot isostatically pressed (b) Y-TZP ceramics after exposure in air for 2000 hours 0 2.0; • 2.5; Δ3.0; ▲ 5.0 mol% Y_2O_3. Adapted from (27).

Lepistö et al. (32) also have found an increase in strength after the exposure. This has been interpreted to be caused by the formation of a porous surface layer in the sample during exposure. If the layer is not too thick and the layer is mechanically weak, it may envelope the surface cracks and prevent them from growing deeper into the material, thus eliminating the effect of surface cracks on the strength. When the layer is thick enough, it will decrease the strength by decreasing the load bearing cross section of the sample. Although the monoclinic phase content in the surface of the aged samples may be the same, the strength after the exposure can be different depending on the phase transformation rate of the material and the thickness of the transformed layer in the surface of the samples.

Fracture Toughness

Chen et al. (33) have reported the decrease of fracture toughness caused by aging-induced phase transformation. The minimum in fracture toughness coincides with the maximum transformed monoclinic content in the surface of the sample, as shown in Figure 17-8.

It has been shown that the materials with high fracture toughness (high instability) are very sensitive to degradation during aging (29). Sato et al. (25) and Tsukuma & Shimada (34) have also reported similar results where decreased fracture toughness of Y-TZP ceramics yielded improved structural stability against aging-induced transformation.

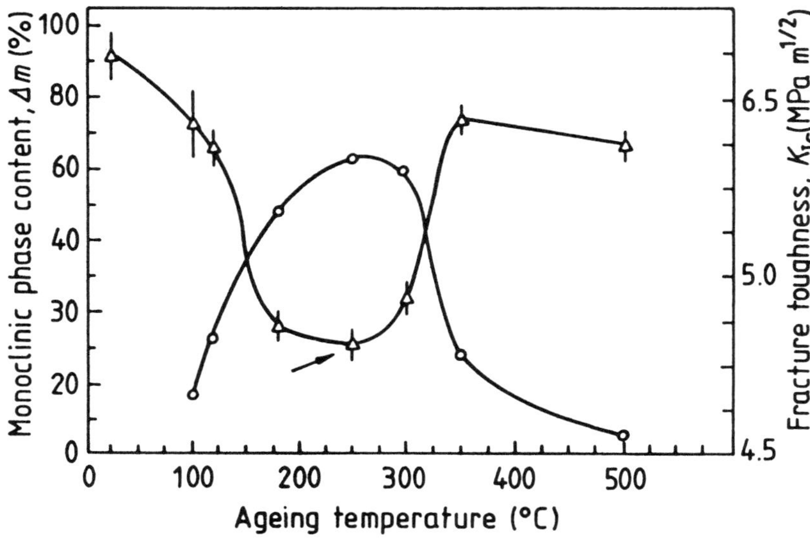

Figure 17-8. The coincidence of the maximum monoclinic phase content (o) with the minimum fracture toughness (Δ) measured after aging 168 hours at 250°C. Adapted from (33).

Wear Resistance

Wear experiments in various environments have shown that the surrounding atmosphere can affect the wear behavior of zirconia ceramics. Fischer et al. (35) have found that polar liquids, such as water or hexadecane can increase the wear by intergranular brittle fracture mechanism. It was suggested that the increased wear may be caused by the environmentally induced stress corrosion cracking.

Reuhkala et al. (36) also found a drastic increase in sliding contact wear of 3Y-TZP ceramic in immersed water at 100°C as compared to dry sliding conditions.

Electrical Conductivity

Kobayashi et al. (14) found that after aging in air at 300°C the electrical conductivity of the ZrO_2-Y_2O_3 ceramics was lower than before the aging. Decreased conductivity coincided with the increased monoclinic phase content after the aging.

Electrical conductivity in zirconia is based on oxygen vacancy movement and vacancy concentration. Alloying of zirconia with lower-valency metal oxides introduces vacancies into the structure and increases the electrical conductivity. Also, the increased vacancy concentration may stabilize the tetragonal structure of zirconia. Therefore, the decreased electrical conductivity of the aged material could be caused by the change in vacancy concentration which can also decrease the stability of the tetragonal structure.

PROPOSED METHODS TO AVOID PHASE TRANSFORMATION

There is a general agreement that aging-induced degradation of TZP ceramics is nucleated on the surface of the specimen with the help of water. Several methods to control the aging-induced surface transformation have been presented:

1) Tetragonal grains may be made smaller than a critical value. This will increase the stability of tetragonal phase both by decreasing the driving force (volume is decreased) and by increasing the surface energy barrier for transformation (surface to volume ratio is increased).

2) The stabilizer content may be made larger than a critical value. By increasing the stabilizer content the chemical free energy difference is increased, thus decreasing the driving force for transformation.

3) The transformable matrix may contain non-transformable second-phase inclusions, such as Al_2O_3 in TZP ceramics (17,24). Alumina has been found to prevent the grain growth of zirconia, and by increasing the elasticity of the matrix the stability of the tetragonal phase is improved.

4) The surfaces may be coated by a non-transformable coating, such as CVD-Al_2O_3 (37). By shielding the tetragonal grains in the surface from the surrounding humid atmosphere it is possible to prevent the transformation to monoclinic structure and the accompanying degradation of mechanical properties. The coating must be fully dense and stable in the atmosphere. Such coatings can be produced by CVD. However, such processing is complicated and expensive.

5) Recrystallization may be induced at the surface (38). Combined surface grinding and subsequent heat treatment has been shown to induce recrystallization in the surface of Y-TZP ceramics. The resultant small grain size region in the surface can stabilize the subsurface large grain structure and prevent the degradation of the structure in humid conditions. By this post-sintering treatment it is possible to combine high fracture toughness of the large-grained bulk material and good stability against water-induced transformation on the surface material.

However, it should be pointed out that the increased stability of the tetragonal phase, as proposed in (1-3), will decrease the transformation toughening effect which means a decrease in mechanical properties.

PROPOSED MECHANISM FOR DEGRADATION

Corrosion Mechanism

Yoshimura et al. (39) have exposed ZrO_2+2.66mol%Y_2O_3 materials to redistilled water under 5 and 100 MPa at 250°C. The aging treatment was found to bring about the formation of monoclinic phase in the surface of the samples and the lattice was found to be expanded. Reheating the aged samples at 400°C in vacuum returned the lattice to its original size with accompanying weight decrease of 0.233%. This reversible expansion and shrinkage of the monoclinic lattice was proposed to be caused by OH^- which entered into the lattice during the aging treatment and was excluded during the subsequent heat treatment. From the weight change it was concluded that 60% of the oxygen vacancies in the material were occupied by OH^- during the exposure. The occurrence of OH^- in the surface of the exposed samples was detected by infrared spectroscopy (IR) and raman spectroscopy (RS).

The degradation of Y-TZP ceramics in humid atmosphere was proposed to proceed according to the following steps (39):

1 - Chemical adsorption of water at the surface.

2 - Formation of Zr-OH and/or Y-OH bonds at the surface with the formation of stressed sites at the surface.

3 - The accumulation of stresses by the migration of OH^- at the surface and in the lattice. This results in the formation of suitable nucleation defects for transformation (40).

4 - Nucleation of monoclinic phase in the tetragonal grains. Transformation of tetragonal phase to monoclinic structure will cause micro- and macrocracking.

The accumulation of stresses at surface sites will increase the free energy of the tetragonal phase, thus decreasing the activation energy for the transformation, and will ultimately bring about the spontaneous transformation to the monoclinic structure.

Sato et al. (20) have also proposed the formation of Zr-OH bonds as a result of the reaction between water and zirconia. The mechanism for degradation is similar to the water-induced stress corrosion of silica proposed by Michalske and Freiman (21). Sato et al. (41) have explained the reaction between water and zirconia on the basis of the studies of activation energies for transformation in both yttria- and ceria-doped zirconias. These studies gave approximately similar values for both alloys according to which Sato et al. proposed that water reacted primarily with Zr-O-Zr bonds and not with the stabilizing agent. In opposition to the mechanism described by Yoshimura et al., Sato et al. proposed that the interaction between water and zirconia released internal stresses which stabilize the tetragonal structure, thus inducing the transformation to monoclinic symmetry. This can be explained by the fact that the release of internal stresses can lower the free energy of monoclinic phase, thus increasing the driving force for transformation, ultimately leading to a spontaneous transformation to monoclinic structure.

On the basis of IR spectroscopy results, it is questionable if water adsorbed into the surface of the aged sample is really dissociated and reacted with ZrO_2 or Y_2O_3 to form the hydroxides mentioned above. Also, the adsorbed water in the surface may give a similar spectrum because of the stretching vibration of O-H in the water molecule. Lepistö et al. (42) have analyzed the surface of ZrO_2+3mol%Y_2O_3 specimens after aging in water vapor at 150°C by infrared spectroscopy and x-ray photoelectron spectroscopy (XPS). IR-spectra revealed that the surface contained either water or OH^-. Also, the O 1s XPS-spectrum of the aged surface confirmed the presence of OH^-. However, the subsurface transformed layer was not found by XPS to contain OH^-. Therefore, it is believed that the surface of the aged specimens may contain OH^- either from the aging treatment or from the air but the subsurface region does not contain OH^-, and the formation of zirconium or yttrium hydroxides cannot be considered to be the principal mechanism for degradation in Y-TZP ceramics.

The role of glassy grain boundary phase was studied by Mecartney (29). He did not find any evidence of water-induced stress corrosion in the glassy phase. However, the glassy phase could increase the stability of the tetragonal phase as discussed earlier. Lange et al. (43) have excluded the effect of grain boundary glassy phase on the degradation of zirconia and do not consider the stress corrosion mechanism in the glassy phase to be important.

Destabilization Theory

Lange et al. (43) have proposed that the degradation is caused by the destabilization of zirconia as a result of reaction with water. TEM analysis of ZrO_2+6.6mol%Y_2O_3 aged in water vapor showed that small crystallites in the size range of 20 to 50nm were formed during aging. Crystallites were formed adjacent to the transformed monoclinic as well as the cubic grains. Energy dispersive x-ray analysis showed that the crystallites were rich in yttrium, and selected area diffraction studies identified the crystallites as α-$Y(OH)_3$.

They also found that aging studies of the materials with approximately equal Y_2O_3 contents but varying silicon contents did not show any differences, indicating that the silicon-rich grain boundary glassy phase does not play a major role in the degradation of zirconia ceramics under humid conditions.

They proposed that, as a consequence of the reaction between yttrium and water, yttrium is drawn out of the surface. When a sufficient amount of yttrium is depleted from a volume element in the surface it will transform to the monoclinic structure. Further depletion of yttrium will result in the growth of the monoclinic nucleus beyond the critical size so that the spontaneous transformation of the tetragonal grains to monoclinic structure can occur. The continuation of the aging-induced degradation will now depend on the grain size of the material. If the grain size is large, the transformation can cause microcracking of the surface and therefore open up the subsurface grains to react with water and the degradation will proceed. If the grain size of the material is small, the transformation of the surface grains does not induce microcracking of the surface, and the subsurface grains are not exposed to the surrounding water-containing atmosphere, thus limiting the transformation.

Yoshimura et al. (39) have criticized the $Y(OH)_3$ crystallite formation by slow diffusion of yttrium in zirconia. For this reason the formation of suitable nuclei for transformation during the exposure is not probable.

Axelsson (18) did not find any chemical changes in the surface of aged specimens by secondary ion mass spectroscopy (SIMS). For example, yttrium content in the surface of $ZrO_2+Y_2O_3$ specimens was unaffected by the aging treatment in humid air at 250°C.

Kagawa et al. (44) have studied the hydration of particle surfaces of various powders (ZrO_2, Y_2O_3, ZrO_2+3mol%Y_2O_3) by measuring electrophoretic mobilities. It was found that the surfaces of ZrO_2 particles were not hydrated by the reaction with water, or at least the amount of $Zr(OH)_x$ formed during the reaction with water was so small that it could not be detected by the mobility measurements. The surfaces of the Y_2O_3 powder were found to be hydrated even though the formation of $Y(OH)_3$ was not found by XRD measurements.

Thus far, it seems that there is not enough analytical evidence for yttrium depletion in the surface, which makes the destabilization theory questionable.

Stress-Induced Transformation

Schmauder et al. (30) have stated that the transformation of the tetragonal phase to monoclinic structure under humid conditions may be controlled by the eigenstresses from thermal expansion anisotropy at grain boundaries. The nucleus for the transformation may arise from the increase of free energy with corresponding decrease in activation energy for transformation in a critical volume element at the edges of grains. Water from the surrounding atmosphere affects the transformation by increasing the free energy of the tetragonal phase. The degradation is believed to be caused by stress-corrosion of the tetragonal phase while the stress corrosion of the grain boundary glassy phase is excluded. However, the principal reason for stresses is the anisotropic thermal expansion of zirconia. As a result of the transformation, microcracks are formed in the surface, thus making it possible for water to penetrate deeper into the material and transformation can proceed. Further transformation can lead to macrocracking which is the main reason for strength reduction. In this mechanism the role of water in inducing the transformation is not clearly described.

Wang et al. (31) have also regarded the stability of Y-TZP ceramics as controlled by the stresses caused mainly by the shear strains related to the transformation. In the surface of the samples shear strains can be removed and the strain energy term can be minimized by the preferential transformation of aligned surface grains with suitable orientation. This will accompany the formation of microcracks, which reduces the strain energy and also allows the randomly aligned surface grains to transform to monoclinic structure. The formation of microcracks allows further exposure of the subsurface grains to the environment and further transformation results. The preferred orientation of the grains found in the aged surface is a consequence of minimizing the strain energy and barrier for transformation to monoclinic structure. However, this model does not explain how water affects the degradation.

SUMMARY

This review summarizes the experimental results published on the aging-induced phase transformation of TZP ceramics under humid conditions at low temperatures. If this transformation takes place in an uncontrolled manner it degrades the mechanical properties of the material. The effects of different material and environmental parameters on the transformation are presented and discussed. Although the factors which affect and control the degradation are known, the actual mechanism of degradation is still under controversy. Many of the proposed mechanisms regard water as playing an important role in the degradation, but no unanimously accepted reaction mechanism has been presented so far. It seems that the research made so far has mainly concentrated on the consequences of the transformation and not so much on the mechanism itself.

The degradation starts from the surface through the interaction of water with the surface structure. Along with the degradation, the actual reaction front is subsequently buried by the resulting destructed structure making it difficult to get information on the microscale processes which take place in the reaction zone. In order to get analytical information on the reaction itself, more surface-sensitive analytical methods should be used to study the aged surface and the changes in its chemical structure which have already occurred after a short period of exposure, when the reactions are still taking place in the very near surface region.

Since the actual mechanisms of degradation are not clear, it is difficult to control and avoid the degradation without losing the beneficial effects of phase transformation toughening. Most of the proposed procedures to control the degradation increase the stability of the structure and thus decrease the mechanical properties of the material. Some of the post-sintering treatments may also lead to more expensive materials.

REFERENCES

1. Subbarao, E.C., *Adv. Cer.* 3: 1 (1981).
2. Bansal, G.K., and Heuer, A.H., *Acta Metall.* 20: 1281-9 (1972).
3. Bansal, G.K., and Heuer, A.H., *Acta Metall.* 22: 409-117 (1974).
4. Kriven, M.K., *J. Am. Cer. Soc.*, 71: 1021-30 (1988).
5. *Science and Technology of Zirconia, Advances in Ceramics* (A.H. Heuer, and L.W. Hobbs ed.), Vol. 3, The American Ceramic Society (1981).
6. *Science and Technology of Zirconia II, Advances in Ceramics* (N. Claussen, M. Ruhle, and A.H. Heuer eds.), Vol. 12, The American Ceramic Society (1984).
7. *Science and Technology of Zirconia III*, Advances in Ceramics (S. Somiya, N. Yamamoto, and H. Hanagida eds.), Vol. 24a and 24b, American Ceramic Society (1988).
8. Garvie, R.C., Hannink, R.H.J., and Pascoe, R.T., *Nature* 258: 703-4 (1975).
9. Evans, A.G., and Cannon, R.M., *Acta Metall.* 34: 701-800 (1986).
10. Chen, I.W., and Chiao, Y.H., *Acta Metall.* 31: 1627-38 (1983).
11. Chen, I.W., and Chiao, Y.H., *Acta Metall.* 33: 1827-45 (1985).
12. Ruhle, M., and Evans, A.G., *Progress in Mat. Sci.* 33: 85-167 (1989).
13. Butler, E.P., *Materials Science and Technology* 1: 417-32 (1985).
14. Kobayashi, K., Kuwajima, H., and Masaki, T., *Solid State Ionics* 3/4: 489-93 (1981).
15. Yoshimura, M., *Cer. Bull.* 67: 1950-5 (1988).
16. Kriven, W.M., *J. Am. Cer. Soc.* 71: 1021-30 (1988).
17. Sato, T., and Shimada, M., *J. Am. Cer. Soc.* 67: C212-C213 (1984).
18. Axelsson, K-O., *Influence of Reactive Gases on the Chemical Composition and Structure of ZrO_2*, Dissertation, Chalmers University of Technology, Gothenburg, 1985, 52 p.
19. Sato, T., and Shimada, M., *J. Mat. Sci.* 20: 3988-92 (1985).
20. Sato, T., and Shimada, M., *J. Am. Cer. Soc.* 68: 356-9 (1985).
21. Michalske, T.A., and Freiman, S.W., *J. Am. Cer. Soc.* 66: 284-8 (1983).
22. Sato, T., Ohtaki, S., Endo, T., and Shimada, M., *J. Mat. Sci. Let.* 5: 1140-2 (1986).
23. Tsukuma, K., *Am. Cer. Soc. Bull.* 65: 1386-9 (1986).
24. Sato, T., Endo, T., and Shimada, M., in: *Ceramic Microstructures '86, Role of Interfaces* (J. A. Pask, and A.E. Evans eds.), pp 215-222, Plenum Press.
25. Sato, T., Ohtaki, S., and Shimada, M., *J. Mat. Sci.* 20: 1466-70 (1985).
26. Watanabe, M., Iio, S., and Fukuma, I., in: *Science and Technology of Zirconia II* (N. Claussen, M. Ruhle, and A.H. Heuer eds.), Vol. 12, pp 391-398, American Ceramic Society (1984).
27. Masaki, T., *Int. High Technology Ceramics* 2: 85-98 (1986).
28. Ruhle, M., Claussen, N., and Heuer, A.H., in: *Science and Technology of*

Zirconia II (N. Claussen, M. Ruhle, and A.H. Heuer eds.) Vol. 12, pp 352-370, American Ceramic Society (1984).
29. Mecartney, M.L., *J. Am. Cer. Soc.* 70: 54-8 (1987).
30. Schmauder, S., and Schubert, H., *J. Am. Cer. Soc.* 69: 534-40 (1986).
31. Wang, J., and Stevens, R., *J. Mat. Sci. Let.*, 8: 1195-8 (1989).
32. Lepistö, T.T., Lintula, P.V., and Mäntylä, T.A., *Cer. Eng. Sci. Proc.* 9: 1517-24 (1988).
33. Chen, S-Y., and Lu, H-Y., *J. Mat. Sci.* 23: 1195-1200 (1988).
34. Tsukuma, K., and Shimada, M., *J. Mat. Sci. Let.* 4: 857-61 (1985).
35. Fischer, T.E., Anderson, M.P., Jahanmir, S., and Salher, R., *ASLE Trans.*
36. Reuhkala, P.J., Lepistö, T.T., and Mäntylä, T.A., in: *Proc. 5th International Congress on Tribology* (K. Holmberg, and I. Nieminen eds.), Vol. 5, pp 339-344, International Tribology Council and the Finnish Society for Tribology (1989).
37. Iio, S., Watanabe M., Kuroda, K., Saka, H., and Imura, T., in: *Science and Technology of Zirconia III* (S. Somiya, N. Yamamoto, and H. Hanagida ed.), Vol. 24a, pp 49-54, American Ceramic Society (1988).
38. Whalen, P.J., Reidinger, F., and Antrim, R.F., *J. Am. Cer. Soc.* 72: 319-21 (1989).
39. Yoshimura, M., Noma, T., Kawabata, K., and Somiya, S., *J. Mat. Sci. Let* 6: 465-7 (1987).
40. Chen, I-W., and Chiao, Y-H., in: *Science and Technology of Zirconia II* (N. Claussen, M. Ruhle, and A.H. Heuer eds.), Vol. 12, pp 33-45, American Ceramic Society (1984).
41. Sato, T., and Shimada, M., *Ceramic Bulletin* 64: 1382-4 (1985).
42. Lepistö, T.T., and Mäntylä, T.A., *Cer. Eng. Sci. Proc.* 10: 658-67 (1989).
43. Lange, F.F., Dunlop, G.L., and Davis, B.I., *J. Am. Cer. Soc.* 69: 237-40 (1986).
44. Kagawa, M., Omori, M., Syono, Y., Imamura, Y., and Usui, S., *J. Am. Cer. Soc.* 70:C212-C213 (1987).

18

High Temperature Corrosion of Engineering Ceramics

Nathan S. Jacobson, James L. Smialek, and Dennis S. Fox

National Aeronautics and Space Administration
Lewis Research Center
Cleveland, Ohio

INTRODUCTION

Current and future applications of silicon-based ceramics include heat engines, heat exchangers, and fuel cells. In such applications the ceramic will encounter temperatures often in excess of 1000°C as well as harsh chemical environments. These environments may contain gases such as oxygen, hydrogen, steam, and sulfur oxides. The environments may also lead to condensed phase deposits, which are often quite corrosive. The deposit may be an alkaline salt such as Na_2SO_4 or a mixture of oxides. The focus of this chapter is on deposit-induced corrosion in silicon-based ceramics.

Silicon-based ceramics encompass a wide range of materials: various types of monolithic silicon carbides and silicon nitrides, as well as composites of these two materials. These materials are unstable in air and rely on the formation of a thin film of SiO_2 to impart stability. The interaction of this film with the corrosive deposit is the primary focus in hot corrosion studies. Table 18-1 lists the specific types of silicon-based ceramics which will be covered in this chapter. In general only the chemically vapor deposited (CVD) materials are highly pure. The other materials are formed from powder compacts and require additional compounds to achieve full densification. These additions may have a major effect on the corrosion resistance of a particular material.

Monolithic silicon carbide and silicon nitride are brittle and their strength is influenced by surface flaws. This is an important fact to consider in a surface

Table 18-1. Types of Silicon-Based Ceramics

Designation	Material	Additives, impurities	Type, manufacturer
CVDSC	Chemically vapor deposited SiC	----	e.g. Thermoelectron, Waltham, MA
SCSC	Single crystal SiC	Fe	
SASC	Sintered α-SiC	B,C	Carborundum Niagara Falls, NY
SBSC	Sintered ß-SiC	B,C	General Electric Schenectady, NY
HPSC	Hot pressed SiC	Al_2O_3	Norton Co. Worcester, MA
CVDSN	Chemically vapor deposited Si_3N_4	----	e.g. Union Carbide Cleveland, OH
HIP RBSN	Hot isostatically pressed reaction bonded Si_3N_4	----	Ref. 23
SSN (Al_2O_3,Y_2O_3)	Sintered Si_3N_4	Al_2O_3, Y_2O_3	AY-6, GTE Products Towanda, PA
SSN (Y_2O_3)	Sintered Si_3N_4	Y_2O_3	PY-6, GTE Products Towanda, PA
Si_3N_4/SiC_w	Hot isostatically pressed Si_3N_4 Tateho SCW#1 whiskers	Y_2O_3, Al_2O_3	GN-10, Garrett Ceramic Components Div., Torrance, CA
Si_3N_4/SiC_f	Reaction bonded Si_3N_4/AVCO SCS-6 fibers	Si,C	NASA Lewis Research Center Cleveland, OH

degradation process such as corrosion. Ceramic composites are a more recent development. Examples are silicon nitride matrices reinforced by silicon carbide fibers or silicon carbide particulates. These reinforcements are intended to limit crack propagation and thus make the strength of the composite less dependent on surface flaws. However, in their current state of development, these composites are generally porous and contain various additives. This can exacerbate corrosion. In summary, silicon-based ceramics have additives and physical characteristics which can have a major influence on the behavior of the material in a corrosive environment.

A detailed literature survey on the deposit-induced corrosion of ceramics has been presented elsewhere (1) and will only be briefly summarized here. The early studies of this phenomena were the etchant studies of SiC (2,3) by basic molten salts, such as Na_2CO_3. These studies were important from the standpoint of corrosion in that they showed that basic molten salts were among the few compounds which would etch SiC and create a rough surface. Most of the corrosion studies tend to emphasize heat engines, since this is where in practice corrosion has been a real problem. Extensive studies of metals and alloys have shown that in this application, Na_2SO_4 is the most corrosive agent. The Na_2SO_4 deposits form from ingestion of Na either in the intake air or in the fuel and subsequent reaction with sulfur oxides from the fuel (4):

$$2NaCl(g) + SO_2(g) + 1/2O_2(g) + H_2O(g) \rightarrow Na_2SO_4(l) + 2HCl(g) \qquad (18\text{-}1)$$

Sodium sulfate-induced corrosion is termed "hot corrosion." Previous work has shown that for silicon-based ceramics the primary mode of attack is dissolution of the silica scale (1,5,6). This will be discussed in more detail later. Another group of studies have centered around ceramic heat exchangers. In this case the deposit varies depending on the application; it may be an oxide mixture or a salt. Oxide dissolution and metal silicide formation are the primary modes of attack (7,8). Other studies center on more exotic applications for ceramics such as fuel cells and magneto-hydrodynamics (9,10). In these cases the ceramics encounter various molten salts. In this chapter, the primary emphasis will be on heat engines and Na_2SO_4 deposits. However, similarities and differences between this and other types of deposit-induced corrosion will also be discussed.

TEST METHODS

Experimental studies of corrosion are generally performed using three techniques, namely, burner rigs, laboratory crucible tests, and laboratory thin film tests. A burner rig is a variable flow velocity flame tunnel with salt impurities carefully injected into the flame. This most closely simulates the situation which may be encountered in a heat exchanger or heat engine. However, burner rig tests are expensive and time-consuming, and it may be difficult to precisely control the various parameters effecting corrosion. Laboratory tests fall into two categories: crucible tests, wherein the specimen is set at the bottom of a crucible containing the salt; and thin film tests, wherein the specimen is coated with a film of the salt and then heated in a furnace. Crucible tests may be appropriate to model certain heat exchangers and other applications where a thick deposit is observed. However, for heat engine applications a crucible test may produce an unrealistically low oxygen pressure at the melt/specimen interface. Therefore, thin film laboratory tests are most often used to simulate the situation observed in a heat engine. It is best to use burner rig tests to develop an overall picture of corrosion, and laboratory thin film tests to examine details of the corrosion mechanism.

THERMODYNAMICS OF CORROSION

Reaction 18-1 shows the route by which sodium sulfate forms in a gas turbine engine. Sodium sulfate is a highly stable molecule and tends to form quite readily and deposit on engine parts. Typically, corrosion occurs between the melting point of sodium sulfate and the dew point for sodium sulfate deposition. Above the melting point the salt tends to spread readily and transport rates through the salt are rapid, allowing for corrosion. The dew point is the highest possible temperature at which a condensed phase deposit can form on a surface in the hot gas path. It can be calculated by means of a thermodynamic free energy minimization program (11,12). Inputs to the program are the combustion reactants, temperature, and pressure. Figure 18-1 shows the dew point as a function of pressure and sulfur and sodium impurity levels. Future

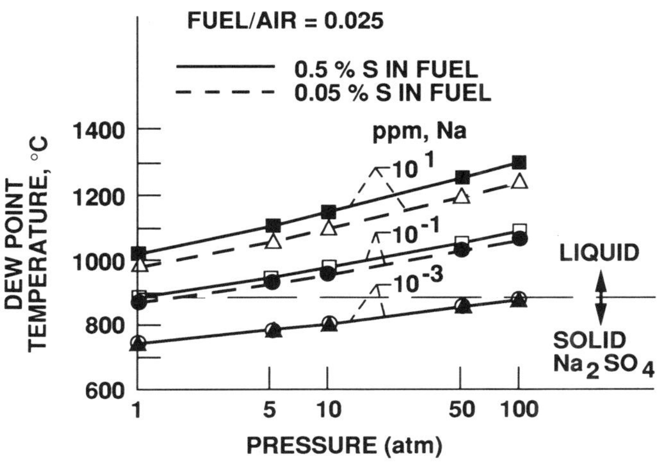

Figure 18-1. Calculated dew points for Na_2SO_4 deposition.

heat engines are expected to operate at higher temperatures and pressures. The high temperatures may limit deposition, since in some cases the dew point will be exceeded. However, the higher pressures may extend the range of corrosion by increasing the dew point. Thus in order to assess the importance of corrosion, the operating parameters of a system must be fully known.

Once deposition has occurred, the next step is the chemical process of corrosion. Figures 18-2(a) and (b) show a coupon of SASC tested in the burner rig with and without injected sodium chloride. The first specimen has only undergone oxidation and looks very similar to the as-received material. The

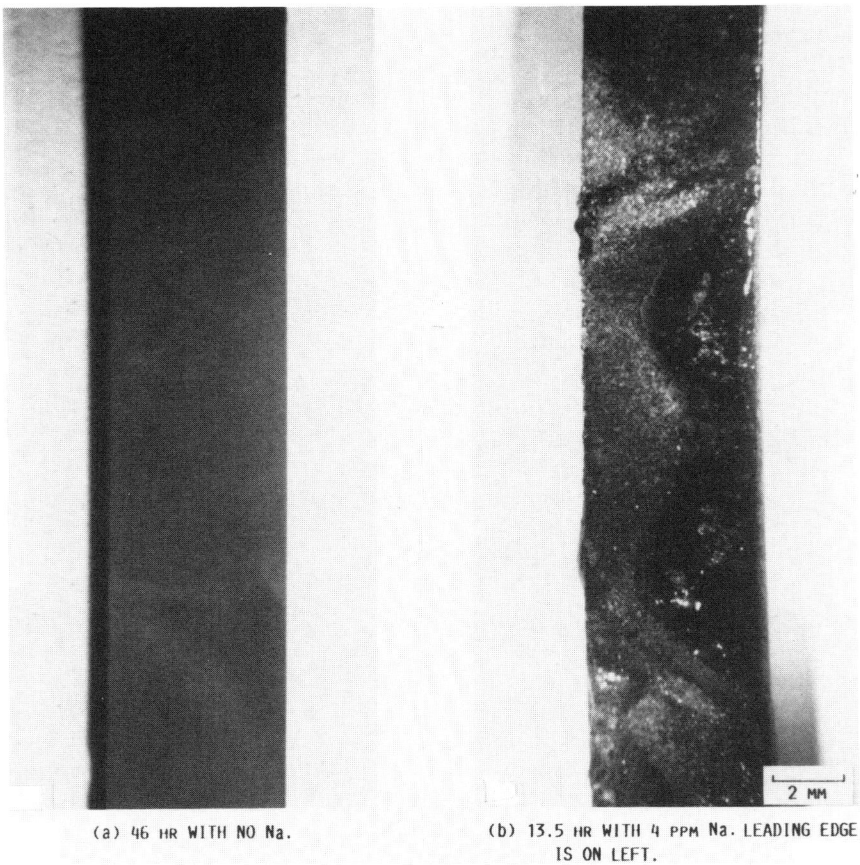

Figure 18-2. Optical micrographs of SASC treated in the burner rig at 1000°C with Jet A fuel.

second specimen shows a substantial amount of glassy corrosion product resulting from the Na_2SO_4 deposit. Figure 18-3 shows a polished cross section

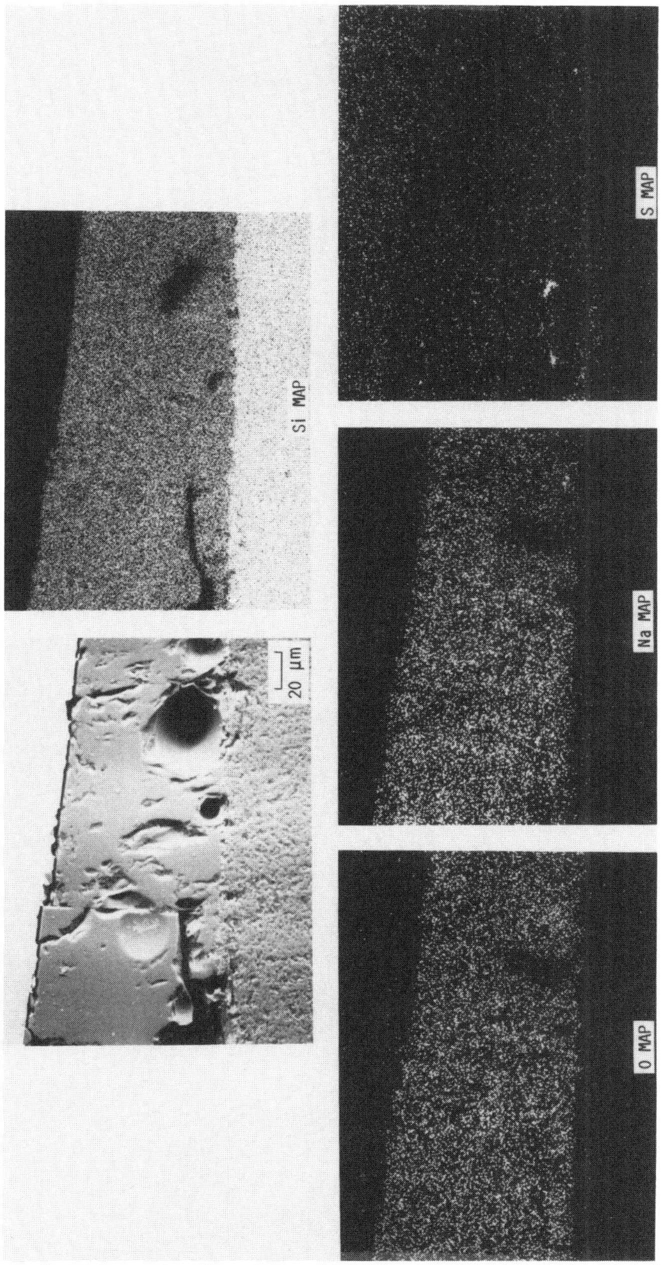

Figure 18-3. Electron microprobe analysis of a polished cross section of glassy products on SASC formed by burner rig corrosion at 1000°C with Jet A fuel for 13.5 hr with 4 ppm Na.

of this corroded specimen and associated elemental maps. The uniform distribution of Na, Si, and O suggest $(Na_2O) \cdot x(SiO_2)$ formation. The reaction scheme is as follows:

$$SiC(s) + 3/2\ O_2(g) \rightarrow SiO_2(s) + CO(g) \tag{18-2}$$

$$Na_2SO_4(l) \rightarrow Na_2O(s) + SO_3(g) \tag{18-3}$$

$$2\ SiO_2(s) + Na_2O(s) \rightarrow (Na_2O) \cdot 2(SiO_2)(l) \tag{18-4}$$

Thus corrosion occurs by oxidation of the SiC and dissolution of the oxide scale. The solid, protective SiO_2 scale is replaced by a liquid, unprotective $Na_2O \cdot 2(SiO_2)$ layer and the ceramic sustains corrosive attack.

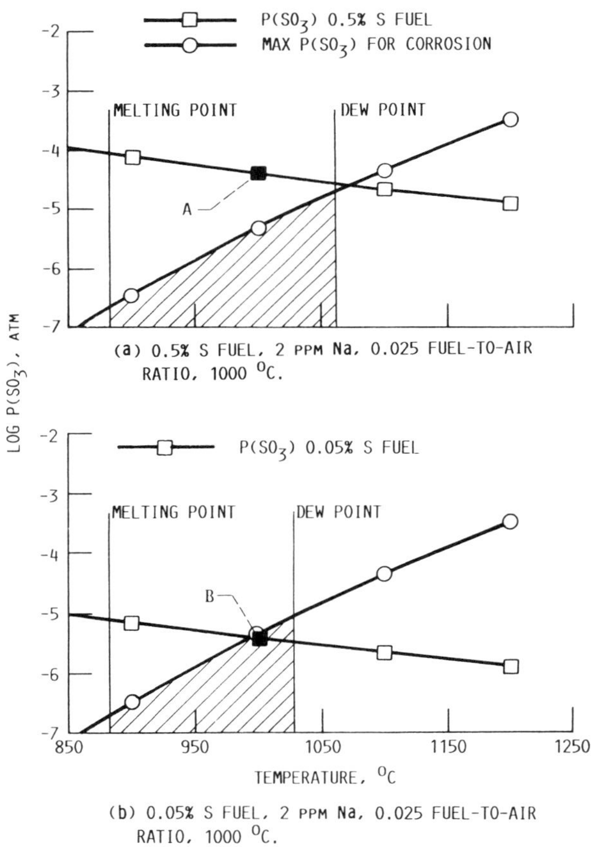

Figure 18-4. Calculated corrosion regimes for SiO_2

The key reactant is the Na_2O, which is released by reaction 18-3. The activity of Na_2O, $a(Na_2O)$, is a simple index of basicity. A large activity is termed a basic molten salt and set by a low $P(SO_3)$ over Na_2SO_4; a smaller activity of

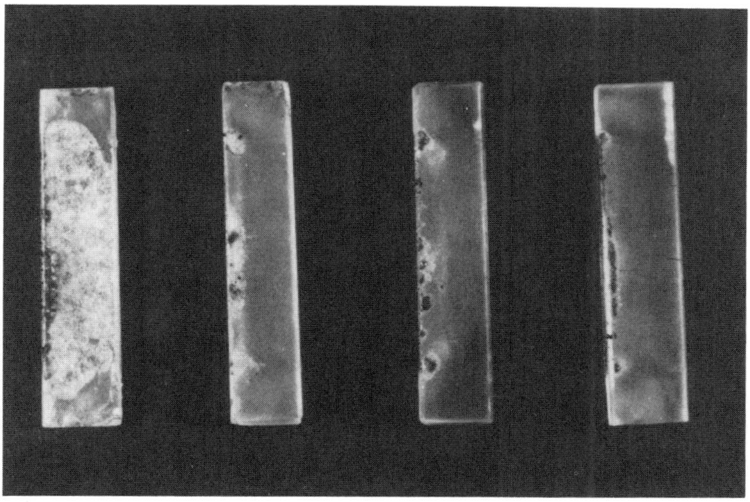

(a) NO. 2 DIESEL FUEL (0.5% S), 5HR.

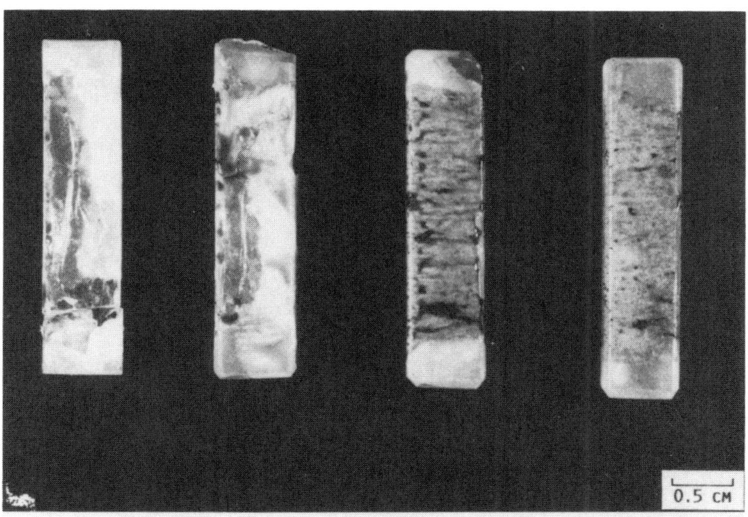

(b) JET A FUEL (0.05%S), 1 HR.

Figure 18-5. SiO_2 coupons treated in the burner rig, 2ppm Na, 1000°C, for high and low $P(SO_3)$.

Na_2O is termed an acidic molten salt and set by a high $P(SO_3)$ over Na_2SO_4. As reaction 18-4 shows, this activity is the key factor in the dissolution reaction. Unlike other protective oxides, SiO_2 is an acidic oxide and can only be dissolved by a basic salt, i.e. a larger activity of Na_2O. From reaction 18-4 a threshold activity of Na_2O for SiO_2 dissolution can be calculated.

On the basis of these thermodynamic arguments, corrosion regimes for these SiO_2-protected materials can be calculated as a function of temperature and $P(SO_3)$ (since it sets $a(Na_2O)$). Two diagrams which correspond to the NASA burner rig are shown in Figure 18-4(a) and (b). Each diagram corresponds to a different sulfur level in the fuel and hence a different basicity in the deposit. The threshold value of $P(SO_3)$ for SiO_2 dissolution remains the same. The lower temperature boundary is the melting point of sodium sulfate. The upper temperature boundary is the dew point, which changes slightly with the addition of more sulfur. However, in the higher sulfur case the $P(SO_3)$ generated does not intersect the dissolution regime--i.e., the salt is too acidic to dissolve the SiO_2. In the lower sulfur case, the $P(SO_3)$ generated does intersect the dissolution regime between 995 and 1025°C and the SiO_2 is dissolved in that range. These predictions have been verified experimentally with quartz coupons, treated under the conditions corresponding to (A) and (B) in Figure 18-4. The results are shown in Figure 18-5(a) and (b). Note the coupons treated in the higher sulfur fuel show less attack. Chemical analysis showed that sodium silicate was only produced on the specimens treated in the lower sulfur fuel.

The type of diagram shown in Figure 18-4 has been calculated for a range of pressures (11) and provides general guidelines for predicting when corrosion would be expected to occur. However, other elements in the ceramics complicate the situation. Table 18-2 shows the amount of corrosion for several ceramics with various additives, which have been corroded in the high sulfur fuel. This is a situation when corrosion is not predicted, but in each case some corrosion occurs. The effect of excess carbon in SiC is particularly pronounced, and Y_2O_3 and Al_2O_3 additions have a lesser, but still significant, effect. These oxides may migrate to the surface and react with the SiO_2 and Na_2O to form a corrosion compound.

Table 18-2. Total Corrosion Products: Burner Rig Results

Material	Amount of Corrosion Product mg/cm^2
SASC	33.77±5.17
HPSC	14.15±0.55
SSN(Y_2O_3)	5.24±1.78

* T=1000°C, 2ppm Na as NaCl, 0.5% S in fuel, 40 hr.

It is important to look at the effect of carbon in more detail, since it is present both intrinsically and as an additive in some types of SiC. As mentioned, the a(Na_2O) is the most important parameter in corrosion reactions. Its value can be measured electrochemically using a zirconia membrane for oxide anions and a mullite membrane for sodium cations (13). The net effect is a direct measurement of a(Na_2O). The results of Na_2SO_4/SO_3 with and without a layer of carbon below it are shown in Figure 18-6. Note that the layer of carbon drives the melt strongly basic. It is quite likely this occurs through the following reaction:

$$Na_2SO_4 + 2\ C \rightarrow Na_2S + 2\ CO_2 \tag{18-5}$$

$$Na_2S + 3\ Na_2SO_4 \rightarrow 4\ Na_2O + 4\ SO_2 \tag{18-6}$$

This explains the observed strong effect of carbon on the corrosion process in an acidic salt.

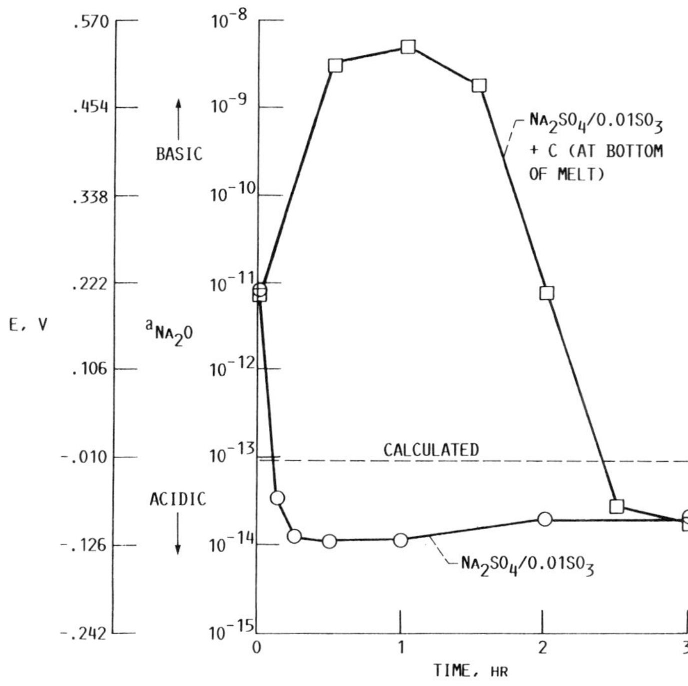

Figure 18-6. Effect of carbon on the basicity of Na_2SO_4. T = 1000°C.

Corrosion in other types of melts exhibits some similarities to and differences from to corrosion by Na_2SO_4. Corrosion by Na_2CO_3 occurs more readily, since Na_2CO_3 is a more basic salt which readily dissociates:

$$Na_2CO_3 \rightarrow Na_2O + CO_2 \tag{18-7}$$

Corrosion by NaCl in the presence of moisture can be quite similar to Na_2SO_4 and Na_2CO_3 corrosion, since Na_2O is formed (14):

$$2\ NaCl + H_2O \rightarrow Na_2O + 2\ HCl \tag{18-8}$$

The discussion up to this point has emphasized dissolution of SiO_2 by basic molten salts. In general SiO_2 is fairly unreactive to acidic molten salts. However, Na may induce small changes in the silica film on SiC and Si_3N_4 and enhance oxidation (15). Sodium may induce devitrification in amorphous SiO_2 and the resultant volume change could lead to cracking of the oxide and more rapid oxidation. Sodium could also dope the silica scale and lead to faster oxidation. These effects must be considered in the presence of an acidic salt deposit.

Another important variable to consider is the oxygen potential at the melt/ceramic interface. At low oxygen partial pressures SiC and Si_3N_4 undergo a transition from passive oxidation-forming SiO_2 to active oxidation-forming SiO (16). Some investigators have proposed that in a deep melt active oxidation occurs (15,16):

$$SiC(s) + O_2(g) \rightarrow SiO(g) + CO(g) \tag{18-9}$$

This may explain the excessive bubbling and attack observed in deep melts.

Corrosion by slags is more complex than salt corrosion. These are complex mixtures of different oxides and there is no simple, measurable index of activity.

Table 18-3. Typical Coal Slag Composition*

	Acidic	Basic
SiO_2	54	46
Fe_2O_3	21	11
Al_2O_3	19	14
CaO	0.1	19
MgO	0.9	7
Na_2O	0.6	1.1
K_2O	1.7	0.5
TiO_2	1.3	0.3
Ratio basic oxides to acidic oxides**	0.29	1.14

*From Ref. 24

** Sum wt% (Fe_2O_3 + CaO + MgO+ Na_2O+ K_2O)/Sum wt%(SiO_2 + Al_2O_3 + TiO_2)

A measure of basicity is the ratio of basic oxides to acidic oxides. Table 18-3 shows typical compositions of laboratory simulated coal slags.

Three mechanisms of corrosion in the presence of coal slags have recently been discussed (7). Two are similar to those observed in the salt corrosion case --formation of a passive SiO_2 film and dissolution of the SiO_2 film. A third is silicide formation, which occurs only in the case of a low oxygen potential:

$$13\ SiC + 5\ Fe(\text{slag matrix}) \rightarrow Fe_5Si_{13} + 13\ C \qquad (18\text{-}10)$$

Clearly there are both similarities and differences between molten salt and slag corrosion.

KINETICS OF CORROSION

Corrosion is a complex process involving several steps--deposition, oxidation, and dissolution. The kinetics of deposition have been studied extensively and the reader is referred to other discussions of this topic (17,18). Again it is important to recognize the important effects of both temperature and pressure.

Laboratory thin film tests are most helpful to study the kinetics of the actual corrosion process. In this case the reaction is followed either by removing the sample from the furnace at incremental times for chemical analysis, or by following the weight change as a function of time (Thermogravimetric Analysis - TGA). The corrosion products are composed of silicate and silica, which can be separated and analyzed by a process based on the water solubility of silicate and HF solubility of silica. This procedure has been described in detail elsewhere (6). Initial studies were done with Na_2CO_3, which is a model basic molten salt and expected to react quite readily with SiO_2. Typical kinetic curves are shown in Figures 18-7(a) and (b), constructed both from thermogravimetric and chemical analysis. Silicon, SiC, and Si_3N_4 all show this general behavior. There appear to be three regimes. Region I is a period of weight loss, which is attributed to coupled oxidation and dissolution:

$$SiC + 3/2\ O_2 \rightarrow SiO_2 + CO \qquad (18\text{-}11)$$

$$SiO_2 + Na_2CO_3 \rightarrow Na_2SiO_3 + CO_2 \qquad (18\text{-}12)$$

For a number of different pure silicon, silicon carbide, and silicon nitride materials the amount of weight loss in this Region I corresponds to these types of reactions (19). Region II, where rapid weight gain is observed, corresponds to a period of rapid oxidation, during which an oxide film grows below the Na_2SiO_3 layer. This film appears to be porous, which explains its rapid growth rate. Finally in Region III, where the weight gain is slow, this layer becomes

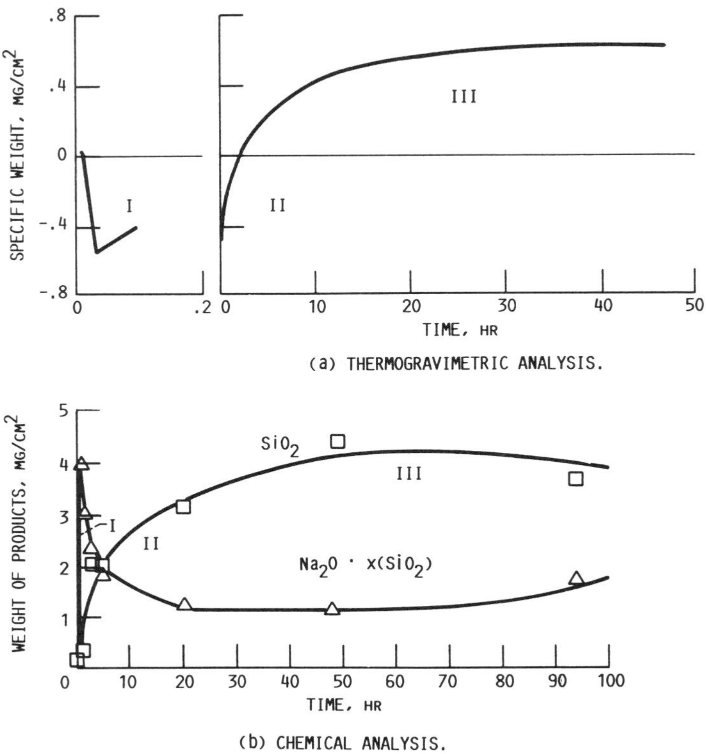

Figure 18-7. Kinetic curves for the reaction SASC and Na_2CO_3 at 1000°C.

dense and the overall reaction slows. Figure 18-8 shows a polished cross section and associated elemental maps in Region III, which indicate a thick scale composed of an outer sodium silicate layer and an inner silica layer.

Next consider Na_2SO_4. Figure 18-9 shows a TGA plot of HIP RBSN reacting with Na_2SO_4 in a flowing oxygen environment. In such an environment, a(Na_2O) is uncontrolled and varies from $5x10^{-14}$ to $1x10^{-12}$ at 1000°C. This is too acidic to dissolve SiO_2. This is reflected in Figure 18-9 - the initial weight loss over 40 hr is the same as that for Na_2SO_4 on a Pt coupon. After 40 hr some limited oxidation and dissolution occur. This is in contrast to Figure 18-10, which shows the reaction of SASC with Na_2SO_4/O_2. Note that this trace is very similar to that obtained for reactions with Na_2CO_3. This is due to the fact that carbon creates basic conditions in the melt bottom, as described earlier.

In an atmosphere of 10^{-4} atm P(SO_3), the activity of Na_2O in Na_2SO_4 becomes equal to $5x10^{-13}$. Table 18-4 shows the amount of corrosion products for several different silicon-based ceramics corroded in this atmosphere. Note extensive corrosion occurred only in the SASC case. In fact this was the most

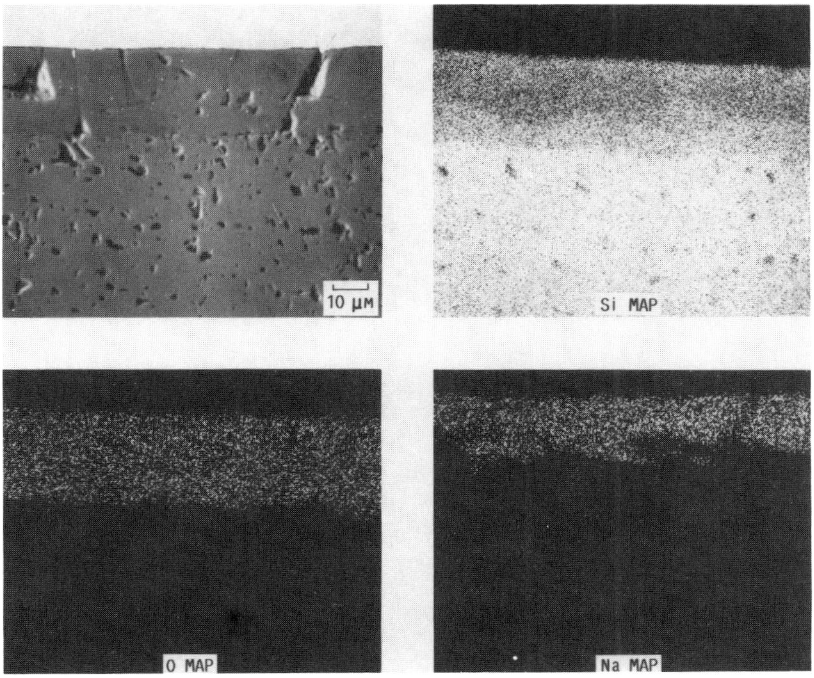

Figure 18-8. Electron microprobe analysis of a polished cross section of corrosion products formed from the reaction of SASC + Na_2CO_3 at 1000°C for 48 hours.

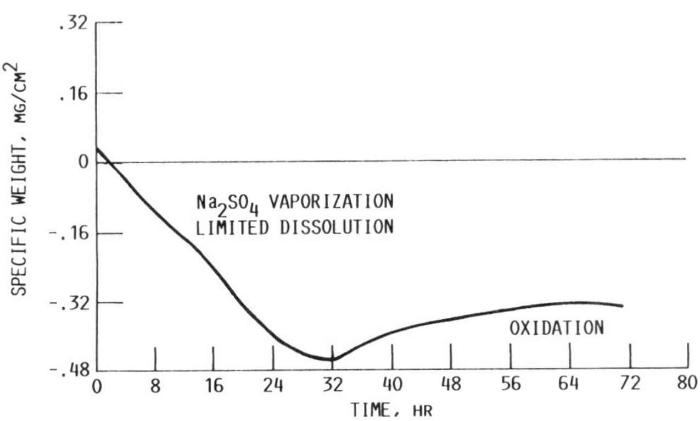

Figure 18-9. Thermogravimetric analysis curve for HIP RBSN + Na_2SO_4/O_2 at 1000°C.

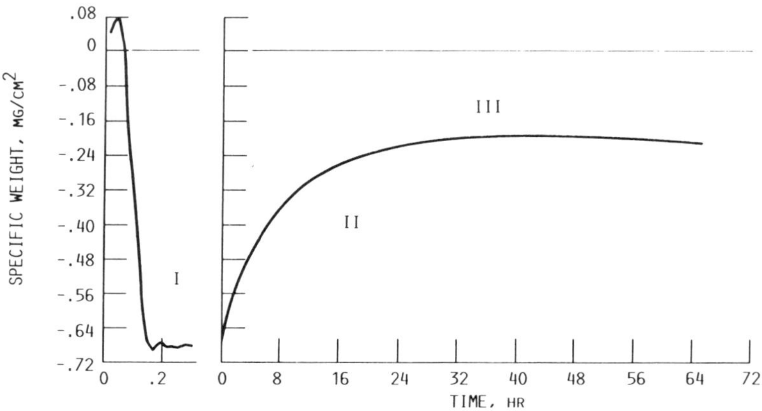

Figure 18-10. Thermogravimetric analysis curve for SASC + Na_2SO_4/O_2 at 1000°C.

Table 18-4. Corrosion Products-Laboratory Tests with Na_2SO_4/SO_3 [SiC, Si_3N_3 + $Na_2SO_4/0.01SO_3$-O_2 -- 48 hours, 1000°C]

Material	Water Soluble: Na_2SO_4, mg/cm^2	Water Soluble: Na_2O x (SiO_2) mg/cm^2	Water Soluble: Na_2O x (SiO_2) x	Water Insoluble SiO_2, mg/cm^2
SCSC	1.64 ± 0.86	<0.01	----	1.82 ± 0.96
HPSC	2.01 ± 0.03	<0.01	----	0.39 ± 0.02
SASC*	0.49 ± 0.06	0.97 ± 0.28	1.6 ± .09	10.50 ± 0.66
SSN**	0.23 ± 0.11	----	----	1.30 ± 0.20

* (B,C)
** (Y_2O_3, Al_2O_3)

extensive corrosion encountered in the laboratory studies. This was attributed to basic conditions at the melt bottom and acidic conditions at the melt top, which create a series of self-sustaining reactions as shown in Figure 18-11. This is termed "fluxing" and is common in the hot corrosion of metals.

Thus far the kinetics of laboratory reactions have been discussed. In a burner rig or engine, where continuous deposition occurs, the situation may be somewhat different. If the salt is sufficiently basic, continuous oxidation and dissolution may occur. This could occur until the ceramic is consumed.

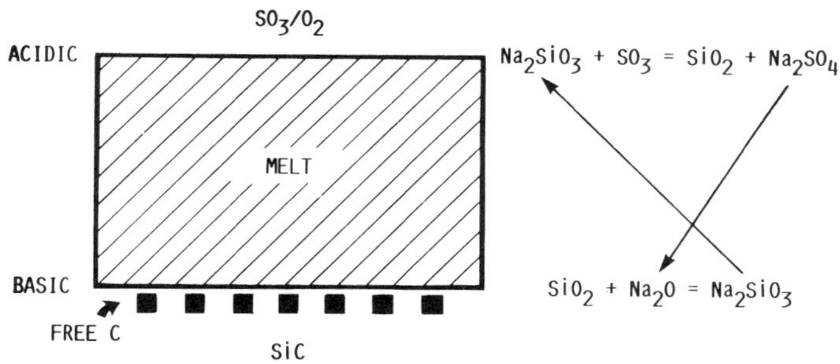

Figure 18-11. Schematic of fluxing mechanism operative in the reaction of SASC + Na_2SO_4/SO_3.

MICROSTRUCTURAL AND MECHANICAL PROPERTY EFFECTS

As the ceramic is converted to silica and silicate, it is being consumed. Clearly the form of surface recession is important, since surface finish is so critical to crack initiation and ultimate strength. For example, uniform surface recession would lead to only limited strength decreases, whereas deep pitting could lead to substantial strength decreases (20).

In the case of SiC, the corrosion products are cleanly removed by a 10 percent HF solution, revealing the attacked surface for microstructural examination. Figure 18-12 shows a sequence of the as-received material, the corroded material with corrosion products, and the corroded material with the corrosion products cleanly removed. Note the extensive pitting and grain boundary attack. Microstructural observation has shown that attack occurs within the grain at structural discontinuities, at the grain boundaries, and in the form of large pits (21). The first relates back to the early studies of molten salts as an etchant for SiC.

The formation of large pits has been correlated with the observations of gas bubbles as seen in Figure 18-13. This shows a series of micrographs taken at exactly the same place in the silicate, silica, and bare substrate. Note that the bubble emerging from the corrosion products corresponds to a pit in the substrate. This suggests that when a gas bubble forms, it exposes a fresh section of the substrate for attack, leading to eventual pit formation. This is shown schematically in Figure 18-14.

One might expect these pits to have a major effect on strength. Figure 18-15 shows the effect of corrosion on four point bend strength for SASC and three different salt systems. Note that Na_2SO_4/SO_3 causes the most severe strength degradation--this is the fluxing situation described earlier of SiC. Figure 18-16 shows the strength degradation for burner rig corrosion. The most severe

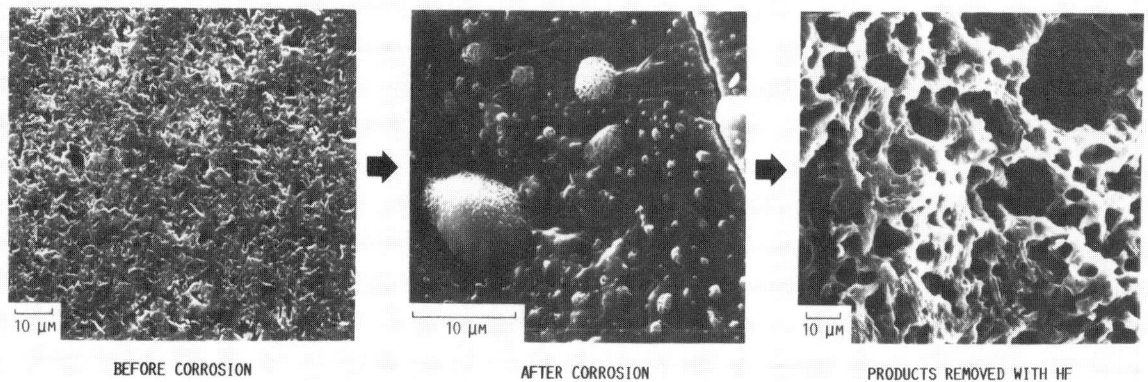

Figure 18-12. Sequence showing SASC before corrosion, after corrosion with Na_2SO_4/SO_3 at 1000°C for 48 hours, and with the corrosion products removed with HF.

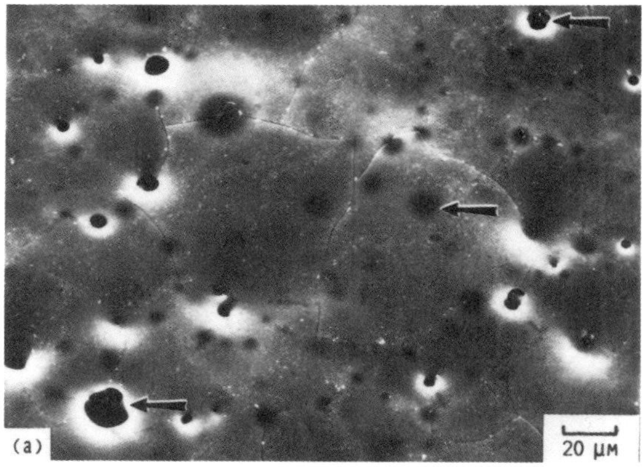

(a) AS-CORRODED SURFACE SHOWS BUBBLES AND PORES IN THE SILICATE LAYER.

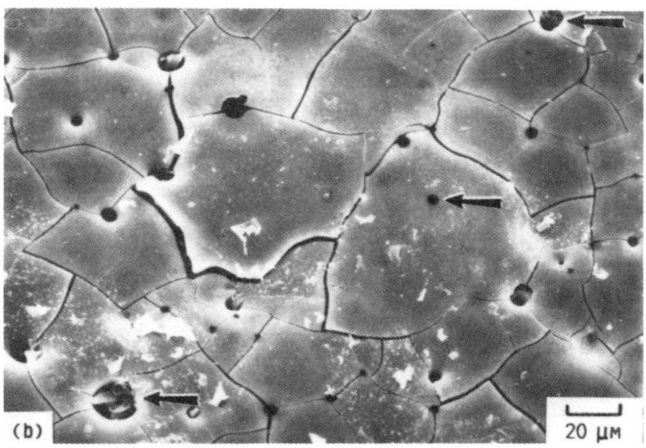

(b) WATER LEACH REVEALS BUBBLES AND PORES IN THE SILICA LAYER.

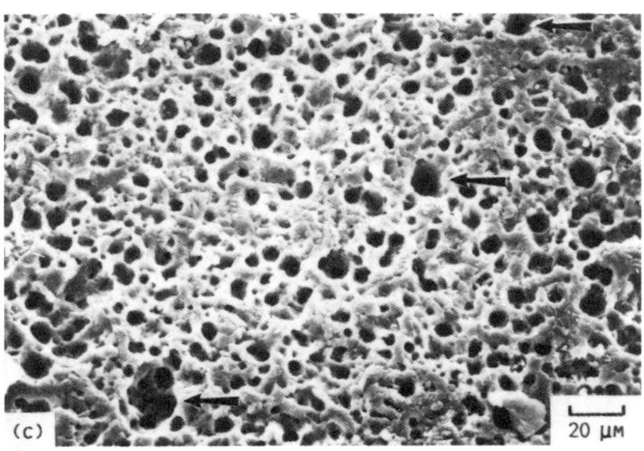

(c) HF LEACH SHOWS THE CORRELATION OF THESE BUBBLES AND PORES TO SOME OF THE PITS IN THE CERAMIC, AS SHOWN BY THE THREE ARROWS.

Figure 18-13. Electron micrographs of SASC + Na_2CO_3 reacted for 1 hour at 1000°C.

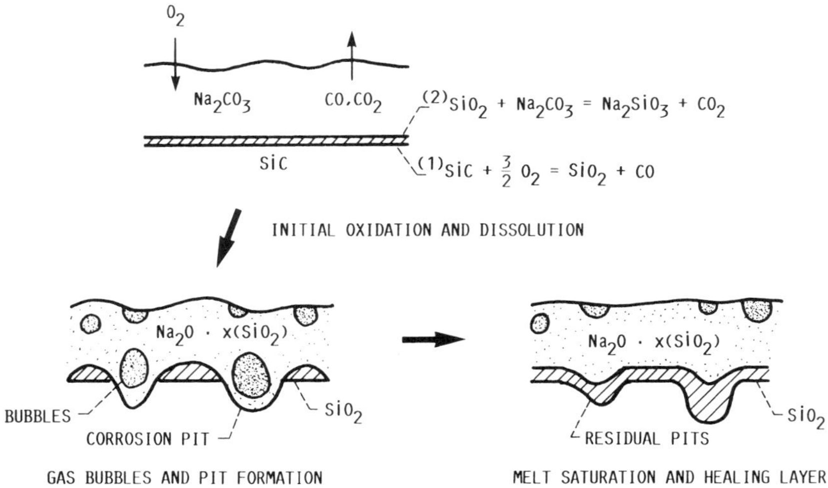

Figure 18-14. Schematic of proposed pitting mechanism in SASC via bubbles.

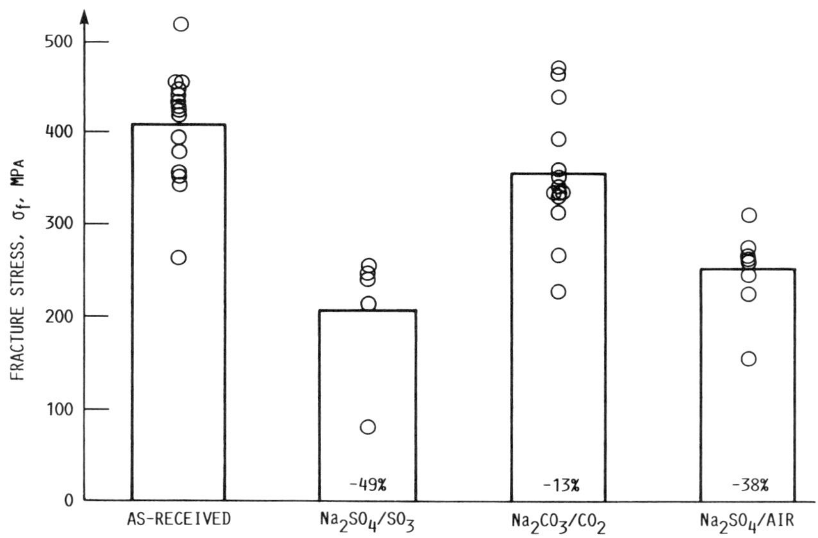

Figure 18-15. Strength degradation of SASC after furnace corrosion test for 48 hours at 1000°C.

strength degradation occurs for the SiC with excess Si. It appears the excess Si corrodes quite readily.

In order to confirm that these strength degradations did indeed come from corrosion effects, fracture origins of these samples were located. In most instances, they were attributed to a surface pit caused by corrosion. Figure 18-17 shows a fracture origin of a SiC sample treated in the laboratory in Na_2SO_4/SO_3. Figure 18-18 shows a fracture origin of a SiC sample treated in the burner rig. The relationship of corrosion pits to fracture strength can be further explored by correlating pit depth to fracture strength. This is given by the following equation:

$$\sigma_f = Z/Y \quad K_{IC}/(\mathbf{a})^{1/2} \tag{18-13}$$

Here σ_f is the fracture strength, K_{IC} is the fracture toughness, Z is a flaw shape parameter, Y is a geometric factor, and **a** is the depth of an atomically sharp crack. Clearly the corrosion pits are not necessarily atomically sharp cracks. However one would expect a plot of σ_f versus $1/(\mathbf{a})^{1/2}$ to cluster about a line with slope similar to that observed from previous fracture toughness measurements. This is shown in Figure 18-19. This further verifies that corrosion pits control strength.

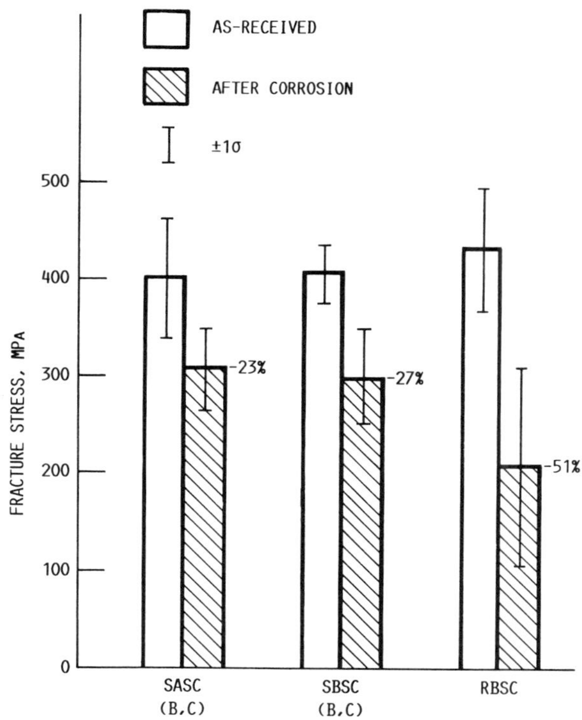

Figure 18-16. Strength degradation of SiC after burner rig corrosion at 1000°C, 4 atm, 2 ppm Na, Jet A fuel, 40 hours.

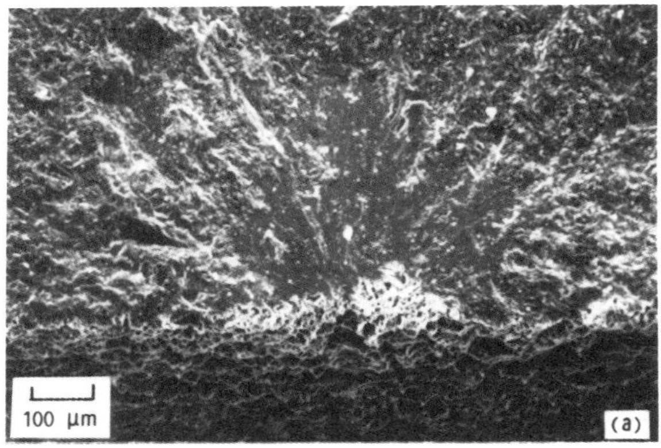

(a) RADIAL CRACK LINES EXTENDING TO THE ORIGIN.

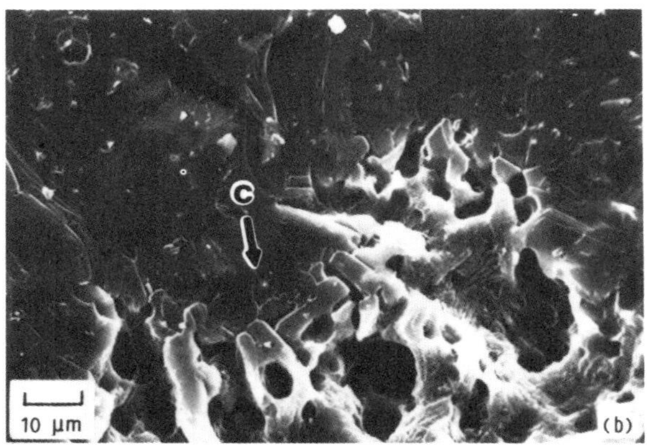

(b) HONEYCOMB CORROSION PIT AT THE ORIGIN.

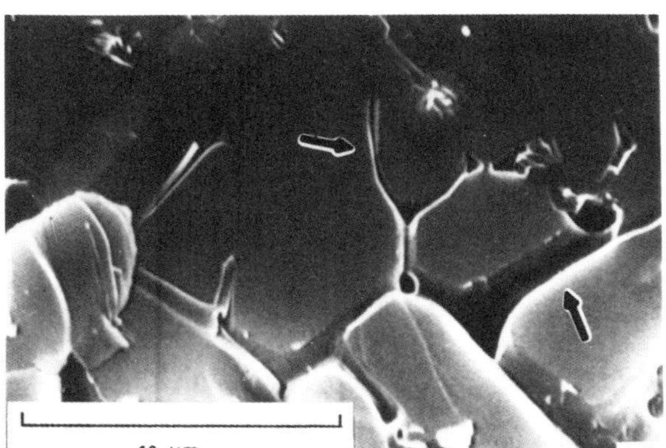

(c) PREFERENTIAL GRAIN BOUNDARY ATTACK AHEAD OF PIT IN (B).

Figure 18-17. Corrosion pit fracture origin for SASC after Na_2SO_4/SO_3 at 1000°C for 48 hours. Scale removed by HF dissolution.

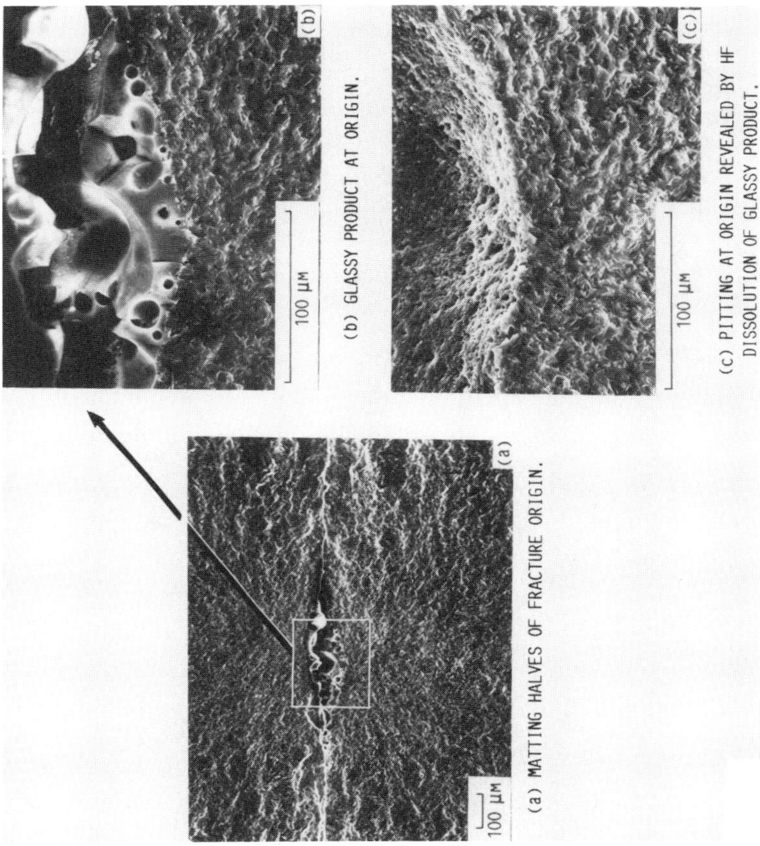

Figure 18-18. Corrosion pit fracture origin for SASC after burner rig corrosion using Jet A fuel, 4 ppm Na, and 1000°C for 15 hours.

Molten salt attack of Si_3N_4 is more subtle. Figure 18-20 shows a sequence of attack--as received, corroded with products present, and the corroded substrate with corrosion products removed by an HF treatment. It should be noted that HF tends to attack the grain boundary phase in Si_3N_4, so only a weak HF treatment was used to remove the products and keep the grain boundary phase intact. Note that the predominant mode of attack is grain boundary attack and that pitting is minimal.

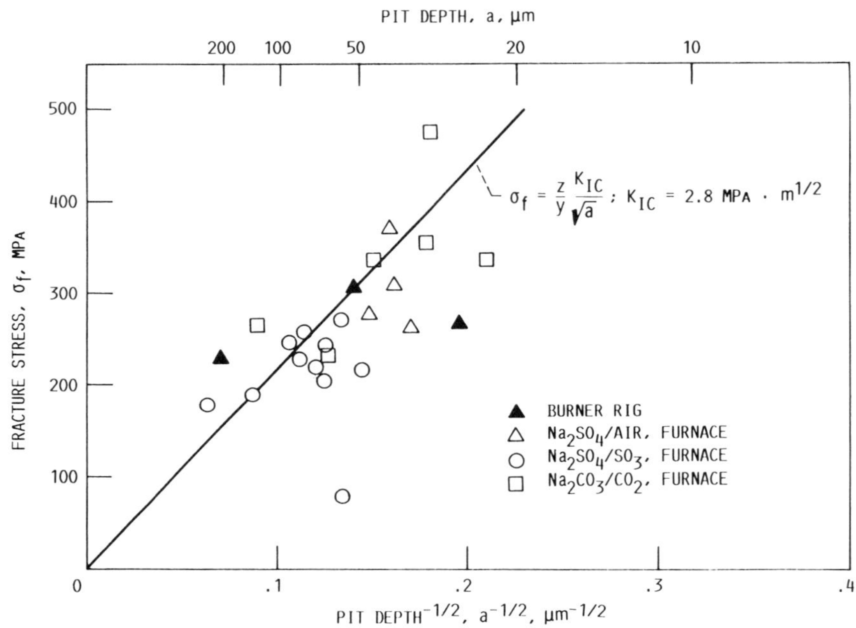

Figure 18-19. Correlation of fracture strength with corrosion pit depth for SASC corroded in the laboratory furnace and burner rig.

As mentioned, Si_3N_4 may contain various oxide additives. In oxidation these tend to migrate to the surface. A polished cross section and associated elemental maps (Figure 18-21) show that a similar effect is occurring in the corrosion case. Note the gap in Y--it has migrated from the substrate into the corrosion product. Figure 18-22 shows a secondary ion mass spectrometry (SIMS) profile into a corroded specimen, starting at the melt/substrate interface. This again shows the Y depletion zone, as well as some Na and S penetration into the sample, very likely along grain boundaries.

The effect of this microstructural change is a loss in room temperature strength as shown in Figure 18-23. Note that the strength decrease is clearly greater than that for pure oxidation. Figure 18-24 shows a typical fracture origin --at the surface-which suggests corrosion as a cause of strength degradation. The strength data at elevated temperatures (Figure 18-25) are more complex to

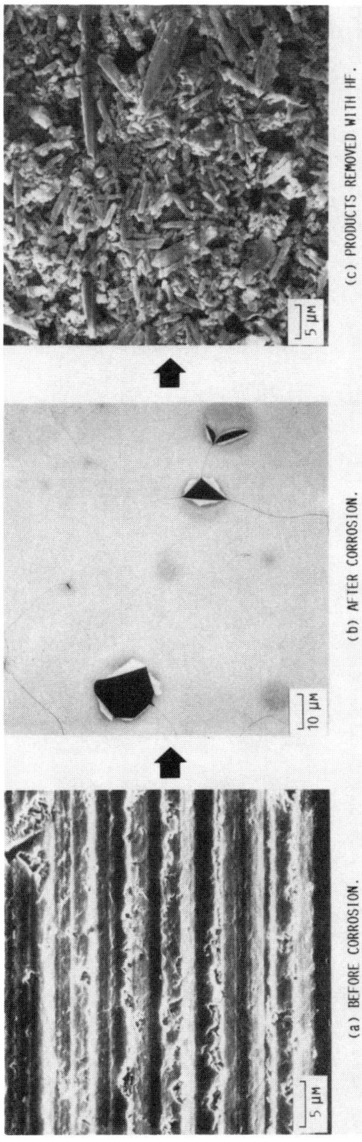

Figure 18-20. Sequence showing SSN (Y_2O_3, Al_2O_3) before corrosion, after corrosion in Na_2SO_4/O_2 for 48 hours at 1000°C, and after corrosion products are removed with a mild HF treatment.

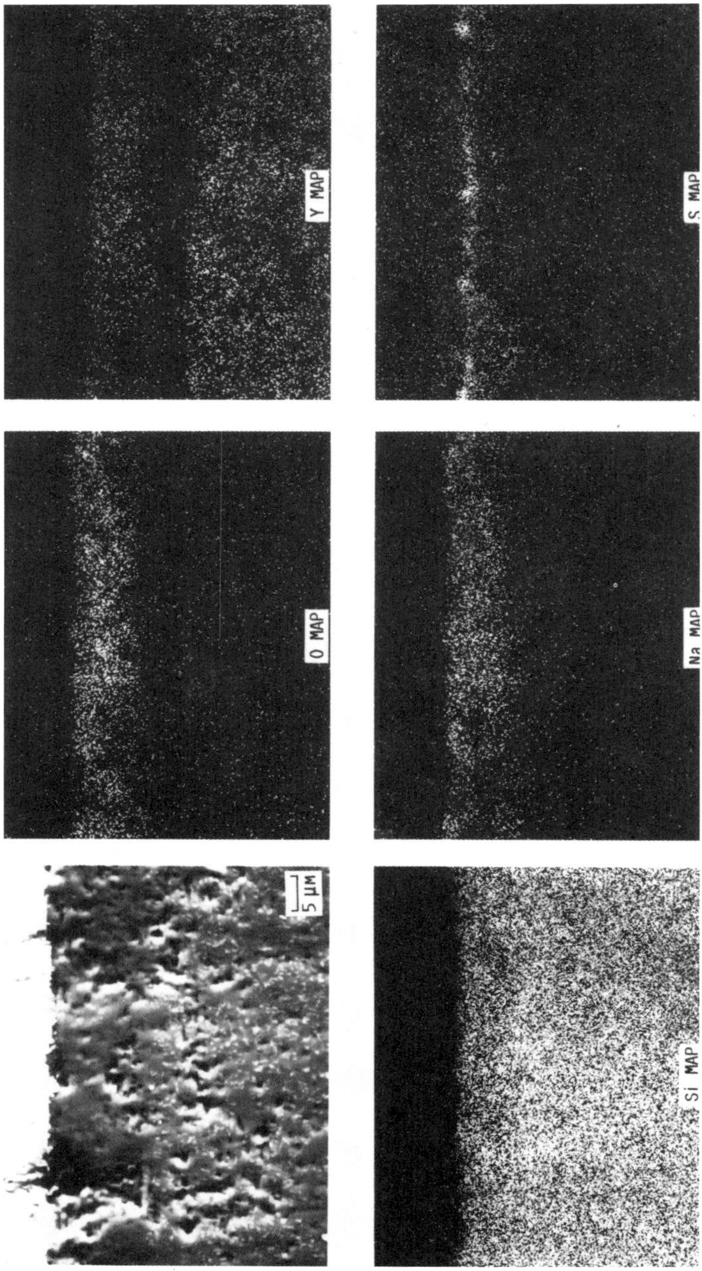

Figure 18-21. Electron microprobe analysis showing Y depletion in the outer region of SSN (Y_2O_3, Al_2O_3) reacted with Na_2SO_4/O_2 for 48 hours at 1000°C.

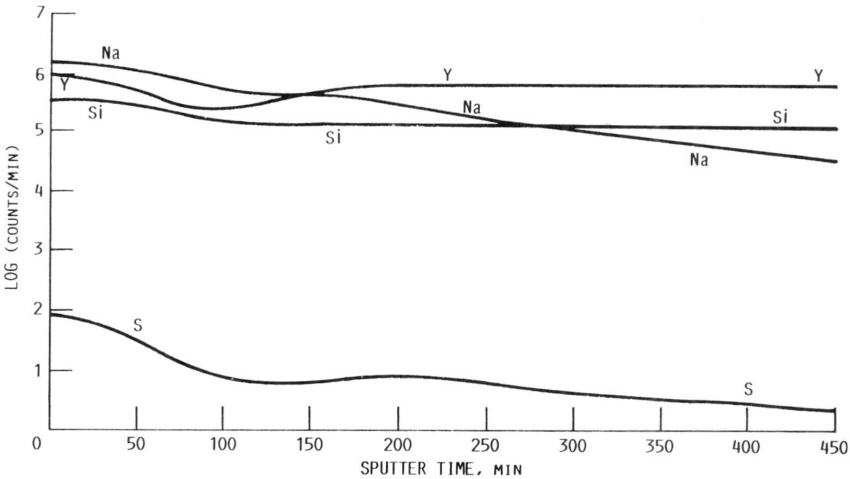

Figure 18-22. SIMS profile for SSN (Y_2O_3, Al_2O_3) reacted with Na_2SO_4/O_2 for 48 hours at 1000°C. Angle lapped specimen. Profiling started at the scale nitride interface for a total depth of 45μm after 450 min. sputtering.

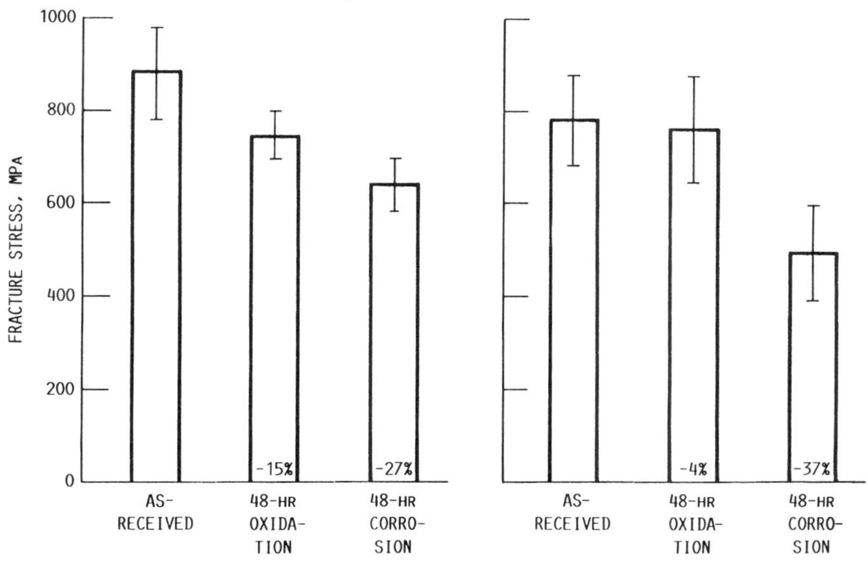

Figure 18-23. Room temperature strengths for (a) SSN (Y_2O_3,Al_2O_3) and (b) SSN (Y_2O_3). Corrosion treatment consisted of laboratory furnace reaction with Na_2SO_4/O_2.

interpret and point toward a fundamental problem with many types of Si_3N_4. That is, the actual testing of these materials at elevated temperatures leads to a strength decrease, regardless of previous exposure. Nonetheless, the room temperature tests clearly indicate that the strength of Si_3N_4 is affected by corrosion.

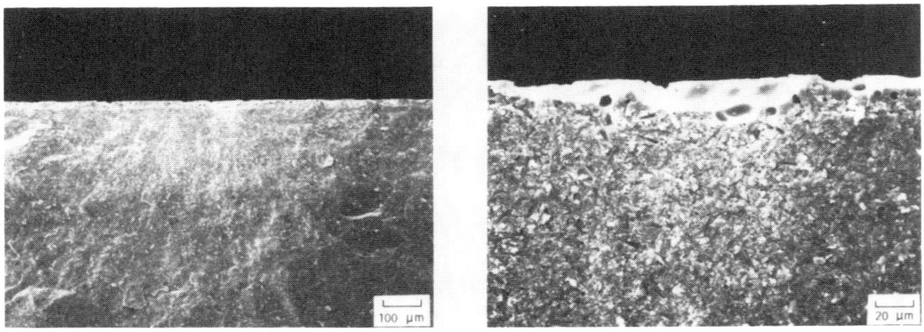

Figure 18-24. Fracture origin for SSN (Y_2O_3, Al_2O_3) reacted with Na_2SO_4/O_2 for 48 hours at 1000°C.

COMPOSITES

As discussed in the previous section, the strength of a monolithic ceramic is quite sensitive to surface finish. Hence corrosion has a major impact on strength. Ceramic matrix composites offer the advantage of being relatively insensitive to surface finish and therefore should show better performance in a corrosive environment.

Composite development is a rapidly moving field. New and improved composites are constantly becoming available. To date we have tested two composite materials (22). One is a Si_3N_4 reinforced with 30 v/o SiC whiskers. The other is a fiber-reinforced Si_3N_4. The first material showed a substantial strength reduction--more than the monolithic matrix alone (Figure 18-26). This is attributed to the selective attack of the SiC whiskers and subsequent pitting.

The fiber reinforced Si_3N_4 was about 30 v/o porous and contained SiC fibers with a carbon coating. Both of these are detrimental to corrosion. A simple oxidation treatment led to extensive cracking, as shown in Figure 18-27(a). A hot corrosion treatment led to extensive attack as shown in Figure 18-27(b). Current efforts are aimed at producing a less porous composite and higher purity SiC fibers. Such a material would be expected to show substantially better corrosion behavior.

PROTECTIVE COATINGS

Several investigators in the field of hot corrosion of ceramics are looking at protective coatings as a solution to the problem. Mullite is a material which looks promising. It shows a good thermal expansion match to SiC, and if no free silica is present, it shows good corrosion resistance. Initial results give coating lifetimes of about 20 hr in the NASA LeRC burner rig. It appears that

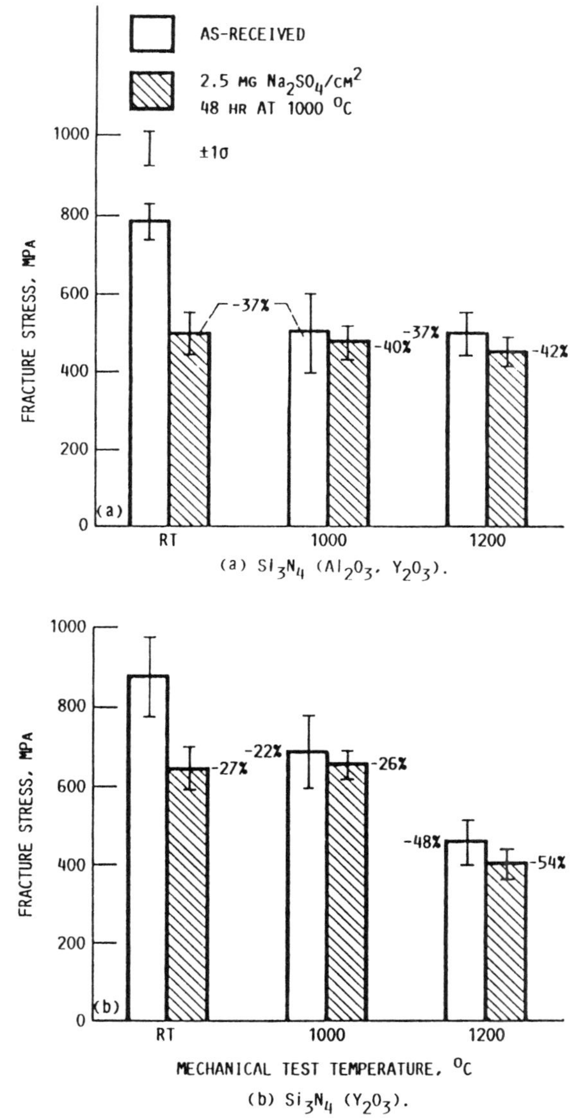

Figure 18-25. Room temperature and high temperature strengths for (a) SSN (Al_2O_3, Y_2O_3) and (b) SSN (Y_2O_3). Corrosion treatment consisted of laboratory furnace reaction with Na_2SO_4/O_2.

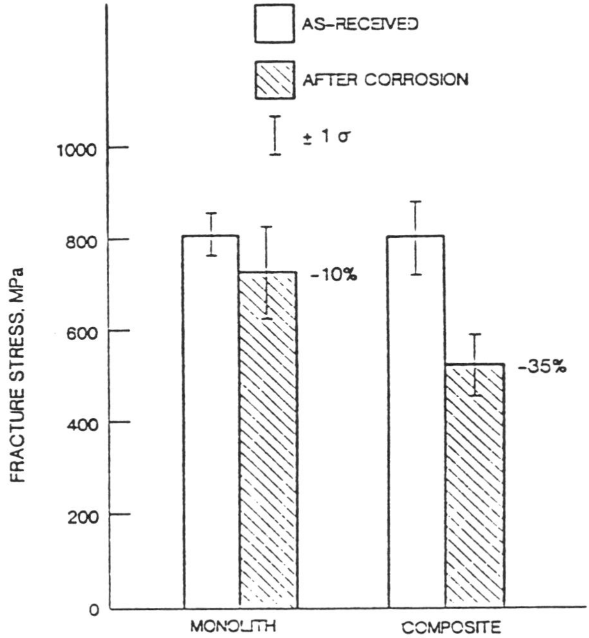

Figure 18-26. Strength of whisker reinforced Si_3N_4 after corrosion in a burner rig. Using Jet A fuel and 2ppm Na at 1000°C for 40 hrs.

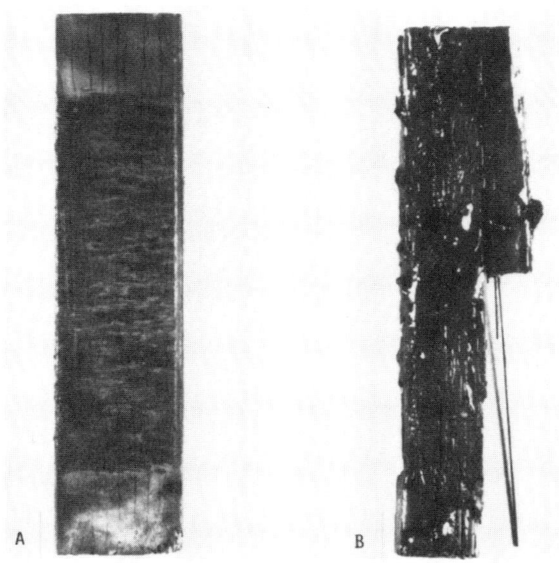

Figure 18-27. Fiber reinforced Si_3N_4 after (a) oxidation (Jet A fuel, 1000°C, 40hrs) and (b) corrosion (same conditions, 2ppm Na added).

the coating eventually cracks and allows salt penetration. Current efforts are directed at solving this cracking problem.

SUMMARY AND CONCLUSIONS

As silicon-based ceramics find more applications in various energy conversion systems, corrosion will continue to become more of an issue. As with metals, corrosion by deposits may be the most severe. The emphasis of this chapter has been on corrosion by molten salts, though corrosion by slags is similar in many ways.

The thermodynamics of molten salt corrosion have been discussed. Sodium sulfate forms by the reaction of ingested sodium with sulfur in the fuel. Corrosion occurs between the melting point of sodium sulfate and the dew point for sodium sulfate deposition. Furthermore, corrosion can only occur when the salt is basic enough to dissolve the silica scale. These facts allow prediction of corrosion regimes, which provide general guidelines for corrosion prediction. However, corrosion is more complex, and additive or impurity elements in the specific ceramic have a major impact on corrosion. Carbon is particularly deleterious, since it tends to drive the salt more basic and promote corrosion.

Kinetics of corrosion are complex, since the process has several steps. For basic molten salts, a rapid oxidation-dissolution step is followed by a region of accelerated oxidation, and an eventual slowing of the reaction. For acidic molten salts, reaction rates are quite slow, with limited attack. However, in a situation with carbon present on the substrate, the bottom of the melt becomes basic, whereas the top of the melt is acidic. This creates a fluxing situation, which leads to extensive corrosion.

This corrosive attack produces some major microstructural changes which adversely affect mechanical properties. In the case of SiC, extensive pitting occurs which leads to a substantial strength decrease. In the case of Si_3N_4, extensive grain boundary attack occurs, as does migration of impurity cations into the scale. This leads to a room temperature strength decrease. Higher temperature effects are complex and require further understanding.

Solutions to the hot corrosion problem lie in flaw-resistant ceramics and/or protective coatings. Ceramic matrix composites offer the potential of flaw resistant strength. However, in their current state of development, they have chemical impurities and porosity which lead to corrosion problems. Refractory oxide coatings show potential for limiting corrosion. Here too, more development is necessary to achieve an adequate protective coating.

REFERENCES

1. Jacobson, N.S., Smialek, J.L., and Fox, D.S., *Molten Salt Corrosion of SiC*

and Si3N4, NASA TM-101346 (1988).

2. Gabor, T., and Stickler, R., *Nature* 199: 1054-1056 (1963).
3. Faust, J.W., Jr., and Liaw, H.M., in: *Silicon Carbide - 1973* (R.C. Marshall, J.W. Faust Jr., and C.E. Ryan ed.), pp 657-667, University of South Carolina Press (1974).
4. Kohl, F.J., Stearns, C.A., and Fryburg, G.C., in: *Metal-Slag-Gas Reactions and Processes* (Z.A. Foroulis and W.W. Smeltzer ed.), pp 649-664, The Electrochemical Society (1975).
5. McKee, D.W., and Chatterji, D., *J. Am. Ceram. Soc.* 59: 441-444 (1976).
6. Jacobson, N.S., and Smialek, J.L., *J. Am. Ceram. Soc.* 68: 432-439 (1985).
7. Ferber, M.K., Ogle, J., Tennery, V.J., and Henson, T., *J. Am. Ceram. Soc.* 68: 191-197 (1985).
8. Federer, J.I., Tiegs, T.N., Kotchick, D.M., and Petrak, D., *Analysis of Candidate SiC Recuperator Materials Exposed to Industrial Furnace Environments*, ORNL/TM-9677 (1985).
9. Sato, T., Kanno, Y., and Shimada, M., *Int. J. High Technol. Ceram.* 2: 279-290 (1986).
10. Tajiri, K., Nishio, T., Asahina, T., and Kosaka, M., in: *Joint International Symposium on Molten Salts* (G. Mamantov, M. Blander, C. Hussey, C. Mamantov, M.L. Saboungi, and J. Wilkes ed.), pp 717-722, The Electrochemical Society (1987).
11. Jacobson, N.S., *Oxid. Met.* 31: 91-103 (1989).
12. Gordon, S., and McBride, B.J., *Computer Program for the Calculation of Complex Chemical Equilibrium Compositions, Rocket Performance, Incident and Reflected Shocks, and Chapman-Jouguet Detonations*, NASA SP-273 (1976).
13. Watt, G.W., Andresen, R.E., and Rapp, R.A., in: *Proceedings of the Second International Symposium on Molten Salts* (J. Braunstein and J.R. Selman ed.), pp 81-109, The Electrochemical Society (1981).
14. McNallan, M., in: *Corrosion of Ceramic Materials Workshop, Proceedings* (B.K. Kennedy ed.), pp 239-251, Center for Advanced Materials, Pennsylvania State University (1987).
15. Blachere, J.R., and Pettit, F.S., *High-Temperature Corrosion of Ceramics*, DOE/ER/45117-1 (1985).
16. Federer, J.I., *Adv. Ceram. Mater.* 3: 293-295 (1988).
17. Santoro, G.J., Kohl, F.J., Stearns, C.A., Gokoglu, S.A., and Rosner, D.E., *Experimental and Theoretical Deposition Rates From Salt-Seeded Combustion Gases of a Mach 0.3 Burner Rig*, NASA TP-2225 (1984).
18. C.A. Stearns, F.J. Kohl, and D.E. Rosner, in: *High Temperature Corrosion* (R.A. Rapp ed.), pp 441-450, National Association of Corrosion Engineers (1983).
19. Fox, D.S., and Jacobson, N.S., *J. Am. Ceram. Soc.* 71: 128-138 (1988).
20. Smialek, J.L., and Jacobson, N.S., *J. Am. Ceram. Soc.* 69: 741-752 (1986).
21. Jacobson, N.S., and Smialek, J.L., *J. Electrochem. Soc.* 133: 2615-2621

(1986).
22. Fox, D.S., *Molten Salt Corrosion of Ceramic Matrix Composites*, unpublished research, NASA Lewis Research Center, Cleveland, OH.
23. Watson, G.K., Moore, T.J., and Millard, M.L., *J. Am. Ceram. Soc.* 67: C208-C210 (1984).
24. Becher, P.F., *J. Mater. Sci.* 19: 2805-2814 (1984).

Section IV.

Corrosion of Ceramic Superconductors

19

Corrosion of Ceramic Superconductors: An Overview

Aaron Barkatt

Department of Chemistry
The Catholic University of America
Washington, D.C. 20064

Hamid Hojaji and Karen A. Michael

Vitreous State Laboratory
The Catholic University of America
Washington, D.C. 20064

INTRODUCTION

The discovery of ceramic superconducting materials (1) with high critical temperatures (T_c) was one of the major developments in materials science during the 1980's. Most of the studies on these materials concern their unique electric and magnetic properties. However, in the course of their becoming the basis for applications and devices, rather than just scientific curiosities, other properties of these materials will have to be characterized and optimized. These include mechanical properties, such as strength and hardness, as well as chemical properties, in particular resistance to corrosion and weathering. These latter properties will have an important role in determining the suitability of these materials for various potential applications and the extent of the need for protective measures in each case. These properties are also important in view of the limitations which they impose on the use of various processing techniques. Furthermore, studies of surface properties of these materials, such as their reactivity toward various liquid and gaseous environments, provide additional insight into their unusual chemical composition and structure.

The unique character of superconducting oxide ceramics in providing very high T_c values, as compared with those observed in other types of superconducting materials discovered so far, is shown in Figure 19-1. At present, several main families of superconducting oxides are known.

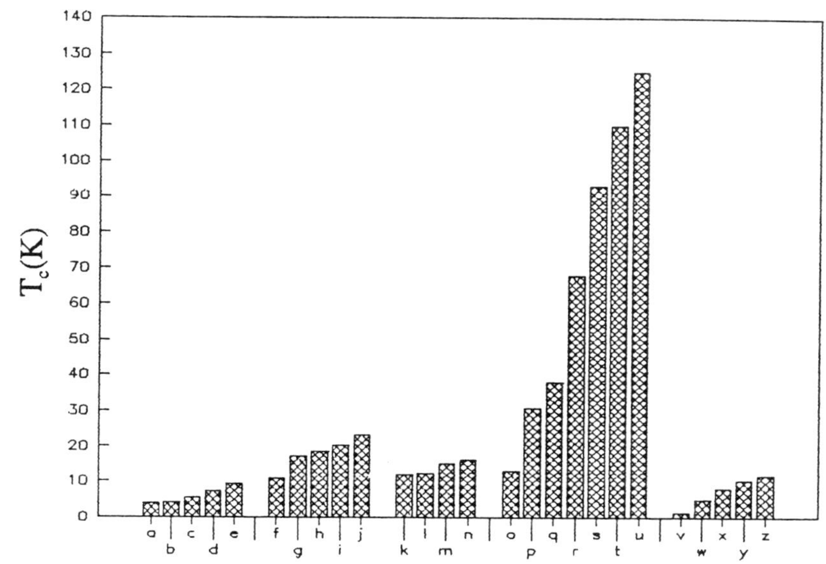

1. Metallic Elements
   a. Sn
   b. Hg
   c. V
   d. Pb
   e. Nb
2. Metal/metal and metal/semimetal compounds
   f. Zr_2Rh
   g. V_3Si
   h. Nb_3Sn
   i. Nb_3Ga
   j. Nb_3Ge
3. Simple metal/nonmetal compounds
   k. MoN
   l. Mo_2C
   m. $PbMo_6S_8$
   n. NbN
4. Oxide Ceramics
   o. $BaPb_{1-x}Bi_xO_3$
   p. $Ba_{1-x}K_xO_3$
   q. $La_{1-x}(Sr,Ba)_xCuO_4$
   r. $Pb_2Sr_2(Y,Ca)Cu_3O_8$
   s. $YBa_2Cu_3O_7$
   t. $Bi_2Sr_2Cu_3O_{10}$
   u. $Tl_2Ba_2Ca_2Cu_3O_{10}$
5. Organic Compounds[130]
   v. $(TMTSF)_2ClO_4$
      TMTSF = tetramethyltetraselenafulvalene
   w. $(BEDT\text{-}TTF)_2AuI_2$
      BEDT-TTF = bis(ethylenedithio)tetrathiafulvalene
   x. $(BEDT\text{-}TTF)_2I_3$ (0.5 kbar)
   y. $(BEDT\text{-}TTF)_2Cu(NCS)_2$
   z. $(BEDT\text{-}TTF)_2Cu[N(CN)_2]Br$

Figure 19-1. Superconductivity materials: Families and superconducting transition temperatures, T_c (K).

Table 19-1. Respresentative Superconducting Compositions.

Typical Composition	Critical Temperature (K)	References
$BaPb_{1-x}Bi_xO_3$	13	2,3
$Ba_{1-x}K_xO_3$	30.5	4
$Nd_{2-2x}Ce_xCuO_{4+x}$	34	5
$La_{1-x}(Sr,Ba)_xCuO_4$	38	1, 6-12
$Pb_2Sr_2(Y,Ca)Cu_3O_8$	68	13-16
$YBa_2Cu_3O_7$	93	17-20
$Bi_2Sr_2Cu_3O_{10}$	110	21,22
$Tl_2Ba_2Ca_2Cu_3O_{10}$	125	23,24

Representative compositions are given for each of them in Table 19-1 together with the highest critical temperature observed for each family. A wide range of substitutions has been carried out within each of these families. For instance, within the $YBa_2Cu_3O_7$ family (often designated 123), all the rare earths (except Pr and Pm) have been substituted wholly or in part for yttrium, Sr for part of the barium, F for part of the oxygen, and Ag for a small fraction of the copper, all these substitutions having no major effects on T_c (25). Intentional changes in stoichiometry, e.g. an increase in the yttrium content above its level in $YBa_2Cu_3O_7$, have also been attempted with a view to improving magnetic flux pinning (26,27). A large remanent magnetization is associated with a high critical current, J_c (28). The resistivity drop and the magnetization hysteresis loop indicative of the extent of flux trapping are shown in Figure 19-2 for the case of a high-J_c melt-based specimen of $YBa_2Cu_3O_7$ prepared at the authors' laboratory.

As shown in Table 19-1, three families of ceramic superconducting compounds exhibit critical temperatures above the normal boiling point of nitrogen (77 K), and are therefore the focus of current technological interest. Among these three families, which are based on the complex cuprates of yttrium (123), bismuth and thallium, respectively, the largest amount of data is available for the 123 materials. In the case of bismuth-based materials it is generally difficult to remove all the impurity phases. In the case of the thallium-based materials, major problems are encountered stemming from the volatility and toxicity of thallium. In addition, flux creep, which limits the current-carrying capability of high-T_c superconductors in most practical applications, has been observed at much lower temperatures in the cases of the Bi- and Tl-based materials than in the case of $YBa_2Cu_3O_7$ (29). As a result, most applications in the near future are likely to be based on the use of 123 materials. Accordingly, most of the studies on processing techniques as well as on mechanical, thermal and chemical properties have been carried out on these materials. The highest values of J_c obtained to date with 123 materials are of the order of 10^7 A/cm^2 for thin films and 10^5 A/cm^2 for bulk specimens prepared using melt-based techniques. Since the J_c values reflect the effectiveness of various processing methods in providing sites for flux trapping and controlling weak links within the specimens (27), the values cited above, unlike the T_c values, cannot be regarded as materials constants. Thus, it can be expected that as the development of the Bi- and Tl-based compositions goes on, materials will be produced which will have higher J_c levels, comparable to those obtained with the yttrium-based materials. Indeed, in the case of wires, the highest J_c values obtained to date, which are of the order of 10^4 A/cm^2 are those measured in Tl- and Bi-based specimens. Tl-based thin films with J_c values on the order of 10^6 A/cm^2 have also been produced. It should also be noted that, particularly in the case of bulk specimens, J_c values given in the literature are only seldom based on direct conduction measurements. Most of the reported values are based on magnetization measurements and involve the use of a characteristic dimension (28) which is subject to some uncertainty, depending on the relative importance of the sizes of individual grains, of aligned domains and of entire specimens, respectively. These values also depend, as shown in Figure 19-2b, on the orientation of the grains in the specimen with respect to the applied field.

WEATHERING IN MOISTURE-CONTAINING ENVIRONMENTS

Because of the prominent role of $YBa_2Cu_3O_7$ among ceramic superconducting materials, most of the studies described below were focused on this material. Wherever results obtained on other materials are discussed, the type of material is indicated.

During the early period of work on ceramic superconducting materials many cases were reported of superconducting behavior, observed in resistivity measurements or in magnetic measurements, which deteriorated and often vanished altogether upon repeating the measurements over periods as short as a few days. Many irreproducible readings could be traced back to problems involving the physical measurements. Others, however, were attributed to chemical instability in the presence of atmospheric water during annealing of the specimens between 200-300°C (30). In general, both room temperature and low temperature resistivity of samples exposed to humid environments are found to increase (31). Two detrimental effects of the attack of water on resistivity measurements are the appearance of an electrochemical EMF at the electrode contact and the formation of insulating barriers at grain boundaries. The presence of carbon dioxide in the environment together with moisture can accelerate the latter effect, although the presence of CO_2 alone does not have a detrimental effect (30). The sharpness of the superconducting transition is affected before its onset temperature. For instance, exposure of $YBa_2Cu_3O_7$ to air with 100% relative humidity at 60°C for 80 minutes causes the temperature

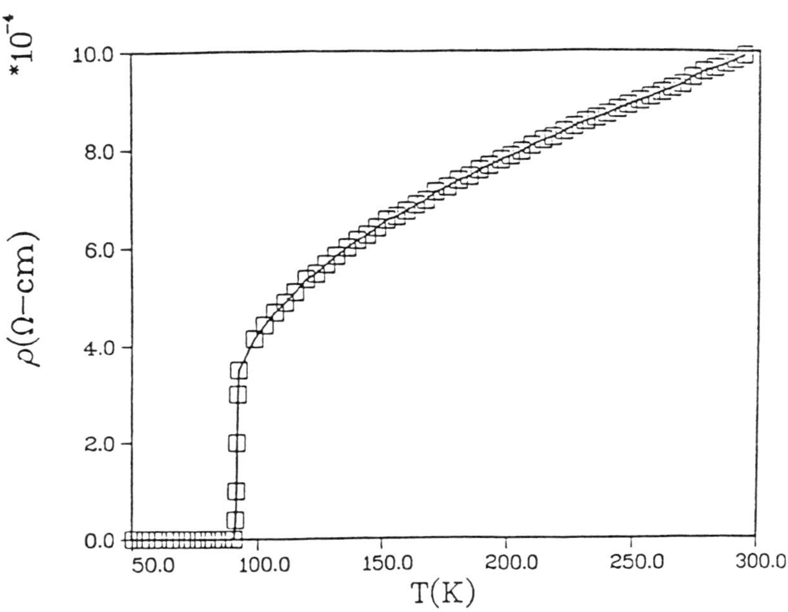

a)

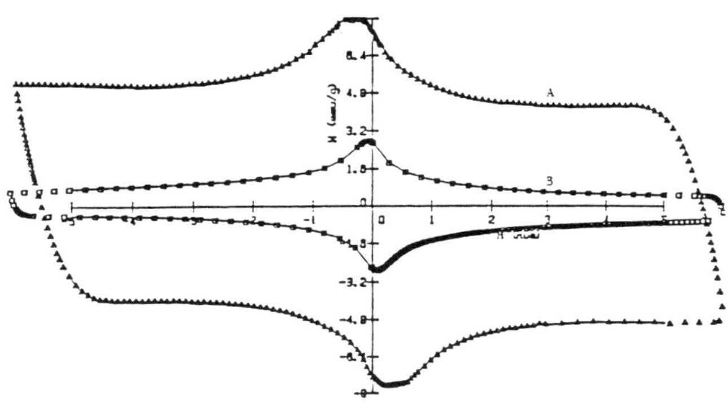

b)

Figure 19-2. Characteristics of superconducting materials: Melt-based $YBa_2Cu_3O_7$.
a) Resistivity drop at the transition temperature; (b) Magnetic flux trapping: Hysteresis of DC magnetization vs. applied magnetic field at 77K.

Surface studies on samples exposed the humid environments have shown evidence of a mostly amorphous structure (33) of segregation of barium onto the surface of the exposed specimen in the form of $BaCO_3$ (34). A mechanism based on product analysis on the surface of samples exposed to liquid water (33) (see below) has been proposed to be also valid in the case of exposure to atmospheric water vapor and CO_2 (35). This mechanism consists of the following reactions:

$$2\ YBa_2Cu_3O_7 + 3\ H_2O = Y_2BaCuO_5 + 3\ Ba(OH)_2 + 5\ CuO + 0.5\ O_2 \qquad (19\text{-}1)$$

or: (36)

$$4YBa_2Cu_3O_7 + 6H_2O = 2Y_2BaCuO_5 + 10CuO + 6Ba(OH)_2 + O_2 \qquad (19\text{-}1')$$

$$2\ Y_2BaCuO_5 = 2\ BaCuO_2 + 2Y_2O_3 \qquad (19\text{-}1a')$$

$$Ba(OH)_2 + CO_2 = BaCO_3 + H_2O \qquad (19\text{-}2)$$

A slightly different formulation proposed for the reaction of $YBa_2Cu_3O_7$ with atmospheric H_2O and CO_2 is: (37)

$$YBa_2Cu_3O_7 + 1.5H_2O + 5CO_2 = 0.5Y_2(CO_3)_3 + 2BaCO_3 + 1.5[(CuCO_3)\cdot Cu(OH)_2] + 0.25O_2 \qquad (19\text{-}3)$$

The corresponding reaction in the case of La_2CuO_4 is:

$$La_2CuO_4(s) + (x+0.5)H_2O(g) + 3.5CO_2(g) = La_2(CO_3)_3.xH_2O(s) + 0.5(CuCO_3).Cu(OH)_2(s) \qquad (19\text{-}4)$$

Another study identified the products of weathering in humid air as $Ba_2Cu(OH)_6$, $BaCO_3$, CuO and $Y(OH)_3$, with the attack starting at the surface, where microcrystals of $BaCO_3$ appear at an early stage (38). In any case, these weathering reactions are highly exothermic, (37) due in a large part to the large negative free energy of hydration of BaO (39). The total free energy of hydration generally correlates with the extent of dissolution of complex oxide solids during a specified period of exposure to an aqueous environment (40-42).

Initially, all samples of $YBa_2Cu_3O_7$ were prepared by solid-state sintering. The elimination of porosity and the attainment of high density require use of starting material in the form of a fine powder (43). However, the degradation of this material in ambient atmospheres accelerates with decreasing particle size because of the larger surface area (37).

The rate of degradation of $YBa_2Cu_3O_7$ in humid environments strongly depends on the temperature. Degradation in humid air at 40°C is quite pronounced (38). Measurements of the increase in normal state resistivity as a function of temperature between 30 and 50°C were reported to indicate that the time required to reduce the conductivity by half is inversely proportional to the vapor pressure, which rises exponentially with increasing temperature (32). The rapid deterioration of $YBa_2Cu_3O_7$ upon exposure to humid air at 60°C, as reflected in a large increase in resistivity and an apparent drop in the observed value of T_c, is illustrated in Figure 19-3.

X-ray photoelectron spectroscopy (XPS) of $YBa_2Cu_3O_7$ films and pellets (36) shows the presence of a surface layer, enriched with respect to Ba and depleted with respect to Cu, due to the formation of $BaCO_3$ and/or $Ba(OH)_2$ in ambient atmospheres (see above and Ref. 44). The thickness of the layer is initially on the order of 1-100 nm, and its depth increases with time. Adsorbed oxygen, water and carbonate are also present at the surface. Annealing in moist oxygen leads to the formation of the green Y_2BaCuO_5 phase at the surface as a result of reaction 19-1′ or 19-1. The Y_2BaCuO_5 phase layer grows thicker with time, resulting in progressive loss of superconductivity (36). The surface layer of $Bi_2Sr_2CaCu_2O_x$ ceramics is enriched with respect to Ca and Sr and depleted with respect to Bi and Cu, again due to migration of Ca and Sr to the surface to form hydroxides and/or carbonates. Further deterioration of Bi-based pellets

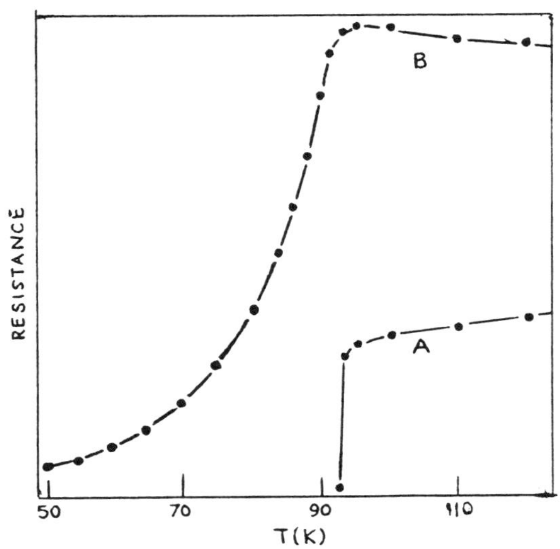

Figure 19-3. Resistance vs. temperature of sintered $YBa_2Cu_3O_7$ samples before and after exposure to humid air (100% humdity at 60°C). (A) at t = 0; (B) at t = 80 min. Adapted from Ref. 54.

upon annealing in oxygen is very slow, showing that these materials are less reactive than the Y-based specimens (36).

Photoelectron emission studies on $YBa_2Cu_3O_7$ samples onto which water was evaporated and cryosorbed at very low temperatures showed that the sorption of water causes irreversible changes in the valence band structure. The number of non-equivalent Ba sites was observed to decrease, and new oxygen 1s core-level components appeared, apparently associated with OH^- ions and H_2O molecules, respectively (45,46). The initial attack of atmospheric water on $(La_{1-x}Sr_x)_2CuO_4$ is particularly rapid since H_2O molecules adsorb dissociatively with a very high sticking coefficient of about 1.0 to form surface hydroxide. CO_2 also adsorbs very effectively, having a sticking coefficient of about 0.3, and forms carbonate surface species. CO has a much lower sticking coefficient (about $3x10^{-4}$) to form similar products to those of CO_2 chemisorption. Oxygen, on the other hand, has little effect at ambient temperatures (47).

DECOMPOSITION UPON CONTACT WITH LIQUID WATER

Attack by liquid water is much more severe. Disc-shaped samples were found to lose about one-third of their diamagnetic susceptibility upon exposure to distilled water at 60°C for 5 minutes and almost all susceptibility within 60 minutes (48). As mentioned above, the reaction of $YBa_2Cu_3O_7$ with liquid water has been concluded to proceed via the mechanism indicated by eqs. 19-1 and 19-2, which is based on the identification of the reaction producrts, viz. $BaCO_3$, CuO and Y_2BaCuO_5 in the x-ray diffraction patterns of the leached samples (33). An insulating thin film of $BaCO_3$ is formed on the surface after drying (33,49-53). Electrochemical polarization is observed in potentiometric measurements (53). Subsequent decomposition of Y_2BaCuO_5 to its constituent oxides may also take place (36,54).

The rapid degradation of $YBa_2Cu_3O_7$ upon contact with liquid water is reflected in dissolution of Ba in the aqueous phase and loss of excess oxygen. The latter, which corresponds to the formal Cu^{3+} oxidation state in the $YBa_2Cu_3O_7$ formula, is evolved as molecular oxygen in pure water (50,54) or consumed in the oxidation of reducing solutes such as iodide if such solutes are present in the aqueous phase (50). The second process proceeds simultaneously with the first one, indicating that as barium ions are leached away, excess oxygen sites in the matrix come in contact with the aqueous medium and are immediately converted to ordinary Cu^{2+} sites. In agreement with the product identification studies mentioned above, Y and Cu remain largely undissolved. The leaching of Ba is also reflected in a rapid rise in pH (36,50,52), which is eventually attenuated as $Ba(OH)_2$ is converted to $BaCO_3$ (50). The maximum levels of dissolved Ba in the aqueous phase at steady state are determined by $BaCO_3$ solubility; the latter is higher in the presence of 0.05 M KI than in pure water. This has been attributed to the Debye-Huckel model, according to which a higher ionic strength lowers the activity coefficients of precipitate-forming ions, resulting in increases in solubility in the presence of added salts (50). On the other hand, the initial rate of attack on $YBa_2Cu_3O_7$ in 1 M NaCl solution at room temperature was observed to be lower than that of pure water under the same conditions (31). This may be attributed to lowering of the activity of H_3O^+ ions, which may constitute the active species during this stage of the interaction (see below). A similar effect of attenuation of leach rates in the presence of high concentrations of inert salts has been observed in the case of leaching of alkali-containing silicate glasses (55).

In liquid water, as well as in water vapor, Ba eventually accumulates on the surface of $YBa_2Cu_3O_7$ as a result of formation of $BaCuO_2$ and $Ba(OH)_2$ deposits. These deposits are non-conducting. Similarly, in the case of the less reactive $Bi_2Sr_2CaCu_2O_x$, Ca and to a lesser extent Sr migrate and accumulate on the surface as a result of the formation of hydroxides and/or carbonates (36).

The rate of the attack of liquid water on $YBa_2Cu_3O_7$, like the corresponding reaction with water vapor (see above), is highly dependent on the temperature. $YBa_2Cu_3O_7$ has been reported to be relatively inert upon exposure to cold water for several hours. However, even slight heating (to 35°C) causes significant decomposition (51). While an overnight soak of $YBa_2Cu_3O_7$ in water at 22°C did not show an effect on its x-ray diffraction pattern, some decrease, accompanied by the appearance of CuO and a little $BaCO_3$, was observed after 30 minutes at 35°C. Most of the $YBa_2Cu_3O_7$ was decomposed upon a 30-minute soak in water at 65°C, while at 90°C decomposition was completed in less than 30 minutes (51). The decomposition products, in agreement with the mechanism described above, (33) are Y_2BaCuO_5, $Ba(OH)_2$ and O_2; and $Ba(OH)_2$ subsequently reacts with CO_2 to form $BaCO_3$. At higher temperatures, aqueous solutions of 10 M NaOH, 1 M NaCl and 1 M KCl, which are less reactive than water at room temperature, cause decomposition of $YBa_2Cu_3O_7$ over short periods of exposure (31).

According to (54), the dependence of the time required for complete hydrolysis on water temperature can be expressed as $t = (1/0.7)\ (\ln(200/T))^5$. However, the data in (54) can also be represented in terms of a simple Arrhenius relationship with an activation energy of (17±2) kcal/mol. This value is similar to the activation energies of the dissolution of other oxide materials such as silicate glasses (57,58). The observed temperature dependence for the attack of aqueous environments on $YBa_2Cu_3O_7$ on one hand, and on a borosilicate silicate on the other, is shown in Figure 19-4.

Since Ba is the most leachable component of the ternary oxide system of Y, Ba and Cu, high-Ba phases are particularly sensitive to attack by water, rapidly undergoing decomposition even at room temperature (51).

INHERENT MECHANISM OF ATTACK BY WATER: IONIC MOBILITY AND MATRIX DISSOLUTION

In highly dense pure-phase materials, predictions of the rate of environmental degradation are based on the inherent dissolution rates of the solid material rather than on the extent of porosity, since mass penetration of water through pores is no longer the rate-controlling mechanism. It was suggested that the mobilities of the constituent ions have an important role (58). However, the dissolution of divalent ions such as Ba^{2+} by diffusion, when these ions are incorporated in less soluble solid matrices, is quite difficult (40). This can account for the slowdown observed once the attack proceeds beyond the original

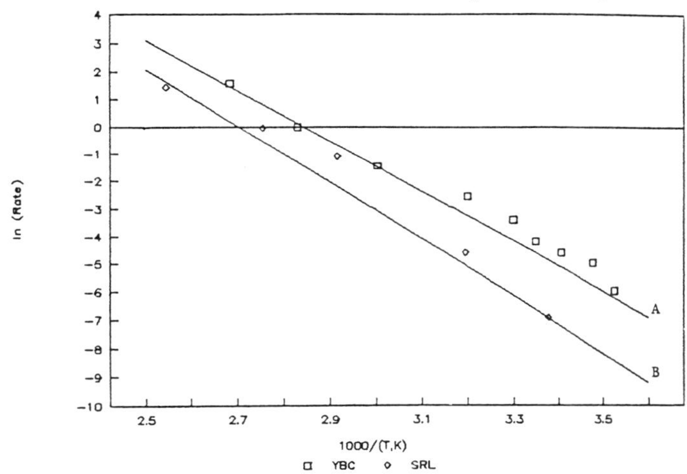

Figure 19-4. Arrhenius plots for corrosion of a ceramic superconductor and a multi-component silicate glass in de-ionized water. A = sintered $YBa_2Cu_3O_7$ (rate - 1/hydrolysis time, hr^{-1}) (Based on Ref. 54). B = SRL TDS-131 borosilicate glass (rate = boron leach rate, $gm^{-2}d^{-1}$) (Based on Ref. 61).

surface (35), which is accompanied by levelling off in the pH rise (52). The pH changes accompanying the leaching of $YBa_2Cu_3O_7$ in water according to Ref. 50 are shown in Figure 19-5. Accordingly, the subsequent rate of dissolution of Ba^{2+} ions corresponds to the rate of formation of a new surface by the relatively slow dissolution of copper and yttrium ions (39). This is also expected to be true for the rate of reduction of excess oxygen sites, which closely follows the rate of dissolution of barium ions (50). The control of barium ion extraction into the solution by the dissolution of copper and yttrium is reflected in the acceleration of the former process in the presence of strong acids or ammonia, which enhance the solubility of both Cu and Y or of Cu alone, respectively (see below) (39). The kinetics of the attack of water on $YBa_2Cu_3O_7$ thus bears some resemblance to the corrosion kinetics of alkali-lime silicate glasses, where an initial stage of rapid preferential leaching of Na^+ or K^+ ions is followed by much slower dissolution, since a new surface has to be created through dissociation and dissolution of Si-O-Si units before further alkali sites can be reached (56,57). Furthermore, in both cases the initial rapid step involves H_3O^+ as an active species. Later stages which are likely to be involved in the corrosion of both types of materials are a rise in the pH of the aqueous phase, formation of depleted surface layers (in the present case, with respect to alkali earth cations), matrix dissolution, deposition of new phases on the surface and diffusion of species through these layers (40,59-62).

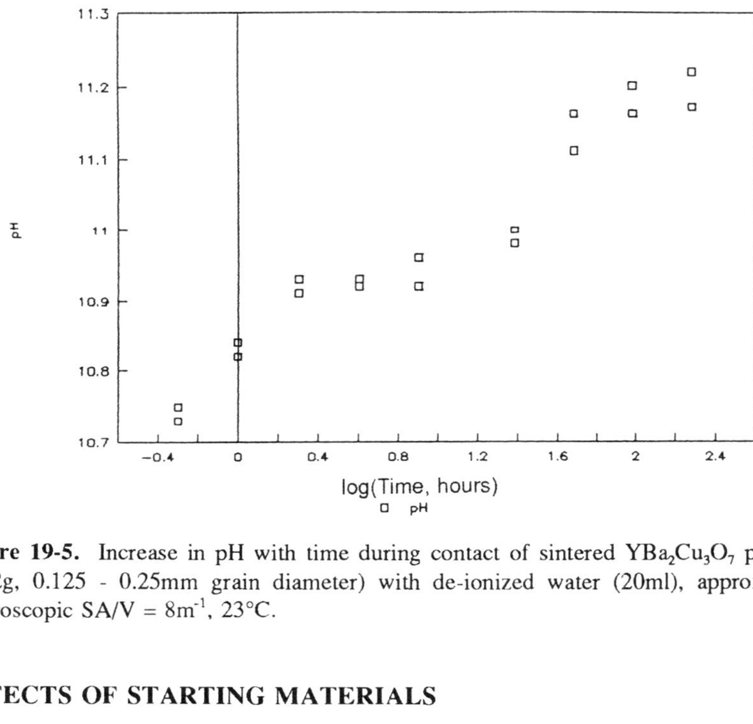

Figure 19-5. Increase in pH with time during contact of sintered $YBa_2Cu_3O_7$ powder (0.02g, 0.125 - 0.25mm grain diameter) with de-ionized water (20ml), approximate macroscopic SA/V = $8m^{-1}$, 23°C.

EFFECTS OF STARTING MATERIALS

It has been reported (63) that films prepared through the use of Y metal, Cu metal and BaF_2 as sources for the deposition of $YBa_2Cu_3O_7$ on substrates such as $SrTiO_3$ using an electron beam co-evaporation method exhibit high resistance toward attack by liquid water as well as water vapor. Their resistance towards attack by water is much higher than that of films prepared through a similar process using Ba metal instead of BaF_2 as an evaporation source. In addition, films prepared using BaF_2 consistently exhibit higher critical temperatures of 85-92K compared with films prepared under the same conditions using Ba metal, which do not have T_c values higher than 60 K. BaF_2-based films also have unusually high values of J_c (10^6 A/cm^2 at 84 K). Films prepared using BaF_2 can survive exposure to humid air for 18 hours or immersion in de-ionized water for 5 minutes before annealing. (Annealing is carried out for 0.5-6 hours at 800-920°C under oxygen.) The better resistance to moisture exhibited by films prepared from BaF_2 may be related to the fact that, as an evaporation source, BaF_2 is much less susceptible to degradation upon exposure to air than Ba metal (63). In addition, it is well-known that the presence of fluoride in starting compositions used to prepare various oxide glasses can lead to a large reduction in the amount of hydroxide incorporated due to the presence of water vapor in the atmosphere during glass preparation. This can be attributed to the replacement of sorbed OH^- by F^- in the oxide structure and the formation of volatile HF in the reaction between absorbed water species and fluorides (64).

ACID/BASE ASPECTS OF THE INTERACTION OF CERAMIC SUPERCONDUCTORS WITH AQUEOUS ENVIRONMENTS

Both $(La_{1-x}Sr_x)_2CuO_4$ and $YBa_2Cu_3O_7$ are basic oxides and dissolve most rapidly in acidic solutions. The corrosion rate of $YBa_2Cu_3O_7$ is extremely high in strongly acidic media, e.g. 0.1 M HCl, where all three constituent oxides (BaO, Y_2O_3 and CuO) are highly soluble and, accordingly, the dissolution process is congruent (39). The accelerated rates of corrosion of $YBa_2Cu_3O_7$ in humid atmospheres which are observed when CO_2 is present (see above) can be attributed to an increase in the acidity of the water, (30) although the stabilization of the reaction products through the formation of $BaCO_3$ (cf. eq. (19-2)) is also significant. The strongly basic BaO has the highest affinity towards water as well as CO_2 (53). Even the attack of pure water on $YBa_2Cu_3O_7$ (cf. eq. (19-1)) can be regarded as an acid/base reaction, with the water behaving as an acid, in view of the basicity of the oxide (30). A similar electrophilic character accounts for the reaction of water with alkali silicate glasses (65,66). Furthermore, the effect of alkalis (e.g., a KOH solution at pH 10) on the surface composition of $YBa_2Cu_3O_7$ is much smaller than that of water (36). Between pH3 and pH13, with NaOH being used in the basic region, the rate of attack is still significant, due to the large free energy of hydration of BaO, but it is lower by more than a factor of 30 than the attack at pH 1, as shown in Figure 19-6 (39). Within this pH range, the concentrations of dissolved barium are much higher than those of dissolved Cu and Y. In addition, the dissolved Ba concentrations slightly decrease with increasing pH. However, as mentioned above, the rate-determining step is the creation of a new surface by the slow dissolution of Cu and Y to expose deeper Ba sites. This is indicated by the increase in Ba dissolution in the highly acidic range, where both Cu and Y become soluble. Furthermore, Ba dissolution is also enhanced by the substitution of NH_3 for NaOH as a basic reagent, which causes a considerable rise in the rate of Cu dissolution (see Figure 19-6) (39). The fact that the leachate composition in the near-neutral and basic range is greatly enriched in barium although dissolution of the other metal

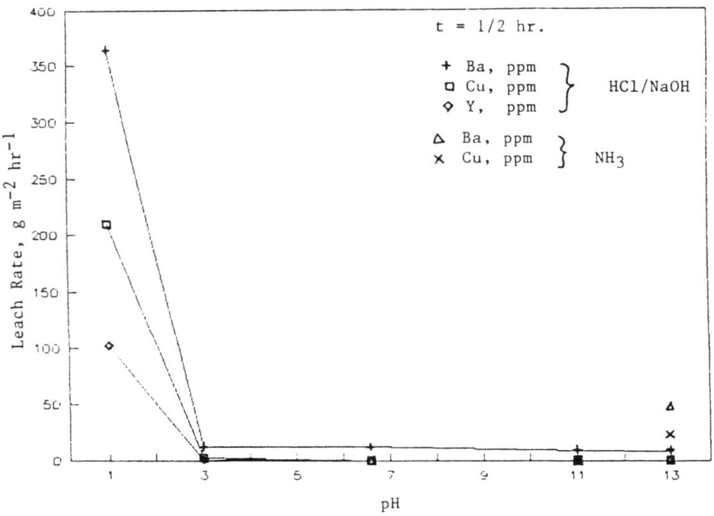

Figure 19-6. Dependence of leach rate of sintered $YBa_2Cu_3O_7$ on leachant pH, 20°C. Based on Ref. 39.

oxides is required to release the barium indicates that the less soluble Y and Cu are reprecipitated as Y_2BaCuO_5 and CuO, in agreement with eq. 19-1. Such a dissolution/re-precipitation mechanism (52) has also been invoked to account for the incongruent dissolution of multi-component oxide glasses (67,68).

Other studies have confirmed the correlation between acidity and rate of attack by aqueous media. In general, the order of reactivity of various aqueous solutions follows the following sequences:

HCl > NH_4OH > H_2O > NaOH (31,32,39,52)
HCl > HNO_3 > CH_3COOH > H_2SO_4 > H_2O > NaOH (52,69,62)
$HClO_4$ >> CF_3COOH >> pH 4 buffer > water > pH 10 buffer
>> dry CH_3CN, epoxy, dry air (70)

The reaction products with $HClO_4$ include a surface deposit of $Ba(ClO_4)_2 \cdot 3H_2O$ while most of the exposed Y and Cu dissolve away (65). The slow action of H_2SO_4 was attributed to the formation of insoluble $BaSO_4$ (69). The relatively high reactivity of ammonia (39,52) compared with the low reactivity of NaOH (31,69) is due to the formation of a highly stable, soluble complex of copper, as manifested by the deep blue color of ammonia solutions after contact with $YBa_2Cu_3O_7$

On the basis of current-voltage measurements on various aqueous solutions it was reported that the electrochemical stability of $YBa_2Cu_3O_7$ in dilute (0.1 M) aqueous solutions is particularly high in the case of H_2SO_3 solutions, while some passivation takes place in oxidizing acids, e.g. HNO_3, H_2SO_4 (which is comparable in electrochemical behavior with NaOH) and H_3PO_4 (which is significantly less corrosive than NH_4OH) (71). More work is required in order to correlate these findings with results of regular immersion tests.

As mentioned above, studies of the temperature dependence of the rate of degradation of $YBa_2Cu_3O_7$, as measured by the decrease in conductivity, have indicated that the reciprocal of the characteristic time of this process is proportional to the vapor pressure of water in the system (32). This implies a second-order reaction, suggesting that the process is not initiated by a direct attack of the H_2O molecule, but through abstraction of Ba by OH^-, followed by stabilization of the activated transition complex by H_3O^+ (32). While this hypothesis requires further experimental confirmation, the attribution of an important role for H_3O^+ is in agreement with the observed low rates of degradation in aqueous solutions containing high concentrations of NaOH (32) or NaCl (see above). Current measurements on $YBa_2Cu_3O_7$ and $La_{1.8}Sr_{0.2}CuO_4$ electrodes have provided direct evidence that the rate of attack by aqueous media on $YBa_2Cu_3O_7$ is limited by proton transport, since this rate linearly increases with increasing hydrogen ion concentration (72). $La_{1.8}Sr_{0.2}CuO_4$ dissolves at roughly half the rate of $YBa_2Cu_3O_7$ with mixed mass transfer and surface rate control (72). Both these materials dissolve at rates higher by more than one order of magnitude than that of CuO.

OXIDATION/REDUCTION ASPECTS OF THE INTERACTION OF CERAMIC SUPERCONDUCTORS WITH AQUEOUS ENVIRONMENTS

All of the types of the novel superconducting oxides characterized to date contain at least one, and often more than one, element which forms cations with more than one oxidation state. These elements include Bi, Tl, Pb, and, in particular, Cu. Thus $(La_{0.9}Sr_{0.1})_2CuO_4$ can be formally regarded as a compound in which 80% of the copper is present in the Cu(II) oxidation state and 20% in the Cu(III) oxidation state, while $YBa_2Cu_3O_7$ formally consists of 67% Cu(II) and 33% Cu(III) (73). Actually, the oxygen content of superconducting materials of the latter type is usually somewhat smaller. The composition of these materials can usually be more accurately described by the formula $YBa_2Cu_3O_x$, with values of x ranging between 6.63 and 7.0. Below 6.63 the non-superconducting tetragonal form supersedes the orthorhombic superconducting phase (74). Usual maximum values of x for materials annealed under oxygen-based atmospheres at ordinary pressures lie between 6.9 and 7.0, corresponding to values of between 27% and 33% for the fraction of Cu(III) out of the total copper content (74,75).

The existence of a strong correlation between the formal Cu(III) content and superconducting properties (9,74-77) has given a strong impetus to studies aimed, on one hand, at accurate determination of the amount of Cu(III) in specimens of superconducting ceramics, and, on the other hand, at understanding the chemical nature of the corresponding species. Measurements of the interaction of superconducting oxides with aqueous solutions have provided a major part of the information now available on these two issues.

The Cu(III):Cu(II) ratio corresponding to the value of x in the formula $YBa_2Cu_3O_x$ can be measured in a combination of two iodometric titrations (78). In the first titration, $YBa_2Cu_3O_x$ powder is dissolved in a mixed solution of 1.0 M KI and 0.7 M HCl in an inert atmosphere, usually consisting of argon (78) or nitrogen (79). Cu(III) behaves as a two-electron oxidant and Cu(II) as a one-electron oxidant:

$$Cu(III) + 3I^- = CuI + I_2 \tag{19-5}$$

$$Cu(II) + 2I^- = CuI + 0.5I_2 \tag{19-6}$$

The reaction is carried out under argon or nitrogen to prevent the oxidation of iodide by atmospheric oxygen. The liberated iodine is titrated with a standard solution of sodium thiosulfate:

$$2\ S_2O_3^{2-} + I_2 = S_4O_6^{2-} + 2\ I^- \tag{19-7}$$

In the second part of the procedure, $YBa_2Cu_3O_x$ is dissolved in aerated 1.0 M HCl to convert all Cu(III) to Cu(II):

$$2Cu(III) + 3H_2O = 2Cu(II) + 0.5O_2 + 2H_3O^+ \tag{19-8}$$

Then 0.7 M I^- is added under argon to liberate iodine according to eq. (19-6) and the iodine titrated with thiosulfate according to eq. (19-7). The difference between the volumes of thiosulfate solution required in the two titrations, respectively, yields the amount of Cu(III) in the sample, while the second titration alone gives the amount of total copper, thus permitting evaluation of the amount of Cu(II). Results of determination of x using this method have been shown to be in very good agreement with results of thermogravimetric measurements under a flowing hydrogen/argon mixture, (78) which are based on the reaction:

$$YBa_2Cu_3O_x + (x-3.5)\ H_2 = 0.5\ Y_2O_3 + 2\ BaO + 3\ Cu + (x-3.5)H_2O \tag{19-9}$$

However, this method requires great care in its use. The inaccuracy resulting from the use of two titration volumes to determine the amount of Cu(III) can be eliminated by carrying out the first titration in the presence of citrate, which forms a complex with Cu(II) and prevents its further reduction (80). This makes it possible to determine the amount of Cu(III) directly without the need for a second titration (81). More seriously, the use of KI in combination with acid concentrations high enough to dissolve $YBa_2Cu_3O_x$ rapidly can cause large errors stemming from Cu-catalyzed oxidation of iodide by traces of atmospheric oxygen (81). These difficulties have led to the development of an improved iodometric method, in which $YBa_2Cu_3O_x$ is dissolved in ice-cold 4.4 M HBr, which is much less air sensitive than HI and does not react with Cu(II):

$$2Cu(III) + 2Br^- = 2Cu(II) + Br_2 \quad (19\text{-}10)$$

The liberated bromine is then reacted with an ice-cold aqueous iodide solution to produce iodine in the presence of a sufficient amount of ammonia to neutralize all but 1 mmol of the acid:

$$Br_2 + 2\ I^- = I_2 + 2\ Br^- \quad (19\text{-}11)$$

A solution of 1 M sodium citrate is added to prevent or reverse oxidation of further iodide by Cu(II), and the iodine is titrated using a standard solution of 0.1 M $Na_2S_2O_3$ according to eq. (19-7) in the presence of a 1% amylose solution as an indicator. This procedure has been successfully applied in the determination of the Cu(III) content of La_2CuO_x and $La_{1.85}Sr_{0.15}CuO_x$ as well as $YBa_2Cu_3O_x$ (81).

Another method which has been used to determine the relative amount of Cu(III) in YBa_2Cu_3Ox is the addition of a known amount of Fe^{2+} to the samples before acid dissolution under inert atmosphere and back-titration of the excess of Fe^{2+} with $K_2Cr_2O_7$:(79)

$$Cu(III) + Fe(II) = Cu(II) + Fe(II) \quad (19\text{-}12)$$

$$6Fe(II) + K_2Cr_2O_7 + 14H_3O^+ = 6\ Fe(III) + 2\ K^+ + 2\ Cr^{3+} + 21\ H_2O \quad (19\text{-}13)$$

The results obtained by this method agree with those obtained by the iodometric method described above.

Specimens of $YBa_2Cu_3O_x$ with $x < 6.5$, which are not superconducting, can be formally considered to consist of mixed Cu(I) and Cu(II) rather than Cu(III) and Cu(II). In such cases the total copper content can be determined by the iodometric technique described above, while Cu(I) is analyzed by acid dissolution of the sample under inert atmosphere in the presence of Fe^{3+}. This additive is reduced by Cu(I) to yield an equivalent amount of Fe^{2+}:

$$Cu(I) + Fe(III) = Cu(II) + Fe(II) \quad (19\text{-}14)$$

The Fe^{2+} formed is titrated with $K_2Cr_2O_7$ according to eq. (19-13). Alternatively, the dissolved Cu(I) can be determined spectrophotometrically, following reaction with the complexing agent 2,9-dimethyl-1,10-phenanthroline (neocuproine). The colored complex may be extracted into 2-propanol (79).

As mentioned above, reactions with solutes in various aqueous media can also be used in studies of the chemical nature of the formal Cu(III) oxidation state by assisting in the evaluation of its reduction potential. As detailed above, the Cu(III) in $YBa_2Cu_3O_x$ readily oxidizes both iodide and bromide in highly acidic solutions (82). However, in neutral solutions, where the reduction potentials of these species are higher, $YBa_2Cu_3O_x$ oxidizes only I^- (standard reduction potential E° = 0.535 V) and not Br^- (E° = 1.087 V). Furthermore, $YBa_2Cu_3O_x$ has been shown to oxidize N,N,N',N'-tetramethyl-p-phenylene diamine (E° = 0.27 V) and $Fe(CN)_6^{4-}$ (E° = 0.46 V), but it is incapable of producing H_2O_2 by the oxidation of water (E° = 1.776 V). These findings indicated that the oxidizing species in $YBa_2Cu_3O_x$ is a rather mild oxidant, with a reduction potential not very far from that of molecular oxygen. This implies that the electron deficiency in this oxide is de-localized rather than concentrated in a well-defined Cu^{3+} cation, which can be expected to be a very strongly oxidizing species. Furthermore, $YBa_2Cu_3O_x$ does not behave as a peroxide, in view of the fact that it does not produce H_2O_2 upon contact with water (82). The assumption of some electron deficiency delocalization on oxygen sites adjacent to the copper atom agrees with the results of x-ray photoelectron spectroscopy and ultraviolet photoemission spectroscopy. These studies show a small density of states at the Fermi level, Cu 2p spectra with only a Cu^{2+} contribution and indication of large O^{2-} vacancies coordinated with Ba^{2+} sites (83) or 2p holes localized mainly on the oxygen atoms on the Cu-O chains (84). Carriers doped into La-Sr-Cu-O and Y-Ba-Cu-O superconducting systems may be viewed as oxygen p-like holes interacting with Cu(II) d^9 states (85). These findings are in agreement with the indications detailed above, which are provided by the studies of the reactions of oxide superconductors with oxidizable solutes in aqueous media.

Electrochemical measurements on $YBa_2Cu_3O_7$, performed against a saturated calomel electrode, indicated high stability of the material in a solution of the reducing acid H_2SO_3, and passivation in oxidizing acid solutions (71) (see above). The rates of attack on $YBa_2Cu_3O_7$ electrodes increase with increasing acid concentration (72). Sintered pellets of $YBa_2Cu_3O_7$, $Bi_2Sr_{2.2}Ca_{0.8}Cu_2O_8$ and $Tl_2Ba_2Ca_2Cu_3O_{10}$, freshly resurfaced by polishing or cleavage, were successfully used to demonstrate room temperature electron transfer between the superconducting phases and redox species such as ferrocene, ferrocene carboxylic acid, bis(pentamethylcyclopentadienyl) iron(II) (Cp_2) or the lithium salt of the radical anion of tetracyanoquinodimethane (LiTCNQ) (70,86). The solvent was 0.1 M of tetraethylammonium perchlorate or acetonitrile or dichloromethane (70,86). (Redox experiments with epoxy-impregnated $YBa_2Cu_3O_7$ electrodes were even attempted in frozen $HClO_4 \cdot 5.5H_2O$ at low temperatures (70). Electron transfer processes were also shown to take place between these superconducting oxides and surface-confined electroactive polymers such as poly-$[Os(bpy)_2(vpy)_2](ClO_4)_2$ or poly-pyrrole films deposited on their surface. Such electron-transfer processes taking place at the surface of cuprate electrodes described above on one hand and at the surface of conventional Pt electrodes on the other are schematically shown in Figure 19-7. Such systems can provide the basis for the development of molecular film/superconductor junctions (86). The major features of electron transfer events which occur at superconducting electrodes as observed in the studies reported in Refs. 70 and 86 are as follows:

1. Because the charge carriers in superconductors are electron pairs, multiple electron transfer events at superconducting electrodes should occur more readily than at metals or semiconductors where the charge carriers are single electrons (or holes).

2. Modulation of faradaic electron transfer events or double layer charging may occur at superconducting surfaces as a result of the superconductor proximity effect.

3. Anomalous magnetic effects associated with superconductors may alter the behavior of superconductor electrodes.

4. Enhanced rates of tunneling might be expected at superconductor surfaces.

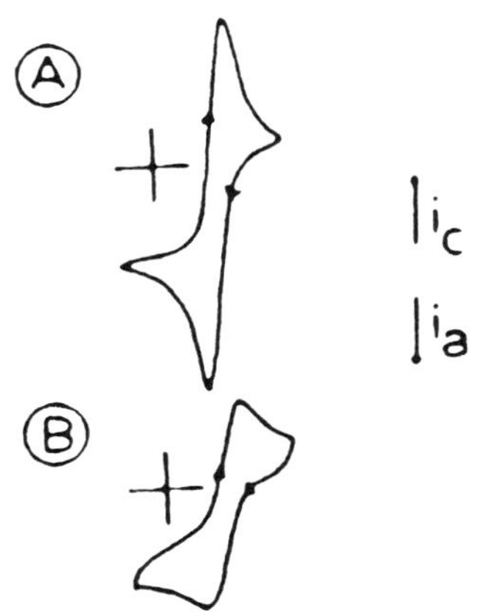

$[(CH_3C)_5]_2Fe \Leftrightarrow [(CH_3C)_5]_2Fe^+ + e^-$

Cyclic voltammetry loops

Solvent: 0.1M $(C_2H_5)_4NClO_4$ in acetonitrile. A = Pt electrode, B = $YBa_2Cu_3O_7$ electrode. Based on Ref. 86.

Figure 19-7. Electron transfer processes involving cuprate and conventional electrodes.

The electronically conducting non-stoichiometric cuprates which exhibit high-T_c superconducting behavior are potentially useful as electrochemical catalysts and oxidation catalysts (58), but further information concerning their chemical properties, in particular ionic diffusion, is required.

It is also interesting to note that perovskite-type materials with oxygen deficiencies such as $BaCe_{0.90}Nd_{0.10}O_{3-\alpha}$ which are structurally related to superconducting oxides, have proven to be promising solid ionic conductors for fuel cell applications (87).

EFFECTS OF PRESENCE OF MOISTURE DURING SYNTHESIS

The detrimental effect of water on superconducting oxides is not limited to cases where these materials, prepared dry, are exposed to moisture or liquid water, resulting in absorption of water into their structure. When water is present in the mix of starting materials used to prepare $(La_{1-x}Sr_x)_2CuO_4$, even firing at 1100°C for 22 hours is insufficient to remove it from the final product (88). Water originating in the starting materials which is not removed during the sintering stage causes a decrease in the volume fraction of the superconducting phase in $(La_{1-x}Sr_x)_2CuO_4$ (88). Likewise, during the preparation of $YBa_2Cu_3O_7$ from $BaCO_3$, Y_2O_3 and CuO, dry milling is found to be best for minimizing moisture degradation, although organic solvents are usually preferred in order to produce a fine, uniform particle size distribution. Direct contact with water is immediately detrimental (89). The presence of water vapor, especially when accompanied by CO_2, during the annealing of $YBa_2Cu_3O_7$ specimens has been observed to result in a large increase in resistivity (30). CO_2 by itself does not increase the resistivity, but it greatly enhances the detrimental effect of water. Thus the resistivity of samples heated at 200-300°C under pure oxygen with a controlled amount of water vapor is much lower than that of samples heated under similar conditions in air containing the same amount of water vapor (30). The presence of water in $(La_{1-x}Ba_x)_2CuO_4$ can be detected by infrared absorption spectroscopy (90).

The sensitivity of superconducting oxides to attack by liquid water and, to a lesser extent, by water vapor precludes the use of many processing methods applicable to the preparation of other ceramic materials. In particular, water is not a promising medium for the growth of crystals of $YBa_2Cu_3O_7$ (91) and of similar cuprates. Attempts to grow $YBa_2Cu_3O_7$ crystals from the corresponding powder hydrothermally (between 150 and 750°C) in a broad range of aqueous media led to decomposition into CuO, Cu_2O, Y_2BaCuO_5 and other products. In the case of $La_{1.7}Sr_{0.3}CuO_4$, the products obtained between 375 and 700°C included CuO, $La(OH)_3$ and $(LaSr)CuO_4$. Even choosing rare earths for which the reaction

$$(RE)_2O_3 + 3\ H_2O = 2\ RE(OH)_3 \tag{19-15}$$

is not favored at the temperature of the experiment, such as Gd, did not prevent decomposition. $Gd_{1.8}Sr_{0.2}CuO_4$ decomposed at 700°C into Gd_2O_3, CuO and other products (91).

Because of the strongly basic character of $YBa_2Cu_3O_7$, some CO_2, like water, tends to remain bound to the product even after calcination, resulting in increased resistance (17,36,89,92,93). Upon calcination, some carbonate groups give rise to carboxylic groups (94).

INTERACTION OF SUPERCONDUCTING OXIDES WITH NON-AQUEOUS SOLVENTS

The study of the stability of ceramic superconducting materials in organic solvents is important because, as shown above, aqueous media are generally too reactive to be used in processing these materials into desired configurations such as single crystals and films. Accordingly, such studies have been mainly pursued with a view to processing applications rather than because of relevance to storage or service conditions.

Early work showed that many organic solvents are much less corrosive towards superconducting oxides than water and aqueous solutions. Washing of $(La_{1-x}Sr_x)_2CuO_4$ powders in ethanol prior to final annealing was found to reduce the water content and improve the conductivity of the final product (88). $YBa_2Cu_3O_7$ powders immersed in hexane and dried by solvent evaporation at 25°C gave identical x-ray patterns and transmission electron micrographs with those obtained with the dry material prior to immersion (52). Similar results were obtained with acetone, ethanol and 2-propanol. However, when the hexane was removed by evaporation at 100°C subsequent to immersion, the splitting of the lines in the x-ray spectrum of the $YBa_2Cu_3O_7$ powder disappeared, indicative of partial reduction. The oxygen content of the material can be restored by annealing at 450°C under oxygen (52). Comparative studies of the effects of methanol and of aqueous solutions, respectively, on $YBa_2Cu_3O_7$ powder showed that at room temperature sample deterioration, accompanied by gas evolution, took place almost immediately in water (and even more rapidly in 1 M HCl), while no reaction was observed in 1 M NaCl or 10 M NaOH over an exposure period of 1 day and in methanol over a period of 5 days (31). At 100°C, decomposition in 10 M NaOH was evident after 2 hours, while in 1 M NaCl the material became slightly corroded in 1 day and in methanol no reaction was observed at the end of 1 day. Only under hydrothermal conditions did the material react with methanol, with complete decomposition taking 4 days at 300°C. The corresponding time in the cases of 10 M NaOH or 1 M NaCl was 2 days. The resulting order of reactivity was: (31)

$$HCl > H_2O > NaOH > NaCl > CH_3OH$$

The decomposition of $YBa_2Cu_3O_7$ powder in methanol and other alcohols under hydrothermal conditions over a period of 2-4 days at temperatures between 150 and 250°C prevented the use of these alcohols as growth media for $YBa_2Cu_3O_7$ crystals (91).

More systematic studies were carried out in order to obtain comparative information on the stability of superconducting oxides in various organic solvents (62,95). Upon placing 0.5-g powdered samples of $YBa_2Cu_3O_7$ in a series of covered test tubes and permitting each of the powders to equilibrate with 15 g of a solvent over a period of 220 hours at a constant temperature of 25°C, it was observed that the dissolution process of $YBa_2Cu_3O_7$ was generally incongruent, except in the case of acidic media (e.g., acetic acid, see above). Ba is preferentially dissolved in solvents with labile hydrogen atoms and a high dielectric constant, such as water and methanol. The dissolution is nearly congruent with respect to Ba and Cu, but not Y, when the dielectric constant is very high (formamide, N-methylformamide). Cu is the predominant species in leachates of $YBa_2Cu_3O_7$ in solvents with a low dielectric constant, such as benzene, toluene, and 2-pentanol as well as in solvents with a higher dielectric constant but no active hydrogen atoms, such as N,N-dimethylformamide. Y dissolves relatively well in solvents with an intermediate dielectric constant of about 20, such as ethanol, 2-propanol and acetone. In addition to the dielectric constant, the water content of a given organic solvent also appears to have an important role, as indicated by the high observed value for the extent of dissolution of Ba in methanol and the significant amounts of dissolved Ba in benzene and toluene (62). The relative extent of corrosion, as reflected in changes in the x-ray patterns of the solvent-exposed powders, appears to follow the sequence:

Acetic acid >> 2-pentanol > formamide >> ethanol > N-methylformamide > 2-propanol > N,N-dimethylformamide > acetone > methanol, benzene, toluene

Alteration phases were observed in the cases of acetic acid, formamide, water and 2-pentanol. Leachate analysis, on the other hand, gives the following order in terms of total dissolved material:

Acetic acid > formamide >> water >> N-methylformamide > methanol >> N,N-dimethylformamide > acetone > 2-propanol > ethanol > toluene > benzene > 2-pentanol

The estimates of the extent of dissolution in the cases of acetic acid, formamide, water and 2-pentanol are too low because of the observed appearance of a solid precipitate (62). This may account, at least in part, for discrepancies between the two series (in particular, in the case of 2-pentanol). Other discrepancies may result from variations in water content of the various solvents throughout the tests. Water content appears to have a large effect on the comparative corrosivity of the various solvents, as indicated by the large effect of 2-pentanol on the x-ray diffraction pattern and the high level of leached Ba in methanol. (These two solvents had the highest water content in the reported work (62)). Allowing for differences in water content, however, the extent of dissolution of $YBa_2Cu_3O_7$ in the various solvents still appears to be related to their dielectric constants.

The practical implication of studies on various solvents is that superconducting oxides such as $YBa_2Cu_3O_7$ are not adversely affected to a significant degree by short exposure (i.e. a few hours or days) to solvents of low acidity and low or moderate dielectric constant such as benzene, toluene, acetone, ethanol, methanol, 2-propanol or N,N-dimethylformamide. This has made it possible to incorporate solvent-assisted procedures in the synthesis and processing of such oxides. In the preparation of $(La_{1-x}Sr_x)_2CuO_4$ powder, for instance, it was observed that ethanol can serve as an effective mixing medium while avoiding the introduction of residual water, which has detrimental effects on product superconductivity (88). Similarly, in preparing $YBa_2Cu_3O_7$ powder it was found that milling in methanol, 2-propanol or xylene for 2 - 4 hours was almost as effective as dry milling in avoiding chemical degradation. The use of such solvents yielded finer particles with a more uniform particle distribution than dry milling. The resulting powders were than processed to give bulk objects or tapes using suitable binders, plasticizers and solvents (89). 0.25 M tetrabutylammonium perchlorate in pyrrole and 0.1 M tetraethylammonium perchlorate in acetonitrile were used as electrolytic media in the electrodeposition of poly-$[Os(bpy)_2(vpy)_2](PF_6)_2$ and poly(pyrrole) on $YBa_2Cu_3O_7$, $Bi_2Sr_{2.2}Ca_{0.8}Cu_2O_8$ and $Tl_2Ba_2Ca_2Cu_3O_{10}$ pellets, which were subsequently used in electron transfer experiments in acetonitrile media (86). Dry acetonitrile does not affect $YBa_2Cu_3O_7$ over long periods of time, and the same is true for epoxy resin used to impregnate the $YBa_2Cu_3O_7$ electrodes in such experiments (70). Soaking in acetonitrile proved useful in removing Tl oxide films from the surfaces of $Tl_2Ba_2Ca_2Cu_3O_{10}$ electrodes.

A major application which requires compatibility between organic media and superconducting oxides is the fabrication of films by screen printing (96). Screen-printed films prepared for electronic device studies are shown in Figure 19-8. In this application the superconducting oxide powder is suspended in a liquid carrier by means of thorough mixing to form a paste, which is subsequently printed through a suitable screen onto a ceramic substrate (e.g., quartz, alumina, yttria-stabilized zirconia, La_2CuO_4 (96-98) or Y_2BaCuO_5 (99-101)) and fired to remove the liquid phase and produce a consolidated film. The carrier liquid phase must have suitable viscosity and wettability with respect to the superconducting oxide powder and the substrate, without reacting with the latter, either during paste preparation and printing at room temperature or during subsequent firing, to produce a superconducting film on the substrate.

Early carrier liquid formulations consisted of a single mono- or bifunctional aliphatic alcohol, such as butanol (102), heptanol (96-98), octanol (96),

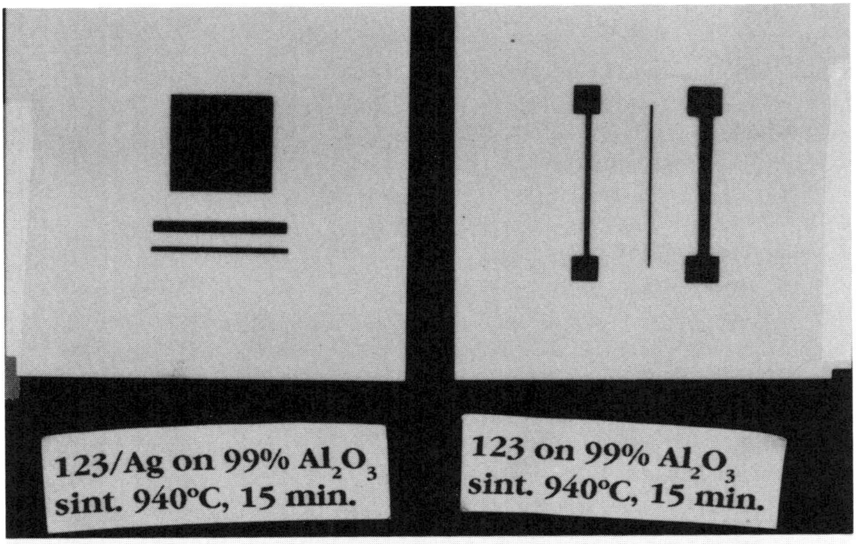

a) Samples prepared at the University of Florida

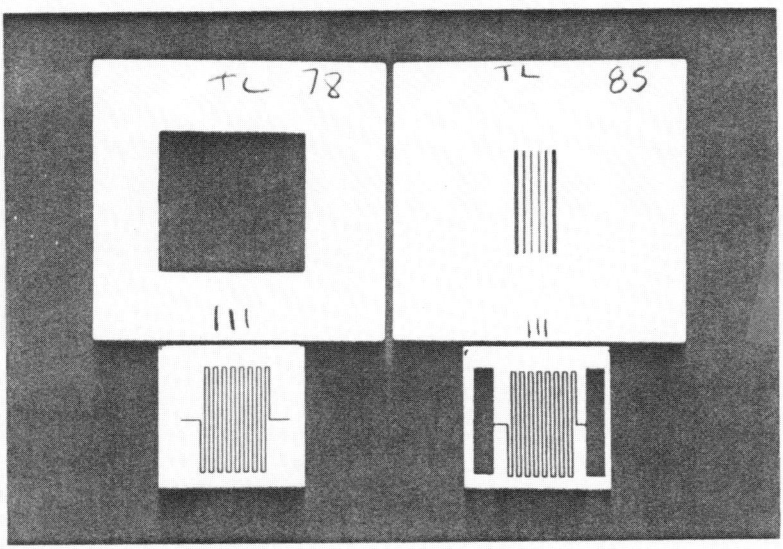

b) Samples prepared at the Catholic University of America

Figure 19-8. Screen-printed $YBa_2Cu_3O_7$ films.

ethanediol (103) or propanediol (96). Since these alcohols have relatively low viscosities, which are also hard to control due to extreme sensitivity to the presence of traces of moisture, more sophisticated mixtures of organic liquids were later developed to prepare superconducting oxide pastes. Such mixtures contain binders such as ethyl cellulose (100,104,105) and 2-(2-butoxyethoxy)ethyl acetate (100,105), highly viscous alicyclic alcohols such as alpha-terpineol (100,104,105) and solvents such as 2-propanol (105) or commercial thinners (104), which are introduced in order to control the viscosity. A small amount of glass powder was added in one case (104) in order to reduce superconducting oxide/substrate interaction during sintering and to improve film adhesion. Mixtures such as those described here have higher and more easily controlled viscosities and are less moisture-sensitive than the simple alcohols used in earlier studies.

Other methods used in the fabrication of films and tapes, such as slip casting, also make use of suitable combinations of organic binders, plasticizers and dispersants (101,106).

EFFECTS OF MICROSTRUCTURE: POROSITY AND GRAIN BOUNDARIES

As described above, early studies on superconducting oxides indicated that these materials are highly sensitive to the moisture in ambient environments. For instance, it was reported that even $YBa_2Cu_3O_7$ strips protected by a 0.5-mm thick epoxy coating can be expected to exhibit significant deterioration under ambient conditions, and that bare samples undergo degradation even more quickly (30). Strips (30) and disc-shaped samples (33) exposed to air at a temperature of 85°C and a relative humidity of 85% for 50-90 minutes exhibited both a large increase in resistivity (30) and disappearance of the x-ray diffraction pattern of $YBa_2Cu_3O_7$ (33, 107). Sintered pellets of $(La_{1-x}Ba_x)_2CuO_4$ were observed to be particularly sensitive to degradation upon atmospheric exposure, forming a white powdery layer and crumbling after an exposure period of only a few days (47). Similar behavior was reported for $YBa_2Cu_3O_7$ (108). At the end of an exposure period of a few days at 40°C and relative humidity of 100%, sintered pellets of $YBa_2Cu_3O_7$ exhibit intense disorder and disruption of the parent structure, indicative of bulk decomposition which begins at the surface, from which microcrystals of $BaCO_3$ are observed to grow (38). Even storage of sintered bars of $YBa_2Cu_3O_7$ in a desiccator with a $CaSO_4$ drying agent reduces the extent of resistivity increase only by a factor of 2, i.e., from 4% to 2% over a period of 49 days (32).

However, these early studies were carried out on polycrystalline, sintered samples which usually had a high degree of porosity (70,86) and a large fraction of impurity phases. More recently, it has been shown that an increase in density from about 75% to about 90% of the theoretical value results in a decrease by more than a factor of 2 in the rate of attack by liquid water (43). (The theoretical density of $YBa_2Cu_3O_7$ is 6.383 g/cm^3 based on x-ray diffraction studies (25).) A similar effect has been observed upon comparing the rates of degradation of samples with densities of 50% and 80% of the theoretical value, respectively, when immersed in de-ionized water at 80°C for 4-5 hours (69). The effect is much more pronounced in the case of weathering in humid environments (109), since rapid attack is limited to surfaces directly exposed to the ambient atmosphere; degradation subsequent to the initial surface reaction is quite slow (35). It should be emphasized that because of the strong dependence of sensitivity to environmental attack on porosity, it is not possible to establish direct quantitative correlation among leachability data obtained in different studies on sintered samples, since the extent of porosity of such samples is determined by the exact conditions of preparation used in each individual study.

High critical currents and large capabilities for magnetic flux trapping in high-T_c superconducting materials are generally observed in specimens prepared using melt-based techniques (110-115) or crystal growth methods (91,116-118). In general, the densities of these materials are much closer to the theoretical values than those of specimens prepared by solid-state sintering. In the absence of extensive porosity, attack by moisture on all parts of the specimen except for the outermost surface region is much slower. Accordingly, resistance to attack by atmospheric moisture is of more immediate concern in applications involving films or wires in uncontrolled environments than in applications based on bulk materials, such as levitation, magnetic energy storage and shielded cavities.

In the cases of polycrystalline samples, the attack by water takes place preferentially at grain boundaries, as is usually the case with ceramic materials in general (30,54). The exposure of $YBa_2Cu_3O_7$ samples to air with 100% relative humidity at 60°C for 20 minutes gave rise to the growth of needle-shaped crystals at the grain boundaries (32,54). The composition and corrosion resistance of grain boundary regions is obviously important. Indeed, some of the problems encountered with early specimens were related to the intergranular phases (e.g. carbonates or cuprates) (109,119).

Again, improved methods of preparation, designed to provide higher J_c, have the additional advantage of having the potential of minimizing preferential leaching at grain boundaries. Stoichiometric polycrystalline specimens of $YBa_2Cu_3O_7$ usually exhibit precipitation of $BaCuO_2$ at grain boundaries (26,27). In addition to being an insulating phase which interrupts the superconducting current flow at the grain boundaries, $BaCuO_2$ forms a likely point of attack because of the high proportion of highly soluble BaO and the absence of Y_2O_3 from its composition. The introduction of excess yttrium eliminates the $BaCuO_2$ phase, while giving rise to the appearance of Y_2BaCuO_5 as a second phase. The presence of the latter at moderate amounts does not have a detrimental effect on J_c (26,27). At the same time, it can be expected to improve the chemical durability of the specimen, since it is a stable product of the reaction described by eq. (19-1). These findings are consistent with the results reported in 107.

PROTECTION AGAINST ENVIRONMENTAL ATTACK

Because of the high sensitivity of sintered specimens to attack by humid environments, encapsulation is required in many applications in order to retard corrosion (91). Several attempts were made to develop coatings to provide protection against such attack. Coating of $YBa_2Cu_3O_7$ strips with a 0.5-mm thick fast-setting epoxy was found to reduce the rate of change of resistance upon exposure to air at 85°C and 85% relative humidity by a factor of 3 (30). Encapsulation of sintered $YBa_2Cu_3O_7$, $Bi_2Sr_{2.2}Ca_{0.8}Cu_2O_8$ and $Tl_2Ba_2Ca_2Cu_3O_{10}$ pellets in an epoxy matrix effectively fills the outer pores and permits the use of such pellets as electrodes for periods of 4, 7 and 14 days, respectively, in water (70,86). Such epoxy encapsulation also diminishes the problem of large capacitive background currents caused by the high porosity of these sintered materials. Epoxy encapsulation is even effective in strongly acidic media such as 5.5 M $HClO_4$ (70). The epoxy itself does not have detrimental effects over long periods of time (70). Even small amounts (<2%) of acrylic resins, for instance, can greatly enhance the durability of $YBa_2Cu_3O_7$, upon exposure to moist air, liquid water and even acidic solutions (120). Attempts to apply protective coatings to the surface of $(La_{1-x}Sr_x)_2CuO_4$ or $YBa_2Cu_3O_7$ using metals such as iron (121-123) and titanium (124) have resulted in deterioration in the electric and magnetic properties of the oxides due to reactions between the metal and the oxide. Gold, on the other hand, is apparently effective to some extent in protecting the surface of $(La_{1-x}Sr_x)_2CuO_4$ specimens (125).

As mentioned above, weathering is of particular concern in the case of films, which have a high ratio of exposed surface area to thickness. In the cases of thin films of Y-Ba-Cu-O materials prepared by using electron beam evaporation to deposit them on MgO substrates, it was observed that the addition of a silver layer to the starting structures improves the conductivity of the films. The use of a gold layer, on the other hand, reduces the temperature at which the resistance falls to zero (126). The addition of the silver layer to such films deposited on ZrO_2-Y_2O_3 or MgO substrates results in a large decrease in the extent of moisture-induced degradation as reflected in film resistance before and after warming and re-cooling under nitrogen (127,128). A large beneficial effect of silver layers on resistance to degradation has also been reported under much more severe conditions of immersion in liquid water at 20°C. In the latter case, films with silver layers on ZrO_2-Y_2O_3 substrates were observed to retain their superconductivity at liquid nitrogen temperature even after 5 hours of immersion, and contact resistance remained low even after 60 hours. In the absence of silver, however, contact resistance rapidly increased with increasing immersion time, making measurements at 77K difficult after only 5 minutes of immersion (128,129).

SUMMARY

The sensitivity of superconducting oxides to moisture is an important consideration affecting their durability in various environments. This factor also precludes the use of many types of ceramic processing methods which involve aqueous environments in the fabrication of solids with a desired shape and size (e.g. films, tapes, wires, single crystals) from superconducting oxide powders. In general, all known types of superconducting oxides are sensitive, to a certain extent, to attack by humid environments, and, to a much larger extent, to attack by liquid water. The severity of the attack depends on the time of exposure and, in particular, on the temperature. The sensitivity of different superconducting oxides to water appears to decrease according to the sequence

$$YBa_2Cu_3O_7 > (La_{1-x}Sr_x)_2CuO_4 > Bi_2Sr_{2.2}Ca_{0.8}Cu_2O_8 > Tl_2Ba_2Ca_2Cu_3O_{10},$$

but the differences among these materials are not very large. Most of the information available at the present has been acquired in studies on $YBa_2Cu_3O_7$.

Early samples of superconducting materials had very poor resistance to water. Fortunately, improvements in preparative methods stemming from the need to obtain high critical current densities have had the added benefit of greatly reducing the rate of attack by water vapor as well as liquid water. The improvement regarding both properties has been due primarily to the minimization of porosity and, in the case of polycrystalline specimens, to elimination of insulating, highly soluble phases such as $BaCuO_2$ from the grain boundaries, for instance by intentional introduction of the more stable Y_2BaCuO_5 as a minority phase.

When accelerated attack through pores and grain boundaries is prevented, superconducting oxide specimens exhibit corrosion mechanisms similar in many respects to those of other multi-component ceramics and glasses which contain highly basic metal oxides. In the present case, the alkaline earth (usually Ba) leaches out rapidly from the surface, but subsequent attack is much slower even with respect to Ba extraction since it necessitates the creation of a new surface. The latter process involves dissolution of other oxide components (those of Y and Cu) which have much lower solubilities. Eventually the corrosion process leads to the formation of altered phases, such as $BaCO_3$, on the surface. This mechanism is compatible with the observation that the overall rate of dissolution rises very sharply with increasing acidity of the aqueous medium, which increases the solubility of Cu and Y, or with the addition of ammonia, which forms a soluble complex with Cu.

The electron deficiency sites (formally represented as Cu(III)) are very labile in the presence of water and are converted to Cu(II) with the evolution of molecular oxygen or reactions with oxidizable solutes in the aqueous medium. Such reactions permit evaluation of their reduction potential, which has a moderate value indicative of some delocalization of the excess positive charge.

The redox properties of superconducting oxides have potential applications in the areas of electrochemistry and catalysis.

Because of the sensitivity of superconducting oxides to attack by aqueous media, processing methods which require mixing of the oxide powder with a liquid phase, such as the preparation of films by screen printing, are generally based on the use of organic solvents and binders. Except for solvents with very high dielectric constants and active hydrogen atoms, such as formamide and N-methylformamide, such solvents do not cause considerable degradation as long as the time of exposure, especially at elevated temperatures, is kept short.

ACKNOWLEDGEMENTS

The authors are very grateful to S. A. Olszowka for preparation of the manuscript and to Drs. S. Alterescu and A.N. Thorpe for consistent encouragement and support. This work was supported by the National Air and Space Administration under Contract No. NAG 5-1017.

REFERENCES

1. Bednorz, J.G., and Muller, K.A., *Z. Phys. B* 64: 189-193 (1986).
2. Sleight, A.N., Gillson, J.L., and Bierstadt, F.E., *Solid State Comm.* 17: 27 (1975).
3. Thanh, T.D., Koma, A., and Tanaka, S., *App. Phys.* 22: 205 (1980).
4. Cava, R.J., and Batlogg, B., *MRS Bull.* 14: 49-52 (1989).
5. Vanderah, T.A., and Osofsky, M.S., in: *High-Temperature Superconductors: Fundamental Properties and Novel Materials* (J. Narayan, C.W. Chu, L.F. Schneemeyer, and D.K. Christen ed.), MRS Symp. Ser. 169, Materials Research Society (1990).
6. Bednorz, J.G., Takashige, M., and Muller, K.A., *Europhys. Lett.* 3: 379 (1987).
7. Uchida, S., Takagi, H., Kitazawa, K., and Tanaka, S., *Jpn. J. Appl. Phys.* L123 (1987).
8. Kishio, K., Kitazawa, K., Kanbe, S., Yasuda, I., Sugii, N., Takagi, H., Uchida, S., Fueki, K., and Tanaka, S., *Chem. Lett.* 429 (1987).
9. Cava, R.J., van Dover, R.B., Batlogg, B., and Rietman, A., *Phys. Rev. Lett.* 58: 408 (1987).
10. Tarascon, J.M., Greene, L.H., McKinnon, W.R., Hull, G.W. and GeBalle, T.H., *Science* 235: 1373 (1987).
11. Chu, C.W., Hor, P.H., Meng, R.L., Gao, L., Huang, Z.J., and Wang, Y.Z., *Phys. Rev. Lett.* 58: 405 (1987).
12. Raveau, B., Michel, C., and Hervieu, M., in: *Chemistry of High-Temperature Superconductors* (D.L. Nelson, M.S. Whittingham, and T.F. George ed.),

ACS Symp. Ser. 351, pp 122-135, American Chemical Society.

13. Cava, R.J., Batlogg, B., Krajewski, J.J., Rupp, L.W., Schneemeyer, L.F., Siegrist, T., vanDover, R.B., Marsh, P., Peck, W.F., Jr., Gallagher, P.K., Glarum, S.H., Marshall, J.H., Farrow, R.C., Waszczak, J.V., Hull, R., and Trevor, P., *Nature* 336: 211-214 (1988).
14. Cava, R.J., Marezio, M., Krajewski, J.J., Peck, W.F., Jr., Santoro, A., and Beech, F., *Physica C* 157: 272 (1989).
15. Hull, R., Bonar, J.M., Schneemeyer, L.F., Cava, R.J., Krajewski, J.J., and Wasczak, J.V., *Phys. Rev. B* 39: 9614 (1989).
16. Subramanian, M.A., Gopalakrishnan, J., Torardi, C.C., Gai, P.L., Boyes, E.D., Askew, T.R., Flippen, R.B.N., Farneth, W.E., and Sleight, A.W., *Physica C* 1573: 124 (1989).
17. Wu, M.K., Ashburn, J.R., Torng, C.J., Hor, P.H., Meng, R.L., Gao, L., Huang, Z.J., Wang, Y.Q., and Chu, C.W., *Phys. Rev. Lett.* 58: 908-910 (1987).
18. Hikami, S., Hirai, T., and Kagoshima, S., *Jpn. J. Appl. Phys.* 26: L314 (1987).
19. Zhao, Z., Chen, L., Yang, Q., Huang, Y., Chen, G., Tang, R., Liu, G., Cui, C., Chen, L., Wang, L., Guo, S., Li, S., and Bi, J., *Kexue Tongbao* no. 6 (1987).
20. Cava, R.J., Batlogg, B., van Dover, R.B., Murphy, D.W., Sunshine, S.A., Siegrist, T., Remeika, J.P., Reitman, R.E., Zahurak, S., and Espinosa, G.P., *Phys. Rev. Lett.* 58: 1676 (1987).
21. Maeda, H., Tanaka, Y., Fukutomi, N., and Asano, T., *Jpn. J. Appl. Phys.* 27: L209 (1988).
22. Chu, C.W., Bechtold, J., Gao, L., Hor, P.H., Huang, Z.J., Meng, R.L, Sun, Y.Y., Wang, Y.Q., and Xue, Y.Y., *Phys. Rev. Lett.* 60: 941-943 (1988).
23. Sheng, Z.Z., Hermann, A.M., El Ali, A., Almasan, C., Estrada, J., Datta, T., and Matson, R.J., *Phys. Rev. Lett.* 60: 937-940 (1988).
24. Hazen, R.M., Finger, L.W., Angel, R.J., Prewitt, C.T., Ross, N.L., Hadidiacos, C.G., Heaney, P.J., Veblen, D.R., Sheng, Z.Z., El Ali, A., and Hermann, A.M., *Phys. Rev. Lett.* 60: 1657-1660 (1988).
25. Poole, Jr., C.P., Datta, T., and Farach, H.A., *Copper Oxide Superconductors*, pp 122-130, Wiley-Interscience (1988).
26. Wadayama, Y., Kudo, K., Nagata, A., Ikeda, K., Hanada, S., and Izumi, O., *Jpn. J. Appl. Phys.* 27: L1221-L1224 (1988).
27. Hojaji, H., Barkatt, A., Michael, K.A., Hu, S., Thorpe, A.N., Ware, M.F., Talmy, I.G., Haught, D.A., and Alterescu, S., *J. Mater. Res.* 5: 721-730 (1990).
28. Bean, C.P., *Phys. Rev. Lett.* 8: 250 (1962).
29. Pool, R., *Science* 244: 914-916 (1989).
30. Kitazawa, K., Kishio, K., Hasegawa, T., Ohtomo, A., Yaegashi, S., Kanbe, S., Park, K., Kuwahara, K., and Fueki, K., in: *High-Temperature Superconductors* (M.B. Brodsky, R.C. Dynes, K. Kitazawa, and

H.L. Tuller ed.), *Mater. Res. Soc. Symp. Proc.*, Vol. 99, pp 33-40 (1988).

31. Barns, R.L., and Laudise, R.A., *Appl. Phys. Lett.* 51: 1373-1375 (1987).
32. Gaier, J.R., Hepp, A.H., Curtis, H.B., Schupp, D.A., Hambourger, P.D., and Blue, J.W., in *High-Temperature Superconductors*, Mater. Res. Soc. Symp., Boston, MA, 1988; NASA TM-101401, NASA Lewis Research Center, Cleveland, OH, 1988.
33. Yan, M.F., Barns, R.L., O'Bryan, H.M., Jr., Gallagher, P.K., Sherwood, R.C., and Jin, S., *Appl. Phys. Lett.* 51: 532-534 (1987).
34. Zhang, J.P., Li, D.J., and Marks, L.D., in: *High-Temperature Superconductors* (M.B. Brodsky, R.C. Dynes, K. Kitazawa, and H.L. Tuller ed.), *Mater. Res. Soc. Symp. Proc.*, Vol. 99, pp 965-968 (1988).
35. Horowitz, H.S., Bordia, R.K., Flippen, R.B., and Johnson, R.E., in: *High-Temperature Superconductors* (M.B. Brodsky, R.C. Dynes, K. Kitazawa, and H.L. Tuller ed.), *Mater. Res. Soc. Symp. Proc.*, Vol. 99, pp 903-906 (1988).
36. Nefedov, V.I., Sokolov, A.N., Tyzykhov, M.A., Oleinikov, N.N., Yeremina, Ye. A., and Kolotyrkina, M.A., *J. Electr. Micr. Rel. Phenom.* 49: 47-60 (1989).
37. Morss, L.R., Sonneberger, D.C., and Thorn, R.J., in: *High-Temperature Superconductors* (M.B. Brodsky, R.C. Dynes, K. Kitazawa, and H.L. Tuller ed.), *Mater. Res. Soc. Symp. Proc.*, Vol. 99, pp 571-574 (1988).
38. Hyde, B.G., Thompson, J.G., Withers, R.L., Fitzgerald, J.G., Stewart, A.M., Bevan, D.J.M., Anderson, J. S., Bitmead, J., and Peterson, M.S., *Nature* 327: 402 (1987).
39. Komori, M., Kozuka, H., and Sakka, S., *J. Mater. Sci.* 24: 1889-1894 (1989).
40. Paul, A., *J. Mater. Sci.* 12: 2246 (1977).
41. Paul, A., and Youssefi, A., *J. Mater. Sci.* 13: 97-107 (1978).
42. Jantzen, C.M., and Plodinec, M.J., *J. Non-Cryst. Solids* 67: 207-223 (1984).
43. Hojaji, H., Barkatt, A., and Hein, R.A., *Mat. Res. Bull.* 23: 869-879 (1988).
44. Simmons, J.H., this volume.
45. Iqbal, Z., Leone, E., Chin, R., Signorelli, A.J., Bose, A., and Eckhardt, H., *J. Mater. Res.* 2: 768-774 (1987).
46. Qiu, S.L., Ruckman, M.W., Brookes, N.B., Johnson, P.D., Chen, J., Lin, C.L., Strongin, M., Sinkovic, B., Crow, J.E., and Jee, C.-S., *Phys. Rev. B* 37: 3747-3750 (1988).
47. Kurtz, R.L., Stockbauer, R., Madey, T.E., Mueller, D., Shih, A., and Toth, L., *Phys. Rev. B* 37: 7936-7939 (1988).
48. Safari, A., Wachtman, J.B., Jr., Parkhe, V., Caracciolo, R., Jeter, D., Rao, A.S., and Sundar, H.G.K., in: *High-Temperature Superconductors* (M.B. Brodsky, R.C. Dynes, K. Kitazawa, and H.L. Tuller ed.), *Mater. Res. Soc. Symp. Proc.*, Vol. 99, pp 269-272 (1988).
49. Rao, C.N.R., Ganguly, P., Gopalakrishnan, J., and Sarma, D.D., *Mat. Res. Bull.* 22: 1159 (1987).

50. Barkatt, A., Hojaji, H., and Michael, K.A., *Adv. Cer. Mater.* 2: 701-709 (1987).
51. Nakada, I., Sato, S., Oda, Y., and Kohara, T., *Jpn. J. Appl. Phys.* 26: L697 (1987).
52. Frase, K.G., Liniger, E.G., and Clarke, D.R., *Adv. Cer. Mater.* 2: 698-700 (1987).
53. Kitazawa, K., Kishio, K., Hasegawa, T., Nakamura, O., Shimoyama, J., Sugii, N., Ohtomo, A., Yaegashi, S., and Fueki, K., *Jpn. J. Appl. Phys.* 26: L1979-L1981 (1987).
54. Dexin, Z., Mingshan, X., Ziqing, Z., Shubin, Y., Huansui, Z., and Shuxia, S., *Solid State Comm.* 65: 339 (1988).
55. Feng, X., and Barkatt, A., in: *Scientific Basis for Nuclear Waste Management X* (J.K. Bates, and W.B. Seefeldt ed.), *Mater. Res. Soc. Symp. Proc.*, Vol. 84, pp 519-529 (1987).
56. Rana, M.A., and Douglas, R.W., *Phys. Chem. Glasses* 2: 179 (1961).
57. Rana, M.A., and Douglas, R.W., *Phys. Chem. Glasses* 2: 196 (1961).
58. Nelson, D.L., Whittingham, M.S., and George, T.F., in: *Chemistry of High Temperature Superconductors* (D.L. Nelson, M.S. Whittingham, and T.F. George ed.), ACS Symp. Ser. 351, pp 308-312, American Chemical Society (1987).
59. Clark, D.E., Pantano, C.G., and Hench, L.L., *J. Am. Ceram. Soc.* 59: 37-39 (1976).
60. Hench, L.L., *J. Non-Cryst. Solids* 25: 343 (1977).
61. Barkatt, Aa., Gibson, B.C., Macedo, P.B., Montrose, C.J., Sousanpour, W., Barkatt, Al., Boroomand, M.A., Rogers, V.L., and Penafiel, L.M., *Nucl. Tech.* 73: 140-164 (1986).
62. Trolier, S.E., Atkinson, S.D., Fuierer, P.A., Adair, J.H., and Newnham, R.E., *Am. Ceram. Soc. Bull.* 67: 759-762 (1988).
63. Mankiewich, P.M., Scofield, J.H., Skocpol, W.J., Howard, R.E., Dayem, A.H., and Good, E., *Appl. Phys. Lett.* 51: 1753-1755 (1987).
64. Adams, R.V., *Phys. Chem. Glasses* 2: 50-54 (1961).
65. Charles, R.J., *J. Appl. Phys.* 29: 1549 (1958).
66. Budd, S.M., *Phys. Chem. Glasses* 2: 111 (1961).
67. Grambow, B., in: *Scientific Basis for Nuclear Waste Management VIII* (C.M. Jantzen, J.A. Stone, and R.C. Ewing ed.), pp 15-27, Materials Research Society (1985).
68. Barkatt, Aa., Saad, E.E., Adiga, R.B., Sousanpour, W.S., Barkatt, Al., Feng, X., O'Keefe, J.A., and Alterescu, S, in: *Materials Stability and Environmental Degradation* (Aa. Barkatt, E.D. Verink, Jr., and L.R. Smith ed.), pp 129-142, Materials Research Society (1988).
69. Morrish, A.H., Zhou, X.Z., Raudsepp, M., Maartense, I., Eaton, J.A., and Luo, Y.L., *Can. J. Phys.* 65: 808-809 (1987).
70. McDevitt, J.T., Longmire, M., Gollmar, R., Jernigan, J.C., Dalton, E.F., McCarley, R., and Murray, R.W., *J. Electroanal. Chem.* 243: 465 (1988).

71. Chin, T.S., Shih, H.C., Lin, C.A., Huang, T.W., Tsai, W.T., and Hung, M.P., in: *High-Temperature Superconductors: Fundamental Properties and Novel Materials Processing* (J. Narayan, C.W. Chu, L.F. Schneemeyer, and D.K. Christen ed.), MRS Symp. Ser. 169, paper M7.99, Materials Research Society (1990).
72. Rosamilia, J.M., Miller, B., Schneemeyer, L.F., Waszczak, J.V., and O'Bryan, H.M., Jr., *J. Electrochem. Soc.* 134: 1863-1864 (1987).
73. Rigney, M.M., Poole, C.P., Jr., and Farach, H.A., *J. Phys. Chem. Solids* 49: xxx (1988).
74. Gallagher, P.K., *Adv. Cer. Mater.* 2: 632-639 (1987).
75. Tarascon, J.M., Barboux, P., Bagley, B.C., Greene, L.H., McKinnon, W.R., and Hull, G.W., in: *Chemistry of High-Temperature Superconductors* (D.L. Nelson, M.S. Whittingham, and T.F. George ed.), ACS Symp. Ser. 351, pp 198-210, American Chemical Society (1987).
76. Alp, E.E., Shenoy, G.K., Hinks, D.G., Capone II, D.W., Soderholm, L., and Schuttler, H.-B., *Phys. Rev. B* 35: 7199 (1987).
77. Jorgensen, J.D., Beno, M.A., Hinks, D.G., Soderholm, L., Vohn, K.J., Hitterman, R.L., Grace, J.D., Schuller, I.K., Segre, C.U., Zhang, K., and Kleefisch, M.S., *Phys. Rev. B* 36: 3608 (1987).
78. Harris, D.C., and Hewston, T.A., *J. Solid State Chem.* 69: 182-185 (1987).
79. Matacotta, F.C., Nobile, G., Serrini, G., Giardina, M.D., and Merlini, A.E., in: *High-Temperature Superconductors* (M.B. Brodsky, R.C. Dynes, K. Kitazawa, and H.L. Tuller ed.), *Mater. Res. Soc. Symp. Proc.*, Vol. 99, pp 561-565, Materials Research Society (1988).
80. Hume, D.N., and Kolthoff, I.M., *Ind. Eng. Chem., Anal. Ed.* 16: 103 (1944).
81. Appelman, E.H., Morss, L.R., Kini, A.M., Geiser, U., Umezawa, A., Crabtree, G.W., and Carlson, K.D., *Inorg. Chem.* 26: 3237-3239 (1987).
82. Barkatt, Aa., Hojaji, H., and Michael, K.A., *Mat. Res. Bull.* 23: 735-742 (1988).
83. Steiner, P., Kinsinger, V., Sander, I., Siegwart, B., Hufner, S., and Politis, C., *Z. Phys. B* 67: 19 (1987).
84. Yarmoff, J.A., Clarke, D.R., Drube, W., Karlsson, U.O., Taleb-Ibrahimi, A., and Himpsel, F.J., *Phys. Rev. B* 36: 3967 (1987).
85. Fujimori, A., Takayama-Muromachi, E., and Uchida, Y., *Solid State Comm.* 63: 857 (1987).
86. McDevitt, J.T., McCarley, R.L., Dalton, E.F., Gollmar, R., Murray, R.W., Collman, J., Yee, G.T., and Little, W.A., in: *Chemistry of High Temperature Superconductors II* (D.L. Nelson, and T.F. George ed.), ACS Symp. Ser. 377, pp 207-222, American Chemical Society (1988).
87. Iwahara, H., Uchida, H., Morimoto, K., *J. Electrochem Soc.* 137:462-465 (1990).
88. Kishio, K., Sugii, N., Kitazawa, K., and Fueki, K., *Jpn. J. Appl. Phys.*, 26: L466 (1987).
89. Poeppel, R.B., Flandermeyer, B.K., Dusek, J.T., and Bloom, I.D., in: *Chemistry of High-Temperature Superconductors* (D.L. Nelson, M.S.

Whittingham, and T.F. George ed.), ACS Symp. Ser. 351, pp 261-265, American Chemical Society (1987).

90. Ohbayashi, K., Ogita, N., Udagawa, M., Aoki, Y., Maeno, Y., and Fujita, T., *Jpn. J. Appl. Phys.* 26: L420 (1987).
91. Laudise, R.A., Schneemeyer, L.F., and Barns, R.L., *J. Crystal Growth* 85: 569-575 (1987).
92. Cima, M.J., and Rhine, W.E., *Adv. Cer. Mater.* 2: 329-336 (1987).
93. Kayser, M.H., Borglum, B., Antony, G., Shyu, S.G., and Buchanan, R.C., in: *High-Temperature Superconductors* (M. B. Brodsky, R.C. Dynes, K. Kitazawa, and H. L. Tuller ed.), *Mater. Res. Soc. Symp. Proc.*, Vol. 99, pp 159-164, Materials Research Society (1988).
94. Parmigiani, F., Chiarello, G., Ripamonti, N., Goretzki, H., and Roll, U., *Phys. Rev. B*, 36: 7148-7150 (1987).
95. McAndrew, T.P., Frase, K.G., and Shaw, R.R., in: *Proc. Am. Vacuum Soc. Symp.* (J.M.E. Harper, R.J. Colton, and L.C. Feldman ed.) (1987). *Am. Inst. Phys. Conf. Proc.*, Vol. 16 (1988).
96. Koinuma, H., Hashimoto, T., Kawasaki, M., and Fueki, K., *Jpn. J. Appl. Phys.* 26: L399-L401 (1987).
97. Kionuma, H., Hashimoto, T., Nakamura, T., Kishio, K., Kitazawa, K., and Fuiki, K., *Jpn. J. Appl. Phys.* 26: L761-L762 (1987).
98. Koinuma, H., Kawasaki, M., Hashimoto, T., Nagata, S., Kitazawa, K., Fueki, K., Masubuchi, K., and Kudo, M., *Jpn. J. Appl. Phys.* 26: L763-L765 (1987).
99. Wang, W.N., Lu, H.B., Lin, W.J., Yao, P.C., Hsu, H.E., Tai, M.F., and Ku, H.C., *Jpn. J. Appl. Phys.* 273: L1268-L1270 (1988).
100. Yoshiara, K., Kagata, K., Yokoyama, S., Hiroki, T., Higuma, H., Yamazaki, T., and Nakahigashi, K., *Jpn. J. Appl. Phys.* 27: L1492-L1494 (1988).
101. Dorris, S.E., Lanagan, M.T., Moffatt, D.M., Leu, H.J., Youngdahl, C.A., Balachandran, U., Cazzato, A., Bloomberg, D.E., and Goretta, K.C., *Jpn. J. Appl. Phys.* 28: L1415-L1416 (1989).
102. Ginley, D.S., Mitchell, M.A., Fu, W., Kwak, J.F., Venturini, E.L., and Baughman, R.J., in: *High-Temperature Superconductors* (M.B. Brodsky, R.C. Dynes, K. Kitazawa and H.L. Tuller ed.), *Mater. Res. Soc. Symp. Proc.*, Vol. 99, pp 673-676 (1988).
103. Kitagawa, T., Shibata, S., Okazaki, H., Kimura, T., and Enomoto, Y., *Jpn. J. Appl. Phys.* 27: L1113-L1115 (1988).
104. Bhattacharya, D., Maiti, C.K., Pramanik, P., Dey, T.K., Ghatak, S.K., Chopra, K.L., Kashyap, S.C., Pandya, D.K., and Gogoi, B., *Thin Solid Films* 164: 115-118 (1988).
105. Sacchi, M., Sirotti, F., Morten, B., and Prudenziati, M., *Appl. Phys.Let. 53* 1110-1112 (1988).
106. Singh, J.P., Balachandran, U., Shi, D., Degener, J.K., and Poeppel, R.B., *Mater. Lett.* 7: 72 (1988).
107. Chandler, G.T., this volume.

108. Thomas, J.H., III, and Labib, M.E. in *High T_c Superconductors, Proc. Am. Vacuum Soc. Symp.*, (J.M.E. Harper, R.J. Colton, and L.C. Feldman ed.) (1987). *Am. Inst. of Phys. Conf. Proc.*, Vol. 165, p 374 (1988).
109. Dou, S.X., Liu, H.K., Bourdillon, A.J., Tan, N.X., Zhou, J.P., Sorrell, C.C., and Easterling, K.E., *Mod. Phys. Lett. B* 1: 363 (1988).
110. McHenry, M.E., McKittrick, J., Sasayama, S., Kwapong, V., O'Handley, R.C., and Kalonji, G., *Phys. Rev. B* 37: 623-626 (1988).
111. Costa, G.A., Ferreti, M., Olcese, G.L., Calzona, V., Cimberle, M.R., Ferdeghini, C., Putti, M., and Siri, A.S., *Solid State Comm.* 68: 923-928 (1988)
112. Venturini, E.L., Ginley, D.S., Baughman, R.J., Morosin, B., and Kwak, J.F., in: *High-Temperature Superconductors* (M.B. Brodsky, R.C. Dynes, K. Kitazawa, and H.L. Tuller ed.), *Mater. Res. Soc. Symp. Proc.*, Vol. 99, pp 639-642 (1988).
113. Jin, S., Tiefel, T.H., Sherwood, R.C., van Dover, R.B., Davis, M.E., Kammlott, G.W., and Fastnacht, R.A., *Phys. Rev. B* 37: 7850-7853 (1988).
114. Hojaji, H., Michael, K.A., Barkatt, A., Thorpe, A.N., Ware, M.F., Talmy, I.G., Haught, D.A., and Alterescu, S., *J. Mater. Res.* 4: 28-32 (1989).
115. Salama, K., Selvamanickam, V., Gao, L., and Sun, K., *Appl. Phys. Lett.* 54: 2352-2357 (1989).
116. Crabtree, G.W., Kwok, W.K., Umezawa, A., Soderholm, L., Morss, L., and Alp, E.E., *Phys. Rev. B* 36: 5258 (1987).
117. Schneemeyer, L.F., Waszczak, J.V., Siegrist, T., van Dover, R.B., Rupp, L.W., Batlogg, B., Cava, R.J., and Murphy, D.W., *Nature* 328: 601 (1987).
118. Worthington, T.K., Gallagher, W.J., and Dinger, T.R., in: *Novel Superconductivity International Workshop* (S.A. Wolf, and V.Z. Kresin ed.), p 781, Plenum (1987).
119. Sahu, D., Langner, A., George, T.F., Weaver, J.H., Meyer, H.M., III, D.L. Nelson, and A. Wold, in: *Chemistry of High-Temperature Superconductors II* (D.L. Nelson, and T.F. George ed.), ACS Symp. Ser. 377, pp 1-15, American Chemical Society (1988).
120. Manning, M.A., and Barkatt, Aa., to be published.
121. Gao, Y., Wagener, T.J., Weaver, J.H., Flandermeyer, B., and Capone, D.W., II, *Appl. Phys. Lett.* 51: 1032 (1987).
122. Hill, D.M., Meyer III, H.M., Weaver, J.H., Flandermeyer, B., and Capone II, D.W., *Phys. Rev. B* 36: 3979 (1987).
123. Weaver, J.H., Gao, Y., Wagener, T.J., Flandermeyer B., and Capone, D.W., II, *Phys. Rev. B* 36: 3975 (1987).
124. Meyer III, H.M., Hill, D.M, Anderson, S.G., Weaver, J.H., and Capone, D.W., II, *Appl. Phys. Lett.* 51: 1750 (1987).
125. Meyer, H.M., III, Wagener, T.J., Hill, D.M., Gao, Y., Anderson, S.G., Krahn, S.D., Weaver, J.H., and Capone II, D.W., *App. Phys. Lett.* 51: 1118 (1987).

126. Chang, C.-A., *App. Phys. Lett.* 52: 924-926 (1988).
127. Chang, C.-A., *App. Phys. Lett.* 53: 1113-1115 (1988).
128. Chang, C.-A., in: *High-T_c Superconductors, Proc. Am. Vacuum Soc. Symp.* (J.M.E. Harper, R.J. Colton and L.C. Feldman ed.) (1987). *Amer. Inst. of Phys. Conf. Proc.*, Vol. 165 (1988).
129. Chang, C.-A., and Tsai, J.A., *App. Phys. Lett.* 53: 1976-1978 (1988).
130. Recently, higher T_c values have been obtained with C_{60} derivatives. In the case of Cs_2RbC_{60}, T_c = 33K according to Tanigaki, K., Ebbesen, T.W., Saito, S., Mizuki, J., Tsai, J.S., Kubo, Y., and Kuroshima, S., *Nature*, 352: 222-223 (1991).

20

Role of Processing in the Corrosion of Ceramic Superconductors

Gregory T. Chandler

Department of Materials Science and Engineering
University of Florida
Gainesville, Florida

INTRODUCTION

Since the discovery by Wu, Chu and co-workers (1,2) in 1987 of superconductivity in the Y-Ba-Cu-O system, most of the research activities in the area of processing of ceramic superconductors have been focused on the $YBa_2Cu_3O_{7-x}$ (123) compound because of the high temperature of its superconducting transition ($\approx$ 94 K). As in any ternary compound, the variables of temperature and composition are important in the synthesis of $YBa_2Cu_3O_{7-x}$. In addition, control of oxygen partial pressure and cooling rate is important in attaining the defect perovskite structure of the $YBa_2Cu_3O_{7-x}$ phase. The superconducting phase has also been found to be highly sensitive to water and water vapor (3-9) which not only contributes to the difficulty of the synthesis but also to the difficulty in handling the material after it has been produced.

Identifying and understanding the mechanisms of the reactions of the $YBa_2Cu_3O_{7-x}$ phase with water and water vapor are important to the future use of ceramic superconductors. This chapter examines the effect of processing conditions on the stability of the $YBa_2Cu_3O_{7-x}$ in a humid environment. Specifically, the effects of particle size, particle size distribution and sintering temperature are discussed.

REACTIVITY WITH WATER

Degradation Reactions

According to Yan et al. (3), the following reactions are expected between the superconducting phase and water vapor:

$$3H_2O + 2YBa_2Cu_3O_{7-x} \rightarrow Y_2BaCuO_5 + 3Ba(OH)_2 + 5CuO + 0.5O_2 \quad (20\text{-}1)$$

If CO_2 is present in the atmosphere $Ba(OH)_2$ reacts further to form $BaCO_3$:

$$Ba(OH)_2 + CO_2 \rightarrow BaCO_3 + H_2O \quad (20\text{-}2)$$

Superconducting samples increase in weight with the adsorption of water and the formation of reaction products. From the degradation reactions (1) and (2) the theoretical weight percent increase after complete reaction is 14.5%.

The effect of degradation on a $YBa_2Cu_3O_{7-x}$ sample is shown in Figure 20-1.

Figure 20-1. Effect of degradation on $YBa_2Cu_3O_{7-x}$.

Figures 20-2 and 20-3 show SEM micrographs of a $YBa_2Cu_3O_{7-x}$ fracture surface before and after exposure to humidity. As shown in Figure 20-3, after reaction with water, the morphology of the grains is severely altered with small crystallites forming on many of the grains. Energy dispersive x-ray spectroscopy (EDS) of an area that is covered by small crystallites (Figure 20-3a) shows a predominance of Ba compared to spectra of a sample that has not been exposed to humidity (Figure 20-2). X-ray diffraction (XRD) and transmission electron microscopy (TEM) have confirmed these crystallites to be $BaCO_3$. EDS analysis of an altered area that is not covered by the crystallites (Figure 20-3b) shows a predominance of Cu which indicates the possibility of CuO in this area.

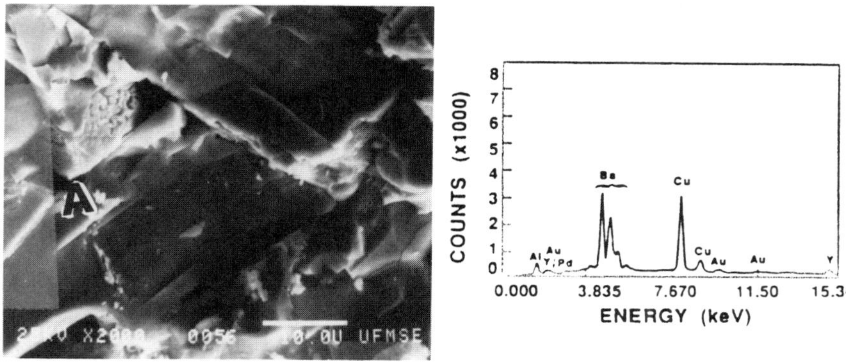

Figure 20-2. Micrograph and energy dispersive x-ray spectrum of a fracture surface of a $YBa_2Cu_3O_{7-x}$ sample before exposure to humidity. (Al, Au and Pd are detected from the sample holder and coating used in SEM).

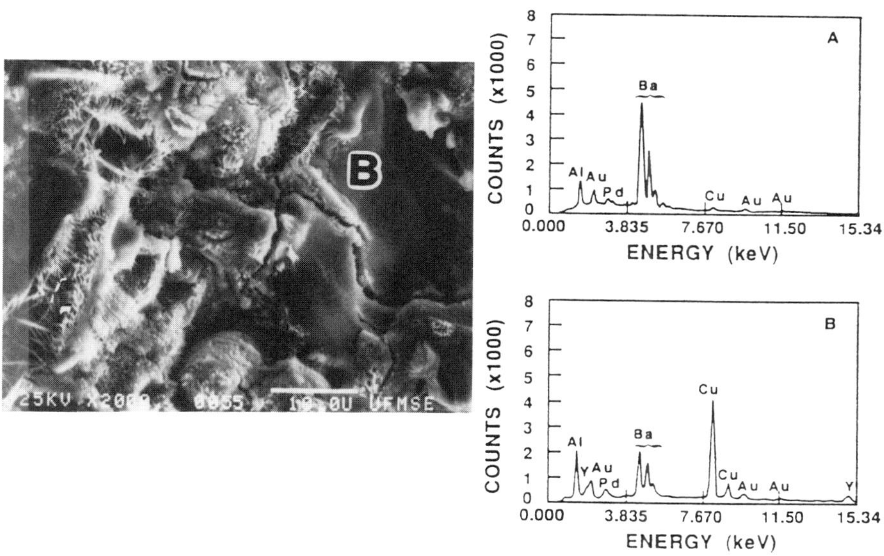

Figure 20-3. Micrograph and energy dispersive x-ray spectrum of a fracture surface of a $YBa_2Cu_3O_{7-x}$ sample after exposure to humidity: A) area where needle-shaped crystals are present; B) area where needle-shaped crystals are absent. (Al, Au and Pd are detected from sample holder and coating used in SEM)

Effect on Superconducting Properties

The sensitivity of the $YBa_2Cu_3O_{7-x}$ to water and water vapor is a major obstacle to the use of these superconductors. After exposing a superconducting

123 sample to air at 85°C and 85% relative humidity for 90 minutes, Yan showed that the x-ray diffraction pattern of the 123 phase essentially disappeared from the surface of the sample. Comparison of the x-ray diffraction peak intensity at a 2θ of 58.3° for successive layers to the original peak intensity showed that water vapor reacts readily to a considerable depth (≈ 320 μm) in a short time.

Yan also showed that the superconductivity of a 123 sample is greatly degraded after the sample has been exposed to boiling water for 15 minutes. While the onset of the transition still occured at the original transition temperature of 94 K, the water-treated sample exhibited a much broader transition and a much reduced susceptibility signal. Garland (4) has shown that the superconducting fractions of powder samples degrade exponentially in the presence of high-humidity atmosphere. For 98% relative humidity at 20°C the time constant is approximately 22 days. Barns and Laudise (5) found that after 70 minutes of exposure to 85°C, 85% relative humidity conditions, superconductivity was no longer present in thin film samples at 77 K.

Effects of Other Solvents

Dissolution behaviors of 123 powders in various solvents besides water have been studied to identify those that may be useful for processing. Trolier et al. (6) have shown that dissolution of 123 powders was quite low in benzene, toluene, acetone, ethanol, methanol, isopropanol, and dimethyl formamide. Table 20-1 shows the solvents studied and the phases present in the powder after

Table 20-1. Initial Water Content of Solvents and Predominant Phase Following $YBa_2Cu_3O_{7-x}$ Exposure to Solvent. Source: Reference 6.

Water Content Solvent	Predominant (wt%)	Y-Ba-Cu Phase
Acetone	0.32	$YBa_2Cu_3O_{7-x}$
Isopropanol	0.11	$YBa_2Cu_3O_{7-x}$
Ethyl alcohol	0.27	$YBa_2Cu_3O_{7-x}$
Benzene	0.021	$YBa_2Cu_3O_{7-x}$
Toluene	0.015	$YBa_2Cu_3O_{7-x}$
Methanol	0.16	$YBa_2Cu_3O_{7-x}$
Isoamyl alcohol	0.31	$BaCO_3$, CuO + additional phase(s)
Formamide	0.05	Multiple phases ?
N-Methyl formamide	0.16	$YBa_2Cu_3O_{7-x}$
N,N-Dimethyl formamide	0.060	$YBa_2Cu_3O_{7-x}$
Acetic acid	0.046	$C_4H_6CuO_4$+$YBa_2Cu_3O_{7-x}$

exposure. Frase et al. (7) have shown that the superconducting material appears to be completely unstable in the presence of strong acids and bases.

SAMPLE PREPARATION AND CHARACTERIZATION

Many different processing techniques for producing $YBa_2Cu_3O_{7-x}$ material have been investigated by researchers throughout the world since 1987. Because of the strong relationship between processing and environmental stability of 123 ceramic superconductors, the following section gives detailed information on the sample preparation and characterization used to assess the effects of particle size, particle size distribution and sintering temperature on the stability of 123 material.

Samples were prepared from the solid state reaction of powders of Y_2O_3 (<10 μm diam.), $BaCO_3$ (1 μm diam.) and CuO (<800 μm diam.) mixed in a Y:Ba:Cu molar ratio of 1:2:3. The powders were calcined in an Al_2O_3 crucible at 950°C in air for three hours, quenched in air and then ground in acetone with a porcelain mortar and pestle. To obtain a more homogeneous mixture, a second calcination step was performed at 950°C in air for six hours. The powders were then quenched again in air and ground with a mortar and pestle. The powders were screened through ASTM standard 200, 230, 270, 325, and 400 mesh brass sieves. The particle sizes that were used to prepare samples were <38 μm, 45-53 μm and 63-75 μm.

Samples were also prepared from commercially available calcined powders with reported average particle sizes of 1 μm, 2 μm and 20 μm, obtained from Superconductive Components, Inc[1]. The 1 μm powders were prepared from a precipitation of 99.999% minimum purity nitrates of Y, Ba and Cu. Powders with an average particle size of 20 μm for hand ground material and 2 μm for mechanically milled material were prepared from a dry mix and calcination of oxides of ACS grade Y and Cu and $BaCO_3$.

Calcined powders were pressed at 430 MPa into pellets each weighing approximately 0.1 g with an average height of 1.35 mm and average diameter of 4.85 mm. The pellets were placed in an Al_2O_3 combustion boat and sintered at 920°C, 950°C or 975°C for one hour in a tube furnace. They were cooled slowly to 580°C and annealed for three hours in flowing oxygen followed by slow cooling to room temperature in flowing oxygen. The sintering and annealing schedules are shown in Figure 20-4. Superconducting samples were characterized by Archimedes density measurements in acetone, SEM-EDS, x-ray diffraction analysis and magnetic susceptibility measurements.

Samples were exposed to air at 50°C and 85% relative humidity in a Blue M humidity chamber for up to 460 hours. Earlier studies indicated that the

---

[1]1145 Chesapeake Avenue, Columbus, OH 43212

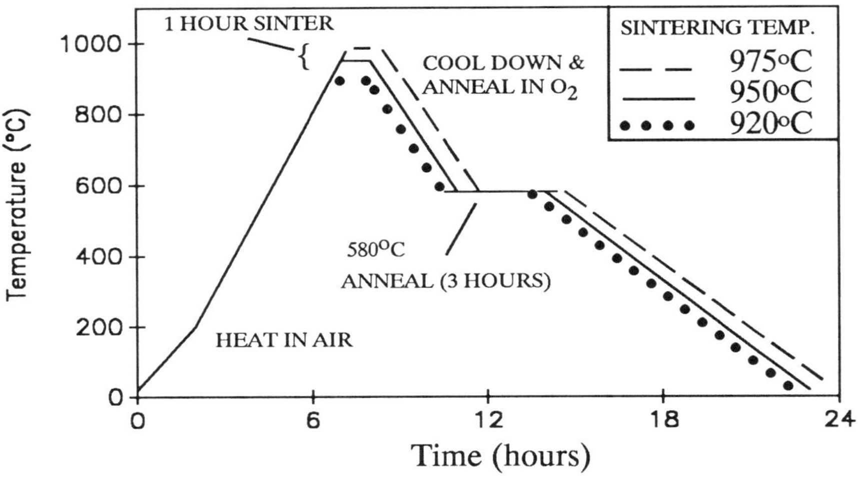

Figure 20-4. Sintering and annealing schedules for the 123 superconductors.

reaction products formed at the above conditions were also formed at ambient conditions and at 50°C and 95% relative humidity. However, the reaction rate was too slow at ambient conditions and too fast at the extreme conditions to obtain useful data within a reasonable time period. The reaction at 50°C and 85% relative humidity proceeded at a rate at which useful data could be obtained in a reasonable time period.

Weight measurements were made on six samples from each set of processing conditions for up to 460 hours of exposure to humidity. Weight measurements on the samples were discontinued when parts of the samples began to exfoliate. SEM was used to examine samples after 36 and 168 hours of exposure. X-ray analysis was performed on powders after 36 hours of exposure. Susceptibility measurements were performed on samples after 18, 36, 84, 132 and 460 hours of exposure.

EFFECT OF PARTICLE SIZE ON THE STABILITY OF $YBa_2Cu_3O_{7-x}$ IN A HUMID ENVIRONMENT

Density and Weight Changes

Samples with the highest bulk density (≈ 87% theoretical) in this particular study were produced from 1 μm and <38 μm powders as shown in Figure 20-5a-c. To better understand the reason why the 1 μm and <38 μm powders produced the higher bulk density samples, a measure of the particle size distributions of

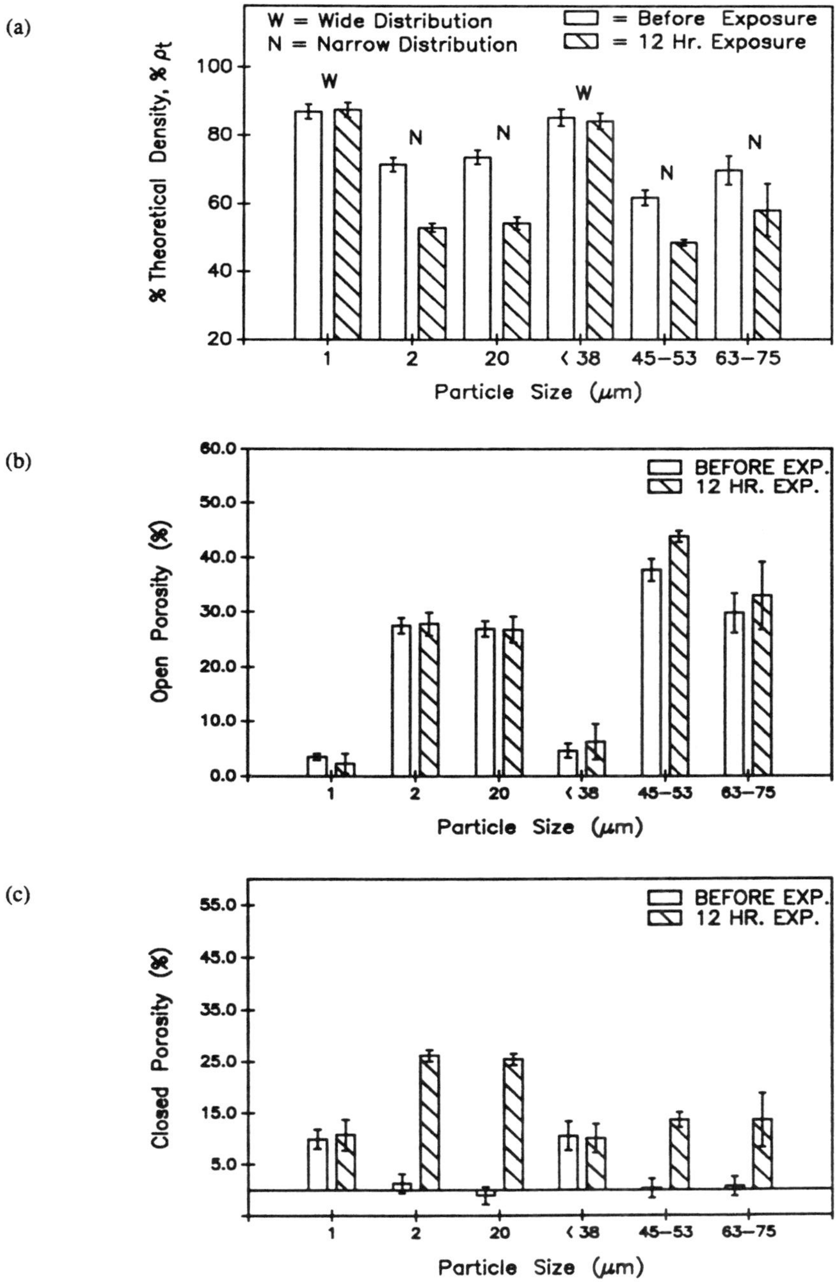

Figure 20-5. a) Bulk density of $YBa_2Cu_3O_{7-x}$ samples before and after exposure to air at 50°C and 85% relative humidity as a function of particle size particle size distribution. b) % open porosity; c) % closed porosity.

as-calcined powders was made as shown in Figure 20-6a-f. The 1 μm and <38 μm powders had wider size distributions compared to the other particle size ranges. A wider particle size distribution produces higher bulk density samples because there is an increased occupancy of void spaces by smaller particles. Because of denser packing due to a wider particle size distribution, enhanced sintering produces less open porosity (Figure 20-5b). The high bulk density samples, however, had larger amounts of closed porosity compared to the samples with narrow particle size distributions (Figure 20-5c). Because of enhanced sintering, there is an increase in the number of pores which are isolated and separated from one another.

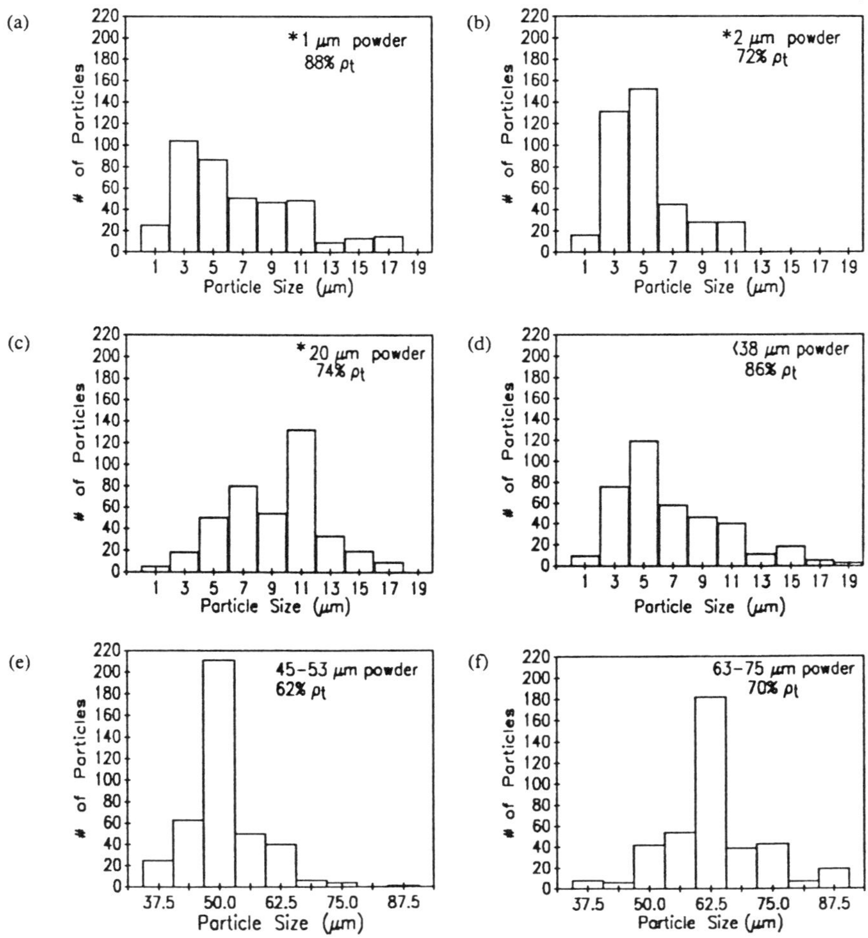

Figure 20-6. Particle size distributions of $YBa_2Cu_3O_{7-x}$ powders used in this study (* denotes average particle sizes reported by Superconductive Components, Inc.).

After 12 hours of exposure to humidity the bulk densities for the 63-75 μm, 45-53 μm, 20 μm and 2 μm samples decreased (Figure 20-5a), due primarily to the formation of $BaCO_3$ which has a density of 4.43 g/cc as compared to 6.38 g/cc for fully dense $YBA_2Cu_3O_{7-x}$. An increase in the percent closed porosity as shown in Figure 20-5c suggests that the reaction products that form fill the channels between particles and close off open spaces. This is plausible since the molar volume of the reaction products is greater than that of the reactants from which they are formed.

The <38 μm and 1 μm samples did not experience a noticeable change in bulk or apparent density after 12 hours of exposure due to minimal formation of reaction products. It appears that reaction with water initiates on the sample surface and progresses into the bulk of the sample when there is a transport of the atmospheric moisture into the superconductor. The 1 μm and <38 μm samples had much lower open porosity which greatly inhibited reaction between the superconducting phase and the water vapor.

The higher bulk density samples also contained a secondary nonsuperconducting Y_2BaCuO_5 phase (or 211 phase) which appeared after sintering. This may be due to liquid phase sintering at or above 950°C in a slightly Y poor region of the sample which may be the result of incomplete mixing (10). The presence of a liquid phase also strongly enhances densification. The samples with the 211 phase present had larger grain sizes, which is consistent with data obtained by Hwang, et al. (11). The effect of the formation of this secondary phase on the superconducting properties and the stability of the material in a humid environment will be discussed later.

Table 20-2 gives a summary of the XRD results from samples before and after exposure to humidity. After 36 hours of exposure to air at 50°C and 85% relative humidity the 2 μm, 20 μm, 45-53 μm and 63-75 μm samples contained the reaction products $BaCO_3$, and CuO as well as the 211 phase. There was no indication of the presence of the superconducting phase. The 1 μm and <38 μm samples, however, still showed presence of the superconducting phase with no indication of reaction products $BaCO_3$ and CuO after 36 hours of exposure.

Figure 20-7 shows the percent weight change versus exposure time to humidity for superconducting samples produced from different particle size ranges. The highest weight gains in early time periods (<50 hours) were in samples produced from 20 μm particle sizes. Samples produced from 63-75 μm, 45-53 μm and 2 μm particle sizes had similar weight increases compared to the 20 μm samples. Samples produced from <38 μm and 1 μm powders had the lowest weight gains in early time periods. The smaller weight gains in the 1 μm and <38 μm samples are due to the diminished formation of reaction products in early time periods.

Also shown in Figure 20-7 are the times at which the samples failed to levitate a samarium/cobalt magnet at liquid nitrogen temperature. The 63-75 μm, 45-53 μm, 20 μm and 2 μm samples failed to levitate the magnet after 12 hours of exposure which corresponded to approximately a 10% weight gain for the 63-

Table 20-2. Summary of X-ray Data for Particle Size Distribution Test

Particle size (µm)	Phases Present prior to exposure	Phases Present after 36 hours at 50°C, 85% RH
1	$YBa_2Cu_3O_7$ $BaCuO_2$ $YBaCuO_5$	$YBa_2Cu_3O_7$ $BaCuO_2$ $YBaCuO_5$
2	$YBa_2Cu_3O_7$	$YBaCuO_5$ $BaCO_3$ CuO
20	$YBa_2Cu_3O_7$	$YBaCuO_5$ $BaCO_3$

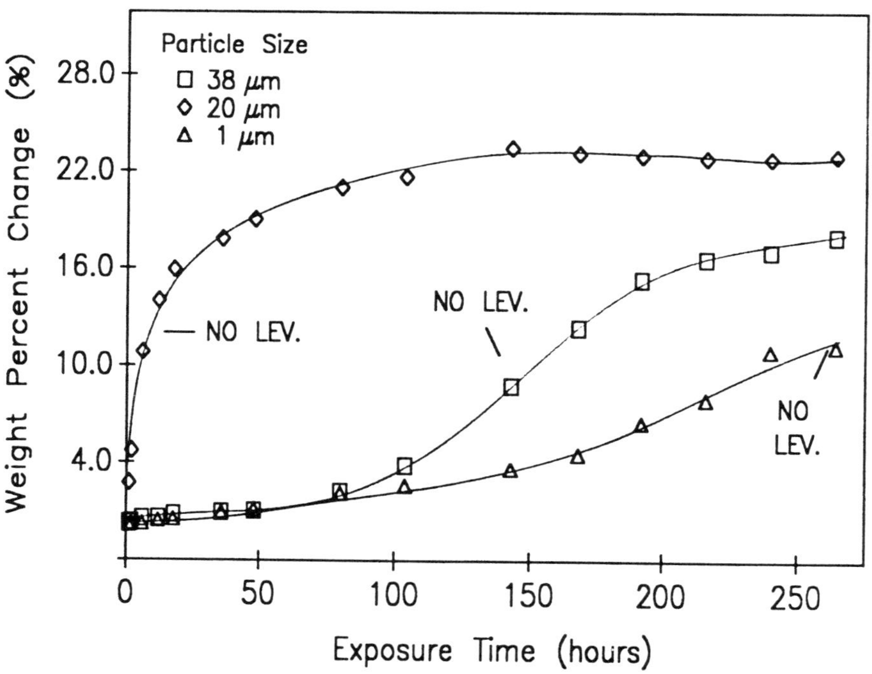

Figure 20-7. Weight percent change of $YBa_2Cu_3O_{7-x}$ samples versus exposure time to air at 50°C and 85% relative humidity as a function of particle size. No Lev. indicates the time when the samples fail to levitate a samarium/cobalt magnet at 77K.

75 μm, 20 μm and 2 μm samples and a 3% weight gain for the 45-53 μm samples. The <38 μm and 1 μm samples failed to levitate the magnet after 148 hours and 268 hours, respectively, which corresponded to a weight increase of approximately 10%. A more quantitative determination of the degradation of superconducting properties will be discussed later in the text.

Change in Microstructure

Figure 20-8 shows SEM micrographs of sample fracture surfaces before exposure and after 36 hours of exposure to humidity. The 1 μm and < 38 μm samples exhibited much larger grain sizes after sintering, suggesting that liquid phase sintering did occur in these samples. It is well established that the presence of a liquid phase during sintering strongly enhances densification of ceramic oxides. The larger grain size can be explained by the fact that the transport of atoms through the liquid is faster than through the solid.

After 36 hours of exposure to humidity the 2 μm, 20 μm, 45-53 μm and 63-75 μm samples showed extensive formation of reaction products. The morphology of the grains had been altered severely with small crystallites present on most of the grains. The 1 μm and <38 μm samples showed little if any reaction product formation after 36 hours of exposure to humidity.

Change in Superconducting Properties

A quantitative determination of the degradation of the superconducting properties is shown in Figure 20-9. Samples were analyzed using a Magnetic Properties Measurement System (MPMS) to determine the superconducting fraction. Samples were cooled to 5 K in zero field, heated to 120 K in a 100 Gauss field and finally cooled to 5 K in the field. The ratio of the susceptibility after cooling in the field to the susceptibility after cooling in zero field gives the superconducting fraction of the sample. Superconducting fractions of samples after exposure to humidity were determined by comparing the susceptibility of the exposed specimen after field cooling to the susceptibility of the unexposed specimen after zero-field cooling.

The high density samples (1 μm and <38 μm) exhibited little change in superconducting fraction after 36 and 84 hours of exposure; however, the low density samples showed a large decrease in superconducting fraction after 18 hours of exposure to humidity due to formation of reaction products. Before exposure to humidity, however, the samples that had high densities and contained the 211 phase had lower superconducting fractions compared to samples with lower densities and no 211 phase present.

An attempt was made to determine the effect of the 211 phase on the stability of the superconducting 123 phase. Small amounts (≈1 g) of single phase 211 powders and single phase 123 powders were placed in separate test tubes containing deionized water. After 24 hours the water was decanted and the

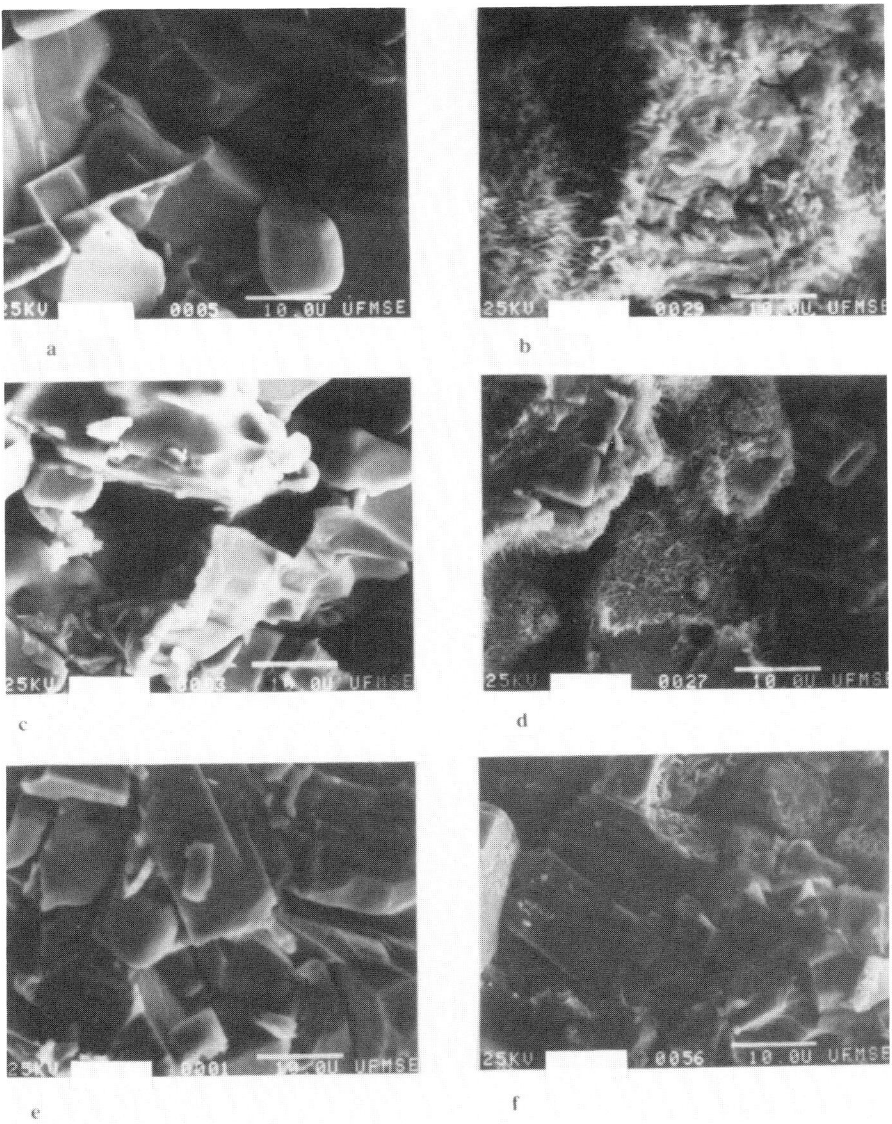

Figure 20-8. SEM micrographs of fracture surfaces of $YBa_2Cu_3O_{7-x}$ samples before and after exposure to humidity. a) 63-75 μm before exposure; b) 63-75 μm after exposure; c) 45-53 μm before exposure; d) 45-53 μm after exposure; e) <38 μm before exposure; f)<38 μm after exposure.

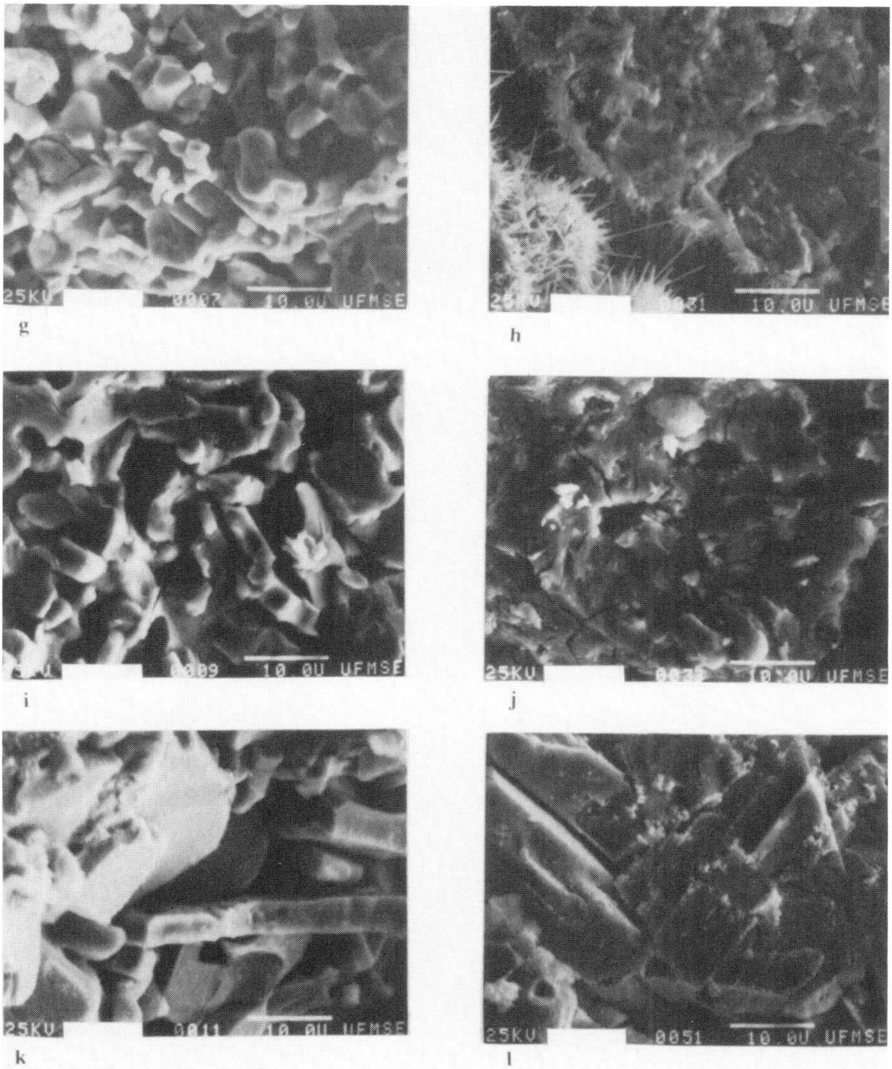

Figure 20-8 cont., g) 20 μm before exposure; h) 20 μm after exposure; i) 2 μm before exposure; j) 2 μm after exposure; k) 1 μm before exposure; l) 1 μm after exposure.

(a)

(b)

(c)

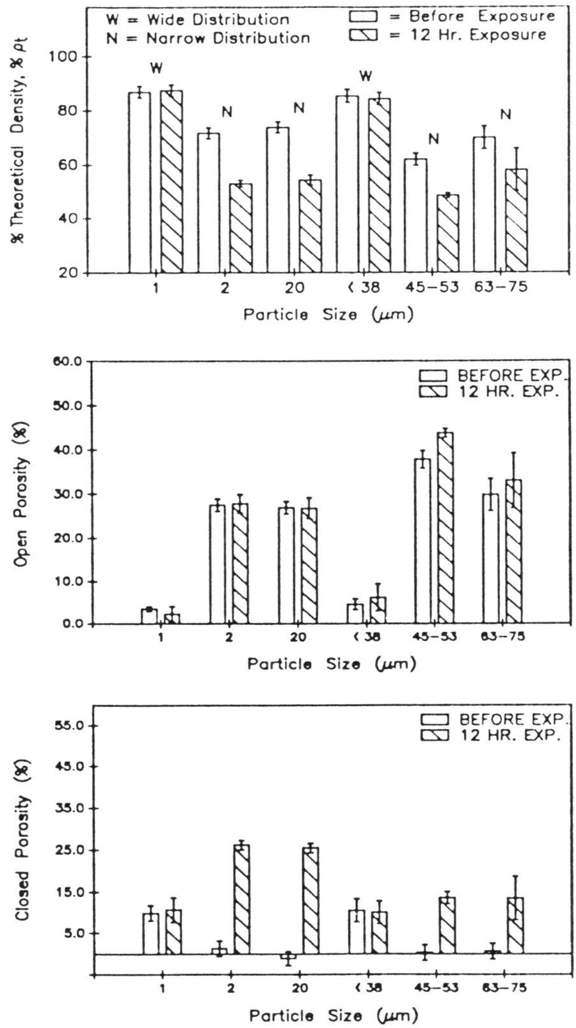

Figure 20-9. Superconducting fraction of $YBa_2Cu_3O_{7-x}$ samples versus exposure time to air at 50°C and 85% relative humidity as a function of particle size.

powders were dried at 110°C for one hour. The powders were then analyzed by x-ray diffraction. The 123 powders had decomposed into Y_2BaCuO_5, $BaCO_3$ and CuO, while the 211 powders contained no measurable reaction products. If the liquid phase coats the grains of the superconducting phase during sintering, then the 211 phase may act as a barrier to water reaction. However, the presence of the 211 phase leads to lower superconducting properties before exposure.

EFFECT OF SINTERING TEMPERATURE ON THE STABILITY OF $YBa_2Cu_3O_{7-x}$ IN A HUMID ENVIRONMENT

For samples with a wide particle size distribution, density increases as sintering temperature increases from 920°C to 950°C as shown in Figure 20-10. Little change in density, however, occured upon increasing sintering temperature from 950°C to 975°C, and small changes in density with increasing sintering temperature occured in samples with narrow particle size distributions.

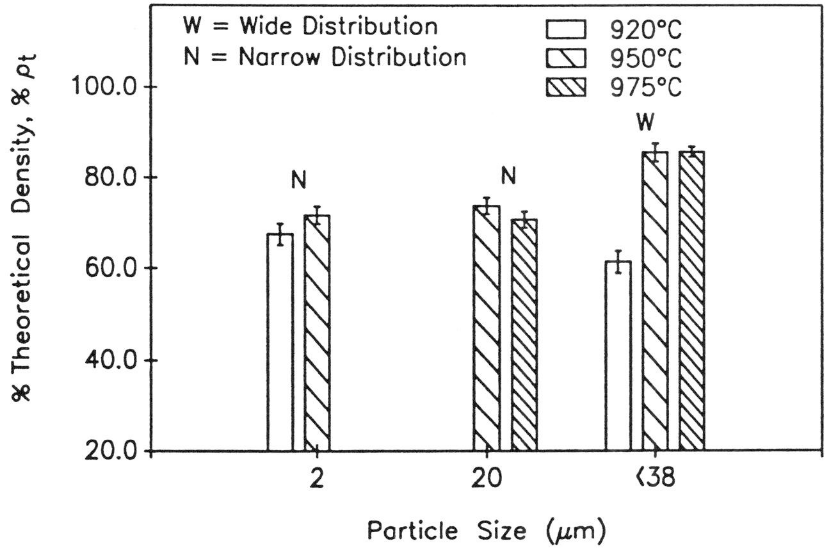

Figure 20-10. Percent theoretical density of $YBa_2Cu_3O_{7-x}$ samples as a function of particle size and sintering temperature.

Figure 20-11 shows the superconducting fraction of the <38 μm samples versus time exposed to humidity for three sintering temperatures. The <38 μm samples sintered at 950°C and 975°C showed less change in superconducting fraction after long exposure times than samples sintered at 920°C, while samples sintered at 975°C had a higher corrosion resistance due to slightly lower porosity and increased 211 phase formation.

PROPOSED MODEL FOR THE ROLE OF PROCESSING IN THE CORROSION OF CERAMIC SUPERCONDUCTORS

Figure 20-12 gives a generalized view of the effects of processing conditions on the superconducting fraction and corrosion resistance of $YBa_2Cu_3O_{7-x}$. Using

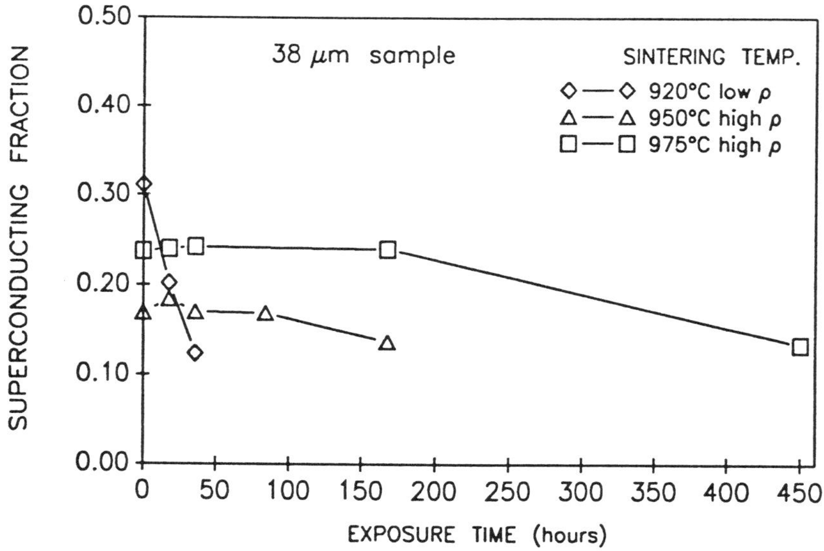

Figure 20-11. Superconducting fraction of $YBa_2Cu_3O_{7-x}$ samples versus exposure time to air at 50°C and 85% relative humdity as a function of sintering temperature.

a narrow particle size distribution to form samples or using a low sintering temperature (Point A) will result in samples with high porosity, small grain size and little or no 211 phase. Superconducting fractions are high but the corrosion resistance is quite low. The low corrosion resistance is due to the high porosity and small grain size.

Using a wide particle size distribution to form samples or a high sintering temperature (Point C) will result in samples with low porosity, large grain size and a high percentage of the 211 phase. Superconducting fractions are low because of the presence of the 211 phase; however, the 211 phase may coat the grains of the superconductor and inhibit their reaction with water. Samples that contain the same amount of the 211 phase as point C but have a smaller grain size and higher porosity (Point C') will have lower corrosion resistance compared to point C. Because of a larger surface area due to smaller grains, the 211 phase does not completely coat the superconducting grains as compared to point C.

Ideal processing conditions would result in a sample with maximum superconducting fraction and maximum corrosion resistance with minimum 211 phase formation (Point B).

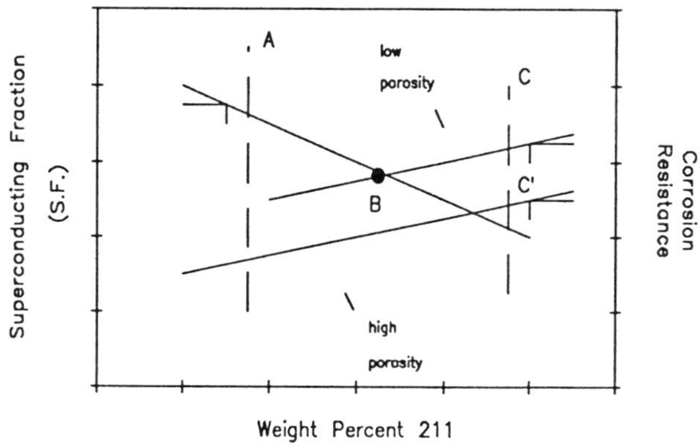

A B C

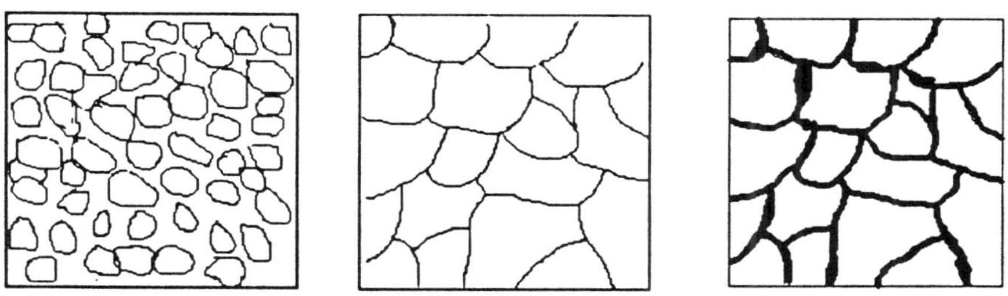

A

Processing conditions:
Narrow particle size distribution or low sintering T. resulting in:
High porosity
Small grain size
Little or no 211
High S.F.
Low corrosion resist.

B

Wide particle size distribution
Low porosity
Max. S.F. and Max. corrosion resistance with minimum 211

C

Wide particle size distribution
Low porosity
Large grain size
High % 211 possibly at grain boundary
Low S.F.
High corrosion resist.

C'

Narrow particle size distribution
High porosity
High % 211
Low S.F.
Lower corrosion resist. Compared to C

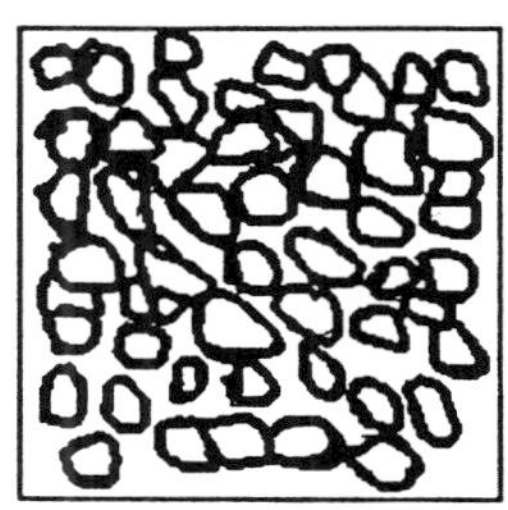

Figure 20-12. Proposed model for the effects of processing conditions on the environmental stability of $YBa_2Cu_3O_{7-x}$ ceramic superconductors.

ACKNOWLEDGEMENT

This work was supported by 3M Company and the Defense Advanced Research Projects Agency (DARPA).

REFERENCES

1. M.K. Wu, J.R. Ashburn, C.J. Torng, P.H. Hor, R.L. Meng, L. Gao, Z.J. Huang, Y.Q. Wang, and C.W. Chu, *Phys. Rev. Lett.*, 58:908 (1987).
2. C.W. Chu, P.H. Hor, R.L. Meng, L. Gao, Z.J. Huang, and Y.Q. Wang, *Phys. Rev. Lett.*, 58: 405 (1987).
3. M.F. Yan, R.L. Barns, H.M. O'Bryan, Jr., P.K. Gallagher, R.C. Sherwood, and S. Jin, *Appl. Phys. Letters*, 51: 532 (1987).
4. M.M. Garland, *J. Mat. Res.*, 3: 830 (1988).
5. R.L. Barns and R.A. Laudise, *Appl. Phys. Lett.* 51: 1373 (1987).
6. S.E. Trolier, S.D. Atkinson, P.A. Fuierer, J.H. Adair, and R.E. Newnham, *Cer. Bull.*, 67: 759 (1988).
7. K.G. Frase, E.G. Liniger, and D.R. Clarke, *Adv. Ceram. Matls.*, 32 [3B]: 698 (1987).
8. N.P. Bansal and A.L. Sandkuhl, *Appl. Phys. Lett.*, 52: 323 (1988).
9. A. Barkatt, H. Hojaji and K.A. Michael, *Adv. Ceram. Matls.*, 32 [3B]: 701 (1987).
10. G. Fisher, *Am. Cer. Soc. Bull.*, 67: 725 (1988).
11. N.M. Hwang, Y.K. Park, H.K. Lee, J.H. Hahn, G.W. Bahng, K.W. Lee, H.G. Moon, and J.C. Park, *J. Am. Ceram. Soc.*, 71: C210 (1988).

21

Corrosion of $YBa_2Cu_3O_{7-x}$ in High and Low Humidity Environments

Yasuro Ikuma
Department of Industrial Chemistry
Kanagawa Institute of Technology
Atsugi, Kanagawa 243-02, Japan

Masahiro Yoshimura
Research Laboratory of Engineering Materials
Tokyo Institute of Technology
Midori, Yokohama, Kanagawa 227, Japan

INTRODUCTION

Highly durable ceramics such as ZrO_2 and SiO_2 degrade when immersed in water. Some investigators (1,2) have recognized that water is sometimes dissolved into these oxides and that OH^- is diffused in the oxide. Furthermore, the oxidation of metal nitrides by the reaction of water produces a film of oxide on the surface. For further oxidation to take place, diffusion of reactant and/or products through the oxide layer becomes a rate determining step (3-6). Thus, the importance of diffusion in the oxide with respect to corrosion cannot be overemphasized.

One of the superconductive materials ($YBa_2Cu_3O_{7-x}$) can be degraded easily by water. Its sensitivity to water poses a serious problem in the general use of the material in the natural environment. Because the material has an alkali-earth element, a reaction of this component with water is the major cause of degradation. On the other hand, the dissolution of OH^- into the material matrix has a different effect on properties and chemical structure of the compound (7,8). This leads to the concept of treating high and low humidity exposure in different ways. If the humidity is high and the atmosphere is saturated, liquid water will condense on the surface of the oxide and will react with it. However, if the humidity is low, water will not condense on the surface. It will only be adsorbed onto the surface and diffused into the bulk of material.

This type of difference in the role of humidity may not have a significant consequence on SiO_2 and ZrO_2 because they do not decompose to form other materials when they are exposed to water. However, $YBa_2Cu_3O_{7-x}$ does decompose into different materials. In this sense we should emphasize the fact that the reaction of $YBa_2Cu_3O_{7-x}$ with a high humidity atmosphere should be distinguished from the reaction with a low humidity atmosphere. In this chapter we will describe mainly the difference in effect of these two types of humidity on $YBa_2Cu_3O_{7-x}$.

$YBa_2Cu_3O_{7-x}$ IN LIQUID WATER

Reaction of $YBa_2Cu_3O_{7-x}$ with liquid water has been reported by Yan et al. (9):

$$2\ YBa_2Cu_3O_{7-x} + 3H_2O \rightarrow 5CuO + 3Ba(OH)_2 + Y_2BaCuO_5 + (1/2)O_2 \qquad (21\text{-}1)$$

Many extensive studies of this problem have been performed by numerous investigators (7, 9-19). The products of reaction they have reported are summarized in Table 21-1. Although there are some differences in experimental

Table 21-1. Products of Reaction Between $YBa_2Cu_3O_{7-x}$ and Water Reported in the Literature. L: liquid water; Sa: saturated water vapor; V: water vapor. Hwang et al. (16) claimed that Y_2BaCuO_5 was detected by x-ray diffraction. However, the diffraction pattern corresponds to $Ba_2Cu(OH)_6$ and $Y(OH)_3$.

Literature (Ref.)	State of H_2O	Treatment Temp./°C	Phases identified				
			$Y(OH)_3$	$Ba(OH)_2$	CuO	Y_2BaCuO_5	$Ba_2Cu(OH)_6$
Yan (9)	L	RT-75		yes	yes	yes	
Nakada (10)	L	22-90		yes	yes		
Frase (11)	L	35		yes	yes		
Kitaguchi (12)	L	RT-550	yes	yes	yes		
Jin (13)	L	0-RT		yes			
Yoshimura (14)	L	RT-60		yes	yes		
"	L	75	yes	yes	yes		
"	L	>90	yes	yes	yes		yes
Hyde (7)	L, Sa	RT-40		yes	yes		yes
Dexin (15)	L, Sa	RT-60	yes	yes	yes		
Hwang (16)	L, Sa	100		yes	yes		yes*
Bansal (17)	L, V	23-55		yes	yes		
Shinozaki (18)	Sa, V	20-200	yes	yes	yes		
Sakai (19)	Sa, V	RT	yes	yes	yes		

conditions, it is a common observation in the table that $YBa_2Cu_3O_{7-x}$ decomposes when it is treated in liquid water. $Ba(OH)_2$ (or $BaCO_3$) has been commonly detected. The existence of CO_2 gas in the system determines whether or not $BaCO_3$ is found. $YBa_2Cu_3O_{7-x}$ will not maintain the original crystal structure if Ba ions are removed due to corrosion. Y or Cu will eventually be converted into yttrium or copper oxides or hydroxides. Although CuO is a common product, some researchers do not observe it, probably because of slow crystallization of CuO near room temperature. $Y(OH)_3$ is less common in the list. The slower crystallization of these compounds, particularly of $Y(OH)_3$, makes their detection difficult with standard wide angle x-ray diffraction. Yoshimura et al (14) observed the crystallization of $Y(OH)_3$ above 75°C, but not below 60°C. They found that the compound $Ba_2Cu(OH)_6$ was produced above 90°C. Therefore, this compound seems to be synthesized from $Ba(OH)_2$ and CuO in hot water. On the other hand, the formation of Y_2BaCuO_3 during decomposition of $YBa_2Cu_3O_{7-x}$ appears to be difficult, because the rearrangement of the lattice from $YBa_2Cu_3O_{7-x}$ to Y_2BaCuO_5 whose crystal structures are not related, should be almost impossible at ambient temperatures. The formation of Y_2BaCuO_5 has not been confirmed after Yan et al. (9). Y_2BaCuO_5 did not decompose at temperature of 75-100°C, but decomposed into $Y(OH)_3$, CuO and $Ba(OH)_2$ at 300°C under pressurized (100MPa) water. We can conclude, therefore, that the reaction of $YBa_2Cu_3O_{7-x}$ with water vapor is expressed by:

$$YBa_2Cu_3O_{7-x} + (7/2)H_2O \rightarrow Y(OH)_3 + 2Ba(OH)_2 + 3CuO + (1/4)O_2 \qquad (21\text{-}2)$$

If CO_2 is present in the atmosphere, the reaction

$$Ba(OH)_2 + CO_2 \rightarrow BaCO_3 + H_2O \qquad (21\text{-}3)$$

will proceed. This latter reaction has been verified by Bansal et al. (17) who performed experiments with and without the presence of CO_2 gas.

The problem with the reaction of either liquid or gaseous water is a simultaneous change in the physical properties of the material. After reaction with water, Dexin et al. (15) have reported that the Tc_o of the material changes to a lower temperature. In some cases, reduction of magnetic susceptibility (9,21) was observed.

REACTION OF $YBa_2Cu_3O_{7-x}$ IN SATURATED OR HIGH HUMIDITY

As Horowitz et al. (22) and Ikuma et al. (23) have shown, saturated or high humidity has a different effect on $YBa_2Cu_3O_{7-x}$ degradation compared to low humidity. In high humidity $YBa_2Cu_3O_{7-x}$ sintered powders are degraded so much that sample integrity is compromised (22). This corresponds to a very large increase in weight of $YBa_2Cu_3O_{7-x}$ observed by Ikuma et al. (23). Low humidity, however, does not affect the properties of the powder (22) even after exposure to a humid environment for over 600 hours. The weight increase of $YBa_2Cu_3O_{7-x}$ powder under similar conditions (23) was several orders of magnitude smaller than the weight increase at high humidity (Figure 21-1). Sakai et al. (19) analyzed the product of reaction between $EuBa_2Cu_3O_{7-x}$ and H_2O vapor at room temperature. $BaCO_3$, CuO and $Eu(OH)_3$ were detected at relative humidity higher than 60%. However, at lower humidity, the behavior was different.

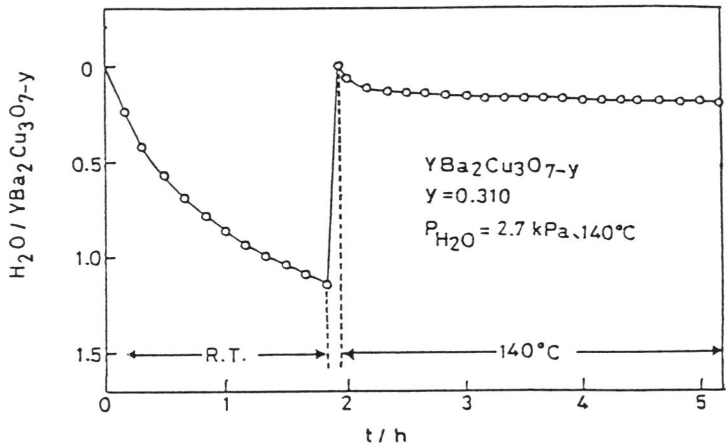

a)

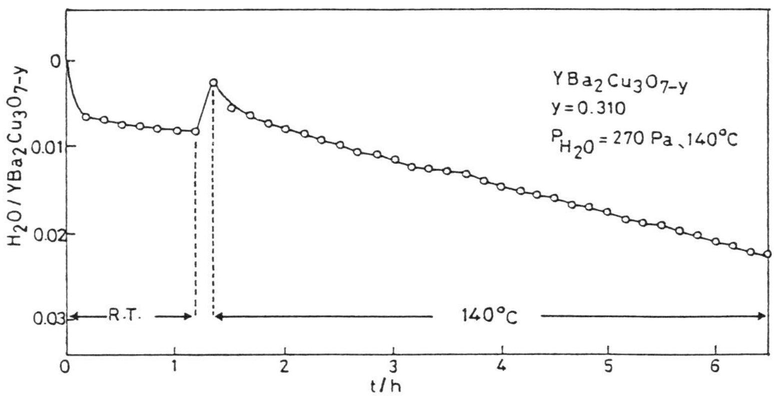

b)

Figure 21-1. The weight gain (Δw_t) as a function of time exposed to water vapor (23): a) at p_{H2O} = 2.7kPa and; b) p_{H2O} = 270Pa. Specimens were pretreated in oxygen at 590°C.

Some products composed of Ba/Cu = 1/1 were seen at the surface. It is natural, therefore, to distinguish between the effects of low humidity and high humidity. The difference is due not only to differences in reaction rates but also to differences in corrosion mechanisms.

It would be the first impression that, the exposure of $YBa_2Cu_3O_{7-x}$ to a high humidity atmosphere is different from submersion in water; submersion in water is more severe than the exposure to moisture. However, the products of reaction are the same as those shown in Table 21-1. The only difference is the time that is needed to get the same amount of reaction products. The reaction at high humidity takes longer to reach the same level of decomposition as the reaction in liquid water.

The same effect of water and high humidity on $YBa_2Cu_3O_{7-x}$ is understood easily if we realize that water will condense on the surface of the oxide when it is exposed to high humidity (24). Sometimes saturation is not required for the condensation of water to take place. For example, vapor pressure of water on a curved surface is different from the value on a flat surface. On the surface of a concave hole, the vapor pressure is lower. Thus, the condensation of water vapor is favored. In high humidity environments, a thin film of liquid water forms on the sample, and the reaction of the samples is similar to what would occur if the sample were immersed. This explains in part why the degradation of $YBa_2Cu_3O_{7-x}$ at high humidity is similar to the degradation in liquid water.

There is another factor that should be taken into account. Figure 21-2 shows the vapor pressure of pure water and the vapor pressure of $Ba(OH)_2$ aqueous

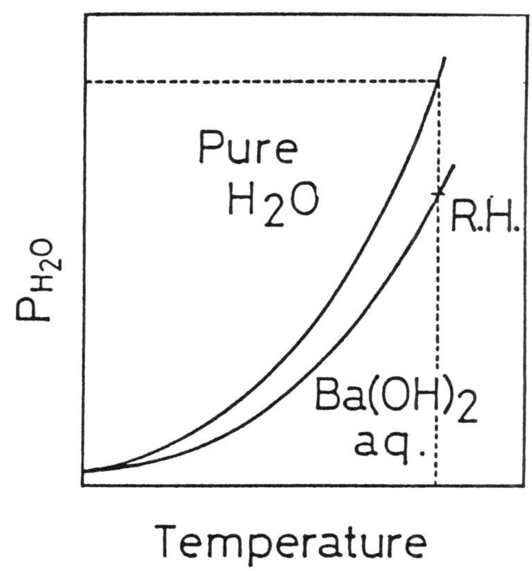

Figure 21-2. Schematic diagram of saturated water vapor pressure of pure H_2O and $Ba(OH)_2$ aqueous solution (25). R.H.: relative humidity.

solution (25). The vapor pressure is lower over $Ba(OH)_2$ aqueous solution than pure water. Consequently, the water can be condensed at the surface of $YBa_2Cu_3O_{7-x}$ from relative humidity lower than saturated, once $Ba(OH)_2$ has formed. As there is no question about the formation of $Ba(OH)_2$ when $YBa_2Cu_3O_{7-x}$ is exposed to water, we would anticipate the rapid condensation of water on $YBa_2Cu_3O_{7-x}$ surface even if the humidity is not saturated. Once the water condenses on the surface, the reaction with $YBa_2Cu_3O_{7-x}$ would be the same as the reaction of $YBa_2Cu_3O_{7-x}$ with liquid water.

REACTION OF $YBa_2Cu_3O_{7-x}$ IN LOW HUMIDITY

In the low humidity regime, the reaction between $YBa_2Cu_3O_{7-x}$ and water is different from the previous two cases. In particular, the diffusion of OH^- becomes important.

Thompson et al. (26) have early proposed a hydration reaction of $YBa_2Cu_3O_{7-x}$:

$$V_O^{\cdot\cdot} + O_O^x + H_2O \rightarrow 2OH_O^{\cdot} \qquad (21\text{-}4)$$

where $V_O^{\cdot\cdot}$ = oxygen vacancy with +2 charge
O_O^x = oxygen in a regular lattice site
$OH_O^{\cdot}$ = hydroxyl in an oxygen site with a single positive charge hole.

Qiu et al. (27) and Kurtz et al. (28) have spectroscopically studied $YBa_2Cu_3O_{7-x}$ when H_2O is present in the system and have observed the existence of OH^- species. Nishihara et al. (29) have studied $YBa_2Cu_3O_{7-x}$ by proton NMR and their results support reaction 21-4. They have suggested that the reaction in the early stage does not destroy the superconductivity of $YBa_2Cu_3O_{7-x}$.

All of these studies lead us to the conclusion that, when $YBa_2Cu_3O_{7-x}$ is exposed to low humidity, OH^- is adsorbed at the surface. At low temperature ($\approx$100°C) and low vapor pressure, the absorption of water is diffusion-controlled as shown in Figure 21-3.

The change in weight of specimen (Δw_t) was monitored during the hydration (23). Δw_t was plotted as a function of the square root of time. The linearity of the curve seen in the figure indicates that the process is controlled by diffusion. When the reaction was examined by measuring the H_2O pressure using a manometer (20), the plot of pressure versus square root of time is linear, again supporting the existence of a diffusion-controlled process. Taking reaction 21-4 into account, it can be concluded that OH^- is the diffusing species.

OH^- will probably occupy a site on the (001) plane. Assuming a spherical geometry of the specimen, we can use equation 21-5 (30) to calculate the diffusion coefficient of OH^- (D_{OH}):

$$\Delta w_t/\Delta w_\infty = 6(D_{OH}t/(\pi a^2))^{0.5} \qquad (21\text{-}5)$$

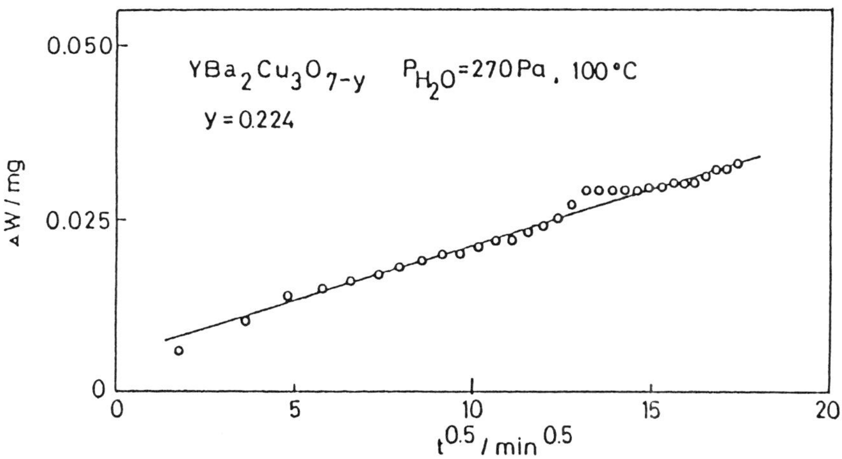

Figure 21-3. The weight gain at 100°C in p_{H2O} = 270Pa as a function of square root of time (23).

where Δw_t is the weight gain at time t, Δw_∞ is the corresponding quantity at infinite time, and a is the radius of the specimen. This equation results from the application of Fick's second law for small times to a problem of diffusion in a sphere when surface concentration is constant. The diffusion coefficients of OH^- (23) are shown in Figure 21-4. Also shown in the figure are the oxygen tracer diffusion coefficients (D_{oxy}) in the system (31-35), the chemical diffusion (36), the diffusion estimated from the conductivity measurement (37), and so on (38--43). Most of the tracer diffusion coefficients in the figure are in the direction perpendicular to the c axis. The diffusion along the c axis would be smaller. The tracer diffusion coefficient measured by one of the present authors (32) is low due to the fact that the average value was determined ignoring the anisotropy of the diffusion coefficient. The value of D_{OH} is also averaged over all the directions. Therefore, it should be compared with those of similar situations; i.e. Ref. (32). D_{OH} is slightly larger than the diffusion coefficient of oxygen. This is expected due to the smaller charge of OH^- compared to O^{2-}. Taking into consideration the uncertainty during the calculation (some assumptions are made) and the necessary extrapolation, the results are in fair accord with the prediction. D_{OH} is rather close to D_{oxy} in the direction perpendicular to the c axis.

Changes in x-ray diffraction spectra of $YBa_2Cu_3O_{7-x}$ during the reaction in low humidity were studied (20) at various values of x. At x = 0.1 and 1.0, H_2O was adsorbed slowly. However the x-ray diffraction pattern did not change much. When $YBa_2Cu_3O_{6.43}$ (x = 0.57) was annealed at various temperatures

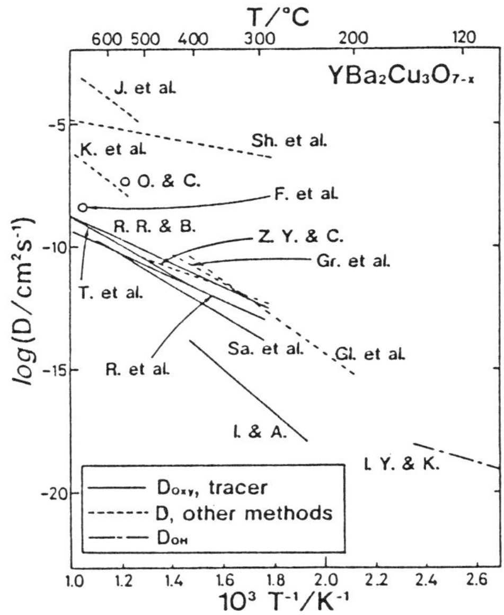

Figure 21-4. Diffusion coefficients in $YBa_2Cu_3O_{7-x}$. D_{OH}: (23); D_{oxy}: (31-35); D:(36-43).

(170-250°C) in p_{H2O} = 2.1KPa, which is far from saturated at these temperatures, H_2O is adsorbed rapidly at a level as high as y = 0.082 - 0.087 ($YBa_2Cu_3O_{7-x}{\cdot}yH_2O$). A similar phenomenon was also observed (20) in $YBa_2Cu_3O_{6.23}$. Figure 21-5 shows that a large amount of H_2O is incorporated into $YBa_2Cu_3O_{6.23}$. In spite of this large concentration of water, x-ray diffraction revealed that the

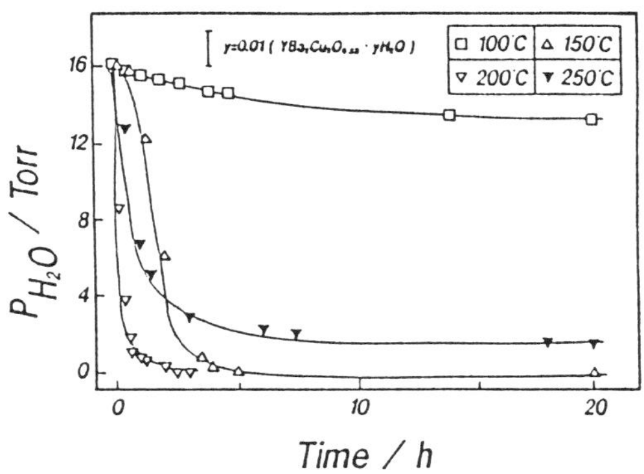

Figure 21-5. Water vapor absorption curve for $YBa_2Cu_3O_{6.23}$ at various temperatures (20).

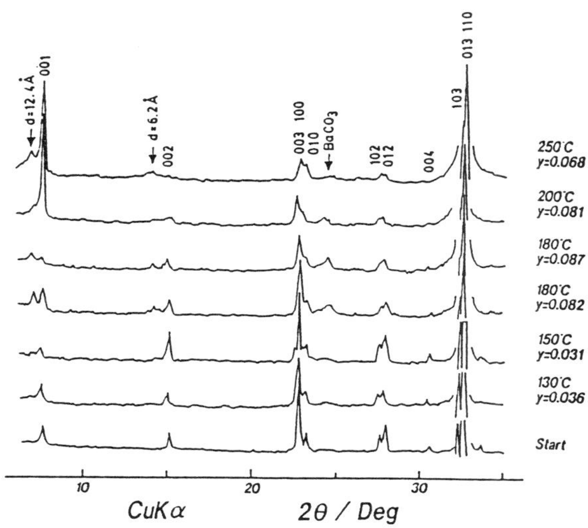

a)

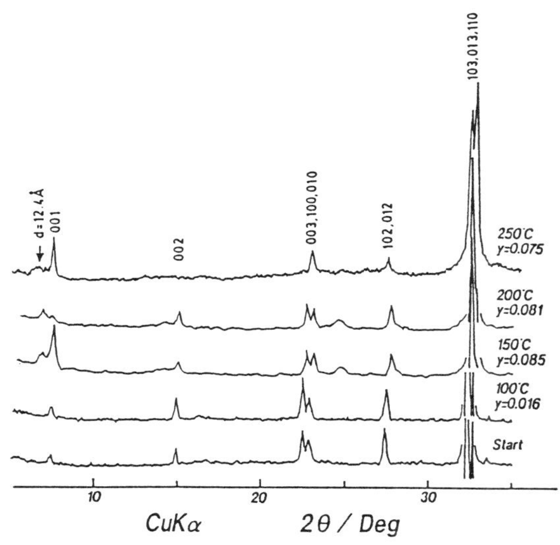

b)

Figure 21-6. X-ray diffraction patterns of (a) $YBa_2Cu_3O_{6.43}$ and (b) $YBa_2Cu_3O_{6.23}$ treated in water vapor. (2.1kPa) (20). "y" is the value in $YBa_2Cu_3O_{7-x} \cdot yH_2O$. These are in the low humidity regime.

crystal structure is not changed (Figure 21-6). X-ray diffraction from the (001) plane became strong and a satellite peak was observed (Figure 21-6). Obviously, the hydration process introduces some kind of planar defect in the lattice. The presence of planar defects was also recognized by Hyde et al (7) and Thompson et al. (26). After aging 60 days at room temperature (Figure 21-7), the peak from the (001) plane of the specimen with y = 0.081 became very weak, while peaks from (013) and (110) remained nearly unchanged. The satellite peak also

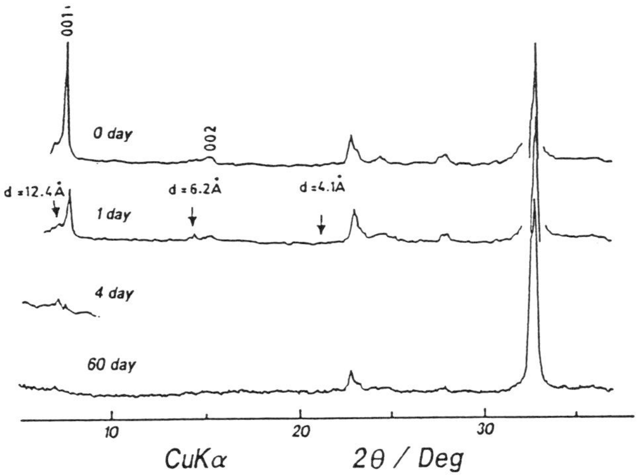

Figure 21-7. Change in x-ray diffraction patterns of $YBa_2Cu_3O_{6.43} \cdot 0.081H_2O$ by leaving in atmospheric air (20).

remained. Further investigation is needed in order to understand how OH^- (or H_2O) is incorporated into the crystal in such a high quantity. These results suggest that diffusion of OH^- could be important in the hydration (or reaction) of $YBa_2Cu_3O_{7-x}$ with water at low pressure.

Summarizing these results, we obtain an adsorption diagram (20) shown in Figure 21-8. At room temperature, water is condensed at the surface because of the saturated humidity (p_{H2O} = 2.1kPa). At temperatures between 100 and 150°C or 100 and 200°C, adsorption is observed, which is controlled by diffusion. The range of temperatures where slow adsorption is observed varies depending on the value of x. At higher temperatures there is a region of fast adsorption. In this case, the mechanism of adsorption is not yet understood. However, the x-ray diffraction revealed some interesting features.

There are several reports (44-48) which claim that large amounts of H_2 gas have been adsorbed by $YBa_2Cu_3O_{7-x}$. We could not repeat these experiments successfully. It is possible there is a crucial factor to their observation that we

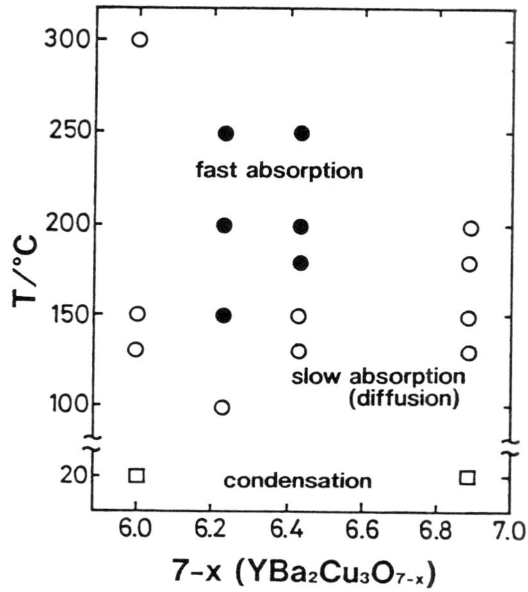

Figure 21-8. Absorption map of interaction between $YBa_2Cu_3O_{7-x}$ and H_2O (20).

have overlooked. If water is formed by a reaction with H_2 as shown by Gallagher et al. (49), adsorption of H_2 by $YBa_2Cu_3O_{7-x}$ is actually a reaction with H_2 to form H_2O. Once water is formed on the surface or within the bulk, the processes discussed in this chapter may occur.

SUMMARY

Reaction of $YBa_2Cu_3O_{7-x}$ with H_2O was categorized into three groups:

1. Reaction with liquid water,
2. Reaction with high or saturated humidity, and
3. Reaction with low humidity.

In the first two groups, the reaction results in the formation of $Ba(OH)_2$ (or $BaCO_3$), $Y(OH)_3$. This is a degradation which involves the decomposition of $YBa_2Cu_3O_{7-x}$. In the third group, H_2O was adsorbed and OH^- diffused into the bulk of the specimen. The diffusion coefficient of OH^- was calculated and found to be the same order of magnitude as the diffusion coefficient of oxygen in the system.

REFERENCES

1. Yoshimura, M., Noma, T., Kawabata, K., and Somiya, S., *J. Mat. Sci. Lett.* 6: 465-67 (1987).
2. Acocella, J., Tomozawa, M., and Watson, E.B., *J. Non-Cryst. Solids* 65: 355-72 (1984).
3. Yoshimura, M., Kase, J., and Somiya, S., *J. Mat. Res.* 1: 100-3 (1986).
4. Sato, T., Haryu, K., Endo, T., and Shimada, M., *J. Mat. Sci.* 22: 2277-80 (1987).
5. Sato, T., Haryu, K., Endo, T., and Shimada, M., *J. Mat. Sci.* 22: 2635-40 (1987).
6. Yoshimura, M., Kase, J., and Somiya, S., *Yogyo-Kyokai-Shi* 94: 129-34 (1986).
7. Hyde, B.G., Thompson, J.G., Withers, R.L., Fitzgerald, J.G., Stewart, A.M., Bevan, D.J.M., Anderson, J.S., Bitmead, J., and Paterson, M.S., *Nature* 327: 402-03 (1987).
8. Kitazawa, K., Kishio, K., Hasegawa, T., Nakamura, O., Shimoyama, J., Sugii, N., Ohtomo, A., Yaegashi, S., and Fueki, K., *Japan. J. Appl. Phys.* 26: L1979-81 (1987).
9. Yan, M.F., Barns, R.L., O'Bryan, H.M., Jr., Gallagher, P.K., Sherwood, R.C., and Jin, S., *Appl. Phys. Lett.* 51: 532-4 (1987).
10. Nakada, I., Sato, S., Oda, Y., and Kohara, T., *Japan. J. Appl. Phys.* 26: L697-8 (1987).
11. Frase, K.G., Liniger, E.G., and Clarke, D.R., *Adv. Cer. Mat.* 2(3B): 698-700 (1987).
12. Kitaguchi, H., Takada, J., Osaka, A., Miura, Y., Yamamoto, N., Oka, Y., Kiyama, M., Unesaki, T., and Tomii, Y., *Funtai Oyobi Funmatsu Yakin* 34: 651-8 (1987).
13. Jin, S.G., Liu, L.G., Zhu, Z.Z., and Huang, Y.L., *Solid State Commun.* 69: 179-82 (1989).
14. Yoshimura, M., Inoue, S., Ogasawara, N., Nakamura, T., Takagi, Y., Liang, L.S., Somiya, S., *Funtai Oyobi Funmatsu Yakin*, 34:659-62 (1987).
15. Dexin, Z., Mingshan, X., Ziqing, Z., Shubin, Y., Huansui, Z., and Shuxia, S., *Solid State Commun.* 65: 339-41 (1988).
16. Hwang, N.M., Bahng, G.W., Park, Y.K., Park, J.C., Moon, H.G., and Yoon, D.N., *J. Mat. Sci. Lett.* 8: 517-19 (1989).
17. Bansal, N., and Sandkuhl, A.L., *Appl. Phys. Lett.* 52: 323-5 (1988).
18. Shinozaki, K., Kurosawa, T., Nanjyo, A., Mizutani, N., and Kato, M., *Thermochimica Acta*, 136: 297-306 (1988).
19. Sakai, K., Suenaga, T., and Sata, T., *J. Cer. Soc. Japan* 97: 677-82 (1989).
20. Yoshimura, M., Inoue, S., Ikuma, Y., Takase. K., and Ishizawa, N., submitted to *Japan. J. Appl. Phys.*
21. Ono, M., Koshimura, M., Abe, M., Matsushima, H., Kurasawa, K., Wada, H., and Takahashi, M., *J. Magnet. Soc. Japan*, 12: 431-6 (1988).

22. Horowitz, H.S., Bordia, R.K., Flippen, R.B., Johnson, R.E., and Chowdhry, U., *Mat. Res. Bull.* 23: 821-30 (1988).
23. Ikuma, Y., Yoshimura, M., and Kabe, S., *J. Mat. Res.* 5: 17-21 (1990).
24. Chang, C.A., *Appl. Phys. Lett.* 53: 1113-5 (1988).
25. Yoshimura, M., Ishikawa, Y., Inoue, S., and Somiya, S., *Proc. MRS on Advanced Materials* 6: 421-26 (1989).
26. Thompson, J.G., Hyde, B.G., Withers, R.L., Anderson, J.S., FitzGerald, J.D., Bitmead, J., Paterson, M.S., and Stewart, A.M., *Mat. Res. Bull.* 22: 1715-24 (1987).
27. Qiu, S.L., Ruckman, M.W., Brookes, N.B., Johnson, P.D., Chen, J., Lin, C.L., Strongin, M., Sinkovic, B., Crow, J.E., and Jee, C.S., *Phys. Rev. B* 37: 3747-50 (1988).
28. Kurtz, R.L., Stokbauer, R., Madey, T.E., Mueller, D., Shih, A., and Toth, L., *Phys. Rev. B* 37: 7936-39 (1988).
29. Nishihara, H., Nishida, N., Takabatake, T., Kishio, K., Ohtomo, A., Hayashi, K., Ishikawa, M., Nakazawa, Y., Koga, K., Tamegai, T., and Kitazawa, K., *Japan. J. Appl. Phys.* 27: 1652-57 (1988).
30. Crank, J., *The Mathematics of Diffusion*, 1970.
31. Routbort, J.L., Rothman, S.J., Nowicki, L.J., and Goretta, K.C., *Mat. Sci. Forum* 34-36: 315-21 (1988).
32. Ikuma, Y., and Akiyoshi, S., *J. Appl. Phys.* 64: 3915-17 (1988).
33. Turrillas, X., Kilner, J.A., Kontoulis, I., and Steele, B.C., *J. Less-Common Metals* 151: 229-36 (1989).
34. Rothman, S.J., Routbort, J.L., and Baker, J.E., *Phys. Rev. B*, 40: 8852-60 (1989).
35. Sabras, J., Dolin, C., Ayache, J., Monty, C., Maury, R., and Fert, A., *Colloq. Phys.*, 51: C1-1035-42 (1990).
36. Kishio, K., Suzuki, K., Hasegawa, T., Yamamoto, T., and Kitazawa, K., *J. Solid State Chem.*, 82: 192-202 (1989).
37. Grader, G.S., Gallagher, P.K., Thomson, J., and Gurvitch, M., *Appl. Phys. A*, 45: 179-83 (1988).
38. Zhang, J., Yang, M., and Chen, T., *Mat. Lett.* 6: 379-84 (1988).
39. O'Sullivan, E.J.M., and Chang, B.P., *Appl. Phys. Lett.* 52: 1441-43 (1988).
40. Fiory, A.T., Martin, S., Schneemeyer, L.F., Fleming, R.M., White, A.E., and Waszczak, J.V., *Phys. Rev. B*, 38: 7129-32 (1988).
41. Glowacki, B.A., Highmore, R.J., Peters, K.F., Greer, A.L., and Evetts, J.E., *Supercond. Sci. Tech.* 1: 7-11 (1988).
42. Shi, D., Krucpzak, J., Tang, M., Chen, N., and Bhadra, R., *J. Appl. Phys.*, 66: 4325-28 (1989).
43. Jantsch, S., Ihringer, J., Maichle, J.K., Prandl, W., Kemmler-Sack, S., Kiemel, R., Losch, S., Schafer, W., Schlichenmaier, M., and Hewat, A.W., *J. Less-Common Metals* 150: 167-75 (1989).
44. Reilly, J.J., Suenaga, M., Johnson, J.R., Thompson, P., and Moodenbaugh,

A.R., *Phys. Rev. B* 36: 5694-97 (1987).

45. Nicolas, M., Daou, J.N., Vedel, I., Vajda, P., Burger, J.P., Lesueur, J., and Dumoulin, L., *Solid State Commun.* 66: 1157-60 (1988).
46. Fujii, H., Kawanaka, H., Ye, W., Orimo, S., and Fukuba, H., *Japan. J. Appl. Phys.*, 27: L525-28 (1988).
47. Kamiyama, T., Tomiyoshi, S., Omori, M., Yamauchi, H., Kajitani, T., Matsunaga, T., and Yamamoto, H., *Physica* 148B: 491-93 (1987).
48. Kato, T., Aihara, K., Kuniya, J., Kamo, T., and Matsuda, S., *Japan. J. Appl. Phys.* 27: L564-66 (1988).
49. Gallagher, P.K., O'Bryan, H.M., Sunshine, S.A., and Murphy, D.W., *Mat. Res. Bull.* 22: 995-1006 (1987).

22

Surface Science Techniques for Analysis of Corrosion of the Ceramic Superconductors

Temel H. Büyüklimanli and Joseph H. Simmons

University of Florida
Gainesville, FL 32611

INTRODUCTION

Since the discovery of the new family of oxide superconducting materials a wide variety of analytical techniques have been employed to determine their structure and the structural features responsible for the superconductivity phenomenon. In these studies, it was quickly discovered that these compounds corrode rapidly when exposed to humid or even ambient atmosphere. X-Ray Photoelectron Spectroscopy (XPS) has been used extensively to study the structure of corroded surfaces, composition and electronic density of states of these materials. Because of the ability of XPS to provide bond structure information much investigation has also been conducted of fracture surfaces to reveal bulk characteristics such as the copper oxidation states, and the possibility of peroxide formation.

X-RAY PHOTOELECTRON SPECTROSCOPY

X-rays XPS analysis relates the bonding state of constituent atoms to the kinetic energy of core and valence electrons emitted from inner and outer shell states when a sample is irradiated with x-rays. The kinetic energy E_k of the emitted photoelectron is given by,

$$E_k = h\nu - E_b - \Phi \tag{22-1}$$

where $h\nu$ equals the energy of incident photons, E_b is the binding energy of electrons (with reference to Fermi level) and Φ is the spectrometer work function. Equation (22-1) only holds for compounds in electrical equilibrium with the spectrometer. Electrical insulators build up a positive charge on the surfaces, causing a shift in the kinetic energy of the emitted electrons. The resulting charging depends strongly on the X-ray flux, the electrical resistivity of the sample, and electron back flux from the X-ray source and other components in the system. For an electrical insulator the kinetic energy of the photoelectron will be given as,

$$E_k = h\nu - E_b - \Phi - Q \qquad (22\text{-}2)$$

The additional Q term takes into account the sample charging effect. However, the relative positions of the photoelectron lines are not influenced in general by the surface charging, and therefore the choice of an appropriate reference line of known energy allows an absolute energy calibration. Since the ceramic superconductors are conductive ($\sigma =$) at room temperature where most of these measurements are made, there is no need for charging compensation. However, the degraded superconductors become insulators and the line for surface adventitious carbon (CH_x) is often used as a source of calibration (E_b=284.6eV). The chemically altered surface film hinders XPS analysis of the bulk structure because the depth resolution of XPS is ~1-4nm (or 3-6 monolayers). Therefore, the extension of results from XPS measurements to the bulk structure requires a careful cleaning of the surface adsorbates. There is no generally accepted method for removing surface contamination. Many methods have been suggested, but each has serious drawbacks. For example, vacuum fracture commonly practiced in sample preparation for XPS studies has been criticized for polycrystalline materials because of the possible enhancement of grain boundaries at the fracture surface leading to a Ba-rich layer; (14,15) low energy ion milling in situ causes differential sputtering of the elements, especially oxygen, and leads to incorrect compound stoichiometries; (14,16,17) surface scraping with a diamond or steel file is the most often used method, yet suffers from the same grain boundary exposure problem as sample fracture (14,16,18). Heating the sample until the contaminants are evaporated seems to be the most reliable method for measurement of the bulk structure (16).

APPLICATION OF XPS TO 123 COMPOUNDS

Evidence of degradation prompted the measurements of surface changes upon exposure to ambient atmosphere and water. XPS was used to compliment other techniques such as x-ray diffraction (XRD) (1) to identify the products resulting from a reaction of these ceramics with different ambient environments.

By purposely exposing the 123 to corrosion-producing conditions it was possible to separate bulk components from contaminants and to study the effect of the corrosion process on superconducting samples. An analysis of corrosion product formation could also be conducted to determine the mechanisms of corrosion in these materials. Results are discussed below.

MEASUREMENT OF SURFACE CONTAMINATION USING XPS

The study reviewed here analyzed corrosion product formation in two types of samples (pressed powder pellets and thin films) subjected to air and 85% relative humidity at 50°C. Surface species identification was conducted by XPS while Scanning Auger Microprobe (SAM) spectroscopy was conducted to measure depth variations in component concentrations.

The pellets were prepared by mixing and sintering powders of Y_2O_3 (<10μm diam.), $BaCO_3$ (1μm diam.) and CuO (<800μm diam.) (19). Pellets with an average height of 1.35mm and average diameter of 4.85mm were pressed from 0.1g of calcined powder at 430MPa in a laboratory press. The pellets were then sintered at 950°C for one hour in an Al_2O_3 combustion boat. They were cooled slowly to 600°C and annealed for three hours in flowing oxygen followed by slow cooling to room temperature in flowing oxygen.

Films several microns thick with no particular orientation were deposited by RF sputtering using Y_2O_3, CuO and BaF_2 targets. They were treated with water vapor followed by a dry oxygen anneal to remove residual fluorine. Details of the preparation are given elsewhere (20).

Two pellet samples from the same batch and two films from the same deposition were prepared and one of each was tested for superconductivity while the surface of the other was examined in the corrosion study. All samples were of the 123 nominal composition and had T_c in the range 80-90 K. Samples were analyzed immediately after preparation, following an ambient air exposure of 18 days and after exposure to 85% relative humidity and 50°C in a humidity chamber. XPS measurements were performed on a Perkin Elmer Phi ESCA 5000 with pass energies at 17.9 and 8.95eV and Mg x-rays operated at 300 watts. Angle resolved measurements were also made on the same instrument between 15°- 80°. The typical binding energy resolution was 1.4eV full width at half max (FWHM). Sputter depth profiling using three point differentiation and peak to peak measurements was conducted on a Perkin Elmer Phi 660 SAM. The beam energy of the electrons and Ar ions was 10keV and 3keV, respectively. The ion beam diameter was 1mm creating a ~100Å/min sputtering yield. Typical vacua in both instruments were near 10^{-9} torr. Comparisons were made between the XPS and SAM data. In order to minimize the irradiation damage during the measurements, fast data acquisitions were performed at the expense of higher resolution. Samples were analyzed without surface cleaning in order to examine the effect of the environment on the actual sample surface.

XPS spectra of Ba 4d and $3d_{5/2}$, O 1s, Cu $2p_{3/2}$, Y 3d and C1s orbitals were collected. For charge referencing the C1s peak was used after curve fitting with 8 peaks (Figure 22-1) and assigning the lowest binding energy peak to hydrocarbon species. This peak was set to 284.6eV.

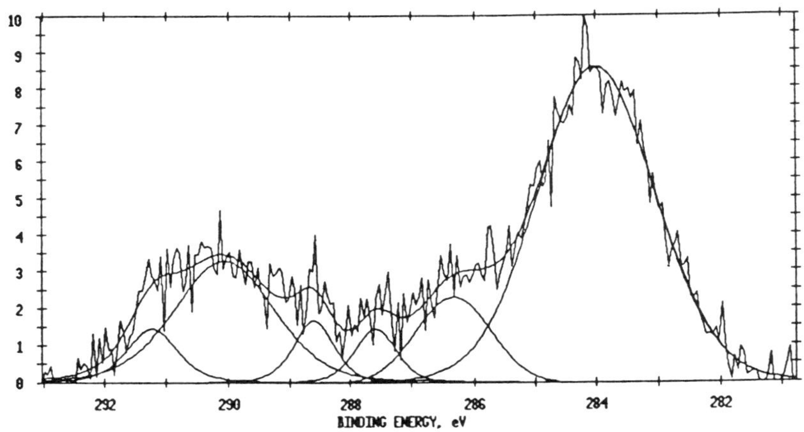

Figure 22-1. C1s XPS spectra from a 123 pellet.

The XPS analyses of fresh and exposed sample surfaces from the pellets and thin films are presented first. Figure 22-2 shows the relative changes in surface Ba and Cu atomic concentrations between various conditions of exposure ((a) freshly prepared, (b) following 18 days in ambient air, and following (c) 5, (d) 12 and (e) 48 hours in a controlled humidity chamber at 50°C with a relative humidity of 85%), on both the bulk sample pellet and the thin film. Barium increases with respect to Cu both under ambient and humid air exposure. After exposure to humid environments the Cu concentration decreases drastically in the surface region of both samples and the superconductivity is quenched. The decrease in Cu concentration occurs more rapidly in the pellet.

Concentration profiles for Cu, Ba, Y, O, and C were measured as a function of distance from the surface by both angle-resolved XPS and argon-ion sputtered scanning Auger measurements (SAM). Both approaches are subject to some limitations and these are discussed below.

Angle resolved XPS measurements have been reported by a variety of authors (18,22,23). The results of this study clearly show an increase in Ba and O content and a decrease in Y and Cu content towards the surface. The Ba increase essentially agrees with other reports where it was measured with different techniques upon exposure to CO_2, air and moisture (3,7,12,22,24,25,26,27). Angle resolved XPS studies require very smooth surfaces (ie. the surface roughness must be less than the escape depth of the electrons) (18). This clearly was not the case in the pellet samples and was

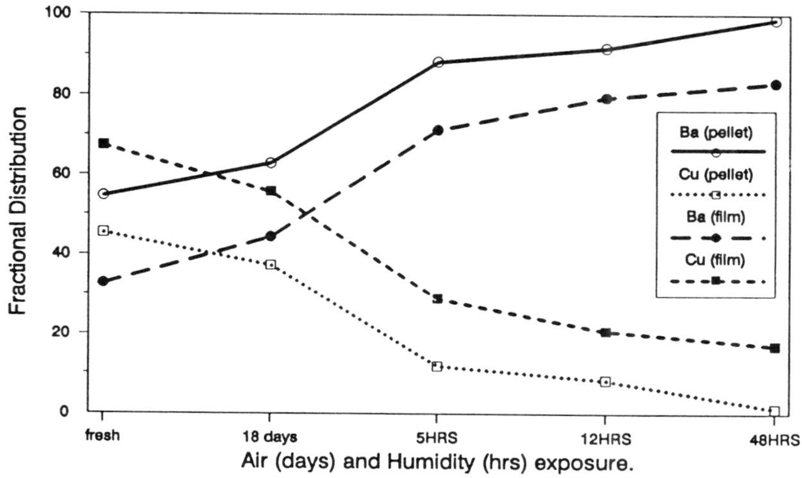

Figure 22-2. Relative change of Ba and Cu atomic concentration during moisture treatment of 123 ceramics measured by XPS.

marginal in the thin film samples. Therefore, only qualitative changes in species concentration are possible with this method. Quantitative measurements of the depth profiles were obtained using argon-ion sputtered SAM analysis. The argon ion beam was adjusted to produce a sputtering rate of about ~ 100Å/min. In order to test for any occurrence of differential oxygen sputtering, data acquisition was continued well into the bulk, at which point, the sputtering yield saturated and no further change in concentration of Ba, O, Y, Cu was observed. This guaranteed that the overall sputtering rates were equivalent for the 4 components. A SAM trace of the depth profile in a freshly prepared pellet sample (Figure 22-3) confirms these results as the relative peak intensities of each of the 4 components remain unchanged at depths greater than 200nm. This is the same depth where large changes are observed in the exposed pellet.

Figures 22-4 and 22-5 show peak to peak depth profiles in both the pellet and thin film samples after a 48 hour exposure to humidity. The data show a clear accumulation of Ba and O at the sample surfaces and a depletion of Cu for pressed powder. The intensity of Y was not sufficient to make any judgement. The depletion and enrichment regions exhibit about the same thickness with Ba, Cu, Y and O approaching their bulk concentrations at about 600nm in the pellet sample. By contrast, the composition of the film was not as severely altered. The depth profile shows that there was no appreciable Ba enrichment and a much thinner Cu depletion layer (Figure 22-5). As shown below (Figures 22-6 and 22-7) it was observed that formation of carbonates and

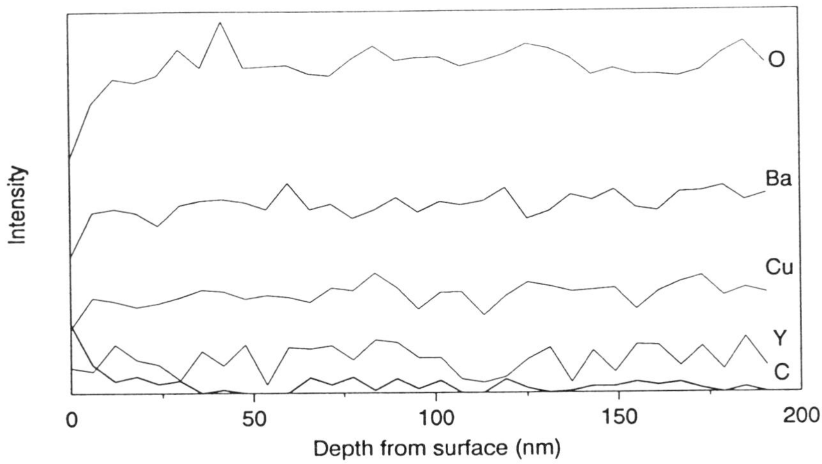

Figure 22-3. Depth profile of fresh 123 pellet measured by SAM.

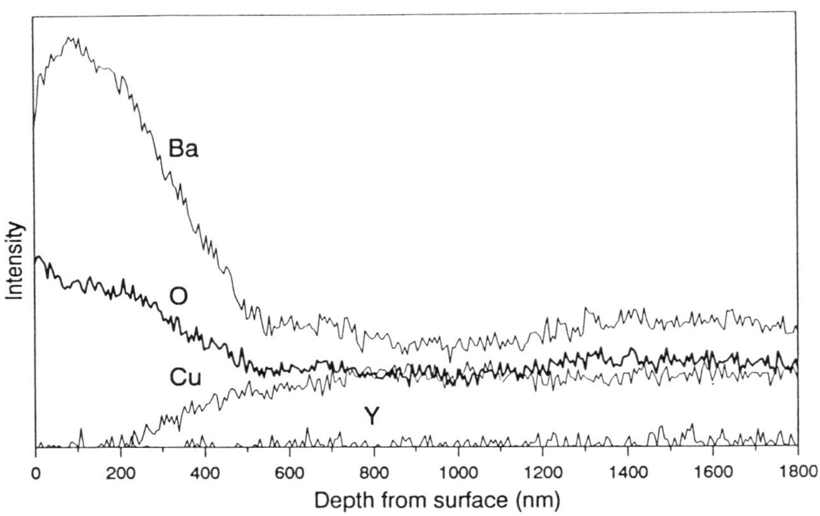

Figure 22-4. Ion assisted depth profile measured by SAM of the sintered pellet after 48 hours at 80% relative humidity and 50°C.

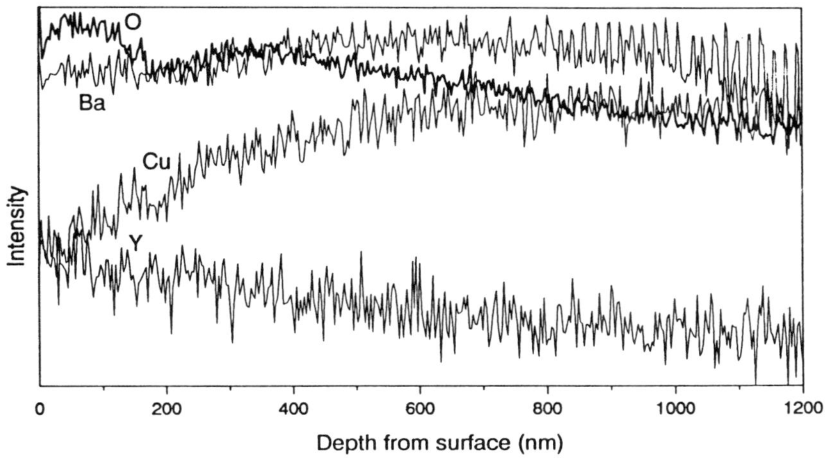

Figure 22-5. Ion assisted depth profile measured by SAM of the RF sputtered film after 48 hours at 80% relative humidity and 50°C.

hydroxyl groups and loss of superconducting component occurs more rapidly in pellets. Such composition variations through mass transport along the grain boundaries have been reported previously (7,8,10,18,15,22,28).

The heats of formation of $Ba(OH)_2.8H_2O$ (-798.8 kcal/mole), $BaO_2.8H_2O$ (-718.8 kcal/mole) and $BaC_2O_4.2H_2O$ (-471 kcal/mole) (29) are strongly negative and consequently favor the formation of such compounds on the sample surface. The high diffusivity of Ba along the grain boundary provides the excess Ba to fuel these reactions. A similar conclusion was obtained by others (3,25,30) based on the narrowing of the XPS Ba peak associated with release of oxygen in vacuum upon heating (16). The result is further supported by the detection of non-superconducting Ba containing phases at the grain boundaries (31). Samples exposed to humidity were normally analyzed immediately. No changes were observed in Ba or O features when some samples were reanalyzed after they had been kept in UHV (10^{-9}torr) overnight. We have also observed that in some cases carbonation of Ba occurs in high stress areas especially in pellets formed from pressed powders. Sequeira et al. (2) also conclude that non-superconducting amorphous phases formed at the grain boundaries were caused by strain fields and partial release of oxygen. Several studies have linked the degradation of 123 to oxygen defects in these materials (5,13,28). In similar tests on higher density pellets we measured a less severe degradation while porous laser ablated films showed much greater degradation in moist air. These results further support the speculation that the structure and size of the grain boundary (or grain) plays an important role in the resistance to degradation in humid air. Zandbergen et al. (5) have demonstrated by means of transmission electron microscopy (TEM) that defect rich regions decomposed more rapidly.

XPS analysis of the surface can also yield information about chemical changes occurring with various degrees of exposure by means of chemical shift analysis of the deconvoluted peaks. The Ba4d O1s and peaks were deconvoluted into a series of components (Figures 22-6 and 22-7), closely matching the binding energy positions reported in Ref. 22, except for the superconducting cuprate species in Ba4d. An additional component which grew after exposure to humidity was identified as due to hydroxyl forms of Ba. The binding energies and component species assignments are shown in Tables 22-1 and 22-2 for the Ba4d and O1s XPS peaks. Studies by Ford et al. (16) in which the Ba and O XPS peaks were analyzed, interpreted the 531eV peak as a combination of surface and bulk species. Most studies (7,22,25) associated the peak with BaO_2 or $BaCuO_2$ formed on the surface after heat-treatment (16) and exposure to atmosphere (3,7). The observations reported here show that the 531 line is still present after Cu is no longer detected on the surface. Therefore, the assignment to BaO_2 only is more appropriate as shown in Table 22-2. The cuprate species have lower binding energies (528 and 529 eV) as shown in Table 22-2. The hydroxide line was found to have a higher binding energy than predicted by previous authors (18,22,23). The 534 and 535.6eV lines increased with exposure to humidity and most probably correspond to $Ba(OH)_2$, $Ba(OH)_2.8H_2O$, $BaC_2O_4.2H_2O$ and /or molecular water (32). The O1s peaks often detected in the 531-534 eV range have been attributed to a multiplicity of forms of chemically and physically bound water on the surface. The chemisorbed and physisorbed oxygen and hydroxyl are also found in this range making a clear interpretation difficult (4,33). The high binding energy O1s peak in $La_2CuO_{4+\delta}$ system was also attributed to water (32). The possibility of the presence of $BaC_XO_YH_Z$ in this study arose from the fact that carbon had a component at 292.1eV, which is much higher than the reported $BaCO_3$ peak at 288.6±0.3eV, (24,30,34,35) also observed in this study (Figure 22-1). This carbon peak is most likely related to a hydroxyl or a water compound since these usually have higher binding energies. In light of these considerations we agree with Qui et al. (3) that O dimerization assumptions (36,37) in HTSCs which result from the observation of O1s peaks in the range 531-535 eV at low temperatures which subsequently disappear at higher temperatures should be carefully examined since increasing temperatures can also lead to the removal of OH species. These different binding energy speculations can only be resolved by water reaction of BaO_2 and $BaCO_3$ and identification by x-ray diffraction (XRD) and XPS. In making the assignments of Tables 22-1 and 22-2, we insured that no impurities were present which would interfere with our interpretations. Curve fitting was performed using the second derivative method (38) to identify the component peaks after subtraction of the x-ray line shape (often called deconvolution). This allowed an accurate measure of the number of components present and their binding energies. While curve fitting different spectra, the binding energy positions were

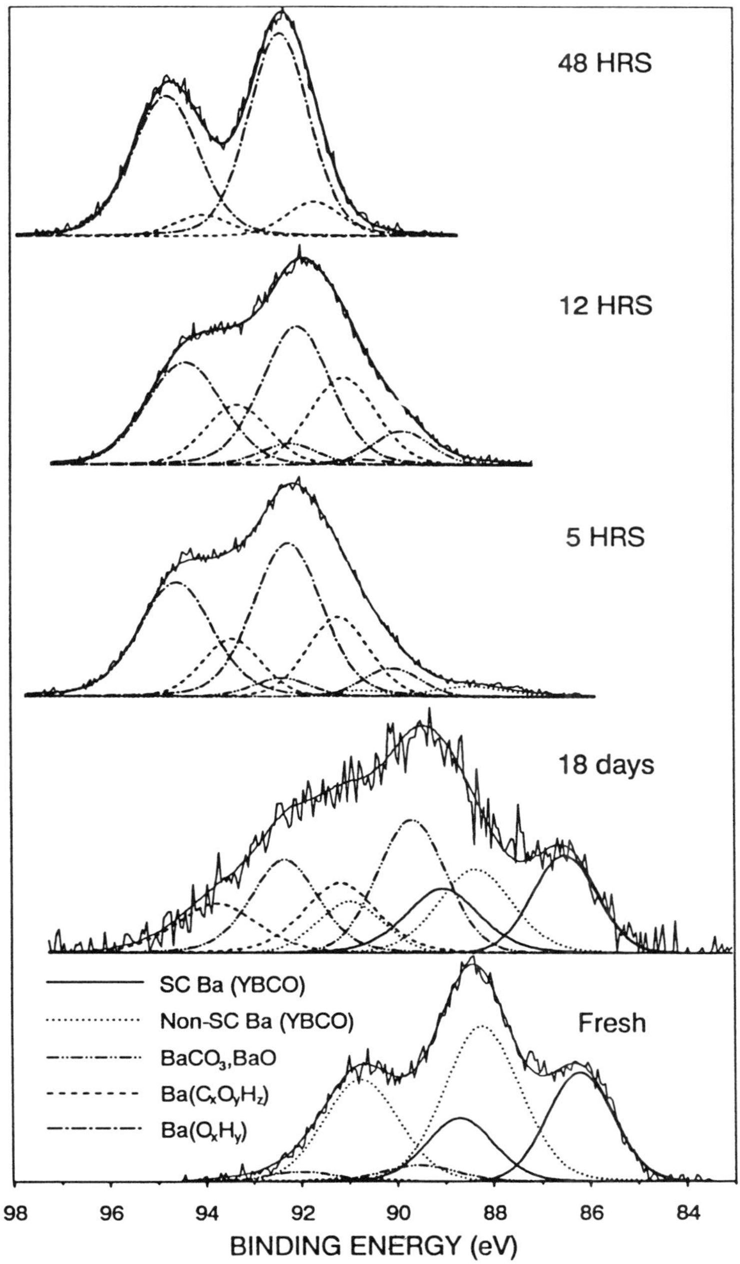

Figure 22-6. Ba4d spectra of the 123 pellet after exposure, slowing the component peaks.

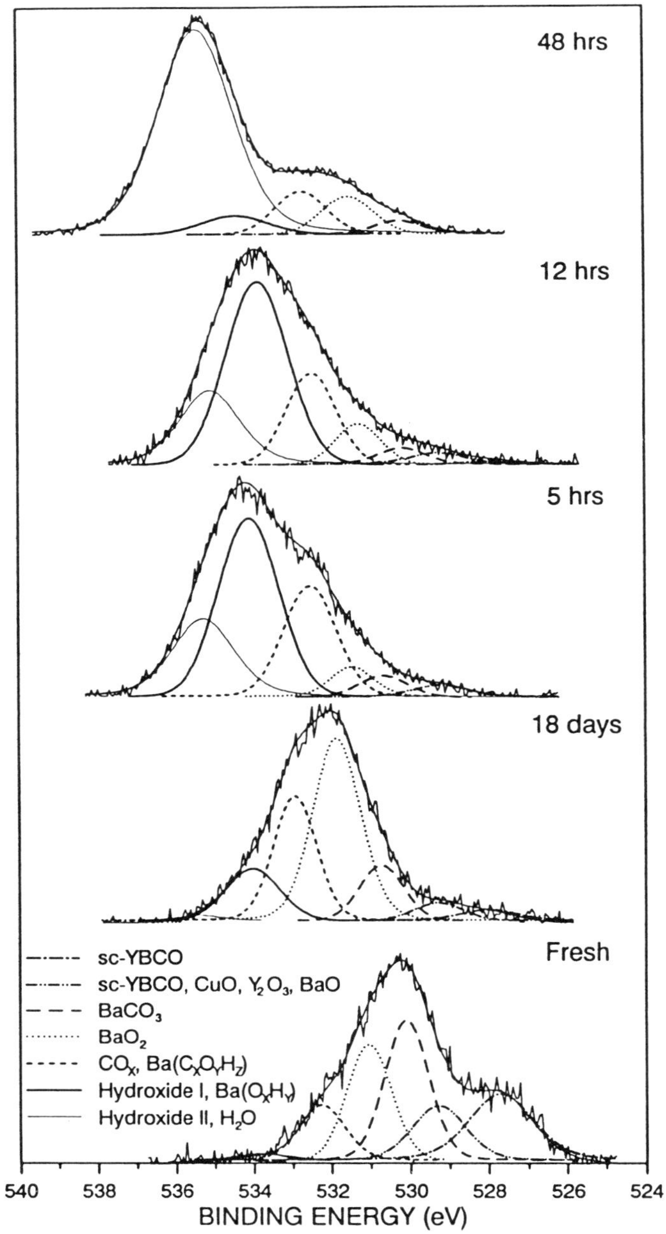

Figure 22-7. O1s spectra of the 123 pellet after exposure, slowing the component peaks.

kept within ±0.2eV of the values given in Tables 22-1 and 22-2. The same FWHM (±0.2eV) was resolved for all peaks. Generally 90% Gaussian and 10% Lorentzian curves were used with integral baseline subtraction. For the Ba analysis, the 4d orbital was used rather than the 3d peak since the chemical shifts of the higher electronic state are greater. We found the 4d spin states (5/2 and 3/2) to be separated by 2.5 eV as measured on BaO and $BaCO_3$ samples after

Table 22-1. Components Used to Curve Fit Ba4d Spectra of 123 Compounds.

B.E.(eV)±0.2	Assignment
86.4	superconducting Ba (YBCO)
88.2	non-superconducting Ba (YBCO)
89.6	$BaCO_3$, BaO
91.0	$Ba(C_xO_yH_y)$
92.1	$Ba(O_xH_y)$

Table 22-2. Components Used to Curve Fit O1s Spectra of 123 Compounds.

B.E.(eV)±0.2	Assignment
528.0	Superconducting O (YBCO)
529.0	CuO, Y_2O_3, BaO, YBCO
530.2	$BaCO_3$
531.3	BaO_2
532.5	CO_x, $Ba(C_xO_yH_z)$
534.0	Hydroxide I, $Ba(O_xH_y)$
535.6	Hydroxide II, H_2O

low energy ion sputter cleaning. This close separation complicates peak deconvolution. For every Ba species it was necessary to use two peaks with 2.5 eV separation and the intensity of the 4d 5/2 peak was taken at 1.5 times the 4d 3/2 peak intensity as measured in the BaO and $BaCO_3$ samples. These considerations are in close agreements with earlier reports (16,22,39). The

$Ba(OH)_X$ species in this study had a larger binding energy than predicted by the same reports. Figures 22-6 and 22-7 show the deconvolution of the Ba 4d and O1s peaks for the bulk sample under all cases of exposure studied. While many of the peak assignments in XPS studies are based on direct and indirect inference both in this paper and in most of the literature, the assignments shown in Tables 22-1 and 22-2 agree well with those of other authors and are consistent with a wide variety of measurements (22,39). Instances where we differ from the literature have been identified and justified by other measurements.

Figures 22-8 and 22-9 show the variation of different surface chemical species identified using Tables 22-1 and 22-2 as a function of exposure condition. In the deconvolution of the Ba 4d peak, 2 barium cuprate peaks have been identified. The first is associated with a superconducting bulk cuprate

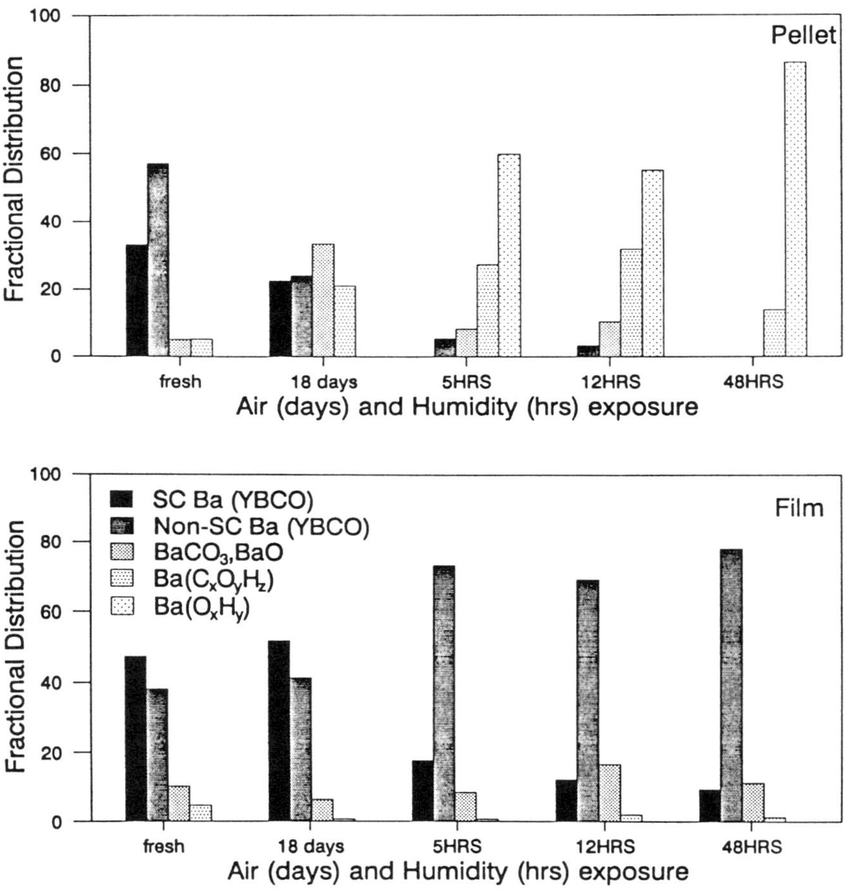

Figure 22-8. Change in the Ba4d XPS peak due to ambient atmosphere and humidity exposure calculated from the peak areas of Figure 22-6.

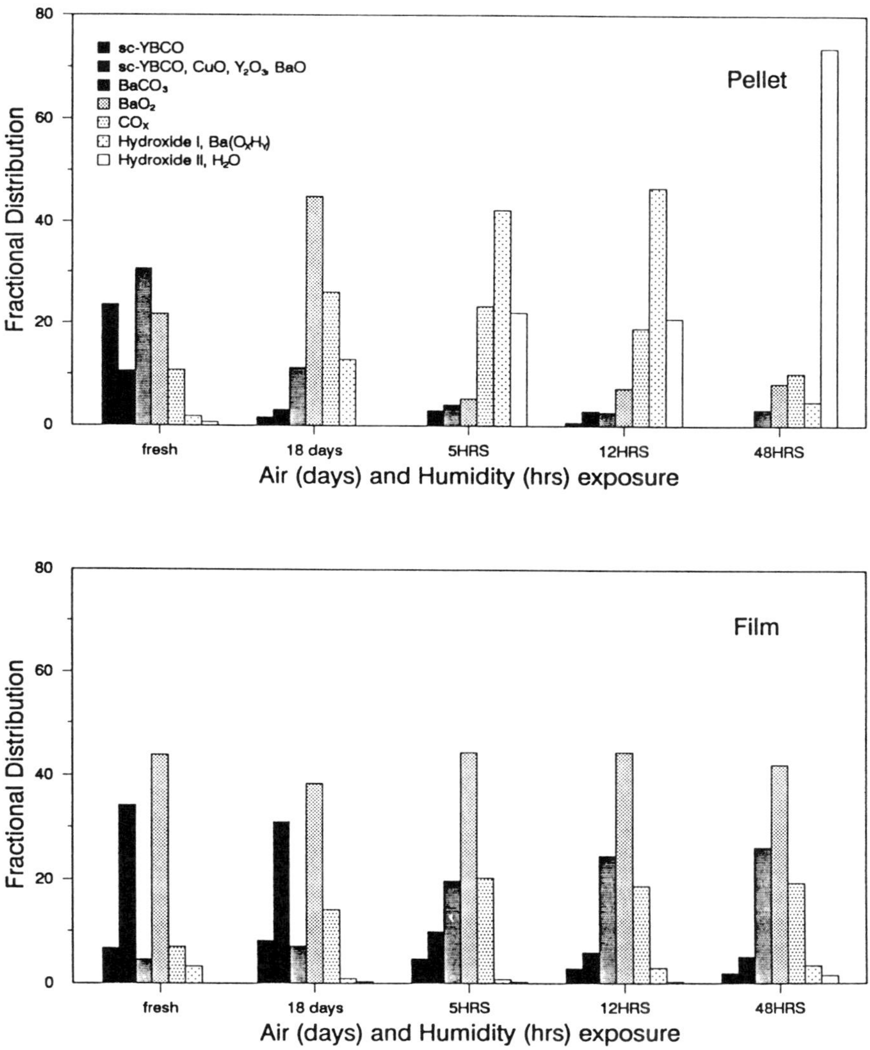

Figure 22-9. Change in the O1s XPS peak due to ambient atmosphere and humidity exposure calculated from the peak areas of Figure 22-7.

structure and the second is a non-superconducting cuprate structure, often associated with a surface species. This assignment is consistent with several reports of XPS analysis on superconducting 123 compounds. Ambient air exposure causes a greater degradation of the two barium cuprate peaks in the pellet than in the film. In fact, the film shows an unexpected resistance to degradation in ambient air. Exposure to humid air, however, has a drastic effect on the pellet and a five hour exposure is sufficient to virtually eliminate the superconducting cuprate peak from the surface, with the rapid formation of barium carbonate and hydroxide. The film still has significant superconducting cuprate at the end of the experiment and the predominant reacted Ba species is the non-superconducting cuprate rather than the hydroxide or carbonate compounds. This shows that the reaction with water remained localized to the surface of the film, possibly due to reduced diffusion of Cu to the interior and Ba to surface. In fact, both the SAM and XPS data show a significant Cu concentration at the film surface in contrast to the pellet which has none. Also, no excess Ba was present on the film surface. The non-superconducting cuprate phase is likely an oxygen deficient structure ($YBa_2Cu_3O_6$) (22,40) which lost an oxygen and a Ba and picked up CO_2 from the atmosphere to form $BaCO_3$.

These results are supported by the peak resolution of the O1s spectra (Figures 22-7 and 22-9) where large hydroxide peaks in the pellet sample become predominant upon exposure to humid environments while the superconducting Ba and O species are still measurable for the films. The carbonate peak is also present in the pellet, and the cuprate peaks vanish from the surface upon exposure to humid air.

SUMMARY AND CONCLUSIONS

YBCO samples exposed to ambient air and humidity exhibit similarities and differences in behavior. The samples which consist of sintered pellets and RF sputtered films, exhibit a loss of superconducting phase with exposure and a tendency to form carbonate species in ambient air and hydrate and hydroxide species in humid air. Both types of samples exhibit a large excess of surface Ba and a depletion in Cu. However, the Cu surface concentration of the pellet vanishes with exposure to 80% relative humidity at 50°C over 48 hours while the film still manages to retain a significant portion of its initial surface Cu concentration. While the pellet exhibits a surface primarily covered with carbonate and hydroxide species, the film is primarily covered by a non-superconducting barium cuprate phase, most likely oxygen deficient. Differences in severity of degradation between these two samples and also between high porosity pellets and laser ablated films which exhibited even more severe degradation indicate that the rate controlling process in the environmental attack of YBCO superconducting perovskites occurs by Ba diffusion to the surface and Cu depletion from the surface through the grain boundary. Thus samples with fine grains and a small grain boundary volume exhibit a good resistance to environmental degradation (i.e. the RF sputtered films), while samples with larger grains and a thicker grain boundary are very susceptible to phase changes and chemical reactions in a thick region below the surface (pellets exhibited a transformed layer about 600nm thick after only 48 hours of exposure to humid air). These results obtained by surface analysis techniques shed a new light to the corrosion of superconducting ceramics as well as complimenting the earlier conclusions made by other experimental methods. They also serve as a standard to accurately identify the bulk related properties in YBCO compounds.

ACKNOWLEDGEMENTS

The authors wish to thank Paul Holloway for valuable discussions, Eric Lambers for help with SAM data, Greg Chandler for fabricating the HTSC pellets, Kelly Truman and Carl Mueller for fabricating the film samples. This research was funded by DARPA under Contract No. MDA972-88-J-1006.

REFERENCES

1. Yan, M.F. Barns,R.L., O'Bryan, Jr., H.M., Gallagher, P.K., Sherwood, R.C. and Jin, S., *Appl. Phys. Lett.* 51: 532 (1987).
2. Sequeira, A., Rajagopal H. and Yakhmi, J.V., *Solid State Commun.* 65: 991 (1988).
3. Qui, S.L., Ruckman, M.W., Brookes, N.B., Johnson, P.D., Chen, J., Lin, C.L., Sinković, B. and Strongin, M., *Phys. Rev. B.* 37: 3747 (1988).
4. Qui, S.L., Lin, C.L., Ruckman, M.W., Chen, J., Chen, D.H., Xu, Y., Moodenbaugh, A.R., Strogin, M.,Nichols, D. and Crow, J.E., *AIP Conf. Proc.* 1988, No.182, (G. Margaritondo, R. Joynt and M. Onellion, eds.) 368 (1989).
5. Zandbergen, H.W., Gronsky, R. and Thomas, G., *Phys. Stat. Sol.* (a) 105: 207 (1988).
6. Thompson, J.G., Hyde, B.G., Withers, R.L., Anderson, J.S., Fitz Gerald, J.G., Bitmead, J., Paterson, M.S. and Steward, A.M., *Mat. Res. Bull.* 22: 1715 (1987).
7. Nefedov, V.I., Sokolov, A.N., Tyzykhov, M.A., Oleinikov, N.N., Yeremina, Ye.A. and Kolotyrkina, M.A., *J. Electron Spec. Relat. Phen.* 49: 47 (1989).
8. Fitch, L.D. and Burdick, V.L., *J. Am. Ceram. Soc.*, 72: 2020 (1989).
9. Hyde, B.G., Thompson, J.G., Withers, R.L., FitzGerald, J.G., Steward, A.M., Bevan, D.J.M., Anderson, J.S., Bitmead, J. and Paterson, M.S., *Nature*, 327: 402 (1987).
10. Hwang, N.M., Bahng, G.W., Park, Y.K., Park, J.C., Moon, H.G. and Yoon, D.N., *J. Mater. Sci. Lett.*, 8: 517 (1989).
11. Harris, L.B. and Nyang, F.K., *Solid State Comm.*, 67: 359 (1988).
12. Liu, H-X, Cocke, D.L., Naugle, D.G. and Pandey, R.K., *Solid State Ionics*, 32/33: 1125 (1989).
13. Ruckman, M.W., Herald, S.M., DiMarzio, D., Chen, H., Moodenbaugh,

A.R., and Yang, C.Y., *AIP Conf. Proc.* 1988, No.182, (G. Margaritondo, R. Joynt and M. Onellion, eds.) 360: (1989).

14. Myhra, S., Riviere, J.C., Steward, A.M. and Healy, P.C., *Z. Phys. B-Condensed Matter*, 72: 413 (1988).
15. Stucki, F., Brüesch, P. and Baumann, Th., *Physica C* 156: 461 (1988).
16. Ford, W.K., Anderson, J., Rubenacker, G.V., Drumheller, J.E., Chen, C.T., Hong, M., Kwo, J. and Liou, S.H., *J. Mater. Res.* 4: 16 (1989).
17. Ford, W.K., Chen, C.T., Anderson, J., Hong, M., Kwo, J., Liou, S.H., Rubenacker, G.V. and Drumheller, J.E., *Phys. Rev. B* 37: 7924 (1988).
18. van der Marel, D., van Elp, J., Sawatzky, G.A and Heitmann, D., *Phys. Rev. B*, 37: 5136 (1988).
19. Stoffel, N.G., Morris, P.A., Bonner, W.A., LaGraffe, D., Tang, M., Chang, Y., Margaritondo, G. and Onellion, M., *Phys. Rev. B*, 38: 213 (1988).
20. Ahmad, I., Chandler, G.T. and Clark, D.E., *Mat. Res. Soc. Symp. Proc.* 124: 239 (1988).
21. Truman, K., Ph.D. dissertation, University of Florida, Dept. of Materials Sci. and Eng. (1990).
22. Herzog, W., Scwarz, M., Sixl, H. and Hoppe, R., *Z.Phys. B*, 72: 19 (1988).
23. Halbritter, J., Walk, P., Mathes, H.J., Hauser, B. and Rogalla, H., *Z. Phys. B-Condensed Matter*, 73: 277 (1988).
24. Steiner, P., Kinsinger, B., Sander, I., Siegwart, B., Hüfner, S. and Politis, C., *Z. Phys. B-Condensed Matter*, 67: 19 (1987).
25. Caracciolo, R., Hegde, M.S., Wachtman,Jr., J.B., Inam, A. and Venketesan, T., *AIP Conf. Proc.* 1988, No.182, (G. Margaritondo, R. Joynt and M. Onellion, eds.) 232 (1989).
26. Wu, X.D., Inam, A., Hegde, M.S., Venkatesan, T., Chang, C.C., Chase, E.W., Wilkens, B. and Tarascon, J.M., *Phys. Rev. B*, 38: 9307 (1988).
27. Jaeger, H., Hofmann, S., Kaiser, G., Schulze, K. and Petzow, P., *Physica C*, 153-155: 133 (1988).
28. Gavaler, J.R. and Braginski, A.I., *Physica C* 153-155: 1435 (1988).
29. Marks, L.D., Li, D.J., Shibahara, H. and Zhang, J.P., *J. Electron Micros. Tech.*, 8: 297 (1988).
30. Wagman, D.D., Evans, W.H., Parker, V.B., Schumm, R.H., Bailey, S.M., Halow, I., Churney, K.L. and Nuttall, R.L., *National Bureau of Standards Tech. Notes*, 270-3, 270-4, 270-5, 270-6, 270-7 and 270-8.
31. Majumdar, D., Chatterjee, D. and Pav-Pujalt, G., *Physica C*, 158: 413, (1989).
32. Stucki, F., Brüesch, P. and Baumann, T., *Physica C*, 153-155: 200 (1988).

33. Rogers, Jr., J.W., Shinn, N.D., Schirber, J.E., Venturini, E.L., Ginley,D.S. and Morosin, B., *Phys. Rev. B*, 38: 5021 (1988).
34. McIntyre, N.S., *Practical Surface Analysis by Auger and X-Ray Photoelectron Spectroscopy*, (D. Briggs and M.P. Seah, eds.) John Wiley & Sons, 410 (1983)
35. Fjellvåg, H., Karen, P., Kjekshus, A. and Grepstad, J.K., *Acta Chem. Scand.*, A42: 171 (1988).
36. Parmigiani, F., Chiarello, G., Ripamoni, N., Goretzki, H. and Roll, U., *Phys. Rev. B*, 36: 7148 (1987).
37. Sarma, D.D., Sreedhar, K., Ganguly, P. and Rao, C.N.R., *Phys. Rev. B*, 36: 2371 (1987), Kohiki, S., Hamada, T. and Wada, T., *Phys. Rev. B*, 36: 2290 (1987).
38. Kachel, T., Sen, P., Dauth, B., and Campagna, M., *Z. Phys. B*, 70: 137 (1988).
39. Sherwood, P.M.A., *Practical Surface Analysis by Auger and X-Ray Photoelectron Spectroscopy,* (D. Briggs and M.P. Seah, eds.) John Wiley & Sons. 454 (1983)
40. Verhoeven, J.A.T. and van Doveren, H., *Appl. Surface Science*, 5: 361 (1980).
41. Liu, R., Olson, C.G., Yang, A.B., Gu, C., Lynch, D.W., Arko, A.J., List, R.S., Bartlett, R.J., Veal, B.W., Liu, J.Z., Paulikas, A.P. and Vandervoort, K. *Phys. Rev. B*, 40: 2650 (1989).

23

Reducing Corrosion of Ceramic Superconductors with Sol-Gel Coatings

S.S. Bayya and V.R.W. Amarakoon

NYS College of Ceramics at Alfred University
Alfred, NY 14802

INTRODUCTION

High temperature superconductivity has been at the center of attention in the ceramic oxide world since the discovery of this property for $La_{2-x}Ba_xCuO_4$ - type compounds by Bednorz and Muller (1) at temperatures two times higher than any previously observed. In the subsequent years, interest has shifted dramatically to the higher T_c materials discovered by Wu et al. (2), in the Y-Ba-Cu-O system, Maeda et al. (3) in the Bi-Sr-Ca-Cu-O system and Sheng et al. (4) in the Tl-Ba-Ca-Cu-O system.

Although we can now reliably prepare phase pure material of known crystal structure, much remains to be learned in order to construct a practical device. The new materials must satisfy a formidable array of physical requirements, depending on the applications. They must be able to carry currents of more than 10^5 amperes per square centimeter at 77K. For many applications the superconductors must be strong and flexible enough to withstand large magnetic, gravitational and centrifugal forces (for example, in spinning turbine generators and motors). They must be easy to shape. Perhaps of even more importance is their environmental stability. This latter property can be substantially improved through the use of coatings. The objective of this chapter is to discuss the use of sol-gel coatings for protecting ceramic superconductors.

MOISTURE DEGRADATION

A major problem is that most superconducting compounds are chemically unstable due to reaction with water. $YBa_2Cu_3O_x$ (123) degrades on exposure to the atmosphere. Hyde et al. (5), using x-ray diffraction, identified the products of its degradation in air as $Ba_2Cu(OH)_6$, $BaCO_3$, CuO and $Y(OH)_3$. Enhanced degradation in air at 40°C saturated with water vapor was shown by this study to begin at the surface, from which microcrystals of $BaCO_3$ grow. Yan et al. (6) have shown that the 123 superconductor decomposes in water with the formation of CuO, $Ba(OH)_2$ and Y_2BaCuO_5 (211), and evolves oxygen. They proposed the following decomposition reactions associated with the reduction of trivalent copper ions:

$$3H_2O + 2YBa_2Cu_3O_7 \rightarrow Y_2BaCuO_5 + 3Ba(OH)_2 + 5CuO + 0.5O_2 \quad (23\text{-}1)$$

$Ba(OH)_2$ can further react with CO_2 present in the atmosphere.

$$Ba(OH)_2 + CO_2 \rightarrow BaCO_3 + H_2O \quad (23\text{-}2)$$

Wang et al. (7) proposed a three stage corrosion mechanism for the degradation of 123 superconductors in water. Using x-ray diffraction, infrared spectroscopy and transmission electron microscopy, they determined microcrack formation along the layers perpendicular to the c-axis. This was followed by lattice dislocation due to the increasing number of microcracks and their subsequent propagation and the formation of corrosion product in the final stage. Frase et al. (8) have shown that in addition to chemical reactions with powders, the presence of humidity promotes slow crack growth. This effect is particularly pronounced in the presence of strong acids and bases.

Bansal and Sandkuhl (9) studied the chemical durability of 123 superconductors in water and 100% humidity at three temperatures using pH measurements, x-ray diffraction, and scanning electron microscopy. They determined that the 123 compound is highly susceptible to attack from moisture and degrades in the ambient atmosphere, particularly in high humidity. The rate of corrosion was determined to be rapid when it comes in contact with liquid water and the dissolution was highly incongruent. Similar results were reported by Barkatt et al. (10)[1].

Komori et al. (11) studied the chemical durability of 123 compound in aqueous solution of varying pH. They found that it was very unstable in high acidic solutions (pH ~ 1) with the cations dissolving in the same ratio as in the compound. They also observed that the leaching rate of cations increased

---

[1]For an extensive discussion of 123 degradation see Barkatt's chapter in this book.

sharply as the pH decreased below 3 and decreased gradually as pH increased above 3. In highly basic solutions (with NaOH) barium ions leached out preferentially and almost no yttrium and copper ions were leached out. When NH_4OH was used as a basic reagent, the amounts of barium ions leached out were higher than in the NaOH solution of the same pH value. Dissolution of Cu ions also increased.

Garland (12) studied the degradation behavior of 123 powder in a high-humidity environment using changes in the a.c. susceptibility. It was found that the fraction of 123 compound in a sample exposed to moisture at 98% relative humidity and 20°C, decreased exponentially with time. Karioris and Vance (13) studied the decomposition of 123 powders in moisture-saturated air at 23°C using x-ray diffraction. They suggested that after the initial decomposition in water, further corrosion was prevented by the development of a protective layer. Kumar et al. (14) studied the stability of 123 compound at 925°C in air, and found that the decomposition products of corrosion ($BaCO_3$, CuO and Y_2O_3) form a boundary layer on the 123 grains, and that this phase limits the grain growth and densification during sintering.

METHODS OF SURFACE PASSIVATION

Barns and Laudise (15) used room-temperature and liquid nitrogen temperature resistance measurements to study the stability of smaller surface area specimens (thin sheets). They found that such specimens react slowly with humid air at room temperature and rapidly in air at 85°C and 85% relative humidity (the standard conditions used for accelerated humidity testing (15)). The resultant decomposition products were non-superconducting phases which formed in hours to days, depending on the conditions. Furthermore, they found that a plastic encapsulant[2] considerably slowed such reactions even though it did not eliminate them.

Chang (16) studied the moisture-induced degradation of the 123 superconducting films using layer structures deposited at low temperatures with and without a silver covering layer, and using films deposited at high temperatures with a silver barrier layer. Upon removing the samples from liquid nitrogen and warming in flowing nitrogen, a large rise in film resistivity was observed for the structures without silver, resulting in reduced transition temperature and reduced critical current density at 77K. Much less degradation was observed, however, with the samples containing a silver layer. For samples deposited at high temperatures, no degradation was observed following similar moisture exposures. Hill et al. (17) used x-ray photoelectron spectroscopy to investigate passivation

---

[2] A fast-setting epoxy No: 04001 - Hardman, Inc., Belleville, NJ

of the high temperature superconductors 123 compound and $Bi_2Ca_{1+x}Sr_{2-x}Cu_2O_{8+y}$, involving overlayer materials deposited in vacuum onto freshly cleaved surfaces. CaF_2 was found to be completely inert and to form uniform overlayers on the surface. Deposition of Bi, Si, and Al in activated oxygen produced oxides and exhibited minimal reaction and disruption of the superconductor surface. In contrast, results for Bi, Si, and Al deposition in nonactivated oxygen or ultrahigh vacuum showed extensive substrate disruption due to surface reactions and oxygen withdrawal. These materials with passivated overlayer films have not been tested for their stability in the presence of moisture.

Vasquez et al. (18) showed that treatment of 123 films with a nonaqueous solution of HF in absolute ethanol results in the formation of a passivating oxyfluoride surface, as determined by x-ray photoelectron spectroscopy. The passivation properties of chemically-treated films were tested by monitoring the growth of non-superconducting species, as a function of air exposure time, for both HF-treated and untreated films. In a separate study, these same authors (19) described wet chemical techniques for the treatment of 123 surfaces resulting in the formation of native compounds such as CuI, $BaSO_4$, CuS, Cu_2S and oxalates with little or no reactivity to water. Treatment with dilute HI resulted in the formation of a native iodide film containing 80-90% CuI with small amounts of YI_3 and BaI_2. Treatment with dilute H_2SO_4 resulted in the formation of a film containing 95% $BaSO_4$ and 5% $Y_2(SO_4)_3$. Cu_2S was formed on the surface with a dilute Na_2S solution. An oxalate film with equal amounts of $Y_2(C_2O_4)_3$ and BaC_2O_4 resulted from treatment with dilute oxalic acid. X-ray photoelectron spectra showed no significant changes when the sulfide, sulfate, or oxalate films were dipped in water, while the iodide film showed evidence of $Ca(OH)_2$ formation.

SOL-GEL COATING OF POWDERS

Recent work done at NYS College of Ceramics (20-24) indicates that sol-gel coating offers a promising method of improving ceramic microstructures and properties. Some of the advantages to this processing technique include the following:

1. Possibility of lowering sintering temperature and thereby avoiding the loss of volatile components. This has historically been a problem in attaining dense electronic ceramics with technical properties.

2. Generally, liquid phase additions tend to dilute the electrical properties. However, by using sol-gel coating procedures, the amount of liquid phase incorporated could be minimized to ~3 mole%.

3. The ability to add dopants and counter dopants (<0.5 atom%) in a

uniform and controlled way by incorporation via the sol allows processing without a costly chemical powder preparation technique.

4. Sol-gel technology is amenable to mass production with a minimum of changes required in manufacturing.

MnZn ferrite powders coated with sol-gel phase containing tetraethylorthosilicate (TEOS) and calcium acetate in ethyl alcohol, resulted in the improvement of microstructures and properties (20). Without $CaSiO_3$ coating, MnZn ferrite exhibited exaggerated grain growth and poor microstructures. The coating resulted in a uniform grain size distribution, with $CaSiO_3$ located preferentially in the grain boundary regions. X-ray diffraction of these powders indicated the power of the coating process in controlling the phase distribution during sintering. The coating process prevented the formation of non-magnetic alpha-Fe_2O_3 in MnZn ferrites.

Sb-doped SnO_2 is difficult to sinter at temperatures lower than 1500°C, limiting its application for gas sensing. Sol-gel coating of powders resulted in a low temperature liquid phase sintering of Sb-doped SnO_2 (21) with uniform microstructures and improved electrical properties.

Semiconducting positive temperature coefficient of resistance (PTCR) ceramics were processed by sol-gel coating of $BaTiO_3$ powders (22). Dopants (Y), counter dopants (Mn, F) and additives (Si) were incorporated homogeneously within the microstructure with improved properties. The development of uniform microstructures indicated the potential ability of the coating process to increase the reproducibility of PTCR devices on an industrial scale. Incorporation of F via sol-gel increased the reliability of devices, making them less sensitive to reducing environments.

Sol-gel coating formation involves following steps:

1) Mixing of alkoxides with water for hydration.

2) If the alkoxide is immiscible with water, then a cosolvent is required which would dissolve both the alkoxide and water so that the two can interact.

3) Hydrolysis of alkoxides.

4) Mixing of powder with sol, which involves the distribution of sol around the particles.

5) Completion of remaining hydrolysis.

6) Condensation of the gel at the surface of the particle.

7) Removal of excess liquid from the system by evaporation.

In this investigation, sol-gel coating of $YBa_2Cu_3O_{7-x}$ powders was used to determine the stability of superconductors in humid environments.

EXPERIMENTAL PROCEDURE

$YBa_2Cu_3O_x$ powders were prepared from $BaCO_3$[3], CuO[4] and Y_2O_3[3] starting materials, using the conventional mixed oxide method. Powders were ball milled in plastic bottles for 12 hours using zirconia balls[5] and acetone. The mixture was dried and calcined at 930°C for 24 hours. Calcined powder was ground and sieved through a 200 mesh sieve and was then divided into 3 parts and used as described below. Powder-1 was not coated with any gel. Powder-2 was coated with 2.5 mole% silica gel and powder-3 was coated with 2.5 mole% titania gel. Sol-gel coating was done in a way similar to that described in reference 21. A schematic flow diagram for this process is shown in Figure 23-1.

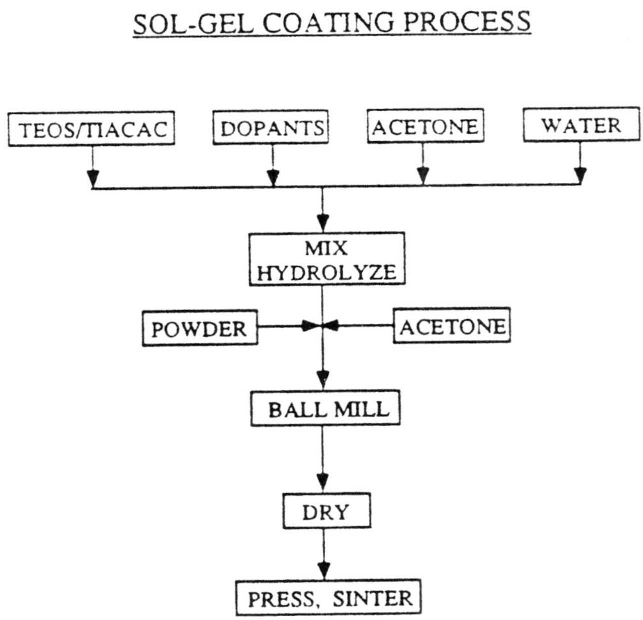

Figure 23-1. Flow diagram of the sol-gel coating process.

---

[3]Fisher Scientific, Rochester, NY
[4]Aldrich Chemical Co., Inc., Milwaukee, WI
[5]Zircoa Inc., Solon, OH

It was found that in order to apply the sol-gel coating to $YBa_2Cu_3O_x$ powders, acetone rather than ethyl alcohol should be used as a cosolvent for mixing H_2O and other alkoxides (23). $YBa_2Cu_3O_x$ phase reacted with alcohol and formed small amounts of Y_2BaCuO_5, $Ba(OH)_2$, $BaCO_3$ and CuO, whereas it was stable in acetone (Figure 23-2).

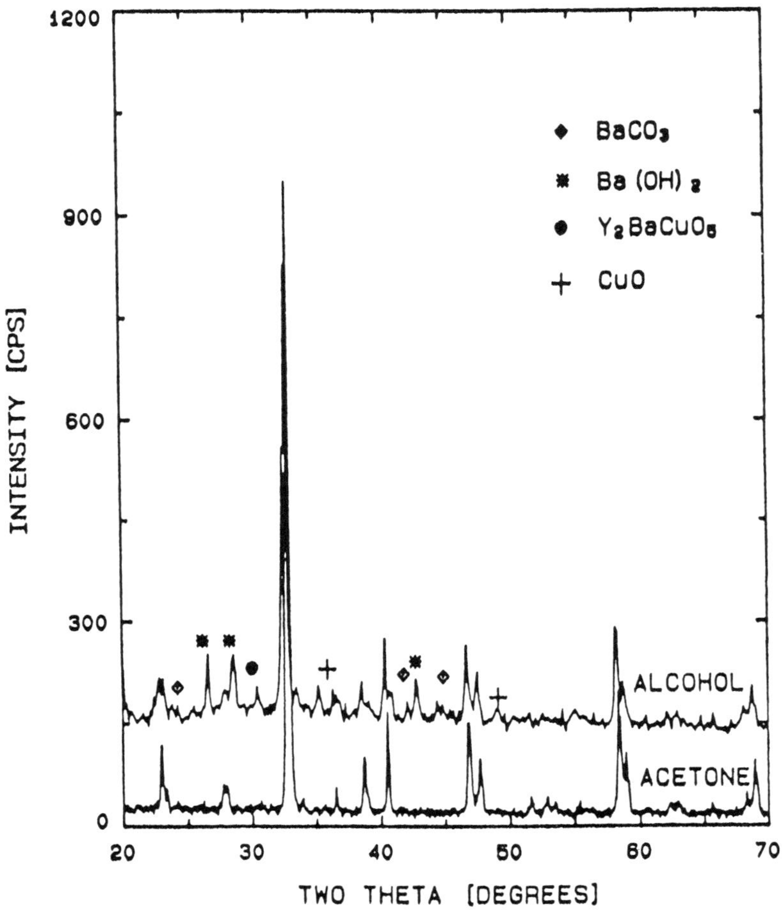

Figure 23-2. XRD pattern of $YBA_2Cu_3O_x$ powder coated with 2.5mol% silica gel using alcohol and acetone as two different cosolvents.

Coated powders were dried and pressed in the shape of bars under a pressure of 13,790 N/cm^2 (20,000 psi). Silica gel-coated samples were sintered at 940°C for 12 hours and annealed at 570°C for 6 hours, whereas samples without coating and samples with titania gel-coating were sintered at 970°C for 6 hours and annealed at 570°C for 6 hours in flowing air.

These sintered bars were placed on a wire mesh in a desiccator containing water. The water level in the desiccator was maintained below the sample level to prevent a direct contact between samples and water. The desiccator was kept in an oven at 95°C. The chamber inside the desiccator was always filled with water vapor. 100% relative humidity was assumed. Samples were taken out of the desiccator at different times (0.0 hour, 1.0 hour, 2.5 hours, 4.0 hours and 7.0 hours from the beginning) for analysis.

CHARACTERIZATION

Direct resistivity measurements were made on uncorroded samples of silica-gel coated, titania-gel coated, and uncoated bars by the standard four point probe resistivity method. X-ray diffraction was done on the powdered bars and not on the solid bar surface in order to see the effect of corrosion reaction on the bulk. Samples 1, 2 and 3 each with different corrosion times were characterized using x-ray diffraction[6]. The uncorroded samples (corrosion time = 0.0 hour) of sample 1, sample 2 and sample 3 were characterized using an Analytical Scanning Transmission Electron Microscope[7] (ASTEM). Microstructure analysis was also performed on corroded samples using a scanning electron microscope with EDS capabilities[8].

RESULTS AND DISCUSSION

The resistivity measurements (Figure 23-3) showed only a slight change (~2-3K) in the transition temperature and also a slight increase in the resistivity above T_c (critical superconducting transition temperature) of coated $YBa_2Cu_3O_{7-x}$ superconductors from that of uncoated samples. Titania-gel coated samples showed a T_c of 86.8K while silica-gel coated samples showed a T_c of 87.5K. Uncoated samples showed a T_c of 89.5K. These transition temperatures were taken as the temperature midway between 5% and 95% of resistive transition. Lower transition temperatures on silica and titania substitution have been reported in the literature. Slightly higher resistance above the transition temperature can be attributed to the increased grain boundary resistivity.

The x-ray analysis (Figures 23-4-6) showed that as the corrosion time increases - the fraction of $YBa_2Cu_3O_x$ decreases, and the amount of Y_2BaCuO_5, $BaCO_3$, CuO, $Ba(OH)_2 \cdot H_2O$ and $CuO \cdot 3H_2O$ increases steadily.

The uncoated samples and the ones coated with silica swelled to about 1.5 times their initial volume after 7 hours corrosion. SEM analysis of the corroded

---

[6]Philips Electronic Instruments, Mount Vernon, NY

[7]JEM 2000-FX JEOL, Japan

[8]ETEC Corp. Hayward, CA

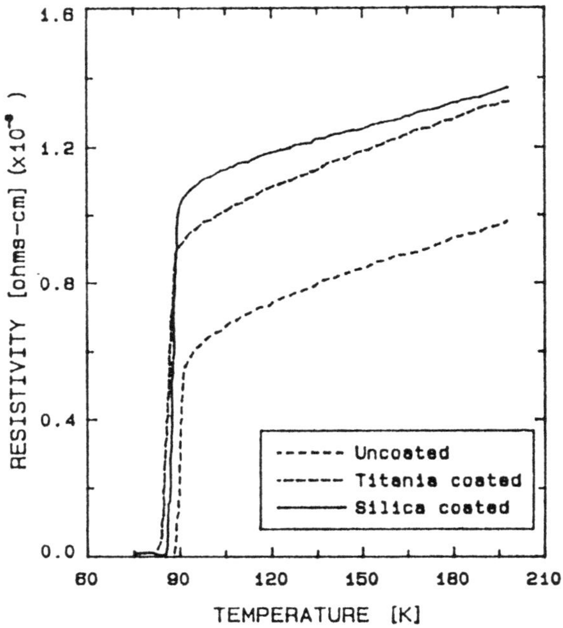

Figure 23-3. 4-Point resistivity data of coated and uncoated $YBa_2Cu_3O_{7-x}$ superconductor samples.

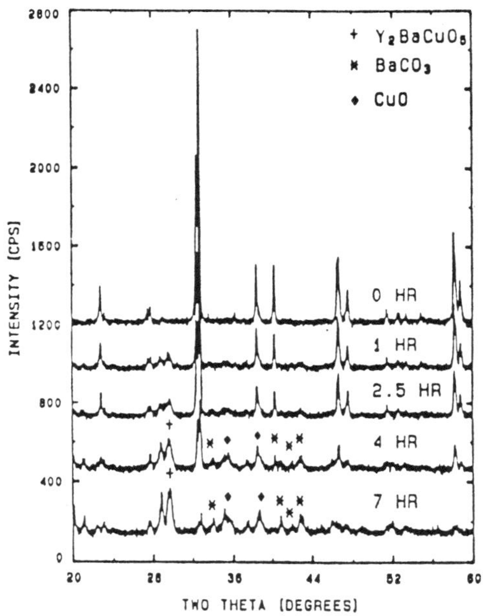

Figure 23-4. XRD pattern of $YBa_2Cu_3O_x$ uncoated samples after corrosion.

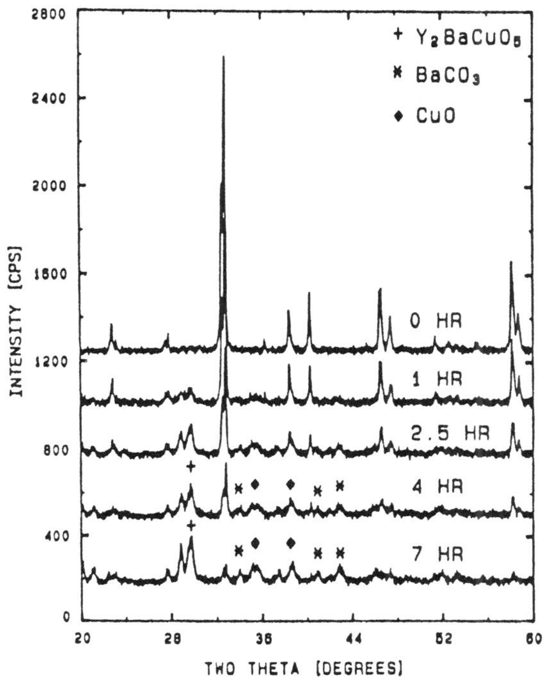

Figure 23-5. XRD pattern of $YBa_2Cu_3O_x$ silica (gel) coated samples after corrosion.

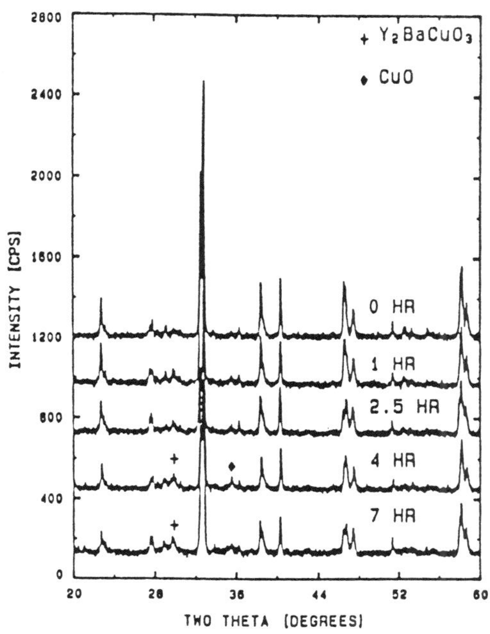

Figure 23-6. XRD pattern of $YBa_2Cu_3O_x$ titania (gel) coated samples after corrosion.

samples showed that with increasing corrosion time the microstructure disintegrated and became more open (Figure 23-7a-c). These results indicated that the dissociation of the $YBa_2Cu_3O_x$ structure occurred from the surface and went deeper with the corrosion time. For this reason x-ray diffraction was not carried out on the bulk surface, but rather on the crushed powder.

The x-ray diffraction pattern of silica-coated samples (Figure 23-5) does not show any improvement over the uncoated samples, but that of titania-coated samples shows a remarkable difference in the degradation behavior (Figure 23-6). Though, these results indicate that the corrosion of $YBa_2Cu_3O_x$ cannot be eliminated completely, they do show that coating with titania can impede the corrosion process substantially. TEM studies of sol-gel coated powders before sintering, done by Brooks (24), have shown that a uniform, very thin gel layer of a few Angstroms thickness was developed on the surface of the powder particles. This layer of coating after sintering can either remain at the grain boundary as a second phase, or go to triple points, or go into the lattice, or even react with the material to form a new phase depending on the reactivity of this coating phase with the material. TEM studies of sintered uncoated, silica-coated, and titania-coated samples are shown in Figure 23-8. Figure 23-8a shows the grain boundary of an uncoated specimen. In studies of silica-coated sintered samples, no layer could be seen on the grain boundary (Figure 23-8b). Figure 23-8c shows a TEM micrograph of the titania-coated sintered sample. Here a second phase was seen at the grain boundary. This phase was not continuous and covered a very small fraction of the grain boundary region. The x-ray beam could not be focused on that small region to prove that it was titania. But this second phase was not sufficient to conclude that the titania coating has created a diffusion barrier at the grain boundary which prevented the reaction of $YBa_2Cu_3O_x$ with moisture. XPS studies done by Brow (25) have shown that the grain boundary of $YBa_2Cu_3O_x$ is very sensitive to the atmosphere and is affected first when exposed to a humid atmosphere.

Rosamilia, et al. (26) have shown that for the reaction:

$$Cu^{+3} + 0.5H_2O \rightarrow H^+ + 0.25O_2 + Cu^{+2} \quad (23\text{-}3)$$

the half-cell voltage versus the standard calomel electrode (SCE) in acid media changes from 0.866 V at a pH of 2 to 0.807 V at a pH of 3. These potentials are above that of the O_2/H_2O electrode. This implies that the reduction of Cu^{+3} will be more energetically favorable than the hydrolysis of water. Thus the water reaction is a concern for high temperature cuperate superconductors. Ti^{+4} has a comparable ionic size to Cu^{+3} ($\pm 15\%$) and so it can substitute at the Cu^{+3} site in the lattice. Since Ti is introduced from sol-gel coating, it is expected to be near the grain boundary at the Cu^{+3} sites. Ti^{+4} can be expected to diffuse into the grain exponentially following the Arrhenius behavior. Thus a Ti-rich skin would be formed on the grains, with Ti occupying Cu sites. As the corrosion reaction starts from the grain surface the electrode potential of Ti^{+4}/Ti^{+3} starts

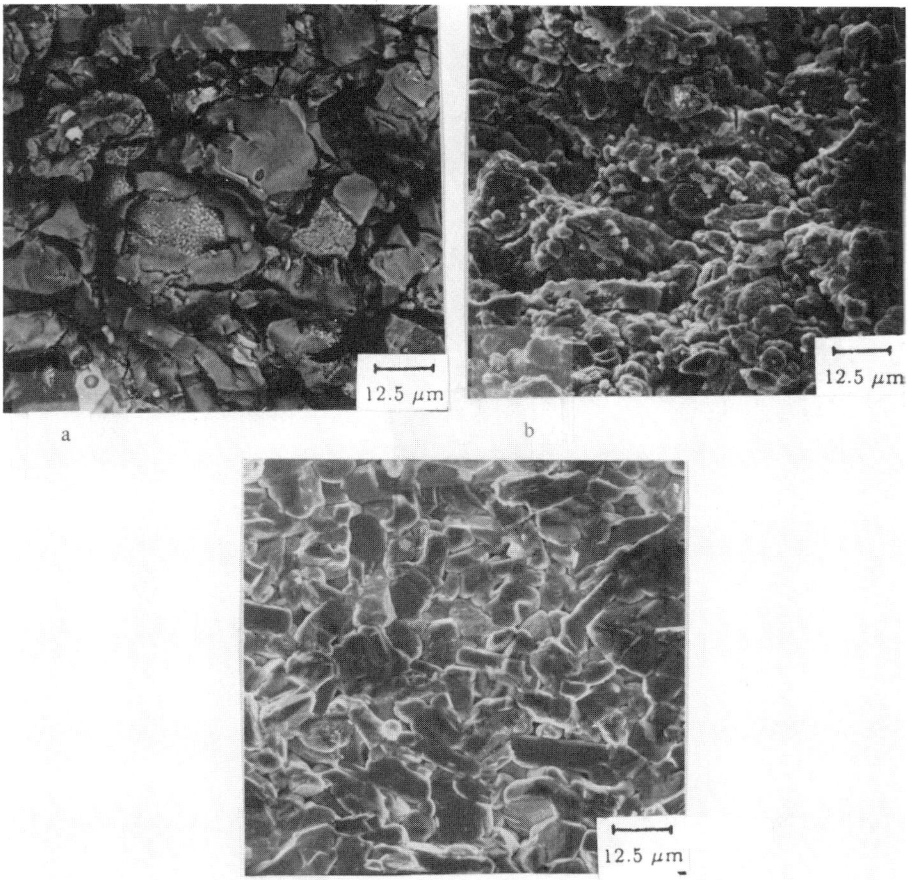

Figure 23-7. SEM micrographs of sintered $YBa_2Cu_3O_x$ samples. a) uncoated sample after 7 hours corrosion, fractured surface, backscattered image at 800X; b) coated sample with 2.5mol% silica (gel) after 7 hours corrosion, fractured surface, secondary image at 800X; c) coated sample with 2.5mol% titania (gel) after 7 hours corrosion, fractured surface, secondary image at 800X.

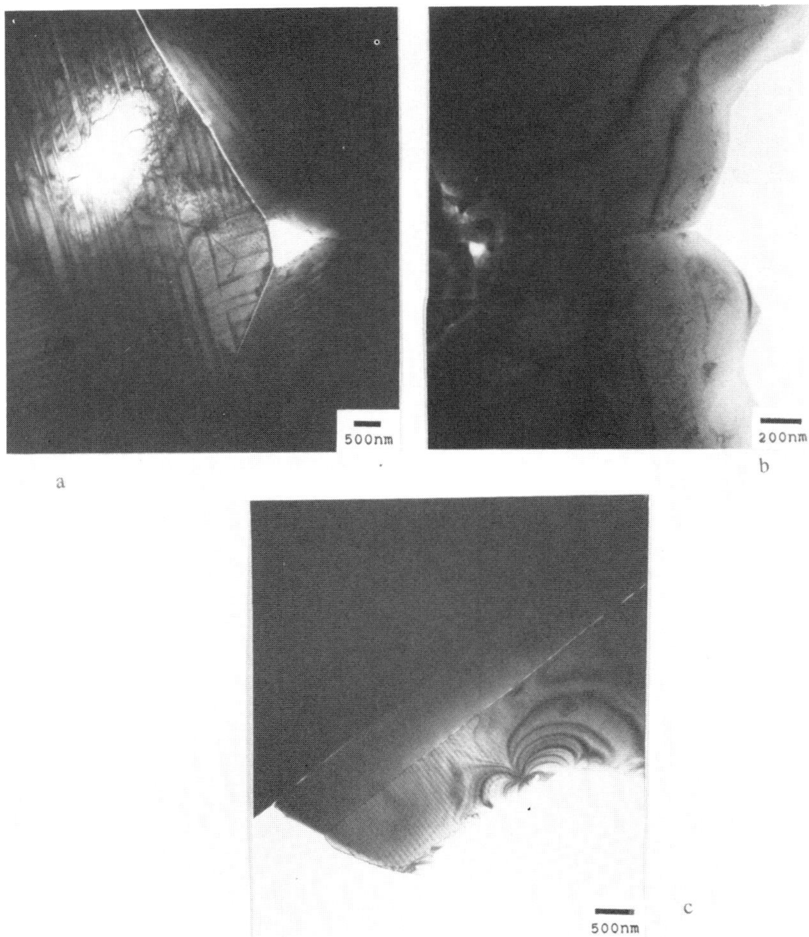

Figure 23-8. TEM micrographs of sintered $YBa_2Cu_3O_x$ samples. a) uncoated sample after 7 hours corrosion, b) coated sample with 2.5mol% silica; c) coated sample with 2.5mol% titania.

competing with O_2/H_2O equilibrium. It turns out that in the 0.4 - 2.6 pH range the half cell potential of Ti^{+4}/Ti^{+3} is lower than that of the O_2/H_2O electrode, giving rise to hydrolysis of water instead of Ti^{+4} reduction. Since all Cu^{+3} are not replaced by Ti^{+4} near the grain surface, there would be some Cu^{+3} present which for the above-mentioned reason would reduce to Cu^{+2}. This explains why we saw a restricted corrosion in titania-coated samples instead of a complete elimination of corrosion. Silica-coated samples do not show any resistance to corrosion due to silica's inability to act like the TiO_2 layer on the surface of $YBa_2Cu_3O_x$.

Similar electrochemical effects may be responsible for the improved stability of 123 compound upon treatment with NF_3 gas. Several authors (27-29) have demonstrated that these fluoride treatments could dramatically increase corrosion resistance. As demonstrated in the case of PTCR thermistors (22), addition of TiO_2 and F via a sol-gel coating process will allow for the improved stability of 123 superconductors against corrosion in a humid environment. Passivating Ag layers may also be incorporated via the sol-gel coating process to improve stability against corrosion and with increased grain orientation for the improvement of critical current density, another major limitation for the application of high T_c superconductors in practical devices.

CONCLUSIONS

A sol-gel coating technique was developed to coat the ceramic superconducting $YBa_2Cu_3O_x$ powder. Acetone was found to be a better cosolvent for mixing alkoxides with water, instead of alcohol. The effect of sol-gel coating on the water corrosion of $YBa_2Cu_3O_x$ samples was studied. Studies showed that the uncoated superconductor was very unstable in a humid atmosphere at high temperatures. Sol-gel coating with silica did not prevent the corrosion of $YBa_2Cu_3O_x$, while sol-gel coating with titania impeded the corrosion quite considerably. Corrosion in humid atmosphere of $YBa_2Cu_3O_x$ superconductors may be explained as being due to the reduction of Cu^{+3}. TiO_2 coating allowed for the formation of a Ti layer on the surface, substituting for Cu in $YBa_2Cu_3O_x$, and result in a significant resistance to corrosion. Silica-coated samples do not show any resistance to corrosion with moisture due to the inability of silica to act in a similar fashion.

ACKNOWLEDGEMENTS

The authors would like to thank the Center for Advanced Ceramic Technology (CACT) at Alfred University for supporting this research.

REFERENCES

1. Bednorz, J.G., Muller, K.A., *Z. Phys.* B 64: 189 (1986).
2. Wu, M.K., Ashburn, J.R., Torng, C.J., Hor, P.H., Meng, R.I., Gao, L., Huang, Z.J., Wang, Q., and Chu, C.W., *Phys. Rev. Lett.* 58(9): 908-910 (1987).
3. Maeda, H., Tanaka, Y., Fukutomi, M., Asano, T., *Jpn. J. Appl. Phys. Lett.* 27(2): L209-L210 (1988).
4. Sheng, Z.Z., and Hermann, A.,M., *Nature* 332: 55-58 (1988).
5. Hyde, B.G., Thompson, J.G., Withers, R.L., FitzGerald, J.G., Stewart, A.M., Bevan, D.J.M., Anderson, J.S., Bitmead, J., and Paterson, M.S., *Nature* 327: 402-403 (1987).
6. Yan, M.F., Barns, R.L., O'Bryan, H.M., Jr., Gallaghar, P.K., Sherwood, R.C., and Jin, S., *Appl. Phys. Lett.* 51(7): 532-534 (1987).
7. Wang, J., Stevens, R., and Bultitude, J., *J. Mat. Sci.*, 23: 3393-3397 (1988).
8. Frase, K.G., Liniger, E.G., and Clarke, D.R., *Adv. Ceram. Mat.* 2(3B): 698-700, Special Issue (1987).
9. Bansal, N.P., and Sandkuhl, A., *Appl. Phys. Lett.* 52(4): 323-325 (1988).
10. Barkatt, A., Hojaji, H., and Michael, K.A., *Adv. Ceram. Mater.* 2(3B): 701-709 (1987).
11. Komori, K., Kozuka, H., and Sakka, S., *J. Mat. Sci.* 24: 1889-1894 (1989).
12. Garland, M.M., *J. Mat. Res.* 3(5): 830-831 (1988).
13. Karioris, F.G., and Vance, E.R., *Mat. Lett.* 6(1-2): 16-18 (1987).
14. Kumar, B., Harmer, R.S., Wittberger, T.N., Cordonnier, S.K., and Rai, A.K., *J. Mat. Sci.* 23: 3879-3883 (1988).
15. Barns, R.L., and Laudise, R.A., *Appl. Phys. Lett.* 51(17): 1373-1375 (1987).
16. Chang, C., *Appl. Phys. Lett.* 53(12): 1113-1115 (1988).
17. Hill, D.M., Meyer, H.M., III, and Weaver, J.H., *Appl. Phys. Lett.* 53(17): 1657-1659, (1988).
18. Vasquez, R.P., Hunt, B.D., and Foote, M.C., *Appl. Phys. Lett.* 54(23): 2373-2375 (1989).
19. Vasquez, R.P., Foote, M.C., and Hunt, B.D., *Appl. Phys. Lett.* 55(17): 1801-1803 (1989).
20. Sainamthip, P., *Microstructure Control and Grain Boundary Engineering of Chemically Derived MnZn Ferrites*, Ph.D. Thesis, Alfred University (1987).
21. Selmi, F.A., Amarakoon, V.R.W., *Ceram. Eng. Soc. Proc* , 8(9-10) 1120-1127 (1987).
22. Selmi, F.A., Amarakoon, V.R.W., *J. Am. Ceram. Soc.* 71(11): 934-937 (1988).
23. Bayya, S.S., *Grain Boundary Engineering of* $YBa_2Cu_3O7\text{-}x$ *Superconductor Via Sol-Gel Coating*, M.S. Thesis, Alfred University, Alfred, NY (1989).
24. Brooks, K.G., *Magnetic Properties of Sol-Gel Coated Li-Zn Ferrite*, presented at the American Ceramic Society 90th Annual Meeting (1988).

25. Brow, R.K. in: *Research Update 1988, Ceramic Superconductors II*, pp 598-606 Am. Ceram. Soc. Pub. (1988).
26. Rosamilia, J.M., Miler, B., Schneemeyer, L.F., Waszczak, J.W., and O'Bryan, H.M., Jr., *J. Electrochem. Soc.* 134(7): 1863-1864 (1987).
27. Tressaud, A., Chevalier, B., Lepine, B., Dance, J.M., Lozano, L., Grannec, J., Etourneau, J., Tournier, R., Sulpice, A., and Lejay, P., *Mod. Phys. Lett.* B 2(10): 1183-1188 (1988).
28. Yoshimura, M., Ishikawa, Y., Inoue, S., and Somiya, S., *MRS Int'l Mtg. on Adv. Mats. Proc.*, Vol. 6, pp 421-426 (1989).
29. Axelson, S., LaGraff, J., and Taylor, J., submitted to *J. Am. Ceram. Soc.* for publication.

24

Summary

D.E. Clark, R.L. Schulz, and B.K. Zoitos

University of Florida
Gainesville, Florida

INTRODUCTION

Glass and ceramics are important engineering materials used in a wide range of demanding applications that include ground and air transportation, hazardous waste disposal, communications, energy generation, medical devices, housing, and food and beverage packaging. Recent advances in the development of high T_c superconductors will result in even more widespread use of ceramics. Products fabricated from these materials will include medical diagnostic equipment, strong magnets for levitating high speed trains, high efficiency motors, zero resistance wires for power transmission and microwave devices. These ceramic materials exhibit their superconducting properties at liquid nitrogen temperatures (instead of the liquid helium temperatures required for metallic superconductors) which makes their use practical and economical. It is worth noting that glass and ceramics are used over a wide range of temperatures (from cryogenic to white-hot) and environmental conditions (air, vacuum, reactive and inert gases). Thus, their environmental stability is of major importance. Indeed, environmental stability is one of the main reasons that glass and ceramics are selected over other materials.

All materials change when subjected to a changing environment (Table 24-1). However, with proper understanding these changes can be controlled and in some cases used to advantage. The objective of this book has been to summarize in detail our current understanding of material/environmental interactions based on theoretical models, experimental results and practical examples. As with any field of science and engineering our breadth and depth of understanding of corrosion is continually improving. Thus it is not possible to cover every aspect of the field in one book. In the previous twenty three chapters we selected topics in which extensive research has been conducted in

Table 24-1. Factors Affecting Glass and Ceramic Corrosion

I. Environmental Factors
1. Temperature
2. Exposure Time (continuous or cycled)
3. Humidity
4. Solution pH
5. External Stresses
6. Radiation
7. Solution Composition (i.e., deionized water vs. ground water; presence of cations, complexing agents, colloids)
8. Saturation and Approach to Equilibrium
9. Pressure
10. Atmosphere Composition (i.e., concentration of CO_2, SO_2, etc.)
11. Solution Eh

II. Physical Factors
1. Weathering vs. Aqueous Corrosion
2. Dynamic (Stirring) vs. Static vs. Flow vs. Cyclic Exposures
3. Exposed Surface Area-to-Solution Volume Ratio (SA/V)
4. Bulk vs. Powders vs. Fibers

III. Specimen State
1. Bulk Composition
2. Thermal History
   a. degree of annealing for stress and density
   b. phase separation (glass in glass)
   c. % crystallization
3. Prior Corrosion Exposure History
4. Surface Features
   a. surface roughness (i.e., as formed vs. polished surfaces)
   b. surface composition
   c. surface flaws
5. Homogeneity
6. Surface Treatment
7. Surface Charge
8. Film Stability
9. Oxidation State of Constituents

the last 10-20 years. This research has provided a strong scientific understanding on which the chapters were based.

In this chapter we briefly summarize some the important corrosion phenomena that should be more thoroughly addressed in future work. These include zeta potential and its role in glass corrosion, surface passivation, stress corrosion, glass and ceramic electrodes, and electrochemical corrosion.

ZETA POTENTIAL AND ITS ROLE IN GLASS CORROSION

Background

When a glass is subjected to an aqueous medium an electrical charge develops on the surface of the glass due to the adsorption and desorption of ions between the surface and surrounding solution. The charge at the surface is balanced by a layer of oppositely charged ions (counter-ions) known as the Stern layer. Additional counter-ions are attracted to the surface, but are repelled by the Stern layer causing the development of a diffuse layer of counter-ions. Together these two layers are referred to as the electrical double layer (1).

The electrical potential at or near the outer limit of the diffuse layer is referred to as the zeta potential and determines the electrostatic interaction force between particles (2). Zeta potential plays an important role in determining colloid stability. The formation of colloids has a direct effect on precipitation layers that may form on glasses during leaching as well as influencing saturation effects of the leaching medium.

Measurement Techniques

Zeta potential values are obtained using a variety of techniques. A brief description of the more commonly used techniques as well as a recently developed technique follows.

Microelectrophoresis. In microelectrophoresis a stereomicroscope is used to observe colloidal particles inside an electrophoresis chamber. Electrodes attached at each end of the cell create an electric field across the chamber. Charged colloids move in the field and by measuring particle velocity and knowing the field strength, it is possible to calculate the electrophoretic mobility, μ, which is a measure of particle velocity (V) in an electric field (E) (μ_E = V/E). The zeta potential is calculated from the Helmholtz-Smoluchowski equation (3-5).

$$\mu_E = (4\pi\varepsilon_o) \cdot \frac{D\zeta}{4\pi\eta} = \frac{\varepsilon\zeta}{\eta} \tag{24-1}$$

where: μ_E = electrophoretic mobility
ε_o = permittivity of free space
D = relative dielectric constant of solution
ε = dielectric constant of solution
ζ = zeta potential
η = viscosity

Streaming potential. In streaming potential measurements the potential is determined by measuring the potential difference between the ends of a porous plug of particles when a liquid is forced through the plug. Equation 24-2 may be used to determine zeta potential from information obtained using this technique (6).

$$\zeta = (4\pi\eta\lambda E)/(\varepsilon P) \tag{24-2}$$

where:

ζ	=	zeta potential
η	=	viscosity
E	=	streaming potential
ε	=	dielectric constant of the liquid
P	=	applied pressure difference
λ	=	specific conductance

Electrokinetic Sonic Amplitude (ESA). ESA measurements are a relatively new means of determining electrophoretic mobility, μ. The premise that sound waves could generate electric fields in a charged particle system was postulated by DeBye in 1933 (4). The ESA instrument developed by Matec[1] works on the following principles. When an alternating field is applied across a colloidal suspension, a pressure wave is generated by the movement of charged colloidal particles. The field causes the charged particles to be displaced and then relax toward their initial positions. When the pulse rates are sufficiently high, the movement of the particles creates an acoustic pressure wave that impinges on a transducer. The transducer then converts the mechanical wave to an electrical impulse, the magnitude of which is termed the electrokinetic sonic amplitude with units of Pascals/volt/meter. The dynamic mobility measured by ESA techniques is not the same as the mobility measured by low frequency or microelectrophoresis methods because of particle inertia effects. Particle velocities are out of phase with the applied field and the higher the frequency of the applied field the more the particle will lag the field. Therefore, dynamic mobility is a function of frequency, particle size and density (7,8). Zeta potential can be calculated from the following relationship provided the suspension contains spherical particles with a thin double layer and low zeta potential (7,8):

$$\mu_a = \varepsilon\zeta/\eta \; G(\omega a^2/\nu) \tag{24-3a}$$

where:		
μ_d	=	dynamic mobility
a	=	particle radius
ε	=	dielectric constant of suspension
ν	=	kinematic viscosity of the liquid (η/ρ)
ω	=	angular frequency
η	=	viscosity of the liquid

[1] Matec Applied Sciences, Inc. Hopkinton, MA

and G(α), an inertial term equal to:

$$G(\alpha) = [1\text{-}1/9i\alpha(3 + 2\Delta\rho/\rho)/\{1 + (1\text{-}i)\sqrt{\alpha/2}\}]^{-1} \qquad (24\text{-}3b)$$

where: $\alpha = \omega a^2/\nu$

$\Delta\rho/\rho = (\text{density}_{particle} - \text{density}_{liquid})/\text{density}_{liquid}$

i = imaginary ($\sqrt{-1}$)

Direct calculation of zeta potential from ESA measurements does not hold for highly concentrated solutions.

Glass Corrosion

Several researchers have investigated changes in zeta potential values during glass corrosion experiments to determine the effect the electrical field has on glass leaching (5,9-15) and while some dispute its importance, (9) others feel it is an important tool in understanding glass corrosion.

Mularie et al. (10) showed that the addition of a surface potential term to first order diffusion equations gave a more realistic representation of leaching kinetics and suggested that the parameters which affect the double layer (i.e., pH, T, impurities) will in turn influence the leach rate.

Horn and Onoda (6) used zeta potential measurements to determine the surface charge of vitreous silica and soda-lime-silicate glasses in various electrolyte solutions. They found it was possible to reverse the sign of the surface charge of vitreous silica by adding aluminum chloride to the solution. Further study indicated the positive surface charge created by the adsorption of Al^+ ions could be masked by exposing the surface to a citrate ion solution.

Lee et al. (11,12) investigated the effect of leaching on zeta potential measurements using borosilicate glasses (SRL-131-29.8% TDS) and proposed the following relationship between glass leaching, zeta potential, surface film and colloid formation. Upon immersion in an aqueous solution a negative surface charge develops due to the release of alkali ions and $\equiv SiO^-$ groups are produced. These groups react with water to form silanol groups. As a result, the solution pH increases and zeta potential decreases. Metal cations (i.e. Al^+,) are absorbed on the glass surface when the solution pH is increased sufficiently to allow hydrolysis of the cation to occur. Charged particles are formed in the leachate due to reactions between the metal cations and monosilicic acid. Figure 24-1 shows a schematic representation of this process. Whether or not the charged particles remain in solution or are absorbed onto the glass surface depends on their charge and the zeta potential. Those with the same charge as the surface and a high zeta potential will be repelled from the surface and may grow in solution by absorbing ions from the leachate. This process prevents saturation and increases the glass leach rate. Those with an opposite charge and low zeta potential will be attracted to the surface. The reversal of sign in zeta potential observed by Lee when $AlCl_3$ is added to the leachate is shown in Figure 24-2a,b. This figure also illustrates the change in zeta potential, pH, and specific conductance as a function of leaching time.

McVay and Buckwalter (15) investigated the effects of iron on glass leaching in various aqueous solutions (deionized water, tuff and basalt groundwaters) under static conditions and noted that the formation of iron silicate colloids have a significant effect on leaching reactions of borosilicate glass wasteforms (PNL 7668). The colloids remove elements from solution via precipitate formation and reduce saturation effects and thus, increase the leach rate. Further, it was found that when iron is present, elements removed from the glass are incorporated in the precipitates. They suggest additional studies are necessary to determine whether or not these precipitates may act as a potential means for radionuclide transport/entrapment within the repository.

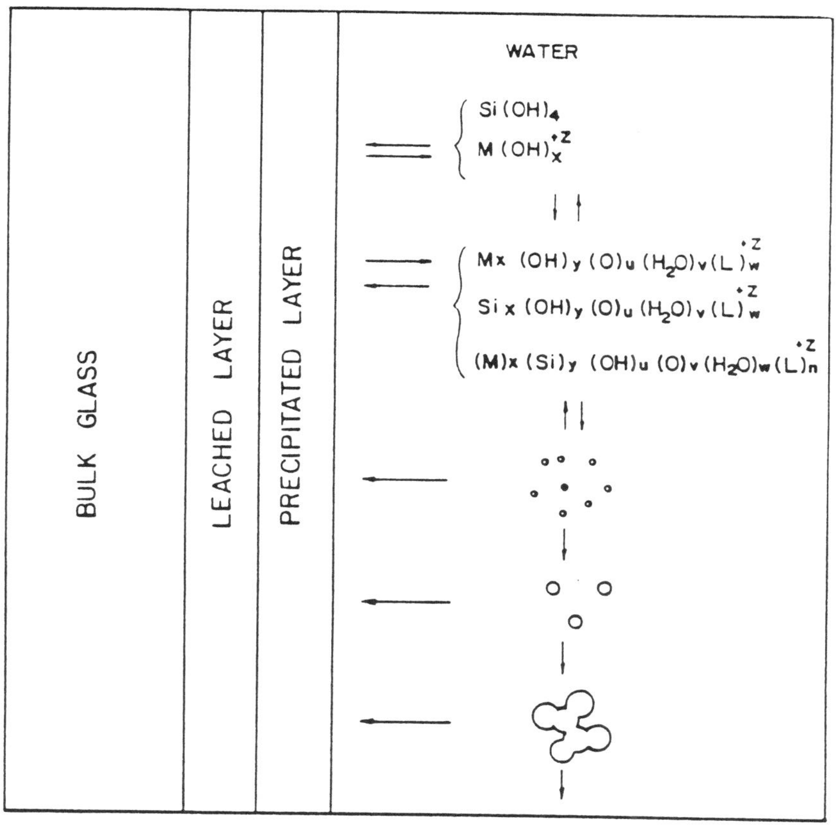

Figure 24-1. Schematic representation of mechanisms of colloid and surface layer formation during waste glass leaching (from Ref. 11).

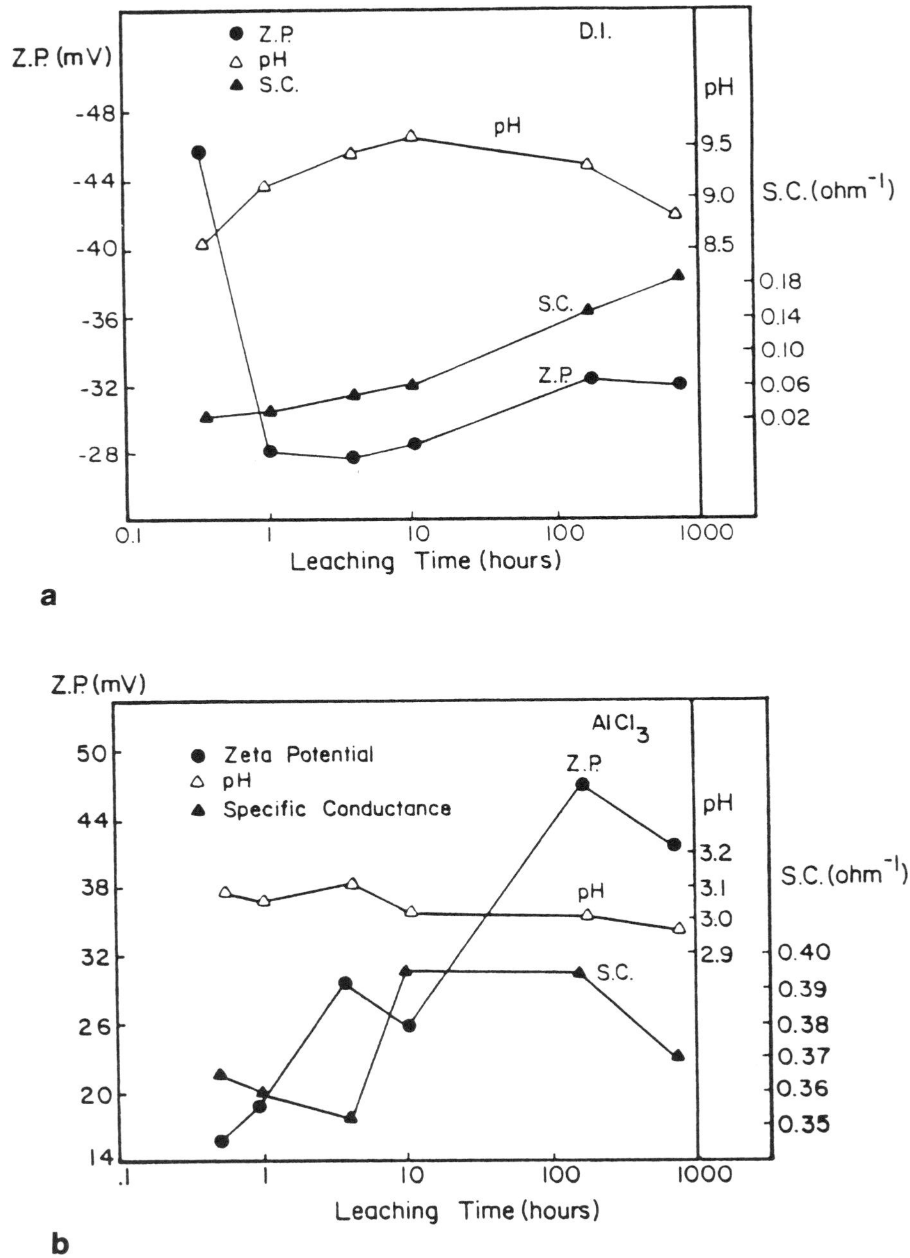

Figure 24-2. Time dependence of the zeta potential, specific conductance and solution pH of SRL 131-29.8% TDS waste glass (static conditions, 90°C). a) Deionized water and b) 1028 ppm $AlCl_3$ solution.

SURFACE PASSIVATION

Corrosion of glasses and ceramics usually results in the formation of a surface layer. At least five types of surface layers have been identified and are discussed in the book (see Hench's chapter). Both the chemistry and the structure of these layers can be complex (see Figure 1-5 White's chapter). Multilayers form under certain exposure conditions and the leachant leach layer can range in thickness from less than one micron to several hundred microns. Further, the densities of these layers vary substantially (16). Thus, it can be stated with certainty that the surfaces of most glasses and ceramics are not representative of the bulk since corrosion processes occur immediately after fabrication. Less certain is to what extent, if any, these surface layers protect the underlying material. However, numerous studies have shown that some passivation can be achieved by altering the surface chemistry (17-29).

Three approaches have been used to passivate glass surfaces. The most common approach has been to manipulate the bulk composition so as to achieve protective layers during normal exposure (25-28). Unfortunately, the bulk compositions that generally yield reduced corrosion are more difficult to melt and form (there are exceptions). Nevertheless, CaO and Al_2O_3 additions to soda silicate glasses can provide substantial improvement in these glasses as shown in Figure 24-3. Both the infrared reflection spectra and scanning electron micrographs reveal considerably less corrosion in the glass with CaO.

A second approach involves exposing the hot glass (during annealing) to reactive gases such as sulfur dioxide and water vapor (29). This results in the extraction of alkali from a few microns depth leaving a nearly pure silica surface that is more durable than the bulk. The alkali collects on the surface in the form of a sulfate salt that can be washed off when the glass is cooled.

A third method of surface passivation is to expose freshly prepared materials to special solutions. Iler (18) has shown that when pure silica is exposed to a solution containing aluminum ions, the solubility of the former is reduced. He attributed this phenomenon to the adsorption of aluminum ions onto the silica surface resulting in the formation of an alumino-silicate layer. The solubility of silica with respect to this layer is less than with respect to the underlying bulk silica. Hudson and Bacon (20) investigated a number of ions including beryllium, zinc, aluminum, strontium, antimony, barium, lead, bismuth, tin, zirconium and copper and found that all of these resulted in surface passivation to some extent with beryllium and zinc being the most effective. Tait and Jensen (21) have also found the presence of zinc ions in solution to be very effective in reducing corrosion of alkali borosilicate glass. Dilmore et al. (17) compared the corrosion mechanisms and kinetics of a lithia-alumina-silica glass to a lithia-silica glass (33L) exposed to solutions containing aluminum ions. They found that 50ppm of aluminum added to solution provided nearly the same protection to the 33L glass as did adding 2.5 mol% Al_2O_3 (33L-2.5A) directly to the glass composition (Figure 24-4). Interestingly, their work also

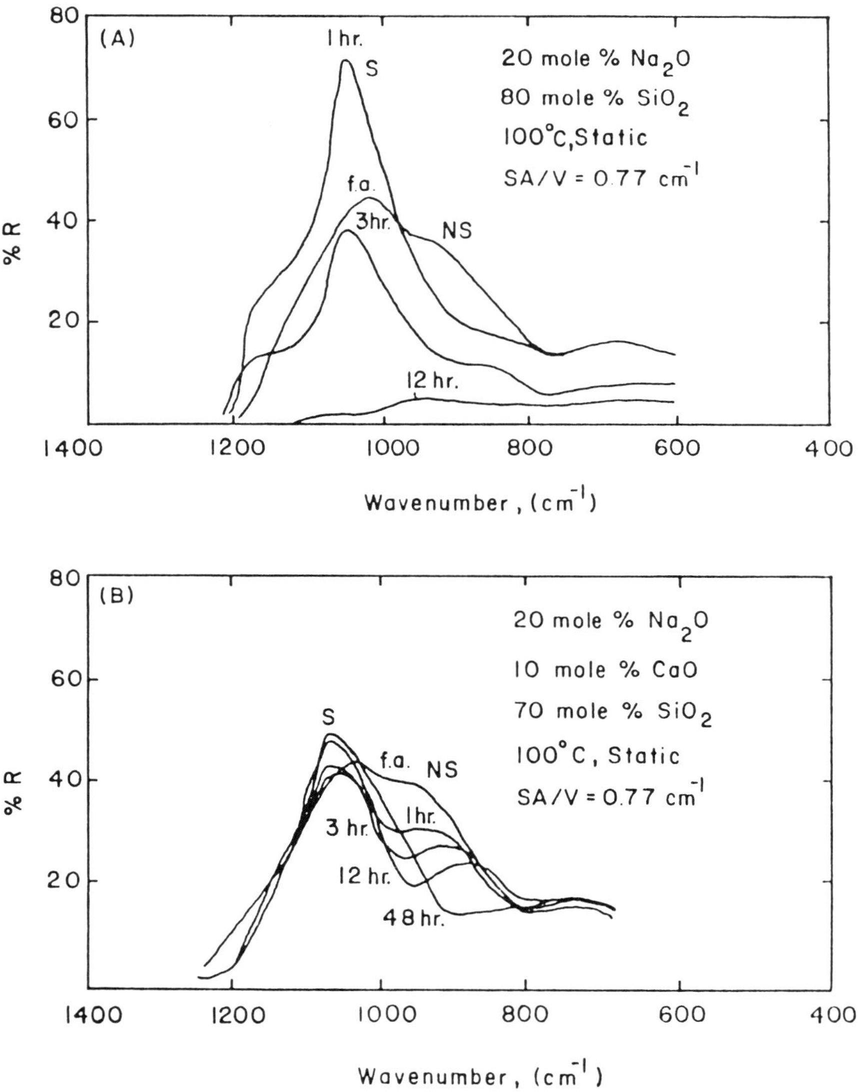

Figure 24-3a. Infrared reflection spectra for glass specimens corroded at 100°C in static aqueous solution. A). 20%Na_2O-80% SiO_2; and B). 20%Na_2O-10CaO-70%SiO_2 (from Ref. 27).

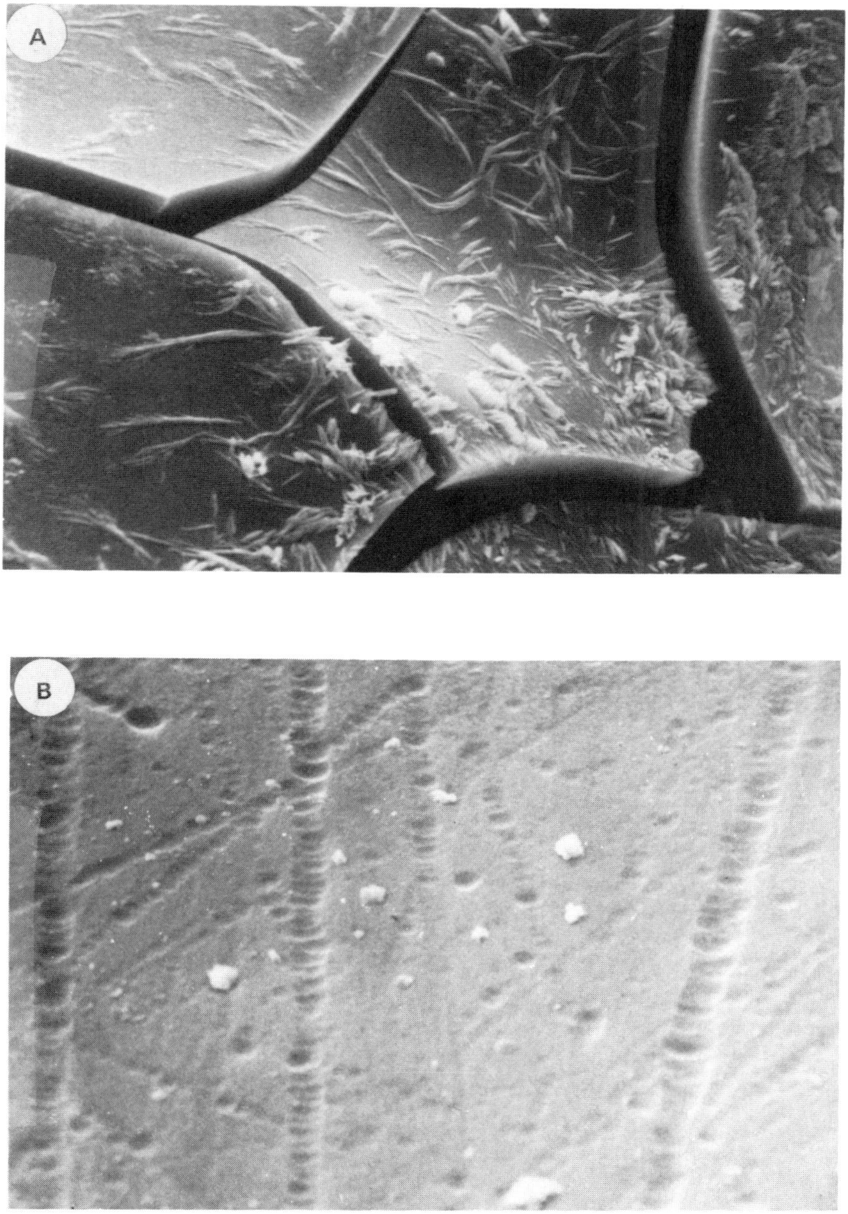

Figure 24-3b. Scanning electron micrographs of surfaces A). 20%Na_2O-80% SiO_2 glass corroded for 12 h and B) 20%Na_2O-10CaO-70%SiO_2 glass corroded for 9 days (from Ref 27).

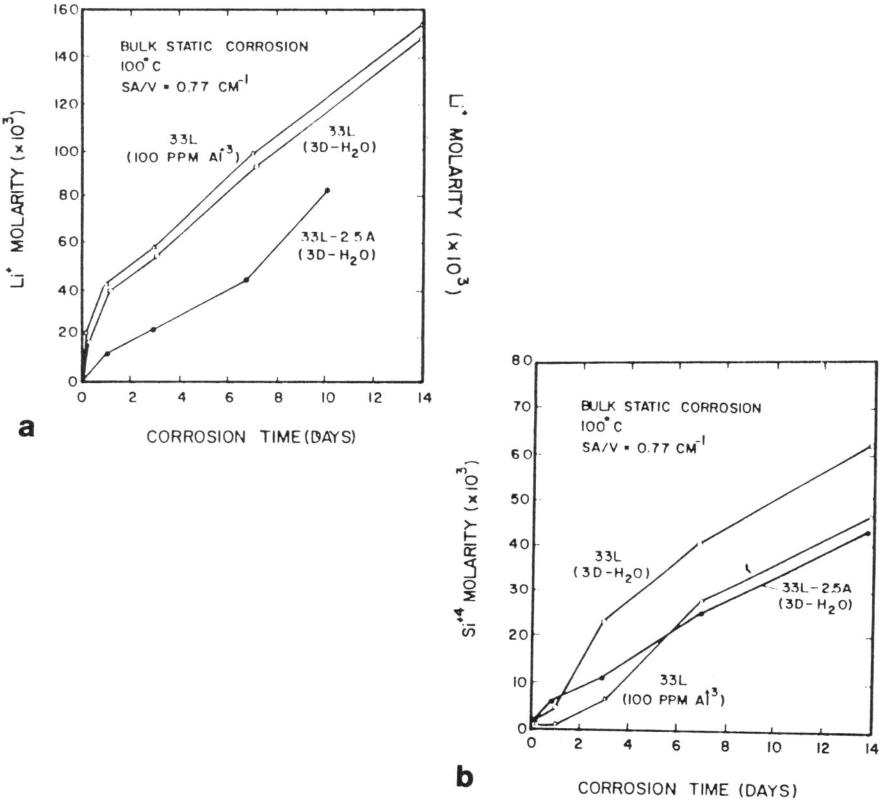

Figure 24-4. Solution data for bulk (33L and 33L-2.5A glasses corroded in triple-distilled water containing 100 ppm Al^{3+} ($AlCl_3$) showing concentrations of A). Li^+ and B). Si^{4+} (from Ref. 15).

demonstrated that the presence of aluminum in solution reduced the silica dissolution but had little effect on the dissolution of lithia. These results are consistent with the formation of an alumino-silicate surface layer that reduces solubility of silica but does not provide a barrier to diffusion of mobile species. That is, the layer behaves as a selective barrier.

Passivation through exposure to tailored solutions is a potentially valuable process that should be explored further. Additionally, these studies should be expanded to include technically important ceramics such as superconductors. Exposure to mixed solutions, stepwise exposure to multiple solutions and exposure of pre-corroded glasses to various tailored solutions can provide flexibility to the process and permit fabrication of surface layers engineered for specific functions.

STRESS CORROSION

When a glass is stressed, its rate of corrosion changes (30-34). Tensile stresses lead to enhanced corrosion rates with respect to the unstressed glass. This phenomenon has been explained in terms of the relative energies required to break the bonds in stressed materials. Since tensile stresses stretch the bonds, the amount of additional activation energy required to break the bonds or transport alkali through the structure during corrosion is reduced and hence the rate of corrosion increases. The same rationale can be used to explain the beneficial effect of compressive stresses.

Stress corrosion is particularly important when sharp cracks are present which is the usual case in brittle materials. It is well known that under load, the stress in the vicinity of the sharp crack tip is substantially greater than in the bulk material. When these materials, in the unstressed state, are exposed to water the sharp tips will become blunted and their mechanical performance will be improved. However, if these same materials are placed in water or a humid environment under tensile stress, the rate of crack growth can be enhanced. Wiederhorn (33) has identified three regions in the crack velocity versus stress diagram corresponding to: I) crack velocity is dependent on the applied load and, is controlled by the rate of reaction between water and the glass at the crack tip; II) crack velocity is independent of the applied load and is controlled by the rate of water transport to the crack tip, and; III) crack velocity is very sensitive to the applied load and is independent of corrosion processes.

Many factors such as the crack length, crack tip radius, surface energy and applied stress affect the crack or corrosion velocity. McColm (35) has quantified the effects of these parameters in the following equation:

$$V = K\exp - [E_o + \gamma (V_m / r) + \sigma_a (1 + (\frac{C}{r})^{1/2}) (\frac{\delta E}{\delta \sigma})_{\sigma=0}] / RT \qquad (24\text{-}4)$$

where:
- V = crack velocity or corrosion rate normal to the surface
- K = constant
- E_o = activation energy for corrosion
- γ = surface energy
- V_m = molar volume
- r = crack tip radius
- C = crack length
- σ_a = applied stress
- T = absolute temperature
- R = gas constant

When $\sigma_a = 0$, the second term in the exponential controls and the corrosion rate normal to the surface at the crack tip is less than the corrosion rate parallel to the surface. As a result, the crack tip becomes blunted.

Once a crack reaches a critical length under an applied load, crack velocity increases substantially and failure occurs. Universal fatigue curves are used to compare the mechanical performance of materials exposed to a wide range of environments. In these diagrams the normalized log time to failure (i.e, for cracks to reach a critical length) is plotted versus the normalized applied load (35).

Glasses and ceramics are being considered for use in many structural applications where both mechanical and chemical properties are important. Although progress has been made in recent years regarding the synergistic interaction between mechanical and chemical processes much research is still needed in order to better predict the performance of complex materials.

GLASS AND CERAMIC ELECTRODES

Glass electrodes have been used for many years to measure the concentration of hydrogen and other species in solution. In practice, the glass serves as an electrolyte or membrane that separates two solutions. One is a reference solution with known chemistry, while the other is the solution to be measured. Reference electrodes placed into these solutions measure the potential difference between them. There are two contributions to the measured potential; the boundary potentials between solutions and the glass and the diffusion potential (36). The boundary potential results from the exchange of ions between the glass surface and solution while the diffusion potential is due to the interdiffusion of ions in the glass. The electrode potential is the sum of these two contributions (36-38):

$$V = V_B + V_D \tag{24-5}$$

There is a V_B and V_D associated with each glass/solution interface (36). Their values depend on the glass composition and the solution chemistries. When the solution chemistry is the same on both sides of the glass membrane, V = 0. In the more typical case, the chemistries of solution are different and the value of V depends on the concentration of the particular ion in the unknown solution.

Doremus (37) provides an excellent overview of the glass electrode used to measure the hydrogen concentration ($-\log[H^+] = pH$). The electrode consists of a glass bulb into which a buffered solution of known hydrogen concentration and a reference electrode are placed. This assembly, together with a second reference electrode, is placed into the solution to be measured. In order for the electrode to function properly, steady state must be quickly established and the glass surface must be saturated with hydrogen ions. Under these conditions (37):

$$V = V_o + (2.3RT/F)\log[H^+] \tag{24-6}$$

where V_o = determined from standard solution

R	=	gas constant
T	=	absolute temperature
F	=	Faraday constant

In this case, the diffusion potential, V_D, is constant because the surface is always saturated with hydrogen ions and its contribution is accounted for in V_o. In modern glass electrodes, both reference electrodes are packaged inside the glass bulb.

Eisenmann (39) found that by altering the glass composition to include Al_2O_3, the glass electrode becomes more responsive to other ions. In fact, through careful tailoring of the glass composition the electrode can be made to be selectively responsive to ions such as Li, Na, and K. The more general equation describing this phenomenon is (39):

$$V = V_o + (2.3RT/F)\log[a_i^{1/n} + (K_{ij}^{Pot} \bullet a_{ij})^{1/n}] \qquad (24\text{-}7)$$

where a_i and a_j correspond to the concentrations of exchangeable ions from solution (e.g H and Na), n is a constant that depends on the glass composition and the particular pair of exchangeable ions, and K_{ij}^{Pot} is the selectivity. A value of 10 for K_{HNa}^{Pot} means that the glass electrode is 10 times more sensitive to Na than to H ions at equivalent concentrations. K_{ij}^{Pot} can be altered and controlled by adding Al_2O_3 to the glass.

Isard (36) provides a historical review of the glass electrode with respect to glass composition. The first commercial glass electrodes were developed at Corning in the late 1920's and were essentially soda-lime-silicates. In the 1940's glass compositions containing lithia as the major alkali oxide were developed. These electrodes were less sensitive to other cations in solution and were more durable.

Glass compositions for electrodes are selected based on two important parameters; electrical resistivity and chemical durability. In order to make accurate potential measurements, the resistivity of the glass should be less than 10^{12} ohm-cm (36). Further, the glass should be chemically durable.

The role of glass corrosion on electrode potential has not been evaluated in depth. Both sources of potential, boundary and diffusion, are a result of surface or near surface reactions. Obviously these reactions are beneficial and essential to the operation of the electrode. However, other types of corrosion reactions, such as network dissolution can be detrimental to the electrode's performance.

Ceramic electrodes of ZrO_2, TiO_2 and ß-Al_2O_3 are currently used in gas sensors and fuel cells. Many of the same principles discussed for glass electrodes can also be applied to ceramic membranes. Their major role is to separate liquids, gases and molten salts so that the reactions that occur at the interfaces can be used for monitoring purposes or for energy generation. These ceramics must be fairly conductive and resistant to corrosion in the environments that they are expected to operate.

There appears to be still much opportunity in the design and development of glass and ceramic electrodes that can operate reliably in harsh environments.

ELECTROCHEMICAL CORROSION

Although the science of electrochemical corrosion is well developed for metals, its application to glass and ceramics has been limited. The reason for this is that these latter materials are electrical insulators and therefore they have little tendency to give up electrons. Further, since they are electrical insulators, measurement of the corrosion current is difficult and requires very sensitive equipment. In a typical electrochemical corrosion experiment, the material to be tested is fabricated into an electrode and immersed into the test solution. If the electrode material is a metal dipped into an acidic solution we might expect the following reactions at the metal/solution interface:

$$m^{\bullet} \rightarrow m^{n+} + ne^{-} \qquad \text{anodic reaction}$$

$$nH^{+} + ne^{-} \rightarrow (n/2)H_2 \qquad \text{cathodic reaction}$$

In this case, the corrosion rate can be measured by periodically weighing the anode.

$$CR = \frac{\Delta m}{t} \tag{24-8}$$

where Δm is the mass loss in time t.

If the metal is connected to an inert electrode and a reference electrode, also immersed into the solution, the electrons generated at the anode can be transferred to the cathode where H_2 will be released. Using Faraday's law, the mass corroded can be derived from the current flowing between the electrodes and the corrosion rate can be determined independent of weight loss. The value of this technique is that it allows corrosion rates to be obtained in short times where weight loss experiments are usually not very sensitive.

Divakar et al. (40) have recently reported the use of electrochemical methods for measuring the corrosion rates of SiC. Corrosion rates were determined using the following equations:

$$CR = \frac{I_{corr} A\,(EW)}{F} \quad \text{g/s} \tag{24-9}$$

$$CR = \frac{0.13 I_{corr}(EW)}{d} \quad \text{mpy} \qquad (24\text{-}10)$$

where

I_{corr}	=	corrosion current density, amps/cm^2
A	=	anode surface area, cm^2
EW	=	equivalent weight = molecular wt/valence
F	=	Faraday constant
d	=	density of anode, g/cm^3

The only unknown in these equations is I_{corr} which was measured using a dc cathodic polarization technique. Although the I_{corr} for these tests was several orders of magnitude lower ($\approx 10^{-8}$amps/cm^2) than for metals, the results were very reproducible and able to discriminate between samples of reaction sintered and pressureless sintered SiC. Corrosion rates varied with materials and test solution but generally were in the range of 10^{-3} mpy.

Electrochemical corrosion techniques appear to offer certain advantages in the study of glass and ceramic corrosion. More of these other studies should be conducted and results from these compared to results from weight loss and solution analyses. Furthermore, as with metals, it may be possible to passivate glass and ceramic surfaces using electrochemical methods. This possibility should also be explored.

ACKNOWLEDGEMENTS

The authors thank Westinghouse Savannah River Company under subcontract AX-715234 for partial financial support during the preparation of this chapter.

REFERENCES

1. Zeta Meter, Inc. Instrumentation pamphlet.
2. Gaudin, A.M., Fuerstenau, D.W., *Mining Eng.*, Transactions AIME, 66-72, (1955).
3. Bikerman, J.J., *Surface Chemistry - Theory and Applications*, Academic Press, NY, pp. 404-433 (1958).
4. O'Brien, R.W., *J. Fluid Mech.* 190, pp. 71-86, (1988).
5. Hunter, R.J., *Zeta Potential in Colloid Science - Principles and Applications*, Academic Press, p. 69 (1981).

6. Horn, Jr., J.M., Onoda, Jr., G.Y., *J. Am. Ceram. Soc.*, 61:11-12, pp. 523-527, Nov-Dec. (1978).
7. Springgate, Mark, Doctoral Student, University of Florida, personal communication (1991).
8. Cannon, D.W., *Conversion of ESA to Mobility and Zeta Potential*, Instrumentation information, Matec Applied Sciences, Hopkinton, MA.
9. Sullivan, T.M., Machiels, A.J., *J. Non-Cryst. Solids*, 55, pp. 269-28 (1983).
10. Mularie, W.M., Furth, W.F., Westwood, A.R.C., *J. of Mat. Sci.*, 14, pp. 2659-2664 (1979).
11. Lee. C.T. PhD Dissertation, *Surface and Solution Chemistry of Glass/Water Interactions*, University of Florida, 1986.
12. Lee, C.T., and Clark, D.E., *Mat. Res. Soc. Symp. Proc.*, Vol. 44 pp. 221-228 (1985).
13. Jednacak, J., Pravdic, V., *J. of Colloid & Interface Sci.*, 49:1, pp. 16-23 (1974).
14. Jednacak-Biscan, J., Mikac-Dadic, V., Pravdic, V., and Haller, W., *J. of Colloid & Interface Sci.*, 70:1, pp. 18-28 (1979).
15. McVay, G.L., Buckwalter, C.Q., "The Effect of Iron on Waste Glass Leaching", PNL-SA-10474 (1982).
16. Zhu, B.F. PhD Dissertation, University of Florida (1987).
17. Dilmore, M.F., Clark, D.E., and Hench, L.L., *Am. Ceram. Soc. Bull.*, 58:11, pp. 1111-1114 (1979).
18. Iler, R.K., *J. Colloid & Interface Sci.*, 43:2, pp. 399-408 (1973).
19. Oka, Y. and Tomozawa, M., *J. Non-Cryst. Solids*, 42, pp. 535-544 (1980).
20. Hudson, G.A. and Bacon, F.R., *Am. Ceram. Soc. Bull.* 37:4 pp. 185-188 (1958).
21. Tait, J.C. and Jensen, C.D., *J. Non-Cryst. Solids*, 49, pp. 363-377 (1982).
22. Persson, H.R., *Glass Tech.*, 3:1 pp. 17-35 (1962).
23. Tomozawa, M., Oka, Y., and Wahl, J.M., *J. Am. Ceram. Soc.*,64:2, pp. C-32-33 (1981).
24. Oka, Y., Ricker, K.S., and Tomozawa, M., *J. Am. Ceram. Soc.*, 62:1 12, pp. 631-632 (1979).
25. Dilmore, M.F., Clark, D.E., and Hench, L.L., *J. Am. Ceram. Soc.*, 57:11 pp. 1040-1045 (1978).
26. Clark, D.E., Dilmore, M.F., Etheridge, E.C., and Hench, L.L., *J. Am. Ceram. Soc.*, 59:1-2, pp. 62-65 (1976).
27. Clark, D.E., Pantano, Jr., C.G., and Hench, L.L. *Glass Corrosion*, Magazines for Industry, Inc., NY (1979).
28. Rana, M.A., and Douglas, R.W., *Phys. Chem. Glasses*, 2:6, pp. 179-195 (1961).
29. Douglas, R.W., Isard, J.O., *J. Soc. Glass Tech.*, 33, pp. 289-335 (1949).
30. Rothermel, D.L., *J. Am. Ceram. Soc.*, 50:11, pp. 574-577 (1967).
31. Charles, R.J., *J. Appl. Phys.*, 29:11, pp. 1549-1553 (1958).
32. Charles, R.J., *J. Appl. Phys.*, 29:11, pp. 1554-1560 (1958).

33. Wiederhorn, S.M., *J. Am. Ceram. Soc.*, 50:8, pp. 407-414 (1967).
34. Wiederhorn, S.M. and Bolz, L.H., *J. Am. Ceram. Soc.*, 53:10, pp. 543-548 (1970).
35. McColm, I.J., *Ceramic Science for Materials Technology*, Leonard Hill, (Distributed in the United States by Chapman and Hall) New York, NY, (1983).
36. Isard, J.O., in: *Glass Electrodes for Hydrogen and Other Cations, Principles and Practice*, (G. Eisenmann, ed.), Marcel Dekker, Inc., NY (1967).
37. Doremus, R.H., ibid.
38. Eisenman, G., ibid.
39. Eisenman, G., ibid.
40. Divakar, R., Seshardri, S.G., and Srinvasan, M., *J. Am. Ceram. Soc.*, 72:5, pp. 780-784 (1989).

Index

A

abrasive wear, 483
acid resistance, 278
acid-base reactions, 12
acidic molten salt, 522
acidproof bricks, 455
acids, 559-561,567,573-574
AEM, 342
aging-induced trans., 496
Al machining, 488,489
Al_2O_3, 461
albite, 13
alkali depletion, 131
alkali extraction, 123,301
alkaline resistance, 283
alumina, 484
aluminum-oxide compos., 488-489
American Optical test, 282
anal. electron micros., 342
analytical approach, 38
ancient glasses, 151
angle resolved XPS, 618
aoyake, 274
aqueous attack, 151
aqueous solution, 606
archaeological ceramics, 393
archaeological glasses, 330,393
Arrhenius equation, 21
ASTEM, 639
ASTM standards, 56,61-64
atomic absorption spect., 91
Auger electron spectroscopy, 88,411
autoclave methods, 275
autoclaving, 399

B

Ba4d spectra, 623
$Ba(OH)_2$, 603
$BaCO_3$, 553-556, 603
bacteria, 293
$BaCuO_2$, 553,572,574
ball-milling, 276
basic molten salt, 521
basicity, 559-561
BET, 87
$Bi_2Sr_2CaCu_2O_x$, 554
$Bi_2Sr_2Cu_3O_7$, 549-574
binary glasses, 93
binders, 571
bioactive ceramics, 298

Bioglass, 299
biomechanical, 311
biomineralization, 311
bitumen adhesives, 422
blooming, 278
bone, 298
borate glasses, 288
borosilicate glasses, 151,218-259
borosilicate glazes, 422
burner rigs, 516

C

cadmium release, 387
calcite, 349
carbide, 484
catalysts, 566,575
ceramic cutting tools, 481
ceramic electrodes, 660
ceramic heat exchangers, 516
ceramic matrix composites, 540
ceramic oxides, 632
ceramic superconductors,548-575,583-599,601-612,632-647
CH_3COOH, 560, 568
charge referencing, 618
chem. deg. (cut tool), 485
chemical diffusion, 607
chemical dissol. (cut tool), 486
chemical durability, 52,232,270
chemical shifts, 625
climatic resistance, 277
cloisonne enamels, 422
closed systems, 19
CO, 555
CO_2, 551,559
coating composition, 380
coatings for HMF, 327
coefficient of linear expansion, 421
cold working (opt. glasses), 272
colorimetry, 79
composites, 298
composition effect, 32
composition profile, 299
compositional analysis, 87
condensation, 605
congruent dissolution, 4,299
conservation science, 393
construction materials, 455
corrosion
 ancient ceramics, 393-427
 ancient glasses, 393-427
 borosilicate glass, 218-258
 ceramic superconductors,548-575,583-599,601-612,632-647
 characterization,78-96,103-119
 construction materials, 455-478
 enamels, 372-390
 engineering ceramics, 514-543
 environment,199
 glass, 111, 218-258,315-330
 glazes, 372-390
 heavy-metal fluoride glass, 399
 kinetics, 374
 mechanisms (TZP), 508
 models, 31,246-249,434
 natural glasses, 335
 optical glasses, 273-295
 prediction, 29-49
 regimes, 522
 resistance tests, 385
 stress, 659
 tests, 52-77
 theory, 2-26
 TZP ceramics, 492-511
coulombic force, 285
crazing, 419
crown glass, 270
Cu(III), 555, 561-564,574
CuO, 603
curve fitting, 622
cutting fluids, 486
cutting tool coatings, 490
cutting tool materials, 483
cutting tool selection, 489

D

deconvolution, 622
defect, 273
degradation factors, 425-426
degradation mech. (cut. tools), 483
degradation of TZP ceramics
  corrosion mechanism, 508
  density, 502,
  destabilization theory, 509
  electrical conductivity, 506
  fracture toughness, 505
  grain boundary phase, 503
  grain shape, 503
  grain size, 502
  stabilizing agent, 500
  stabilizer content, 500
  strength, 504
  wear resistance, 506
density, 502
deposit-induced corrosion, 514
depth profiles (SAM), 619
devitrification, 431
dew point, 517
differential thermal analysis, 423
diffusion, 557
diffusion coefficient, 606
diffusion wear (cut tool), 486
dimming, 273
dispersion, 270
durability testing, 274

E

Egyptian faience, 422
Eh, 21
Eh effects, 160
electrical conductivity, 506
electrical double layer, 650
electrochemical corrosion, 662
electrochemistry, 551
electrodes, 564-566
electrokinetic sonic amplitude, 651
electron micro., 85
electron microprobe, 88
electron spec. chem. an., 82
electron transfer,561-566
element distributions, 111
element profiling, 104
enamels, 372,422
end user, 39
environ. conditions, 199,499
environmental stability, 632
ESCA, 82
etchant studies SiC, 516
EXAFS, 84
experimental tech. (gl. dur.), 172

F

factors affecting corrosion, 649
Fe_2NiO_4 spinel, 10
fiber waveguides, 315
Fick's law, 13
field tests, 75,234
films, 551, 554, 558, 569-571, 573
fireclay, 455
first order reactions, 9
flint glass, 270
fluorozirconate glasses, 319
fracture (cut. tool), 485
fracture toughness, 505
free drift, 20
free energy hyd., 163
FSZ, 495
FTIR (HMF), 324
fully stabilized zirconia, 495
fundamental studies, 47
fungi, 291

G

gel layer, 15
general rate equation (gl. ds), 132
geochemical, 122
geochemical models, 122
geochemistry/groundwater, 363
geochemistry/mineral dep., 363

geochemistry/seawater, 363
geologic systems, 334
geological glasses, 330
glass coatings, 372
glass corrosion rate, 135
glass dissolution, 122
glass durability, 151
(*see also corrosion*)
glass electrodes, 660
glass hydration mech., 154
glass/seawater int., 330
glass/water reaction, 129
glass-ceramic, 298,431,432
GLASSOL, 123
GLASSOL modeling, 140
glazes, 372,415
grain boundaries, 571,572,574,621
grain boundary phase, 503
grain shape, 503
grain size, 502
Grambow equation, 16
Guan celadon, 419
gypsum, 398

H

H_2SO_4, 458
H_3PO_4, 455
HCl, 458
health/safety, 44
heat engines, 516
heats of formation, 621
heavy-metal fluoride glass, 315
HF, 458
high alumina, 455
high silica glass, 355
HLW, 218
HMF glasses
  aqueous corrosion, 317
  composition, 316
  effect of pH, 318
  surface condition, 320
  test methods, 317
HNO_3, 458
humidity, 551-555
 561,571,573,574,584,602,619
hydration, 299
hydroxyl attack, 373

I

ICP, 79
impurities, 115
incongruent dissolution, 5
inductively coupled plasma, 79
industrial glasses, 151
infrared reflection (gl-cer), 443-445
infrared spectroscopy, 81,508
in-situ tests, 184
interdiffusion theory, 124
interfacial strength, 310
interferences, 36
ion exchange, 6,158,301,373
ionic field strength, 168
ionic potential, 168
ionic radii, 286
IR lenses, 315

K

KI, 555, 563
kinetic approach gl. dur., 153
kinetic modification, 298
Korean celadon, 418
Koryo dynasty, 417

L

$La_{1-x}(Sr,Ba)_xCuO_4$, 549,550
laboratory tests, 184
lanthanum, 284
lanthanum borate, 289
laser windows, 315
latent scratch, 274
leach rate, 13,69
leaching, 233
leaching environments, 189

lead-containing glasses, 284
lead release, 387
leaded glass, 424
lime, 398
linear expansion, 475
lithia-alumino-silicates, 434
lithia-disilicates, 434
Longquan celadon, 419
low silica glass, 337

M

macrocracking, 508
matrix dissolution, 123,158
MCC-1, 172
MCC-3, 174
medieval stained glass, 424
methodology, 39
MgF_2, 4
MgO, 4
microanalysis, 103
microcracking, 508
microelectrophoresis, 650
micro-organisms, 291
microphase separation, 290
microstructural char., 84
MIIT, 250
mixed alkali effect, 286
modeling, 29
modified random network, 165
modulus of rupture, 421
moisture barrier, 424
moisture expansion, 421

N

Na_2O x SiO_2, 520
$NaAlSi_3O_8$, 13
NaCl, 556
NaOH, 556
natural glasses, 331
Neolithic period, 398
Nernst theory, 7
network dissol., 301
NH_4OH, 559-561
nickel-bas. alloys (mach.), 488,489
nickel ferrite, 11
nonaqueous solvents, 498
non-bridging oxygen bonds, 163
non-metal. comp. (mach), 488,489
nonsuperconducting phase, 591
nuclear magnetic resonance, 82
nuclear reaction analysis, 90
nuclear waste, 51,151,218,358

O

O1s, 624
O_2, 555
open systems, 19
optical glasses, 269
organic solvents, 560, 567-571
Ostwald ripening, 399
oxidation, 561-566
oxide effects, 32
oxygen tracer, 607

P

palagonite, 338
Paleolithic period, 395
Parthian period, 415
partially stabilized zirconia, 494
particle size, 276
particle size (SC), 583
particle size dist. (SC), 583
passivation, 634
PCT, 175
pH, 557-561
pH corrosion effect, 73,160
pH electrodes, 91
pH stat, 20
phase transformation, 495
phosphate glasses, 269,289
PHREEQE, 123
pinholes, 419
pitting, 529
pitting corrosion, 417

planetary geology, 360
plastic deformation, 484
plutonic rock, 331
polycrystalline solids, 431
porosity, 557, 571-574
pottery, 415
Pourbaix diagrams, 162
Pourbaix stability fields, 201
powdered glass, 276
precision/accuracy, 38
predicting corrosion, 29
preservation, 394
pristine glass, 424
processing conditions (SC), 583
protective coatings, 540,573
protocol, 34
PSZ, 494
Pyrex®, 269
Pyroceram®, 431

Q

quartz, 10

R

radioactivity, 218
Raman spectroscopy, 81
reaction path modeling, 138
red shale, 456
redox species, 160
reduction potential, 564, 574
refractive index, 270
Roman period, 404
restoration, 393
Rutherford backscattering, 89

S

SAM (cer. sup), 618
SA/V, 66,190
SAXS, 87
screen printing, 569, 575
secondary ion mass spec. 103-121,253
secondary mineralization, 349
secondary minerals, 351-352
secondary spectrum, 288
selective dissol., 299
SEM, 436-439
shiroyake, 274
SiC, 456
silica, 455
silica transport, 135
silicate glasses, 122,285
silicon-based ceramics, 514
silicon carbide, 455
silicon nitride, 484
SIMS, 88,103
sintering temp. (SC), 583
slags, 524
small angle x-ray scattering, 87
smectites, 349
sodium aluminosilicate, 6
sodium tri-phosphate, 284
sol-gel coatings, 632
sol-gel coatings cor, 633
sol-gel corrosion, 633
solution analysis, 79, 91
spalling, 419
specialty materials, 455
spectroscopy, 81
spherulites, 411
stabilizer content, 500
stabilizing agent, 500
staining, 274
standard acetate soln., 278
steel machining, 488, 489
stoichiometry, 115
strain point, 421
streaming potential, 650
strength, 504
strength degradation, 532
stress corrosion, 508,659
stress-induced transf., 510
superconducting fraction, 593
superconductivity, 583,632
superconductor (J_c), 550-551, 572
superconductor (melt-derived), 551-552, 572

superconductor (sintered), 564, 571-572
superconductor (synthesis), 558,
superconductor (T_c), 548-550, 552-554, 558, 582
superconductor (types), 549, 574
superconductors (ceramic), 548-575
surface analysis techn. cer sup., 615
surface chemical species, 626
surface layer form., 158,186
surface layers, 125
surface modification, 298
surface passivation, 655

T

target environment conditions, 37
Taylor equation, 487
TEM, 85,342,642
temperature effects, 93, 192, 496,556
test protocol, 37
tetragonal zirconia polycrystals, 492
therm. equilibria, 177
thermal expansion, 422
thermod. model (gl. cor), 152,153
Thermodyne test, 277
titanium alloys (mach), 488,489
$Tl_2Ba_2Ca_2Cu_3O_{10}$, 549
transport-controlled reaction, 7
tungsten carbide, 484
tungsten carbide/cobalt, 487
TZP, 492

U

unfired clay, 423
use-wear, 394

V

vapor phase, 25
vapor pressure, 605
vitrification, 222
volcanic rock, 331
volume expansion, 474

W

Wallace-Wicks equation, 16
waste management, 219
water, 555-575,602
water solutions, 498
water vapor, 551-555, 561,
water/water vapor, 498
wear resistance, 506
weathering, 151,275,393,551-555

X

x-ray abs. fine-edge struct., 84
x-ray photoelectron spect., 82,509
XPS, 82,88,326,509,615
XRD, 622,636

Y

$Y(OH)_3$, 603
Y_2BaCuO_5, 550, 553-556,560,566,569,572,574,603
$YBa_2Cu_3O_7$, 549-574
$YBa_2Cu_3O_{7-x}$, 583,601
Yue celadon, 419

Z

zeolites, 349
zero-order kinetics, 9
zeta potential, 650
zirconia ceramics, 493